The Evolutionary Roots of Human Brain Diseases

The Evolutionary Roots of Human Brain Diseases

EDITED BY

Nico J. Diederich
Consultant, Department of Neurology, Centre Hospitalier de Luxembourg, Luxembourg
Ausserplanmässiger Professor für Neurologie, University of Cologne, Germany

Martin Brüne
Professor of Psychiatry & Head of the Division of Social Neuropsychiatry and Evolutionary Medicine
LWL University Hospital Bochum, Department of Psychiatry, Psychotherapy and Preventive Medicine, Ruhr-University Bochum, Germany

Katrin Amunts
Professor and Director, Institute of Neuroscience and Medicine (INM-1), Research Center Jülich, Germany
C. and O. Vogt Institute for Brain Research, University Hospital Düsseldorf, Germany

Christopher G. Goetz
Professor, Department of Neurological Sciences
Professor, Department of Pharmacology
Rush University Medical Center, USA

OXFORD
UNIVERSITY PRESS

Oxford University Press is a department of the University of Oxford. It furthers the University's objective of excellence in research, scholarship, and education by publishing worldwide. Oxford is a registered trade mark of Oxford University Press in the UK and certain other countries.

Published in the United States of America by Oxford University Press
198 Madison Avenue, New York, NY 10016, United States of America.

© Oxford University Press 2024

All rights reserved. No part of this publication may be reproduced, stored in a retrieval system, or transmitted, in any form or by any means, without the prior permission in writing of Oxford University Press, or as expressly permitted by law, by license, or under terms agreed with the appropriate reproduction rights organization. Inquiries concerning reproduction outside the scope of the above should be sent to the Rights Department, Oxford University Press, at the address above.

You must not circulate this work in any other form
and you must impose this same condition on any acquirer.

Library of Congress Cataloging-in-Publication Data
Names: Diederich, Nico J., editor. | Brüne, Martin, editor. |
Amunts, Katrin, editor. | Goetz, Christopher G., editor.
Title: The evolutionary roots of human brain diseases / edited by Nico J. Diederich,
Martin Brüne, Katrin Amunts, and Christopher G. Goetz.
Description: New York, NY : Oxford University Press, [2024] |
Includes bibliographical references and index.
Identifiers: LCCN 2024013899 (print) | LCCN 2024013900 (ebook) |
ISBN 9780197676592 (hardback) | ISBN 9780197676615 (epub) | ISBN 9780197676622
Subjects: MESH: Brain Diseases—physiopathology | Brain Diseases—etiology |
Biological Evolution | Brain—anatomy & histology | Neuropsychology
Classification: LCC RC346 (print) | LCC RC346 (ebook) | NLM WL 348 |
DDC 616.8—dc23/eng/20240430
LC record available at https://lccn.loc.gov/2024013899
LC ebook record available at https://lccn.loc.gov/2024013900

DOI: 10.1093/med/9780197676592.001.0001

This material is not intended to be, and should not be considered, a substitute for medical or other professional advice. Treatment for the conditions described in this material is highly dependent on the individual circumstances. And, while this material is designed to offer accurate information with respect to the subject matter covered and to be current as of the time it was written, research and knowledge about medical and health issues is constantly evolving and dose schedules for medications are being revised continually, with new side effects recognized and accounted for regularly. Readers must therefore always check the product information and clinical procedures with the most up-to-date published product information and data sheets provided by the manufacturers and the most recent codes of conduct and safety regulation. The publisher and the authors make no representations or warranties to readers, express or implied, as to the accuracy or completeness of this material. Without limiting the foregoing, the publisher and the authors make no representations or warranties as to the accuracy or efficacy of the drug dosages mentioned in the material. The authors and the publisher do not accept, and expressly disclaim, any responsibility for any liability, loss, or risk that may be claimed or incurred as a consequence of the use and/or application of any of the contents of this material.

Printed by Integrated Books International, United States of America

Contents

Foreword   ix
Contributors   xi

Introduction   1
   Nico J. Diederich, Martin Brüne, Katrin Amunts, and Christopher G. Goetz

Part 1   **Human Brain Evolution: From Anatomy to Function**

1. Human Telencephalization   9
   Felix Ströckens and Katrin Amunts

2. Evolutionary Aspects of Glial Cells: From Physiology to Pathology   38
   Corrado Calì, Nicole L. Ackermans, Patrick R. Hof, and Pierre J. Magistretti

3. The Contribution of Mitochondrial Evolution and Dysfunction to Neurodegeneration   58
   Kobi Wasner, Sandro L. Pereira, and Anne Grünewald

4. Intrinsic Templates for Neurodegenerations Featuring Disease-Specific Axonal or Dendritic Vulnerability   81
   Toshiki Uchihara

5. Humans and Nonhuman Primates   105
   Geneviève Konopka and Emre Caglayan

6. Adaptive Archaic Introgression   131
   Olga Dolgova and Oscar Lao

7. Goal-Directed and Habitual Behaviors: Anatomical and Functional
   Circuits in Health and Neurological Disease   154
   Ledia F. Hernandez and Ignacio Obeso

Part 2   How Human Brain Diseases Are Impacted by Human Evolution

8. Alzheimer's Disease, the Parietal Lobes, and the Evolution of the
   Human Genus   181
   Emiliano Bruner and Heidi I. L. Jacobs

9. Parkinson's Disease: Overstrain Focused on Basal Ganglia and
   Brainstem Nuclei   205
   Nico J. Diederich and Christopher G. Goetz

10. Brain Diseases Associated with Unstable Repeats   228
    Katharine E. Shelly, Emily G. Allen, and Peng Jin

11. The Properties of Cortico-motoneuronal Connections and Their
    Evolutionary Significance for Amyotrophic Lateral Sclerosis   252
    Roger N. Lemon

12. REM Sleep Behavior Disorder: Nocturnal Replay of "Fight or Flight"?   270
    Nico J. Diederich and Isabelle Arnulf

13. An Evolutionary Psychoneuroimmunological Approach to Major
    Depressive Disorder   294
    Markus J. Rantala and Javier I. Borráz-León

14. Schizophrenia: Embracing the Spectrum   314
    John S. Allen

15. Williams Syndrome and Autism: Dysfunction of Frontal Networks   334
    Isabel August and Katerina Semendeferi

16. Attention-Deficit Hyperactivity Disorder: An Evolutionary View   360
    Annie Swanepoel

17. Addiction: Diverted Reward and Motivation Principles   373
    Roger Sullivan and Edward Hagen

Part 3   Consequences and Perspectives on Research and Clinical Sciences

18. Conditions of Comparative Brain Connectomics   399
    Kathleen S. Rockland, Daniel Zachlod, and Katrin Amunts

19.	Are Evolutionary Concepts Helpful in Designing Preventive Strategies for Brain Diseases? *Gilberto Levy and Bruce Levin*	426
20.	Evolutionary Aspects of Neuropsychopharmacology *Martin Brüne, Riadh Abed, and Paul St. John-Smith*	449
21.	Ongoing Human Evolution? *Nicole Bender, Frank Rühli, and Maciej Henneberg*	472
22.	Human Cultural Evolution Outpaces Biological Evolution: A Brain Connectomic Approach *Jean-Pierre Changeux*	491
23.	Concluding Remarks and Future Directions *Martin Brüne, Nico J. Diederich, Christopher G. Goetz, and Katrin Amunts*	510

Glossary   521
Index   533

Foreword

"Nothing in biology makes sense except in the light of evolution," a famous quote from the 1970s by Theodosius Dobzhansky, a Ukrainian American researcher, can undoubtedly be applied also to many, but not all, aspects of medicine. In *The Evolutionary Roots of Human Brain Diseases*, Nico Diederich together with Christopher Goetz, Martin Brüne, and Katrin Amunts have collected 22 very interesting chapters written by experts in different areas to form an insightful overview. The chapters extend from which parts of the human brain that have expanded during evolution, and equally important the underlying microstructure, to the many different disorders of the brain. They include the basal ganglia disorders, cognitive symptoms such as Williams syndrome, autism, attention-deficit syndrome, schizophrenia, depression, and more.

Biological evolution during the last billion years did expand gradually. By the Cambrian explosion some 600–500 million years ago the eukaryotic cell had evolved. This was a major step with the amazing design of the cell nucleus; the energy production with mitochondria; active membrane transport; ion channels for sodium, calcium, and potassium; ionotropic and metabotropic receptors; and most signaling molecules including neuropeptides, monoamines, glutamate, and gamma-aminobutyric acid. With these building blocks, nervous systems could be built with action potentials for rapid transmission of information and chemical synaptic transmission. With this cellular infrastructure, advanced multicellular organisms could then evolve. All the different invertebrate groups and the vertebrates took off at this period, and then each followed its own evolutionary path.

The nervous system of vertebrates would very early converge on a common bauplan with the basal ganglia, cortex/pallium, hypothalamus, a midbrain, brainstem, and spinal cord. This basic organization had evolved already some 500 million years ago, and it has been maintained ever since. This includes also the many different microcircuits of the brain. An important difference, however, is the number of cells; and, of course, the larger the number of cells, the greater the potential for elaborate processing. Take, for instance, the cortical columns. They are designed in a similar way in mice and humans, but the number of cortical columns is orders of magnitude greater in humans, which then will have a much larger potential for complex processing. However, the size of the brain does not automatically translate into advanced cognitive ability. Elephants have a much larger brain than humans, and some whales have brains even weighing up to 7.8 kg (!!). Conversely, corvids (crows, ravens, and jays) with relatively small brains have an advanced cognitive ability considered similar to that of apes.

Let me take one example of a very evolutionarily conserved system, that of the dopamine neurons in the substantia nigra pars compacta (SNc). In all groups of vertebrates the SNc has a similar organization with regard to its projection pattern and input from other parts of the nervous system. Moreover, experimental degeneration of the dopamine system leads to very similar types of motor symptoms in all groups of vertebrates. However, only in humans do the clinical symptoms of Parkinson's occur. The motor symptoms occur primarily due to a lack of dopamine innervation of the putamen. In humans, the dopamine axonal arbor in the putamen is very extensive, with nearly a million synaptic varicosities per axon that must be supplied with energy from a single cell body (!), which thus becomes vulnerable. In other animals, as well as for other parts of the ventral and dorsal striatum in humans (including the caudate nucleus), the number of varicosities per neuron is smaller. This serves as one of many examples provided within this interesting volume on how evolution can contribute to our understanding of the brain and its many disorders.

Sten Grillner
Department of Neuroscience
Karolinska Institutet
Stockholm, Sweden
August 21, 2023

Contributors

Riadh Abed
Evolutionary Psychiatry Special Interest Group
Royal College of Psychiatrists
London, United Kingdom

Nicole L. Ackermans
Department of Biological Sciences
College of Arts and Sciences
University of Alabama
Tuscaloosa, AL, USA

Emily G. Allen
Department of Human Genetics
Emory University
Atlanta, GA, USA

John S. Allen
Department of Anthropology
Indiana University
Bloomington, IN, USA

Katrin Amunts
Cecile & Oskar Vogt Institute for Brain Research
Medical Faculty
University Hospital Düsseldorf
Heinrich-Heine University Düsseldorf
Düsseldorf, Germany
Institute of Neurosciences and Medicine (INM-1)
Research Centre Jülich
Jülich, Germany

Isabelle Arnulf
Sleep Disorder Unit
Hôpital Pitié-Salpêtrière and Sorbonne University
Paris, France

Isabel August
Department of Anthropology
University of California, San Diego
La Jolla, CA, USA

Nicole Bender
Clinical Evolutionary Medicine Group
Institute of Evolutionary Medicine
Medical Faculty
University of Zurich
Zurich, Switzerland

Contributors

Javier I. Borráz-León
Institute for Mind and Biology
The University of Chicago
Chicago, IL, USA

Martin Brüne
Professor of Psychiatry & Head of the Division of Social Neuropsychiatry and Evolutionary Medicine
LWL University Hospital Bochum, Department of Psychiatry, Psychotherapy and Preventive Medicine, Ruhr-University Bochum
Germany

Emiliano Bruner
Centro Nacional de Investigación sobre la Evolución Humana
Burgos, Spain
Centro de Investigación Enfermedades Neurológicas
Madrid, Spain

Emre Caglayan
Department of Neuroscience
UT Southwestern Medical Center
Dallas, TX, USA

Corrado Calì
Department of Neuroscience
University of Torino
Torino, Italy
Neuroscience Institute Cavalieri Ottolenghi
Orbassano, Italy

Jean-Pierre Changeux
CNRS UMR 3571
Institut Pasteur
Communications Cellulaires
Collège de France
Paris, France

Nico J. Diederich
Consultant, Department of Neurology
Centre Hospitalier de Luxembourg
Luxembourg
Ausserplanmässiger Professor für Neurologie
University of Cologne
Germany

Olga Dolgova
Integrative Genomics Lab
CIC bioGUNE
Derio (Bizkaia), Spain

Christopher G. Goetz
Professor, Department of Neurological Sciences
Professor, Department of Pharmacology
Rush University Medical Center
Chicago, USA

Anne Grünewald
Luxembourg Centre for Systems Biomedicine
University of Luxembourg
Esch-sur-Alzette, Luxembourg
Institute of Neurogenetics
University o Lübeck
Lübeck, Germany

Edward Hagen
Department of Anthropology
Washington State University
Vancouver, WA, USA

Maciej Henneberg
School of Biomedicine and Adelaide Nursing School
The University of Adelaide
Adelaide, Australia
Institute of Evolutionary Medicine
University of Zurich
Zurich, Switzerland
Unit for Biocultural Variation in Obesity
University of Oxford
Oxford, United Kingdom

Ledia F. Hernandez
Facultad HM de Ciencias de la Salud
Universidad Camilo José Cela
Madrid, Spain

Patrick R. Hof
Nash Family Department of Neuroscience
Icahn School of Medicine at Mount Sinai
New York, NY, USA

Heidi I. L. Jacobs
Gordon Center for Medical Imaging
Massachusetts General Hospital/Harvard Medical School
Boston, MA, USA

Peng Jin
Department of Human Genetics
Emory University
Atlanta, GA, USA

Geneviève Konopka
Department of Neuroscience
UT Southwestern Medical Center
Dallas, TX, USA

Oscar Lao
Algorithms for Population Genomics Group
Institut de Biologia Evolutiva
Barcelona, Spain

Roger N. Lemon
Department of Clinical and Movement Neurosciences
UCL Queen Square Institute of Neurology
London, United Kingdom

Bruce Levin
Department of Biostatistics
Mailman School of Public Health
Columbia University
New York, NY, USA

Gilberto Levy
Independent researcher
Rio de Janeiro, Brazil

Pierre J. Magistretti
Biological and Environmental Sciences and Engineering Department
King Abdullah University Of Science And Technology
Thuwal, Saudi Arabia

Ignacio Obeso
HM-CINAC (Centro Integral de Neurociencias Abarca Campal)
Fundación de Investigación HM
HM Hospitales
Psychobiology and Methodology in Behavioral Sciences Department
Complutense University of Madrid
Madrid, Spain

Sandro L. Pereira
Luxembourg Centre for Systems Biomedicine
University of Luxembourg
Esch-sur-Alzette, Luxembourg

Markus J. Rantala
Department of Biology
University of Turku
Turku, Finland

Kathleen S. Rockland
Department of Anatomy & Neurobiology
Chobanian & Avedisian School of Medicine
Boston University
Boston, MA, USA

Frank Rühli
Institute of Evolutionary Medicine
Medical Faculty
University of Zurich
Zurich, Switzerland

Katerina Semendeferi
Department of Anthropology
University of California, San Diego
Center for Academic Research and Training in Anthropogeny
La Jolla, CA, USA

Katharine E. Shelly
Department of Human Genetics
Emory University
Atlanta, GA, USA

Paul St. John-Smith
Evolutionary Psychiatry Special
   Interest Group
Royal College of Psychiatrists
London, United Kingdom

Felix Ströckens
Cecile & Oskar Vogt Institute for Brain
   Research
Medical Faculty
University Hospital Düsseldorf
Heinrich-Heine University Düsseldorf
Düsseldorf, Germany

Roger Sullivan
Department of Anthropology
California State University
Sacramento, CA, USA

Annie Swanepoel
North East London NHS Foundation Trust
Chelmsford, United Kingdom

Toshiki Uchihara
Clinical Professor, Department of
   Neurology, Tokyo Medical and
   Dental University, 1-5-45 Yushima,
   Bunkyo-ku, Tokyo, Japan

Kobi Wasner
Luxembourg Centre for Systems
   Biomedicine
University of Luxembourg
Esch-sur-Alzette, Luxembourg

Daniel Zachlod
Institute of Neurosciences and
   Medicine (INM-1)
Research Centre Jülich
Jülich, Germany

Introduction

Nico J. Diederich, Martin Brüne, Katrin Amunts, and Christopher G. Goetz

Why This Book?

Traditionally, studies and textbooks in neurology or psychiatry, as well as allied disciplines, deal with proximate causes of diseases and therapies but remain mute or minimally interested in their ultimate causes including the phylogeny and adaptive significance of disease manifestations. Yet, as clinicians or basic researchers, we are conscious of potential evolutionary roots of neurological and psychiatric symptoms, often offering a rudimentary explanation but never delving deeply into the current role of evolutionary science as it relates to health and disease. We may miss appreciation of the role of adaptive properties, evolutionarily based neuronal circuitries, unbalanced cellular energy demands, and the potential health consequences of residual syndromic behaviors that were possibly useful in early times of human development but presently are obsolete and pathological. The problem is amplified because there is often no interdisciplinary dialogue between anthropology and evolutionary biology, on one side, and clinical sciences, on the other side. However, the evolutionary tracing back of disease pathways may disclose unexpected insights and trigger the design of innovative research as well as propel the development of new therapeutic interventions. There could also be a better apprehension of compensatory behaviors, at both the cellular level as well as the systemic and behavioral levels, that could be the expected fruits of such collaborations. So far scientists fall short in modeling the complexity of human (social) life, human language, manual dexterity, and mental or emotional behaviors that typify human neurological or psychological function and dysfunction. Finally, there remain obstacles in the form of poor animal modeling for human brain diseases and for human longevity. The present book aims to fill these gaps by presenting an evolutionary view of neurological and psychiatric conditions that is meant to complement and enrich existing medical perspectives.

Placement in Time

By exploring the evolutionary roots of human brain diseases, are we a lone voice in the wilderness? From a historical perspective, this is not the case. *Alfred Russel Wallace* and *Charles Darwin* set the path, the latter by his detection of similarities between human and animal behavior in *The Expression of the Emotions in Man and Animals* (Darwin and Prodger 1872/1998). From the neurological perspective, *John Hughlings Jackson* (1874, pp. 80–86) first proposed an evolutionarily grounded hierarchical organization of the nervous system, further morphologically conceptualized by *Paul D. MacLean* (1990) when proposing different evolutionary cerebral layers in his "triune brain." We presently know that such simplistic parallelism between ontogeny and phylogeny does not exist. From the ethological perspective, *Konrad Lorenz* (1950) and others dissected human instinctive behaviors and resulting pathological syndromes. *Ernst Mayr* (1961) and *Nikolaas Tinbergen* (1963) coined the concepts "proximate" and "ultimate" causes of diseases. In the psychiatric domain, several authors have emphasized that psychiatric disorders could, in part, reflect maladaptive extremes of evolved cognitive, emotional, and behavioral processes. Despite these admirable works, in-depth genetic, subcellular, supracellular, or connectomic conceptualizations have not yet been fully developed in the context of neurological and psychiatric medicine. Meeting this goal requires a rigorous and comprehensive approach to be uniformly tested and validated. Indeed, the evolution coin has to be considered from both sides. Thus, the holistic analysis of human brain evolution reveals that "advantageous" evolutionary changes that favorably matched with former environments may have turned over time into disadvantages, mismatching with a new environment. The present book aims to fill in this lacuna.

What Are New Insights?

Since 2010, numerous studies have dissected various aspects of human brain evolution. They range from innovative morphometric comparisons between the brains of humans and other primates such as allometry or connectomics, to genetic analyses. The genome has been compared in different hominids. Differing genetic expression and mosaicism in various brain areas have been reported. There has been a broad discussion on energy-based stress of cellular organelles. Evolutionary anthropology has presented arguments on how brain development paralleled the acquisition of bipedal gait, sophistication of manual skills, acquisition of language, widening of social networks, and optimization of theory of mind. We begin to understand the impact of adaptive archaic introgression from Neanderthals or Denisovans to *Homo sapiens*. These new insights allow the definition of evolutionary mechanisms and principles such as trade-offs, exaptation, mismatch, human-accelerated regions of genetic changes, etc. They have now entered the realm of medical disease.

A Few Examples of What Will Be Discussed

Provocatively, it can be questioned if neurological and psychiatric diseases should be the price to pay for evolutionary adaptations. But is the term "price to pay" adequate? The chapters in this book try to give pondered responses to this question. For instance, they allude to pleiotropic antagonism, when early favorable developmental processes as seen in Huntington's disease later turn on to trigger neurodegeneration after the reproductive phase. Gene changes inducing novelty-seeking as seen in attention-deficit hyperactivity syndrome could have been advantageous in early times but constitute a serious handicap for the regulated modern social life. The constraints of the exceptionally long human life on cellular organelles and metabolism will be considered: could they explain why, for instance, Parkinson's disease naturally manifests itself only in humans but not in closely related primates? Is there maladaptive stressing of basal ganglia nuclei and circuitries, considering the exceptional telencephalization in humans? When applying allometric comparisons, do we just perform as expected? Is deviating neuronal connectivity as seen in schizophrenia or Alzheimer's dementia part of the problem or a way to delay the inescapable disease trajectory? Genetic and brain connectivity differences between humans and other primates have been highlighted in the context of schizophrenia and autism. From an ethological point of view, the sickness behavior seen in depression will be presented as an evolutionary sequela linked to combating microbial and neuroinflammation associated with hunting or fighting behaviors. Similarly, rapid eye movement sleep behavior disorder could be reminiscent of a nocturnal replay of a "fight or flight" program in response to danger. Other topics to be addressed will be a critical revisiting of Western lifestyle diseases, the tracing back of neurotransmitters to ancient biomodulators, and the advocation that in humanity cultural evolution already outpaces biological evolution. Such themes are treated as a special focus in individual chapters but also transcend into other chapters throughout the volume.

The broadness of the topics with their eclecticism is deliberate. Thus, the authors and the editors come from various scientific backgrounds. Gender and cultural balances are respected, and this is not a book from one school of thinking, meaning that different perspectives, levels of discussion, and often controversial standpoints will be presented. We strongly believe that light emerges from the clash of ideas.

How the Book Is Organized

The format of the chapters is purposefully organized so that researchers and clinicians from disparate disciplines juxtapose findings from different domains. In the first part of the book, entitled "Human Brain Evolution: From Anatomy to Function," from Chapters 1–7 there will be a short presentation of essential anatomical, physiological, and genetic findings underlying human brain evolution and the potential for human brain diseases. We consider these chapters as indispensable to the understanding of the clinical chapters. The second part, entitled "How Human Brain Diseases Are Impacted by Human Evolution," from Chapters 8–17 constitutes the core of the book as there will be detailed presentations of the evolutionary impact on selected, but very common, diseases in clinical neuroscience. The applicability to therapeutic strategies will be proposed. The last part of the book, entitled "Consequences and Perspectives on Research and Clinical Sciences," from Chapters 18–23 discusses the consequences and perspectives on research and clinical sciences. It will be the most critical part of the book and will present potential directions for future research as well as cultural implications. Whereas all authors have been encouraged to develop their arguments in their own way, the editorial team has proposed a common chapter organization with the sections: historical background; evolutionary evidence for human specificities (including genetic aspects if available); counterarguments and evidence against evolutionary influences; clinical and therapeutic implications; unanswered questions and future work; and, finally, concluding bullet points presented in special boxes. In this way, readers can access individual chapters with the confidence of new information but organized in a systematic presentation style.

Biased Choice of Diseases?

We openly acknowledge that the range of topics discussed here is limited. The disease choices are common diseases, such as Alzheimer's dementia, depression, schizophrenia, Parkinson's disease, and autism, to name only a few. We do not cover most rare diseases with a known mono- or oligogenetic origin. We have also consciously left aside primary immunological diseases impacting the brain, even though we are well aware that bacteriological and viral agents considered "former friends" have impacted our innate neuroimmune systems. We consider in-depth analysis of the evolutionary roots of neuroimmunological processes to be a worthy topic for a future volume.

Evolutionary Rooting as an All-Encompassing Straitjacket?

We do not propose evolutionary rooting as an all-encompassing straitjacket for understanding brain diseases. Such a reductionist concept would not adequately reflect the complexity of neurological and, even more so, psychiatric diseases during a long human life with multiple exposures, triggers, and life events. Consequently, although we focus on evolution and a highly longitudinal view of pathophysiology, we do not mean to displace or exclude "proximate" influences or current environmental risk factors on disease. As a metaphor, our text is a discussion that searches underground for the "ultimate" roots of diseases so that each disease or tree in the forest is more clearly understood in its full manifestation, individually and collectively within the forest. We are digging downward to consider the depth of the roots or the time-base of diseases and studying how far each tree's roots extend laterally as one disease's history may relate to the roots of another disease. With this foundation, we hope that scientists and clinicians will better understand neurological and psychiatric diseases and be better equipped to interpret the "aboveground" or clinical manifestations of diseases and the environmental influences impacting health.

Tackling Controversies?

The authors have been encouraged to embrace controversies, ambiguities, and unknowns. Throughout the text, we have asked for balanced and undiluted discussions where evolutionary explanation advances our understanding but also where such explanations pose conflicting or negative arguments. Some of the presented hypotheses remain highly speculative, even when only applicable to one human brain disease. The attentive reader may criticize that some suggestions have been elaborated in the splendid isolation of the ivory tower of pure science, without extensive resonance or testing by colleagues from other disciplines. But by bringing these ideas into one book and appealing to a wide readership we hope to promote interdisciplinary dialogue and confrontation that will help us to leave trodden paths and allow new thinking. Domains that will be enriched by such cross-fertilization should be numerous and may include inquiries of involved neuronal circuitries, analyses of induced cellular energy demands, and scrutiny of potential health consequences of residual syndromic behaviors left behind from earlier evolutionary periods.

Will Evolutionary Insights Impact Clinical Practice Today?

Our ultimate goal is to present to readers concepts and observations that impact their own views of clinical medicine. This effort is not *l'art pour l'art* but clearly the tenet that looking backward actually allows a forward vision empowered by new insights and clarity. Evolutionary considerations open new research avenues, drawing attention to comparative connectomics, both between neurologic and psychiatric diseases as well as between humans and other primates. These discussions should facilitate dialog between different disciplines, particularly drawing neurologists, psychiatrists, clinical psychologists, and other health providers together. For instance, neuropsychologists will take advantage of the disclosure of archaic and deeply rooted behavioral functions or dysfunctions. Also, through understanding how exaptation is used by evolution to set up "new" networks, neuroradiologists exploring brain networks by new technologies (functional magnetic resonance imaging, etc.), may gain better insight into brain connectivity. The readers will themselves detect numerous other applications to their field of interest.

As an editorial team we humbly admit that our endeavor is ambitious, eclectic, and often only surface-scratching in regard to some diseases. However, we know that science proceeds by asking new questions instead of answering old ones as proposed by evolutionary psychologist Robin Dunbar (2014). With this principle in mind, we gained confidence that this book will foster new collaborations, incite new discussions, and advance clinical neurosciences in a direction heretofore uncharted and that you, as reader, may have new insight and pleasure when scrolling through the chapters.

References

Darwin, C., & Prodger, P. (1998). *The expression of the emotions in man and animals*. Oxford University Press. (Original work published 1872)

Dunbar, R. (2014). *Human evolution: A Pelican introduction*. Penguin.

Jackson, J. H. (1874). On the nature of the duality of the brain. *Medical Press and Circular, 1*, 19, 80–86.

Lorenz, K. Z. (1950). The comparative method in studying innate behavior patterns. In *Society for Experimental Biology: Physiological mechanisms in animal behavior* (Symposium IV, pp. 221–268). Academic Press.

MacLean, P. D. (1990). *The triune brain in evolution: Role in paleocerebral functions*. Springer Science & Business Media.

Mayr, E. (1961). Cause and effect in biology: Kinds of causes, predictability, and teleology are viewed by a practicing biologist. *Science, 134*(3489), 1501–1506.

Tinbergen, N. (1963). On aims and methods of ethology. *Zeitschrift für tierpsychologie, 20*(4), 410–433.

PART 1

Human Brain Evolution

From Anatomy to Function

Human Telencephalization

Felix Ströckens and Katrin Amunts

Introduction

In 1963, Leon C. Megginson, professor of business management at Louisiana State University, published an article on the economic interactions between the United States and Europe (Megginson 1963). In this article, he used concepts of evolutionary biology to describe competition in business and explained the ideas of Darwinism to his readers with a now famous quote: "It is not the most intellectual of the species that survives; it is not the strongest that survives; but the species survives is the one that is able best to adapt and adjust to the changing environment in which it finds itself." This statement summarizes perfectly one of the core principles of evolution, which is highly relevant when studying brain structure, function, and disease state.

During the course of evolution, an immense selection pressure occurred with consequences for neuroanatomical and physiological features. The challenges occurring in the world of a mouse are certainly different from the ones in the world of a monkey, and bodies, brains, and behavior of a species are optimized to survive in their individual worlds. The human brain is no exception and has adapted to cope with ecological and sociocultural challenges unfolding in the environment in which humans evolved (Simpson and Belsky 2008). The telencephalon (or endbrain), which is the largest part of the brain, containing the cerebral cortex, some subcortical nuclei, and the hippocampus, changed significantly in size and shape during evolution, with extant avian and mammalian species possessing on average a much larger telencephalic mass in absolute and relative terms in comparison to other vertebrates (Güntürkün et al. 2017; Tsuboi et al. 2018). This trend, called "telencephalization," can also be observed in primate brain evolution. However, a larger telencephalon is by far not the only difference between humans and other mammalian species.

Within this chapter, we give a brief overview of the anatomy of the human brain and then highlight some of the most important differences in macro- and microstructure in comparison to rodents and monkeys—the major mammalian model species used to investigate brain diseases (Box 1.1). We focus on the mammalian brain and its changes

> **BOX 1.1 Mammalian Model Species in Neuroscience**
>
> A large part of brain research comes from animal model species, predominantly mice and rats. The development of transgenic mouse models in conjunction with techniques like optogenetics and in vivo genome editing (e.g., CRISPR/Cas) led to a further surge in mouse studies (e.g., as animal models for brain diseases). Limitations arise because of considerable differences in brain structure, function, behavior, and the respective environment (e.g., nocturnal lifestyle, role of olfactory and somatosensory stimuli). In addition, rodents have a much shorter life span in comparison to humans, do not naturally suffer from several neuropathologies common in humans (e.g., Parkinson's), and/or show different symptoms (e.g., amyotrophic lateral sclerosis).
>
> Research in primates, mostly rhesus monkeys and marmosets, has been performed as well (e.g., studies on face and object recognition). The absolute number of such studies is currently relatively low (i.e., 2%–3% compared to rodent publications). In 2001, the first transgenic rhesus monkey was created. Today, monkey strains with mutant genes of human origin known to play a role in diseases like Parkinson's, Huntington's, and schizophrenia exist, with some of the strains showing disease-like symptoms (Liu et al. 2016). While some see this approach as a new avenue to potential cures, others perceive such experiments as critical due to the transfer of human genetic material to a host species (Feng et al. 2020). In the same line fall experiments with human organoids. These are three-dimensional cell cultures derived from human stem cells, mimicking organs or parts of organs, including structures of the human brain. Recently, human cortical organoids were transplanted into the cortex of rats. After transplantation, the organoids integrated to some extent into the rats' neuronal network, showed responses when the animal was stimulated, and could even affect the rats' behavior (Revah et al. 2022).
>
> Studies in other mammalian species are less numerous and include, for example, those in guinea pigs (auditory system), cats (visual system), and dogs (associative learning principles, relevant for, e.g., development and treatment of anxiety disorders). Furthermore, studies in marine animals like cetaceans (dolphins and whales) with very large brains, but also small shrew species with brain weights of less than 100 mg, have been performed. These species show specific adaptations, for example, a highly gyrified but very thin and mostly agranular neocortex in cetaceans or a massive reduction in the number of cortical areas in shrew species.
>
> (For more information, see, e.g., Catania et al. 1999; Butti et al. 2011; Perrin 2014; Bolker 2019.)

> **BOX 1.2 Lessons from Nonmammalian Species**
>
> A large amount of knowledge on the nervous system stems from nonmammalian species. For example, the small nematode worm *Caenorhabditis elegans* has 302 neurons, and the full connectome of these neurons has been established. This allowed researchers to link the activity of single sensory neurons and motor neurons with behavior (Cook et al. 2019). Scientists of the *Drosophila* connectome project try to model the connectivity of every single neuron in the brain of the fruit fly (about 135,000 neurons), including interindividual variance. As of 2021, the project has managed to model about half of the neurons. At its end, the project expects to have a tool to fully model and understand neuronal circuits consisting of several hundreds of neurons responsible for complex behaviors like feeding, locomotion, and sexual behavior (Hulse et al. 2021).
>
> Songbirds have evolved a complex vocalization system. The structure of songs of passerines like zebra finches or robins shows a remarkable similarity to the structure of human spoken language. Bird songs consists of several notes, which are combined into syllables, which again can be combined into motifs. The components can be more or less freely combined by the bird to create an individual song. How the components are combined is learned during a critical phase from the parents and can be further adapted by the bird based on personal experience (e.g., songs of other birds or the sounds of a regular-appearing traffic jam close to the bird's habitat). Also, the underlying neuronal network bears some remarkable similarities to the human language system, with distinct areas for song perception and song production resembling, at least to some extent, Broca's and Wernicke's areas in humans, including function, structure, and network asymmetries.
>
> (For further information on bird song, see Aamodt et al., 2020; Güntürkün et al., 2020.)

during evolution for practical reasons, although, from a research point of view, data from other vertebrate clades or species beyond the vertebrate taxon (e.g., the fruit fly where gene–behavior mechanisms have been studied; Box 1.2) are increasingly important for our understanding of the evolution of brain pathologies. Last, we discuss some implications that adaptive changes on the neuroanatomical level during evolution might have had on the occurrence of neural disorders.

A Short Summary of Mammalian and Human Evolution

When exploring possible evolutionary roots of human brain diseases, it is necessary to understand the evolutionary trajectory of the mammalian clade and the phylogenetic relationship between modern humans and the model species used to investigate these diseases. The ancestor of all recent mammals diverged from its sister clade—comprising modern reptiles and birds—about 310 million years ago (mya), giving rise to several extinct clades as well as to the mammalian lineage (Rowe 2017). The mammalian lineage branched into several further clades, including branches leading to modern rodents (about 80–60 mya) and primates (about 77 mya; Pozzi et al. 2014; Upham et al. 2019; see Figure 1.1 for an overview). The primate branch gave rise to several clades, including, for example, New- and Old-World monkeys, lesser apes, and great apes. Molecular clock data indicate that about 8 mya a new branch split from the other great apes—the hominins. Fossil records indicate that several species evolved from this lineage, with the oldest one being more than 7 million years old (*Sahelanthropus tchadensis*). However, it took another 5 million years before the first species of the *Homo* genus (*H. habilis*, 2.4–1.6 mya) appeared. Around 200,000 years ago, three new *Homo* species appeared, of which two (*H. neanderthalensis* and the Denisova human) went extinct 50,000–30,000 years ago. The remaining species is *Homo sapiens*, the modern human (Pozzi et al. 2014; Beaudet et al. 2019), who, even though being the last extant *Homo* species, still carries genetic information from Neanderthals and Denisovans, gained by cross-breeding events, which might affect our brain structure, function, and associated diseases even today (Kuhlwilm et al. 2016; see also Chapters 5 and 6).

The Bauplan of the Human Brain

The human brain has changed during the course of evolution but still follows the bauplan of the vertebrate brain in general and the mammalian brain layout in particular. Like all vertebrate brains, the human brain can be subdivided into five main subdivisions as defined by their appearance during ontogeny, folding pattern, and gene expression (Ishikawa et al. 2012). The most caudal part is the myelencephalon or medulla oblongata, connecting the brain to the spinal cord and containing, besides some nuclei involved in autonomic functions like breathing and heart rate control, mostly ascending and descending fibers of sensory or motor structures, respectively. The metencephalon joins rostrally to the myelencephalon and consists of the pons, parts of the brainstem, and the cerebellum. The pons contains many fibers (e.g., connections to the cerebellum), and its rostral margin marks the border to the mesencephalon. The brainstem of the metencephalon contains the main nuclei of the serotonergic (parts of the raphe nuclei, with the others being located in myelin and mesencephalon) and noradrenergic (locus coeruleus) transmitter systems, involved, for example, in the pathomechanisms of schizophrenia, depression, and Alzheimer's disease. The cerebellum is involved in motor coordination, learning, and fine-tuning. Although it takes only approximately 20% of the total brain volume, it comprises about 80% of the 86 billion cerebral neurons (Herculano-Houzel et al. 2015; Table 1.1). The mesencephalon contains, with the substantia nigra and the ventral tegmental area, core elements of the dopaminergic system, with high

CHAPTER 1 HUMAN TELENCEPHALIZATION | 13

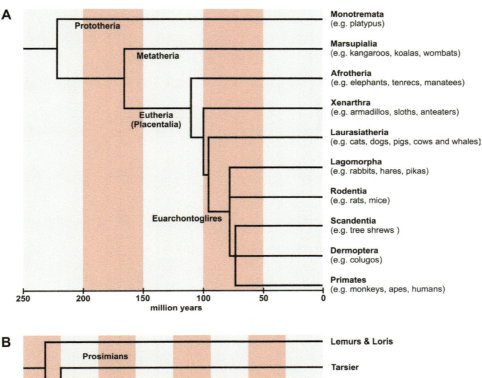

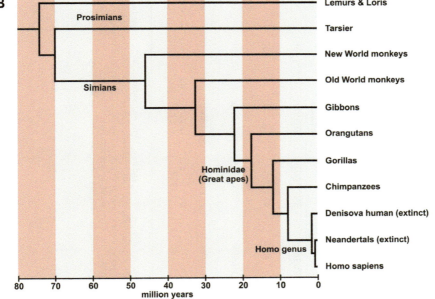

Figure 1.1 Simplified phylogenetic trees for the mammalian class (A) and the primate order (B). Trees are based on data from Pozzi et al. (2014), Kaas (2017b), Rowe (2017), Upham et al. (2019), and Beaudet et al. (2019). Note that there is some uncertainty on the exact time frame for each radiation and that the positioning of some of the shown clades (like Scandentia and Demoptera in [A]) is still under debate.

TABLE 1.1 Selection of Neuroanatomical Measures for Common Mammalian Model Species

Species	Body mass (g)	Brain volume (mm³)	Neuron numbers in whole brain (million)	Cortical volume (mm³)	Cortical surface (mm²)	Average cortical thickness (mm)	Gyrification index[a]	Neuron numbers in cortex (million)	Neuron densities cortex (n/mg)
Human (*Homo sapiens*)	70,000	1,395,000	86,000	980,000	1,890,000	2.7	2.99	16,340	13,500
Chimpanzee (*Pan troglodytes*)	55,400	404,000	28,000	238,000	634,000	2.5	2.3	7,400	19,300
Rhesus macaque/monkey (*Macaca mulatta*)	10,400	90,000	6,376	55,000	212,000	2.1	1.75	1,710	24,470
Rat (*Rattus norvegicus*)	292	1,970	188.9	716	644	1.2	1.02	31.0	41,000
Mouse (*Mus musculus*)	40	508	67.9	170	348	0.9	1.03	13.6	79,000

Nonhuman primate and human brain/cortex volumes: (Rilling and Insel 1999; Hayashi et al. 2021); mouse brain/cortex volumes: (Badea et al. 2007); rat brain/cortex volumes: (Vetreno et al. 2017); mouse brain/cortex volume: (Welniak-Kaminska et al. 2019); human, chimpanzee, rhesus monkey cortical surface: (Hayashi et al. 2021); mouse cortical surface: (Badea et al. 2007); human, chimpanzee, rhesus monkey cortical thickness: (Hayashi et al. 2021); rat cortical thickness: (Vetreno et al. 2017); mouse cortical thickness: (Lerch et al. 2008); human, rhesus monkey, mouse rat neuron numbers: (Herculano-Houzel et al. 2015); chimpanzee neuron numbers: (Collins et al. 2016); chimpanzee neuron densities: (Young et al., 2013); all gyrification index data: (Pillay and Manger 2007). Note that the gyrification data is based on the calculation method of Zilles et al. (1988).

relevance for many neurological disorders, including Parkinson's (compare Chapter 9). Furthermore, it contains the superior and inferior colliculi as visual and auditory relay nuclei, respectively. Rostrally, the mesencephalon merges with the diencephalon, comprising the thalamus, epithalamus, and hypothalamus. While the thalamus is the central relay station for the cortex, the epi- and hypothalamus mostly serve vegetative functions.

The telencephalon is by far the largest part of the human brain, encasing most of the diencephalon and parts of the mesencephalon. It can be further subdivided into the pallium (mantle) and subpallium. The largest structure of the subpallium is the striatum, consisting of the nucleus caudatus and putamen, followed by the globus pallidus, nucleus accumbens, olfactory tubercle,[1] and ventral pallidum. These structures are often, together with the mesencephalic substantia nigra and diencephalic nucleus subthalamicus, summarized as basal ganglia. The basal ganglia serve as a hub region for the control and adjustment of the majority of all motor functions and are also involved in learning and memory processes like conditioning, procedural learning, and memory, as well as reinforcement learning and decision-making. Many neural disorders with motor symptoms (e.g., Parkinson's, amyotrophic lateral sclerosis) affect parts of the basal ganglia (Cox and Witten 2019).

In mammals, the pallium consists of the layered cerebral cortex and pallial nuclei like the claustrum and amygdala.[2] The latter is involved in the processing of emotions like fear, anxiety, and aggression, as well as sexual behavior. Many anxiety and sexual disorders, as well as depression and Parkinson's disease, are related to impaired or altered amygdala functions, rendering it a prime target for neuropathology research (Baird et al. 2007; Janak and Tye 2015; Diederich et al. 2016; Palomero-Gallagher and Amunts 2022) The olfactory bulb, the primary olfactory region, receives input from the olfactory mucosa and sends efferents to many pallial areas and the amygdala. Recent studies indicate that SARS-CoV-2 can enter the brain via the olfactory mucosa, infecting olfactory bulb neurons, and from there spreads over wide parts of the brain, likely causing short- and long-term symptoms (Meinhardt et al. 2021).

Based on cytoarchitecture, the cerebral cortex can be subdivided into the allocortex and isocortex. The isocortex forms, with about 90%, the largest part of the cortical sheet, consisting, with few exceptions, of six cell layers with radially arranged neurons and is considered to be phylogenetically younger (Zilles and Amunts 2012). The isocortex, evolved with the onset of early mammals, is also called the "neocortex" (Rakic 2009). In contrast, the allocortex has a different number of layers (e.g., three in the hippocampus and up to nine in the entorhinal cortex) and was already present in the reptilian ancestor of mammals, with homolog structures (e.g., hippocampus) in modern reptiles and birds. It can be further subdivided into the paleocortex, containing mostly areas related to olfaction (like the olfactory bulb, anterior olfactory nucleus, and piriform cortex), and the archicortex, containing the hippocampus. The hippocampus is crucial for encoding and retrieval of declarative memories and is highly involved in orientation and navigation in space, with hippocampal lesions often causing amnesia and impairment of spatial memory (Knierim 2015). The hippocampus borders the subicular complex with the pre- and parasubiculum, the entorhinal cortex, the retrosplenial cortex, and parts of the cingulate cortex (see Zilles and Amunts 2012 for a more detailed review on different cortex types). The entorhinal cortex is a relay region connecting the neocortex and the hippocampus, and it is involved in memory processing and the analysis of spatial representations. It is one of the first locations revealing neurofibrillary tangles during early stages of Alzheimer's disease (Braak stage I), before tangles spread to other parts of the cortex and can thus be used as a diagnostic criterion (Braak and Braak 1991; Garcia and Buffalo 2020; see also Chapter 8).

The cerebral cortex is subdivided into areas; 43 were listed in the historical map of Brodmann (1909), who also emphasized that cortical areas can be found in all mammals. He analyzed cortical parcellation schemes across species, including prosimians, Chiroptera, Carnivora, Rodentia, insectivores, and marsupials, and identified similarities and dissimilarities in cytoarchitectonics underlying functional systems (Brodmann 1909). In addition to cytoarchitectonic differences, he found differences in the number of areas. For example, the frontal lobe of Hapalidae (New-World monkeys, e.g., marmosets) consists only of three areas, while the same region in humans has five areas more (Brodmann 1909). Brodmann was convinced that each area serves a certain function and that cytoarchitecture and myeloarchitecture, as investigated by Cécile Vogt and Oskar Vogt (1919), go hand in hand (see Chapter 18).

Today, it is widely acknowledged that even more areas exist. A recent classification scheme based on multimodal Magnetic Resonance Imaging (MRI) proposes a range of 180–200 areas (Glasser et al. 2016). The number of areas defined by the Julich-Brain cytoarchitectonic maps is in the same range (Amunts et al. 2020). Going beyond Brodmann's map, the Julich-Brain cytoarchitectonic maps also capture differences between different brains (i.e., consider intersubject variability) and provide true three-dimensional information. That makes the maps a powerful tool to interpret findings of in vivo neuroimaging studies in healthy subjects and patients on a sound microstructural basis (Amunts et al. 2020). Recent cytoarchitectonic studies also showed that the borders of the majority of areas do not coincide with gyri/sulci, including borders of higher associative areas, which renders brain macroscopy alone an imprecise predictor of microstructural organization. Cytoarchitectonic maps have been applied to elucidate the progression of disease in patients with Parkinson's, extending findings of postmortem studies and our knowledge on Braak stages (Pieperhoff et al. 2022; compare Chapter 9). A comparison of interindividual variability shows differences in localization and extent between the areas (Amunts et al. 2020). Interestingly, interindividual variability seems to be lower in areas considered phylogenetically older, like the hippocampus, amygdala, and entorhinal cortex, as well as in subcortical nuclei, while neocortical areas show a larger degree of variance (Amunts et al. 2005).

Macroscopically, the cerebral cortex (including both allo- and neocortical areas) can be subdivided into six lobes: the frontal, parietal, temporal, and occipital lobes, visible from the outside, as well as the insular and limbic lobes, encased by the frontal, parietal, and temporal lobes. The lobes differ between mammalian species in size and gyrification. The macroscopic structure is helpful for general orientation and is, especially in a comparative context, necessary due to the lack of detailed cytoarchitectonic maps in most mammals. For a more detailed review of the macroscopical structures of the human cortex, see, for example, ten Donkelaar et al. (2018). In the following, the basic layout and most important functions of the human cerebral lobes will be described.

The **frontal lobe** is separated by the central sulcus from the parietal and by the lateral sulcus (or Sylvian fissure) from the temporal lobe. The areas of the frontal lobe can be roughly subdivided into two functional units: the posterior areas are involved in motor functions. The primary motor cortex occupies the anterior wall of the central sulcus and posterior aspects of the precentral gyrus and is responsible for the execution of voluntary movements. Rostrally, premotor (at the lateral surface) and supplementary (at the mesial surface) motor cortices are found, which are involved in planning of movements, selection of action sequences, and motor learning (Luppino and Rizzolatti 2000; Nachev et al. 2008). More anteriorly, areas of the prefrontal cortex (PFC) are situated, which serve a large variety of higher cognitive functions. Areas for language processing and speech production are located in the ventral PFC, at the inferior frontal gyrus and neighboring regions, and are often termed "Broca's region" (lateralized to the left side in most cases). Areas of the dorsolateral PFC participate in functions like social behavior and personality formation, executive functions, planning, and decision-making, as well as working and long-term memory (Fuster 2015).

Posterior to the central sulcus lies the **parietal lobe**, separated by the parieto-occipital sulcus from the occipital lobe. The anterior parietal lobe is mainly involved in somatosensory functions: the primary somatosensory area (S1) receives information on tactile stimuli from the body surface via the thalamic ventral posterolateral nucleus. Adjacent to S1, within the parietal operculum (parietal part of the wall of the lateral sulcus, covering the insular cortex), lies the secondary somatosensory area (S2), receiving projections from S1 and possibly direct input from the thalamus. Besides processing of tactile information, S2 seems also to be involved in cognitive processing of touch and somatosensory memory. In contrast, the posterior parietal cortical regions serve as associative integration areas combining visual, motor, and touch information (Ackerley and Kavounoudias 2015).

The **occipital lobe** can be separated at the mesial surface by the parieto-occipital sulcus from the parietal lobe, while the transitions from occipital to temporal/parietal lobes on the superolateral surface follow an imaginary line between the superior aspects of the parieto-occipital and the preoccipital notch. The occipital lobe is mostly associated with visual processing. The wall of the calcarine sulcus and adjacent parts of the cuneus and lingual gyrus contain the primary visual cortex (V1), receiving input from the retina via the thalamic lateral geniculate nucleus. V1 is surrounded on both superior and inferior sides by secondary visual areas (V2–V4), further processing visual stimuli and relaying them to visual association areas in the parietal and temporal lobes (Freud et al. 2016).

The **temporal lobe** contains the transverse/Heschl's gyri with the primary auditory cortex. It receives auditory input from the inner ear via the medial geniculate nucleus of the thalamus. This core region is surrounded by further auditory responsive regions, forming belt-shaped areas around the core. Higher auditory and associative areas cover the planum temporale as well as the posterior superior temporal sulcus and gyrus. These areas process specific qualities of acoustic stimuli and relay them to associative auditory areas in the temporal lobe (Moerel et al. 2014; Zachlod et al. 2020). Especially noteworthy in humans is Wernicke's region in the posterior superior temporal gyrus, which is involved, among other things, in comprehension of spoken and written language (Friederici 2011). Anatomically, Wernicke's region is not sharply defined. Further examples of functions associated with the temporal lobe comprise identification and categorization of colors, objects, gestures, and faces (Lafer-Sousa et al. 2016; Papeo et al. 2019).

The **limbic lobe** forms a ring around the corpus callosum and the brainstem and is visible from the mesial side of the brain. The large cingulate gyrus surrounds the corpus callosum. Clockwise follow the parahippocampal gyrus and the uncus, consisting of the ambient, semilunar, and uncinate gyri. These gyri contain the entorhinal, piriform, perirhinal, and retrosplenial cortices as well as the hippocampal formation. The ring is closed by the subcallosal gyrus, next to the anterior aspects of the cingulate cortex. The structures of the limbic lobe form together, with other cortical areas, subpallial structures (most notably the amygdala), nuclei of the diencephalon, and the olfactory bulb functional networks called "limbic system(s)." These networks partly overlap but can be roughly divided into networks centered around (1) olfaction, memory, and spatial information; (2) integration of visceral sensation and emotion with semantic memory and behavior; and (3) autobiographical memories and introspective, self-directed thinking (Catani et al. 2013).

In the depth of the Sylvian fissure lies the **insular lobe** (insula), separated from the other lobes by the circular sulcus. The insula contains parts of the primary gustatory cortex and is a cortical hub, maintaining reciprocal connections to sensory areas of all modalities, as well as parts of the limbic system. It receives neuromodulatory input from the brainstem. Insular functions are even more heterogenous than functions in the other cortical lobes, but there is evidence of insula involvement in salience detection, risk and reward assessment, and emotional valence (Gogolla 2017).

Differences in Pallial Layout and Architecture between Human and Nonhuman Model Species

Even though the brain of all mammalian species follows a general bauplan (Figure 1.2), areas or even the whole brain can vary considerably between species. Furthermore, functions dedicated to a specific area or network in one mammalian species might have a different function in another. And, vice versa, similar functions may rely on different

CHAPTER 1 HUMAN TELENCEPHALIZATION | 19

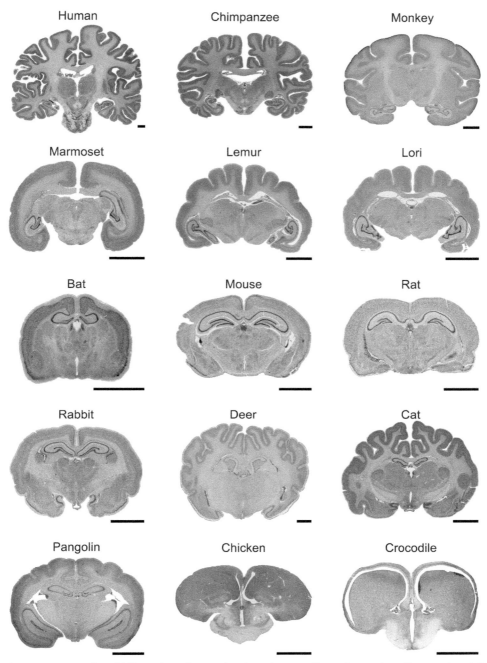

Figure 1.2 Examples of different brain layouts in selected mammalian, avian, and reptilian species. Cell body–stained sections with either Cresyl violet or Merker silver staining. Magnification of brains in images differs between species to highlight organizational differences. Note the differences in size (compare scale bars, indicating 5 mm for all species except for bat and mouse, where they indicate 2.5 mm) and gyrification between different brains and the complete lack of gyrification in avian and reptilian species with a lack of a neocortex. Species are arranged by increasing phylogenetic distance from humans: human (*Homo sapiens*), chimpanzee (*Pan troglodytes*), monkey (*Macaca fascicularis*), marmoset (*Callithrix jacchus*), lemur (*Lemur catta*), loris (*Loris tardigradus*), bat (*Miniopterus schreibersii*), mouse

(networks of) areas. In the following, some interspecies differences and similarities will be introduced to give a flavor of interspecies variation (see Table 1.1 for a summary of basic neuroanatomic differences).

Brain Size and Number of Neurons

One of the most striking differences between primate and other mammalian brains is the higher brain volume in relation to body size. On average, primate brains are about twice as large as the brain of a nonprimate species with a comparable body size (van Dongen 1998). Also, within the primate order, brain/body scaling varies. While the phylogenetically older prosimians (e.g., lemurs, lorises) possess brain volumes about two times larger than rodents of similar body weights, simians (i.e., New-World monkeys, Old-World monkeys, and apes) show relative brain volumes that are even 1.5 times larger than those of prosimians. The human brain is up to four times larger than would be expected for a brain of a similar-sized monkey or nonhuman ape (Halley and Deacon 2017; Table 1.1). This expansion of volume is especially visible in the neocortex. Prosimians have larger neocortices than rodents, while simians have larger neocortices than prosimians (Halley and Deacon 2017). Interestingly, the human neocortex/rest of brain volume ratio differs only marginally from other simians. However, there are considerable differences in individual neocortical areas between nonprimates, nonhuman primates, and humans.

Differences in neuron numbers likely indicate differences in processing capacities, especially when compared to brain weight. Over all mammals, the general distribution of neurons within the brain is relatively similar. The majority of neurons are found in the cerebellum (75%–80%) and the telencephalon (15%–20%; Herculano-Houzel et al. 2015). In comparison to other mammalian species (including rodents), primate brains gain more neurons when increasing in size. Thus, a primate has higher neuron densities than a nonprimate with a similar-sized brain.[3] In contrast to relative brain volume measures, the scaling of neuron numbers and brain volume seems to be constant among primates (i.e., the human brain is "just" a normal upscaled primate brain regarding neuron number). Still, given the enlarged cerebral and cortical volumes, humans hold the highest absolute number of neurons among all primates for both the whole brain and the neocortex (Herculano-Houzel 2012).

Taken together, the brains of primates are more encephalized and more telencephalized and contain more neurons in comparison to rodents. Humans differ from nonhuman primates in an even larger encephalization but show, at least in comparison

---

Figure 1.2 (Continued)

(*Mus musculus*), rat (*Rattus norvegicus*), rabbit (*Oryctolagus cuniculus*), deer (*Capreolus capreolus*), cat (*Felis silvestris*), pangolin (*Phataginus tricuspis*), chicken (*Gallus gallus domesticus*), and crocodile (*Crocodylus niloticus*). Brain sections are part of the Vogt Collection and Stephan Collection of the Cécile and Oskar Vogt Institute for Brain Research, Heinrich Heine University Düsseldorf. We would like to thank Dr. Manuel Marx and René Hübbers for their help retrieving and scanning the sections.

to monkeys and apes, neither a larger neocortex nor larger neuron numbers as would be expected from their brain size.

Features of Cells and Cellular Composition

Comparative analysis of features like neuron and glia morphology, neuron physiology, and local neuronal circuit layout, as well as gene or protein expression patterns (see Chapter 5), have resulted in new insights into brain evolution (for reviews, see Vasile et al. 2017; Schmidt and Polleux 2021). As an example, pyramidal neurons represent one of the most abundant cell types of the cortex and are found in all mammalian species. However, their precise morphology does differ. Human pyramidal neurons are on average bigger, have larger and more branched dendritic trees, and form up to 40% more synapses in comparison to both mice and monkeys (Elston et al. 2001; Benavides-Piccione et al. 2002; Sherwood et al. 2020). While this could be partially explained by the large thickness of the human cortex (Sherwood et al. 2020), they also have some unique features regarding dendritic morphology (Mohan et al. 2015; Figure 1.3). These morphological differences are accompanied and possibly even related to differences in dendritic physiology. Human layer II/III pyramidal neurons have a lower specific membrane capacitance than mice, resulting in a lower depolarization charge required for a somatic potential (Eyal et al. 2016). Furthermore, it was recently shown that a subset of these neurons can generate graded,

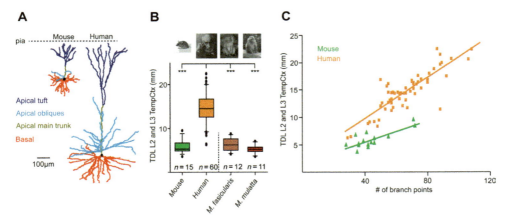

Figure 1.3 Even though mouse and human pyramidal neurons seem to appear, besides size differences, rather similar, human pyramidal neurons are not just upscaled mouse neurons. Human pyramidal neurons in layers II and III of the temporal cortex have larger and more branched dendritic trees, especially when comparing different sections of the dendritic tree (A). A difference in the size of the dendritic tree can even be observed when comparing human pyramidal neurons with neurons from monkey species, even though the monkey cortex is much thicker than the mouse one (B; compare Table 1.1). With increasing length, dendrites in humans also tend to branch more in comparison to those in mice, visible from the steeper regression line for humans (C). These dendritic differences coincide with further structural differences, like higher synapse densities in humans, and functional differences, like a higher membrane capacitance in humans in comparison to mice (see main text). TDL = total dendritic length; TempCtx = temporal cortex. Figure reproduced from Mohan et al. (2015) with permission of the publisher and author.

calcium-dependent dendritic action potentials, which have not yet been described in any other species. These graded potentials enable coding properties on a dendrite level, which were thought to be only possible in multilayer neural networks (Gidon et al. 2020). Since dendrite morphology and physiology have been associated with cognitive performance in humans (Genç et al. 2018; Goriounova et al. 2018), it is likely that these human-specific evolutionary adaptations are relevant for the high cognitive capabilities of *Homo sapiens*.

Another neuron type, which has been claimed to be unique for humans, are von Economo neurons (VENs). First described by Santiago Ramón y Cajal and Constantin von Economo, these neurons, with a very thin and elongated cell body and a distinct axon and dendrite morphology, only appear in layer Vb of the anterior cingulate cortex and frontal insular cortex (Banovac et al. 2021). While the function of VENs and their projection targets are still unknown, their morphology is altered in neuropathologies mostly linked to humans like Alzheimer's disease, autism spectrum disorder, or Parkinson's disease. However, comparative studies have found VENs in great apes and monkeys and cells highly resembling VENs in other mammalian species. Further, a recent transcriptomic analysis found a high homology to extratelencephalic projection neurons in mice, making it likely that VENs are not uniquely human but a "morphological diversification of an evolutionarily conserved cell type" (Hodge et al. 2020).

Not only neurons differ in morphology and physiology between species but also glial cell types, for example, protoplasmatic astrocytes. Protoplasmatic astrocytes are involved, among other functions, in the formation of the blood–brain barrier, metabolic support of neurons, and extracellular ion homeostasis. Furthermore, they are excitable by neurotransmitters and are connected via gap junctions, allowing for calcium-dependent propagation of signals within astrocyte networks (i.e., calcium waves). In comparison to rodents, human protoplasmatic astrocytes are threefold larger (i.e., longer and more branched) and cover more neuronal synapses and blood vessels. In addition, calcium waves of human astrocyte networks propagate more than five times faster than in rodents. The implications of these differences are currently not well understood, and further research is needed on how these differences could contribute to interspecies differences in physiology and behavior (Vasile et al. 2017; compare Chapter 2).

Gyrification

One of the most characteristic features of the human brain is the highly convoluted surface with dozens of gyri and sulci. The role of gyrification is discussed in the context of different neuropsychiatric diseases, for example, schizophrenia or bipolar disorder (Sasabayashi et al. 2021), as well as in cortical malformation in neurodevelopmental disorders (see Chapters 14 and 15).

Gyrification increases the surface area of the cortical sheet and can be found in many mammalian species. However, the strength of gyrification varies among mammals (Figure 1.4). The gyrification index (GI) of mice (outer cortical surface and sulcal surface divided by outer cortical surface), as a measure for the degree of gyrification, is about

1.03, while the GI of rhesus monkeys is 1.75. Chimpanzee and human brains are even more gyrified, reaching GIs of 2.3 and 2.99, respectively (Pillay and Manger 2007). While the GI in humans does weakly correlate with cognitive performance, especially in frontal cortical areas (Gautam et al. 2015), a general association of GI to cognitive capabilities over all mammals is questionable. Instead, data from several studies indicate a strong correlation of GI to brain size, culminating in respective GIs of approximately 3.8 and 5.7 in the large-brained African elephant and killer whale (Manger et al. 2012). Still, a comparison of the gyrification pattern between closely related species can be informative. As an example, the lateral temporal lobe of rhesus monkeys consists only of two gyri separated by a longitudinal sulcus. While the superior gyrus is mostly involved in auditory functions, the inferior gyrus comprises associative visual areas. In chimpanzees, an additional longitudinal sulcus separates the temporal lobe in three rather straight gyri, while the three lateral temporal gyri in humans are strongly convoluted, resulting in a temporal surface area much larger in absolute terms but also in relation to the rest of the cortex even in comparison to chimpanzees. Although there has been a certain level of reorganization and relocalization of temporal areas in human evolution, the separation into a dorsal auditory and a ventral visual-associative temporal lobe present in monkeys persists. However, the newly gained cortical surface area, for example, the middle temporal gyrus, seems to be involved in functions which either are human-specific or provide humans a certain advantage over other primates, like semantic and social processing, including aspects of language and theory of mind (Braunsdorf et al. 2021).

Cortical Parcellation

Human brains also differ in the layout, volume, and number of cortical areas from both rodents and nonhuman primates. While there are also certain differences in brainstem and cerebellar structures (see, e.g., Baizer 2014), we here focus on interspecies differences of the neocortex.

Visual Areas

A major difference between rodent and primate species is the reliance on different sensory systems to guide behavior and corresponding adaptive changes of the neural substrate as visible in the visual system. While many vertebrate species have four cone opsin types, the majority of mammals lost two cone opsins during the course of evolution and are, therefore, dichromatic, resulting in a reduced ability to perceive color (Jacobs 2009). Thus, especially nocturnal mammals, like rats and mice, rely more on their olfactory and somatosensory systems. Primates, on the other hand, (re)gained an additional cone pigment around 30–40 mya. Therefore, many extant diurnal primate species are trichromatic and have good color vision. This and other adaptations of the peripheral visual system, like more frontally placed eyes with a resulting larger binocular field of view or a single, deep fovea densely packed with photoreceptors, allowing high spatial resolutions, are accompanied by distinct changes in cortical and subcortical areas. In comparison to many mammalian species, including rodents, larger portions of the primate neocortex are involved in visual processing, spanning from areas in the occipital lobe to temporal and parietal areas. These areas are not only larger in size but also larger in number and show a higher complexity and specialization in comparison to rodent visual areas (Kaas 2012; Kaas and Balaram 2014). Among others, areas like the middle temporal area (MT/V5) or medial superior temporal area, known to play a major role in motion perception, seem to be unique for primates. Furthermore, primates exhibit a larger posterior parietal cortex, and frontal areas involved in visuomotor functions (e.g., frontal eye field) are more strongly developed (Kaas and Balaram 2014).

The comparison of primate brains also reveals considerable differences between visual cortical areas. Besides adaptations to specific living environments, like a reduction of visual area size and complexity in nocturnal primates, prosimians have, in general, fewer visual cortical areas than Old-World monkeys, which have a lower number of visual cortex areas than humans (Kaas 2017). The latter difference is relevant for clinical neuroscience since the rhesus monkey (and to a lesser degree the marmoset) serves as the model species (e.g., to study cortical blindness; Goebel et al. 2001). While most primary and secondary visual areas like V1–3 and V5/MT are present in both monkeys and humans (Vanduffel et al. 2001) and are thought to be homologs (with notable exceptions like V4), the systems become less similar when entering the parietal lobes. The posterior parietal lobe contains visual association areas which are, in conjunction with parts of the occipital lobe, summarized under the term "dorsal visual stream" ("where" stream). The areas of this stream are responsible for executing movements under visual control, to process object locations, and to perceive three-dimensional objects (Freud et al. 2016). Among other things, the dorsal stream comprises caudal parts of the inferior parietal lobule, a region which is much larger and more parcellated in human brains in comparison to monkeys. It is involved in several functions thought to be better developed in humans like numerosity and tool use (Van Essen and Dierker 2007; Caspers et al. 2013; Niu et al. 2021).

Several areas of the human temporal lobe and adjacent occipital regions form the ventral visual stream, involved in identification and recognition of objects ("what" stream). The ventral stream is organized in functional strips and patches with stimulus preferences such as color, shape, places, body parts, or faces (de Haas et al. 2021). The relative position and functional characteristics of these strips and patches are highly similar between humans and monkeys. However, due to the emergence of additional temporal auditory and language regions during human evolution (see below, "Auditory/Vocalization Areas"), these areas moved from the lateral temporal surface in monkeys (e.g., inferior occipital and inferior temporal gyrus) to more basal aspects of the temporal lobe (e.g., fusiform and parahippocampal gyrus) in humans (Lafer-Sousa et al. 2016).

Auditory/Vocalization Areas

The general layout of the primary and secondary auditory areas in rodents and primates is similar, and these areas are considered homologs. The central core region receives auditory inputs from the thalamic corpus geniculatum mediale. The core region can be subdivided into different areas, which can vary even between species of the same clade, depending on the range of sound frequencies perceived (e.g., rodents have an ultrasound field, which is not present in primates). The core region in both rodents and primates is surrounded by areas forming the so-called belt and parabelt regions. These areas are tonotopically and hierarchically organized, with areas on higher hierarchy levels processing more complex stimulus properties (Romanski and Averbeck 2009; Geissler et al. 2016). However, the belt and parabelt organization of the human brain is an ongoing topic of research, and its exact homologies to the monkey brain are still a matter of discussion (Zachlod et al. 2022).

Auditory association areas on the adjacent superior and middle temporal gyrus are enlarged in humans compared to monkeys (about 30 times larger) and chimpanzees (about three times larger; Van Essen and Dierker 2007). It is commonly accepted that this increase in size is correlated to the appearance of human language since functional studies indicate that these newly gained areas are involved in the processing of not only spoken but also written language. This includes analysis of spectrotemporal features of sounds to identify phonemes and syllables and of pitch changes in sounds to identify prosody (Bhaya-Grossman and Chang 2022). These areas form, together with the presumed Wernicke's region in the posterior superior temporal gyrus and Broca's region in the inferior frontal gyrus, the core network of the human language. Cytoarchitectonics and nonlinguistic functional properties indicate that nonhuman primates possess smaller, homolog regions to Broca's and Wernicke's regions. Moreover, the connectivity also differs between species: the arcuate fasciculus connecting Broca's with Wernicke's region is much larger in humans and comprises more prominent connections to the superior, middle, and inferior temporal gyrus (Rilling et al. 2008; compare Chapter 18).

Olfactory Areas

The olfactory system of primates is well developed and sensitive for many different odorants but differs from that of rodents, which have a much higher reliance on olfactory cues in their life. In mice, the olfactory bulb makes up about 2% of the total brain volume, compared to only 0.01% in humans (McGann 2017). It is debated whether primates possess fewer genes coding for olfactory receptors, possibly resulting in weaker abilities to discriminate odors. Primates also seem to lack an accessory olfactory system, including a functional vomeronasal organ, which is present in rodents to detect liquid, low-volatility odorants, like some rodent pheromones (McGann 2017). While many studies have investigated olfactory areas in rodents, data in primates are relatively rare; but available data indicate that areas are relatively similar in terms of connectivity, function, and structure, even though absolute size and localization might have changed due to reorganization of the brain during evolution (Zhou et al. 2019). A major difference between the two species groups is the lack of adult neurogenesis in primates in the olfactory bulb. While both species groups show a constant renewal of olfactory receptors embedded in the nasal mucosa, only in rodents do newly proliferated neurons migrate from the subventricular zone of the lateral ventricles into the olfactory bulb to constantly replace gamma-aminobutyric acid-ergic interneurons. It has been shown that suppression of this neurogenesis alters the receptor profile, and thus possibly also impairs olfactory guided behavior (Lothmann et al. 2021). Since the earliest symptoms of several neurological pathologies (e.g., schizophrenia) are characterized by a loss of function or alteration of olfactory discrimination and since rodents are the standard model to investigate the causes of these diseases, this difference in neurogenesis might be of more importance than visible at first glance.

Somatosensory Areas

Mammalian species possess a primary sensory neocortical area (S1), which seems to be homologous among them. At the same time, size and layout of S1 are more variable than the layout of primary areas of the other sensory modalities. The reason is that S1 contains a direct representation of cutaneous receptors and thus is also a representation of the body surface, which differs between species. In rodents, S1 occupies a relatively larger area of the cortical surface than in primates, and about 50% of S1 is involved in processing input from vibrissae and perioral structures, underlining the importance of facial somatosensation in these nocturnal animals. In primates, S1 is relatively smaller, with large portions devoted to the skin of the palms and lips. Furthermore, primates newly evolved several additional primary somatosensory areas adjacent to S1 within the anterior parietal lobe, processing cutaneous and proprioceptive inputs. Additionally, associative somatosensory areas differ between primates and rodents. Within the posterior parietal cortex, visual, auditory, and tactile inputs converge with afferents from the motor cortex. One function of the posterior parietal cortex is to guide hand movements under the control of sensory, and here mostly visual, input (O'Connor et al. 2021). Given the sophisticated manipulation properties of the primate hand, it is not surprising to find a larger cortical area devoted to this task in this clade. Even though rodents do possess a comparable, possibly homologous area, this area is rather small and seems not to receive input from sensory modalities outside somatosensation.

Prefrontal Areas

The prefrontal cortex (PFC) includes associative areas that are involved in functions often summarized as executive functions or cognitive control (e.g., attention, working memory, and decision-making; for a review, see Fuster 2015). Anatomically, PFC areas in human brains are defined by a distinct granular layer IV, input from the mediodorsal nucleus of the thalamus, and a relatively strong dopaminergic innervation. However, these criteria do not fully apply to other species. In monkeys, for example, the mediodorsal thalamus projects also into motor areas, while other cortical structures receive much higher dopaminergic input than the PFC. In mice, cortical areas rostral to the motor cortex lack even the defining cytoarchitectonic feature of a granular layer IV. This finding gave rise to the debate over whether mice actually have a PFC, even though they possess cognitive traits that would require an intact (granular) PFC in humans and even though these traits are processed by frontal areas rostral to the motor cortex (Carlén 2017; Preuss and Wise 2022; Vogel et al. 2022). Recent functional MRI data indicate that the PFC in both species possesses a very similar functional connectivity, supporting the idea of two different, and possibly analogous, areas serving identical, or at least similar, functions (Schaeffer et al. 2020).[4] Indeed, evidence has been provided that the granular PFC is a unique feature in primates, which possibly took over functions from the agranular PFC in nonprimate species (Preuss and Wise 2022). The comparison of monkey and human brains shows that the human PFC is much larger in absolute terms but also in relation to the rest of the frontal lobe (Figure 1.4). In monkeys, the PFC makes up about 50% of the frontal lobe, while the human PFC covers about 80% of the frontal lobe volume. A major hotspot of evolution seems to be the lateral PFC, which is, in addition to the middle temporal gyrus and inferior parietal lobule, one of the regions with the largest expansion (Van Essen and Dierker 2007). The dorsolateral PFC is involved in working memory, executive control, mood regulation, decision-making under conflicting information, and cognitive aspects of theory of mind (see Hertrich et al. 2021 for a recent review on dorsolateral PFC functions). It can be delineated from the ventrolateral PFC by the inferior frontal sulcus, a landmark that has been conceptualized as a border between two different functional systems. However, a recent study has shown that this sulcus itself encompasses a number of distinct cortical areas that may form a third domain of the PFC, in addition to the dorsolateral and ventral PFC (Ruland et al. 2022). The ventral PFC of the inferior frontal gyrus is involved in language-related functions (Broca's region). Given the importance of these processes in social interactions, it is arguable that the expansion of the lateral PFC in humans occurred in interplay with the development of more and more complex social relationships, likely also driven by the emergence of language.

Brain Asymmetries

Several behaviors and brain functions in humans are lateralized, meaning that one task or function is predominantly executed or processed by either the left or the right brain hemisphere. Most notable examples are handedness (with about 80%–90% of the population being right-handed) and language production/processing, which is lateralized in around 95% of the population to the left hemisphere (Güntürkün et al. 2020). Brain asymmetries are not unique to humans. Instead, they occur in all vertebrate clades and even in invertebrates, including, for example, limb preferences, although usually not as strong or persistent as in humans (for review, see Ocklenburg et al. 2013; Ströckens et al. 2013). For instance, chimpanzees exhibit handedness, and their brains show a higher cortical thickness in the right precentral gyrus on the population level (Xiang et al. 2020).

In most humans, the right frontal and the left occipital lobes are protruded as seen in horizontal MR images, giving the impression that the brain has a counterclockwise torque. At the microscopical level, left–right differences have been demonstrated (e.g., for area 44 of Broca's region; for a review, see Amunts 2010). However, the association

28 | Part 1 Human Brain Evolution: From Anatomy to Function

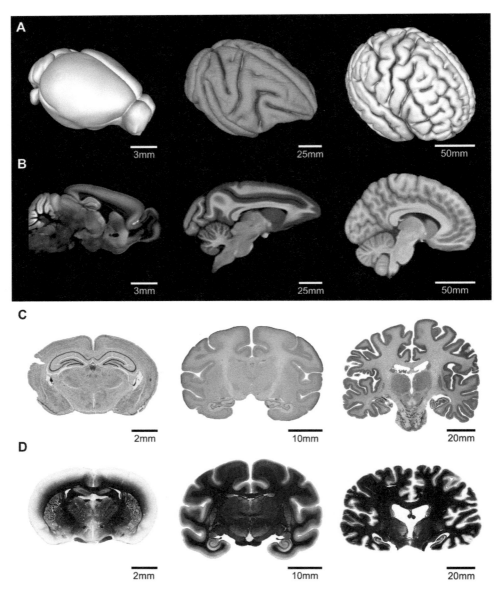

Figure 1.4 Digital reconstruction (A), sagittal magnetic resonance (MR) images (B), histological sections stained for cell bodies with Cresyl violet/Merker silver stain (C), and histological sections stained with Gallyas silver staining for fibers (D) of the mouse (first column), monkey (second column), and human brain (last column). Pictures in (A) and (B) taken from the EBRAINS atlas database of the Human Brain Project (www.ebrains.eu). Brain sections shown in (C) and (D) are from the Vogt Collection of the Cécile and Oskar Vogt Institute for Brain Research, Heinrich Heine University Düsseldorf.

between such measures and functional asymmetries is not well understood (Kong et al. 2018).

Analyses of brain volume growth within the primate lineage showed a disproportional increase in the left PFC in the evolution of primates and a stronger coupling in volumetric growth between left frontal motor areas and the right posterior cerebellar hemisphere in comparison to the contralateral side. The latter is only visible in the phylogenetic branch leading to humans and chimpanzees, and it has been suggested that this could be correlated to the evolution of a more pronounced/distinct handedness and possibly even to the evolution of the lateralized human language system (Smaers et al. 2013). In addition, differences in structural asymmetry on the fiber tract level between humans and nonhuman primates have been reported.

Conclusion

This chapter gave a brief overview of the bauplan of the mammalian brain and its changes in telencephalization. While these changes allowed humans to evolve their sophisticated cognitive capabilities, they also formed the basis for neurological disorders not present, or at least less common, in nonhuman species. At the end of this chapter, we would like to mention a few examples, of which some will be outlined in more detail in the following chapters. One of the consequences of a higher telencephalization is higher metabolic costs, mostly driven by the upkeep costs of neurons. Humans have the largest number of neurons among primates, and, accordingly, their brain maintenance costs are also highest among primates (Herculano-Houzel 2012). The size of the inner carotid artery (ICA) and its blood flow increased during evolution (by a factor of about six compared to our early hominin ancestors). It is likely that these adaptations of brain blood supply happened to cope with the increasing energy demands of the brain (Seymour et al. 2016). However, the sizes of the ICA and associated vessels of the circle of Willis still are relatively small and show a high risk for stenosis (prevalence for ICA stenosis about 4%; Rockman et al. 2013). It could be argued that due to the increased energy/blood supply demands, also the risk of stroke increased during human evolution, especially in combination with human longevity (see next paragraph). Furthermore, also other, more indirect factors need to be considered. To cover the high neuronal and other costs of the human-specific lifestyle, humans, and, before, our primate and hominin ancestors, constantly developed new foraging strategies based on complex social interactions and social learning, including the development of cultural habits (as can be observed in nonhuman primates; Davis et al. 2016). While the obtained sociocultural traits turned out to be highly efficient, they possibly also favored the development of culturally induced mental disorders (see Chapter 22).

Furthermore, large brains require long maturation times; thus, human longevity increases (González-Lagos et al. 2010). Since high age is a risk factor for many neurological diseases, it could be argued that one of the disadvantages of high encephalization and telencephalization is an increased risk for certain neurological diseases (see Chapters 8–17). Furthermore, an increased life span also increases the chance for genetic alterations and, thus, the chance of, for instance, glio- and astrocytomas (even though species with a high longevity evolved some mechanisms to counter this; Gorbunova et al. 2014).

The addition of cortical areas during human evolution (e.g., in the parietal and temporal cortex), enabling human traits like language or complex tool use, and the addition of new circuits and increasing complexity of networks (see Chapter 18) may also contribute to neurological disorders (e.g., schizophrenia or autism spectrum disorders), as will be outlined in Chapters 14, 15, and 17.

Taken together, the human brain underwent significant adaptive changes during the course of evolution; with some of these adaptations likely came costs. Although these costs were outweighed by the benefits in reproductive fitness on the population average, they are still visible in the form of human-specific neurological disorders in certain individuals.

Key Points to Remember

- The ancestors of modern mammals emerged about 310 million years ago, giving rise to rodent (80–60 mya), primate (77 mya), and early human species (8 mya). The modern human (*Homo sapiens*) evolved from these early human species about 200,000 years ago.
- Even though the basic bauplan of the brain is identical in all mammals, neuroanatomical macro- and microstructure can differ vastly between mammalian clades and species depending on the species lifestyle, living environment, and phylogenetic ancestry.
- Primate species possess, in relation to body size, larger brains and more neurons than rodent species, while humans possess larger brains and more neurons than nonhuman primates.
- Most of this increase in brain volume and neuron number can be attributed to the telencephalic neocortex, which is in relative terms larger in primates in comparison to rodents and larger in humans in comparison to nonhuman primates.
- The largest neocortical volume expanse during human evolution occurred in the lateral PFC, the middle temporal gyrus, and the inferior parietal lobule—areas which are related to higher cognitive functions, language processing, and hand coordination.
- Microstructural differences comprise larger pyramidal neurons with larger and more branched dendritic trees in humans compared to rodents, which might be related to differences in cell physiology.
- While evolutionary adaptations of the human brain look at first glance highly advantageous, they could also be correlated to several neurological diseases present in humans but mostly not in other mammalian species.

Notes

1. Cells of the olfactory tubercle originate from both pallial and subpallial proliferation zones; and thus, it is sometimes considered part of the paleocortex. Due to similarities in receptor expression to other subpallial areas (Lothmann et al. 2021), we list it here as a subpallial structure.

2. Also, the amygdala and olfactory bulb contain cells from different proliferation zones. Amygdalar subnuclei differ in their receptor expression pattern, allowing a segregation into subpallial, pallial, and even cortical fractions, while cells of pallial and subpallial origin in the olfactory bulb are intermingled. For the sake of simplicity, we list amygdala and olfactory bulb here as pallial nuclear structures and refer to Kedo et al. (2018) and Tufo et al. (2022) for further information.
3. As a side note, we as primates should not feel too proud about our high neuron densities since avian species like crows and parrots have even more favorable scaling rules endowing them with neuron densities double those in primates (Olkowicz et al. 2016; Ströckens et al. 2022).
4. The phenomenon of analogous brain areas taking over similar executive or higher cognitive functions is not unique to mammalian species. Avian species do not have a neocortex but possess pallial areas with strikingly similar functional properties to the mammalian PFC and a comparable behavioral output (Güntürkün et al. 2017).

References

Aamodt, C. M., Farias-Virgens, M., & White, S. A. (2020). Birdsong as a window into language origins and evolutionary neuroscience. *Philosophical Transactions of the Royal Society of London, Series B, Biological Sciences*, *375*(1789), Article 20190060. https://doi.org/10.1098/rstb.2019.0060

Ackerley, R., & Kavounoudias, A. (2015). The role of tactile afference in shaping motor behaviour and implications for prosthetic innovation. *Neuropsychologia*, *79*(Pt. B), 192–205. https://doi.org/10.1016/j.neuropsychologia.2015.06.024

Amunts, K. (2010). Structural indices of asymmetry. In K. Hugdahl, R. Westerhausen, & R. Westerhausen (Eds.), *Two halves of the brain: Information processing in the cerebral hemispheres* (pp. 145–176). MIT Press.

Amunts, K., Kedo, O., Kindler, M., Pieperhoff, P., Mohlberg, H., Shah, N. J., Habel, U., Schneider, F., & Zilles, K. (2005). Cytoarchitectonic mapping of the human amygdala, hippocampal region and entorhinal cortex: Intersubject variability and probability maps. *Anatomy and Embryology*, *210*(5–6), 343–352. https://doi.org/10.1007/s00429-005-0025-5

Amunts, K., Mohlberg, H., Bludau, S., & Zilles, K. (2020). Julich-Brain: A 3D probabilistic atlas of the human brain's cytoarchitecture. *Science*, *369*(6506), 988–992. https://doi.org/10.1126/science.abb4588

Badea, A., Ali-Sharief, A. A., & Johnson, G. A. (2007). Morphometric analysis of the C57BL/6J mouse brain. *NeuroImage*, *37*(3), 683–693. https://doi.org/10.1016/j.neuroimage.2007.05.046

Baird, A. D., Wilson, S. J., Bladin, P. F., Saling, M. M., & Reutens, D. C. (2007). Neurological control of human sexual behaviour: Insights from lesion studies. *Journal of Neurology, Neurosurgery and Psychiatry*, *78*(10), 1042–1049. https://doi.org/10.1136/jnnp.2006.107193

Baizer, J. S. (2014). Unique features of the human brainstem and cerebellum. *Frontiers in Human Neuroscience*, *8*, Article 202. https://doi.org/10.3389/fnhum.2014.00202

Banovac, I., Sedmak, D., Judaš, M., & Petanjek, Z. (2021). Von Economo neurons—Primate-specific or commonplace in the mammalian brain? *Frontiers in Neural Circuits*, *15*, Article 714611. https://doi.org/10.3389/fncir.2021.714611

Beaudet, A., Du, A., & Wood, B. (2019). Evolution of the modern human brain. *Progress in Brain Research*, *250*, 219–250. https://doi.org/10.1016/bs.pbr.2019.01.004

Benavides-Piccione, R., Ballesteros-Yáñez, I., DeFelipe, J., & Yuste, R. (2002). Cortical area and species differences in dendritic spine morphology. *Journal of Neurocytology*, *31*(3–5), 337–346. https://doi.org/10.1023/a:1024134312173

Bhaya-Grossman, I., & Chang, E. F. (2022). Speech computations of the human superior temporal gyrus. *Annual Review of Psychology*, *73*, 79–102. https://doi.org/10.1146/annurev-psych-022321-035256

Bolker, J. A. (2019). Selection of models: Evolution and the choice of species for translational research. *Brain, Behavior and Evolution*, *93*(2–3), 82–91. https://doi.org/10.1159/000500317

Braak, H., & Braak, E. (1991). Neuropathological staging of Alzheimer-related changes. *Acta Neuropathologica*, *82*(4), 239–259. https://doi.org/10.1007/BF00308809

Braunsdorf, M., Blazquez Freches, G., Roumazeilles, L., Eichert, N., Schurz, M., Uithol, S., Bryant, K. L., & Mars, R. B. (2021). Does the temporal cortex make us human? A review of structural and functional diversity of the primate temporal lobe. *Neuroscience and Biobehavioral Reviews*, *131*, 400–410. https://doi.org/10.1016/j.neubiorev.2021.08.032

Brodmann, K. (1909). *Vergleichende lokalisationslehre der Grosshirnrinde in ihren Prinzipien dargestellt auf Grund des Zellenbaues*. Verlag von Johann Ambrosius Barth.

Butti, C., Raghanti, M. A., Sherwood, C. C., and Hof, P. R. (2011). The neocortex of cetaceans: Cytoarchitecture and comparison with other aquatic and terrestrial species. *Annals of the New York Academy of Sciences*, *1225*, 47–58. https://doi.org/10.1111/j.1749-6632.2011.05980.x

Carlén, M. (2017). What constitutes the prefrontal cortex? *Science*, *358*(6362), 478–482. https://doi.org/10.1126/science.aan8868

Caspers, S., Schleicher, A., Bacha-Trams, M., Palomero-Gallagher, N., Amunts, K., & Zilles, K. (2013). Organization of the human inferior parietal lobule based on receptor architectonics. *Cerebral Cortex*, *23*(3), 615–628. https://doi.org/10.1093/cercor/bhs048

Catani, M., Dell'acqua, F., & Thiebaut de Schotten, M. (2013). A revised limbic system model for memory, emotion and behaviour. *Neuroscience and Biobehavioral Reviews*, *37*(8), 1724–1737. https://doi.org/10.1016/j.neubiorev.2013.07.001

Catania, K. C., Lyon, D. C., Mock, O. B., & Kaas, J. H. (1999). Cortical organization in shrews: Evidence from five species. *The Journal of Comparative Neurology*, *410*(1), 55–72. https://doi.org/10.1002/(SICI)1096-9861(19990719)410:1<55::AID-CNE6>3.0.CO;2-2

Collins, C. E., Turner, E. C., Sawyer, E. K., Reed, J. L., Young, N. A., Flaherty, D. K., & Kaas, J. H. (2016). Cortical cell and neuron density estimates in one chimpanzee hemisphere. *Proceedings of the National Academy of Sciences of the United States of America*, *113*(3), 740–745. https://doi.org/10.1073/pnas.1524208113

Cook, S. J., Jarrell, T. A., Brittin, C. A., Wang, Y., Bloniarz, A. E., Yakovlev, M. A., Nguyen, K. C. Q., Tang, L. T. H., Bayer, E. A., Duerr, J. S., Bülow, H. E., Hobert, O., Hall, D. H., & Emmons, S. W. (2019). Whole-animal connectomes of both *Caenorhabditis elegans* sexes. *Nature*, *571*(7763), 63–71. https://doi.org/10.1038/s41586-019-1352-7

Cox, J., & Witten, I. B. (2019). Striatal circuits for reward learning and decision-making. *Nature Reviews Neuroscience*, *20*(8), 482–494. https://doi.org/10.1038/s41583-019-0189-2

Davis, S. J., Vale, G. L., Schapiro, S. J., Lambeth, S. P., & Whiten, A. (2016). Foundations of cumulative culture in apes: Improved foraging efficiency through relinquishing and combining witnessed behaviours in chimpanzees (*Pan troglodytes*). *Scientific Reports*, *6*, Article 35953. https://doi.org/10.1038/srep35953

de Haas, B., Sereno, M. I., & Schwarzkopf, D. S. (2021). Inferior occipital gyrus is organized along common gradients of spatial and face-part selectivity. *The Journal of Neuroscience*, *41*(25), 5511–5521. https://doi.org/10.1523/JNEUROSCI.2415-20.2021

Diederich, N. J., Goldman, J. G., Stebbins, G. T., & Goetz, C. G. (2016). Failing as doorman and disc jockey at the same time: Amygdalar dysfunction in Parkinson's disease. *Movement Disorders*, *31*(1), 11–22. https://doi.org/10.1002/mds.26460

Elston, G. N., Benavides-Piccione, R., & DeFelipe, J. (2001). The pyramidal cell in cognition: A comparative study in human and monkey. *The Journal of Neuroscience*, *21*(17), Article RC163. https://doi.org/10.1523/JNEUROSCI.21-17-j0002.2001

Eyal, G., Verhoog, M. B., Testa-Silva, G., Deitcher, Y., Lodder, J. C., Benavides-Piccione, R., Morales, J., DeFelipe, J., de Kock, C. P. J., Mansvelder, H. D., & Segev, I. (2016). Unique membrane properties and enhanced signal processing in human neocortical neurons. *eLife*, *5*, Article e16553. https://doi.org/10.7554/eLife.16553

Feng, G., Jensen, F. E., Greely, H. T., Okano, H., Treue, S., Roberts, A. C., Fox, J. G., Caddick, S., Poo, M. M. Newsome, W. T., & Morrison, J. H. (2020). Opportunities and limitations of genetically modified non-human primate models for neuroscience research. *Proceedings of the National Academy of Sciences of the United States of America*, *117*(39), 24022–24031. https://doi.org/10.1073/pnas.2006515117

Freud, E., Plaut, D. C., & Behrmann, M. (2016). "What" is happening in the dorsal visual pathway. *Trends in Cognitive Sciences*, *20*(10), 773–784. https://doi.org/10.1016/j.tics.2016.08.003

Friederici, A. D. (2011). The brain basis of language processing: From structure to function. *Physiological Reviews*, *91*(4), 1357–1392. https://doi.org/10.1152/physrev.00006.2011

Fuster, J. M. (2015). *The prefrontal cortex* (5th ed.). Academic Press.

Garcia, A. D., & Buffalo, E. A. (2020). Anatomy and function of the primate entorhinal cortex. *Annual Review of Vision Science*, *6*, 411–432. https://doi.org/10.1146/annurev-vision-030320-041115

Gautam, P., Anstey, K. J., Wen, W., Sachdev, P. S., & Cherbuin, N. (2015). Cortical gyrification and its relationships with cortical volume, cortical thickness, and cognitive performance in healthy mid-life adults. *Behavioural Brain Research*, *287*, 331–339. https://doi.org/10.1016/j.bbr.2015.03.018

Geissler, D. B., Schmidt, H. S., and Ehret, G. (2016). Knowledge about sounds—Context-specific meaning differently activates cortical hemispheres, auditory cortical fields, and layers in house mice. *Frontiers in Neuroscience*, *10*, Article 98. https://doi.org/10.3389/fnins.2016.00098

Genç, E., Fraenz, C., Schlüter, C., Friedrich, P., Hossiep, R., Voelkle, M. C., Ling, J. M., Güntürkün, O., & Jung, R. E. (2018). Diffusion markers of dendritic density and arborization in gray matter predict differences in intelligence. *Nature Communications*, *9*(1), Article 1905. https://doi.org/10.1038/s41467-018-04268-8

Gidon, A., Zolnik, T. A., Fidzinski, P., Bolduan, F., Papoutsi, A., Poirazi, P., Holtkamp, M., Vida, I., & Larkum, M. E. (2020). Dendritic action potentials and computation in human layer 2/3 cortical neurons. *Science*, *367*(6473), 83–87. https://doi.org/10.1126/science.aax6239

Glasser, M. F., Coalson, T. S., Robinson, E. C., Hacker, C. D., Harwell, J., Yacoub, E., Ugurbil, K., Andersson, J., Beckmann, C. F., Jenkinson, M., Smith, S. M., & Van Essen, D. C. (2016). A multi-modal parcellation of human cerebral cortex. *Nature*, *536*(7615), 171–178. https://doi.org/10.1038/nature18933

Goebel, R., Muckli, L., Zanella, F. E., Singer, W., & Stoerig, P. (2001). Sustained extrastriate cortical activation without visual awareness revealed by fMRI studies of hemianopic patients. *Vision Research*, *41*(10–11), 1459–1474. https://doi.org/10.1016/s0042-6989(01)00069-4

Gogolla, N. (2017). The insular cortex. *Current Biology*, *27*(12), R580–R586. https://doi.org/10.1016/j.cub.2017.05.010

González-Lagos, C., Sol, D., & Reader, S. M. (2010). Large-brained mammals live longer. *Journal of Evolutionary Biology*, *23*(5), 1064–1074. https://doi.org/10.1111/j.1420-9101.2010.01976.x

Gorbunova, V., Seluanov, A., Zhang, Z., Gladyshev, V. N., & Vijg, J. (2014). Comparative genetics of longevity and cancer: Insights from long-lived rodents. *Nature Reviews Genetics*, *15*(8), 531–540. https://doi.org/10.1038/nrg3728

Goriounova, N. A., Heyer, D. B., Wilbers, R., Verhoog, M. B., Giugliano, M., Verbist, C., Obermayer, J., Kerkhofs, A., Smeding, H., Verberne, M., Idema, S., Baayen, J. C., Pieneman, A. W., de Kock, C. P. J., Klein, M., & Mansvelder, H. D. (2018). Large and fast human pyramidal neurons associate with intelligence. *eLife*, *7*, Article e41714. https://doi.org/10.7554/eLife.41714

Güntürkün, O., Stacho, M., & Ströckens, F. (2017). The brains of reptiles and birds. In J. H. Kaas (Ed.), *Evolution of nervous systems* (2nd ed., pp. 172–210). Elsevier Science.

Güntürkün, O., Ströckens, F., & Ocklenburg, S. (2020). Brain lateralization: A comparative perspective. *Physiological Reviews*, *100*(3), 1019–1063. https://doi.org/10.1152/physrev.00006.2019

Halley, A. C., & Deacon, T. W. (2017). The developmental basis of evolutionary trends in primate encephalization. In J. H. Kaas (Ed.), *Evolution of nervous systems* (2nd ed., pp. 149–162). Elsevier Science.

Hayashi, T., Hou, Y., Glasser, M. F., Autio, J. A., Knoblauch, K., Inoue-Murayama, M., Coalson, T., Yacoub, E., Smith, S., Kennedy, H., & Van Essen, D. C. (2021). The nonhuman primate neuroimaging and neuroanatomy project. *NeuroImage*, *229*, Article 117726. https://doi.org/10.1016/j.neuroimage.2021.117726

Herculano-Houzel, S. (2012). The remarkable, yet not extraordinary, human brain as a scaled-up primate brain and its associated cost. *Proceedings of the National Academy of Sciences of the United States of America*, *109*(S1), 10661–10668. https://doi.org/10.1073/pnas.1201895109

Herculano-Houzel, S., Catania, K., Manger, P. R., & Kaas, J. H. (2015). Mammalian brains are made of these: A dataset of the numbers and densities of neuronal and nonneuronal cells in the brain of glires, primates, scandentia, eulipotyphlans, afrotherians and artiodactyls, and their relationship with body mass. *Brain, Behavior and Evolution*, *86*(3–4), 145–163. https://doi.org/10.1159/000437413

Hertrich, I., Dietrich, S., Blum, C., & Ackermann, H. (2021). The role of the dorsolateral prefrontal cortex for speech and language processing. *Frontiers in Human Neuroscience, 15*, Article 645209. https://doi.org/10.3389/fnhum.2021.645209

Hodge, R. D., Miller, J. A., Novotny, M., Kalmbach, B. E., Ting, J. T., Bakken, T. E., Aevermann, B. D., Barkan, E. R., Berkowitz-Cerasano, M. L., Cobbs, C., Diez-Fuertes, F., Ding, S. L., McCorrison, J., Schork, N. J., Shehata, S. I., Smith, K. A., Sunkin, S. M., Tran, D. N., Venepally, P., . . . Lein, E. S. (2020). Transcriptomic evidence that von Economo neurons are regionally specialized extratelencephalic-projecting excitatory neurons. *Nature Communications, 11*(1), Article 1172. https://doi.org/10.1038/s41467-020-14952-3

Hulse, B. K., Haberkern, H., Franconville, R., Turner-Evans, D., Takemura, S. Y., Wolff, T., Noorman, M., Dreher, M., Dan, C., Parekh, R., Hermundstad, A. M., Rubin, G. M., & Jayaraman, V. (2021). A connectome of the *Drosophila* central complex reveals network motifs suitable for flexible navigation and context-dependent action selection. *eLife, 10*, Article e66039. https://doi.org/10.7554/eLife.66039

Ishikawa, Y., Yamamoto, N., Yoshimoto, M., & Ito, H. (2012). The primary brain vesicles revisited: Are the three primary vesicles (forebrain/midbrain/hindbrain) universal in vertebrates? *Brain, Behavior and Evolution, 79*(2), 75–83. https://doi.org/10.1159/000334842

Jacobs, G. H. (2009). Evolution of colour vision in mammals. *Philosophical Transactions of the Royal Society B: Biological Sciences, 364*(1531), 2957–2967. https://doi.org/10.1098/rstb.2009.0039

Janak, P. H., and Tye, K. M. (2015). From circuits to behaviour in the amygdala. *Nature, 517*(7534), 284–292. https://doi.org/10.1038/nature14188

Kaas, J. H. (2012). The evolution of neocortex in primates. *Progress in Brain Research, 195*, 91–102. https://doi.org/10.1016/B978-0-444-53860-4.00005-2

Kaas, J. H. (2017). Evolution of visual cortex in primates. In J. H. Kaas (Ed.), *Evolution of nervous systems* (2nd ed., pp. 187–201). Elsevier Science.

Kaas, J. H., & Balaram, P. (2014). Current research on the organization and function of the visual system in primates. *Eye and Brain, 6*, 1–4. https://doi.org/10.2147/EB.S64016

Kedo, O., Zilles, K., Palomero-Gallagher, N., Schleicher, A., Mohlberg, H., Bludau, S., & Amunts, K. (2018). Receptor-driven, multimodal mapping of the human amygdala. *Brain Structure & Function, 223*(4), 1637–1666. https://doi.org/10.1007/s00429-017-1577-x

Knierim, J. J. (2015). The hippocampus. *Current Biology, 25*(23), R1116–R1121. https://doi.org/10.1016/j.cub.2015.10.049

Kong, X. Z., Mathias, S. R., Guadalupe, T., ENIGMA Laterality Working Group, Glahn, D. C., Franke, B., Crivello, F., Tzourio-Mazoyer, N., Fisher, S. E., Thompson, P. M., & Francks, C. (2018). Mapping cortical brain asymmetry in 17,141 healthy individuals worldwide via the ENIGMA Consortium. *Proceedings of the National Academy of Sciences of the United States of America, 115*(22), E5154–E5163. https://doi.org/10.1073/pnas.1718418115

Kuhlwilm, M., Gronau, I., Hubisz, M. J., de Filippo, C., Prado-Martinez, J., Kircher, M., Fu, Q., Burbano, H. A., Lalueza-Fox, C., de La Rasilla, M., Rosas, A., Rudan, P., Brajkovic, D., Kucan, Ž., Gušic, I., Marques-Bonet, T., Andrés, A. M., Viola, B., Pääbo, S., . . . Castellano, S. (2016). Ancient gene flow from early modern humans into eastern Neanderthals. *Nature, 530*(7591), 429–433. https://doi.org/10.1038/nature16544

Lafer-Sousa, R., Conway, B. R., & Kanwisher, N. G. (2016). Color-biased regions of the ventral visual pathway lie between face- and place-selective regions in humans, as in macaques. *The Journal of Neuroscience, 36*(5), 1682–1697. https://doi.org/10.1523/JNEUROSCI.3164-15.2016

Lerch, J. P., Carroll, J. B., Dorr, A., Spring, S., Evans, A. C., Hayden, M. R., Sled, J. G., & Henkelman, R. M. (2008). Cortical thickness measured from MRI in the YAC128 mouse model of Huntington's disease. *NeuroImage, 41*(2), 243–251. https://doi.org/10.1016/j.neuroimage.2008.02.019

Liu, Z., Li, X., Zhang, J. T., Cai, Y. J., Cheng, T. L., Cheng, C., Wang, Y., Zhang, C. C., Nie, Y. H., Chen, Z. F., Bian, W. J., Zhang, L., Xiao, J., Lu, B., Zhang, Y. F., Zhang, X. D., Sang, X., Wu, J. J., Xu, X., . . . Qiu, Z. (2016). Autism-like behaviours and germline transmission in transgenic monkeys overexpressing MeCP2. *Nature, 530*(7588), 98–102. https://doi.org/10.1038/nature16533

Lothmann, K., Amunts, K., & Herold, C. (2021). The neurotransmitter receptor architecture of the mouse olfactory system. *Frontiers in Neuroanatomy, 15*, Article 632549. https://doi.org/10.3389/fnana.2021.632549

Luppino, G., & Rizzolatti, G. (2000). The organization of the frontal motor cortex. *News in Physiological Sciences, 15*(5), 219–224. https://doi.org/10.1152/physiologyonline.2000.15.5.219

Manger, P. R., Prowse, M., Haagensen, M., & Hemingway, J. (2012). Quantitative analysis of neocortical gyrencephaly in African elephants (*Loxodonta africana*) and six species of cetaceans: Comparison with other mammals. *The Journal of Comparative Neurology, 520*(11), 2430–2439. https://doi.org/10.1002/cne.23046

McGann, J. P. (2017). Poor human olfaction is a 19th-century myth. *Science, 356*(6338), Article eaam7263. https://doi.org/10.1126/science.aam7263

Megginson, L. C. (1963). Lessons from Europe for American business. *The Southwestern Social Science Quarterly, 44*, 3–13.

Meinhardt, J., Radke, J., Dittmayer, C., Franz, J., Thomas, C., Mothes, R., Laue, M., Schneider, J., Brünink, S., Greuel, S., Lehmann, M., Hassan, O., Aschman, T., Schumann, E., Chua, R. L., Conrad, C., Eils, R., Stenzel, W., Windgassen, M., . . . Heppner, F. L. (2021). Olfactory transmucosal SARS-CoV-2 invasion as a port of central nervous system entry in individuals with COVID-19. *Nature Neuroscience, 24*(2), 168–175. https://doi.org/10.1038/s41593-020-00758-5

Moerel, M., de Martino, F., & Formisano, E. (2014). An anatomical and functional topography of human auditory cortical areas. *Frontiers in Neuroscience, 8*, Article 225. https://doi.org/10.3389/fnins.2014.00225

Mohan, H., Verhoog, M. B., Doreswamy, K. K., Eyal, G., Aardse, R., Lodder, B. N., Goriounova, N. A., Asamoah, B., Brakspear, A. B., Groot, C., van der Sluis, S., Testa-Silva, G., Obermayer, J., Boudewijns, Z. S., Narayanan, R. T., Baayen, J. C., Segev, I., Mansvelder, H. D., & de Kock, C. P. J. (2015). Dendritic and axonal architecture of individual pyramidal neurons across layers of adult human neocortex. *Cerebral Cortex, 25*(12), 4839–4853. https://doi.org/10.1093/cercor/bhv188

Nachev, P., Kennard, C., & Husain, M. (2008). Functional role of the supplementary and pre-supplementary motor areas. *Nature Reviews Neuroscience, 9*(11), 856–869. https://doi.org/10.1038/nrn2478

Niu, M., Rapan, L., Funck, T., Froudist-Walsh, S., Zhao, L., Zilles, K., & Palomero-Gallagher, N. (2021). Organization of the macaque monkey inferior parietal lobule based on multimodal receptor architectonics. *NeuroImage, 231*, Article 117843. https://doi.org/10.1016/j.neuroimage.2021.117843

Ocklenburg, S., Ströckens, F., & Güntürkün, O. (2013). Lateralisation of conspecific vocalisation in non-human vertebrates. *Laterality, 18*(1), 1–31. https://doi.org/10.1080/1357650X.2011.626561

O'Connor, D. H., Krubitzer, L., & Bensmaia, S. (2021). Of mice and monkeys: Somatosensory processing in two prominent animal models. *Progress in Neurobiology, 201*, Article 102008. https://doi.org/10.1016/j.pneurobio.2021.102008

Olkowicz, S., Kocourek, M., Lučan, R. K., Porteš, M., Fitch, W. T., Herculano-Houzel, S., & Němec, P. (2016). Birds have primate-like numbers of neurons in the forebrain. *Proceedings of the National Academy of Sciences of the United States of America, 113*(26), 7255–7260. https://doi.org/10.1073/pnas.1517131113

Palomero-Gallagher, N., & Amunts, K. (2022). A short review on emotion processing: A lateralized network of neuronal networks. *Brain Structure & Function, 227*(2), 673–684. https://doi.org/10.1007/s00429-021-02331-7

Papeo, L., Agostini, B., & Lingnau, A. (2019). The large-scale organization of gestures and words in the middle temporal gyrus. *The Journal of Neuroscience, 39*(30), 5966–5974. https://doi.org/10.1523/JNEUROSCI.2668-18.2019

Perrin, S. (2014). Preclinical research: Make mouse studies work. *Nature, 507*(7493), 423–425. https://doi.org/10.1038/507423a

Pieperhoff, P., Südmeyer, M., Dinkelbach, L., Hartmann, C. J., Ferrea, S., Moldovan, A. S., Minnerop, M., Diaz-Pier, S., Schnitzler, A., & Amunts, K. (2022). Regional changes of brain structure during progression of idiopathic Parkinson's disease—A longitudinal study using deformation based morphometry. *Cortex, 151*, 188–210. https://doi.org/10.1016/j.cortex.2022.03.009

Pillay, P., & Manger, P. R. (2007). Order-specific quantitative patterns of cortical gyrification. *European Journal of Neuroscience, 25*(9), 2705–2712. https://doi.org/10.1111/j.1460-9568.2007.05524.x

Pozzi, L., Hodgson, J. A., Burrell, A. S., Sterner, K. N., Raaum, R. L., & Disotell, T. R. (2014). Primate phylogenetic relationships and divergence dates inferred from complete mitochondrial genomes. *Molecular Phylogenetics and Evolution, 75*, 165–183. https://doi.org/10.1016/j.ympev.2014.02.023

Preuss, T. M., and Wise, S. P. (2022). Evolution of prefrontal cortex. *Neuropsychopharmacology, 47*(1), 3–19. https://doi.org/10.1038/s41386-021-01076-5

Rakic, P. (2009). Evolution of the neocortex: A perspective from developmental biology. *Nature Reviews Neuroscience, 10*, 724–735. https://doi.org/10.1038/nrn2719

Revah, O., Gore, F., Kelley, K. W., Andersen, J., Sakai, N., Chen, X., Li, M. Y., Birey, F., Yang, X., Saw, N. L., Baker, S. W., Amin, N. D., Kulkarni, S., Mudipalli, R., Cui, B., Nishino, S., Grant, G. A., Knowles, J. K., Shamloo, M., . . . Paşca, S. P. (2022). Maturation and circuit integration of transplanted human cortical organoids. *Nature, 610*(7931), 319–326. https://doi.org/10.1038/s41586-022-05277-w

Rilling, J. K., Glasser, M. F., Preuss, T. M., Ma, X., Zhao, T., Hu, X., & Behrens, T. E. J. (2008). The evolution of the arcuate fasciculus revealed with comparative DTI. *Nature Neuroscience, 11*(4), 426–428. https://doi.org/10.1038/nn2072

Rilling, J. K., and Insel, T. R. (1999). The primate neocortex in comparative perspective using magnetic resonance imaging. *Journal of Human Evolution, 37*(2), 191–223. https://doi.org/10.1006/jhev.1999.0313

Rockman, C. B., Hoang, H., Guo, Y., Maldonado, T. S., Jacobowitz, G. R., Talishinskiy, T., Riles, T. S., and Berger, J. S. (2013). The prevalence of carotid artery stenosis varies significantly by race. *Journal of Vascular Surgery, 57*(2), 327–337. https://doi.org/10.1016/j.jvs.2012.08.118

Romanski, L. M., and Averbeck, B. B. (2009). The primate cortical auditory system and neural representation of conspecific vocalizations. *Annual Review of Neuroscience, 32*, 315–346. https://doi.org/10.1146/annurev.neuro.051508.135431

Rowe, T. B. (2017). The emergence of mammals. In J. H. Kaas (Ed.), *Evolution of nervous systems* (2nd ed., pp. 1–47). Elsevier Science.

Ruland, S. H., Palomero-Gallagher, N., Hoffstaedter, F., Eickhoff, S. B., Mohlberg, H., & Amunts, K. (2022). The inferior frontal sulcus: Cortical segregation, molecular architecture and function. *Cortex, 153*, 235–256. https://doi.org/10.1016/j.cortex.2022.03.019

Sasabayashi, D., Takahashi, T., Takayanagi, Y., & Suzuki, M. (2021). Anomalous brain gyrification patterns in major psychiatric disorders: A systematic review and transdiagnostic integration. *Translational Psychiatry, 11*(1), Article 176. https://doi.org/10.1038/s41398-021-01297-8

Schaeffer, D. J., Hori, Y., Gilbert, K. M., Gati, J. S., Menon, R. S., & Everling, S. (2020). Divergence of rodent and primate medial frontal cortex functional connectivity. *Proceedings of the National Academy of Sciences of the United States of America, 117*(35), 21681–21689. https://doi.org/10.1073/pnas.2003181117

Schmidt, E. R. E., and Polleux, F. (2021). Genetic mechanisms underlying the evolution of connectivity in the human cortex. *Frontiers in Neural Circuits, 15*, Article 787164. https://doi.org/10.3389/fncir.2021.787164

Seymour, R. S., Bosiocic, V., & Snelling, E. P. (2016). Fossil skulls reveal that blood flow rate to the brain increased faster than brain volume during human evolution. *Royal Society Open Science, 3*(8), Article 160305. https://doi.org/10.1098/rsos.160305

Sherwood, C. C., Miller, S. B., Karl, M., Stimpson, C. D., Phillips, K. A., Jacobs, B., Hof, P. R., Raghanti, M. A., & Smaers, J. B. (2020). Invariant synapse density and neuronal connectivity scaling in primate neocortical evolution. *Cerebral Cortex, 30*(10), 5604–5615. https://doi.org/10.1093/cercor/bhaa149

Simpson, J. A., & Belsky, J. (2008). Attachment theory within a modern evolutionary framework. In J. Cassidy & P. R. Shaver (Eds.), *Handbook of attachment: Theory, research, and clinical applications* (3rd ed., pp. 131–157). Guilford Press.

Smaers, J. B., Steele, J., Case, C. R., & Amunts, K. (2013). Laterality and the evolution of the prefronto-cerebellar system in anthropoids. *Annals of the New York Academy of Sciences, 1288*(1), 59–69. https://doi.org/10.1111/nyas.12047

Ströckens, F., Güntürkün, O., & Ocklenburg, S. (2013). Limb preferences in non-human vertebrates. *Laterality, 18*(5), 536–575. https://doi.org/10.1080/1357650X.2012.723008

Ströckens, F., Neves, K., Kirchem, S., Schwab, C., Herculano-Houzel, S., & Güntürkün, O. (2022). High associative neuron numbers could drive cognitive performance in corvid species. *The Journal of Comparative Neurology, 530*(10), 1588–1605. https://doi.org/10.1002/cne.25298

ten Donkelaar, H. J., Tzourio-Mazoyer, N., & Mai, J. K. (2018). Toward a common terminology for the gyri and sulci of the human cerebral cortex. *Frontiers in Neuroanatomy, 12*, Article 93. https://doi.org/10.3389/fnana.2018.00093

Tsuboi, M., van der Bijl, W., Kopperud, B. T., Erritzøe, J., Voje, K. L., Kotrschal, A., Yopak, K. E., Collin, S. P., Iwaniuk, A. N., & Kolm, N. (2018). Breakdown of brain–body allometry and the encephalization of birds and mammals. *Nature Ecology & Evolution, 2*(9), 1492–1500. https://doi.org/10.1038/s41559-018-0632-1

Tufo, C., Poopalasundaram, S., Dorrego-Rivas, A., Ford, M. C., Graham, A., & Grubb, M. S. (2022). Development of the mammalian main olfactory bulb. *Development, 149*(3), Article dev200210. https://doi.org/10.1242/dev.200210

Upham, N. S., Esselstyn, J. A., & Jetz, W. (2019). Inferring the mammal tree: Species-level sets of phylogenies for questions in ecology, evolution, and conservation. *PLOS Biology, 17*(12), Article e3000494. https://doi.org/10.1371/journal.pbio.3000494

van Dongen, P. A. M. (1998). Brain size in vertebrates. In R. Nieuwenhuys, H. J. ten Donkelaar, & C. Nicholson (Eds.), *The central nervous system of vertebrates* (pp. 2099–2134). Springer.

Vanduffel, W., Fize, D., Mandeville, J. B., Nelissen, K., Van Hecke, P., Rosen, B. R., Tootell, R. B., & Orban, G. A. (2001). Visual motion processing investigated using contrast agent-enhanced fMRI in awake behaving monkeys. *Neuron, 32*(4), 565–577. https://doi.org/10.1016/s0896-6273(01)00502-5

Van Essen, D. C., & Dierker, D. L. (2007). Surface-based and probabilistic atlases of primate cerebral cortex. *Neuron, 56*(2), 209–225. https://doi.org/10.1016/j.neuron.2007.10.015

Vasile, F., Dossi, E., & Rouach, N. (2017). Human astrocytes: Structure and functions in the healthy brain. *Brain Structure & Function, 222*(5), 2017–2029. https://doi.org/10.1007/s00429-017-1383-5

Vetreno, R. P., Yaxley, R., Paniagua, B., Johnson, G. A., & Crews, F. T. (2017). Adult rat cortical thickness changes across age and following adolescent intermittent ethanol treatment. *Addiction Biology, 22*(3), 712–723. https://doi.org/10.1111/adb.12364

Vogel, P., Hahn, J., Duvarci, S., & Sigurdsson, T. (2022). Prefrontal pyramidal neurons are critical for all phases of working memory. *Cell Reports, 39*(2), Article 110659. https://doi.org/10.1016/j.celrep.2022.110659

Vogt, C., & Vogt, O. (1919). *Allgemeine Ergebnisse unserer Hirnforschung*. Verlag von Johann Ambrosius Barth.

Welniak-Kaminska, M., Fiedorowicz, M., Orzel, J., Bogorodzki, P., Modlinska, K., Stryjek, R., Chrzanowska, A., Pisula, W., & Grieb, P. (2019). Volumes of brain structures in captive wild-type and laboratory rats: 7T magnetic resonance in vivo automatic atlas-based study. *PLOS ONE, 14*(4), Article e0215348. https://doi.org/10.1371/journal.pone.0215348

Xiang, L., Crow, T. J., Hopkins, W. D., & Roberts, N. (2020). Comparison of surface area and cortical thickness asymmetry in the human and chimpanzee brain. *Cerebral Cortex, 34*(2), Article bhaa202. https://doi.org/10.1093/cercor/bhaa202

Young, N. A., Collins, C. E., & Kaas, J. H. (2013). Cell and neuron densities in the primary motor cortex of primates. *Frontiers in Neural Circuits, 7*, Article 30. https://doi.org/10.3389/fncir.2013.00030

Zachlod, D., Kedo, O., & Amunts, K. (2022). Anatomy of the temporal lobe: From macro to micro. In G. Miceli, P. Bartolomeo, & V. Navarro (Eds.), *Handbook of clinical neurology: Vol. 187. The temporal lobe* (pp. 17–51). Elsevier. https://doi.org/10.1016/B978-0-12-823493-8.00009-2

Zachlod, D., Rüttgers, B., Bludau, S., Mohlberg, H., Langner, R., Zilles, K., & Amunts, K. (2020). Four new cytoarchitectonic areas surrounding the primary and early auditory cortex in human brains. *Cortex, 128*, 1–21. https://doi.org/10.1016/j.cortex.2020.02.021

Zhou, G., Lane, G., Cooper, S. L., Kahnt, T., & Zelano, C. (2019). Characterizing functional pathways of the human olfactory system. *eLife, 8*, Article e47177. https://doi.org/10.7554/eLife.47177

Zilles, K., & Amunts, K. (2012). Architecture of the human cerebral cortex. In J. K. Mai & G. Paxinos (Eds.), *The human nervous system* (3rd ed., pp. 826–885). Elsevier.

Zilles, K., Armstrong, E., Schleicher, A., & Kretschmann, H. J. (1988). The human pattern of gyrification in the cerebral cortex. *Anatomy and Embryology, 179*(2), 173–179. https://doi.org/10.1007/BF00304699

Evolutionary Aspects of Glial Cells

From Physiology to Pathology

Corrado Calì, Nicole L. Ackermans, Patrick R. Hof, and Pierre J. Magistretti

Historical Background

Glial cells, also known as "neuroglia" or simply "glia," are the non-neuronal cells that make up the supportive tissue of the nervous system. Although glial cells were first described in the mid-19th century, their functions and importance in the nervous system were largely ignored until relatively recently. To date, we have a more substantial understanding of the role of these cells as essential players in the functioning of the nervous system.

The first recorded observations of glial cells were made by the French anatomist Xavier Bichat in the early 19th century. He described the nervous system as being composed of two types of tissues: the gray matter, which he thought was responsible for sensation and perception, and the white matter, which he believed was responsible for motion. Bichat noted that the gray matter contained large, irregularly shaped cells with numerous processes. However, Bichat did not yet recognize the importance of these cells in the nervous system, and their role remained largely unknown for many years.

In the 1850s, the German pathologist Rudolf Virchow also observed the presence of small, star-shaped cells in the brain and spinal cord, which he called "stellate cells." These cells were later identified as astrocytes, a type of glial cell that is now known to be important in the regulation of the extracellular environment of the nervous system (Calì

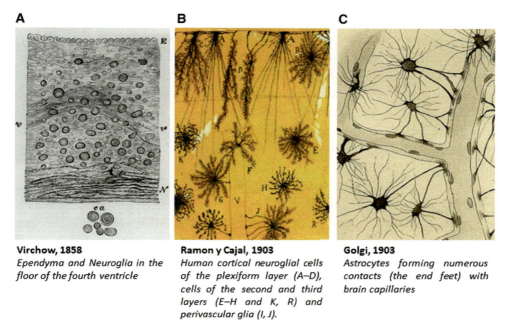

Figure 2.1 Early observers of astrocytes. (A) Drawing of neuroglia from Virchow (1860). Reproduced from Virchow (1860). (B) Drawing from Ramon y Cajal (1903). Adapted from Verkhratsky et al. (2018). (C) Drawing adapted from Camillo Golgi (1903).

2017). However, like Bichat, Virchow still failed to recognize the significance of these cells (Figure 2.1A).

It was not until the end of the 19th century that glial cells began to receive more attention from researchers. One of the key figures in this shift was the Spanish neuroscientist Santiago Ramón y Cajal, who is often considered the father of modern neuroscience. Ramón y Cajal used a staining technique (Figure 2.1B) developed by the Italian physician Camillo Golgi (Figure 2.1C) to produce highly detailed images of the nervous system, including the structure of glial cells (Magistretti and Ransom 2002).

The school of Ramón y Cajal was able to identify two main types of glial cells: astrocytes and oligodendrocytes. He deemed the former "protoplasmic astrocytes," as the larger and more numerous of the two types. They had multiple branches that extended out to surround neurons and blood vessels, forming a complex network of support. Oligodendrocytes, first referred to as "mesoglia," were smaller and had fewer branches (Robertson 1899). However, it was Ramón y Cajal's scholar Pio del Rio-Hortega who first suggested that oligodendrocytes were responsible for producing the myelin sheath surrounding axons (Pérez-Cerdá et al. 2015).

Through his work, del Río-Hortega identified a third type of glial cell: the microglia. Microglia are the immune cells of the nervous system, responsible for defending against infections and clearing away cellular debris. They are derived from the same stem cells as macrophages, which are the immune cells found in the rest of the body (Paolicelli et al.

2022). The name "astrocyte" was coined by the German scientist Michel von Lenhossek from the Greek words *astron*, meaning "star," and *kytos*, meaning "cell." In 1921, del Río-Hortega published a paper describing the morphology of glial cells in the central nervous system (CNS). In this paper, he used the term "astrocyte" to describe a specific type of glial cell that had numerous radiating processes resembling the rays of a star. In the years following del Río-Hortega's discovery of microglia, research on glial cells began to accelerate. In the 1930s, the American neuroscientist Harold Saxton Burr proposed a theory of "neural fields" in which glial cells played an essential role as regulators of electrical potential (Fields 2009).

More recent discoveries have shed light on the diverse functions of glial cells and their intricate interactions with neurons. It has become evident that glial cells participate in synaptic transmission, regulate synaptic plasticity, and contribute to information processing in the CNS. Moreover, studies have highlighted the involvement of glial cells in neuroinflammation and the release of pro-inflammatory cytokines, which can have both detrimental and protective effects on the brain (Araque et al. 2014; Calì et al. 2009, 2014; de Oliveira Figueiredo et al. 2022; Petrelli et al. 2020, 2023).

The history of glial cell research has been shaped by various experimental techniques and approaches, including histological staining, electron microscopy, genetic manipulation, and advanced imaging techniques. These tools have allowed researchers to unravel the complex biology of glial cells and their roles in normal brain function and disease. The history of the main discoveries surrounding glial cells has revealed their crucial roles in CNS development, function, and pathology. From the initial characterization of glia as structural elements to their recognition as active participants in neuronal signaling and immune responses, these cells have emerged as key players in understanding the complexities of the brain and its associated disorders (Fan and Agid 2018). Further research in this field holds great promise for developing novel therapeutic strategies targeting glial cells to treat a wide variety of neurological conditions.

Evolutionary Evidence

A lack of fossil evidence leads to the evolutionary origin of glial cells being relatively unknown. To understand glial origins, we observe developmental pathways for adult glial cells in combination with phylogenetic invertebrate neuroanatomy (Hartline 2011). For instance, the gene glial cell missing (*gcm*) controlling the differentiation of glial cells is not expressed in the mammalian CNS (Hosoya et al. 1995; Jones et al. 1995; Kim et al. 1998). The origin of glial cells can be traced back to the emergence of complex nervous systems in early metazoans, and the emergence of astrocytes can be traced back to the early chordates, where primitive forms of glial cells are present. As vertebrates evolved, they developed larger and more complex bodies requiring longer axons. Supposedly, glial cells evolved to support an increasingly demanding CNS (Figure 2.2). Specialized glial cells were necessary to insulate large axons at regular intervals, in the form of myelin.

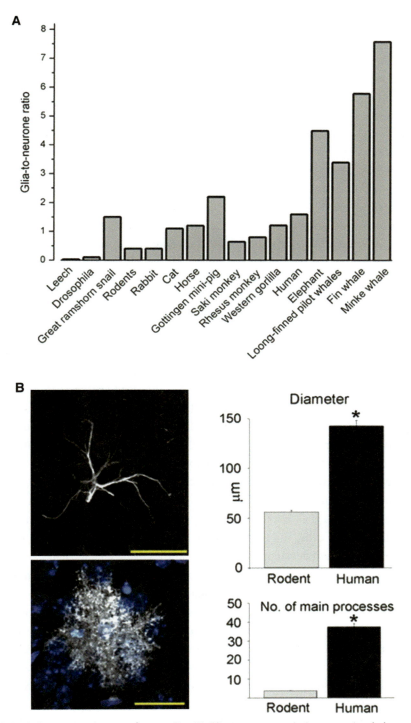

Figure 2.2 Phylogenetic advance of neuroglia. (A) Glia-to-neuron ratio increases in phylogeny, from invertebrates to mammals. Interesting to notice some outliers. Reproduced from Verkhratsky and Butt (2018). (B) Comparison between murine (top left) and human (bottom left) astrocytes. Graphs on the right show an increase in size and complexity. Adapted from Oberheim et al. (2009).

Thus, Schwann cells in the peripheral nervous system and oligodendrocytes in the CNS are specialized in myelin secretion, promoting rapid nerve conduction and velocity and ensuring the evolutionary success of vertebrates (Ackerman and Monk 2016). The evolution of astrocytes is closely linked to the development of radial glia, which serve as neural stem cells during embryonic neurogenesis. Radial glia eventually differentiated into astrocytes in more advanced organisms, and the complexity and diversity of astrocytes increased, leading to the specialized functions they exhibit today. In fish and amphibians, radial cells are the predominant astroglia-like cell, whereas most radial glia are replaced by astrocytes in adult mammals. Astrocytes are present in lizards, snakes, and caimans but absent in turtles. Despite astrocyte appearance in evolution possibly being an apomorphic trait, the combined presence of radial glia and astrocytes in reptiles may suggest astrocyte evolution as a contributor to increased brain size and complexity (Figure 2.2B; Falcone 2022). Astrocyte-like cells have even been observed in various invertebrate species, including *Hydra*, mollusks, *Drosophila*, and *Caenorhabditis elegans*, suggesting their early evolutionary appearance (Sharma et al. 2021). However, invertebrates' astrocyte-like cells are structurally and functionally distinct from those of vertebrates. While the astrocyte-like cells of invertebrates contribute to diverse functions maintaining neural circuits and overall CNS homeostasis, their roles differ from those of vertebrate astrocytes. For instance, in *Hydra*, astrocyte-like cells support interneuronal communication and play a role in neuronal patterning and regeneration (Ghysen 2003). In *Drosophila*, astrocyte-like glia provide metabolic support to neurons and regulate synaptic transmission (He et al. 2023; Scott et al. 2023). These findings suggest that even in simpler organisms, astrocyte-like cells fulfill essential roles in neuronal function and neural network integrity. Nonetheless, the phylogenetic origin of glial cells and the existence of a relationship between invertebrates and vertebrates is debated.

Developmentally, glia derive from the ectoderm. More specifically, astrocytes and oligodendroglia are neuroectoderm-derived, while microglia derive from embryonic yolk-sac cells (Sharma et al. 2021). The development of a CNS in vertebrates requires supportive cells like glial cells, and astrocytes are heterogeneous across species, throughout the CNS and at different developmental stages. Studies using glial fibrillary acidic protein immunohistochemistry revealed ependymal cells as abundant in fish and amphibians, variable in reptiles, and scarce in birds and mammals (Falcone et al. 2022).

In vertebrates, astrocytes exhibit increased morphological and functional diversity but a conserved genetic program of orthologous genes in rodents and primates, including humans (Geirsdottir et al. 2019; Bakken et al. 2021; Falcone, 2022; Falcone et al. 2021; Falcone et al. 2022). Astrocytes in the vertebrate CNS are characterized by their unique stellate morphology, extensive processes, and close association with neurons and blood vessels (Calì, Agus, et al. 2019). In fish and amphibians, astrocytes show increased complexity and structural organization compared to invertebrate glial cells. These astrocytes contribute to ion homeostasis, metabolic support, and neurotransmitter recycling (Jurisch-Yaksi et al. 2020).

In reptiles, birds, and mammals, astrocytes further evolved into diverse subtypes with specialized functions. Mammalian astrocytes, in particular, demonstrate remarkable complexity and heterogeneity. They display a star-shaped morphology with extensive branching processes and possess unique properties that contribute to synaptic regulation, neurovascular coupling, and modulation of neuronal excitability. Overall, glial morphology, including soma size, branching, and territorial occupancy, varies greatly among species (Geirsdottir et al. 2019; Falcone et al. 2019).

Recent research has revealed the existence of distinct astrocyte subtypes with specialized functions in the vertebrate CNS. These subtypes include, among others, protoplasmic astrocytes in the gray matter and fibrous astrocytes in the white matter. Protoplasmic astrocytes are involved in synapse formation, neurotransmitter uptake, and metabolic support, while fibrous astrocytes provide structural support and contribute to myelination (Bayraktar et al. 2020; Escartin et al. 2021; Bezzi et al. 2023).

Astrocytes in mammals, particularly humans and great apes, exhibit remarkable functional complexity and have been extensively studied (Figure 2.2B; Colombo et al. 2004; Falcone et al. 2019, 2022; Munger et al. 2022). There is evidence that, among primates, astrocytes are crucially involved in brain expansion and evolution and that cortical layer I astrocytes likely play a major function in human brain metabolic regulation and cognition. Interestingly, a particular type of astrocyte, the intralaminar astrocyte (ILA), resides in cortical layer I in most therian mammal orders. ILAs can be classified into two morphological types based on the length of their processes, with pial ILAs being confined to layer I and subpial ILAs showing processes extending to the deep layers of the cortex (Colombo et al. 2004; Falcone, 2022; Falcone et al. 2022). It is interesting to note that while pial ILAs are broadly distributed in all therian mammal species examined, subpial ILAs are found only in Primates and exhibit an increased morphologic complexity in great apes and humans compared to the rest of anthropoid species. The processes of ILAs contact neuronal blood vessels and meninges, implying evolutionary differences in the composition of the blood–brain barrier and meningeal interfaces in hominid primates including humans.

Glial Metabolism Supports Brain Evolution

Astrocytes also play essential roles in maintaining the blood–brain barrier, regulating extracellular ion concentrations, and modulating synaptic activity. Astrocytes are also involved in neurogenesis, neuroinflammation, and the formation and maintenance of tripartite synapses, where they actively participate in bidirectional communication with neurons and other glial cells by sensing neurotransmitters and releasing gliotransmitters from lamelliform perisynaptic processes in juxtaposition to synapses (Bezzi et al. 1998; Calì, Tauffenberger, and Magistretti 2019; Veloz Castillo et al. 2021). In humans, studies have revealed that astrocytes regulate extracellular ion concentrations, such as potassium and calcium, ensuring optimal neuronal excitability. Astrocytes also participate in the formation and maintenance of the blood–brain barrier, which protects the brain from harmful substances. Furthermore, astrocytes provide metabolic support to neurons by taking up glucose and lactate from the bloodstream and metabolizing them into energy sources (Bittar et al. 1996; Carteron et al. 2018).

The utilization of energy by the brain varies significantly across different species (Herculano-Houzel 2011). In rodents like rats and mice, glucose utilization ranges from 0.3 to 1.2 μmol/min, while in nonhuman primates, it increases to 30–60 μmol/min. Humans have the highest glucose utilization at 450 μmol/min. Similarly, the percentage of energy resources allocated to the brain also varies, with rodents using 2% of whole-body energy consumption, nonhuman primates using 9%–12%, and humans using 20% (Mink et al. 1981). Interestingly, the increase in energy utilization by the brain in relation to body size is greater than what would be expected. In humans, the neocortex, which accounts for over 40% of whole-brain energy utilization, has undergone a massive development (Martin 1981; Karbowski 2007). This expansion of the neocortex is associated with higher cognitive functions in humans compared to other nonhuman primate species (Lennie 2003).

The increase in brain size and neocortex expansion in humans is accompanied by a significant increase in the number of neurons (Figure 2.2). Humans have approximately 16 billion cortical neurons, while nonhuman primates have 1.7–5.5 billion and rodents have 10–31 million (Herculano-Houzel 2011). Moreover, the complexity of dendritic arborization, including the number of spines and synapses, increases more than linearly in humans. These factors lead to a lower density of neurons per unit volume in humans compared to other species and a more efficiently connected brain network. Consequently, the energy utilization per unit volume, measured by glucose use, is considerably lower in humans (Herculano-Houzel 2011). The question of whether energy consumption per neuron remains constant across species or if neurons with larger arborizations consume more energy is still debated.In addition to neurons, the contribution of glial cells, particularly astrocytes, to brain energy metabolism evolution is noteworthy. The glia/neuron ratio increases across evolution, with a ratio of 1.6 in the human neocortex. The glia/neuron ratio in mice and rats, the most commonly used models in neuroscience research, is about 0.4 and, interestingly for some large mammals like elephants and whales, can reach values from 5 to 8 (Figure 2.2A). Hence, the hypothesis that the number of glial cells might be linearly linked to higher cognitive capabilities falls under this evidence. However, astrocytes, the cells that are mainly responsible for glucose uptake, are more abundant in the human cortex than in other species and are larger and more complex compared to rodents (Figure 2.2B), pointing to the fact that the processing power of the human neocortex might be linked to energy management. Indeed, each human neocortical neuron consumes three times more adenosine triphosphate (ATP) than neurons in other species, primarily for maintaining resting membrane potential and firing (Lennie 2003).

Gene-profiling studies have shown that genes involved in energy production, particularly oxidative phosphorylation, are highly expressed in the human neocortex compared to nonhuman primates (Cáceres et al. 2003; Uddin et al. 2004; Fu et al. 2011; Bakken et al. 2021). This suggests that the human neocortex has evolved to prioritize energy production through an energy-efficient metabolic pathway (Grossman et al. 2004; Khaitovich et al. 2006). However, this pathway also carries the risk of producing more potentially damaging reactive oxygen species (ROS). Metabolomic analysis indicates specific metabolic changes in the human prefrontal cortex, reflecting a fourfold increase compared to nonhuman primates. Notably, humans show remarkable differences in metabolites related to learning, memory, and synaptic transmission compared to nonhuman primates (Fu et al. 2011), possibly representing a frailty for the onset of neurological and psychiatric disease specific to humans. For instance, the higher turnover of glutamate observed in newborns (Fu et al. 2011) might represent a ground for dysregulating synaptogenesis during development, which might result in the development of psychiatric disorders such as schizophrenia or autism (Petrelli et al. 2020, 2023; see Chapters 14 and 15, respectively) or aberrant ROS accumulation resulting in oxidative stress, which is a common denominator of many chronic diseases and pathological states like Alzheimer's disease (Chapter 8) and Parkinson's disease (Chapter 9; Jomova et al. 2023). There is an accelerated metabolome evolution in the human prefrontal cortex and skeletal muscle, two organs sharing an energy management system, such as the accumulation of glycogen and the use of glycolysis when a power boost is needed, such as during learning and memory (Veloz Castillo et al. 2021). Such evolution indicates a preferential allocation of energy production to the brain in humans (Bozek et al. 2014).

The metabolic coupling between astrocytes and neurons has been formalized based on extensive experimental evidence as the astrocyte–neuron–lactate shuttle (ANLS) model (Pellerin and Magistretti 1994; Magistretti and Allaman 2018). The ANLS puts at center stage astrocytes as the main gateway of glucose into the brain via glucose transporter 1 and the site of aerobic glycolysis resulting in lactate release for the use of neurons (Magistretti and Allaman 2018). Lactate can also be produced from glycogen, which is exclusively localized in astrocytes under physiological conditions (Calì, Tauffenberger, and Magistretti 2019; Veloz Castillo et al. 2021). The ANLS has been demonstrated in invertebrates such as the *Drosophila* nervous system (Volkenhoff et al. 2015) and the honeybee drone retina (Tsacopoulos and Magistretti 1996).

Comparisons of lactate dehydrogenase (LDH) isoform expression levels across species reveal that lactate is the metabolic substrate for the tricarboxylic acid cycle and electron transport chain in the human neocortex (Duka et al. 2014). Observations in rodents show that LDHA (the form present in lactate-producing cells in peripheral tissues) is expressed exclusively in astrocytes, while neurons express only LDHB (Bittar et al. 1996). The predominant form of LDH in the human neocortex is LDHB, supporting the idea that neurons primarily utilize lactate as an energy substrate.

In addition to being an energy substrate, lactate released by astrocytes acts as a signaling molecule promoting expression of plasticity and neuroprotective genes in neurons (Margineanu et al. 2018). For instance, it has been shown that the lactate cycle stimulated by the glutamate–glutamine cycle can act on *N*-methyl-D-aspartate (NMDA) receptors by activating the transduction of immediate early genes, which are relevant to plasticity (e.g., *ARC*, *c-FOS*, and *Zif268*). Lactate potentiates NMDA receptor signaling and modifies the intracellular pH by promoting the increase of nicotinamide adenine dinucleotide (Yang et al. 2014); furthermore, the transfer of lactate from astrocytes to neurons is necessary for memory consolidation (Suzuki et al. 2011; Gao et al. 2016). Interestingly, analyses at the ultrastructural level showed that glycogen granules in astrocytes are predominantly localized close to neuronal dendritic spines (Figure 2.3; Calì et al.

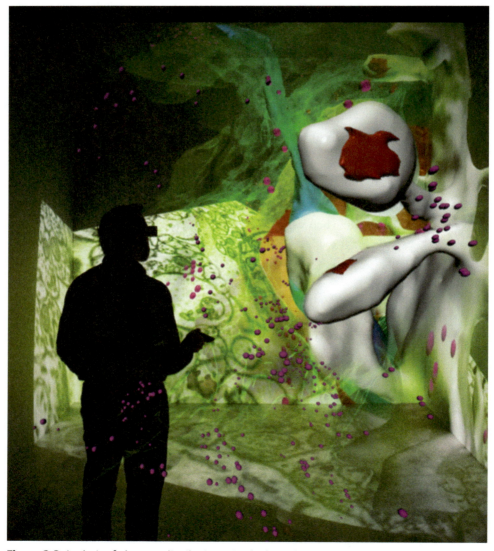

Figure 2.3 Analysis of glycogen distribution using high-resolution three-dimensional electron microscopy, three-dimensional models of astrocytes, and immersive virtual reality tools reveals that glycogen is compartmentalized around spines. Adapted from Calì et al. (2016).

2016; Agus et al. 2018). These astrocytic glycogen granules are subject to plasticity: their size and numbers increase in the hippocampus after learning paradigms such as novel object recognition (Vezzoli et al. 2019).

Data from gene expression analysis in the whole human brain indicate that expression of monocarboxylate transporter 2, which is exclusively expressed in neurons where it mediates the entry of lactate, is correlated to glucose utilization as monitored by fluorodeoxyglucose positron emission tomography (^{18}F-FDG PET) and to functional magnetic resonance imaging (MRI) images in brain regions associated with attention, arousal, and stress function (Medel et al. 2022).

Overall, these findings demonstrate that energy utilization in the brain varies across species, with humans exhibiting unique characteristics such as higher energy consumption, increased neuron and glia counts, and specific gene expression patterns related to energy production. These factors contribute to the distinctive higher cognitive functions observed in humans compared to other nonhuman primate species (Bittar et al. 1996; Duka et al. 2014).

The phylogenesis of astrocytes highlights their evolutionary significance and their integral role in supporting the function and development of the nervous system. From their early appearance in invertebrates to their highly specialized roles in mammals, astrocytes have undergone significant diversification and functional adaptation. Further research is needed to unravel the molecular mechanisms underlying astrocyte evolution and to fully understand their contributions to brain function and neurological disorders.

Glia in Health and Disease

Neurological disorders pose significant challenges to individuals and society, and there is a pressing need for effective treatment options. In recent years, there have been exciting advancements in the field of neurology, particularly in the area of clinical treatment targeting glial cells.

Glial cells have important roles in the realm of brain disease. Astrocytic swelling can be a helpful component in regulating homeostasis; by physically increasing their volume and retracting some of the smaller processes. Astrocytes can limit the flow of neurotransmitters that might non-specifically activate synapses and aid in sealing the blood–brain barrier. Astrocytes have also been suggested to aid in detoxification by swelling via osmotic forces. On the contrary, reactive gliosis resulting in dramatic swelling can be harmful in acute injuries such as stroke and trauma. This can lead to glial scarring or induce pain or even epilepsy.

Myelinating glia are essential participants in nervous cell function and specifically participate in cell repair and regeneration. Perturbation of this critical process creates nervous system deregulation, as is the case for diseases like multiple sclerosis, for example, which is caused by failed myelin repair. Other neurodegenerative disorders present neuropathologically with a prominent form of glial tauopathy. While astrogliosis and tau accumulation can be observed in Alzheimer's disease, this is particularly evident in globular glial tauopathy, a rare neurodegenerative disease characterized by globular phosphorylated tau inclusions in astrocytes and oligodendrocytes (Murray et al. 2014; Kovacs 2015) and chronic traumatic encephalopathy (McKee et al. 2009, 2013, 2015) that are associated with distinct forms of neuronal and glial hyperphosphorylated tau accumulation. Recent reports documented the presence of astrogliosis, but not of tau gliopathy, in the frontal and parietal cortex of combative bovid species (Ackermans et al. 2022), as well as abnormalities of perivascular cortical astrocytes in rats subjected to repeated mild traumatic brain injury, a known cause of neuropsychiatric comorbidities like depression (Chapter 13; Sosa et al. 2013; Gama Sosa et al. 2019; Dickstein et al. 2021).

In addition, astrocytes containing abnormally phosphorylated tau protein have been described in postmortem brains form elderly patients and characterized as "aging-related tau astrogliopathy" to refer to a spectrum of pathological presentations (Kovacs et al. 2016; 2017). In the context of primate brain evolution, it is worth noting that significant alterations in glial cell populations have been reported during aging in several primate species including mice, lemurs, marmosets, tamarins, squirrel monkeys, macaques, baboons, vervets, and guenons, as well as great apes (for a comprehensive review, see Freire-Cobo et al. 2021, 2023). All species examined show the presence of reactive astrocytes and microglia, frequently in the vicinity of, or in direct contact with, age-related amyloid deposits. An age-related decline in oligodendrocyte and myelinated axons is also well documented (Peters 2009; Peters and Kemper 2012). The case of great apes (hominids) is of particular relevance in the context of age-related neurodegenerative pathologies. Whereas most nonhuman primates do not show extensive tauopathy in aging but do show a certain degree of amyloid deposition, great apes consistently exhibit Alzheimer's disease–type degenerative changes in the cerebral cortex (Perez et al. 2013; 2016; Edler et al. 2018; Munger et al. 2019; Freire-Cobo et al. 2021). Quantitative analyses in chimpanzees have shown the presence of inflammatory response around such lesions with severe astrogliosis and microglial activation with particular involvement of subpial ILAs, indicating a certain degree of involvement of glial responses in brain aging in great apes, although these changes are not as marked as in humans with Alzheimer's disease (Edler et al. 2018; 2023; Munger et al. 2019). This suggests that chimpanzees' cognitive capabilities may be somewhat better preserved during aging compared to humans, while not being entirely immune to neurodegenerative changes.

A significant breakthrough in neurological research has been the utilization of induced pluripotent stem cells (iPSCs) to model and investigate various neurological disorders. iPSC-derived neurons have enabled researchers to study disease mechanisms, identify cellular and molecular deficits, and develop new treatment strategies. Considering the success of the approach, iPSC-derived glia are more and more used to highlight their aberrant phenotypes in a plethora of models of disorders such as Alzheimer's disease, Parkinson's disease, fragile X syndrome, Rett syndrome, and autism spectrum disorder. Indeed, in most of them, astrocytes and microglia respond to chemical stimuli that make them reactive and active, respectively, a similar response to acute injury or inflammation. The use of iPSCs offers the chance to evaluate the properties of these cells *in vitro* using omics approaches, under controlled culture environments, or even transplanting them to develop pathological models. The use of brain organoids, three-dimensional models that mimic aspects of human brain development, has also proven valuable in understanding neurodevelopmental abnormalities and exploring potential therapeutic interventions (Cordella et al. 2022).Unanswered Questions

Despite remarkable advances in understanding the diverse functions of glial cells including astrocytes, oligodendrocytes, and microglia in brain health, synaptic transmission, and neurological disorders, several unanswered questions remain in the field of glial cell biology, offering exciting avenues for future research and promising perspectives for future discoveries.

Heterogeneity

The heterogeneity and functional diversity of glial cells are intriguing features. More and more studies are showing how structural and functional diversity of astrocytes is linked to neuronal regional subtypes, implying that specialized functions are adapted to fulfill the needs of specific brain regions or circuits. Moreover, their heterogeneity appears at multiple scales, across different areas, within neuropil of the same region, or in subdomains of the same cell: for instance, electron microscopic studies are revealing the diversity of astrocytic processes in the neuropil. Understanding the molecular and physiological mechanisms driving this diversity, the structural heterogeneity, as well as the functional implications of different glial subtypes is an important and unresolved subject.

Neuron–Glia Interactions

Glial cells actively engage in brain development by forming synapses, performing myelination, and regulating blood flow. However, not all methods through which glial cells regulate and interact with neurons are entirely recognized. To this regard, metabolic interaction between glial cells, in particular astrocytes, and neurons is of the outmost importance. The significance of glial cells in these fundamental elements of brain biology can give important insights into brain development and plasticity.

Disease

While glial cells have been linked to a variety of neurological diseases such as Alzheimer's disease, multiple sclerosis, amyotrophic lateral sclerosis (Chapter 11), schizophrenia, autism spectrum disorder, and ADHD (Chapter 16), to name a few, the precise contributions and underlying processes remain poorly understood. Unveiling the role of glial cells in disease etiology, their interactions with neurons, and their influence on synaptic function and connectivity offers considerable potential for the development of innovative treatment techniques. At present, only few glia-specific drugs are available to treat neurological diseases (Finsterwald et al. 2015; Beard et al. 2021). For instance, l-lactate during brain reperfusion has been successfully used to limit the effects of ischemia (Magistretti and Allaman 2018) or to treat clinical depression (Carrard et al. 2021); and inhibitors of LDH, glycogen synthase, or phosphorylase have been used to treat epilepsy and its comorbidities, such as stripendiol (Sada et al. 2015) and SKF83959 (Guo et al. 2022).

Enteric Nervous System

Beyond the CNS, there is a specialized population of glial cells known as "enteric glia" in the enteric nervous system (ENS). The involvement of enteric glia in gastrointestinal illnesses such as enteric neuropathies and intestinal motility problems is still being studied, with the ENS thought to be able to adjust the activity of the CNS via the gut–brain axis, most likely via the vagus nerve. Understanding intercellular communication within the ENS, as well as the role of enteric glial cells in gut homeostasis and disease development, is an attractive topic for future study.

Future Perspectives

The field of glial cell biology holds tremendous potential for advancing our understanding of brain function and the development of novel therapeutic approaches. Future research efforts should focus on addressing the unanswered questions and exploring the following avenues.

> *Glial Biomarkers*: Identification of glial cells and their different subtypes to allow histochemical studies and genetic manipulation suffers from the problem of specificity to identify these cells. The development of specialized biomarkers formatted to detect astrocyte or microglial inflammation in blood, cerebrospinal fluid, and saliva is an emerging technique that shows promise in the early detection of neurodegenerative disease. These techniques will also broaden our longitudinal understanding of glial physiology in vivo and how it changes over time.
>
> *Advanced Imaging Techniques*: Advancements in imaging technologies, such as super-resolution microscopy and live-cell imaging, will enable researchers to study glial cells with enhanced spatial and temporal resolution. Advanced imaging will facilitate the investigation of glial cell dynamics, intercellular communication, and their interactions with neurons in real time. PET-MRI using targets of glial inflammation is another technique currently expanding our understanding of pathological glial interactions in vivo throughout the brain.
>
> *Single-Cell Transcriptomics*: The application of single-cell RNA sequencing to glial cells will provide comprehensive insights into their heterogeneity, molecular profiles, and functional states. This approach will help uncover new glial subtypes and signaling pathways, shedding light on their diverse functions and potential therapeutic targets. In particular, large transcriptomics databases will allow the identification of novel specific markers.

Genetic Manipulation: Glia-specific genetic tools, such as cell-specific knockout and gene-editing techniques, will allow researchers to manipulate glial cells selectively. This will enable precise investigation of their contributions to brain development, synaptic plasticity, and disease pathogenesis.

Functional Studies: Advancing our understanding of glial cell function will require integrating experimental approaches, including electrophysiology, optogenetics, and calcium imaging, to decipher the dynamic interactions between glial cells and neurons in both healthy and diseased states.

Therapeutic Interventions: With a deeper understanding of glial cell biology, the development of targeted therapies aimed at modulating glial functions to treat neurological disorders becomes a realistic possibility. Glia-specific drugs, gene therapies, and stem cell–based approaches may emerge as potential strategies for restoring brain homeostasis and promoting repair (Zhang et al. 2023).

In conclusion, while significant progress has been made in unraveling the roles of glial cells in the brain, numerous unanswered questions remain. Exploring the heterogeneity, functional diversity, contributions to brain development and disease, and interactions with neurons will fuel further advancements in glial cell biology. The future perspectives on this field are promising as they hold the potential to uncover fundamental mechanisms underlying brain function and provide novel therapeutic strategies for neurological disorders.

Key Points to Remember

- Glial cells make up the supportive system of the nervous system and have a variety of functions essential to maintaining nervous system regulation.
- Vertebrates and invertebrates have similar "glia-type cells" with different functions. As brains increase in complexity, so do glial cells, especially astrocyte-type cells.
- Primates show a specific subset of astrocytes known as "intralaminar astrocytes."
- Glial cells are heterogeneous among brain regions, depending on environmental molecules, but also ultrastructurally due to their high compartmentalization.
- Glial dysregulation can result in a number of neurological disorders.
- Glial morphological and functional diversity remains to be investigated and possibly depends on specific molecular markers.

References

Ackerman, S. D., & Monk, K. R. (2016). The scales and tales of myelination: Using zebrafish and mouse to study myelinating glia. *Brain Research*, *1641*, 79–91. https://doi.org/10.1016/j.brainres.2015.10.011

Ackermans, N. L., Varghese, M., Williams, T. M., Grimaldi, N., Selmanovic, E., Alipour, A., Balchandani, P., Reidenberg, J. S., & Hof, P. R. (2022). Evidence of traumatic brain injury in headbutting bovids. *Acta Neuropathologica*, *144*, 5–26. https://doi.org/10.1007/s00401-022-02427-2

Agus, M., Boges, D., Gagnon, N., Magistretti, P. J., Hadwiger, M., & Cali, C. (2018). GLAM: Glycogen-derived Lactate Absorption Map for visual analysis of dense and sparse surface reconstructions of rodent brain structures on desktop systems and virtual environments. *Computers & Graphics, 74*, 85–98. https://doi.org/10.1016/j.cag.2018.04.007

Araque, A., Carmignoto, G., Haydon, P. G., Oliet, S. H. R., Robitaille, R., & Volterra, A. (2014). Gliotransmitters travel in time and space. *Neuron, 81*, 728–739. https://doi.org/10.1016/j.neuron.2014.02.007

Bakken, T. E., Jorstad, N. L., Hu, Q., Lake, B. B., Tian, W., Kalmbach, B. E., Crow, M., Hodge, R. D., Krienen, F. M., Sorensen, S. A., Eggermont, J., Yao, Z., Aevermann, B. D., Aldridge, A. I., Bartlett, A., Bertagnolli, D., Casper, T., Castanon, R. G., Crichton, K., . . . Lein, E. S. (2021). Comparative cellular analysis of motor cortex in human, marmoset and mouse. *Nature, 598*, 111–119. https://doi.org/10.1038/s41586-021-03465-8

Bayraktar, O. A., Bartels, T., Holmqvist, S., Kleshchevnikov, V., Martirosyan, A., Polioudakis, D., Ben Haim, L., Young, A. M. H., Batiuk, M. Y., Prakash, K., Brown, A., Roberts, K., Paredes, M. F., Kawaguchi, R., Stockley, J. H., Sabeur, K., Chang, S. M., Huang, E., Hutchinson, P., . . . Rowitch, D. H. (2020). Astrocyte layers in the mammalian cerebral cortex revealed by a single-cell in situ transcriptomic map. *Nature Neuroscience, 23*, 500–509. https://doi.org/10.1038/s41593-020-0602-1

Beard, E., Lengacher, S., Dias, S., Magistretti, P. J., & Finsterwald, C. (2021). Astrocytes as key regulators of brain energy metabolism: New therapeutic perspectives. *Frontiers in Physiology, 12*, Article 825816. https://doi.org/10.3389/fphys.2021.825816

Bezzi, P., Carmignoto, G., Pasti, L., Vesce, S., Rossi, D., Rizzini, B. L., Pozzan, T., & Volterra, A. (1998). Prostaglandins stimulate calcium-dependent glutamate release in astrocytes. *Nature, 391*, 281–285. https://doi.org/10.1038/34651

Bezzi, P., Magnaghi, V., Paolicelli, R. C., & Hornung, J.-P. (2023). Glial heterogeneity: Impact on neuronal function and dysfunction [Editorial]. *Frontiers in Neuroanatomy, 17*, Article 1249919. https://doi.org/10.3389/fnana.2023.1249919

Bittar, P. G., Charnay, Y., Pellerin, L., Bouras, C., & Magistretti, P. J. (1996). Selective distribution of lactate dehydrogenase isoenzymes in neurons and astrocytes of human brain. *Journal of Cerebral Blood Flow & Metabolism, 16*, 1079–1089. https://doi.org/10.1097/00004647-199611000-00001

Bozek, K., Wei, Y., Yan, Z., Liu, X., Xiong, J., Sugimoto, M., Tomita, M., Pääbo, S., Pieszek, R., Sherwood, C. C., Hof, P. R., Ely, J. J., Steinhauser, D., Willmitzer, L., Bangsbo, J., Hansson, O., Call, J., Giavalisco, P., & Khaitovich, P. (2014). Exceptional evolutionary divergence of human muscle and brain metabolomes parallels human cognitive and physical uniqueness. *PLOS Biology, 12*, Article e1001871. https://doi.org/10.1371/journal.pbio.1001871

Cáceres, M., Lachuer, J., Zapala, M. A., Redmond, J. C., Kudo, L., Geschwind, D. H., Lockhart, D. J., Preuss, T. M., & Barlow, C. (2003). Elevated gene expression levels distinguish human from non-human primate brains. *Proceedings of the National Academy of Sciences of the United States of America, 100*, 13030–13035. https://doi.org/10.1073/pnas.2135499100

Calì, C. (2017). Astroglial anatomy in the times of connectomics. *Journal of Translational Neuroscience, 2*, 31–40. https://doi.org/10.3868/j.issn.2096-0689.2017.04.004

Calì, C., Agus, M., Kare, K., Boges, D. J., Lehvaslaiho, H., Hadwiger, M., & Magistretti, P. J. (2019). 3D cellular reconstruction of cortical glia and parenchymal morphometric analysis from serial block-face electron microscopy of juvenile rat. *Progress in Neurobiology, 183*, Article 101696. https://doi.org/10.1016/j.pneurobio.2019.101696

Calì, C., Baghabra, J., Boges, D. J., Holst, G. R., Kreshuk, A., Hamprecht, F. A., Srinivasan, M., Lehvaslaiho, H., & Magistretti, P. J. (2016). Three-dimensional immersive virtual reality for studying cellular compartments in 3D models from EM preparations of neural tissues. *Journal of Comparative Neurology, 524*, 23–38. https://doi.org/10.1002/cne.23852

Cali, C., Lopatar, J., Petrelli, F., Pucci, L., & Bezzi, P. (2014). G-protein coupled receptor–evoked glutamate exocytosis from astrocytes: role of prostaglandins. *Neural Plasticity, 2014*, Article 254574. https://doi.org/10.1155/2014/254574

Calì, C., Marchaland, J., Spagnuolo, P., Gremion, J., & Bezzi, P. (2009). Regulated exocytosis from astrocytes physiological and pathological related aspects. *International Review of Neurobiology, 85*, 261–293. https://doi.org/10.1016/S0074-7742(09)85020-4

Calì, C., Tauffenberger, A., & Magistretti, P. (2019). The strategic location of glycogen and lactate: From body energy reserve to brain plasticity. *Frontiers in Cellular Neuroscience, 13*, Article 82. https://doi.org/10.3389/fncel.2019.00082

Carrard, A., Cassé, F., Carron, C., Burlet-Godinot, S., Toni, N., Magistretti, P. J., & Martin, J.-L. (2021). Role of adult hippocampal neurogenesis in the antidepressant actions of lactate. *Molecular Psychiatry, 26*, 6723–6735. https://doi.org/10.1038/s41380-021-01122-0

Carteron, L., Solari, D., Patet, C., Quintard, H., Miroz, J.-P., Bloch, J., Daniel, R. T., Hirt, L., Eckert, P., Magistretti, P. J., & Oddo, M. (2018). Hypertonic lactate to improve cerebral perfusion and glucose availability after acute brain injury. *Critical Care Medicine, 46*, 1649–1655. https://doi.org/10.1097/CCM.0000000000003274

Colombo, J. A., Sherwood, C. C., & Hof, P. R. (2004). Interlaminar astroglial processes in the cerebral cortex of great apes. *Anatomy and Embryology, 208*, 215–218. https://doi.org/10.1007/s00429-004-0391-4

Cordella, F., Ferrucci, L., D'Antoni, C., Ghirga, S., Brighi, C., Soloperto, A., Gigante, Y., Ragozzino, D., Bezzi, P., & Di Angelantonio, S. (2022). Human iPSC-derived cortical neurons display homeostatic plasticity. *Life, 12*, Article 1884. https://doi.org/10.3390/life12111884

de Oliveira Figueiredo, E. C., Calì, C., Petrelli, F., & Bezzi, P. (2022). Emerging evidence for astrocyte dysfunction in schizophrenia. *Glia, 70*, 1585–1604. https://doi.org/10.1002/glia.24221

Dickstein, D. L., De Gasperi, R., Gama Sosa, M. A., Perez-Garcia, G., Short, J. A., Sosa, H., Perez, G. M., Tschiffely, A. E., Dams-O'Connor, K., Pullman, M. Y., Knesaurek, K., Knutsen, A., Pham, D. L., Soleimani, L., Jordan, B. D., Gordon, W. A., Delman, B. N., Shumyatsky, G., Shahim, P.-P., . . . Elder, G. A. (2021). Brain and blood biomarkers of tauopathy and neuronal injury in humans and rats with neurobehavioral syndromes following blast exposure. *Molecular Psychiatry, 26*, 5940–5954. https://doi.org/10.1038/s41380-020-0674-z

Duka, T., Anderson, S. M., Collins, Z., Raghanti, M. A., Ely, J. J., Hof, P. R., Wildman, D. E., Goodman, M., Grossman, L. I., & Sherwood, C. C. (2014). Synaptosomal lactate dehydrogenase isoenzyme composition is shifted toward aerobic forms in primate brain evolution. *Brain, Behavior and Evolution, 83*, 216–230. https://doi.org/10.1159/000358581

Edler, M. K., Munger, E. L., Maycon, H., Hopkins, W. D., Hof, P. R., Sherwood, C. C., & Raghanti, M. A. (2023). The association of astrogliosis and microglial activation with aging and Alzheimer's disease pathology in the chimpanzee brain. *Journal of Neuroscience Research, 101*, 881–900. https://doi.org/10.1002/jnr.25167

Edler, M. K., Sherwood, C. C., Meindl, R. S., Munger, E., Hopkins, W. D., Ely, J. J., Erwin, J. M., Perl, D. P., Mufson, E. J., Hof, P. R., & Raghanti, M. A. (2018). Microglia changes associated to Alzheimer's disease pathology in aged chimpanzees. *Journal of Comparative Neurology, 526*, 2921–2936. https://doi.org/10.1002/cne.24484

Escartin, C., Galea, E., Lakatos, A., O'Callaghan, J. P., Petzold, G. C., Serrano-Pozo, A., Steinhäuser, C., Volterra, A., Carmignoto, G., Agarwal, A., Allen, N. J., Araque, A., Barbeito, L., Barzilai, A., Bergles, D. E., Bonvento, G., Butt, A. M., Chen, W.-T., Cohen-Salmon, M., . . . Verkhratsky, A. (2021). Reactive astrocyte nomenclature, definitions, and future directions. *Nature Neuroscience, 24*, 312–325. https://doi.org/10.1038/s41593-020-00783-4

Falcone, C. (2022). Evolution of astrocytes: From invertebrates to vertebrates. *Frontiers in Cell and Developmental Biology, 10*, Article 931311. https://doi.org/10.3389/fcell.2022.931311

Falcone, C., Penna, E., Hong, T., Tarantal, A. F., Hof, P. R., Hopkins, W. D., Sherwood, C. C., Noctor, S. C., & Martínez-Cerdeño, V. (2021). Cortical interlaminar astrocytes are generated prenatally, mature postnatally, and express unique markers in human and nonhuman primates. *Cerebral Cortex, 31*(1), 379–395.

Falcone, C., Wolf-Ochoa, M., Amina, S., Hong, T., Vakilzadeh, G., Hopkins, W. D., Hof, P. R., Sherwood, C. C., Manger, P. R., Noctor, S. C., & Martínez-Cerdeño, V. (2019). Cortical interlaminar astrocytes across the therian mammal radiation. *Journal of Comparative Neurology, 527*(10), 1654–1674. doi:10.1002/cne.24605

Fan, X., & Agid, Y. (2018). At the origin of the history of glia. *Neuroscience, 385*, 255–271. https://doi.org/10.1016/j.neuroscience.2018.05.050

Fields, R. D. (2009). *The other brain: From dementia to schizophrenia, how new discoveries about the brain are revolutionizing medicine and science*. Simon & Schuster.

Finsterwald, C., Magistretti, P. J., & Lengacher, S. (2015). Astrocytes: New targets for the treatment of neurodegenerative diseases. *Current Pharmaceutical Design, 21*, 3570–3581. https://doi.org/10.2174/1381612821666150710144502

Freire-Cobo, C., Edler, M. K., Varghese, M., Munger, E., Laffey, J., Raia, S., In, S. S., Wicinski, B., Medalla, M., Perez, S. E., Mufson, E. J., Erwin, J. M., Guevara, E. E., Sherwood, C. C., Luebke, J. I., Lacreuse, A., Raghanti, M. A., & Hof, P. R. (2021). Comparative neuropathology in aging primates: A perspective. *American Journal of Primatology, 83*, Article e23299. https://doi.org/10.1002/ajp.23299

Freire-Cobo, C., Rothwell, E. S., Varghese, M., Edwards, M., Janssen, W. G. M., Lacreuse, A., & Hof, P. R. (2023). Neuronal vulnerability to brain aging and neurodegeneration in cognitively impaired marmoset monkeys (*Callithrix jacchus*). *Neurobiology of Aging, 123*, 49–62. https://doi.org/10.1016/j.neurobiolaging.2022.12.001

Fu, X., Giavalisco, P., Liu, X., Catchpole, G., Fu, N., Ning, Z.-B., Guo, S., Yan, Z., Somel, M., Pääbo, S., Zeng, R., Willmitzer, L., & Khaitovich, P. (2011). Rapid metabolic evolution in human prefrontal cortex. *Proceedings of the National Academy of Sciences of the United States of America, 108*, 6181–6186. https://doi.org/10.1073/pnas.1019164108

Gama Sosa, M. A., De Gasperi, R., Perez Garcia, G. S., Perez, G. M., Searcy, C., Vargas, D., Spencer, A., Janssen, P. L., Tschiffely, A. E., McCarron, R. M., Ache, B., Manoharan, R., Janssen, W. G., Tappan, S. J., Hanson, R. W., Gandy, S., Hof, P. R., Ahlers, S. T., & Elder, G. A. (2019). Low-level blast exposure disrupts gliovascular and neurovascular connections and induces a chronic vascular pathology in rat brain. *Acta Neuropathologica Communications, 7*, Article 6. https://doi.org/10.1186/s40478-018-0647-5

Gao, V., Suzuki, A., Magistretti, P. J., Lengacher, S., Pollonini, G., Steinman, M. Q., & Alberini, C. M. (2016). Astrocytic β2-adrenergic receptors mediate hippocampal long-term memory consolidation. *Proceedings of the National Academy of Sciences of the United States of America, 113*, 8526–8531. https://doi.org/10.1073/pnas.1605063113

Geirsdottir, L., David, E., Keren-Shaul, H., Weiner, A., Bohlen, S. C., Neuber, J., Balic, A., Giladi, A., Sheban, F., Dutertre, C.-A., Pfeifle, C., Peri, F., Raffo-Romero, A., Vizioli, J., Matiasek, K., Scheiwe, C., Meckel, S., Mätz-Rensing, K., van der Meer, F., . . . Prinz, M. (2019). Cross-species single-cell analysis reveals divergence of the primate microglia program. *Cell, 179*, 1609–1622.e16. https://doi.org/10.1016/j.cell.2019.11.010

Ghysen, A. (2003). The origin and evolution of the nervous system. *International Journal of Developmental Biology, 47*, 555–562.

Golgi, C. (1903). *Opera omnia* (Vol. 3). Ulrico Hoepli.

Grossman, L. I., Wildman, D. E., Schmidt, T. R., & Goodman, M. (2004). Accelerated evolution of the electron transport chain in anthropoid primates. *Trends in Genetics, 20*, 578–585. https://doi.org/10.1016/j.tig.2004.09.002

Guo, L., Gao, T., Jia, X., Gao, C., Tian, H., Wei, Y., Lu, W., Liu, Z., & Wang, Y. (2022). SKF83959 attenuates memory impairment and depressive-like behavior during the latent period of epilepsy via allosteric activation of the sigma-1 receptor. *ACS Chemical Neuroscience, 13*, 3198–3209. https://doi.org/10.1021/acschemneuro.2c00629

Hartline, D. K. (2011). The evolutionary origins of glia. *Glia, 59*, 1215–1236. https://doi.org/10.1002/glia.21149

He, L., Wu, B., Shi, J., Du, J., & Zhao, Z. (2023). Regulation of feeding and energy homeostasis by clock mediated *Gart* in *Drosophila*. *Cell Reports, 42*, Article 112912. https://doi.org/10.1016/j.celrep.2023.112912

Herculano-Houzel, S. (2011). Scaling of brain metabolism with a fixed energy budget per neuron: Implications for neuronal activity, plasticity and evolution. *PLOS ONE, 6*, Article e17514. https://doi.org/10.1371/journal.pone.0017514

Hosoya, T., Takizawa, K., Nitta, K., & Hotta, Y. (1995). Glial cells missing: A binary switch between neuronal and glial determination in *Drosophila*. *Cell, 82*, 1025–1036. https://doi.org/10.1016/0092-8674(95)90281-3

Jomova, K., Raptova, R., Alomar, S. Y., Alwasel, S. H., Nepovimova, E., Kuca, K., & Valko, M. (2023). Reactive oxygen species, toxicity, oxidative stress, and antioxidants: Chronic diseases and aging. *Archives of Toxicology, 97*, 2499–2574. https://doi.org/10.1007/s00204-023-03562-9

Jones, B. W., Fetter, R. D., Tear, G., & Goodman, C. S. (1995). Glial cells missing: A genetic switch that controls glial versus neuronal fate. *Cell, 82*, 1013–1023. https://doi.org/10.1016/0092-8674(95)90280-5

Jurisch-Yaksi, N., Yaksi, E., & Kizil, C. (2020). Radial glia in the zebrafish brain: Functional, structural, and physiological comparison with the mammalian glia. *Glia, 68*, 2451–2470. https://doi.org/10.1002/glia.23849

Karbowski, J. (2007). Global and regional brain metabolic scaling and its functional consequences. *BMC Biology, 5*, Article 18. https://doi.org/10.1186/1741-7007-5-18

Khaitovich, P., Tang, K., Franz, H., Kelso, J., Hellmann, I., Enard, W., Lachmann, M., & Pääbo, S. (2006). Positive selection on gene expression in the human brain. *Current Biology, 16*, R356–R358. https://doi.org/10.1016/j.cub.2006.03.082

Kim, J., Jones, B. W., Zock, C., Chen, Z., Wang, H., Goodman, C. S., & Anderson, D. J. (1998). Isolation and characterization of mammalian homologs of the *Drosophila* gene glial cells missing. *Proceedings of the National Academy of Sciences of the United States of America, 95*, 12364–12369. https://doi.org/10.1073/pnas.95.21.12364

Kovacs, G. G. (2015). Neuropathology of tauopathies: Principles and practice [Review]. *Neuropathology and Applied Neurobiology, 41*, 3–23. https://doi.org/10.1111/nan.12208

Kovacs, G. G., Ferrer, I., Grinberg, L. T., Alafuzoff, I., Attems, J., Budka, H., Cairns, N. J., Crary, J. F., Duyckaerts, C., Ghetti, B., Halliday, G. M., Ironside, J. W., Love, S., Mackenzie, I. R., Munoz, D. G., Murray, M. E., Nelson, P. T., Takahashi, H., Trojanowski, J. Q., . . . Dickson, D. W. (2016). Aging-related tau astrogliopathy (ARTAG): Harmonized evaluation strategy. *Acta Neuropathologica, 131*, 87–102. https://doi.org/10.1007/s00401-015-1509-x

Kovacs, G. G., Xie, S. X., Lee, E. B., Robinson, J. L., Caswell, C., Irwin, D. J., Toledo, J. B., Johnson, V. E., Smith, D. H., Alafuzoff, I., Attems, J., Bencze, J., Bieniek, K. F., Bigio, E. H., Bodi, I., Budka, H., Dickson, D. W., Dugger, B. N., Duyckaerts, C., . . . Trojanowski, J. Q. (2017). Multisite assessment of aging-related tau astrogliopathy (ARTAG). *Journal of Neuropathology & Experimental Neurology, 76*, 605–619. https://doi.org/10.1093/jnen/nlx041

Lennie, P. (2003). The cost of cortical computation. *Current Biology, 13*, 493–497. https://doi.org/10.1016/S0960-9822(03)00135-0

Magistretti, P. J., & Allaman, I. (2018). Lactate in the brain: From metabolic end-product to signalling molecule. *Nature Reviews Neuroscience, 19*, 235–249. https://doi.org/10.1038/nrn.2018.19

Magistretti, P. J. & Ransom, B. R. (2002). *Astrocytes*. In Davis, K.L. (ed) Neuropsychopharmacology: the fifth generation of progress: an official publication of the American College of Neuropsychopharmacology. Lippincott Williams & Wilkins. ISBN: 978-0-78-172837-9

Margineanu, M. B., Mahmood, H., Fiumelli, H., & Magistretti, P. J. (2018). L-Lactate regulates the expression of synaptic plasticity and neuroprotection genes in cortical neurons: A transcriptome analysis. *Frontiers in Molecular Neuroscience, 11*, Article 375. https://doi.org/10.3389/fnmol.2018.00375

Martin, R. D. (1981). Relative brain size and basal metabolic rate in terrestrial vertebrates. *Nature, 293*, 57–60. https://doi.org/10.1038/293057a0

McKee, A. C., Cantu, R. C., Nowinski, C. J., Hedley-Whyte, E. T., Gavett, B. E., Budson, A. E., Santini, V. E., Lee, H.-S., Kubilus, C. A., & Stern, R. A. (2009). Chronic traumatic encephalopathy in athletes: Progressive tauopathy after repetitive head injury. *Journal of Neuropathology & Experimental Neurology, 68*, 709–735. https://doi.org/10.1097/NEN.0b013e3181a9d503

McKee, A. C., Stein, T. D., Kiernan, P. T., & Alvarez, V. E. (2015). The neuropathology of chronic traumatic encephalopathy. *Brain Pathology, 25*, 350–364. https://doi.org/10.1111/bpa.12248

McKee, A. C., Stern, R. A., Nowinski, C. J., Stein, T. D., Alvarez, V. E., Daneshvar, D. H., Lee, H.-S., Wojtowicz, S. M., Hall, G., Baugh, C. M., Riley, D. O., Kubilus, C. A., Cormier, K. A., Jacobs, M. A., Martin, B. R., Abraham, C. R., Ikezu, T., Reichard, R. R., Wolozin, B. L., . . . Cantu, R. C. (2013). The spectrum of disease in chronic traumatic encephalopathy. *Brain, 136*, 43–64. https://doi.org/10.1093/brain/aws307

Medel, V., Crossley, N., Gajardo, I., Muller, E., Barros, L. F., Shine, J. M., & Sierralta, J. (2022). Whole-brain neuronal MCT2 lactate transporter expression links metabolism to human brain structure and function. *Proceedings of the National Academy of Sciences of the United States of America, 119*, Article e2204619119. https://doi.org/10.1073/pnas.2204619119

Mink, J. W., Blumenschine, R. J., & Adams, D. B. (1981). Ratio of central nervous system to body metabolism in vertebrates: Its constancy and functional basis. *American Journal of Physiology: Regulatory, Integrative and Comparative Physiology, 241*, R203–R212. https://doi.org/10.1152/ajpregu.1981.241.3.R203

Munger, E. L., Edler, M. K., Hopkins, W. D., Ely, J. J., Erwin, J. M., Perl, D. P., Mufson, E. J., Hof, P. R., Sherwood, C. C., & Raghanti, M. A. (2019). Astrocytic changes with aging and Alzheimer's disease-type pathology in chimpanzees. *Journal of Comparative Neurology, 527*, 1179–1195. https://doi.org/10.1002/cne.24610

Munger, E. L., Edler, M. K., Hopkins, W. D., Hof, P. R., Sherwood, C. C., & Raghanti, M. A. (2022). Comparative analysis of astrocytes in the prefrontal cortex of primates: Insights into the evolution of human brain energetics. *Journal of Comparative Neurology, 530*, 3106–3125. https://doi.org/10.1002/cne.25387

Murray, M. E., Kouri, N., Lin, W.-L., Jack, C. R., Dickson, D. W., & Vemuri, P. (2014). Clinicopathologic assessment and imaging of tauopathies in neurodegenerative dementias. *Alzheimer's Research & Therapy, 6*, Article 1. https://doi.org/10.1186/alzrt231

Oberheim, N. A., Takano, T., Han, X., He, W., Lin, J. H., Wang, F., Xu, Q., Wyatt, J. D., Pilcher, W., Ojemann, J. G., & Ransom, B. R. (2009). Uniquely hominid features of adult human astrocytes. *Journal of Neuroscience, 29*(10), 3276–3287.

Paolicelli, R. C., Sierra, A., Stevens, B., Tremblay, M.-E., Aguzzi, A., Ajami, B., Amit, I., Audinat, E., Bechmann, I., Bennett, M., Bennett, F., Bessis, A., Biber, K., Bilbo, S., Blurton-Jones, M., Boddeke, E., Brites, D., Brône, B., Brown, G. C., . . . Wyss-Coray, T. (2022). Microglia states and nomenclature: A field at its crossroads. *Neuron, 110*, 3458–3483. https://doi.org/10.1016/j.neuron.2022.10.020

Parpura, V. and Verkhratsky, A. (2012). Astrocytes revisited: concise historic outlook on glutamate homeostasis and signaling. *Croatian medical journal, 53*, 518–528. doi: HYPERLINK "https://doi.org/10.3325%2Fcmj.2012.53.518"10.3325/cmj.2012.53.518

Pellerin, L., & Magistretti, P. J. (1994). Glutamate uptake into astrocytes stimulates aerobic glycolysis: A mechanism coupling neuronal activity to glucose utilization. *Proceedings of the National Academy of Sciences of the United States of America, 91*, 10625–10629. https://doi.org/10.1073/pnas.91.22.10625

Perez, S. E., Raghanti, M. A., Hof, P. R., Kramer, L., Ikonomovic, M. D., Lacor, P. N., Erwin, J. M., Sherwood, C. C., & Mufson, E. J. (2013). Alzheimer's disease pathology in the neocortex and hippocampus of the western lowland gorilla (*Gorilla gorilla gorilla*). *Journal of Comparative Neurology, 521*, 4318–4338. https://doi.org/10.1002/cne.23428

Perez, S. E., Sherwood, C. C., Cranfield, M. R., Erwin, J. M., Mudakikwa, A., Hof, P. R., & Mufson, E. J. (2016). Early Alzheimer's disease–type pathology in the frontal cortex of wild mountain gorillas (*Gorilla beringei beringei*). *Neurobiology of Aging, 39*, 195–201. https://doi.org/10.1016/j.neurobiolaging.2015.12.017

Pérez-Cerdá, F., Sánchez-Gómez, M. V., & Matute, C. (2015). Pío del Río Hortega and the discovery of the oligodendrocytes. *Frontiers in Neuroanatomy, 9*, Article 92. https://doi.org/10.3389/fnana.2015.00092

Peters, A. (2009). The effects of normal aging on myelinated nerve fibers in monkey central nervous system. *Frontiers in Neuroanatomy, 3*, Article 11. https://doi.org/10.3389/neuro.05.011.2009

Peters, A., & Kemper, T. (2012). A review of the structural alterations in the cerebral hemispheres of the aging rhesus monkey. *Neurobiology of Aging, 33*, 2357–2372. https://doi.org/10.1016/j.neurobiolaging.2011.11.015

Petrelli, F., Dallérac, G., Pucci, L., Calì, C., Zehnder, T., Sultan, S., Lecca, S., Chicca, A., Ivanov, A., Asensio, C. S., Gundersen, V., Toni, N., Knott, G. W., Magara, F., Gertsch, J., Kirchhoff, F., Déglon, N., Giros, B., Edwards, R. H., . . . Bezzi, P. (2020). Dysfunction of homeostatic control of dopamine by astrocytes in the developing prefrontal cortex leads to cognitive impairments. *Molecular Psychiatry, 25*, 732–749. https://doi.org/10.1038/s41380-018-0226-y

Petrelli, F., Zehnder, T., Laugeray, A., Mondoloni, S., Calì, C., Pucci, L., Molinero Perez, A., Bondiolotti, B. M., De Oliveira Figueiredo, E., Dallerac, G., Déglon, N., Giros, B., Magrassi, L., Mothet, J.-P., Mameli, M., Simmler, L. D., & Bezzi, P. (2023). Disruption of astrocyte-dependent dopamine control in the developing medial prefrontal cortex leads to excessive grooming in mice. *Biological Psychiatry, 93*, 966–975. https://doi.org/10.1016/j.biopsych.2022.11.018

Robertson, W. (1899). On a new method of obtaining a black reaction in certain tissue-elements of the central nervous system (platinum method). *Scottish Medical and Surgical Journal, 4*, 23–30.

Sada, N., Lee, S., Katsu, T., Otsuki, T., & Inoue, T. (2015). Epilepsy treatment. Targeting LDH enzymes with a stiripentol analog to treat epilepsy. *Science, 347*, 1362–1367. https://doi.org/10.1126/science.aaa1299

Scott, H., Novikov, B., Ugur, B., Allen, B., Mertsalov, I., Monagas-Valentin, P., Koff, M., Baas Robinson, S., Aoki, K., Veizaj, R., Lefeber, D. J., Tiemeyer, M., Bellen, H., & Panin, V. (2023). Glia–neuron coupling via a bipartite sialylation pathway promotes neural transmission and stress tolerance in *Drosophila*. *eLife, 12*, Article e78280. https://doi.org/10.7554/eLife.78280

Sharma, K., Kanchan, B., & Ukpong, B. E. (2021). A comparative biology of microglia across species. *Frontiers in Cell and Developmental Biology, 9*, Article 652748. https://doi.org/10.3389/fcell.2021.652748

Sosa, M. A. G., De Gasperi, R., Paulino, A. J., Pricop, P. E., Shaughness, M. C., Maudlin-Jeronimo, E., Hall, A. A., Janssen, W. G. M., Yuk, F. J., Dorr, N. P., Dickstein, D. L., McCarron, R. M., Chavko, M., Hof, P. R., Ahlers, S. T., & Elder, G. A. (2013). Blast overpressure induces shear-related injuries in the brain of rats exposed to a mild traumatic brain injury. Acta Neuropathologica Communications, 1, Article 51. https://doi.org/10.1186/2051-5960-1-51

Suzuki, A., Stern, S. A., Bozdagi, O., Huntley, G. W., Walker, R. H., Magistretti, P. J., & Alberini, C. M. (2011). Astrocyte–neuron lactate transport is required for long-term memory formation. *Cell, 144*, 810–823. https://doi.org/10.1016/j.cell.2011.02.018

Tsacopoulos, M., & Magistretti, P. J. (1996). Metabolic coupling between glia and neurons. *Journal of Neuroscience, 16*, 877–885. https://doi.org/10.1523/JNEUROSCI.16-03-00877.1996

Uddin, M., Wildman, D. E., Liu, G., Xu, W., Johnson, R. M., Hof, P. R., Kapatos, G., Grossman, L. I., & Goodman, M. (2004). Sister grouping of chimpanzees and humans as revealed by genome-wide phylogenetic analysis of brain gene expression profiles. *Proceedings of the National Academy of Sciences of the United States of America, 101*, 2957–2962. https://doi.org/10.1073/pnas.0308725100

Veloz Castillo, M. F., Magistretti, P. J., & Calì, C. (2021). l-Lactate: Food for thoughts, memory and behavior. *Metabolites, 11*, Article 548. https://doi.org/10.3390/metabo11080548

Verkhratsky, A., Bush, N. A. O., Nedergaard, M., & Butt, A. (2018). The special case of human astrocytes. *Neuroglia, 1*, 21–29. https://doi.org/10.3390/neuroglia1010004

Verkhratsky, A., & Butt, A. M. (2018). The history of the decline and fall of the glial numbers legend. *Neuroglia, 1*, 188–192. https://doi.org/10.3390/neuroglia1010013

Vezzoli, E., Cali, C., De Roo, M., Ponzoni, L., Sogne, E., Gagnon, N., Francolini, M., Braida, D., Sala, M., Muller, D., Falqui, A., & Magistretti, P. J. (2019). Ultrastructural evidence for a role of astrocytes and glycogen-derived lactate in learning-dependent synaptic stabilization. *Cerebral Cortex, 30*, 2114–2127. https://doi.org/10.1093/cercor/bhz226

Virchow, R. (1860). *Cellular pathology: As based upon physiological and pathological histology. Twenty lectures delivered in the Pathological institute of Berlin during the months of February, March and April, 1858*. RM De Witt.

Volkenhoff, A., Weiler, A., Letzel, M., Stehling, M., Klämbt, C., & Schirmeier, S. (2015). Glial glycolysis is essential for neuronal survival in *Drosophila*. *Cell Metabolism, 22*, 437–447. https://doi.org/10.1016/j.cmet.2015.07.006

Yang, J., Ruchti, E., Petit, J.-M., Jourdain, P., Grenningloh, G., Allaman, I., & Magistretti, P. J. (2014). Lactate promotes plasticity gene expression by potentiating NMDA signaling in neurons. *Proceedings of the National Academy of Sciences of the United States of America, 111*, 12228–12233. https://doi.org/10.1073/pnas.1322912111

Zhang, W.-J., Wu, C.-L., & Liu, J.-P. (2023). Schwann cells as a target cell for the treatment of cancer pain. *Glia, 71*, 2309–2322. https://doi.org/10.1002/glia.24391

3

The Contribution of Mitochondrial Evolution and Dysfunction to Neurodegeneration

Kobi Wasner, Sandro L. Pereira, and Anne Grünewald

Mitochondrial Description, Endosymbiosis, and Nuclear Involvement

Within the trillions of cells giving life to the human body, a single cell comprises a world in itself; thousands of cellular organelles work in harmony to create proteins, chemicals, and signals responsible for all biological and chemical processes that allow the life of the organism to flourish. While the largest percentage of the code that dictates biological processes lies within nuclear DNA, a small fraction of the code lies within one particular organelle: the mitochondrion.

According to the endosymbiotic theory, mitochondria were once individual, free-living bacteria at a time when the earth was composed of single-celled organisms, roughly 2 billion years ago (Gray et al. 1999). Around that time, the alpha-proteobacteria that became mitochondria were engulfed by a proto-eukaryotic cell; but rather than being digested or killing the host cell, both cells survived, and together they thrived. The host cell provided an extra layer of protection for the mitochondrion, and the mitochondrion provided energy—in the form of adenosine triphosphate (ATP)—that the host cell could use to propagate its own existence. This partnership is the essence of a symbiotic relationship that, without which, may not have led to the world as we know it today (Gray et al. 1999).

The so-called powerhouses of the cell, mitochondria execute vital functions including the Krebs cycle, calcium signaling, lipid metabolism, and, most importantly, the generation of ATP—the source of energy found in all known forms of life necessary to drive and support many processes in living cells (Perier and Vila 2012). Mitochondria create ATP through the electron transport chain (ETC): a series of protein complexes (complexes I–IV) and molecules that transfer electrons through redox reactions across the inner mitochondrial membrane. The reducing equivalent molecules nicotinamide adenine dinucleotide (NADH) and flavin adenine dinucleotide donate electrons to respiratory complex I and complex II, respectively. Electrons are then transferred through the other ETC components according to a gradient of increasing reduction potential, ultimately being accepted by O_2. The energy released by the flow of electrons through the ETC is used to pump H^+ ions from the mitochondrial matrix to the intermembrane space, generating an electrochemical H^+ gradient across these two compartments. The proton motive force associated with the H^+ gradient is used as potential energy to sustain the activity of ATP synthase (also referred to as "complex V"), which, as the name indicates, catalyzes the synthesis of ATP.

Given that they were once independent organisms, mitochondria have their own DNA, which contains the code to help sustain its existence and multiply for proliferation. Mitochondrial DNA (mtDNA) does not encode all of the ETC protein complexes we observe at present, however; complex II of the ETC, for example, is entirely encoded by nuclear DNA, suggesting either the transfer of mitochondrial genes to the nucleus, the evolutionary development of nuclear genes to support mitochondria, or both. Indeed, mitochondria multiply and exist to tens or even thousands per cell depending on the cell's energy demand. At present and no longer totally independent, mitochondrial biogenesis requires coordinated nuclear–mitochondrial signaling to ensure a healthy mitochondrial network. Until recently, the common scientific understanding was that the mitochondrial genome is exclusively inherited via the maternal lineage. By contrast, a pedigree evidencing biparental mtDNA inheritance called this concept into question (Luo et al. 2018).

Surrounded by a double-membrane system, the mitochondrion can be divided into four compartments: the outer mitochondrial membrane, the inner mitochondrial membrane, the intermembrane space (which separates the latter two), and the matrix—the space within the inner mitochondrial membrane (Perier and Vila 2012). The inner membrane is highly folded, forming what are termed "cristae," which extend into the matrix and increase the inner membrane surface area, and thus ETC efficiency (Perier and Vila 2012). The matrix holds mtDNA, which is packaged into small, protein-rich structures known as "nucleoids" that exist in multiple copies per mitochondrion (Perier and Vila 2012). The small, circular mitochondrial genome encodes 13 mitochondrial proteins that form components of the ETC, as well as machinery for synthesizing RNA and proteins. The greater part of proteins necessary to create and maintain mitochondria are encoded in nuclear DNA, which are then synthesized in the cytosol and imported into mitochondria (Perier and Vila 2012).

Mitochondria are highly dynamic organelles and form interconnected networks whose biogenesis and structure are immensely dependent upon the needs of the cell (Lackner 2014). Mitochondria are continually generated and degraded, and their shape and distribution within the cell are maintained and regulated by several mitochondrial activities including fission, fusion, motility, and tethering (Lackner 2014). Mitochondrial fission, or division, involves the transport, distribution, and regulated degradation of the organelle. When adding to the mitochondrial network, mitochondrial fusion allows the sharing of contents between mitochondrial compartments, which facilitates communication between the organelles (Chen and Chan 2010). Motility and tethering coordinate mitochondrial cellular distribution to ensure that they are trafficked to and maintained at the locations within the cell where they are needed (Lackner 2014). Should any of these processes be disturbed, detrimental consequences can take place that result in cell death and the onset of diseases.

During evolution, the human brain grew dramatically in size and functional complexity to account for an increasing skill set that would, for instance, enable communication and social interaction between individuals (Gonçalves et al. 2015). This development was regulated by pathways that we are only just beginning to understand in more detail. One hypothesis is that the same pathways are also associated with neurodegenerative and mental disorders as they are human-specific diseases (Gonçalves et al. 2015). Given that increasing brain size goes hand in hand with higher energy requirements, enhanced mitochondrial signaling is likely a key component of human evolution and thus also critical in the pathogenesis of psychiatric and age-related neurological diseases.

Mitochondrial Contribution to Aging

The cellular mechanisms that mediate aging are still some of the greatest mysteries that puzzle scientists today. Several theories have been proposed, such as antagonistic pleiotropy, which stipulates that the upregulation of genes in the older organism have an antagonistic function during development (Sun et al. 2016). That is, genes that positively coordinate growth during youth may serve to accelerate cellular senescence later in life (Sun et al. 2016). As early as the 1920s, it was discovered that metabolic rates appeared to inversely correlate with life span (Jang et al. 2018), raising the notion that mitochondria contribute to aging.

Without mitochondria, existing animal cells would depend on anaerobic glycolysis for ATP as a tiny fraction of ATP (a net yield of two molecules in total) is produced upon the conversion of glucose to pyruvate. Oxidative phosphorylation (the metabolic pathway by which mitochondria produce ATP), on the other hand, fully oxidizes glucose, resulting in a net production of 30–32 molecules of ATP per molecule of glucose. Despite this overwhelmingly profitable system of converting oxygen into energy, about 0.1%–2% of electrons passing through the ETC are prematurely and incompletely reduced to the superoxide radical O_2^-, a toxic byproduct and precursor to most other reactive oxygen species (ROS). ROS are extremely reactive chemicals derived from diatomic oxygen and are consequences of normal metabolism by mitochondria. Intrinsic to cell signaling and homeostasis, ROS are present at low and stationary levels in healthy cells (Herb et al. 2021). Antioxidants, including superoxide dismutase, consist of proteins and enzymes that mitigate the harmful effects of free radicals such as damage to proteins, lipids, and DNA (Hughes and Reynolds 2005; Herb et al. 2021).

Free-Radical Aging Theory

Denham Harmon first proposed the free-radical aging theory in the 1950s, which proposed that cellular damage caused by ROS results in cellular senescence (Herb et al. 2021). This theory later expanded to include age-related diseases and was then modified and specified to mitochondria; that is, with mitochondria being the major source of ROS production, mitochondria-generated ROS causes irreversible damage to lipids, proteins, and mtDNA (Jang and Van Remmen 2009). Located in the matrix, and therefore in close proximity to the sites of ROS production, mtDNA is highly susceptible to oxidative damage. Moreover, lacking introns, histones, and other typical proteins surrounding DNA, ROS-induced mtDNA damage is more persistent, extensive, and rapid than in nuclear DNA (Huang and Manton 2004; Jang and Van Remmen 2009). ROS-promoted mtDNA damage includes fragmentation and deletions that likely interfere with the creation of the ETC, thus perturbing oxidative phosphorylation and cellular respiration (Huang and Manton 2004).

In addition to its proximity to the ETC and its lack of protection, mtDNA lacks extensive repair mechanisms as seen in nuclear DNA. Several studies have shown that mtDNA mutations increase with age, especially in tissues with high-energy demands (Huang and Manton 2004; Jang et al. 2018). A single mitochondrion contains several copies of mtDNA, yielding thousands of mtDNA molecules per cell (Perier and Vila 2012). Hence, not all mtDNA copies are identical, giving rise to differences in mutational loads between cells, mitochondria, and mtDNA copies—a phenomenon termed "mtDNA heteroplasmy." In other words, it is a state where wild-type mtDNA molecules coexist with mutated mtDNA molecules within the same tissue (Perier and Vila 2012; Stewart and Chinnery 2015). It is estimated that, in order for pathogenic mtDNA mutations to lead to measurable bioenergetic effects, a mutational threshold of 60%–90% in a given cell must be reached (Jang et al. 2018). The most common somatic (i.e., non-inherited) mtDNA alteration is the formation of large deletions in the major arc. These deletions may arise from replication errors (Perier and Vila 2012; Stewart and Chinnery 2015; Nido et al. 2018) and vary in length but often affect genes encoding subunits of complex I or IV (Rygiel et al. 2015).

Mitochondria-Derived Damage-Associated Molecular Patterns

Aging is associated with a chronic, low-grade activation of the immune system—a condition known as "inflammaging" (Jang et al. 2018). In addition to circulating pro-inflammatory cytokines from the immune system, molecules released from senescent or dying cells—known as damage-associated molecular patterns (DAMPs)—can trigger an immune response. Of endosymbiotic origin, mitochondria hold the potential to be large players in DAMP-induced inflammaging. For example, from an evolutionary standpoint, mtDNA is of bacterial origin, which may "confuse" the immune system. Certain mitochondrial subunits, like cytochrome C, are known to induce apoptosis. Inflammaging also results in aberrant mitochondrial degradation (mitophagy), leading to a buildup of damaged mitochondria that release large quantities of ROS, pro-apoptotic factors, and pro-inflammatory DAMPs like mtDNA.

As mentioned above, mitochondrial dysfunction is observed in a variety of disorders, especially age-related diseases. These include Alzheimer's disease (AD), amyotrophic lateral sclerosis, Huntington's disease, and Parkinson's disease (PD), to name a few (Lezi and Swerdlow 2012). In addition, there are young-onset mental disorders such as schizophrenia that have been linked to mitochondrial dysfunction. In this chapter, we will primarily focus on the manifold roles of mitochondria in the pathogenesis of PD (Figure 3.1) but will highlight parallels in other neurological and psychiatric diseases.

Mitochondria as Central Factor in the Pathogenesis of PD

Over 200 years after the initial description of PD, the exact cause of the disease has yet to be identified. This is due to its multifactorial nature, with both environmental and genetic factors contributing to its onset and progression. Aging is by far the most critical risk factor for developing PD as the majority of idiopathic PD patients develop symptoms after reaching 65 years of age (Riess and Krüger 1999). Other factors are also implicated in the onset of idiopathic PD, namely environmental factors such as long-term exposure to pesticides and heavy metals as seen in farmers, manufacturers, and welders. Genetics can also play a role in PD pathogenesis as known familial mutations result in development of the disease including ~20 monogenic forms and risk factors. A wealth of biological factors have been investigated to determine the underlying mechanisms leading to dopaminergic (DA) neuron death. Roughly 40 years ago, however, scientists shed light on mitochondria as a major player in PD pathogenesis.

Mitochondria first became of particular interest in the pathology of PD in the 1980s, after drug abusers ingested botched batches of synthetic heroin and presented with Parkinson-like symptoms. Neurologist J. William Langston, in collaboration with the National Institutes of Health, investigated the fabricated opioids and discovered a

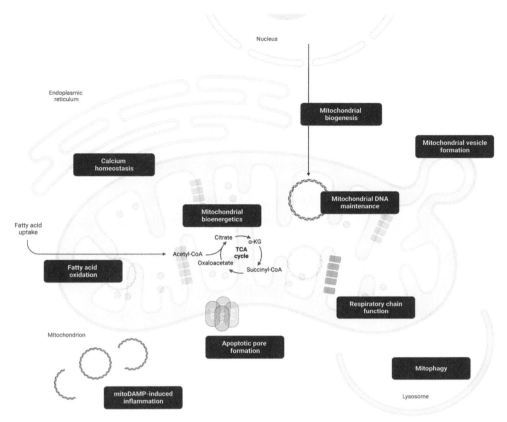

Figure 3.1 Mitochondrial signaling impairments in Parkinson's disease. Scheme summarizing the manifold processes involving mitochondria that are disrupted in Parkinson's disease. CoA = coenzyme A; DAMP = damage-associated molecular pattern; α-KG = alpha-ketoglutarate; TCA = tricarboxylic acid.

toxic, undesired byproduct: 1-methyl-4-phenyl-1,2,5,6-tetrahydropyridine (MPTP). Research in squirrel monkeys confirmed that MPTP exposure resulted in Parkinsonism, which was reversible by administration of levodopa—a PD treatment. Soon thereafter, the cellular mechanisms of MPTP were unraveled, leading to mitochondrial dysfunction as a key player in PD.

Although not toxic itself, MPTP can cross the blood–brain barrier, where it is metabolized into 1-methyl-4-phenylpyridinium (MPP$^+$; Frim et al. 1994). MPP$^+$ is taken up by DA neurons and blocks complex I of the ETC, leading to aberrant oxidative phosphorylation, a buildup of ROS, and cell death (Storch et al. 2004). A hallmark of PD, death of DA neurons of the substantia nigra is evident in postmortem brain tissue, and complex I activity has similarly been shown to be reduced in brain sections (Schapira et al. 1989). Even at present, injection of MPTP is commonly used to study PD in rodents.

The largest portion of PD cases (~90%) are sporadic, originating from environmental factors and aging, among other causes. Roughly 10% of cases are caused by genetic mutations passed down through the family line. Interestingly, about half of the genetic mutations lay within proteins that control mitochondrial function, giving further support for mitochondrial dysfunction in PD. This list includes alpha-synuclein, leucine-rich repeat kinase 2, vacuolar protein sorting-associated protein 35, coiled-coil-helix-coiled-coil-helix domain containing 2, Parkin, phosphatase and tensin homolog–induced kinase 1 (PINK1), and DJ-1 (Borsche et al. 2021).

Mitochondrial Dysfunction and Alpha-Synuclein Aggregation

In addition to alpha-synuclein, Lewy bodies are filled with organellar components including mitochondria, which may indicate a pathophysiological link between the protein and the powerhouses of the cell (Shahmoradian et al. 2019). The upregulation of alpha-synuclein levels or the presence of PD-associated mutations increases its propensity to form pathological oligomers and fibrils (Miraglia et al. 2018). Several lines of research have revealed that alpha-synuclein binds to curved membranes including the outer and inner mitochondrial membrane (Miraglia et al. 2018). By doing so at the level of the translocase of the outer membrane 20 (TOM20), it can block protein import into mitochondria through the translocase of the inner membrane/TOM complex: a protein complex allowing the translocation of nuclear DNA–encoded proteins into mitochondria for the production of ATP. In turn, this disrupts various mitochondrial processes including oxidative phosphorylation and maintenance of mitochondrial membrane potential (Di Maio et al. 2016; Miraglia et al. 2018). In addition, there is evidence suggesting that alpha-synuclein itself can be imported into mitochondria, where it was shown to interfere with complex I function (Devi et al. 2008). Similarly, the main component of the plaques that accumulate in brains of AD patients, amyloid beta (Aβ), can interfere with mitochondrial permeability. In this respect, pathological pores, known as "mitochondrial permeability transition pores," may form in the inner mitochondrial membrane and cause mitochondrial swelling, outer mitochondrial membrane rupture, and cell death. In the presence of Aβ, the threshold for pathological pore opening has been shown to decrease, triggering downstream effects such as enhanced ROS generation, loss of mitochondrial membrane potential, and disrupted calcium homeostasis (Du and Yan 2010). Of note, the mitochondrial permeability transition pore is also formed under other pathophysiological conditions including PD, traumatic brain injury, and stroke.

While these data suggest that mitochondrial dysfunction develops secondarily to protein aggregation, there is also some literature that places mitochondrial dysfunction upstream of alpha-synuclein accumulation (Grünewald et al. 2019). In rodents exposed to paraquat, rotenone, or MPTP, alpha-synuclein aggregation has been observed. These pesticides/drugs are well-established mitochondrial toxins. By contrast, there is no evidence for a direct impact of paraquat, rotenone, or MPTP on alpha-synuclein function or aggregation behavior, which would imply that in these models mitochondrial dysfunction is the trigger for alpha-synuclein accumulation (Grünewald et al. 2019). Along similar lines, exposure of SH-SY5Y (neuroblastoma) cells expressing wild-type amyloid precursor protein (APP) or APP with the Swedish mutation to rotenone revealed that the latter cultures develop higher ROS levels and were more prone to cell death. These findings suggest that pesticide exposure can accelerate protein toxicity via the inhibition of mitochondrial function within AD pathophysiology (Joh and Choi 2017). Further evidence for an interplay between the environment and mitochondrial dysfunction in neurodegenerative disease comes from a recent genetic PD study of ours: we explored the interaction of mitochondrial polygenic risk (i.e., the genetic variability in genes involved in mitochondrial processes such as mitophagy, mitochondrial bioenergetics, and proteostasis) and lifestyle or environmental factors in sporadic PD patients. This analysis revealed that patients with a high mitochondrial risk score who did not smoke had an earlier age at onset than non-smokers with low mitochondrial polygenic risk. A similar association was not found for smokers (Lüth et al. 2023).

Regulation of Mitochondrial Quality Control

Among the aforementioned PD proteins, the most established mitochondrial regulators are PINK1 and Parkin. After the mitochondrial localization of both proteins was established (Grünewald et al. 2019), Narendra and colleagues (2008) first revealed that the kinase PINK1 recruits Parkin to the outer mitochondrial membrane, where the E3 ligase ubiquitinates targets such as mitofusin 1 and 2 (MFN1/2) (Pereira et al. 2023). The resulting poly-ubiquitin chains are then phosphorylated by PINK1, which triggers the recruitment of the autophagy machinery to clear the mitochondrion (Pereira et al. 2023). Together, this process is known as "mitophagy": the selective degradation of defective mitochondria by autophagy. Mitophagy is crucial in keeping cells healthy as it prevents the accumulation of defective, apoptosis-harboring molecules.

While Parkin deficiency is found in familial PD patients harboring mutations in the E3 ubiquitin ligase, loss of Parkin function has also been observed in sporadic PD cases (Dawson and Dawson 2010). In sporadic AD cases, elevated mitochondrial abundance was documented together with increased levels of PINK1 and Parkin, indicative of successful mitophagy initiation but impaired lysosomal degradation of the organelles (Wang et al. 2020). Interestingly, overexpression of *PRKN* in fibroblasts from sporadic AD patients was sufficient to rescue the mitophagy phenotype (Martín-Maestro et al. 2016). Of note, *PRKN* not only has been linked to neurodegenerative disorders, but may also function as a tumor suppressor in the context of several cancers (Wahabi et al. 2018). According to the theory of antagonistic pleiotropy, some genes, which may have beneficial functions during early life, can become harmful in later stages of life. This may also be the underlying cause of Parkin deficiency–mediated neurodegeneration. To prevent the development of cancers, mutated forms of Parkin that contribute to the progression of melanoma by inhibiting mitophagy may have been selected against (Bernardini et al. 2017; Fox 2018).

For the mitochondrial pool to remain balanced, the removal of damaged mitochondria needs to be closely interconnected with the generation of new organelles. The master regulator of mitochondrial biogenesis is peroxisome proliferator–activated receptor gamma (PPARG) C1A PPARG coactivator 1 alpha (PGC-1α), which induces the transcription of nuclear respiratory factors 1 and 2 that in turn regulate the expression of components of the mitochondrial transcription machinery including mitochondrial transcription factor A, (TFAM); mitochondrial transcription factor B1; and mitochondrial transcription factor B2(TFB2M; Gureev et al. 2019). Interestingly, this pathway appears to be disrupted in both sporadic PD and AD (Grünewald et al. 2019; Wang et al. 2020). Quintessentially, in DA neurons isolated from postmortem nigral tissue of PD patients, we observed a depletion of mitochondrial transcription factors TFAM and TFB2M together with a reduction in mtDNA copy number. This phenotype coincided with respiratory chain complex I deficiency in these nerve cells. By contrast, there was no increase in the abundance of mtDNA major arc deletions (Grünewald et al. 2014). These results were later confirmed in an independent study that assessed components of the mitochondrial biogenesis pathway in PD patients alongside mitochondrial disease cases. Interestingly, in AD, several studies reported a reduction in the levels of PGC-1α, which was associated with altered insulin-mediated metabolism (Wang et al. 2020)—a topic that will be further discussed in the next section.

Mitochondrial Bioenergetics

Contrary to most other organs in which the energy expenditure is correlated to their relative size, the brain accounts for only 2% of the body mass and yet stands in the top three most metabolically expensive tissues of the human body, consuming 20% of the total energy budget (Herculano-Houzel 2011). These findings portray the highly demanding energy processes that sustain brain activity, which depend upon efficient provision of energy substrates. Regulatory mechanisms such as neurovascular coupling, through which regional hemodynamics is modulated to match neuronal activity, contribute to the energetic sustenance of the brain (Pereira et al. 2023).

The tight orchestration of brain bioenergetics starts failing during the development of pathologies like neurodegenerative conditions. Glucose is the prime energy substrate of the brain, and its metabolization declines with normal aging, where reductions of almost 30% (between ages 18 and 78) were described (Marcus et al. 2014). Yet, fluorodeoxyglucose-positron emission tomographic studies have illustrated that brain glucose hypometabolism is further pronounced in AD and PD.

PD patients present steep shifts in glucose metabolism in divergent brain regions, which change as a function of disease progression. Brain areas with increased burdens of tau in preclinical AD subjects have lowered levels of aerobic glycolysis—linked to biosynthetic and antioxidant roles—resulting in loss of neuroplasticity and neuroprotection (Vlassenko et al. 2018). Additional proxies of energy status, such as the phosphorylated adenosine-5′-monophosphate-activated protein kinase (AMPK)/AMPK ratio, decline in aging and PD (Hang et al. 2019). These bioenergetic imbalances may be caused by (1) hypoperfusion of the brain driven by structural and functional degeneration of the cerebrovascular system, which causes insufficient input of nutrients and O_2, and faulty cellular processes such as (2) insulin resistance and (3) dysfunctional mitochondrial metabolism (Borghammer et al. 2010; Grünewald et al. 2019).

Hypoperfusion (Reduced Blood Flow) of the Brain

Aβ amyloid angiopathy and ischemic parenchymal alterations are recurrent instances in AD (Mulica et al. 2021). Additional to these effects, Aβ has vasoactive properties and, along with other molecules such as vasoconstrictor endothelin-1, contributes to cerebral hypoperfusion in AD. Neurovascular coupling is equally impaired in preclinical stages of AD (Mulica et al. 2021). In the context of PD, decreased cerebral blood flow and regional deterioration of the vascular network were reported (Mulica et al. 2021).

Insulin Resistance

Epidemiological evidence increasingly supports an association between diabetes and neurodegenerative conditions such as AD and PD (Fox 2018; Han et al. 2023). These findings are supported by studies showing that these maladies share dysregulated molecular mechanisms, particularly insulin resistance. The brain is an insulin-sensitive organ and presents marked expression of the insulin receptor (IR) and insulin-like growth factor receptors (IGF-Rs) 1 and 2 (Ferreira et al. 2018). The mechanism of insulin resistance has been better explored in AD, where impaired signaling through IR and IGF-R and changes to other molecules in the insulin signaling pathway have been established. Exemplary, the insulin-degrading enzyme, which degrades insulin but also Aβ, is saturated under hyperinsulinemia conditions that arise under insulin resistance, reducing its capacity to clear Aβ (Ferreira et al. 2018). Remarkably, insulin resistance has been used as an argument in support of the "mismatched environments hypothesis for AD," which tries to explain the prevalence of this condition under the scope of evolutionary medicine (Fox 2018). Following this hypothesis, features that characterize life in our contemporary societies but were not prevalent in the premodern world, are current triggers or enhancers of AD. In this context, and when qualitatively and quantitatively comparing the diet and the energy expenditure of our hunter–gatherer antecedents with those from contemporary people in modern societies, it is clear that premodern humans were unlikely to develop insulin resistance so frequently, justifying in part their lower susceptibility to AD (Fox 2018).

Dysfunctional Mitochondrial Metabolism

Recent work substantiates the notion that mitochondrial and metabolic impairments are early drivers of PD, rather than secondary actors of other contributing molecular mechanisms. Indeed, proteomic assessment of early to mid-stage Parkinson's brains suggests that bioenergetics and mitochondrial redox state perturbations precede the occurrence of alpha-synuclein pathology and neuronal degeneration (Toomey et al. 2022). Curiously, increased glycolysis and mitochondrial dysfunction are also denoted as markers of early peripheral PD pathology as they are detected in blood mononuclear cells of early idiopathic PD patients and subjects with rapid eye movement–sleep behavior disorder, a prodromal feature of PD (Smith et al. 2018). Changes to glucose metabolism are not restricted to altered rates of glucose uptake and utilization, thus implicating changes in gluconeogenesis—a set of metabolic reactions entailing the de novo synthesis of glucose from non-carbohydrate sources—which has been shown to be elevated in idiopathic PD (Borsche et al. 2023). Increased glucose production and glycolysis could be speculated to be compensatory effects to dysfunctional mitochondrial oxidative phosphorylation (OXPHOS), which is now regarded as a hallmark of PD due to its transversality to many forms of familial and idiopathic PD (Giannoccaro et al. 2017). Adding to evidence discussed in the previous section referring to complex I inhibition by MPTP, epidemiological studies and subsequent confirmatory work established a link between exposure to environmental toxins, such as pesticide, and OXPHOS disruption (Giannoccaro et al. 2017). These findings clearly demonstrate that primary mitochondrial dysfunction leading to bioenergetics deficits operates in a subset of idiopathic PD patients. Interestingly, changes to mtDNA dynamics and copy number can prime the OXPHOS impairment reported in individuals with idiopathic PD (Grünewald et al. 2016).

Mitochondrial dysfunction is also a converging point for many of the familial forms of PD, with PD-associated proteins directly influencing mitochondrial bioenergetics. Parkin has particularly been associated with the regulation of metabolic processes. In the context of tumor biology, Parkin was identified as a downstream target of p53 (Zhang et al. 2011), being capable of enhancing OXPHOS while suppressing glycolysis by means of direct regulation of glycolytic enzymes (Liu et al. 2016; Pereira et al. 2023). Parkin modulates the activity of metabolic enzymes in neuronal models (Bogetofte et al. 2019), and studies in neurons derived from *PRKN*-PD patients show increased lactate production and accumulation of Krebs cycle intermediates (Okarmus et al. 2021; Wasner et al. 2022). These phenotypes were associated with reduced nicotinamide adenine dinucleotide (NAD$^+$):NADH ratios, which conceivably impact the activities of enzymes using these molecules as cofactors, namely the sirtuin 1 (SIRT1) deacetylase. SIRT1 deacetylates PGC-1α, leading to its activation and initiating a process that entails mitochondrial biogenesis. Inhibition of respiratory chain complex I (NADH:ubiquinone oxidoreductase), which utilizes NADH as a substrate, can contribute to the unbalance of NAD$^+$:NADH ratios (Narendra and Thayer 2022). Interestingly, complex I inhibition was observed in induced pluripotent stem cell (iPSC)–derived neuronal cultures from *PRKN*-PD patients (Zanon et al. 2017; Wasner et al. 2022). Finally, Parkin exerts its functions through the control of lipid homeostasis. Parkin ubiquitinates CD36, leading to the stabilization of this fatty acid transporter and ultimately defining the rates of fatty acid uptake and lipid metabolism (Zanon et al. 2017). In accordance with lowered lipid uptake, *PRKN*-PD patients have elevated fatty acid metabolites and oxidized lipids in serum (Okuzumi et al. 2019).

Altered glucose metabolism and mitochondrial physiology also underlie the pathomechanism of schizophrenia. Indeed, higher lactate levels in the cerebrospinal fluid (CSF) is a historical feature of this condition and has been regarded as a consequence of mitochondrial dysfunction (Rajasekaran et al. 2015). Dysregulation of glycolytic enzymes such as hexokinase 1 or of pyruvate dehydrogenase, the prime enzymatic complex regulating the communication between glycolysis and the mitochondrial Krebs cycle, accounts for the mechanism of lactate elevation. And, interestingly, CSF lactate levels correlate with the neurological manifestations of the disease (Rajasekaran et al. 2015).

Mitochondrial Calcium Dyshomeostasis

The regulation of cellular calcium levels occurs at contact sites between mitochondria and the endoplasmic reticulum (MERCS). This connection is maintained by the endoplasmic reticulum (ER)–mitochondria encounter structure complex as well as by proteins such as voltage-dependent anion channel 1 (VDAC1), 75-kDa glucose-regulated protein, vesicle-associated membrane protein-associated protein B (VAPB), protein tyrosine phosphatase–interacting protein 51 (PTPIP51), mitochondrial fission 1, and the mitofusins MFN1 and 2. The area in the ER linking to mitochondria is also denominated as "mitochondria-associated membranes" (MAMs). Given that adult DA neurons in the substantia nigra heavily rely on calcium with respect to their pacemaking, healthy MERCS function is critical for the survival of these cells (Chan et al. 2007). In addition, calcium oscillations in nigral DA neurons may have a bioenergetic function. Intracellular calcium released as a result of these oscillations is taken up by mitochondria, where it drives the tricarboxylic acid cycle, OXPHOS, and thus ATP production. In this fashion, elevated ATP requirements during neuronal activity are accounted for—a feed-forward control mechanism that is phylogenetically preserved (Diederich et al. 2019).

Some research suggests that by mono-ubiquitinating the MERCS tether MFN2, Parkin regulates the stability of these contact sites and thus modulates mitochondrial calcium handling (Pereira et al. 2023). Conversely, other studies have shown that MFN2 is phospho-ubiquitinated by the PINK1/Parkin signaling pathway, which initiates the degradation of the protein and leads to the untethering of MERCS (Pereira et al. 2023). Strengthening these seemingly contradictory results, recent work in *Drosophila* showed that both VDAC mono- and poly-ubiquitination depend on the PINK1/Parkin pathway. In this manner, Parkin regulates the initiation of mitophagy and the suppression of apoptosis. In addition, the authors uncovered that mono-ubiquitination of this pore-forming protein is crucial for the control of mitochondrial calcium uptake (Pereira et al. 2023).

In addition to PINK1 and Parkin, alpha-synuclein was found to localize to the MAM, where it disrupts calcium homeostasis (Calì et al. 2012; Guardia-Laguarta et al. 2014). Modulating the abundance of alpha-synuclein by knockdown or overexpression experiments in HeLa cells suggested that the protein impacts calcium uptake (Calì et al.

2012). In the presence of PD-causing mutations in alpha-synuclein, the protein's affinity to MAMs is reduced, which results in a shift between ER and mitochondria, fragmentation of mitochondria, and a decline in MAM function (Guardia-Laguarta et al. 2014). This deficiency is likely due to the interaction of mutant alpha-synuclein with VAPB–PTPIP51 tethers that leads to loosening of ER–mitochondria associations and in turn to reduced inter-organellar calcium exchange and mitochondrial ATP production (Guardia-Laguarta et al. 2014; Paillusson et al. 2017).

Mitochondrial Genome Alterations

First hints of an involvement of mitochondrial genome alterations in PD came from cytoplasmic hybrid or short "cybrid" experiments. In this setup, patient-derived mtDNA was combined with a control nuclear DNA background. For this purpose, cells devoid of nuclear DNA such as platelets were fused with cells that were, for instance, treated with the intercalating agent ethidium bromide (EtBr). EtBr interferes with mtDNA replication and thus allows for the creation of "rho 0" cells that are completely depleted of mtDNA (Wilkins et al. 2014). Using this approach, Swerdlow and colleagues showed that respiratory chain complex I and IV deficiencies are transmittable from patient to control cells in PD and AD, respectively, suggesting that changes in the mitochondrial genome are the underlying cause of the OXPHOS dysfunction previously observed in postmortem nigral tissue of sporadic patients (Schapira et al. 1989; Swerdlow et al. 1996). However, given the lack of evidence in the literature for strictly maternally inherited forms of PD or AD, the nature of these mutations is likely somatic (Grünewald et al. 2019).

By using a combination of cloning and Sanger sequencing, mtDNA derived from complex IV–deficient nigral neurons of postmortem midbrain tissue from idiopathic PD patients was shown to harbor large major arc deletions (Bender et al. 2006). The deletion spectrum of nigral neurons from PD patients was later further refined using an ultra-deep-sequencing approach. This work revealed that multiple mtDNA species with different deletion characteristics coexist in one cell. Moreover, these mtDNA subtypes differ in the composition of their sequence, which is in line with the clonal expansion hypothesis for mtDNA deletion formation (Nido et al. 2018). Similar work in AD uncovered a drastic increase of the 5-kb deletion in AD patients below the age of 75 (Corral-Debrinski et al. 1994). By contrast, several studies exploring the contribution of the common mtDNA deletion to mitochondrial dysfunction in schizophrenia patients did not find an accumulation in various tested brain regions (Roberts 2021).

Somatic changes in the mitochondrial genome are not only found in the brain, however. A deep-sequencing study using blood-derived mtDNA recently explored mtDNA alterations as possible progression (or penetrance) markers of idiopathic and genetic PD. Excitingly, this work revealed that in heterozygous carriers of mutations in *PINK1* or *PRKN*, the somatic mutational load correlates with PD status. In postmortem and iPSC-derived DA neurons from biallelic *PRKN* mutation carriers, elevated numbers of somatic mtDNA mutations were detected, which further strengthens the role of the E3 ubiquitin ligase in mtDNA maintenance processes. Moreover, the authors uncovered a correlation between the somatic mtDNA mutational burden and blood interleukin-6 (IL-6) cytokine levels in affected and unaffected carriers of disease-causing variants in *PRKN* or *PINK1* (Nido et al. 2018; Trinh et al. 2023). These results connect to a previous study that was performed in a partially overlapping cohort. Here, IL-6 enzyme-linked immunosorbent assay analysis and quantification of circulating cell-free mtDNA in serum samples revealed an increase of both markers in patients with biallelic mutations in *PRKN* or *PINK1*. Moreover, cytokine levels correlated with disease duration in PINK1/PRKN-PD patients. This was not true for idiopathic PD patients (Borsche et al. 2020). Taken together, this work suggests that an axis linking mitochondrial metabolism, mtDNA integrity, and inflammation exists at least in *PRKN*-PD. This topic will be further discussed in the next section.

Mitochondrial Haplotypes and Neurodegeneration

Not only mutations in the mitochondrial genome have been linked to PD and AD. mtDNA is characterized by different single nucleotide polymorphisms (SNPs), denominated as "haplotypes," that can be used to determine the phylogenetic origins of maternal lineages (Lee et al. 2017). Individuals with similar haplotypes share the same "haplogroup" and are likely to stem from a common ancestor—the underlying hypothesis being that the mitochondrial genome accumulated sequence alterations as it was inherited over generations during evolution (Lee et al. 2017). While initial work suggested that the haplogroups J, K, UK, and UKJT are associated with an increased PD risk, more recent studies involving larger data sets did not support such a link (Grünewald et al. 2019). In AD, the UK haplogroup has been linked to elevated disease risk, while haplogroup T was found to be protective (Wang et al. 2020). However, if SNPs that define a certain haplotype occur "out of place" in individuals who are descendants of a different maternal lineage, they may have pathogenic potential. A study exploring the contribution of "out-of-place" mtDNA SNPs in African ancestry PD cases and controls found evidence for a contribution of such variants to PD risk (Müller-Nedebock et al. 2022). By contrast, cybrid studies investigating the functional consequences of such variants in the context of PD are still scarce, precluding firm conclusions about the relevance of inherited mtDNA variations in the pathogenesis of PD (Lang et al. 2022).

mtDNA Epigenetic Modulation

Another underexplored field in mitochondrial PD and AD research as well as aging is the potential epigenetic modulation of the mitochondrial genome. Mostly due to technical limitation, the existence of mtDNA modifications such as methylation (5-methylcytosine [5mC]) and hydroxymethylation (5-hydroxymethylcytosine) remains debatable. While some studies detected mtDNA methylation peaks especially in the D-loop region, others failed to reproduce these findings (Stoccoro and Coppedè 2021). There is evidence suggesting a global hypermethylation of the mtDNA in aged individuals (Mongelli et al. 2023), but in neurodegenerative diseases the situation appears to be less clear. Pyrosequencing revealed a loss of mitochondrial 5mC levels in the D-loop region in nigral tissue from PD patients and an increase in cortical tissue from AD patients (Blanch et al. 2016). In addition, mtDNA methylation patterns appear to be tissue-specific. Contrary to the situation in the brain, in peripheral blood–derived mtDNA, no difference in the methylation pattern of idiopathic PD patients versus controls was observed (Stoccoro et al. 2021).

Mitochondria and Inflammation

The last few years brought increasing excitement to the field of mitochondria-driven inflammation, with implications spanning to numerous human disorders such as infectious diseases, autoimmune conditions, cancer, and neurodegeneration (Marchi et al. 2023). In the context of neurodegenerative conditions and particularly PD, the focus has been put on Parkin and PINK1 proteins since the dysregulation of mitochondrial quality control processes, notably mitophagy, was associated with increased inflammatory outcomes. A link between inflammation and either acutely or chronically induced mitochondrial stress with the backdrop of Parkin and PINK1 deficiency was established in 2018 (Sliter et al. 2018). In this study, an elevation of pro-inflammatory cytokines in the serum of those challenged mice was correlated to increased mtDNA levels circulating in the blood. The inflammatory response was specifically mediated by the cyclic guanosine monophosphate–adenosine monophosphate synthetase (cGAS)–stimulator of interferon genes (STING) signaling pathway (Sliter et al. 2018). Various mitochondrial metabolites and elements can elicit inflammatory responses when released from the mitochondrial compartment and hence are considered DAMPs (Marchi et al. 2023). Depending on the nature of the molecule and the compartment into which the DAMP is released (i.e., cytosol or extracellular milieu), diverse inflammatory pathways can be triggered. mtDNA stress, for example, as the form of depletion of TFAM—a transcription factor instrumental for both mtDNA replication and the ultrastructure of the mtDNA nucleoids—leads to mtDNA efflux into the cytosol. Here, mtDNA activates the cytosolic DNA sensor cGAS, resulting in the triggering of the cGAS–STING immune response (West et al. 2015). Interestingly, inflammation can drive neurodegeneration since STING disruption could rescue the neurodegenerative phenotypes and motor impairments observed in $PRKN^{-/-}$:POLG-deficient mice (Sliter et al. 2018). As discussed in the previous section, *PRKN*-PD patients also present enriched levels of circulating cell-free mtDNA and inflammatory cytokines in the serum (Sliter et al. 2018; Borsche et al. 2020).

In addition, the NLR family pyrin domain containing 3 (NLRP3) inflammasome participates in the pathogenesis of PD, and more recently it has been further implicated in *PRKN* deficiency–mediated pathology. Parkin ubiquitinates NLRP3, tagging it for proteasomal degradation similarly to Parkin interacting substrate (PARIS), as mentioned above. When Parkin is absent, both PARIS and NLRP3 accumulate. PARIS negatively impacts mitochondrial function, resulting in the generation of excessive mitochondria-derived ROS, which in turn activates NLRP3, leading to DA neuron degeneration (Panicker et al. 2022).

PD-Associated Proteins Involved in the Resolution of Apoptosis

Mitochondria are indispensable organelles that are essential for life. Paradoxically, mitochondria are also essential for cell death (Bock and Tait 2020). Programmed cell death, or apoptosis, holds central roles in many processes from embryonic development to immune homeostasis (Bock and Tait 2020). This self-destructive mechanism seen in multicellular organisms allows the removal of damaged, infected, and unwanted cells so that the whole has a greater chance to survive (Wang and Youle 2009). Excessive cell death, however, contributes to neurodegenerative diseases, including PD.

Two major apoptotic pathways exist in vertebrates that differ in their response to different types of stimuli: the extrinsic and intrinsic apoptotic pathways (Wang and Youle 2009). The intrinsic apoptotic pathway, also known as the "mitochondrial apoptotic pathway," involves activation of caspases—proteases that cleave cellular proteins such as DNA repair enzymes and cytoskeletal proteins—by internal mitochondrial substrates (Bock and Tait 2020). This is made possible by permeation of the outer mitochondrial membrane, leading to the release of intramitochondrial proteins. For example, cytochrome C, a crucial molecule of the ETC, is a key activating component of the caspase-9 apoptosome complex upon its release from mitochondria (Bock and Tait 2020). Cytochrome C is released from mitochondria via the mitochondrial permeability transition pore with the apoptotic proteins BAX and BAK acting as outer membrane components (Karch et al. 2013). Interestingly, the PD protein Parkin was shown to ubiquitinate BAK, thereby preventing the formation of BAK oligomers, which facilitate mitochondrial outer membrane permeabilization as the initial step of apoptosis (Bernardini et al. 2019). Moreover, the above-described PINK1/Parkin pathway, which mediates mitochondrial quality control through the regulation of autophagic degradation of malfunctioning mitochondria, is vital for the prevention of mitochondria-induced apoptosis (Pereira et al. 2023). In addition, PINK1 was shown to regulate the switch from autophagy to apoptosis via its interaction with Beclin 1, which prevents the cleavage of the pro-apoptotic protein (Brunelli et al. 2022).

Outlook

During the decades that passed since the identification of mitochondrial dysfunction as an etiological factor for the diseases discussed here, a tremendous leap forward in our understanding of the mechanisms underlying this dysfunction was operated. Mitochondria evolved as a central organelle for the physiology of the eukaryotic cell, in such a way that fine regulation of its function is strictly mandatory for the homeostasis of the whole organism. Still, millions of years into this obligatory symbiotic relationship have not erased the traits of a prokaryotic past. Certain mitochondrial elements such as its DNA can serve as DAMPs, hijacking this harmonious relationship and feeding an inflammatory response that is more and more recognized as part of the etiopathology of PD and AD.

Throughout this chapter, we discussed the current knowledge on the multiple facets of mitochondrial dysfunction and detailed their individual contribution to brain pathologies. Overall, disruption to any of the described mitochondrial activities may lead to bioenergetic and redox unbalance and eventually trigger the release of pro-inflammatory DAMPs. These pathological mechanisms impact neuronal physiology at diverse levels including development, maturation, plasticity, and function and ultimately contribute to neuronal degeneration. Therefore, mitochondrial dysfunction is transversal to many neurological diseases, spanning from early-onset conditions (such as schizophrenia and familial cases of PD) to disorders manifesting later in life (AD and sporadic PD).

In the context of PD, the identification of a heritable component and subsequent studies thereof showed that defective mitochondrial function is not restricted to sporadic patients exposed to viruses or environmental toxins but is rather a fundamental pathological process shared with many of the familial forms of the disease. Yet, one of the ongoing challenges is to unveil the relative contribution of compromised mitochondrial function (and that of each of its facets) to each subtype of the disease. Such knowledge will instrumentally instruct the stratification of PD patients into mechanistically consistent subgroups with defined therapeutic targets, a necessary approach to improve the success of clinical trials targeting mitochondrial dysfunction.

We are just starting to unveil the intricate connections between diverse mitochondrial activities, for example, the newly identified interdependency between mitochondrial bioenergetics and mtDNA homeostasis which, when altered, results in inflammation. Nevertheless, a recent trial aiming at replenishing systemic NAD levels in PD patients already reported promising results with the boost of the brain NAD metabolome, which correlated to diminished inflammatory markers and mild clinical improvement of the patients studied (Brakedal et al. 2022). Further trials on the topic are awaited, with much expectation and hope to preserve the integrity of our oldest symbiotic relationship and to thrive as a whole together later in life.

Key Points to Remember

- During evolution, mitochondria were engulfed by prokaryotic cells, where they serve as energy converters.

- Human brain development led to a rapid increase in size and complexity, demanding increased energy provision by mitochondria.
- Mitochondrial dysfunction is central to human-specific disorders such as PD and AD as well as schizophrenia, implicating a phylogenetic component.
- Mitochondrial impairments in these diseases affect processes such as mitophagy, calcium homeostasis, metabolism, mtDNA maintenance, apoptosis, and inflammatory signaling.
- In concert with the prokaryotic origin of mitochondria, some of its components such as mtDNA can serve as DAMPs and elicit pro-inflammatory responses when released into the cytosol or extracellular space.

Acknowledgment

K. W., S. L. P., and A. G. were supported by the Luxembourg National Research Fund within the ATTRACT program (Model-IPD, FNR9631103).

References

Bender, A., Krishnan, K. J., Morris, C. M., Taylor, G. A., Reeve, A. K., Perry, R. H., Jaros, E., Hersheson, J. S., Betts, J., Klopstock, T., Taylor, R. W., & Turnbull, D. M. (2006). High levels of mitochondrial DNA deletions in substantia nigra neurons in aging and Parkinson disease. *Nature Genetics*, *38*(5), 515–517. https://doi.org/10.1038/ng1769

Bernardini, J. P., Brouwer, J. M., Tan, I. K., Sandow, J. J., Huang, S., Stafford, C. A., Bankovacki, A., Riffkin, C. D., Wardak, A. Z., Czabotar, P. E., Lazarou, M., & Dewson, G. (2019). Parkin inhibits BAK and BAX apoptotic function by distinct mechanisms during mitophagy. *EMBO Journal*, *38*(2), Article e99916. https://doi.org/10.15252/embj.201899916

Bernardini, J.P., Lazarou, M., & Dewson, G. (2017). Parkin and mitophagy in cancer. *Oncogene*, *10*, 1315–1327. https://doi.org/10.1038/onc.2016.302

Blanch, M., Mosquera, J. L., Ansoleaga, B., Ferrer, I., & Barrachina, M. (2016). Altered mitochondrial DNA methylation pattern in Alzheimer disease-related pathology and in Parkinson disease. *American Journal of Pathology*, *186*, 385–397. https://doi.org/10.1016/j.ajpath.2015.10.004

Bock, F. J., & Tait, S. W. G. (2020). Mitochondria as multifaceted regulators of cell death. *Nature Reviews: Molecular Cell Biology*, *21*, 85–100. https://doi.org/10.1038/s41580-019-0173-8

Bogetofte, H., Jensen, P., Ryding, M., Schmidt, S. I., Okarmus, J., Ritter, L., Worm, C. S., Hohnholt, M. C., Azevedo, C., Roybon, L., Bak, L. K., Waagepetersen, H., Ryan, B. J., Wade-Martins, R., Larsen, M. R., & Meyer, M. (2019). *PARK2* mutation causes metabolic disturbances and impaired survival of human iPSC-derived neurons. *Frontiers in Cellular Neuroscience*, *13*, Article 297. https://doi.org/10.3389/fncel.2019.00297

Borghammer, P., Chakravarty, M., Jonsdottir, K. Y., Sato, N., Matsuda, H., Ito, K., Arahata, Y., Kato, T., & Gjedde, A. (2010). Cortical hypometabolism and hypoperfusion in Parkinson's disease is extensive: Probably even at early disease stages. *Brain Structure & Function*, *214*(4), 303–317. https://doi.org/10.1007/s00429-010-0246-0

Borsche, M., König, I. R., Delcambre, S., Petrucci, S., Balck, A., Brüggemann, N., Zimprich, A., Wasner, K., Pereira, S. L., Avenali, M., Deuschle, C., Badanjak, K., Ghelfi, J., Gasser, T., Kasten, M., Rosenstiel, P., Lohmann, K., Brockmann, K., Valente, E. M., . . . Klein, C. (2020). Mitochondrial damage–associated inflammation highlights biomarkers in PRKN/PINK1 parkinsonism. *Brain*, *143*(10), 3041–3051. https://doi.org/10.1093/brain/awaa246

Borsche, M., Märtens, A., Hörmann, P., Brückmann, T., Lohmann, K., Tunc, S., Klein, C., Hiller, K., & Balck, A. (2023). In vivo investigation of glucose metabolism in idiopathic and *PRKN*-related Parkinson's disease. *Movement Disorders, 38*(4), 697–702. https://doi.org/10.1002/mds.29333

Borsche, M., Pereira, S. L., Klein, C., & Grünewald, A. (2021). Mitochondria and Parkinson's disease: Clinical, molecular, and translational aspects. *Journal of Parkinson's Disease, 11*, 45–60. https://doi.org/10.3233/JPD-201981

Brakedal, B., Dölle, C., Riemer, F., Ma, Y., Nido, G. S., Skeie, G. O., Craven, A. R., Schwarzlmüller, T., Brekke, N., Diab, J., & Sverkeli, L. (2022). The NADPARK study: A randomized phase I trial of nicotinamide riboside supplementation in Parkinson's disease. *Cell Metabolism, 34*(3), 396–407. https://doi.org/10.1016/j.cmet.2022.02.001

Brunelli, F., Torosantucci, L., Gelmetti, V., Franzone, D., Grünewald, A., Krüger, R., Arena, G., & Valente, E. M. (2022). PINK1 protects against staurosporine-induced apoptosis by interacting with Beclin1 and impairing its pro-apoptotic cleavage. *Cells, 11*(4), Article 678. https://doi.org/10.3390/cells11040678

Calì, T., Ottolini, D., Negro, A., & Brini, M. (2012). α-Synuclein controls mitochondrial calcium homeostasis by enhancing endoplasmic reticulum–mitochondria interactions. *Journal of Biological Chemistry, 287*, 17914–17929. https://doi.org/10.1074/jbc.M111.302794

Chan, C. S., Guzman, J. N., Ilijic, E., Mercer, J. N., Rick, C., Tkatch, T., Meredith, G. E., & Surmeier, D. J. (2007). "Rejuvenation" protects neurons in mouse models of Parkinson's disease. *Nature, 447*(7148), 1081–1086. https://doi.org/10.1038/nature05865

Chen, H., & Chan, D. C. (2010). Physiological functions of mitochondrial fusion. *Annals of the New York Academy of Sciences, 1201*, 21–25. https://doi.org/10.1111/j.1749-6632.2010.05615.x

Corral-Debrinski, M., Horton, T., Lott, M. T., Shoffner, J. M., McKee, A. C., Beal, M. F., Graham, B. H., & Wallace, D. C. (1994). Marked changes in mitochondrial DNA deletion levels in Alzheimer brains. *Genomics, 23*(2), 471–476. https://doi.org/10.1006/geno.1994.1525

Dawson, T. M., & Dawson, V. L. (2010). The role of parkin in familial and sporadic Parkinson's disease. *Movement Disorders, 25*(S1), S32–S39. https://doi.org/10.1002/mds.22798

Devi, L., Raghavendran, V., Prabhu, B. M., Avadhani, N. G., & Anandatheerthavarada, H. K. (2008). Mitochondrial import and accumulation of alpha-synuclein impair complex I in human dopaminergic neuronal cultures and Parkinson disease brain. *Journal of Biological Chemistry, 283*, 9089–9100. https://doi.org/10.1074/jbc.M710012200

Diederich, N. J., James Surmeier, D., Uchihara, T., Grillner, S., & Goetz, C. G. (2019). Parkinson's disease: Is it a consequence of human brain evolution? *Movement Disorders, 34*, 453–459. https://doi.org/10.1002/mds.27628

Di Maio, R., Barrett, P. J., Hoffman, E. K., Barrett, C. W., Zharikov, A., Borah, A., Hu, X., McCoy, J., Chu, C. T., Burton, E. A., & Hastings, T. G. (2016). α-Synuclein binds to TOM20 and inhibits mitochondrial protein import in Parkinson's disease. *Science: Translational Medicine, 8*(342), Article 342ra78. https://doi.org/10.1126/scitranslmed.aaf3634

Du, H., & Yan, S. S. (2010). Mitochondrial permeability transition pore in Alzheimer's disease: Cyclophilin D and amyloid beta. *Biochimica et Biophysica Acta, 1802*, 198–204. https://doi.org/10.1016/j.bbadis.2009.07.005

Ferreira, L. S. S., Fernandes, C. S., Vieira, M. N. N., & De Felice, F. G. (2018). Insulin resistance in Alzheimer's disease. *Frontiers in Neuroscience, 12*, Article 830. https://doi.org/10.3389/fnins.2018.00830

Fox, M. (2018). "Evolutionary medicine" perspectives on Alzheimer's disease: Review and new directions. *Ageing Research Reviews, 47*, 140–148. https://doi.org/10.1016/j.arr.2018.07.008

Frim, D. M., Uhler, T. A., Galpern, W. R., Beal, M. F., Breakefield, X. O., & Isacson, O. (1994). Implanted fibroblasts genetically engineered to produce brain-derived neurotrophic factor prevent 1-methyl-4-phenylpyridinium toxicity to dopaminergic neurons in the rat. *Proceedings of the National Academy of Sciences of the United States of America, 91*, 5104–5108. https://doi.org/10.1073/pnas.91.11.5104

Giannoccaro, M. P., La Morgia, C., Rizzo, G., & Carelli, V. (2017). Mitochondrial DNA and primary mitochondrial dysfunction in Parkinson's disease. *Movement Disorders, 32*, 346–363. https://doi.org/10.1002/mds.26966

Gonçalves, V. F., Andreazza, A. C., & Kennedy, J. L. (2015). Mitochondrial dysfunction in schizophrenia: An evolutionary perspective. *Human Genetics, 134*, 13–21. https://doi.org/10.1007/s00439-014-1491-8

Gray, M. W., Burger, G., & Lang, B. F. (1999). Mitochondrial evolution. *Science, 283*, 1476–1481. https://doi.org/10.1126/science.283.5407.1476

Grünewald, A., Kumar, K. R., & Sue, C. M. (2019). New insights into the complex role of mitochondria in Parkinson's disease. *Progress in Neurobiology, 177*, 73–93. https://doi.org/10.1016/j.pneurobio.2018.09.003

Grünewald, A., Lax, N. Z., Rocha, M. C., Reeve, A. K., Hepplewhite, P. D., Rygiel, K. A., Taylor, R. W., & Turnbull, D. M. (2014). Quantitative quadruple-label immunofluorescence of mitochondrial and cytoplasmic proteins in single neurons from human midbrain tissue. *Journal of Neuroscience Methods, 232*, 143–149. https://doi.org/10.1016/j.jneumeth.2014.05.026

Grünewald, A., Rygiel, K. A., Hepplewhite, P. D., Morris, C. M., Picard, M., & Turnbull, D. M. (2016). Mitochondrial DNA depletion in respiratory chain–deficient Parkinson disease neurons. *Annals of Neurology, 79*, 366–378. https://doi.org/10.1002/ana.24571

Guardia-Laguarta, C., Area-Gomez, E., Rüb, C., Liu, Y., Magrané, J., Becker, D., Voos, W., Schon, E. A., & Przedborski, S. (2014). α-Synuclein is localized to mitochondria-associated ER membranes. *Journal of Neuroscience, 34*(1), 249–259. https://doi.org/10.1523/JNEUROSCI.2507-13.2014

Gureev, A. P., Shaforostova, E. A., & Popov, V. N. (2019). Regulation of mitochondrial biogenesis as a way for active longevity: Interaction between the Nrf2 and PGC-1α signaling pathways. *Frontiers in Genetics, 10*, Article 435. https://doi.org/10.3389/fgene.2019.00435

Han, K., Kim, B., Lee, S. H., & Kim, M. K. (2023). A nationwide cohort study on diabetes severity and risk of Parkinson disease. *NPJ Parkinson's Disease, 9*, Article 11. https://doi.org/10.1038/s41531-023-00462-8

Hang, L., Thundyil, J., Goh, G. W. Y., & Lim, K.-L. (2019). AMP kinase activation is selectively disrupted in the ventral midbrain of mice deficient in Parkin or PINK1 expression. *Neuromolecular Medicine, 21*, 25–32. https://doi.org/10.1007/s12017-018-8517-7

Herb, M., Gluschko, A., & Schramm, M. (2021). Reactive oxygen species: Not omnipresent but important in many locations. *Frontiers in Cell and Developmental Biology, 9*, Article 716406. https://doi.org/10.3389/fcell.2021.716406

Herculano-Houzel, S. (2011). Scaling of brain metabolism with a fixed energy budget per neuron: Implications for neuronal activity, plasticity, and evolution. *PLOS ONE, 6*, Article e17514. https://doi.org/10.1371/journal.pone.0017514

Huang, H., & Manton, K. G. (2004). The role of oxidative damage in mitochondria during aging: A review. *Frontiers in Bioscience, 9*, 1100–1117. https://doi.org/10.2741/1298

Hughes, K. A., & Reynolds, R. M. (2005). Evolutionary and mechanistic theories of aging. *Annual Review of Entomology, 50*, 421–445. https://doi.org/10.1146/annurev.ento.50.071803.130409

Jang, J. Y., Blum, A., Liu, J., & Finkel, T. (2018). The role of mitochondria in aging. *Journal of Clinical Investigation, 128*, 3662–3670. https://doi.org/10.1172/JCI120842

Jang, Y. C., & Van Remmen, H. (2009). The mitochondrial theory of aging: Insight from transgenic and knockout mouse models. *Experimental Gerontology, 44*, 256–260. https://doi.org/10.1016/j.exger.2008.12.006

Joh, Y., & Choi, W.-S. (2017). Mitochondrial complex I inhibition accelerates amyloid toxicity. *Development & Reproduction, 21*, 417–424. https://doi.org/10.12717/DR.2017.21.4.417

Karch, J., Kwong, J. Q., Burr, A. R., Sargent, M. A., Elrod, J. W., Peixoto, P. M., Martinez-Caballero, S., Osinska, H., Cheng, E. H., Robbins, J., Kinnally, K. W., & Molkentin, J. D. (2013). Bax and Bak function as the outer membrane component of the mitochondrial permeability pore in regulating necrotic cell death in mice. *eLife, 2*, Article e00772. https://doi.org/10.7554/eLife.00772

Lackner, L. L. (2014). Shaping the dynamic mitochondrial network. *BMC Biology, 12*, Article 35. https://doi.org/10.1186/1741-7007-12-35

Lang, M., Grünewald, A., Pramstaller, P. P., Hicks, A. A., & Pichler, I. (2022). A genome on shaky ground: Exploring the impact of mitochondrial DNA integrity on Parkinson's disease by highlighting the use of cybrid models. *Cellular and Molecular Life Sciences, 79*, Article 283. https://doi.org/10.1007/s00018-022-04304-3

Lee, W. T., Sun, X., Tsai, T. S., Johnson, J. L., Gould, J. A., Garama, D. J., Gough, D. J., McKenzie, M., Trounce, I. A., & St. John, J. C. (2017). Mitochondrial DNA haplotypes induce differential patterns of

DNA methylation that result in differential chromosomal gene expression patterns. *Cell Death Discovery*, 3, Article 17062. https://doi.org/10.1038/cddiscovery.2017.62

Lezi, E., & Swerdlow, R. H. (2012). Mitochondria in neurodegeneration. *Advances in Experimental Medicine and Biology*, *942*, 269–286. https://doi.org/10.1007/978-94-007-2869-1_12

Liu, K., Li, F., Han, H., Chen, Y., Mao, Z., Luo, J., Zhao, Y., Zheng, B., Gu, W., & Zhao, W. (2016). Parkin regulates the activity of pyruvate kinase M2. *Journal of Biological Chemistry*, *291*(19), 10307–10317. https://doi.org/10.1074/jbc.M115.703066

Luo, S., Valencia, C. A., Zhang, J., Lee, N. C., Slone, J., Gui, B., Wang, X., Li, Z., Dell, S., Brown, J., Chen, S. M., Chien, Y. H., Hwu, W. L., Fan, P. C., Wong, L. J., Atwal, P. S., & Huang, T. (2018). Biparental inheritance of mitochondrial DNA in humans. *Proceedings of the National Academy of Sciences of the United States of America*, *115*(51), 13039–13044. https://doi.org/10.1073/pnas.1810946115

Lüth, T., Gabbert, C., Koch, S., König, I. R., Caliebe, A., Laabs, B. H., Hentati, F., Sassi, S. B., Amouri, R., Spielmann, M., Klein, C., Grünewald, A., Farrer, M. J., & Trinh, J. (2023). Interaction of mitochondrial polygenic score and lifestyle factors in LRRK2 p. Gly2019Ser Parkinsonism. *Movement Disorders*, *38*(10), 1837–1849. https://doi.org/10.1002/mds.29563

Marchi, S., Guilbaud, E., Tait, S. W. G., Yamazaki, T., & Galluzzi, L. (2023). Mitochondrial control of inflammation. *Nature Reviews: Immunology*, *23*, 159–173. https://doi.org/10.1038/s41577-022-00760-x

Marcus, C., Mena, E., & Subramaniam, R. M. (2014). Brain PET in the diagnosis of Alzheimer's disease. *Clinical Nuclear Medicine*, *39*, e413–e422; quiz e423–e426. https://doi.org/10.1097/RLU.0000000000000547

Martín-Maestro, P., Gargini, R., Perry, G., Avila, J., & García-Escudero, V. (2016). PARK2 enhancement is able to compensate mitophagy alterations found in sporadic Alzheimer's disease. *Human Molecular Genetics*, *25*, 792–806. https://doi.org/10.1093/hmg/ddv616

Miraglia, F., Ricci, A., Rota, L., & Colla, E. (2018). Subcellular localization of alpha-synuclein aggregates and their interaction with membranes. *Neural Regeneration Research*, *13*, 1136–1144. https://doi.org/10.4103/1673-5374.235013

Mongelli, A., Mengozzi, A., Geiger, M., Gorica, E., Mohammed, S. A., Paneni, F., Ruschitzka, F., & Costantino, S. (2023). Mitochondrial epigenetics in aging and cardiovascular diseases. *Frontiers in Cardiovascular Medicine*, *10*, Article 1204483. https://doi.org/10.3389/fcvm.2023.1204483

Mulica, P., Grünewald, A., & Pereira, S. L. (2021). Astrocyte–neuron metabolic crosstalk in neurodegeneration: A mitochondrial perspective. *Frontiers in Endocrinology*, *12*, Article 668517. https://doi.org/10.3389/fendo.2021.668517

Müller-Nedebock, A. C., Pfaff, A. L., Pienaar, I. S., Kõks, S., van der Westhuizen, F. H., Elson, J. L., & Bardien, S. (2022). Mitochondrial DNA variation in Parkinson's disease: Analysis of "out-of-place" population variants as a risk factor. *Frontiers in Aging Neuroscience*, *14*, Article 921412. https://doi.org/10.3389/fnagi.2022.921412

Narendra, D., Tanaka, A., Suen, D.-F., & Youle, R. J. (2008). Parkin is recruited selectively to impaired mitochondria and promotes their autophagy. *Journal of Cell Biology*, *183*, 795–803. https://doi.org/10.1083/jcb.200809125

Narendra, D. P., & Thayer, J. A. (2022). Midbrain on fire: mtDNA ignites neuroinflammation in PRKN-P. *Movement Disorders*, *37*, 1332–1334. https://doi.org/10.1002/mds.29073

Nido, G. S., Dölle, C., Flønes, I., Tuppen, H. A., Alves, G., Tysnes, O. B., Haugarvoll, K., & Tzoulis, C. (2018). Ultradeep mapping of neuronal mitochondrial deletions in Parkinson's disease. *Neurobiology of Aging*, *63*, 120–127. https://doi.org/10.1016/j.neurobiolaging.2017.10.024

Okarmus, J., Havelund, J. F., Ryding, M., Schmidt, S. I., Bogetofte, H., Heon-Roberts, R., Wade-Martins, R., Cowley, S. A., Ryan, B. J., Færgeman, N. J., Hyttel, P., & Meyer, M. (2021). Identification of bioactive metabolites in human iPSC-derived dopaminergic neurons with PARK2 mutation: Altered mitochondrial and energy metabolism. *Stem Cell Reports*, *16*(6), 1510–1526. https://doi.org/10.1016/j.stemcr.2021.04.022

Okuzumi, A., Hatano, T., Ueno, S. I., Ogawa, T., Saiki, S., Mori, A., Koinuma, T., Oji, Y., Ishikawa, K. I., Fujimaki, M., Sato, S., Ramamoorthy, S., Mohney, R. P., & Hattori, N. (2019). Metabolomics-based identification of metabolic alterations in PARK2. *Annals of Clinical and Translational Neurology*, *6*(3), 525–536. https://doi.org/10.1002/acn3.724

Paillusson, S., Gomez-Suaga, P., Stoica, R., Little, D., Gissen, P., Devine, M. J., Noble, W., Hanger, D. P., & Miller, C. C. J. (2017). α-Synuclein binds to the ER–mitochondria tethering protein VAPB to disrupt Ca^{2+} homeostasis and mitochondrial ATP production. *Acta Neuropathologica, 134*(1), 129–149. https://doi.org/10.1007/s00401-017-1704-z

Panicker, N., Kam, T. I., Wang, H., Neifert, S., Chou, S. C., Kumar, M., Brahmachari, S., Jhaldiyal, A., Hinkle, J. T., Akkentli, F., Mao, X., Xu, E., Karuppagounder, S. S., Hsu, E. T., Kang, S. U., Pletnikova, O., Troncoso, J., Dawson, V. L., & Dawson, T. M. (2022). Neuronal NLRP3 is a parkin substrate that drives neurodegeneration in Parkinson's disease. *Neuron, 110*(15), 2422–2437.e9. https://doi.org/10.1016/j.neuron.2022.05.009

Pereira, S. L., Grossmann, D., Delcambre, S., Hermann, A., & Grünewald, A. (2023). Novel insights into Parkin-mediated mitochondrial dysfunction and neuroinflammation in Parkinson's disease. *Current Opinion in Neurobiology, 80*, Article 102720. https://doi.org/10.1016/j.conb.2023.102720

Perier, C., & Vila, M. (2012). Mitochondrial biology and Parkinson's disease. *Cold Spring Harbor Perspectives in Medicine, 2*, Article a009332. https://doi.org/10.1101/cshperspect.a009332

Rajasekaran, A., Venkatasubramanian, G., Berk, M., & Debnath, M. (2015). Mitochondrial dysfunction in schizophrenia: Pathways, mechanisms, and implications. *Neuroscience and Biobehavioral Reviews, 48*, 10–21. https://doi.org/10.1016/j.neubiorev.2014.11.005

Riess, O., & Krüger, R. (1999). Parkinson's disease—A multifactorial neurodegenerative disorder. In H. Przuntek & T. Müller (Eds.), *Journal of neural transmission. Supplementa: Vol. 56. Diagnosis and treatment of Parkinson's disease—State of the art* (pp. 113–125). Springer. https://doi.org/10.1007/978-3-7091-6360-3_6

Roberts, R. C. (2021). Mitochondrial dysfunction in schizophrenia: With a focus on postmortem studies. *Mitochondrion, 56*, 91–101. https://doi.org/10.1016/j.mito.2020.11.009

Rygiel, K. A., Grady, J. P., Taylor, R. W., Tuppen, H. A. L., & Turnbull, D. M. (2015). Triplex real-time PCR—An improved method to detect a wide spectrum of mitochondrial DNA deletions in single cells. *Scientific Reports, 5*, Article 9906. https://doi.org/10.1038/srep09906

Schapira, A. H., Cooper, J. M., Dexter, D., Jenner, P., Clark, J. B., & Marsden, C. D. (1989). Mitochondrial complex I deficiency in Parkinson's disease. *Lancet, 1*, 1269. https://doi.org/10.1016/s0140-6736(89)92366-0

Shahmoradian, S. H., Lewis, A. J., Genoud, C., Hench, J., Moors, T. E., Navarro, P. P., Castaño-Díez, D., Schweighauser, G., Graff-Meyer, A., Goldie, K. N., Sütterlin, R., Huisman, E., Ingrassia, A., Gier, Y., Rozemuller, A. J. M., Wang, J., Paepe, A., Erny, J., Staempfli, A., . . . Lauer, M. E. (2019). Lewy pathology in Parkinson's disease consists of crowded organelles and lipid membranes. *Nature Neuroscience, 22*(7), 1099–1109. https://doi.org/10.1038/s41593-019-0423-2

Sliter, D. A., Martinez, J., Hao, L., Chen, X., Sun, N., Fischer, T. D., Burman, J. L., Li, Y., Zhang, Z., Narendra, D. P., Cai, H., Borsche, M., Klein, C., & Youle, R. J. (2018). Parkin and PINK1 mitigate STING-induced inflammation. *Nature, 561*(7722), 258–262. https://doi.org/10.1038/s41586-018-0448-9

Smith, A. M., Depp, C., Ryan, B. J., Johnston, G. I., Alegre-Abarrategui, J., Evetts, S., Rolinski, M., Baig, F., Ruffmann, C., Simon, A. K., Hu, M. T. M., & Wade-Martins, R. (2018). Mitochondrial dysfunction and increased glycolysis in prodromal and early Parkinson's blood cells. *Movement Disorders, 33*(10), 1580–1590. https://doi.org/10.1002/mds.104

Stewart, J. B., & Chinnery, P. F. (2015). The dynamics of mitochondrial DNA heteroplasmy: Implications for human health and disease. *Nature Reviews: Genetics, 16*, 530–542. https://doi.org/10.1038/nrg3966

Stoccoro, A., & Coppedè, F. (2021). Mitochondrial DNA methylation and human diseases. *International Journal of Molecular Sciences, 22*, Article 4594. https://doi.org/10.3390/ijms22094594

Stoccoro, A., Smith, A. R., Baldacci, F., Del Gamba, C., Lo Gerfo, A., Ceravolo, R., Lunnon, K., Migliore, L., & Coppedè, F. (2021). Mitochondrial D-loop region methylation and copy number in peripheral blood DNA of Parkinson's disease patients. *Genes, 12*(5), Article 720. https://doi.org/10.3390/genes12050720

Storch, A., Ludolph, A. C., & Schwarz, J. (2004). Dopamine transporter: Involvement in selective dopaminergic neurotoxicity and degeneration. *Journal of Neural Transmission, 111*, 1267–1286. https://doi.org/10.1007/s00702-004-0203-2

Sun, N., Youle, R. J., & Finkel, T. (2016). The mitochondrial basis of aging. *Molecular Cell, 61*, 654–666. https://doi.org/10.1016/j.molcel.2016.01.028

Swerdlow, R. H., Parks, J. K., Miller, S. W., Tuttle, J. B., Trimmer, P. A., Sheehan, J. P., Bennett, J. P., Jr., Davis, R. E., & Parker, W. D., Jr. (1996). Origin and functional consequences of the complex I defect in Parkinson's disease. *Annals of Neurology, 40*(4), 663–671. https://doi.org/10.1002/ana.410400417

Toomey, C. E., Heywood, W. E., Evans, J. R., Lachica, J., Pressey, S. N., Foti, S. C., Al Shahrani, M., D'Sa, K., Hargreaves, I. P., Heales, S., Orford, M., Troakes, C., Attems, J., Gelpi, E., Palkovits, M., Lashley, T., Gentleman, S. M., Revesz, T., Mills, K., & Gandhi, S. (2022). Mitochondrial dysfunction is a key pathological driver of early stage Parkinson's. *Acta Neuropathologica Communications, 10*(1), Article 134. https://doi.org/10.1186/s40478-022-01424-6

Trinh, J., Hicks, A. A., König, I. R., Delcambre, S., Lüth, T., Schaake, S., Wasner, K., Ghelfi, J., Borsche, M., Vilariño-Güell, C., Hentati, F., Germer, E. L., Bauer, P., Takanashi, M., Kostić, V., Lang, A. E., Brüggemann, N., Pramstaller, P. P., Pichler, I., . . . Grünewald, A. (2023). Mitochondrial DNA heteroplasmy distinguishes disease manifestation in PINK1/PRKN-linked Parkinson's disease. *Brain, 146*(7), 2753–2765. https://doi.org/10.1093/brain/awac464

Vlassenko, A. G., Gordon, B. A., Goyal, M. S., Su, Y., Blazey, T. M., Durbin, T. J., Couture, L. E., Christensen, J. J., Jafri, H., Morris, J. C., Raichle, M. E., & Benzinger, T. L. (2018). Aerobic glycolysis and tau deposition in preclinical Alzheimer's disease. *Neurobiology of Aging, 67*, 95–98. https://doi.org/10.1016/j.neurobiolaging.2018.03.014

Wahabi, K., Perwez, A., & Rizvi, M. A. (2018). Parkin in Parkinson's disease and cancer: A double-edged sword. *Molecular Neurobiology, 55*, 6788–6800. https://doi.org/10.1007/s12035-018-0879-1

Wang, C., & Youle, R. J. (2009). The role of mitochondria in apoptosis. *Annual Review of Genetics, 43*, 95–118. https://doi.org/10.1146/annurev-genet-102108-134850

Wang, W., Zhao, F., Ma, X., Perry, G., & Zhu, X. (2020). Mitochondria dysfunction in the pathogenesis of Alzheimer's disease: Recent advances. *Molecular Neurodegeneration, 15*, Article 30. https://doi.org/10.1186/s13024-020-00376-6

Wasner, K., Smajic, S., Ghelfi, J., Delcambre, S., Prada-Medina, C. A., Knappe, E., Arena, G., Mulica, P., Agyeah, G., Rakovic, A., Boussaad, I., Badanjak, K., Ohnmacht, J., Gérardy, J. J., Takanashi, M., Trinh, J., Mittelbronn, M., Hattori, N., Klein, C., . . . Grünewald, A. (2022). Parkin deficiency impairs mitochondrial DNA dynamics and propagates inflammation. *Movement Disorders, 37*(7), 1405–1415. https://doi.org/10.1002/mds.29025

West, A. P., Khoury-Hanold, W., Staron, M., Tal, M. C., Pineda, C. M., Lang, S. M., Bestwick, M., Duguay, B. A., Raimundo, N., MacDuff, D. A., Kaech, S. M., Smiley, J. R., Means, R. E., Iwasaki, A., & Shadel, G. S. (2015). Mitochondrial DNA stress primes the antiviral innate immune response. *Nature, 520*(7548), 553–557. https://doi.org/10.1038/nature14156

Wilkins, H. M., Carl, S. M., & Swerdlow, R. H. (2014). Cytoplasmic hybrid (cybrid) cell lines as a practical model for mitochondriopathies. *Redox Biology, 2*, 619–631. https://doi.org/10.1016/j.redox.2014.03.006

Zanon, A., Kalvakuri, S., Rakovic, A., Foco, L., Guida, M., Schwienbacher, C., Serafin, A., Rudolph, F., Trilck, M., Grünewald, A., Stanslowsky, N., Wegner, F., Giorgio, V., Lavdas, A. A., Bodmer, R., Pramstaller, P. P., Klein, C., Hicks, A. A., Pichler, I., & Seibler, P. (2017). SLP-2 interacts with Parkin in mitochondria and prevents mitochondrial dysfunction in Parkin-deficient human iPSC-derived neurons and *Drosophila*. *Human Molecular Genetics, 26*(13), 2412–2425. https://doi.org/10.1093/hmg/ddx132

Zhang, C., Lin, M., Wu, R., Wang, X., Yang, B., Levine, A. J., Hu, W., & Feng, Z. (2011). Parkin, a p53 target gene, mediates the role of p53 in glucose metabolism and the Warburg effect. *Proceedings of the National Academy of Sciences of the United States of America, 108*(39), 16259–16264. https://doi.org/10.1073/pnas.1113884108

4

Intrinsic Templates for Neurodegenerations Featuring Disease-Specific Axonal or Dendritic Vulnerability

Toshiki Uchihara

Introduction

Human neurodegenerative disorders, such as Parkinson's disease (PD), amyotrophic lateral sclerosis (ALS), and Alzheimer's disease (AD), are characterized by selective vulnerability of disease-specific neuronal groups. They contain aggregates of disease-related proteins such as alpha-synuclein (αS; Del Tredici et al. 2002), TAR DNA-binding protein of 43 kDa (TDP-43; Neumann et al. 2006), and tau proteins (Roussarie et al. 2020). However, underlying mechanisms of disease-specific spatiotemporal progression, namely how disease-related proteins form disease-specific lesions in disease-specific distributions along a disease-specific time course, remain a black box (Fu et al. 2018). Although most of these diseases are sporadic, their co-segregation as autosomal-dominant traits in some families led to the identification of disease-related or even disease-causing mutations. Clinicopathological similarity between sporadic cases and genetic counterparts suggests that this shared phenotype is not ascribable to the relevant genetic mutations. Moreover, because disease phenotypes were only partly recapitulated even in animals carrying these mutations, the mere presence of these mutations was found to be insufficient to cause or recapitulate disease phenotypes (Uchihara and Paulus 2008). Therefore, additional conditions or circumstances other than these disease-related mutations may be necessary to generate disease-specific phenotypes. It has been speculated that some inherent "machineries" are an integral part of the human brain and may recapitulate disease phenotypes of neurodegeneration (Ghika 2008). We may call such inherent "machineries" **intrinsic templates**. They may be triggered by some genetic mutations but also by other factors, such as aging or environmental toxins. Once triggered, this inherent system produces specific phenotypes even in the absence of mutations.

Therefore, details of the phenotype are largely dependent on the host through this intrinsic template. Such intrinsic templates are probably the consequence of human evolution (Cookson 2012). However, the precise cellular constructs remain unknown. If an intrinsic template is specific to a disease, each template is expected to trigger the formation of disease-specific lesions and spreading, thus also explaining clinicopathological manifestations of the disease.

This review tries to substantiate what could be the specific architecture of intrinsic templates. For this purpose, our attention is focused on neurite architectures because neurites are differently affected by PD, ALS, and AD (Warren et al. 2013). Moreover, disease-specific neurite lesions, known as "nexopathies," are more tightly related to disease-specific manifestations than the aggregates in the neuronal soma. For example, highly branched unmyelinated axons of nigral dopaminergic (DA) neurons facilitate Lewy body formation and are related to the non-somatotopic nature of symptoms (Uchihara 2017). These DA neurons also cope with the future through reward estimation to assist decision-making (Keiflin and Janak 2015). In contrast, straight myelinated axons of upper and lower motor neurons are necessary for precise and rapid execution of voluntary movements through near real-time feedback (Schwartz 2016). Here, there can be accumulation of phosphorylated TDP-43. Upper and lower motor neurons and their control system interact with the cerebellum and the parietal lobe. They constitute an internal model to reach a goal as machinery, dealing with the present (Shadmehr and Krakauer 2008). Dealing with the past is mediated by formation of memory through dendritic synapses (Cochran et al. 2014). Here, tau accumulates at an early disease stage of AD. Thus, although neurites provide functional connections between neurons in physiological states (Yuste 2015), they are disrupted at an early disease stage, generating disease-specific lesions and manifestations (Warren et al. 2013). Moreover, these neural connections may provide conduits for expanding lesions along disease-specific patterns.

Emergence of "Self" and Its Constitution

According to Damasio (2018), emergence of the self is related to "feelings" or "emotions" and is engendered through complex interactions between the homeostatic core and environments inside (*milieu intérieur*) or outside the body. In this scheme, brain interactions with the body or with the surrounding environment are not direct but mediated through this homeostatic core to generate feelings or emotions that qualitatively impact the mental status or attitude of the individual (Gissis 2020). The homeostatic core is composed of brainstem nuclei and the limbic and autonomic systems, all equipped with highly branched axons and Lewy-prone neurons (Diederich and Parent 2012). The core is structurally interdigitated with the cerebral cortices via neuronal connections, on the one hand, and with the body through spinal cord and peripheral nerves, on the other hand. In addition, this homeostatic core communicates chemically with distant body parts through corticosympathetic loops as proposed by Cannon (Goldstein 2021). Importantly, this homeostatic core is not totally under the control of the cerebral cortex. Behaviorally, its networks regulate subroutine duties for daily survival, and it sometimes activates unexpected "fight or flight" reactions associated with "emotions" or other feelings (Damasio 2018). The cerebral cortex is now "liberated" to concentrate on higher functions such as voluntary movements, visual recognition, communications, etc. It is possible that this "burden-free" circumstance of cerebral cortices promoted its exponential development into a variety of domains with enormous flexibility that allowed rapid adaptation toward a wide range of situations, even if unpredictable. This enormous flexibility may be unique to the human neocortex because it has never been achieved even in any other subhuman primate (Gómez-Robles et al. 2015). Evolution of the nervous system is usually interpreted as a process of progressive increase in its size and complexity, dependent on genomic DNA, at least partly. It is questionable whether the performance of larger brains, such as in the dolphin or elephant, is greater than that of human brains; and flexibility of the cerebral cortex, rather than its size, is likely the more important influential factor for performance.

Highly Branched Axons

Axon-First Pathogenesis Templated by Highly Branched Axons in Lewy Body Disease

In circuit diagrams from neurophysiology textbooks, wiring of brain structure is usually represented by a single straight line that connects one nucleus to another. However, this simplified representation does not take into account the number of neurons in each nucleus, and it remains silent on how axons and dendrites are organized around each neuron. For example, a single axon from a DA neuron of the substantia nigra (SN) undertakes excessive axonal branching to gain contact with 75,000 striatal neurons in the rat brain (Matsuda et al. 2009). Because medial DA axons from the ventral tegmental area project to the frontal cortex, the firing of DA neurons influences not only the extrapyramidal system but also "extra-extrapyramidal" systems such as the reward estimation, as discussed later.

Metabolic demands of the axons are supported by intra-axonal supplies, which become more compromised at their distal ends where metabolic turnover is maximal to support synaptic functions (Cheng et al. 2010). This distal-dominant vulnerability is exaggerated when axons are small in caliber and hyperbranched as seen in nigrostriatal

DA axons (Bolam and Pissadaki 2012). Indeed, selective toxicity of 1-methyl-4-phenyl-1,2,3,6-tetrahydropyridine (MPTP) is mediated by uptake of its toxic metabolite, 1-methyl-4-phenylpyridinium (MPP+), through axon terminals (McCormack et al. 2008). Similarly, innumerable axon terminals may be a portal of entry of potential environmental toxins, such as paraquat or rotenone (McCormack et al. 2008). Their toxic targets are mitochondria, abundant in axon terminals, where energy demand is maximal (Subramaniam and Chesselet 2013). Targeted disruption of the mitochondrial complex 1 in nigral DA neurons results in degeneration of nigrostriatal axons, leading to motor deficits at an early stage (González-Rodríguez et al. 2021). This degeneration is followed by progressive loss of affected DA neurons, leading to overt levodopa-responsive parkinsonism in later disease stages. Moreover, αS is enriched in axon terminals where αS-related pathogenesis is initially triggered and subsequently spreads intra-axonally toward the neuronal soma, as demonstrated in the cardiac sympathetic axons in humans (Orimo et al. 2008). Axonal initiation of αS deposition is followed by somatic extension of the deposits (Orimo et al. 2008), and it has been recapitulated by experimental induction of αS fibrils in cultured cells or in experimental animals (Awa et al. 2022) after exposure to preformed fibrils. Selective down-regulation of autophagy in nigrostriatal neurons (Sato et al. 2018) also leads to intra-axonal deposition of endogenous αS. Although molecular events involving αS in PD and in MPTP, paraquat and rotenone toxicities are quite different, their pathogeneses are similarly mediated by highly branched axons as one of the mechanistic explanations for selective vulnerability of the nigrostriatal system. It is conceivable that highly branched axons serve as "the intrinsic structural template" to accelerate the distal-dominant degeneration. However, this pathogenetic conditioning is not necessarily associated with Lewy pathology as seen in MPTP exposure or in some genetic forms of parkinsonism with nigral degeneration (Deng et al. 2018). This highly divergent nature of axons of Lewy-prone neurons is in sharp contrast with the somatic motor system with straight axons. In contrast, the highly somatotopic organization of the motor system is strengthened by myelination of straight axons so that this structural organization is suitable for speedy voluntary movements with accuracy in terms of space and time. If myelinated axons have less chance to undertake branching, it is likely that myelination plays a pivotal role in organizing highly somatotopic motor systems, advantageous for survival during natural selection. In contrast, it is difficult to execute such rapid movements with accuracy through highly divergent axons in Lewy-prone neurons. This is one of the major reasons why physiological functions of Lewy-prone neurons and their clinical manifestations are barely somatotopic (Figure 4.1).

Multifocal Lewy Body Disease with Capricious Spread: Serial or Parallel?

According to Damasio (2018), the brainstem, the limbic system, and the autonomic system constitute the neuroanatomical basis for the homeostasis core; but its exact boundaries remain obscure (Goldstein 2021). Moreover, most brainstem and limbic

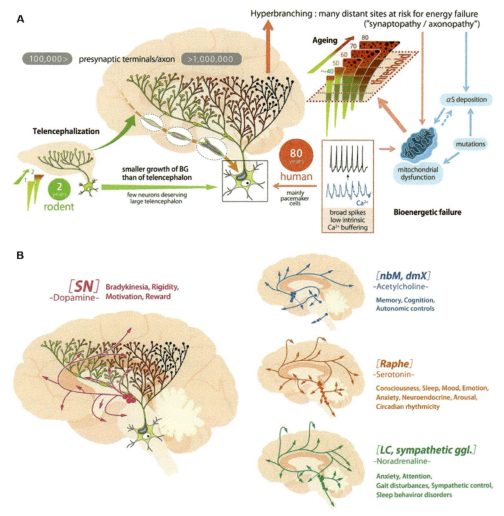

Figure 4.1 (A) Hyperbranching axons as a structural template to facilitate Lewy pathology. Axons from midbrain dopaminergic neurons undertake excessive branching to project to striatum and cerebral cortex, where presynaptic terminals/axon are numerous but less than 100,000 in rat. After preferential enlargement of cerebral hemisphere relative to the brainstem (telencephalization) in humans, midbrain dopaminergic axons undertake further branching (presynaptic terminals/axon >1,000,000), which facilitates distal-dominant degeneration due to energy and metabolic failure. As alpha-synuclein (αS) is enriched in axon terminals, axonal hyperbranching provides an additional load of αS, facilitating αS-related neurodegeneration, which gradually exceeds the threshold to develop clinical manifestations, initially insidiously and later more apparently. These neurons with hyperbranching axons can be pacemaker cells, and their autonomous cell activity relies on feed forward bioenergetics control. (B) Parallel involvement of Lewy-prone systems. Lewy-prone systems are similarly characterized by hyperbranching axons, regardless of the neurotransmitters. This structure templates distal-dominant, αS-related axonal degeneration. Because each system projects to wide areas in the brain, clinical manifestations may be vague, and not focally accentuated. Notably, it is difficult to adjust the output from hyperbranching axon by feedback (see text), which may also explain the instability or fluctuating nature of clinical manifestations of Lewy body disease. ggl. = ganglia; LC = locus coeruleus; SN = substantia nigra; BG = basal ganglia; dmX = dorsal motor nucleus of vagal nerve. Reproduced from Diederich et al. (2019).

systems are characterized by highly branched axons where somatotopic delineation in relation to their functions is poorly defined. Importantly, these brainstem nuclei with highly branched axons preferentially develop Lewy pathology (Uchihara 2017) (Figure 4.1B). One of the prevailing hypotheses is that these Lewy-prone nuclei are interconnected so that Lewy pathology spreads along interconnected wiring (Del Tredici et al. 2002). Numerous experimental studies have tried to demonstrate wiring-based spread of αS deposits after inoculation of aggregated αS (Dehay et al. 2016). However, the quantitative evaluation of αS deposits in human brain failed to demonstrate the presumed correlation between αS deposits and the degree of connection among neuronal groups harboring or not harboring αS deposits (Surmeier et al. 2017a). Moreover, if the functions of these Lewy-prone nuclei are quite diverse and not mechanistically linked, each nucleus may function independently of the others, as demonstrated by transmitter-selective propagation of αS seed (Henrich et al. 2020). Despite this independent construction of Lewy-prone neurons, it has been claimed that Lewy pathology is initiated at the lower brainstem, such as the dorsal motor nucleus of the vagal nerve, and that it spreads sequentially to the locus coeruleus (LC) and SN (upward spreading) as proposed by Del Tredici et al. (2002). Because pathological αS aggregates have the potential to recruit normal αS to make it pathological like a prion protein, Lewy pathology may propagate through neural connections. This elegant scheme is now recognized as the Braak-prion hypothesis (Uchihara et al. 2020), where a local molecular mechanism to form Lewy pathology is combined to explain the spreading of the lesions, leading to disease-specific distribution. It is true that Lewy pathology is most frequent and abundant in the lower brainstem, followed by the pons and the SN. However, this distribution just reflects an overall propensity to develop Lewy pathology, which may be applicable to some patients with PD. In fact, the distribution of Lewy pathology in the human brain is heterogeneous (Berg et al. 2014), and only a minority of Lewy body disease autopsy cases (7%–25%; Adler et al. 2019; Attems et al. 2021) are compatible with the scheme of upward spread. Moreover, rating of Lewy pathology is highly dependent on the examiner (Attems et al. 2021). It is necessary to reconsider the possible mechanism and standardize the operational way of estimating Lewy pathology to reach a sharable consensus to understand Lewy pathology in the human brain in relation to clinical manifestations.

Furthermore, there are numerous pathology reports on isolated emergence of Lewy pathology, all to be grouped under the umbrella of "focal Lewy body disease" (Uchihara 2017). Examples include Lewy bodies only at peripheral sites (Minguez-Castellanos et al. 2007; Probst et al. 2008), the LC (Louis et al. 2005), the olfactory bulb (Beach et al. 2010), the cardiac autonomic system (Miki et al. 2009), and visceral organs (Minguez-Castellanos et al. 2007). Typical PD and dementia with Lewy bodies may be collectively called "multifocal Lewy body disease." Recent development of functional imaging methods has identified two contrasting patterns of progression: brain-first versus body-first progression (Horsager et al. 2020). However, this distinction may just indicate the relative abundance of Lewy pathology, and it does not necessarily indicate if there has been serial progression along neural connections. While serial unidirectional spread of Lewy pathology, like prions along neural connections, is a beautiful and attractive scheme, in the human brain, however, Lewy-prone neurons develop Lewy pathology at *multiple sites* regardless of the connections (Uchihara et al. 2020). Such emergence is hardly explainable by only serial and unidirectional propagation. Random and "capricious" emergence of Lewy pathology in multiple neuronal groups in parallel may be better explained by cell autonomous mechanisms, such as an intrinsic structural template of highly branched axons, as proposed here.

Straight Axons for ALS versus Highly Branched Axons for Lewy Pathology

Neuronal groups preferentially affected in ALS include lower and upper motor neurons (Duyckaerts et al. 2021), neurons in the cerebral cortex, and the neurons receiving direct projection from the cerebral cortex, as shown in Figure 4.2 (Braak et al. 2013). Highly topographical organization is mandatory to achieve fine, individual movements independently or cooperatively upon request. Therefore, axons in the motor system are usually myelinated (ten Donkelaar et al. 2004) so that rapid conduction necessary for rapid and precise movements is executed (Firmin et al. 2014). Several studies have demonstrated that deposition of TDP-43 is found in axons in biopsied specimens of muscles not only in patients with ALS (Mori et al. 2019) but remarkably also in patients who only later developed ALS after the muscle biopsy, suggesting that this TDP pathology is initiated at the distal end of motor axon terminals at least in some patients (Kurashige et al. 2022). In addition to rapid conduction, precise motor control involving multiple inner feedback loops is necessary, as shown in Figure 4.3 (Roth et al. 2014) in a simplified diagram. Detailed discussion of electrophysiological mechanisms for motor control is beyond the scope of this short review. Although this schematic diagram of motor control does not tell how motor commands for voluntary movements are initiated (Granit 1980), once this process starts, the motor command at the motor cortex is forwarded to the cerebellum, which predicts the expected consequences of the movement. Then, the parietal cortex combines this expectation with actual sensory inputs (visual, proprioceptive, and others) to compute an estimate about the current proprioceptive and visual state of the body. This calculated estimate is forwarded back to the motor cortex, where a single motor command is generated for execution based on its policy (Shadmehr and Krakauer 2008). It is worth noting that the genesis of "a single" command is mandatory to initialize and execute an action. This singularity of action is a mandatory constraint, and it is tightly related to the functional polarization of neurons initially idealized by Ramon y Cajal (Berlucchi 1999), whose "neuron doctrine" proposes a fundamental functional unit and "somatotopic organization" of the entire brain. The real-time feedback of motor control is possible only when the necessary feedback loops are available, and the output is calculated as quantified data. Such computational and quantitative characterization of internal feedback models is one of the key fields of neuroscience, which may explain their

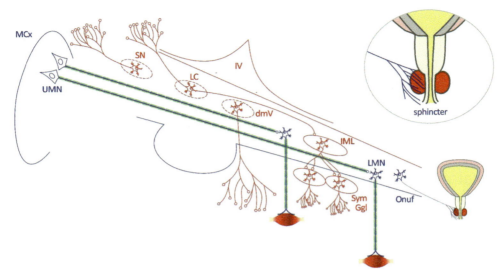

Figure 4.2 Alpha-synuclein (αS) in branched non-myelinated axons of Lewy body disease versus TAR DNA-binding protein (TDP) in straight myelinated axons of amyotrophic lateral sclerosis (ALS). Lewy-prone neurons (red) located in the dorsal neuroaxis are equipped with hyperbranched, non-myelinated axons, which project to wide areas of the brains and to the entire body. αS-related degeneration is initiated at distal axons. Lewy pathology in these nuclei may occur independently as "focal Lewy body disease," not necessarily accompanied by Parkinsonism/nigral degeneration. Parallel emergence of focal Lewy body diseases may constitute "multifocal Lewy body disease," one of the most frequent presentations involving the substantia nigra, as "Parkinson disease." In contrast, ALS-affected motor neurons (blue), located in the ventral neuroaxis are equipped with straight, myelinated axons. Highly somatotopic organization is necessary to perform rapid voluntary movements with high precision in space and time. Initial TDP deposits are seen in distal axons in muscle fibers. Motor neurons in the Onufrowicz nucleus, projecting to the external sphincter around the anus and the urethra (circled), are different from other alpha-motor neurons in terms of their simple function without fine somatotopy. This difference may be one of the conditions that spares Onufrowicz nucleus in ALS. dmV = dorsal motor nucleus of vagal nerve; dmX = dorsal motor nucleus of vagal nerve; IML = intermediolateral nucleus; IV = fourth ventricle; LC = locus coeruleus; LMN = lower motor neuron; MCx = motor cortex; Onuf = Onfrowicz nucleus; SymGgl = sympathetic ganglia; SN = substantia nigra; UMN = upper motor neuron. This tracing was performed by Ms. Noriko Higa at Neuromorphomics Laboratory, Nitobe Memorial Nakano General Hospital, Tokyo, Japan.

algorithmic nature (Rosenblueth et al. 1943). Nevertheless, it is not clear either "how" or "where" motor commands are initiated in the brain.

This axonal architecture is quite different from that of Lewy-prone neurons with poorly myelinated, highly branched axons. As shown in Figure 4.2, the distribution of ALS-affected neurons (Braak et al. 2013) and that of Lewy-prone neurons never overlap but rather are complementary to one another. It is possible that neurons with highly branched unmyelinated axons are liable to develop Lewy pathology, while neurons with straight myelinated axons are preferentially affected in ALS. One of the exceptions is the Onufrowicz nucleus, where motor neurons control the external anal sphincter (Konishi et al. 1978). Given that the function of the sphincter is so simple in contrast to the detailed somatotopic organization of complex hand muscles, branching axons may constitute this

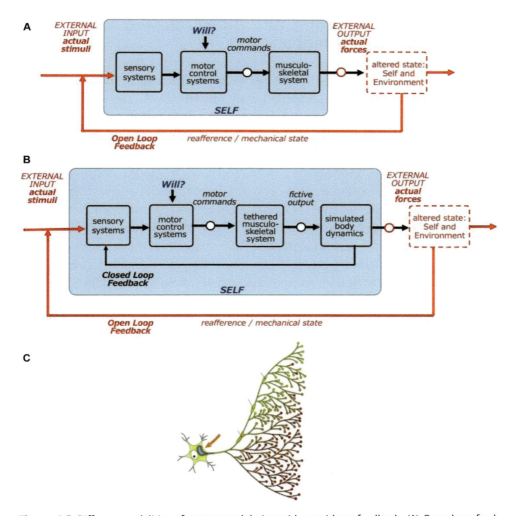

Figure 4.3 Different modalities of output modulation with or without feedback. (A) Open-loop feedback. Motor commands are transferred to musculoskeletal system, then transformed into external output (actual force) for execution. Feedback is possible only after the action is completed. (B) Closed-loop feedback. Motor commands are transferred to the musculoskeletal system to generate fictive output for simulation, usable for closed-loop feedback to modulate output even before execution of the action to improve the temporospatial precision. These feedback systems are based on serially connected units with unified output signals for execution and feedback. (C) Output through hyperbranching axon without feedback. Wiring-based feedback either open-loop (A) or closed-loop (B) is not applicable to the hyperbranching axon. In the absence of such feedback, its activities are inevitably autonomous, unstable, or fluctuating. (A, B) Modified from Roth et al. (2014); (C) modified from Uchihara and Giasson (2016).

system, harboring dual-nature functions (somatic and autonomic; Schellino et al. 2020). It is well known that the Onufrowicz nucleus and anal sphincter function are relatively spared until the terminal stage of ALS (Mannen et al. 1977). This is corroborated by relative preservation of the size and number of the Onufrowicz nucleus neurons in ALS (Takeda et al. 2014).

Fluctuation of Symptoms of Highly Branched Axon without Real-Time Feedback

If the axonal branching is extreme, as seen with Lewy-prone systems, it is not possible to adjust their axonal output easily through neural connections as in motor systems (Figure 4.3C). One of the strategies for output adjustment includes two antagonistic, independent systems: sympathetic versus parasympathetic. In such systems, adjustments are possible only after the consequence of output is achieved at least partly. Such result-oriented modifications are quite different from the motor system with target-oriented modifications guided by near real-time internal feedback to execute smooth voluntary movements with error minimization in order to achieve the expected goal (Figure 4.3A, B). Even if the estimation of errors is calculated in advance, precise modulation of the output as expected may be difficult to achieve, partly because real-time feedback through wiring is not possible for systems with highly branched axons. The contrasts in axonal construction are highlighted Table 4.1. They explain why precise smooth movements are executable with rapidity in the motor system with real-time feedback through myelinated axons. In contrast, output from systems with highly branched axons are more unstable and less precise because real-time feedback, as in the motor system, is lacking. This may explain why clinical manifestations of Lewy body disease can also be unstable and fluctuating (Matar et al. 2020). Fluctuation of cognitive function is one of the most remarkable diagnostic criteria for dementia with Lewy bodies (DLB, McKeith et al. 2017).

The depletion of DA neurons in the SN is the substrate for motor symptoms of PD. However, only secondarily dysfunctional striatal neurons are in fact responsible for the PD symptoms. Similarly, cognitive deficits or hallucinations are related to a decline in the cholinergic drive originated from the nucleus basalis of Meynert through highly branched axons (Yarnall et al. 2011). Dysfunction of the LC may manifest itself as the summation of different cortical and subcortical structures, all driven by highly branched axons originating in the LC (Benarroch 2018). Therefore, clinical manifestations related to such systems with highly branched axons are generally subtle and diffuse, without clear somatotopy (Figure 4.1). Because of this fluctuating nature, detection of early clinical manifestations of DLBD is often challenging, especially before development of characteristic motor symptoms of PD (Berg et al. 2013).

Although the origin of voluntary actions remains to be proven, some of the voluntary actions or their combination as a behavior are driven by estimation of rewards or "desire," which are predictable; but their exact outcome is not unveiled until the behavior is completed. This situation is different from the goal-directed framework of real-time feedback of motor performance with repeated calculations of predicted consequences and corrections during execution of rapid motor executions (Shadmehr et al. 2010).

Muscle movements are executed by rapid data conduction through myelinated fibers that activate neurons as phasic stimuli in motor systems (Figure 4.3). In contrast, reward estimation is performed in systems composed of unmyelinated fibers as in nigrostriatal or in amygdalar systems. Interestingly, these are not included in the feedback loops of

TABLE 4.1 Three Distinct Features of Motor, Nigrostriatal, and Limbic Neurons as Internal Templates for Amyotrophic Lateral Sclerosis (left column), Parkinson's disease (middle column), and Alzheimer's disease D (right column)

Category	Motor control	Reward/neuromodulation	Memory/recognition
Objective	Skill: How to do?	Decision: What to do?	Knowing: What is (was) it?
Direction	Efferent, prediction of self	Efferent, prediction of self/environment	Afferent, from past to present
Performance	Error prediction for real-time feedback correction	Predict possible consequences of an action toward value-oriented choice	Consolidate internal and external data for future retrieval
Goal	Given	To be generated	Not applicable
Performance	Reach the predefined goal by minimizing error (closed loop)	Maximize the value and minimize the risk of an action (open loop)	Plastic consolidation
Strategy	Teleological operation based on internal model to reach a goal	In advance value estimation	Anatomical or functional changes
Prediction error	Correctable during performance	Only after completion of an act	Not applicable
Operating system	Cybernetics: serially connected computing units, each with quantitative input/output	Dopaminergic neurons Autonomic centers	Synaptic functions Sprouting Myelination
Neurite structure	Straight axon	Hyperbranching axon	Axo-dendritic synapse
Calculation	Model-based prediction error	Ongoing without predefined model	
Autonomy	Automatic, dependent, passive	Autonomous, independent, active	Possibly influenced by the self-status
Conduction	Rapid signaling along myelinated axon	Slow signaling along unmyelinated axon	Both myelinated and unmyelinated
Anatomical sites	Motor cortex–anterior horn Cerebellum, parietal cortex	Nigrostriatal dopaminergic neurons Limbic systems	Entorhinal neurons Limbic neurons
Somatotopy	Clear	Obscure	Possibly present
Functional module	Clear	Unclear	Present
Pathological proteins	TAR DNA-binding protein	Alpha synuclein/Lewy pathology	Tau/neurofibrillary pathology

motor execution diagrams (Table 4.1). Furthermore, their tonic firing due to autonomous cellular activity (Surmeier et al. 2017b) is strikingly different from motor neurons or cerebellar neurons with only phasic firing during motor execution. This autonomic nature is probably essential for neurons in reward estimation and decision-making. Indeed, the net output of the motor system is calculated internally along an internal model with feedback loops, which allows near real-time correction before completion of each movement so that the discrepancy between the intended movement and the actual movement is minimized. This discrepancy between the estimated reward and the actual result could be sometimes highly significant, probably because actual results do not come to the scene until a set of actions are executed completely, which hampers the real-time feedback as seen in motor control systems. Under this unsteady expectation/results paradigm, a probabilistic approach rather than the deterministic prediction may be more practical or feasible to maximize the benefit and minimize the cost. However, it remains to be clarified whether the probabilistic decision entertained by artificial intelligence, for example, is more efficient, rational, and, finally, ethically reliable than that of human beings, especially in the field of health care (Schwalbe and Wahl 2020).

Goal-Directed Motor Control System versus Goal-Searching Reward Estimation System

The brain is requested to deal with a changing environment (Schwartz 2016). Difficulties may be enhanced because the environment is changing, sometimes in an unexpected range. The situation may be more difficult if the living organism and its brain are also changing. When an individual organism takes an action in response to external stimuli or internal desire, at least two questions may arise: "How to do?" and "What to do?" "How?" is related to the motor control system, which is highly sophisticated to execute actions to reach a predetermined goal. Although it is not yet clear how brain evolution has been successful in establishing a system of such harmony (Tanaka and Sejnowski 2015), computational neuroscience is successful in demonstrating how each component is connected and interacts during the ongoing action. In this sense, the motor control system is coping with "the present." However, only if the current goal is available, such a sophisticated system of motor control becomes meaningful, as stated by Denny-Brown (cited in Granit 1980, p. 157) in the following: "The cerebral cortex is responsible for goal-directed behaviour." Efficiency is dependent on rapidity, precision, energy cost, etc., which may influence natural selection as a possible motif of evolution. However, this Darwinian interpretation of natural selection is neither meaningful before such a system is established nor sufficient to explain how such sophisticated systems became available prior to natural selection (Pilpel and Rechavi 2015).

In contrast with this predetermined automatic, machine-like feature of the motor control system coping with "the present" to achieve a goal, the reward estimation system is trying to deal with "the future" by estimating expected rewards to choose an action that may "best" balance reward/cost (McShea 2016). This requires autonomous, independent processes to give birth to an aim: "What to do?" Such a creative nature is not affordable by cybernetic systems with precise feedback as in the motor control system, which requires a predefined goal. If such an operation is undertaken in midbrain DA neurons with highly branched axons, the real-time feedback through neural connections, as in the motor control system, is quite unrealistic. It may be related to the independence and the freedom or arbitrariness of the decision as a representative of voluntariness of one's free will. Nevertheless, the ultimate origin of such decisions or voluntariness, either purely internal or induced by external stimuli, remains to be clarified.

The outcome of the motor control is readily apparent based on the discrepancy between the initial plan and its actual result noticed as an error. Internal models of motor control are designed to work within the physical constraints of the actual body size, muscular strength, etc. The reward estimation is similarly under the constraints of the actual body. For example, appetite is satisfied as satiety after some meals, and sexual desire is satisfied usually after sexual activity. In contrast, some desire for financial profit or pursuit of political power could grow so unlimitedly that it may disrupt social systems. It remains to be clarified whether some intrinsic mechanisms to adjust reward within an appropriate range are at work in the human brain or not. Probably, some extrinsic social systems—family, art, ethics, law, education, religion, or even medicine—have been constructed to counteract such unlimited expansion of reward or desire. This could be a kind of interaction between individual human beings and the society. It may be considered as a kind of evolution of the brain in relation to the society.

Memory and Its Disruption by Tau Deposition Initiated at Dendritic Spines

Motor control deals with the current status (the present) by modulating ongoing motor performance through closed-loop feedback, while reward estimation deals with the future by estimating expected rewards and thereby adjusting future activities. In contrast, memory/recognition deals with "the past" by consolidating various episodes or information through plastic alteration of the brain components, as shown in Table 4.1. One of the hypotheses for memory consolidation discusses plastic changes of the axodendritic synapses on the dendritic spine (Boros et al. 2017). In human brains with AD, tau deposition is initiated in the dendrites of parahippocampal neurons (Dorostkar et al. 2015). This dendrite-first cytopathology is in sharp contrast with the axon-first cytopathology of Lewy body disease or ALS. In the human brain with AD, dendritic tau deposits spread into the neuronal soma, then probably into axons in an anterograde direction (Braak et al. 1994; Braak and Braak 1997; Figure 4.4B), which is in sharp contrast with Lewy pathology with retrograde spread from distal axons to neuronal soma. Because tau is primarily an axonal protein, it remains to be explained how deposition of tau is initiated in dendrites followed by expansion into neuronal soma. If tau lesions spread from the parahippocampal area to the hippocampus, then to the limbic cortex, and finally to the neocortex (Braak and Del Tredici 2018), it is conceivable that this spreading of tau is mediated by anterograde spread in neurons connected serially through interneuronal wiring

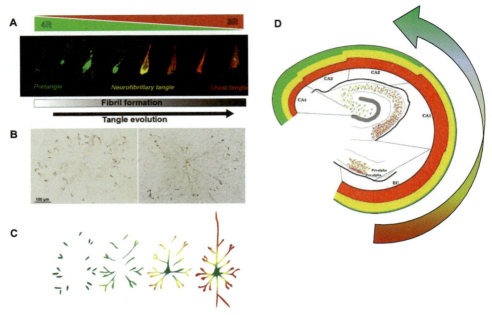

Figure 4.4 Profile shift from four-repeat (4R) to three-repeat (3R) tau is in parallel with tangle evolution, intracellular, and regional extension in Alzheimer's disease. (A) Profile shift from 4R-positive pretangles to 3R-positive ghost tangles. Early tau deposits as pretangles are positive for 4R tau (green). Progressive fibril formation leads to neurofibrillary tangles, positive for both 3R and 4R tau (yellow), then ghost tangles, positive for 3R tau (red). Modified from Uchihara (2020). (B) Tau deposition is initiated at distal dendrites followed by extension toward soma; 100-micron-thick sections immunostained for tau. From Braak and Del Tredici (2018). (C) Intraneuronal spread from distal dendrites to neuronal soma with profile shift from 4R (green) to 3R (red) tau. Modified from Braak and Del Tredici (2018). (D) Regional extension of neurofibrillary tangle from entorhinal cortex (EC) to CA1–4 is accompanied by the same profile shift from 4R (green) to 3R tau (red). Modified from Hara et al. (2013).

via axons to dendrites of downstream neurons (de Calignon et al. 2012). Of note, it is not only that these structures are anatomically connected but this serial connection is also functionally relevant to achieve a hierarchy from memory (hippocampus) to recognition (limbic areas) to finally decision (neocortex). These structural characteristics may serve as a structural template to guide the disease-specific pattern of AD-tau spread, which is tightly related to clinical development of cognitive decline.

An elegant explanation for this disease-specific spread is based on the so-called Braak-prion hypothesis. Pathologically, modified tau molecules may serve as a seed to transform intrinsic tau into a pathological state, as in prion disease. This prion-like pathological transformation continues to provide new seeds that spread along neural connections to form a hierarchical spread of tau lesion, as proposed by Braak (Uchihara et al. 2020). This simple and clear explanation is possible if the tau molecule is a single molecular species that undertakes pathological transformation like prions. However, in the human brain, tau species are divided into three-repeat (3R) and four-repeat (4R) isoforms according to the number of tandem repeats of microtubule-binding domains.

Both 3R and 4R isoforms are involved in the formation of neurofibrillary tangles (NFTs). Double labeling studies demonstrated that pretangles at early stages of NFT formation are positive only for 4R tau, while mature fibrillar NFTs are positive for both 3R and 4R tau, and ghost tangles are positive only for 3R isoforms (Figure 4.4A; Uchihara et al. 2012). This profile shift may be represented during intraneuronal spreading from dendrites to soma (Figure 4.4C). Because 4R tau never transforms itself into 3R tau, this chronologically fixed transition of tau isoforms from 4R to 3R tau is poorly explained by the prion hypothesis. Curiously, mRNA of tangle-bearing neurons in the nucleus basalis of Meynert exhibits a progressive decrease in RNA coding for 3R tau relative to 4R counterparts during tangle evolution from pretangles to ghost tangles (Ginsberg et al. 2006). Moreover, quantification of the number of 4R-positive (4R+) NFTs and 3R-positive (3R+) NFTs along the regional spreading from parahippocampus–subiculum–CA1, CA2/3, and CA4 sectors demonstrated that 3R+ neurons are most frequent in the upstream area, such as the parahippocampal gyrus, while they are progressively replaced by 4R+ neurons in downstream areas (Hara et al. 2013). This indicates that this profile shift from 4R tau to 3R tau, shared between intraneuronal alterations and regional spread, is characteristic of AD. Even if this profile shift represents intraneuronal processes, how it spreads trans-synaptically across neurons is more than a mystery. Although the underlying mechanism to explain this spread of tau lesion with isoform shifting remains unknown, neither 3R nor 4R tau alone is sufficient to explain this profile shift. One may then suppose that some alternative factors other than tau itself (Kara et al. 2018) or some events, such as neuronal firing, may be necessary to explain the trans-synaptic spread of lesions (Wu et al. 2016).

Experimental models trying to reproduce 3R/4R tau-positive AD-NFT in the mouse brain have encountered great difficulties, mainly because of the predominant 4R tau background. Genetically engineered mice expressing both 3R and 4R tau successfully reproduce tau isoform dominance when inoculated with extract from AD patients (3R and 4R), corticobasal degeneration (CBD) patients (4R only), and Pick body disease patients (3R only; Hosokawa et al. 2022), indicating prion-like replication of disease-specific tau isoforms. However, the morphology of tau-positive lesions in these animals remains to be demonstrated for their possible similarity to the original lesions in human brains. Because of such difficulties in recapitulating AD pathology in mouse brains, non-human primate brains have become increasingly attractive as a potential model of AD/aging, especially because both Aβ-positive lesions and tau-positive lesions are found in their aged brains (Rosen et al. 2008). In our series of cynomolgus monkeys (7–36 years old) the followings were found: (1) extensive deposition of Aβ in brains over 25 years of age, (2) selective deposition of 4R tau as pretangles in neurons and as coiled body–like structures in oligodendroglia-like cells and astrocytes, (3) preferential distribution of tau in the basal ganglia and frontal neocortex rather than the hippocampus, and (4) age-associated increases in 30–34-kDa AT8- and RD4-positive tau fragments in sarkosyl-insoluble fractions (Uchihara et al. 2016). These biochemical and morphological features in cynomolgus monkeys are like those of human progressive supranuclear palsy (PSP)/

CBD but quite different from those of AD, suggesting that the intrinsic templates activated along the brain aging process are different between cynomolgus monkeys and humans.

Some of the mutations of the tau gene are related to the generation of pathological tau by the preferential splicing of exon 10 into the tau transcript so that 4R tau deposits similar to those in PSP/CBD are generated. Therefore, the presence or absence of exon 10 may be pivotal to differentiate isoform-specific tau lesions in the human brain (Stancu et al. 2019). Although the adult mouse brain contains only 4R tau, 3R tau is present only at its fetal stage (Takuma et al. 2003). Therefore, splicing of exon 10 is related to species differences (mouse/human), developmental stage (fetus/adult), and disease status. Although it is highly controversial whether "ontogeny recapitulates phylogeny" as postulated by Ernst Haeckel (Elinson and Kezmoh 2010), it is probable that a similar mechanism, such as splicing of exon 10, is related to both the difference in animal species (phylogeny), along with the development of individual organism (ontogeny) and tau pathogenesis. It remains to be clarified how tau molecules are at work in these different contexts.

Why Is Lewy Pathology Only Formed in the Human Brain?

Brain evolution is characterized by a disproportionate increase in the size and complexity of the cerebral cortex known as "telencephalization." Because the brainstem structures are left relatively unchanged, a single DA axon in the human brain has synaptic contacts with more than 1,000,000 striatal neurons by undertaking further axonal branching and elongation (Bolam and Pissadaki 2012). This process enhances the distal-dominant vulnerability. Even with increased metabolic demand, DA neurons are resistant enough to survive and function for decades until the development of PD around age 60 or later. One of the difficulties in establishing animal models of PD with Lewy pathology may be related to these highly branched axons, which are hardly reproducible in animal models. For example, cynomolgus monkey brains develop Aβ deposits after age 20 and tau deposits after age 30 (Uchihara et al. 2016), while αS deposits are absent in the same species up to age 35, suggesting that additional pathogenic conditions or their combination are necessary for Lewy pathology to develop in the brain. Neither of these conditions—highly branched axons, high energy expenditure, or longevity—is achieved in animal brains other than the human brain (Diederich et al. 2019). If the formation of Lewy pathology is dependent on such combined settings, as seen in the human brain, rather than αS itself, it is quite a challenge to replicate Lewy pathology in the brains of experimental animals. In addition to the engineering of αS through genetic and molecular tools, a comparison of different experimental conditions in reference to the human brain may clarify which conditions or molecules are essential in developing Lewy pathology. Such an approach may allow identification of the essential constituents of the intrinsic template (structural, molecular, chronological, etc.) that characterize the selective vulnerability of Lewy body disease in humans.

The Past, the Present, and the Future

Three major neurodegenerative disorders—PD, ALS, and AD—have been compared in terms of disease-specific pathogenesis from the viewpoint of distinctive neurite architectures as a consequence of evolution.

DA neurons in the SN are characterized by highly branched, unmyelinated axons as a basis for distal-dominant degeneration. This selective vulnerability is further enhanced by telencephalization as it enforces axonal stretching and branching, resulting in clinical manifestations without somatotopic accentuation. Abundant αS at each axon terminal facilitates deposition of αS, spreading retrogradely from the axon terminal toward the neuronal soma in these neurons. Highly branched axons seen in other Lewy-prone neurons, such as the LC, the dorsal motor nucleus of the vagal nerve, the nucleus basalis of Meynert, and autonomic neurons, including cardiac sympathetic neurons, similarly facilitate Lewy pathology. Therefore, highly branched axons, as a consequence of telencephalization, may serve as a structural template to facilitate Lewy pathology. This highly branched axon architecture may also explain the generalized nature of the clinical manifestations and missing somatotopy, and it may serve as a functional template for their manifestations.

In contrast, the final motor pathways with rapid signal conduction through myelinated axons are straight. They are related to the highly somatotopic nature of motor control. Although the ultimate origin of voluntary initiation (will, desire) remains to be defined, voluntary movements, once initiated, are under the control of near real-time feedback along the internal models to reach the goal. The motor control system has been progressively sophisticated by integrating long- and short-loop feedbacks involving the cerebellum and other structures. This automatic nature of dealing with the present is in sharp contrast with the reward estimation to deal with the future by midbrain DA neurons.

Autonomic independence of the reward estimation may represent the intrinsic and spontaneous nature of the "free will," and it may mechanistically be related to the inability of real-time feedback on highly branched axons as an output. It may explain the unstable and fluctuating nature of clinical manifestations of Lewy body disease, including dopamine dysregulation syndrome, pathological gambling, etc.

Dealing with the past is mediated by memory, presumably consolidated through synaptic transmission mainly at the dendritic spine. Deterioration of memory in AD may be correlated with tau deposition initially in dendrites, followed by expansion into the neuronal cytoplasm, then into the axon. This anterograde spread may explain how the stereotyped spread of tau along connections is related to clinical progression of AD from amnesia to recognition failure and, finally, to dementia, thus suggesting a functional hierarchy. The brain has evolved to develop different strategies to deal with the past, the present, and the future. Each system has a distinctive neuronal architecture: dendrites for memory (the past/tau), straight axons for motor control (the present/phosphorylated TDP), and highly branched axons for reward estimation (the future/αS). The dysfunction of each system is correlated with the deposition of disease-related proteins, which is facilitated by an underlying neuronal structure under evolution that features disease-specific intrinsic templates and corresponding clinical symptomatology.

Is it the disease-specific protein that causes brain disease? Although such an explanation is simple and straightforward, the protein molecules become pathogenic only after templated by the underlying structure, such as highly branched axons for Lewy pathology, as a consequence of brain evolution. If evolution is the designer of the brain, is it the designer of brain diseases as well? Indeed, brain evolution provides disease-specific intrinsic templates on neurons, which may engender a condition that may facilitate the formation of lesions in a disease-specific fashion. Moreover, clinical manifestations are templated by underlying neuronal architecture leading to disease-specific manifestations. Therefore, the identity of a disease should include at least two aspects: the disease-related protein and the cellular environment (intrinsic template) around the molecule as a consequence of evolution. For example, humans are obligate, habitual, and diverse in their bipedalism and hold the carriage of their bodies upright in a spinally erect posture (Skoyles 2006). The bipedalism and bipedal gait are associated with the progressive increase in the neocortex, which resulted in telencephalization when already hyperbranched axons were stretched and underwent further branching, as in the human brain. This highly branched axon serves as a structural intrinsic template to facilitate the development of PD, as discussed above. In this schema, bipedalism may have seemed initially advantageous, but it led to telencephalization, which unexpectedly facilitated the development of PD and resulted in gait disturbance. Such a complex story is hardly explained by the simple cause-and-results paradigm (Causality). Multiple conditions are combined to result in subsequent consequence (Conditionality). Therefore, the preferential placement of αS as the pathogenic cause may be an oversimplification to understand how Lewy pathology is formed and leads to clinical manifestations. According to Claude Bernard (1865/1927),

> When we know that water, with all its properties, results from combining oxygen and hydrogen in certain proportions, we know everything we can know about it; and that corresponds to the how and not to the why of things. We know how water can be made; but why does the combination of one volume of oxygen with two volumes of hydrogen produce water? We have no idea.

Even if satisfactory answers will never be available to scientists, such "why" questions are everywhere in the field of evolution and medicine. Such "why" questions may be absurd, according to Claude Bernard, but they can be particularly useful to activating our thinking.

Intrinsic templates for AD, ALS, and PD are related to the past, the present, and the future, respectively. It is not possible to explain "why" these intrinsic templates are generated during brain evolution, but it may be possible to understand "how" these intrinsic templates are related to pathogenesis and clinical manifestations in a disease-specific fashion. While this synthetic explanation by itself does not provide straightforward molecular mechanisms, this may highlight disease-specific differences related to both pathogenesis and clinical manifestations. This comprehensive viewpoint may be helpful to establish disease-specific models and strategies for diagnosis and therapy.

Key Points to Remember

- It remains to be explained how disease-related proteins feature specific patterning of neurodegeneration in the human brain. We propose an "intrinsic template" as a novel concept linking disease-specific proteins with selective axonal or dendritic vulnerability unique to the human brain.
- Distal-dominant axonal degeneration of Lewy body disease is templated by highly branched, unmyelinated axons, which may well explain barely somatotopic and generalized clinical features. These highly branched, DA axons are also concerned with reward prediction or decision-making to deal with the future.
- In contrast, straight myelinated axons of upper and lower motor neurons, affected in ALS with TDP-43 deposition, execute highly somatotopic movements with real-time feedback to deal with the present (ongoing processes).
- Tau accumulation in dendritic spines and subsequent spread may account for cognitive and memory deficits of AD, dealing with the past.

References

Adler, C. H., Beach, T. G., Zhang, N., Shill, H. A. Driver-Dunckeley, E., Caviness, J. N., Mehta, S. H., Sabbagh, M. N., Serrano, G. E., Sue, L. I., Belden, C. M., Powell, J., Jacobson, S. A., Zamrini, E., Shprecher, D., Davis, K. J., Dugger, B. N., & Hentz, J. G. (2019). Unified staging system for Lewy body disorders: Clinicopathologic correlations and comparison to Braak staging. *Journal of Neuropathology & Experimental Neurology*, 78(10), 891–899. https://doi.org/10.1093/jnen/nlz080

Attems, J., Toledo, J. B., Walker, L., Gelpi, E., Gentleman, S., Halliday, G., Hortobagyi, T., Jellinger, K., Kovacs, G. G., Lee, E. B., Love, S., McAleese, K. E., Nelson, P. T., Neumann, M., Parkkinen, L., Polvikoski, T., Sikorska, B., Smith, C., Tenenholz Grinberg, L., . . . McKeith, I. G. (2021). Neuropathological consensus criteria for the evaluation of Lewy pathology in post-mortem brains: A multi-centre study. *Acta Neuropathologica*, 141(2), 159–172. https://doi.org/10.1007/s00401-020-02255-2

Awa, S., Suzuki, G., Masuda-Suzukake, M., Nonaka, T., Saito, M., & Hasegawa, M. (2022). Phosphorylation of endogenous alpha-synuclein induced by extracellular seeds initiates at the pre-synaptic region and spreads to the cell body. *Scientific Reports*, 12(1), Article 1163. https://doi.org/10.1038/s41598-022-04780-4

Beach, T. G., Adler, C. H., Sue, L. I., Vedders, L., Lue, L., White, C. L., III, Akiyama, H., Caviness, J. N., Shill, H. A., Sabbagh, M. N., Walker, D. G., & Arizona Parkinson's Disease Consortium. (2010). Multi-organ distribution of phosphorylated alpha-synuclein histopathology in subjects with Lewy body disorders. *Acta Neuropathologica, 119*(6), 689–702. https://doi.org/10.1007/s00401-010-0664-3

Benarroch, E. E. (2018). Locus coeruleus. *Cell and Tissue Research, 373*(1), 221–232. https://doi.org/10.1007/s00441-017-2649-1

Berg, D., Lang, A. E., Postuma, R. B., Maetzler, W., Deuschl, G., Gasser, T., Siderowf, A., Schapira, A. H., Oertel, W., Obeso, J. A., Olanow, C. W., Poewe, W., & Stern, M. (2013). Changing the research criteria for the diagnosis of Parkinson's disease: Obstacles and opportunities. *The Lancet Neurology, 12*(5), 514–524. https://doi.org/10.1016/S1474-4422(13)70047-4

Berg, D., Postuma, R. B., Bloem, B., Chan, P., Dubois, B., Gasser, T., Goetz, C. G., Halliday, G. M., Hardy, J., Lang, A. E., Litvan, I., Marek, K., Obeso, J., Oertel, W., Olanow, C. W., Poewe, W., Stern, M., & Deuschl, G. (2014). Time to redefine PD? Introductory statement of the MDS task force on the definition of Parkinson's disease. *Movement Disorders, 29*(4), 454–462. https://doi.org/10.1002/mds.25844

Berlucchi, G. (1999). Some aspects of the history of the law of dynamic polarization of the neuron. From William James to Sherrington, from Cajal and van Gehuchten to Golgi. *Journal of the History of the Neurosciences, 8*(2), 191–201. https://doi.org/10.1076/jhin.8.2.191.1844

Bernard, C. (1927). *Introduction à l'étude de la médecine expérimentale* [An introduction to the study of experimental medicine]. Henry Schuman. (Original work published 1865)

Bolam, J. P., & Pissadaki, E. K. (2012). Living on the edge with too many mouths to feed: Why dopamine neurons die. *Movement Disorders, 27*(12), 1478–1483. https://doi.org/10.1002/mds.25135

Boros, B. D., Greathouse, K. M., Gentry, E. G., Curtis, K. A., Birchall, E. L., Gearing, M., & Herskowitz, J. H. (2017). Dendritic spines provide cognitive resilience against Alzheimer's disease. *Annals of Neurology, 82*(4), 602–614. https://doi.org/10.1002/ana.25049

Braak, E., & Braak, H. (1997). Alzheimer's disease: Transiently developing dendritic changes in pyramidal cells of sector CA1 of the Ammon's horn. *Acta Neuropathologica, 93*(4), 323–325. https://doi.org/10.1007/s004010050622

Braak, E., Braak, H., & Mandelkow, E. M. (1994). A sequence of cytoskeleton changes related to the formation of neurofibrillary tangles and neuropil threads. *Acta Neuropathologica, 87*(6), 554–567. https://doi.org/10.1007/BF00293315

Braak, H., Brettschneider, J., Ludolph, A. C., Lee, V. M., Trojanowski, J. Q., & Del Tredici, K. (2013). Amyotrophic lateral sclerosis—A model of corticofugal axonal spread. *Nature Reviews Neurology, 9*(12), 708–714. https://doi.org/10.1038/nrneurol.2013.221

Braak, H., & Del Tredici, K. (2018). Spreading of tau pathology in sporadic Alzheimer's disease along cortico-cortical top-down connections. *Cerebral Cortex, 28*(9), 3372–3384. https://doi.org/10.1093/cercor/bhy152

Cheng, H. C., Ulane, C. M., & Burke, R. E. (2010). Clinical progression in Parkinson disease and the neurobiology of axons. *Annals of Neurology, 67*(6), 715–725. https://doi.org/10.1002/ana.21995

Cochran, J. N., Hall, A. M., & Roberson, E. D. (2014). The dendritic hypothesis for Alzheimer's disease pathophysiology. *Brain Research Bulletin, 103*, 18–28. https://doi.org/10.1016/j.brainresbull.2013.12.004

Cookson, M. R. (2012). Evolution of neurodegeneration. *Current Biology, 22*(17), R753–R761. https://doi.org/10.1016/j.cub.2012.07.008

Damasio, A. (2018). *The strange order of things: Life, feeling, and the making of cultures*. Pantheon Books.

de Calignon, A., Polydoro, M., Suarez-Calvet, M., William, C., Adamowicz, D. H., Kopeikina, K. J., Pitstick, R., Sahara, N., Ashe, K. H., Carlson, G. A., Spires-Jones, T. L., & Hyman, B. T. (2012). Propagation of tau pathology in a model of early Alzheimer's disease. *Neuron, 73*(4), 685–697. https://doi.org/10.1016/j.neuron.2011.11.033

Dehay, B., Vila, M., Bezard, E., Brundin, P., & Kordower, J. H. (2016). Alpha-synuclein propagation: New insights from animal models. *Movement Disorders, 31*(2), 161–168. https://doi.org/10.1002/mds.26370

Del Tredici, K., Rüb, U., De Vos, R. A., Bohl, J. R., & Braak, H. (2002). Where does Parkinson disease pathology begin in the brain? *Journal of Neuropathology & Experimental Neurology, 61*(5), 413–426. https://doi.org/10.1093/jnen/61.5.413

Deng, H., Wang, P., & Jankovic, J. (2018). The genetics of Parkinson disease. *Ageing Research Reviews, 42*, 72–85. https://doi.org/10.1016/j.arr.2017.12.007

Diederich, N. J., & Parent, A. (2012). Parkinson's disease: Acquired frailty of archaic neural networks? *Journal of the Neurological Sciences, 314*(1–2), 143–151. https://doi.org/10.1016/j.jns.2011.10.003

Diederich, N. J., Surmeier, D. J., Uchihara, T., Grillner, S., & Goetz, C. G. (2019). Parkinson's disease: Is it a consequence of human brain evolution. *Movement Disorders, 34*(4), 453–459. https://doi.org/10.1002/mds.27628

Diederich, N. J., Uchihara, T., Grillner, S., & Goetz, C. G. (2020). The evolution-driven signature of Parkinson's disease. *Trends in Neurosciences, 43*(7), 475–492. https://doi.org/10.1016/j.tins.2020.05.001

Dorostkar, M. M., Zou, C., Blazquez-Llorca, L., & Herms, J. (2015). Analyzing dendritic spine pathology in Alzheimer's disease: Problems and opportunities. *Acta Neuropathologica, 130*(1), 1–19. https://doi.org/10.1007/s00401-015-1449-5

Duyckaerts, C., Maisonobe, T., Hauw, J. J., & Seilhean, D. (2021). Charcot identifies and illustrates amyotrophic lateral sclerosis. *Free Neuropathology, 2*, Article 12. https://doi.org/10.17879/freeneuropathology-2021-3323

Elinson, R. P., & Kezmoh, L. (2010). Molecular Haeckel. *Developmental Dynamics, 239*(7), 1905–1918. https://doi.org/10.1002/dvdy.22337

Firmin, L., Field, P., Maier, M. A., Kraskov, A., Kirkwood, P. A., Nakajima, K., Lemon, R. N., & Glickstein, M. (2014). Axon diameters and conduction velocities in the macaque pyramidal tract. *Journal of Neurophysiology, 112*(6), 1229–1240. https://doi.org/10.1152/jn.00720.2013

Fu, H., Hardy, J., & Duff, K. E. (2018). Selective vulnerability in neurodegenerative diseases. *Nature Neuroscience, 21*(10), 1350–1358. https://doi.org/10.1038/s41593-018-0221-2

Ghika, J. (2008). Paleoneurology: Neurodegenerative diseases are age-related diseases of specific brain regions recently developed by *Homo sapiens*. *Medical Hypotheses, 71*(5), 788–801. https://doi.org/10.1016/j.mehy.2008.05.034

Ginsberg, S. D., Che, S., Counts, S. E., & Mufson, E. J. (2006). Shift in the ratio of three-repeat tau and four-repeat tau mRNAs in individual cholinergic basal forebrain neurons in mild cognitive impairment and Alzheimer's disease. *Journal of Neurochemistry, 96*(5), 1401–1408. https://doi.org/10.1111/j.1471-4159.2005.03641.x

Gissis, S. B. (2020). Transfer of Lamarckisms and emerging "scientific" psychologies: 19th–early 20th centuries Britain and France. *Studies in History and Philosophy of Biological and Biomedical Sciences, 83*, Article 101146. https://doi.org/10.1016/j.shpsc.2018.08.001

Goldstein, D. S. (2021). Stress and the "extended" autonomic system. *Autonomic Neuroscience, 236*, Article 102889. https://doi.org/10.1016/j.autneu.2021.102889

Gómez-Robles, A., Hopkins, W. D., Schapiro, S. J., & Sherwood, C. C. (2015). Relaxed genetic control of cortical organization in human brains compared with chimpanzees. *Proceedings of the National Academy of Sciences of the United States of America, 112*(48), 14799–14804. https://doi.org/10.1073/pnas.1512646112

González-Rodríguez, P., Zampese, E., Stout, K. A., Guzman, J. N., Ilijic, E., Yang, B., Tkatch, T., Stavarache, M. A., Wokosin, D. L., Gao, L., Kaplitt, M. G., López-Barneo, J., Schumacker, P. T., & Surmeier, D. J. (2021). Disruption of mitochondrial complex I induces progressive parkinsonism. *Nature, 599*, 650–656. https://doi.org/10.1038/s41586-021-04059-0

Granit, R. (1980). *The purposive brain*. MIT Press.

Hara, M., Hirokawa, K., Kamei, S., & Uchihara, T. (2013). Isoform transition from four-repeat to three-repeat tau underlies dendrosomatic and regional progression of neurofibrillary pathology. *Acta Neuropathologica, 125*(4), 565–579. https://doi.org/10.1007/s00401-013-1097-6

Henrich, M. T., Geibl, F. F., Lakshminarasimhan, H., Stegmann, A., Giasson, B. I., Mao, X., Dawson, V. L., Dawson, T. M., Oertel, W. H., & Surmeier, D. J. (2020). Determinants of seeding and spreading of α-synuclein pathology in the brain. *Science Advances, 6*(46), Article eabc2487. https://doi.org/10.1126/sciadv.abc2487

Horsager, J., Andersen, K. B., Knudsen, K., Skjærbæk, C., Fedorova, T. D., Okkels, N., Schaeffer, E., Bonkat, S. K., Geday, J., Otto, M., Sommerauer, M., Danielsen, E. H., Bech, E., Kraft, J., Munk, O. L., Hansen, S. D., Pavese, N., Göder, R., Brooks, D. J., . . . Borghammer, P. (2020). Brain-first versus body-first Parkinson's disease: A multimodal imaging case–control study. *Brain*, *143*(10), 3077–3088. https://doi.org/10.1093/brain/awaa238

Hosokawa, M., Masuda-Suzukake, M., Shitara, H., Shimozawa, A., Suzuki, G., Kondo, H., Nonaka, T., Campbell, W., Arai, T., & Hasegawa, M. (2022). Development of a novel tau propagation mouse model endogenously expressing 3 and 4 repeat tau isoforms. *Brain*, *145*(1), 349–361. https://doi.org/10.1093/brain/awab289

Kara, E., Marks, J. D., & Aguzzi, A. (2018). Toxic protein spread in neurodegeneration: Reality versus fantasy. *Trends in Molecular Medicine*, *24*(12), 1007–1020. https://doi.org/10.1016/j.molmed.2018.09.004

Keiflin, R., & Janak, P. H. (2015). Dopamine prediction errors in reward learning and addiction: From theory to neural circuitry. *Neuron*, *88*(2), 247–263. https://doi.org/10.1016/j.neuron.2015.08.037

Konishi, A., Sato, M., Mizuno, N., Itoh, K., Nomura, S., & Sugimoto, T. (1978). An electron microscope study of the areas of the Onuf's nucleus in the cat. *Brain Research*, *156*(2), 333–338. https://doi.org/10.1016/0006-8993(78)90514-0

Kurashige, T., Morino, H., Murao, T., Izumi, Y., Sugiura, T., Kuraoka, K., Kawakami, H., Torii, T., & Maruyama, H. (2022). TDP-43 accumulation within intramuscular nerve bundles of patients with amyotrophic lateral sclerosis. *JAMA Neurology*, *79*(7), 693–701. https://doi.org/10.1001/jamaneurol.2022.1113

Louis, E. D., Honig, L. S., Vonsattel, J. P., Maraganore, D. M., Borden, S., & Moskowitz, C. B. (2005). Essential tremor associated with focal nonnigral Lewy bodies: A clinicopathologic study. *Archives of Neurology*, *62*(6), 1004–1007. https://doi.org/10.1001/archneur.62.6.1004

Mannen, T., Iwata, M., Toyokura, Y., & Nagashima, K. (1977). Preservation of a certain motoneurone group of the sacral cord in amyotrophic lateral sclerosis: Its clinical significance. *Journal of Neurology, Neurosurgery, and Psychiatry*, *40*(5), 464–469. https://doi.org/10.1136/jnnp.40.5.464

Matar, E., Shine, J. M., Halliday, G. M., & Lewis, S. J. G. (2020). Cognitive fluctuations in Lewy body dementia: Towards a pathophysiological framework. *Brain*, *143*(1), 31–46. https://doi.org/10.1093/brain/awz311

Matsuda, W., Furuta, T., Nakamura, K. C., Hioki, H., Fujiyama, F., Arai, R., & Kaneko, T. (2009). Single nigrostriatal dopaminergic neurons form widely spread and highly dense axonal arborizations in the neostriatum. *Journal of Neuroscience*, *29*(2), 444–453. https://doi.org/10.1523/JNEUROSCI.4029-08.2009

McCormack, A. L., Mak, S. K., Shenasa, M., Langston, W. J., Forno, L. S., & Di Monte, D. A. (2008). Pathologic modifications of alpha-synuclein in 1-methyl-4-phenyl-1,2,3,6-tetrahydropyridine (MPTP)–treated squirrel monkeys. *Journal of Neuropathology & Experimental Neurology*, *67*(8), 793–802. https://doi.org/10.1097/NEN.0b013e318180f0bd

McKeith, I. G., Boeve, B. F., Dickson, D. W., Halliday, G., Taylor, J. P., Weintraub, D., Aarsland, D., Galvin, J., Attems, J., Ballard, C. G., Bayston, A., Beach, T. G., Blanc, F., Bohnen, N., Bonanni, L., Bras, J., Brundin, P., Burn, D., Chen-Plotkin, A., . . . Kosaka, K. (2017). Diagnosis and management of dementia with Lewy bodies: Fourth consensus report of the DLB Consortium. *Neurology*, *89*(1), 88–100. https://doi.org/10.1212/WNL.0000000000004058

McShea, D. W. (2016). Freedom and purpose in biology. *Studies in History and Philosophy of Biological and Biomedical Sciences*, *58*, 64–72. https://doi.org/10.1016/j.shpsc.2015.12.002

Miki, Y., Mori, F., Wakabayashi, K., Kuroda, N., & Orimo, S. (2009). Incidental Lewy body disease restricted to the heart and stellate ganglia. *Movement Disorders*, *24*(15), 2299–2301. https://doi.org/10.1002/mds.22775

Minguez-Castellanos, A., Chamorro, C. E., Escamilla-Sevilla, F., Ortega-Moreno, A., Rebollo, A. C., Gomez-Rio, M., Concha, A., & Munoz, D. G. (2007). Do alpha-synuclein aggregates in autonomic plexuses predate Lewy body disorders? A cohort study. *Neurology*, *68*(23), 2012–2018. https://doi.org/10.1212/01.wnl.0000264429.59379.d9

Mori, F., Tada, M., Kon, T., Miki, Y., Tanji, K., Kurotaki, H., Tomiyama, M., Ishihara, T., Onodera, O., Kakita, A., & Wakabayashi, K. (2019). Phosphorylated TDP-43 aggregates in skeletal and cardiac muscle are a marker of myogenic degeneration in amyotrophic lateral sclerosis and various conditions. *Acta Neuropathologica Communications*, *7*(1), Article 165. https://doi.org/10.1186/s40478-019-0824-1

Neumann, M., Sampathu, D. M., Kwong, L. K., Truax, A. C., Micsenyi, M. C., Chou, T. T., Bruce, J., Schuck, T., Grossman, M., Clark, C. M., McCluskey, L. F., Miller, B. L., Masliah, E., Mackenzie, I. R., Feldman, H., Feiden, W., Kretzschmar, H. A., Trojanowski, J. Q., & Lee, V. M. (2006). Ubiquitinated TDP-43 in frontotemporal lobar degeneration and amyotrophic lateral sclerosis. *Science*, *314*(5796), 130–133. https://doi.org/10.1126/science.1134108

Orimo, S., Uchihara, T., Nakamura, A., Mori, F., Kakita, A., Wakabayashi, K., & Takahashi, H. (2008). Axonal alpha-synuclein aggregates herald centripetal degeneration of cardiac sympathetic nerve in Parkinson's disease. *Brain*, *131*(Pt. 3), 642–650. https://doi.org/10.1093/brain/awm302

Pilpel, Y., & Rechavi, O. (2015). The Lamarckian chicken and the Darwinian egg. *Biology Direct*, *10*, Article 34. https://doi.org/10.1186/s13062-015-0062-9

Probst, A., Bloch, A., & Tolnay, M. (2008). New insights into the pathology of Parkinson's disease: Does the peripheral autonomic system become central? *European Journal of Neurology*, *15*(S1), 1–4. https://doi.org/10.1111/j.1468-1331.2008.02057.x

Rosen, R. F., Farberg, A. S., Gearing, M., Dooyema, J., Long, P. M., Anderson, D. C., Davis-Turak, J., Coppola, G., Geschwind, D. H., Paré, J. F., Duong, T. Q., Hopkins, W. D., Preuss, T. M., & Walker, L. C. (2008). Tauopathy with paired helical filaments in an aged chimpanzee. *Journal of Comparative Neurology*, *509*(3), 259–270. https://doi.org/10.1002/cne.21744

Rosenblueth, R., Wiener, N., & Bigelow, J. (1943). Behavior, purpose and teleology. *Philosophy of Science*, *10*(1), 18–24.

Roth, E., Sponberg, S., & Cowan, N. J. (2014). A comparative approach to closed-loop computation. *Current Opinion in Neurobiology*, *25*, 54–62. https://doi.org/10.1016/j.conb.2013.11.005

Roussarie, J. P., Yao, V., Rodriguez-Rodriguez, P., Oughtred, R., Rust, J., Plautz, Z., Kasturia, S., Albornoz, C., Wang, W., Schmidt, E. F., Dannenfelser, R., Tadych, A., Brichta, L., Barnea-Cramer, A., Heintz, N., Hof, P. R., Heiman, M., Dolinski, K., Flajolet, M., . . . Greengard, P. (2020). Selective neuronal vulnerability in Alzheimer's disease: A network-based analysis. *Neuron*, *107*(5), 821–835.e12. https://doi.org/10.1016/j.neuron.2020.06.010

Sato, S., Uchihara, T., Fukuda, T., Noda, S., Kondo, H., Saiki, S., Komatsu, M., Uchiyama, Y., Tanaka, K., & Hattori, N. (2018). Loss of autophagy in dopaminergic neurons causes Lewy pathology and motor dysfunction in aged mice. *Scientific Reports*, *8*(1), Article 2813. https://doi.org/10.1038/s41598-018-21325-w

Schellino, R., Boido, M., & Vercelli, A. (2020). The dual nature of Onuf's nucleus: Neuroanatomical features and peculiarities, in health and disease. *Frontiers in Neuroanatomy*, *14*, Article 572013. https://doi.org/10.3389/fnana.2020.572013

Schwalbe, N., & Wahl, B. (2020). Artificial intelligence and the future of global health. *Lancet*, *395*(10236), 1579–1586. https://doi.org/10.1016/S0140-6736(20)30226-9

Schwartz, A. B. (2016). Movement: How the brain communicates with the world. *Cell*, *164*(6), 1122–1135. https://doi.org/10.1016/j.cell.2016.02.038

Shadmehr, R., & Krakauer, J. W. (2008). A computational neuroanatomy for motor control. *Experimental Brain Research*, *185*(3), 359–381. https://doi.org/10.1007/s00221-008-1280-5

Shadmehr, R., Smith, M. A., & Krakauer, J. W. (2010). Error correction, sensory prediction, and adaptation in motor control. *Annual Review of Neuroscience*, *33*, 89–108. https://doi.org/10.1146/annurev-neuro-060909-153135

Skoyles, J. R. (2006). Human balance, the evolution of bipedalism and dysequilibrium syndrome. *Medical Hypotheses*, *66*, 1060–1068. https://doi.org/10.1016/j.mehy.2006.01.042

Stancu, I. C., Ferraiolo, M., Terwel, D., & Dewachter, I. (2019). Tau interacting proteins: Gaining insight into the roles of tau in health and disease. *Advances in Experimental Medicine and Biology*, *1184*, 145–166. https://doi.org/10.1007/978-981-32-9358-8_13

Subramaniam, S. R., & Chesselet, M. F. (2013). Mitochondrial dysfunction and oxidative stress in Parkinson's disease. *Progress in Neurobiology*, *106–107*, 17–32. https://doi.org/10.1016/j.pneurobio.2013.04.004

Surmeier, D. J., Obeso, J. A., & Halliday, G. M. (2017a). Parkinson's disease is not simply a prion disorder. *Journal of Neuroscience*, *37*(41), 9799–9807. https://doi.org/10.1523/JNEUROSCI.1787-16.2017

Surmeier, D. J., Obeso, J. A., & Halliday, G. M. (2017b). Selective neuronal vulnerability in Parkinson disease. *Nature Reviews Neuroscience*, *18*(2), 101–113. https://doi.org/10.1038/nrn.2016.178

Takeda, T., Uchihara, T., Nakayama, Y., Nakamura, A., Sasaki, S., Kakei, S., Uchiyama, S., Duyckaerts, C., & Yoshida, M. (2014). Dendritic retraction, but not atrophy, is consistent in amyotrophic lateral sclerosis—Comparison between Onuf's neurons and other sacral motor neurons. *Acta Neuropathologica Communications*, *2*(1), Article 11. https://doi.org/10.1186/2051-5960-2-11

Takuma, H., Arawaka, S., & Mori, H. (2003). Isoform changes of tau protein during development in various species. *Developmental Brain Research*, *142*(2), 121–127. https://doi.org/10.1016/s0165-3806(03)00056-7

Tanaka, H., & Sejnowski, T. J. (2015). Motor adaptation and generalization of reaching movements using motor primitives based on spatial coordinates. *Journal of Neurophysiology*, *113*(4), 1217–1233. https://doi.org/10.1152/jn.00002.2014

ten Donkelaar, H. J., Lammens, M., Wesseling, P., Hori, A., Keyser, A., & Rotteveel, J. (2004). Development and malformations of the human pyramidal tract. *Journal of Neurology*, *251*(12), 1429–1442. https://doi.org/10.1007/s00415-004-0653-3

Uchihara, T. (2017). An order in Lewy body disorders: Retrograde degeneration in hyperbranching axons as a fundamental structural template accounting for focal/multifocal Lewy body disease. *Neuropathology*, *37*(2), 129–149. https://doi.org/10.1111/neup.12348

Uchihara, T. (2020). Neurofibrillary changes undergoing morphological and biochemical changes—How does tau with the profile shift of from four repeat to three repeat spread in Alzheimer brain? *Neuropathology*, *40*(5), 450–459. https://doi.org/10.1111/neup.12669

Uchihara, T., Endo, K., Kondo, H., Okabayashi, S., Shimozawa, N., Yasutomi, Y., Adachi, E., & Kimura, N. (2016). Tau pathology in aged cynomolgus monkey is progressive supranuclear palsy/corticobasal degeneration—but not Alzheimer disease-like—ultrastructural mapping of tau by EDX. *Acta Neuropathologica Communications*, *4*, Article 118. https://doi.org/10.1186/s40478-016-0385-5

Uchihara, T., & Giasson, B. I. (2016). Propagation of alpha-synuclein pathology: Hypotheses, discoveries, and yet unresolved questions from experimental and human brain studies. *Acta Neuropathologica*, *131*(1), 49–73. https://doi.org/10.1007/s00401-015-1485-1

Uchihara, T., Hara, M., Nakamura, A., & Hirokawa, K. (2012). Tangle evolution linked to differential 3- and 4-repeat tau isoform deposition: A double immunofluorolabeling study using two monoclonal antibodies. *Histochemistry and Cell Biology*, *137*(2), 261–267. https://doi.org/10.1007/s00418-011-0891-2

Uchihara, T., & Paulus, W. (2008). Research into neurodegenerative disease: An entangled web of mice and men. *Acta Neuropathologica*, *115*(1), 1–4. https://doi.org/10.1007/s00401-007-0319-1

Uchihara, T., Shibata, N., & Yoshida, M. (2020). Reconsidering the Braak-prion hypothesis: Truths or realities. *Neuropathology*, *40*(5), 413–414. https://doi.org/10.1111/neup.12704

Warren, J. D., Rohrer, J. D., Schott, J. M., Fox, N. C., Hardy, J., & Rossor, M. N. (2013). Molecular nexopathies: A new paradigm of neurodegenerative disease. *Trends in Neurosciences*, *36*(10), 561–569. https://doi.org/10.1016/j.tins.2013.06.007

Wu, J. W., Hussaini, S. A., Bastille, I. M., Rodriguez, G. A., Mrejeru, A., Rilett, K., Sanders, D. W., Cook, C., Fu, H., Boonen, R. A., Herman, M., Nahmani, E., Emrani, S., Figueroa, Y. H., Diamond, M. I., Clelland, C. L., Wray, S., & Duff, K. E. (2016). Neuronal activity enhances tau propagation and tau pathology in vivo. *Nature Neuroscience*, *19*(8), 1085–1092. https://doi.org/10.1038/nn.4328

Yarnall, A., Rochester, L., & Burn, D. J. (2011). The interplay of cholinergic function, attention, and falls in Parkinson's disease. *Movement Disorders*, *26*(14), 2496–2503. https://doi.org/10.1002/mds.23932

Yuste, R. (2015). From the neuron doctrine to neural networks. *Nature Reviews Neuroscience*, *16*(8), 487–497. https://doi.org/10.1038/nrn3962

5

Humans and Nonhuman Primates

Geneviève Konopka and Emre Caglayan

Human evolution is marked by higher cognitive capabilities including an increased ability to interpret the mental state of others (theory of mind), an increased capacity for abstraction, and an increased social closeness in children (Tomasello 2020). Interestingly, some common brain diseases present with alterations in these behavioral traits. For example, individuals with either autism spectrum disorder (ASD) or schizophrenia may have decreased social skills and/or a decreased ability to interpret others (Bianchi et al. 2013a; Kahn et al. 2015). These features of the evolved human brain and common brain disorders also extend to cellular anatomy and function. Human neurons display longer dendrites and an increased number of dendritic spines compared to those of nonhuman primates (NHPs; Bianchi et al. 2013a). Alterations in neuronal morphology are also frequently observed in postmortem brain tissue from individuals with schizophrenia (Moyer et al. 2015) or ASD (Martinez-Cerdeno 2017). This intriguing correspondence in the phenotypic alterations of both human brain evolution and human brain disorders is also accompanied by the possibility that these disorders are unique to humans. Despite years of observation, there has not been compelling evidence of the presence of schizophrenia, ASD, or Alzheimer's disease (AD) in NHPs, although the lack of evidence could be due to several factors, including limitations in population size or species-specific manifestations of the diseases that are challenging to observe in nonhuman species. As genomic tools have advanced, these phenotypic and cellular observations have motivated comparisons of molecular features that may be altered in human brain evolution as well as in human brain disorders.

Early comparisons of human and chimpanzee proteins revealed striking similarity, indicating that molecular evolution on the human lineage is not marked by novel proteins (King and Wilson 1975). Corroborating this prediction, decades of research have identified only a handful of human-specific genes (Florio et al. 2018), while the vast majority of the protein-coding genes are shared between humans and NHPs. In contrast, there are millions of human-specific single-nucleotide changes in both the coding (e.g., synonymous changes) and non-coding genomic regions that are otherwise conserved across species (Chimpanzee Sequencing and Analysis Consortium 2005). The potential functional consequences of such variants are mostly unknown. Through comparative analyses of both synonymous and nonsynonymous coding variants, ~500 genes have been identified as under positive selection in humans (Geschwind and Rakic 2013). Similarly, comparative analyses of conserved non-coding genomic regions with higher substitution rates in humans (compared to a large panel of mammalian species) have identified ~3,000 genomic sequences that exhibit accelerated evolution in humans (also known as "human accelerated regions" [HARs]; Franchini and Pollard 2017). Together, these studies have provided a curated list of genomic features that have been altered specifically on the human lineage, and the identification of these features has fueled additional studies that aim to illuminate the functional consequences of these human-specific changes.

While identification of genomic sequences that are altered in human evolution provides a robust first step for follow-up functional characterization, the majority of these initial studies were low-throughput and do not provide insights into how such sequence changes affect functional molecular networks in the human brain, let alone how they might be altered in comparison to the NHP brain. In contrast, the advent of high-throughput transcriptomic profiling assays (first microarrays, then RNA-sequencing [RNA-seq]) enabled functional comparisons of human and NHP brains, resulting in many discoveries that identified human-specific gene expression networks (Konopka et al. 2012; Liu et al. 2012; Geschwind and Rakic 2013; Sousa et al. 2017; Zhu et al. 2018; Berto et al. 2019). More recently, single-cell transcriptome profiling (e.g., single-nucleus RNA-seq [snRNA-seq]) of postmortem brain tissue became possible, and studies using this approach have revealed the incredible cellular complexity of many brain regions with dozens of cell types identified and found reproducibly across independent studies (Hodge et al. 2019; Bakken et al. 2021), unlocking the functional investigation of cell-type alterations in human brain evolution. Technical challenges in high-throughput epigenomic profiling have also been overcome with techniques such as chromatin immunoprecipitation followed by sequencing (ChIP-seq) on histone markers (e.g., H3K27ac for active enhancers) and open-chromatin profiling with assay for transposase-accessible chromatin using sequencing (ATAC-seq), which is also adapted for single-cell resolution. These tools allowed the discovery of human-specific functions of regulatory elements. In contrast to the initial studies that only relied upon comparisons of DNA sequences, these mRNA and chromatin assays can identify human-specific functions of genomic features at spatial and cellular resolution. However, such studies are in their infancy with respect to the breadth of the comparisons being made. To date, these high-throughput, cellularly resolved functional comparisons have been limited to only a few brain regions and age groups, indicating that the field is still far from uncovering the complete list of human-specific functions of genomic elements in the brain. This is in contrast to DNA comparisons, where recent updates to previous versions of the human genome have expanded coverage of sequencing across the entire length of chromosomes (telomere to telomere sequencing) and have included an increased number of samples contributed from a diverse cross section of the human population (Nurk et al. 2022). This significantly updated and "complete" version of the human genome is now being compared to genomes of other species and will soon provide a nearly complete list of human-specific DNA sequences. One caveat to this catalog of human-specific DNA variants is that much work will remain to be done to characterize them and understand the functional consequences (if any) of these DNA changes. Thus, these resources are complementary, and current research in molecular mechanisms of human brain evolution needs to utilize both resources (Girskis et al. 2021; Ma et al. 2022).

In comparison to identifying human-specific genomic features, identification of brain disease–associated genetic variants has been more challenging due to the small prediction power of a given variant to dissociate individuals with disease from healthy individuals (Manolio et al. 2009). Genome-wide association studies (GWAS) require sequencing of thousands of individuals to yield robust results for common brain disorders (Sullivan and Geschwind 2019). Transcriptomic comparisons have also required large sample sizes (hundreds of samples) to identify dysregulated genes in schizophrenia, ASD, or bipolar disorder (Gandal et al. 2018). Similar to cross-species comparisons, transcriptomic comparisons across tissues, cell types, and developmental time points are currently limited; and the functional characterization of genomic changes identified through GWAS are also incomplete, although both of these limitations should improve in the near future as more researchers are expanding both of these avenues of research. Therefore, the fields of evolutionary and disease neurogenomics have advanced with similar limitations and opportunities, although to date evolutionary comparisons have been more definitive and less vulnerable to statistical power limitation. Since there has been an empirical overlap between evolutionary changes and disease-associated changes in the human brain, many studies have focused on this association at the molecular level, investigating the evidence for direct connections between these processes.

To review both fields and their intersection in more detail, the next sections will provide an overview of (1) human-specific genomic changes and their potential functions, (2) human-specific regulatory changes in bulk tissue with respect to transcriptome and epigenome in developing, and (3) adult brain. The evidence for and against an association with brain disorders will be weighed at each step. Finally, advantages and limitations of understanding any correspondence between human evolution and disease association, as well as prospects for a more definitive understanding, will be outlined.

Evolutionary Evidence Specifically Related to Human Health and Disease

Evolution and Function of Human-Specific DNA Sequences

Human-specific DNA sequence changes are, directly or indirectly, connected to human-specific gene regulatory changes, human-specific novel genes, and human-specific alterations of the protein structures. Genes encoding for proteins expressed in the brain are largely conserved in the human genome (Dumas et al. 2021). Notable exceptions have been an active area of research since altering amino acid composition of a single gene is more amenable to genetic modification in experimental systems than sequence changes in the non-coding genome (Pinson and Huttner 2021). However, in this chapter, we will focus on the gene regulatory effect of DNA sequence changes that comprise the millions of human-specific substitutions, insertions, and deletions (Chimpanzee Sequencing and Analysis Consortium 2005). Since ~98% of the human genome is non-coding, most of these changes are in the non-coding genome, and understanding their functional role is an exciting and ongoing challenge.

Human-specific substitutions can be either functional genomic changes or neutral changes without a functional consequence. To determine functional genomic sequences with increased divergence specifically in human evolution, one approach is to identify the genomic sequences that are highly conserved across many vertebrate species with low substitution rates indicating functionality and then retain the regions that accumulated significantly more human-specific substitutions on this constrained background, indicating accelerated evolution of the conserved elements on the human lineage. The accumulation of studies utilizing this approach has yielded ~3,000 HARs (Franchini and Pollard 2017). While some studies have exclusively focused on non-coding sequences (Prabhakar et al. 2006), other studies have carried out genome-wide analyses and found >90% of the accelerated regions to be non-coding (Pollard et al. 2006), indicating that most HARs are likely to affect gene regulation rather than protein sequences.

Associating HARs to nearby genes for the identification of their potential functions revealed enrichments for genes involved in neuronal functioning, specifically in neurodevelopment (Capra et al. 2013). In their analysis of 2,649 non-coding HARs, Capra et al. (2013) found that 773 are predicted to be developmental enhancers and 251 are predicted to be active in brain. Predicted enhancer activity in development motivated functional characterization of HARs with reporter assays through injection into mouse embryos (Capra et al. 2013; Gittelman et al. 2015). These studies revealed activity of HARs in the developing brain, with the HARs driving different patterns of reporter activity compared to the ancestral state of the genomic region (Capra et al. 2013). Even with ways to narrow down the regions to the most promising candidates, low-throughput methodology, such as reporter assays in mouse embryos, is a major roadblock for the functional characterization of HARs. More recently, studies have been able to parallelize the delivery of genomic constructs into cultured cells, allowing the high-throughput characterization of the effects of sequence changes on regulatory function using massively parallel reporter assays (MPRAs; Girskis et al. 2021; Uebbing et al. 2021; Whalen et al. 2022). MPRAs function as reporter assays that produce RNA as a readout instead of fluorescence. Combined with RNA-seq, this technique allows parallelized screening of the activity of regulatory regions of interest (Klein et al. 2020). Studies utilized MPRAs by delivering HARs and their corresponding chimpanzee sequences into neural stem cells or neural cells in culture. Two studies found that 50%–60% of active HARs displayed significantly altered activity compared to chimpanzee sequences (Girskis et al. 2021; Whalen et al. 2022). Another study found this to be 27.5% (Uebbing et al. 2021). Importantly, the lack of differential activity could be due to the limitations of the culture systems or cell types being utilized, indicating that, even at ~50%, this is likely an underestimation of functional activity of human-specific sequence changes in HARs. Moreover, the functionality of the HARs to drive reporter expression was also similar between the same cell types from human/mouse (Girskis et al. 2021) and human/chimpanzee (Whalen et al. 2022), indicating that HAR functionality is primarily driven by the human-specific sequence changes and not by the *trans* effects of the cellular environment. In addition to uncovering the pattern of HAR activity, these studies further characterized some HARs for their role in human neural stem cells. An example was a HAR-regulated gene, *PPP1R17*, that slows cell cycle progression in neural progenitor cells (Girskis et al. 2021). *PPP1R17* and other genes with human-specific functions (e.g., *SRGAP2C* that promotes radial glia migration and increases spine density [Charrier et al. 2012]) could be molecular factors responsible for neoteny in human brain development.

HARs comprise only a small portion of all human-specific genomic changes (Uebbing et al. 2021). The sequencing of ancient human genomes has also allowed identification of human-specific substitutions that were ancestral in Neanderthals and Denisovans, pointing out regions that were likely changed more recently in modern human evolution (Prüfer et al. 2014). Genes carrying modern human-specific amino acid substitutions are enriched in neurodevelopmental function, similar to HARs (Prüfer et al. 2014). Notably, three of these genes are associated with kinetochore of the mitotic spindle (*CASC5*, *KIF18A*, *SPAG5*; Prüfer et al. 2014). Both human-specific substitutions not linked to HARs and modern human-specific substitutions were also recently characterized by MPRAs (Weiss et al. 2021; Uebbing et al. 2021). Interestingly, human-gained enhancers (HGEs) that have human-specific substitutions that are not necessarily characterized as HARs caused differential activity in 33.9% of the active HGEs (Uebbing et al. 2021). Similarly, ~23% of active modern human-specific substitutions were differentially active compared to the ancestral sequences (Weiss et al. 2021), and the genes associated with the loci of differentially active sequences are enriched for brain anatomy and function (Weiss et al. 2021). These results indicate that modern human-specific substitutions and human-specific substitutions in the non-HAR enhancers are also important sources of molecular evolution in human cells.

Relevance of Human-Specific DNA Evolution in Human Brain Disorders

Disorders negatively affect organismal fitness, yet they are prevalent across animal species, indicating a trade-off for evolving novel adaptive traits (Benton et al. 2021). Interestingly, several brain disorders are either unique to humans or display human-specific characteristics—including schizophrenia, ASD, and AD—supporting the concept that these diseases may be connected to recent evolutionary changes in humans (O'Bleness et al. 2012). The genome-wide identification of both human-specific and disease-relevant genomic changes has allowed studies to rigorously assess the existence and extent of this association (Figure 5.1).

Schizophrenia

Focusing on human evolution and schizophrenia, Xu et al. (2015) found significant enrichment of schizophrenia-related variants in genes near HARs. Interestingly, this enrichment was observed strongly for HARs identified by conservation in primates (pHARs)

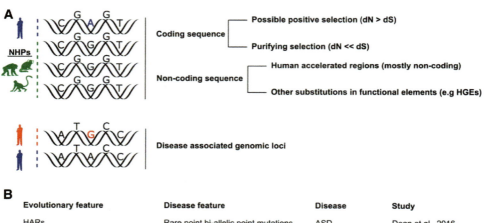

Figure 5.1 Overview of human-specific genomic changes and studies that reported on their association with disease data sets. (A) Diagram of a human-specific substitution. Millions of such substitutions as well as insertions and deletions are found in the human genome. Understanding whether and what functional change they are associated with is an arduous task. Right panel provides a brief overview of the efforts to categorize human-specific genomic changes based on the amino acid sequence changes (positive/purifying selection) and non-coding sequence changes. (B) Table of the studies that provided a statistical test on the overlap of human-specific genomic changes and disease-associated genomic changes. ASD = autism spectrum disorder; dN = nonsynonymous substitution; dS = synonymous substitution; GWAS = genome-wide association study; HAR = human accelerated region; HGE = human-gained enhancer; NHP = nonhuman primate; ORF = open reading frame (a stretch of DNA sequence that does not have stop codons); SNP = single-nucleotide polymorphism.

but not observed for HARs identified by conservation in mammals overall, supporting a unique association of schizophrenia with recent changes in the human lineage (Xu et al. 2015). The authors also found a stronger selective pressure (defined by a higher ratio of substitutions that change amino acid sequence [i.e., nonsynonymous substitutions] to substitutions that do not change the amino acid sequence [i.e., synonymous substitutions]) in the coding sequence of genes near pHARs that distinguished human and chimpanzee genomes. Xu et al. (2015) further showed that pHAR-related genes are enriched for genes that are selectively more expressed in the gamma-aminobutyric acid-ergic (GABAergic) neuronal lineage as well as genes associated with synapse formation. Thus, the functional examination of these particular cellular features, especially in GABAergic neurons, may provide insight into human-specific molecular adaptations that may be important to maintain as trade-offs to susceptibility to schizophrenia. A more recent study also showed that novel open reading frames may be previously uncharacterized potential proteins that have disrupted expression in schizophrenia and are enriched in genomic regions near HARs (Erady et al. 2022). Another study found significant correlations between schizophrenia-associated genetic variants and regions with positive selection in modern humans in comparison to the Neanderthal genome (Srinivasan et al. 2016).

Autism

Similar to HAR–schizophrenia comparisons, Doan et al. (2016) focused on the association of HARs with ASD at the genomic level. The authors found an excess of ASD-related de novo copy-number variants and biallelic point mutations in HARs compared to other regions. They further characterized the effect of HARs on the expression patterns of critical ASD-relevant genes including *CUX1*, *PTBP2*, and *GPC4* by performing in vivo and in vitro reporter assays as well as overexpression of HARs in cultured neurons. For example, the authors showed that a rare autism-linked mutation in HAR426 has a distal interaction with the *CUX1* promoter and causes an increase in the promoter activity of *CUX1* in all layers of the neocortex (Doan et al. 2016). These analyses suggest that genomic changes in human evolution may have contributed to human susceptibility to ASD and schizophrenia specifically but also potentially to other brain diseases.

Alzheimer's Disease

AD is a late age–onset polygenic disease that causes neurodegeneration and memory decline. To date, there is no clear evidence of AD in NHPs, making it a potentially human-specific disease (Finch and Austad 2015). Unlike schizophrenia and ASD, however, research on the overlap between AD risk regions and human-specific genomic changes is lacking with few exceptions (Zhou et al. 2015; Nitsche et al. 2021). Interestingly, one of these studies showed that genes implicated in AD have fewer conserved splice sites than the background, consistently across multiple human and NHP pairs (Nitsche et al. 2021), indicating that more comprehensive studies in the future can illuminate this association further.

Human-Specific Gene Regulatory Changes in Brain Tissues and Cell Types

While genomic changes provide the ultimate resource for identifying the genomic underpinnings of human evolution and disease, the functional consequences of human-specific genomic changes are highly complex. Even if a genomic change is correlated with a detectable phenotype, uncovering the activity of this genomic change in different organs, tissues, and cell types is an arduous task. Moreover, the majority of the genomic changes are non-coding, and their interactions with gene promoters are mostly unknown. Genomic changes can also exert a functional impact indirectly by altering the expression of a gene that subsequently differentially alters the expression patterns or function of other genes. Methodologies have been insufficient to capture such a complex interaction since this would require high-throughput screening of molecular function throughout development to adult stages and at high cellular and regulatory (from DNA to protein) resolution. However, gene expression and chromatin architecture can be investigated in whole tissues and, more recently, at single-cell resolution per tissue. Such breakthroughs have allowed investigators to understand human-specific gene regulatory novelties, especially in brain tissues. In this section, we will summarize the major findings from these studies that span more than a decade and discuss their relevance to the understanding of human brain disorders.

Single-Gene Comparisons

As mentioned above, human-specific changes are largely non-coding but can also rarely be among coding sequences. Nonsynonymous changes in the coding sequences can potentially alter the function of the given protein, which can have indirect effects on the molecular landscape of the cell. Regulatory proteins, such as transcription factors and RNA-binding proteins, can be prioritized to test this hypothesis and uncover potential human-specific gene expression changes caused by these human-specific evolutionary novelties in regulatory proteins. Variants in the coding region of the transcription factor FOXP2 are associated with both language disorders (Lai et al. 2001) and human evolution through positive selection (Enard et al. 2002). Humanized FOXP2 mice that express the two human-specific amino acids show altered ultrasonic vocalizations (~30–~100 kHz), indicating that human-specific FOXP2 sequence functionality may have been altered in brain circuits that underlie motor-relevant behaviors in human evolution (Enard et al. 2009). Another study tested whether human FOXP2 has differential transcriptional targets compared to chimpanzee FOXP2 (FOXP2^{chimp}) in cultured neuronal cells (Konopka et al. 2009). Interestingly, ~100 genes were differentially regulated by FOXP2 compared to FOXP2^{chimp}, indicating that many gene expression changes in humans can be driven by differential *trans* activity of a single regulatory protein. These differentially expressed genes were enriched for genes that are also differentially expressed between human and chimpanzee brain tissues, underscoring the relevance of this finding for in vivo functions (Konopka et al. 2009). A more recent study characterized the modern human-specific coding sequence change in NOVA1, an RNA-binding protein that regulates alternative splicing and is associated with neurological disorders (Trujillo et al. 2021). Despite NOVA1's regulation of alternative splicing, which may not directly affect gene expression levels, the authors found 277 differentially expressed genes in brain organoids expressing the modern-humanized *NOVA1* compared to the ancestral *NOVA1* (Trujillo et al. 2021). Another recent study using overexpression in mouse and ferret cortex as well as human brain organoids found that the modern-human version of *TKTL1* (*hTKTL1*), which codes for an enzyme in the glycolysis pathway, increases the production of basal radial glia and neurons compared to the ancestral variant (Pinson et al. 2022). While this study did not investigate whether there are gene regulatory changes associated with hTKTL1, the authors' phenotypic observations suggest that hTKTL1 evolution likely affects the wiring of the gene regulatory programs. Taken together, these studies have shown that human-specific changes in the function of a single regulatory protein can affect the expression patterns of many other genes, leading to phenotypic alterations and indicating that the molecular networks of a given human brain cell can be very different than a comparable chimpanzee (or other NHP) cell.

Transcriptomic Comparisons in the Developing Brain

The effects of human-specific coding sequence changes can be considered as *trans* effects as they alter gene expression through a diffusible molecule, whereas an HAR regulating a nearby gene's expression is considered a *cis* effect. Strikingly, the studies of protein coding evolution show that functional changes, even in a single protein, can affect the expression patterns of many other genes. It is not feasible, if not currently impossible, to predict the complex regulatory landscape of human brain cells compared to NHP brain cells by DNA sequence data alone. Human-specific phenotypes at different developmental and cellular levels are also challenging to delineate using DNA sequence alone without comprehension of the regulatory landscape of each gene in humans and NHPs in a given biological context. To overcome these challenges, many studies have adopted a more direct approach to understanding the human-specific molecular functionality by comparing the transcriptomes of humans and NHPs. Importantly, these studies use brain tissue from developing and adult humans and NHPs. While initial studies utilized microarray technology, more recent studies have adopted RNA-seq as it is not prone to biases in the predetermined sequences central to hybridization-based microarray technology. Here, we outline these comparative transcriptomics studies and discuss key findings.

The human brain is larger than an NHP brain and has unique cognitive capabilities. These phenotypes are proposed to be due to the heterochronous development (i.e., altered developmental rate or timing) of human features, including brain cell types. To understand whether there is molecular support for these changes and to characterize human-specific molecular alterations related to heterochrony, studies have compared the transcriptomes of the brains of humans and NHPs in early development.

An early study focused on the early postnatal development of the dorsolateral prefrontal cortex (DLPFC) using tissue from humans, chimpanzees, and rhesus macaques and found an excess number of genes that show delayed expression in humans relative to chimpanzee and rhesus macaque (Somel et al. 2009). Later studies also found heterochronous transcriptomic changes in human brain tissues. One study showed that the peak expression of synaptic genes in the prefrontal cortex was delayed until 5 years in humans, whereas this was achieved at ~1 year in chimpanzees and rhesus macaques (Liu et al. 2012). However, a morphological study showed prolonged synaptic maturation in chimpanzees until 5 years old, arguing that the original study may have suffered from low sample sizes in chimpanzees (Bianchi et al. 2013b). A more recent study compared the transcriptomes of human and rhesus macaque brains across prenatal and postnatal development by matching the chronological ages of humans and NHPs according to their transcriptomic profile (Zhu et al. 2018). Comparing the heterochrony in the transcriptomic signatures of five major biological processes (neurogenesis, neuronal differentiation, astrogliogenesis, synaptogenesis, myelination), the authors found that synaptogenesis-related genes were not delayed but accelerated in human neocortex (Zhu et al. 2018). These studies may have yielded different results due to variabilities in the readout (gene expression versus neuronal morphology) and analytical approaches to match the chronological age between species.

In contrast to tissue-based comparisons, in vitro studies using induced pluripotent stem cell (iPSC)–derived neurons offer a more controlled setting to study heterochrony in human and NHP neuronal development. Studies differentiating human and chimpanzee neurons from iPSCs consistently reported slower maturation in human neurons both in terms of neuronal morphology (e.g., dendritic length) and in terms of neuronal function (e.g., synaptic firing; Marchetto et al. 2019; Kanton et al. 2019; Schornig et al. 2021). Transcriptomic comparisons also revealed that genes related to neuronal maturation were differentially expressed in human compared to NHP neurons (Kanton et al. 2019). While this lends support for the initial observation of delayed transcriptomic upregulation of neuronal development in humans (Liu et al. 2012), these studies did not explicitly test whether the heterochronous genes in in vivo development matched the heterochronous genes in in vitro development. Importantly, transcriptomic profiles of monolayer and organoid culture systems have been shown to largely correspond to prenatal development unless cultured for ~1 year (Gordon et al. 2021), while comparative transcriptomic studies targeting neurodevelopment were often from postnatal tissues older than 1 year. Nevertheless, in vitro studies have shown that human neuronal maturation is slower than chimpanzee neuronal maturation outside of their tissue environment, indicating that this property is an intrinsic feature of human neurons.

Another heterochronic biological process associated with human development is myelination. Myelination in the central nervous system is mediated through oligodendrocytes that mature postnatally. Comparisons of myelination levels in human and chimpanzee cortical gray matter throughout postnatal development showed that myelination is prolonged in the human brain, extending beyond late adolescence, whereas it peaked before sexual maturation in chimpanzees (Miller et al. 2012). A carbon dating study of human oligodendrogenesis also showed that oligodendrocyte generation is prolonged in the gray matter of the cortex until ~40 years old, whereas oligodendrocyte generation peaked at ~5 years old in white matter, indicating that the observation of prolonged myelination in humans might be specific to the gray matter of the cortex (Yeung et al. 2014). Transcriptomic comparisons have also shown delayed increased expression of myelination-related genes in humans compared to rhesus macaque (Zhu et al. 2018). A recent study found that the cortical gray matter of the adult human brain has a higher ratio of oligodendrocyte progenitor cells (OPCs) and a lower ratio of mature oligodendrocytes compared to NHPs (Caglayan et al. 2023), indicating that humans may retain a larger oligodendrocyte progenitor pool, and therefore a likely higher progenitor capacity, compared to NHPs even in late adulthood.

The developing brain has many different cell types progressing through various stages of maturation at a given time point. Most investigations that focused on development have been at the tissue level, thus human-specific regulatory patterns throughout development are currently being investigated at the cell-type level. One study compared fetal human and rhesus macaque brains at single-cell resolution in the DLPFC but noted the high variability of expression patterns at single-cell resolution; it utilized bulk transcriptomes to identify differentially expressed genes between the species and investigated their expression across cell types (Zhu et al. 2018). The authors found 14 differentially expressed genes in humans including *TRIM54*—a gene encoding a protein important in axonal growth—expressed in excitatory neurons at lower levels in humans than rhesus macaque (Zhu et al. 2018). Given that even a single regulatory gene can be responsible for over 100 gene expression changes (Konopka et al. 2009; Trujillo et al. 2021), this number is likely an underestimation, and future studies are needed to elucidate the human-specific regulatory changes more accurately.

Transcriptomic Comparisons in the Adult Brain

The differential maturation of the human brain indicates that the molecular architecture of the mature human brain is likely vastly different from that of an NHP brain. Postmortem tissues from the adult brain are more readily accessible to researchers than those from the developing brain, especially for endangered NHPs such as chimpanzees. Therefore, comparative transcriptomic studies on the adult brain have been more frequent and thus more insightful than the limited studies from the developing brain. Additionally, these studies have uncovered the immense cellular heterogeneity in the adult human brain, which is largely reflected in NHP brains (Ma et al. 2022; Caglayan et al. 2023). This section aims to outline these efforts and provide an overview of human-specific molecular features in the adult human brain.

Tissue-level comparisons of human-specific gene expression changes were identified by comparing anatomically matched tissues typically from human, chimpanzee, and rhesus macaque brains, although some studies included more species including bonobo (Khrameeva et al. 2020) and gorilla (Jorstad et al. 2022) as well as a New-World monkey marmoset as an outgroup to rhesus macaque (an Old-World monkey; Castelijns et al. 2020b; Ma et al. 2022). Since gene expression differences between humans and chimpanzees can be due to a change in either humans or chimpanzees, studies have used another NHP (often rhesus macaque) as an outgroup species and identified human-specific gene expression changes that are consistently up-/down-regulated in both human–chimpanzee and human–rhesus macaque comparisons (Konopka et al. 2012; Sousa et al. 2017). Focusing on multiple brain regions from cortical and subcortical regions, these studies have identified hundreds of human-specific gene expression differences. While human-specific gene expression alterations are often reproducible across studies (Konopka et al. 2012), quantitative comparisons have not always yielded the same conclusion. For example, focusing on three brain regions, frontal pole from neocortex, caudate nucleus, and hippocampus, one study found more human-specific gene expression differences in the frontal pole compared to the caudate nucleus or hippocampus (Konopka et al. 2012). However, another study identified a greater number of human-specific gene expression changes in the striatum and thalamus compared to cortical regions (Sousa et al. 2017). A more recent study conducted a comparative survey of 33 anatomical brain regions and found a greater number of human-specific gene expression changes in the cerebral cortex, hypothalamus, and cerebellar gray and white matter regions compared to striatal regions (Khrameeva et al. 2020). Variability across such studies could be explained by differences in the exact anatomical regions used (e.g., striatum encompasses caudate nucleus, putamen, and nucleus accumbens), as well as differences in the analytical and experimental approaches. A meta-analysis that starts from the raw data across these and other studies could be instructive with respect to any potential differences in analytical methods.

Several studies have sought to prioritize specific differentially expressed genes in a number of ways. One approach is to carry out assessment of coexpression networks (coexpressed genes that are likely coregulated). Multiple studies using this approach have found that these gene modules are often not conserved between humans and NHPs (Konopka et al. 2012; Sousa et al. 2017). Notably, one study identified a gene module that contained *FOXP2* as a hub gene (a gene with the highest level of correlation with the other genes in the module), providing further support that *FOXP2* may have important human-specific functions (Konopka et al. 2012). Another hub gene in a human-specific gene module was *CLOCK*, which is implicated in psychiatric diseases in addition to its role in circadian rhythms (Konopka et al. 2012). Human-specific molecular changes in the striatum also revealed that genes involved in dopamine biosynthesis (tyrosine hydroxylase [*TH*] and DOPA decarboxylase [*DDC*]) were human-specifically up-regulated (Sousa et al. 2017). Interestingly, *TH* was down-regulated in several great apes but not in humans in the neocortex, indicating that it was likely down-regulated in the great ape divergence but up-regulated again in the human lineage (Sousa et al. 2017).

A recent single-cell comparative study also found that *TH* expressing human SST (somatostatin) neurons are present in chimpanzees but lack *TH* expression. This is in contrast to human, rhesus macaque, and marmoset SST neurons that express *TH*, although at lower levels in marmosets (Ma et al. 2022). Thus, these studies have uncovered human-specific molecular features at the tissue level.

Brain tissues are notable for containing multiple cell types. The major cell types include neurons, astrocytes, microglia, oligodendrocytes, and OPCs, with many subtypes within each major cell type (especially neurons). Tissue-level comparisons combine transcripts from all cell types in a tissue, potentially masking any cell type–specific effects. To circumvent this problem, some studies have isolated nuclei from postmortem tissue and used flow cytometry to sort several major cell types per tissue sample across species (Berto et al. 2019; Kozlenkov et al. 2020). Comparing differential gene expression in oligodendrocytes, one study found that previous tissue-level comparisons failed to capture the human-specific gene expression changes in oligodendrocytes (Berto et al. 2019). The authors also showed evidence for an increased number of human-specific gene expression changes in oligodendrocytes compared to neurons (Berto et al. 2019). Another study distinguished excitatory and inhibitory neurons from each other and found that many genes with human-specific expression in one cell type were not altered in the other (Kozlenkov et al. 2020). Broad cell-type categories (excitatory neurons, inhibitory neurons, astrocytes, oligodendrocytes, OPCs, microglia) have been further examined across species using snRNA-seq, finding that glial gene expression changes may be more human-specific compared to neuronal gene expression changes despite having similar evolutionary rates of change when all branches are considered (Khrameeva et al. 2020). These results could indicate that glial functions may have been altered in a more human-specific manner. Given the recent discoveries on neuron–glia interactions (e.g., both OPCs and microglia can engulf neuronal processes and affect the specificity and turnover of neuronal connections; Paolicelli et al. 2011; Buchanan et al. 2022), glial function could also indirectly alter the neuronal and neural function of the human brain.

snRNA-seq of postmortem brains has facilitated cellular characterization of brain regions. High-quality data sets have shown that each of the broad cell-type categories of the human cortex (explained above) contain further transcriptomically distinct subtypes (Hodge et al. 2019). Neurons are especially heterogeneous, with more than a dozen distinct subtypes for both excitatory and inhibitory neurons (Hodge et al. 2019). Comparisons between human and NHP brains have revealed that the complex and heterogeneous diversification of neuronal subtypes are largely conserved across species (Bakken et al. 2021; Ma et al. 2022; Caglayan et al. 2023). However, the abundances of subtypes are not always uniform across species. For example, upper-layer excitatory neurons are more abundant in human and chimpanzee compared to other NHPs (Bakken et al. 2021; Ma et al. 2022). Several less abundant subtypes of excitatory neurons, inhibitory neurons, astrocytes, and microglia are also absent in certain species, including a human-specific microglia subtype (Ma et al. 2022). Transcriptomic comparisons of conserved cell types have shown that cell-type identity is more conserved among inhibitory neurons compared to excitatory neurons across species (Bakken et al. 2021). Another study showed that most human-specific gene expression changes are only observed in one or a few subtypes (in both excitatory and inhibitory neurons), and failure to disentangle neuronal subtypes masked these changes (Caglayan et al. 2023). One example is human-specific up-regulation of *FOXP2* in only two (out of 14 detected) excitatory subtypes in the posterior cingulate cortex (PCC; Caglayan et al. 2023). Strikingly, this up-regulation was not observed in a similar comparative transcriptomic study from the DLPFC, and comparisons of the PCC with other cortical regions showed higher *FOXP2* levels in the PCC, indicating that subtype-specific *FOXP2* up-regulation in humans is also region-specific (Caglayan et al. 2023). In addition to its neuronal subtype-specific and region-specific up-regulation, *FOXP2* is human-specifically up-regulated in microglia (Ma et al. 2022). These results revealed novel human-specific changes in the levels of critical regulatory genes, motivating future studies to characterize the functional consequences of these novelties in human evolution.

Epigenomic Comparisons in the Adult Brain

Non-coding genomic elements function to regulate gene expression. These genomic elements are also referred to as "gene regulatory elements" (GREs), and they can be further classified based upon their precise functioning (e.g., enhancers, promoters, silencers). GRE function can be detected by the presence of histone markers, chromatin state (open or closed), or the level of DNA methylation. It is possible to profile these markers through various high-throughput assays and compare the level of epigenomic readout across species. While epigenomic profiling may not be as informative as transcriptomic profiling—since little is known about the functions of non-coding regions—the results of such profiling can help to pinpoint the GREs that function human-specifically and provide further mechanistic insight into the regulatory evolution of the human brain.

Histone proteins are physically associated with DNA and can be modified to facilitate or obstruct the function of a given DNA sequence. Functionally active DNA sequences (such as enhancers and promoters) are free of histone proteins, and the histone proteins flanking these regions are typically modified with H3K27ac or H3K4me1 (Calo and Wysocka 2013). Several comparative studies have captured the active GREs with an antibody against H3K27ac, and these DNA sequences were profiled in postmortem human and NHP brain tissues using ChIP-seq (Vermunt et al. 2016; Castelijns et al. 2020b; Kozlenkov et al. 2020). Similar to genes, GREs are also largely conserved across species at the sequence level; however, their activity can vary between species, indicating that the conserved GREs were modulated during evolution to create novel molecular networks (Vermunt et al. 2016). One study showed that regulatory gains in adult hominins (common in human and chimpanzee) are enriched in elements that regulate oligodendrocyte gene expression, further implicating human-specific changes in the oligodendrocyte lineage (Castelijns et al. 2020b). Another study compared the epigenomes of prenatal human and rhesus macaque brains by profiling enhancers (H3K27ac) and promoters (H3K27ac and H3K4me2) and found enrichments for elements that are linked to genes involved in neuronal proliferation and migration among the human-gained enhancers (Reilly et al. 2015).

Another, more stable epigenetic modification that can change gene expression is DNA methylation. The majority of DNA methylation occurs on CpG sites; however, non-CG methylation (CH methylation) can also modulate gene expression. Multiple studies have shown that CH methylation is enriched in brain tissue, in particular in neurons, and accumulates during the development of neural circuitry (He and Ecker 2015; Jeong et al. 2021). Interestingly, human neurons contain more CH methylation compared to chimpanzee neurons, whereas CG methylation levels are similar between the two species (Jeong et al. 2021). These studies reveal another layer of human-specific gene expression regulation in the brain.

Multiple modalities of the epigenome can be profiled with high-throughput assays at the tissue level; however, achieving cellular resolution in epigenomics has been a technical challenge. Recently, a highly efficient assay for profiling open-chromatin genomic regions, namely ATAC-seq, has been optimized for single-cell sequencing. A recent study compared the transcriptome and epigenome of human, chimpanzee, and rhesus macaque brains at cellular resolution (Caglayan et al. 2023). Focusing on the GREs with human-specific accessibility changes, the authors found that elements that had gains in accessibility specifically in human upper-layer excitatory neurons are enriched for FOS/JUN transcription factor motifs. Since FOS/JUN are immediately transcribed upon neuronal depolarization and target hundreds of genes (Yap and Greenberg 2018), altered accessibility of putative FOS/JUN targets indicates that gene regulation upon neuronal depolarization has likely undergone human-specific modifications, specifically in upper-layer excitatory neurons that are important for higher-order cognition (Toma et al. 2014). The authors also found that human-specific chromatin accessibility gains in deep-layer excitatory neurons are enriched for FOX transcription factor motifs, including FOXP2, that are factors consistently implicated in neurodevelopment and cognitive functions (Golson and Kaestner 2016). Utilizing the comparative genomic sequence data sets, the authors further showed that HARs and modern human-specific variants are enriched within human-specific chromatin changes. Interestingly, while HAR enrichment was observed in all cell types, modern variant enrichment was specific to an upper-layer excitatory subtype, indicating potentially more cell-type specificity in recent human evolution (Caglayan et al. 2023).

Implications of Human-Specific Gene Regulatory Changes in Brain Diseases

Due to the similarities between novelties in human brain evolution and disrupted features in human brain diseases, evolutionary comparative studies also assessed the relevance of human-specific evolutionary novelties for associations with human brain diseases. Many of the key human-specific gene regulatory changes have also been implicated in brain disorders. For example, FOXP2 is important for speech and language (Lai et al. 2001), and CLOCK, TH, and dopamine biology are associated with schizophrenia, ASD, Parkinson's disease, and several other disorders (Schuch et al. 2018; Kosillo and Bateup 2021). More cases have been reported in other comparative transcriptomic studies (Zhu et al. 2018; Khrameeva et al. 2020; Ma et al. 2022).

While individual genes provide entry points for understanding either brain evolution or brain disease in terms of functional characterization, the overlap between gene regulatory changes in evolution and disease needs to be statistically tested with rigor. With this validation, a molecular association between human evolution and human brain disease can be established. Several studies have attempted to show such a statistical association. For example, one study showed that gene coexpression modules enriched in human-specific gene expression changes in oligodendrocytes were also enriched in genomic variants associated with schizophrenia as well as for gene expression changes in oligodendrocytes from individuals with schizophrenia (Berto et al. 2019). Another study found that enhancers enriched in oligodendrocytes and evolved before human–chimpanzee divergence are enriched in ASD-associated epigenomic regions (Castelijns et al. 2020b). A comparative DNA methylation study also showed significant enrichment of human-specific CG hypomethylation in schizophrenia-associated variants. Interestingly, human-specific CH hypermethylation was depleted and CH hypomethylation was not enriched in schizophrenia variants (Jeong et al. 2021). Taken together, these results indicate an association between human brain evolution and human brain disorders, both likely linked to the unique functioning of the human brain. However, future studies need to take into account the fact that disease risk/pathology may be specific to particular brain regions that are not necessarily profiled in evolutionary comparisons (e.g., brainstem). Thus, direct comparisons of brain tissue expression, chromatin state, and/or methylation across species and disease tissue in a brain region–specific manner are needed for support for correlation (Figure 5.2).

Counterevidence of Evolutionary Novelty in Human Brain Disorders

Associations between evolution and disease in brain tissues have often been analyzed in the context of evolutionarily divergent molecular features. While this is informative for establishing the connection between evolution and disease—especially in human brain evolution since some disorders have not been detected in other species—it does not directly test whether the evolutionary novelties are more inclined to be affected in disease. The majority of these studies also do not test whether evolutionarily conserved elements are also affected in disease. Intriguingly, a few studies show that disease-associated changes can be evolutionarily conserved. For example, one study showed that gene

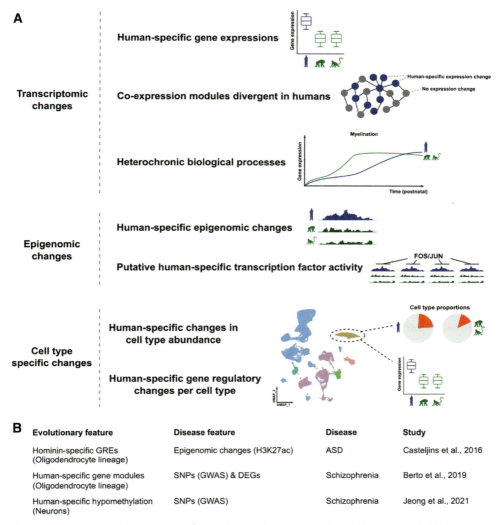

Figure 5.2 Overview of human-specific regulatory changes and studies that reported on their association with disease data sets. (A) Summary of approaches to identify human-specific gene regulatory changes. (B) Table of the studies that provided a statistical test on the overlap of human-specific regulatory changes and disease-associated genomic/regulatory changes. ASD = autism spectrum disorder; DEG = differentially expressed gene; GRE = gene regulatory element; GWAS = genome-wide association study; SNP = single-nucleotide polymorphism.

expression changes in a mouse model (CK-p25) of AD are enriched in the postmortem transcriptomic comparisons between AD and healthy controls (Gjoneska et al. 2015), indicating that the AD-related gene regulatory landscape has been at least partially conserved between humans and mice. A systematic comparison of protein evolution (measured by nonsynonymous to synonymous substitution rate (dN/dS ratio)) also showed that genes associated with psychiatric disorders were evolutionarily conserved (Ogawa and Vallender 2014). However, another study argued an evolutionary effect in AD by showing that the intron–exon structure of long-non-coding RNAs associated with AD is evolutionarily less conserved than that of other long non-coding RNAs (Nitsche et al. 2021). These studies show that disease susceptibility arises in both evolutionarily conserved and divergent molecular features.

Are evolutionarily divergent molecular features associated with brain diseases less present than evolutionarily conserved molecular features? Despite the abundance of comparative genomics studies, to our knowledge, very few studies have provided direct analysis of this question. One study combined genomic variant analysis results from 41 complex disorders—including psychiatric disorders (ASD, schizophrenia) and other brain disorders—and traits (Hujoel et al. 2019). The authors then divided enhancer and promoter regions based on their conservation and annotated the most conserved regions as ancient regions since they are found across mammals. They found that ancient regions have higher enrichment with the disease variants than less conserved regions. Through epigenomic comparisons of humans and marmosets across multiple organs, another study showed that disease-associated enhancers are often conserved between human and marmoset and that newly evolved enhancers show high interindividual variability (Castelijns et al. 2020a), possibly explaining why their disruption does not cause a conspicuous phenotype such as disease symptoms.

Analyses of the differential and conserved regions with respect to DNA methylation profile across human, chimpanzee, and rhesus macaque cortical brain also showed that conserved regions were enriched in variants from several brain disorders including schizophrenia, bipolar disorder, and AD (Jeong et al. 2021). Interestingly, human-specific hypomethylated regions were only enriched in schizophrenia, albeit to a lesser degree than conserved regions, while chimpanzee-specific differential methylations were not enriched in disease variants. Taken together, while human-specific genomic changes may be more associated with human brain disease than changes specific to other species, genomic elements with evolutionarily conserved functions may be more associated with brain disorders than genomic elements with recent evolutionary alteration. Thus, one interpretation of these results is that the evolutionarily conserved elements may be more instrumental to organismal function and that their disruption can lead to more pronounced phenotypes such as disease symptoms. Since genomic features altered in human evolution can also be associated with brain diseases, this could mean that some human-specific changes may have already been incorporated into proper organismal function and should motivate future studies for better understanding of interplay between our recent evolution and brain diseases.

Unanswered Questions and Future Work

Identification of the human-specific molecular regulatory landscape has been instructive but incomplete in several dimensions. For example, nearly all studies compared a few NHP species, mostly including only chimpanzees as the only close relatives to humans.

However, analyses of other great apes have shown that some DNA sequences can be more similar with human/chimpanzee (e.g., between human and gorilla) than the homologous human–chimpanzee comparison due to incomplete lineage sorting (Scally et al. 2012; Prüfer et al. 2012). These sequences are not merely exceptions since ~30% of the gorilla genome has been shown to be more similar with the human/chimpanzee genome compared to their human/chimpanzee homolog sequences (Scally et al. 2012). This indicates that many human-specific regulatory changes may also not have been altered on the human lineage and could be shared with a non-chimpanzee close relative. In addition to incomplete lineage sorting, relying on one species to represent an outgroup (often rhesus macaque) or a close relative species can lead to many false-positive and -negative human-specific regulatory changes. Future studies across many NHPs, particularly other great apes, will further illuminate and refine the extent of human-specific regulatory changes.

Efforts to date have also been limited in the brain regions analyzed at cellular resolution. Bulk transcriptomic comparisons across 33 regions have identified previously overlooked regions to have very high human specificity, such as cerebellar white matter (Khrameeva et al. 2020), that need to be characterized at cellular resolution for a better understanding. Importantly, cellular profiling uncovers not only regulatory changes but also proportional changes, as recently shown with the human-specific shift in the proportion of cortical oligodendrocytes (Caglayan et al. 2023). In addition to greater spatial coverage at cellular resolution, greater temporal coverage at cellular resolution can be very insightful since human development is known to be protracted, although the cellular and molecular features that are associated with neoteny are still poorly understood. Since the human life span is also substantially longer than that of NHPs, studies that increase temporal resolution in mid- and late adulthood may also illuminate how neuronal and glial cells respond to aging in humans and whether this is different than in NHPs. Thus, a comparative outlook involving humans and NHPs should also be insightful to human aging research.

Proteins are the building blocks of cells; however, their high-throughput characterization is currently challenging, especially at cellular resolution. This is especially limiting in human evolution studies since both human and NHP tissues can be scarce, and efficient methodologies that do not require large amounts of starting material are needed. Comparative studies at the proteome level will be highly valuable to further refine the changes we observe at the transcriptome level. Future studies of proteomic comparisons may also reveal changes that occur post-transcriptionally/post-translationally that may not be detectable at the transcript level. In addition to protein quantification, differences in the subcellular protein localization can be informative for putative differential protein function between the species.

Most of these limitations are related to technology and apply to disease comparisons. Since data sets with greater spatial, temporal, and cellular resolution are being generated in disease and across species, future studies will be better positioned to assess the association of human brain evolution and human brain diseases. Based on this review of the literature, such assessments should include the disease associations of not only human-specific changes but also other species–specific changes and conserved elements. This will be important to determine whether the disease association of human-specific changes is unique to humans or whether it is generally enriched in evolutionarily divergent molecular features or mostly in conserved molecular features. However, regardless of the answer, illuminating human-specific alterations will be insightful for disease treatment since the research goal of any therapeutic approach is to understand their function in human brain that is extensively rewired on the human lineage.

Key Points to Remember

- There are millions of human-specific genomic changes, and most of these changes are in the non-coding genome (~98% of the genome).
- Genomic changes in human evolution may have contributed to human susceptibility to ASD and schizophrenia, specifically, but also potentially to other brain diseases.
- iPSC-derived human and chimpanzee neurons consistently reported slower maturation in human neurons.
- Humans may have also retained a larger oligodendrocyte progenitor pool, and therefore a likely higher progenitor capacity, compared to NHPs even in late adulthood.
- *FOXP2* is a transcription factor that both is linked to a human-specific phenotype (language) and has human-specific expression patterns in the cortical brain.
- Characterization of gene regulatory changes at cellular resolution in human brain evolution/diseases is still at its infancy, especially considering the temporal and spatial complexity of the human brain.

Acknowledgments

G. K. is a Jon Heighten Scholar in Autism Research and Townsend Distinguished Chair in Research on Autism Spectrum Disorders at UT Southwestern. E. C. is a Neural Scientist Training Program Fellow in the Peter O'Donnell Brain Institute at UT Southwestern. This work was partially supported by the James S. McDonnell Foundation 21st Century Science Initiative in Understanding Human Cognition Scholar Award to G. K., the National Human Genome Research Institute (HG011641) to G. K., and the National Institute of Mental Health (MH103517) to G. K.

References

Bakken, T. E., Jorstad, N. L., Hu, Q., Lake, B. B., Tian, W., Kalmbach, B. E., Crow, M., Hodge, R. D., Krienen, F. M., Sorensen, S. A., Eggermont, J., Yao, Z., Aevermann, B. D., Aldridge, A. I., Bartlett, A., Bertagnolli, D., Casper, T., Castanon, R. G., Crichton, K., . . . Lein, E. S. (2021). Comparative cellular analysis of motor cortex in human, marmoset and mouse. *Nature, 598*(7879), 111–119. https://doi.org/10.1038/s41 586-021-03465-8

Benton, M. L., Abraham, A., LaBella, A. L., Abbot, P., Rokas, A., & Capra, J. A. (2021). The influence of evolutionary history on human health and disease. *Nature Reviews: Genetics, 22*(5), 269–283. https://doi.org/10.1038/s41576-020-00305-9

Berto, S., Mendizabal, I., Usui, N., Toriumi, K., Chatterjee, P., Douglas, C., Tamminga, C. A., Preuss, T. M., Yi, S. V., & Konopka, G. (2019). Accelerated evolution of oligodendrocytes in the human brain. *Proceedings of the National Academy of Sciences of the United States of America, 116*(48), 24334–24342. https://doi.org/10.1073/pnas.1907982116

Bianchi, S., Stimpson, C. D., Bauernfeind, A. L., Schapiro, S. J., Baze, W. B., McArthur, M. J., Bronson, E., Hopkins, W. D., Semendeferi, K., Jacobs, B., Hof, P. R., & Sherwood, C. C. (2013a). Dendritic morphology of pyramidal neurons in the chimpanzee neocortex: Regional specializations and comparison to humans. *Cerebral Cortex, 23*(10), 2429–2436. https://doi.org/10.1093/cercor/bhs239

Bianchi, S., Stimpson, C. D., Duka, T., Larsen, M. D., Janssen, W. G., Collins, Z., Bauernfeind, A. L., Schapiro, S. J., Baze, W. B., McArthur, M. J., Hopkins, W. D., Wildman, D. E., Lipovich, L., Kuzawa, C. W., Jacobs, B., Hof, P. R., & Sherwood, C. C. (2013b). Synaptogenesis and development of pyramidal neuron dendritic morphology in the chimpanzee neocortex resembles humans. *Proceedings of the National Academy of Sciences of the United States of America, 110*(S2), 10395–10401. https://doi.org/10.1073/pnas.1301224110

Buchanan, J., Elabbady, L., Collman, F., Jorstad, N. L., Bakken, T. E., Ott, C., Glatzer, J., Bleckert, A. A., Bodor, A. L., Brittain, D., Bumbarger, D. J., Mahalingam, G., Seshamani, S., Schneider-Mizell, C., Takeno, M. M., Torres, R., Yin, W., Hodge, R. D., Castro, M., . . . da Costa, N. M. (2022). Oligodendrocyte precursor cells ingest axons in the mouse neocortex. *Proceedings of the National Academy of Sciences of the United States of America, 119*(48), Article e2202580119. https://doi.org/10.1073/pnas.2202580119

Caglayan, E., Ayhan, F., Liu, Y., Vollmer, R., Oh, E., Sherwood, C. C., Preuss, T. M., Yi, S., & Konopka, G. (2023). Molecular features driving cellular and regulatory complexity of human brain evolution. *Nature, 620*, 145–153. https://doi.org/10.1038/s41586-023-06338-4

Calo, E., & Wysocka, J. (2013). Modification of enhancer chromatin: What, how, and why? *Molecular Cell, 49*(5), 825–837. https://doi.org/10.1016/j.molcel.2013.01.038

Capra, J. A., Erwin, G. D., McKinsey, G., Rubenstein, J. L., & Pollard, K. S. (2013). Many human accelerated regions are developmental enhancers. *Philosophical Transactions of the Royal Society B, 368*(1632), Article 20130025. https://doi.org/10.1098/rstb.2013.0025

Castelijns, B., Baak, M. L., Geeven, G., Vermunt, M. W., Wiggers, C. R. M., Timpanaro, I. S., Kondova, I., de Laat, W., & Creyghton, M. P. (2020a). Recently evolved enhancers emerge with high interindividual variability and less frequently associate with disease. *Cell Reports, 31*(12), Article 107799. https://doi.org/10.1016/j.celrep.2020.107799

Castelijns, B., Baak, M. L., Timpanaro, I. S., Wiggers, C. R. M., Vermunt, M. W., Shang, P., Kondova, I., Geeven, G., Bianchi, V., de Laat, W., Geijsen, N., & Creyghton, M. P. (2020b). Hominin-specific regulatory elements selectively emerged in oligodendrocytes and are disrupted in autism patients. *Nature Communications, 11*(1), Article 301. https://doi.org/10.1038/s41467-019-14269-w

Charrier, C., Joshi, K., Coutinho-Budd, J., Kim, J. E., Lambert, N., de Marchena, J., Jin, W. L., Vanderhaeghen, P., Ghosh, A., Sassa, T., & Polleux, F. (2012). Inhibition of SRGAP2 function by its human-specific paralogs induces neoteny during spine maturation. *Cell, 149*(4), 923–935. https://doi.org/10.1016/j.cell.2012.03.034

Chimpanzee Sequencing and Analysis Consortium. (2005). Initial sequence of the chimpanzee genome and comparison with the human genome. *Nature, 437*(7055), 69–87. https://doi.org/10.1038/nature04072

Doan, R. N., Bae, B. I., Cubelos, B., Chang, C., Hossain, A. A., Al-Saad, S., Mukaddes, N. M., Oner, O., Al-Saffar, M., Balkhy, S., Gascon, G. G., Homozygosity Mapping Consortium for Autism, Nieto, M., & Walsh, C. A. (2016). Mutations in human accelerated regions disrupt cognition and social behavior. *Cell, 167*(2), 341–354.e12. https://doi.org/10.1016/j.cell.2016.08.071

Dumas, G., Malesys, S., & Bourgeron, T. (2021). Systematic detection of brain protein-coding genes under positive selection during primate evolution and their roles in cognition. *Genome Research, 31*(3), 484–496. https://doi.org/10.1101/gr.262113.120

Enard, W., Gehre, S., Hammerschmidt, K., Hölter, S. M., Blass, T., Somel, M., Brückner, M. K., Schreiweis, C., Winter, C., Sohr, R., Becker, L., Wiebe, V., Nickel, B., Giger, T., Müller, U., Groszer, M., Adler, T., Aguilar, A., Bolle, I., . . . Pääbo, S. (2009). A humanized version of Foxp2 affects cortico-basal ganglia circuits in mice. *Cell, 137*(5), 961–971. https://doi.org/10.1016/j.cell.2009.03.041

Enard, W., Przeworski, M., Fisher, S. E., Lai, C. S., Wiebe, V., Kitano, T., Monaco, A. P., & Pääbo, S. (2002). Molecular evolution of FOXP2, a gene involved in speech and language. *Nature, 418*(6900), 869–872. https://doi.org/10.1038/nature01025

Erady, C., Amin, K., Onilogbo, T. O. A. E., Tomasik, J., Jukes-Jones, R., Umrania, Y., Bahn, S., & Prabakaran, S. (2022). Novel open reading frames in human accelerated regions and transposable elements reveal new leads to understand schizophrenia and bipolar disorder. *Molecular Psychiatry, 27*(3), 1455–1468. https://doi.org/10.1038/s41380-021-01405-6

Finch, C. E., & Austad, S. N. (2015). Commentary: Is Alzheimer's disease uniquely human? *Neurobiology of Aging, 36*(2), 553–555. https://doi.org/10.1016/j.neurobiolaging.2014.10.025

Florio, M., Heide, M., Pinson, A., Brandl, H., Albert, M., Winkler, S., Wimberger, P., Huttner, W. B., & Hiller, M. (2018). Evolution and cell-type specificity of human-specific genes preferentially expressed in progenitors of fetal neocortex. *eLife, 7,* Article e32332. https://doi.org/10.7554/eLife.32332

Franchini, L. F., & Pollard, K. S. (2017). Human evolution: The non-coding revolution. *BMC Biology, 15*(1), Article 89. https://doi.org/10.1186/s12915-017-0428-9

Gandal, M. J., Zhang, P., Hadjimichael, E., Walker, R. L., Chen, C., Liu, S., Won, H., van Bakel, H., Varghese, M., Wang, Y., Shieh, A. W., Haney, J., Parhami, S., Belmont, J., Kim, M., Moran Losada, P., Khan, Z., Mleczko, J., Xia, Y., . . . Geschwind, D. H. (2018). Transcriptome-wide isoform-level dysregulation in ASD, schizophrenia, and bipolar disorder. *Science, 362*(6420), Article eaat8127. https://doi.org/10.1126/science.aat8127

Geschwind, D. H., & Rakic, P. (2013). Cortical evolution: Judge the brain by its cover. *Neuron, 80*(3), 633–647. https://doi.org/10.1016/j.neuron.2013.10.045

Girskis, K. M., Stergachis, A. B., DeGennaro, E. M., Doan, R. N., Qian, X., Johnson, M. B., Wang, P. P., Sejourne, G. M., Nagy, M. A., Pollina, E. A., Sousa, A. M. M., Shin, T., Kenny, C. J., Scotellaro, J. L., Debo, B. M., Gonzalez, D. M., Rento, L. M., Yeh, R. C., Song, J. H. T., . . . Walsh, C. A. (2021). Rewiring of human neurodevelopmental gene regulatory programs by human accelerated regions. *Neuron, 109*(20), 3239–3251.e7. https://doi.org/10.1016/j.neuron.2021.08.005

Gittelman, R. M., Hun, E., Ay, F., Madeoy, J., Pennacchio, L., Noble, W. S., Hawkins, R. D., & Akey, J. M. (2015). Comprehensive identification and analysis of human accelerated regulatory DNA. *Genome Research, 25*(9), 1245–1255. https://doi.org/10.1101/gr.192591.115

Gjoneska, E., Pfenning, A. R., Mathys, H., Quon, G., Kundaje, A., Tsai, L. H., & Kellis, M. (2015). Conserved epigenomic signals in mice and humans reveal immune basis of Alzheimer's disease. *Nature, 518*(7539), 365–369. https://doi.org/10.1038/nature14252

Golson, M. L., & Kaestner, K. H. (2016). Fox transcription factors: From development to disease. *Development, 143*(24), 4558–4570. https://doi.org/10.1242/dev.112672

Gordon, A., Yoon, S. J., Tran, S. S., Makinson, C. D., Park, J. Y., Andersen, J., Valencia, A. M., Horvath, S., Xiao, X., Huguenard, J. R., Paşca, S. P., & Geschwind, D. H. (2021). Long-term maturation of human cortical organoids matches key early postnatal transitions. *Nature Neuroscience, 24*(3), 331–342. https://doi.org/10.1038/s41593-021-00802-y

He, Y., & Ecker, J. R. (2015). Non-CG methylation in the human genome. *Annual Review of Genomics and Human Genetics, 16,* 55–77. https://doi.org/10.1146/annurev-genom-090413-025437

Hodge, R. D., Bakken, T. E., Miller, J. A., Smith, K. A., Barkan, E. R., Graybuck, L. T., Close, J. L., Long, B., Johansen, N., Penn, O., Yao, Z., Eggermont, J., Höllt, T., Levi, B. P., Shehata, S. I., Aevermann, B., Beller, A., Bertagnolli, D., Brouner, K., . . . Lein, E. S. (2019). Conserved cell types with divergent features in human versus mouse cortex. *Nature, 573*(7772), 61–68. https://doi.org/10.1038/s41586-019-1506-7

Hujoel, M. L. A., Gazal, S., Hormozdiari, F., van de Geijn, B., & Price, A. L. (2019). Disease heritability enrichment of regulatory elements is concentrated in elements with ancient sequence age and conserved function across species. *American Journal of Human Genetics, 104*(4), 611–624. https://doi.org/10.1016/j.ajhg.2019.02.008

Jeong, H., Mendizabal, I., Berto, S., Chatterjee, P., Layman, T., Usui, N., Toriumi, K., Douglas, C., Singh, D., Huh, I., Preuss, T. M., Konopka, G., & Yi, S. V. (2021). Evolution of DNA methylation in the human brain. *Nature Communications, 12*(1), Article 2021. https://doi.org/10.1038/s41467-021-21917-7

Jorstad, N. L., Song, J. H. T., Exposito-Alonso, D., Suresh, H., Castro, N., Krienen, F. M., Yanny, A. M., Close, J., Gelfand, E., Travaglini, K. J., Basu, S., Beaudin, M., Bertagnolli, D., Crow, M., Ding, S. L., Eggermont, J., Glandon, A., Goldy, J., Kroes, T., . . . Bakken, T. E. (2022). Comparative transcriptomics reveals human-specific cortical features. bioRxiv, Article 508480. https://doi.org/10.1101/2022.09.19.508480

Kahn, R. S., Sommer, I. E., Murray, R. M., Meyer-Lindenberg, A., Weinberger, D. R., Cannon, T. D., O'Donovan, M., Correll, C. U., Kane, J. M., van Os, J., & Insel, T. R. (2015). Schizophrenia. *Nature Reviews: Disease Primers, 1*, Article 15067. https://doi.org/10.1038/nrdp.2015.67

Kanton, S., Boyle, M. J., He, Z., Santel, M., Weigert, A., Sanchís-Calleja, F., Guijarro, P., Sidow, L., Fleck, J. S., Han, D., Qian, Z., Heide, M., Huttner, W. B., Khaitovich, P., Pääbo, S., Treutlein, B., & Camp, J. G. (2019). Organoid single-cell genomic atlas uncovers human-specific features of brain development. *Nature, 574*(7778), 418–422. https://doi.org/10.1038/s41586-019-1654-9

Khrameeva, E., Kurochkin, I., Han, D., Guijarro, P., Kanton, S., Santel, M., Qian, Z., Rong, S., Mazin, P., Sabirov, M., Bulat, M., Efimova, O., Tkachev, A., Guo, S., Sherwood, C. C., Camp, J. G., Pääbo, S., Treutlein, B., & Khaitovich, P. (2020). Single-cell-resolution transcriptome map of human, chimpanzee, bonobo, and macaque brains. *Genome Research, 30*(5), 776–789. https://doi.org/10.1101/gr.256958.119

King, M. C., & Wilson, A. C. (1975). Evolution at two levels in humans and chimpanzees. *Science, 188*(4184), 107–116. https://doi.org/10.1126/science.1090005

Klein, J. C., Agarwal, V., Inoue, F., Keith, A., Martin, B., Kircher, M., Ahituv, N., & Shendure, J. (2020). A systematic evaluation of the design and context dependencies of massively parallel reporter assays. *Nature Methods, 17*(11), 1083–1091. https://doi.org/10.1038/s41592-020-0965-y

Konopka, G., Bomar, J. M., Winden, K., Coppola, G., Jonsson, Z. O., Gao, F., Peng, S., Preuss, T. M., Wohlschlegel, J. A., & Geschwind, D. H. (2009). Human-specific transcriptional regulation of CNS development genes by FOXP2. *Nature, 462*(7270), 213–217. https://doi.org/10.1038/nature08549

Konopka, G., Friedrich, T., Davis-Turak, J., Winden, K., Oldham, M. C., Gao, F., Chen, L., Wang, G. Z., Luo, R., Preuss, T. M., & Geschwind, D. H. (2012). Human-specific transcriptional networks in the brain. *Neuron, 75*(4), 601–617. https://doi.org/10.1016/j.neuron.2012.05.034

Kosillo, P., & Bateup, H. S. (2021). Dopaminergic dysregulation in syndromic autism spectrum disorders: Insights from genetic mouse models. *Frontiers in Neural Circuits, 15*, Article 700968. https://doi.org/10.3389/fncir.2021.700968

Kozlenkov, A., Vermunt, M. W., Apontes, P., Li, J., Hao, K., Sherwood, C. C., Hof, P. R., Ely, J. J., Wegner, M., Mukamel, E. A., Creyghton, M. P., Koonin, E. V., & Dracheva, S. (2020). Evolution of regulatory signatures in primate cortical neurons at cell-type resolution. *Proceedings of the National Academy of Sciences of the United States of America, 117*(45), 28422–28432. https://doi.org/10.1073/pnas.2011884117

Lai, C. S., Fisher, S. E., Hurst, J. A., Vargha-Khadem, F., & Monaco, A. P. (2001). A forkhead-domain gene is mutated in a severe speech and language disorder. *Nature, 413*(6855), 519–523. https://doi.org/10.1038/35097076

Liu, X., Somel, M., Tang, L., Yan, Z., Jiang, X., Guo, S., Yuan, Y., He, L., Oleksiak, A., Zhang, Y., Li, N., Hu, Y., Chen, W., Qiu, Z., Paabo, S., & Khaitovich, P. (2012). Extension of cortical synaptic development distinguishes humans from chimpanzees and macaques. *Genome Research, 22*(4), 611–622. https://doi.org/10.1101/gr.127324.111

Ma, S., Skarica, M., Li, Q., Xu, C., Risgaard, R. D., Tebbenkamp, A. T. N., Mato-Blanco, X., Kovner, R., Krsnik, Ž., de Martin, X., Luria, V., Martí-Pérez, X., Liang, D., Karger, A., Schmidt, D. K., Gomez-Sanchez, Z., Qi, C., Gobeske, K. T., Pochareddy, S., . . . Sestan, N. (2022). Molecular and cellular evolution of the primate dorsolateral prefrontal cortex. *Science, 377*(6614), Article eabo7257. https://doi.org/10.1126/science.abo7257

Manolio, T. A., Collins, F. S., Cox, N. J., Goldstein, D. B., Hindorff, L. A., Hunter, D. J., McCarthy, M. I., Ramos, E. M., Cardon, L. R., Chakravarti, A., Cho, J. H., Guttmacher, A. E., Kong, A., Kruglyak, L., Mardis, E., Rotimi, C. N., Slatkin, M., Valle, D., Whittemore, A. S., . . . Visscher, P. M. (2009). Finding the missing heritability of complex diseases. *Nature, 461*(7265), 747–753. https://doi.org/10.1038/nature08494

Marchetto, M. C., Hrvoj-Mihic, B., Kerman, B. E., Yu, D. X., Vadodaria, K. C., Linker, S. B., Narvaiza, I., Santos, R., Denli, A. M., Mendes, A. P., Oefner, R., Cook, J., McHenry, L., Grasmick, J. M., Heard, K., Fredlender, C., Randolph-Moore, L., Kshirsagar, R., Xenitopoulos, R., . . . Gage, F. H. (2019). Species-specific maturation profiles of human, chimpanzee and bonobo neural cells. *eLife*, *8*, Article e37527. https://doi.org/10.7554/eLife.37527

Martínez-Cerdeño, V. (2017). Dendrite and spine modifications in autism and related neurodevelopmental disorders in patients and animal models. *Developmental Neurobiology*, *77*(4), 393–404. https://doi.org/10.1002/dneu.22417

Miller, D. J., Duka, T., Stimpson, C. D., Schapiro, S. J., Baze, W. B., McArthur, M. J., Fobbs, A. J., Sousa, A. M., Sestan, N., Wildman, D. E., Lipovich, L., Kuzawa, C. W., Hof, P. R., & Sherwood, C. C. (2012). Prolonged myelination in human neocortical evolution. *Proceedings of the National Academy of Sciences of the United States of America*, *109*(41), 16480–16485. https://doi.org/10.1073/pnas.1117943109

Moyer, C. E., Shelton, M. A., & Sweet, R. A. (2015). Dendritic spine alterations in schizophrenia. *Neuroscience Letters*, *601*, 46–53. https://doi.org/10.1016/j.neulet.2014.11.042

Nitsche, A., Arnold, C., Ueberham, U., Reiche, K., Fallmann, J., Hackermuller, J., Horn, F., Stadler, P. F., & Arendt, T. (2021). Alzheimer-related genes show accelerated evolution. *Molecular Psychiatry*, *26*(10), 5790–5796. https://doi.org/10.1038/s41380-020-0680-1

Nurk, S., Koren, S., Rhie, A., Rautiainen, M., Bzikadze, A. V., Mikheenko, A., Vollger, M. R., Altemose, N., Uralsky, L., Gershman, A., Aganezov, S., Hoyt, S. J., Diekhans, M., Logsdon, G. A., Alonge, M., Antonarakis, S. E., Borchers, M., Bouffard, G. G., Brooks, S. Y., . . . Phillippy, A. M. (2022). The complete sequence of a human genome. *Science*, *376*(6588), 44–53. https://doi.org/10.1126/science.abj6987

O'Bleness, M., Searles, V. B., Varki, A., Gagneux, P., & Sikela, J. M. (2012). Evolution of genetic and genomic features unique to the human lineage. *Nature Reviews: Genetics*, *13*(12), 853–866. https://doi.org/10.1038/nrg3336

Ogawa, L. M., & Vallender, E. J. (2014). Evolutionary conservation in genes underlying human psychiatric disorders. *Frontiers in Human Neuroscience*, *8*, Article 283. https://doi.org/10.3389/fnhum.2014.00283

Paolicelli, R. C., Bolasco, G., Pagani, F., Maggi, L., Scianni, M., Panzanelli, P., Giustetto, M., Ferreira, T. A., Guiducci, E., Dumas, L., Ragozzino, D., & Gross, C. T. (2011). Synaptic pruning by microglia is necessary for normal brain development. *Science*, *333*(6048), 1456–1458. https://doi.org/10.1126/science.1202529

Pinson, A., & Huttner, W. B. (2021). Neocortex expansion in development and evolution—From genes to progenitor cell biology. *Current Opinion in Cell Biology*, *73*, 9–18. https://doi.org/10.1016/j.ceb.2021.04.008

Pinson, A., Xing, L., Namba, T., Kalebic, N., Peters, J., Oegema, C. E., Traikov, S., Reppe, K., Riesenberg, S., Maricic, T., Derihaci, R., Wimberger, P., Pääbo, S., & Huttner, W. B. (2022). Human TKTL1 implies greater neurogenesis in frontal neocortex of modern humans than Neanderthals. *Science*, *377*(6611), Article eabl6422. https://doi.org/10.1126/science.abl6422

Pollard, K. S., Salama, S. R., Lambert, N., Lambot, M. A., Coppens, S., Pedersen, J. S., Katzman, S., King, B., Onodera, C., Siepel, A., Kern, A. D., Dehay, C., Igel, H., Ares, M., Jr., Vanderhaeghen, P., & Haussler, D. (2006). An RNA gene expressed during cortical development evolved rapidly in humans. *Nature*, *443*(7108), 167–172. https://doi.org/10.1038/nature05113

Prabhakar, S., Noonan, J. P., Pääbo, S., & Rubin, E. M. (2006). Accelerated evolution of conserved noncoding sequences in humans. *Science*, *314*(5800), 786. https://doi.org/10.1126/science.1130738

Prüfer, K., Munch, K., Hellmann, I., Akagi, K., Miller, J. R., Walenz, B., Koren, S., Sutton, G., Kodira, C., Winer, R., Knight, J. R., Mullikin, J. C., Meader, S. J., Ponting, C. P., Lunter, G., Higashino, S., Hobolth, A., Dutheil, J., Karakoç, E., . . . Pääbo, S. (2012). The bonobo genome compared with the chimpanzee and human genomes. *Nature*, *486*(7404), 527–531. https://doi.org/10.1038/nature11128

Prüfer, K., Racimo, F., Patterson, N., Jay, F., Sankararaman, S., Sawyer, S., Heinze, A., Renaud, G., Sudmant, P. H., de Filippo, C., Li, H., Mallick, S., Dannemann, M., Fu, Q., Kircher, M., Kuhlwilm, M., Lachmann, M., Meyer, M., Ongyerth, M., . . . Pääbo, S. (2014). The complete genome sequence of a Neanderthal from the Altai Mountains. *Nature*, *505*(7481), 43–49. https://doi.org/10.1038/nature12886

Reilly, S. K., Yin, J., Ayoub, A. E., Emera, D., Leng, J., Cotney, J., Sarro, R., Rakic, P., & Noonan, J. P. (2015). Evolutionary changes in promoter and enhancer activity during human corticogenesis. *Science*, *347*(6226), 1155–1159. https://doi.org/10.1126/science.1260943

Scally, A., Dutheil, J. Y., Hillier, L. W., Jordan, G. E., Goodhead, I., Herrero, J., Hobolth, A., Lappalainen, T., Mailund, T., Marques-Bonet, T., McCarthy, S., Montgomery, S. H., Schwalie, P. C., Tang, Y. A., Ward, M. C., Xue, Y., Yngvadottir, B., Alkan, C., Andersen, L. N., . . . Durbin, R. (2012). Insights into hominid evolution from the gorilla genome sequence. *Nature, 483*(7388), 169–175. https://doi.org/10.1038/nature10842

Schörnig, M., Ju, X., Fast, L., Ebert, S., Weigert, A., Kanton, S., Schaffer, T., Nadif Kasri, N., Treutlein, B., Peter, B. M., Hevers, W., & Taverna, E. (2021). Comparison of induced neurons reveals slower structural and functional maturation in humans than in apes. *eLife, 10*, Article e59323. https://doi.org/10.7554/eLife.59323

Schuch, J. B., Genro, J. P., Bastos, C. R., Ghisleni, G., & Tovo-Rodrigues, L. (2018). The role of *CLOCK* gene in psychiatric disorders: Evidence from human and animal research. *American Journal of Medical Genetics. Part B, Neuropsychiatric Genetics, 177*(2), 181–198. https://doi.org/10.1002/ajmg.b.32599

Somel, M., Franz, H., Yan, Z., Lorenc, A., Guo, S., Giger, T., Kelso, J., Nickel, B., Dannemann, M., Bahn, S., Webster, M. J., Weickert, C. S., Lachmann, M., Pääbo, S., & Khaitovich, P. (2009). Transcriptional neoteny in the human brain. *Proceedings of the National Academy of Sciences of the United States of America, 106*(14), 5743–5748. https://doi.org/10.1073/pnas.0900544106

Sousa, A. M. M., Zhu, Y., Raghanti, M. A., Kitchen, R. R., Onorati, M., Tebbenkamp, A. T. N., Stutz, B., Meyer, K. A., Li, M., Kawasawa, Y. I., Liu, F., Perez, R. G., Mele, M., Carvalho, T., Skarica, M., Gulden, F. O., Pletikos, M., Shibata, A., Stephenson, A. R., . . . Sestan, N. (2017). Molecular and cellular reorganization of neural circuits in the human lineage. *Science, 358*(6366), 1027–1032. https://doi.org/10.1126/science.aan3456

Srinivasan, S., Bettella, F., Mattingsdal, M., Wang, Y., Witoelar, A., Schork, A. J., Thompson, W. K., Zuber, V., Schizophrenia Working Group of the Psychiatric Genomics Consortium, International Headache Genetics Consortium, Winsvold, B. S., Zwart, J. A., Collier, D. A., Desikan, R. S., Melle, I., Werge, T., Dale, A. M., Djurovic, S., & Andreassen, O. A. (2016). Genetic markers of human evolution are enriched in schizophrenia. *Biological Psychiatry, 80*(4), 284–292. https://doi.org/10.1016/j.biopsych.2015.10.009

Sullivan, P. F., & Geschwind, D. H. (2019). Defining the genetic, genomic, cellular, and diagnostic architectures of psychiatric disorders. *Cell, 177*(1), 162–183. https://doi.org/10.1016/j.cell.2019.01.015

Toma, K., Kumamoto, T., & Hanashima, C. (2014). The timing of upper-layer neurogenesis is conferred by sequential derepression and negative feedback from deep-layer neurons. *Journal of Neuroscience, 34*(39), 13259–13276. https://doi.org/10.1523/JNEUROSCI.2334-14.2014

Tomasello, M. (2020). The adaptive origins of uniquely human sociality. *Philosophical Transactions of the Royal Society of London, 375*(1803), Article 20190493. https://doi.org/10.1098/rstb.2019.0493

Trujillo, C. A., Rice, E. S., Schaefer, N. K., Chaim, I. A., Wheeler, E. C., Madrigal, A. A., Buchanan, J., Preissl, S., Wang, A., Negraes, P. D., Szeto, R. A., Herai, R. H., Huseynov, A., Ferraz, M. S. A., Borges, F. S., Kihara, A. H., Byrne, A., Marin, M., Vollmers, C., . . . Muotri, A. R. (2021). Reintroduction of the archaic variant of *NOVA1* in cortical organoids alters neurodevelopment. *Science, 371*(6530), Article eaax2537. https://doi.org/10.1126/science.aax2537

Uebbing, S., Gockley, J., Reilly, S. K., Kocher, A. A., Geller, E., Gandotra, N., Scharfe, C., Cotney, J., & Noonan, J. P. (2021). Massively parallel discovery of human-specific substitutions that alter enhancer activity. *Proceedings of the National Academy of Sciences of the United States of America, 118*(2), Article e2007049118. https://doi.org/10.1073/pnas.2007049118

Vermunt, M. W., Tan, S. C., Casteljins, B., Geeven, G., Reinink, P., de Bruijn, E., Kondova, I., Persengiev, S., Netherlands Brain Bank, Bontrop, R., Cuppen, E., de Laat, W., & Creyghton, M. P. (2016). Epigenomic annotation of gene regulatory alterations during evolution of the primate brain. *Nature Neuroscience, 19*(3), 494–503. https://doi.org/10.1038/nn.4229

Weiss, C. V., Harshman, L., Inoue, F., Fraser, H. B., Petrov, D. A., Ahituv, N., & Gokhman, D. (2021). The cis-regulatory effects of modern human-specific variants. *eLife, 10*, Article e63713. https://doi.org/10.7554/eLife.63713

Whalen, S., Inoue, F., Ryu, H., Fair, T., Markenscoff-Papadimitriou, E., Keough, K., Kircher, M., Martin, B., Alvarado, B., Elor, O., Laboy Cintron, D., Williams, A., Hassan Samee, M. A., Thomas, S., Krencik, R., Ullian, E. M., Kriegstein, A., Rubenstein, J. L., Shendure, J., . . . Pollard, K. S. (2022). Machine-learning dissection of human accelerated regions in primate neurodevelopment. bioRxiv, Article 256313. https://doi.org/10.1101/256313

Xu, K., Schadt, E. E., Pollard, K. S., Roussos, P., & Dudley, J. T. (2015). Genomic and network patterns of schizophrenia genetic variation in human evolutionary accelerated regions. *Molecular Biology and Evolution, 32*(5), 1148–1160. https://doi.org/10.1093/molbev/msv031

Yap, E. L., & Greenberg, M. E. (2018). Activity-regulated transcription: Bridging the gap between neural activity and behavior. *Neuron, 100*(2), 330–348. https://doi.org/10.1016/j.neuron.2018.10.013

Yeung, M. S., Zdunek, S., Bergmann, O., Bernard, S., Salehpour, M., Alkass, K., Perl, S., Tisdale, J., Possnert, G., Brundin, L., Druid, H., & Frisén, J. (2014). Dynamics of oligodendrocyte generation and myelination in the human brain. *Cell, 159*(4), 766–774. https://doi.org/10.1016/j.cell.2014.10.011

Zhou, H., Hu, S., Matveev, R., Yu, Q., Li, J., Khaitovich, P., Jin, L., Lachmann, M., Stoneking, M., Fu, Q., & Tang, K. (2015). A chronological atlas of natural selection in the human genome during the past half-million years. bioRxiv, Article 018929. https://doi.org/10.1101/018929

Zhu, Y., Sousa, A. M. M., Gao, T., Skarica, M., Li, M., Santpere, G., Esteller-Cucala, P., Juan, D., Ferrández-Peral, L., Gulden, F. O., Yang, M., Miller, D. J., Marques-Bonet, T., Imamura Kawasawa, Y., Zhao, H., & Sestan, N. (2018). Spatiotemporal transcriptomic divergence across human and macaque brain development. *Science, 362*(6420), Article eaat8077. https://doi.org/10.1126/science.aat8077

6

Adaptive Archaic Introgression

Olga Dolgova and Oscar Lao

Human Evolutionary History
Human Origin

The emergence of *Homo sapiens* out of the primate lineage involves the gradual development of human traits, multiple human dispersals all over the world from Africa, and population extinctions. Moreover, human evolutionary history includes interbreeding with other hominins, an essential creative force in the emergence of modern humans (Ahlquist et al. 2021).

Therefore, understanding the complex evolutionary history of humans requires integrative approaches from different disciplines, including genetics. The latest is particularly interesting when archaeological evidence is scarce or not available. Genetic studies suggest that primates diverged from other mammals about 85 million years ago (MYA), in the Late Cretaceous period (Figure 6.1). Within the superfamily Hominoidea, the family Hominidae (great apes) diverged from the family Hylobatidae (gibbons) some 15–20 MYA; the subfamily Homininae (African apes) diverged from Ponginae (orangutans) about 14 MYA; the tribe Hominini (including humans, *Australopithecus*, and chimpanzees) parted from the tribe Gorillini (gorillas) 8–9 MYA; in turn, the subtribes Hominina (humans and extinct bipedal ancestors) and Panina (chimpanzees) separated 4–7 MYA (Prüfer et al. 2012; Figure 6.1).

The earliest documented representative of the genus *Homo* is *Homo habilis*, which evolved around 2.8 MYA in Africa (Wohns et al. 2022; Figure 6.1). According to current evidence, *Homo erectus* and *Homo ergaster* were the first hominins that left Africa, spreading throughout Asia and Europe between 1.3 and 1.8 MYA (Wohns et al. 2022).

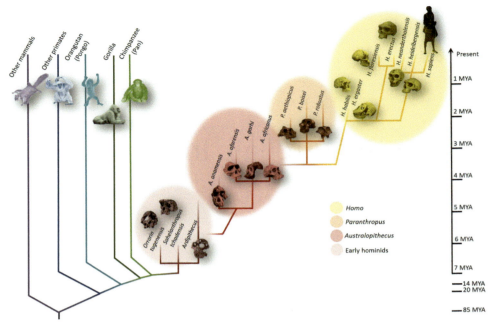

Figure 6.1 Evolutionary tree showing the phylogenetic relationships between major groups of archaic and modern hominins, other primates, and mammals. The representatives of the extinct groups are illustrated with skulls. Time is represented vertically (but not to scale) on the right, with the present time on top. MYA = million years ago.

Human Dispersal

Since the early 1980s, the most accepted demographic scenario for explaining the recent evolution of anatomically modern humans (AMH) has been the out of Africa model (OOA; Stringer and Andrews 1988). According to the OOA, AMH first appeared around 100–200 thousand years ago (KYA) in East Africa, possibly from *Homo heidelbergensis*, *H. rhodesiensis*, or *H. antecessor*. Then, AMH spread throughout the world around 50–60 KYA, leading to the nearly complete replacement of local populations of *Homo erectus*, Denisova hominins, *H. floresiensis*, *H. luzonensis*, and *H. neanderthalensis*, who putatively had been evolving in Eurasia from the first *H. erectus* diaspora (Wohns et al. 2022; Figure 6.2).

Recently, the identification of fossils compatible with AMH anatomy dated ~300 KYA found at Jebel Irhoud in Morocco (Hublin et al. 2017) and a maxilla of 177–194 KYA in Misliya Cave (Hershkovitz et al. 2018) have challenged the widely accepted dating of *H. sapiens* emergence. These findings could suggest that the members of the *H. sapiens* clade left Africa sooner than was previously thought, probably in several waves of OOA migration at different stages of its demographic history. Nevertheless, recent genetic evidence suggests that each modern non-African population descends from the same wave that left Africa between 65 and 50 KYA somewhere in northeast Africa (Wohns et al. 2022).

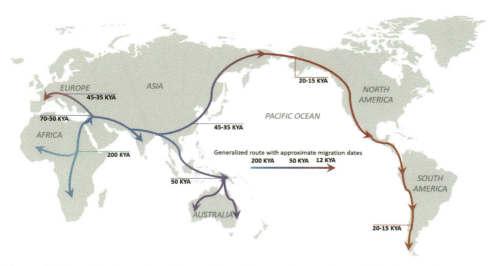

Figure 6.2 First migration paths of humans from Africa to other continents. KYA = thousand years ago.

The geographic paths of the OOA diaspora are also contentious. According to the Southern Dispersal theory (Lahr et al. 2010), around 70 KYA there was a coastal dispersal of modern humans from the Horn of Africa crossing the Bab-el-Mandeb to Yemen at a lower sea level. Supposedly, those early modern humans were dependent upon marine resources for their survival and populated Southeast Asia and Oceania. This would explain the discovery of early human sites in these areas much earlier than those in the Levant (Lahr et al. 2010). The second wave of humans could have later migrated through the Persian Gulf oasis and the Zagros Mountains into the Middle East. Alternatively, they could have moved across the Sinai Peninsula into Asia, shortly after 50 KYA, originating the majority of the human populations in Eurasia (Lahr et al. 2010).

Interbreeding between Archaic and Modern Humans

The hypothesis of interbreeding of AMH with early modern humans, also known as "hybridization," "admixture," or "hybrid-origin theory," has been discussed ever since the discovery of Neanderthal remains in the 19th century (Huxley 1890). In the 21st century, with the advent of molecular biology techniques and computerization, it has been revealed that AMH interbred on several occasions during the Middle Paleolithic and early Upper Paleolithic with other hominid lineages present in Eurasia from 500 KYA up to 50–30 KYA (Green et al. 2010; Figure 6.3).

Interbreeding between Neanderthals and Modern Humans

Homo neanderthalensis (alternatively, *H. sapiens neanderthalensis*) diverged from the last common ancestor with AHM no earlier than ~500 KYA, inhabited Eurasia until ~41–39 KYA, and was gradually replaced by AMH (Green et al. 2010). The reason for such

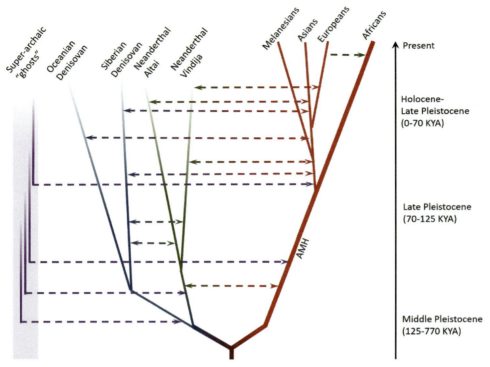

Figure 6.3 Inferred pulses of archaic introgression between anatomically modern humans and archaic humans (see Ahlquist et al. 2021 and Dolgova and Lao 2018 for review). Time is represented vertically (but not to scale) on the right. Anatomically modern human populations are represented in red, two Denisovan populations in green, two Neanderthal populations in blue, and super-archaic "ghosts" (unknown populations of hominins that may have contributed to the genome ancestry of modern humans) in purple. Horizontal lines indicate gene flow between two populations but may represent single or multiple gene flow events between the same two populations. AMH = anatomically modern human; KYA = thousand years ago.

disappearance is contentious. New genomic evidence suggests that the later Neanderthals were perhaps not replaced by incoming modern humans but gradually absorbed by them (Lalueza-Fox 2021).

To date, three high-coverage Neanderthal genomes, as well as multiple low-coverage Neanderthal genomes, have been sequenced (Ahlquist et al. 2021). The draft sequence of the Neanderthal genome revealed that Neanderthals shared more alleles with Eurasian populations (e.g., Spanish, Han Chinese, and Australian) than with sub-Saharan African populations (e.g., San and Yoruba; Green et al. 2010). Further analyses have allowed us to identify that the Neanderthal component in non-African modern humans is more closely related to the Vindija and Mezmaiskaya Neanderthals than to the Altai Neanderthal. These results suggest that the majority of the admixture into modern humans came from Neanderthal populations that diverged about 80–100 KYA from the Vindija (Croatia) and Mezmaiskaya (Russia, region Krasnodar) Neanderthal lineages before the latter two diverged from each other (Prüfer et al. 2017). Furthermore, it was shown that out of

Altai (Siberia), only El Sidrón (Spain) and Vindija Neanderthal lineages display significant rates of gene flow (0.3%–2.6%) into modern humans. This suggests that these two lineages are more closely related to the Neanderthals that interbred with modern humans about 47–65 KYA somewhere in western Eurasia, possibly the Middle East (Kuhlwilm et al. 2016).

Overall, it has been estimated that about 20% of the Neanderthal genome is assimilated into the genome of the modern human populations of East Asians and Europeans. However, some specific genomic regions under positive selection may have degrees of Neanderthal ancestry as high as 64% in Europeans and 62% in Asians (Sankararaman et al. 2014). The specific individual amount of estimated Neanderthal ancestry depends on which methodology is considered and which population is assessed, ranging between 1.5% and 7.3% (Nielsen et al. 2017). East Asian individuals carry somewhat more Neanderthal DNA (2.3%–2.6%) than people in western Eurasia (1.8%–2.4%; Nielsen et al. 2017). Within the African continent, the Neanderthal inferred admixture is the highest among the North African populations with maximal autochthonous North African ancestry, such as Tunisian Berbers, where it was at the same level as or even higher than that of Eurasian populations (Sánchez-Quinto et al. 2012). Low significant rates of Neanderthal admixture have also been observed in the Maasai of East Africa. In contrast, sub-Saharan African groups are the only modern human populations that generally did not experience Neanderthal admixture (Sánchez-Quinto et al. 2012). However, recently Durvasula and Sankararaman (2020) challenged this interpretation, suggesting that Neanderthal admixture also occurred in sub-Saharan Africans.

In contrast to the signal of archaic introgression observed in autosomes, no evidence of Neanderthal maternally inherited mitochondrial DNA (mtDNA) and Y chromosome has been found in modern humans (Green et al. 2010).

Interbreeding with Denisovans

In contrast to the profusion of fossils of Neanderthals, so far five small and highly fragmented fossils discovered at Denisova Cave (Russian Altai, Siberia, Russia) have been identified as another archaic population, the Denisovans (Reich et al. 2010). Nevertheless, despite the location of the Denisovan remains, the greatest proportion of Denisovan admixture in current AMH has been found in Melanesian genomes, accounting for 4%–6% (Vernot et al. 2016). Aboriginal Australians also show an increased allele-sharing with Denisovans, indicating that interbreeding took place 44–54 KYA east of the Wallace Line that divides Southeast Asia and before their common ancestors entered into Sahul (Pleistocene New Guinea and Australia, at least 44 KYA; Vernot et al. 2016). Recently, an independent admixture event with Denisovans was detected when analyzing 118 Philippine ethnic groups, reaching the Ayta Magbukon population with up to ~30%–40% more Denisovan ancestry than in Australo-Melanesians (Larena et al. 2021).

In contrast, Europeans and Africans do not display contributions of the Denisovan genes (Vernot et al. 2016). The presence of Denisovan ancestry in East Asian populations is more contentious. The skeletal remains of an early modern human from the Tianyuan cave (near Zhoukoudian, China) that lived 40 KYA showed a Neanderthal contribution within the range of today's Eurasian modern humans, but it had no discernible Denisovan contribution (Vernot et al. 2016). Thus, the genetic contribution had been always scarce on the mainland, consistent with the lack of Denisovan admixture (Vernot et al. 2016). Nevertheless, based on patterns of Denisovan relatedness, Browning et al. (2018) concluded that there have been at least two separate episodes of Denisovan admixture, one in East Asians (e.g., Japanese and Han Chinese) and another in South Asians (e.g., Telugu and Punjabi) and Oceanians (e.g., Papuans).

Interbreeding with Archaic "Ghost" Populations

The fossil human record is scarce and fragmented, and DNA availability from ancient remains is limited or non-existent. Nevertheless, the presence of known archaic admixture in our genome suggests that genomic approaches could be taken to identify the existence of other archaic—"ghost"—populations that admixed with the first AMH, for which we do not have fossil records or they have been incorrectly assigned to already described archaic populations, as originally happened with the Denisovan remains (Figure 6.3). Within Africa, Hammer et al. (2011) found that three candidate regions introgressed approximately 35 KYA in the human genome of African sub-Saharan populations and concluded that roughly 2% of the genetic material was from archaic hominins separated from the ancestors of the modern human lineage around 700 KYA and inhabited central Africa (Hammer et al. 2011). According to another study published in 2020, 2%–19% of the DNA of four West African populations may have come from an unknown archaic hominin that split from the ancestor of humans and Neanderthals between 1.02 MYA and 360 KYA (Durvasula and Sankararaman 2020). Yet another study in 2019 using deep learning inferred a model where current sub-Saharan populations had received a substantial amount of archaic introgression from a "ghost" population related to AMH (Lorente-Galdos et al. 2019).

Similar to observations within Africa, there is growing evidence that "ghost" archaic populations could have interbred with AMH out of Africa. Additional genetic contributions in Eurasians from an "unknown" ancestral population potentially related to the Neanderthal–Denisovan lineage were identified (Mondal et al. 2019). Interestingly, the authors proposed the existence of a hybrid Neanderthal–Denisovan population contemporary to the discovery of a first-generation Neanderthal–Denisovan hybrid (Slon et al. 2018).

The Functional Value of Introgressed Regions

The identification of archaic introgression in AMH has shown that its distribution is not random in their genomes. This non-random distribution has been explained in terms of

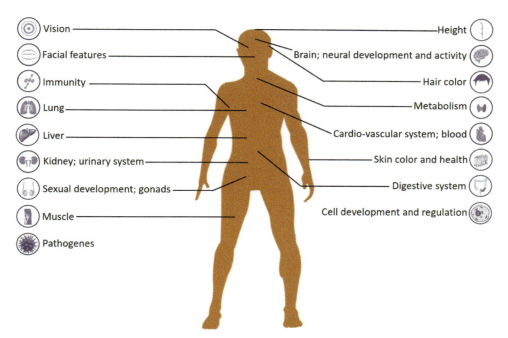

Figure 6.4 The phenotypic impact of introgressed variants from ancient genomes on modern human organs and systems. Updated from Dolgova and Lao (2018, fig. 2).

how often interbreeding occurred and how successful it was. Introgressive hybridization can have both positive and negative effects, affecting different organs and systems (Figure 6.4), summarized in the next sections.

Selection against Introgressed Regions at the Level of Genomes and Individual Loci

The footprint of purifying selection against archaic alleles is ubiquitous in the genomes of non-African humans, which have sequences devoid of introgressed variation (also called "introgression deserts") from Neanderthals and Denisovans. The genome of Upper Paleolithic Eurasian modern humans (such as the Tianyuan) carry more Neanderthal DNA (about 4%–5%) than present-day Eurasians (about 1%–2%), supporting the hypothesis that purifying selection has reduced the Neanderthal contribution in present-day human genomes (Ahlquist et al. 2021). Moreover, selection rates against Neanderthal sequences varied for European and Asian populations (Ahlquist et al. 2021).

The genomic regions affected by introgression deserts are of a high density of functionally important elements, possibly maintained by selection against introgression (Sankararaman et al. 2016). Moreover, it was shown that modern humans' introgression deserts free from Neanderthal and Denisovan DNA are largely overlapping, suggesting repeated loss of archaic DNA at specific loci (Sankararaman et al. 2016). Genes transcribed in meiotic germ cells are particularly sensible for archaic introgression (Dolgova and Lao 2018). Selection has also depleted Neanderthal alleles primarily in promoters, coding regions, conserved non-coding regions, and biological pathways, such as RNA processing (Sankararaman et al. 2016). This phenomenon extends to sex-related chromosomes. In fact, the disproportionate role of sex chromosomes in species differences and hybrid incompatibility constitutes a consistent pattern in speciation (Dolgova and Lao 2018). Following this observation, large regions of low Neanderthal contribution are most pronounced on the X chromosome—with fivefold lower Neanderthal ancestry compared to autosomes (Sankararaman et al. 2016). Similarly, a significant reduction in admixture associated with genes showing testes-specific expression suggests that admixture may have led to reduced male fertility. This would support evidence of reduced introgression on sex chromosomes that may be partly affected by hemizygosity of the X chromosome in males (Dolgova and Lao 2018).

The presence of regions strongly depleted of archaic ancestry has been used as a strategy to identify functional regions contributing to the uniqueness of some modern human traits (Vernot et al. 2016). For example, no Neanderthal ancestry has been detected around the forkhead box protein P2 (*FOXP2*) gene, whose mutations lead to language disorders (Dolgova and Lao 2018). Furthermore, regions where both Neanderthal and Denisova ancestry are lacking show enrichment for genes expressed in specific brain regions (e.g., the striatum in adults and the ventral frontal cortex–ventrolateral prefrontal cortex in infants; Vernot et al. 2016).

Even so, this genomic evidence must be interpreted with caution. When selection against introgression occurs at a big number of genomic loci, its combined effects can leave detectable patterns, while selection on any individual locus may be weak (Dolgova and Lao 2018). Moreover, weaker signals of archaic admixture have been detected in parts of the genome with high gene density and/or low recombination. This confirms the theoretical assumption that the strength of selection against introgressed alleles depends on the density of selected sites and the recombination rate (Skov et al. 2020). Unannotated structural variation in archaic genomes may also explain some of the observed depletion in archaically introgressed alleles (Ahlquist et al. 2021). Depletion of Neanderthal segments in the modern human genome may also be caused by genetic drift, involving intense bottlenecks in the modern human population and strong background selection against deleterious Neanderthal alleles (Ahlquist et al. 2021).

Genomic Signatures of Adaptive Introgression from Archaic to AMH

Despite the generally deleterious effects of archaic hybridization observed in humans, interbreeding between AMH and archaic species in Eurasia provided a source of advantageous archaic genetic variants adapted to local environments and a reservoir for extra genetic variation (Dolgova and Lao 2018). This evolutionary process, called "adaptive introgression," could bring variants at a higher frequency than de novo mutations by the formation of linked blocks of sequences with multiple functional mutations, potentially including co-adapted alleles (Dolgova and Lao 2018). Recent studies have used a variety of methods to identify candidate alleles and regions underlying phenotypic impacts of archaic introgression (Ahlquist et al. 2021).

Adaptive introgression is especially abundant around immune-related genes, suggesting that many alleles adaptive to local pathogens were segregating in Neanderthals and Denisovans and were positively selected after interbreeding with AMH. Since innate immunity genes have evolved under stronger purifying selection than the rest of the genome, this enrichment of introgressed alleles suggests the presence of strong positive selection in the immune system (Ahlquist et al. 2021). In particular, Enard and Petrov (2018) showed that introgressed adaptive haplotypes are enriched for proteins that interact with RNA viruses. Similarly, polygenic adaptive introgression has been identified in pathways associated with immunity (Gouy and Excoffier 2020). Furthermore, population transcriptome studies of immune reactions to viral and bacterial pathogens in a large number of cell cultures from individuals of European versus African ancestry have detected many differences in gene expression and splicing that were driven by Neanderthal introgressed alleles, further suggesting their regulatory impact on immunity (Rotival et al. 2019).

Additionally, multiple studies have used large genotypic data sets with phenotypic data to link archaic alleles with specific physiological phenotypes and traits (Dolgova and Lao 2018; McArthur et al. 2021). While many associations between archaic alleles and phenotypes have been discovered, recent reanalysis has found that quite a few of those associations are the result of the linkage between non-archaic variants and archaic haplotypes (Skov et al. 2020). Many of them might be non-archaic alleles shared between modern and archaic humans, still found in Africans today, that were lost due to the OOA bottleneck and subsequently reintroduced into non-Africans by admixture with archaic humans (Rinker et al. 2020). This highlights the complications in associating phenotypic effects to archaic introgressed alleles. To avoid such biases, McArthur et al. (2021) studied specific Neanderthal introgressed variants across individuals that are statistically associated with phenotypic variation in modern humans. McArthur et al. (2021) showed that Neanderthal introgressed genomic regions are depleted of heritability for all 400 traits considered in their study, except those related to skin and hair. Moreover, cognitive traits were found to be the most depleted for Neanderthal ancestry in this study (McArthur et al. 2021).

In addition to introgressed protein-based variants, increasing evidence suggests that regulatory variants play a central role in adaptive processes (Silvert et al. 2019). In particular, Neanderthal introgressed alleles contribute proportionally more to variation in expression than non-archaic alleles (Ahlquist et al. 2021). They are enriched in variants associated with gene expression, providing evidence that Neanderthal introgression more often functionally affects gene regulation than coding changes (Silvert et al. 2019). In single-nucleotide polymorphisms (SNPs) heterozygous for the Neanderthal allele, the allele-specific expression of introgressed alleles was significantly lower in the brain and testes relative to other tissues (McCoy et al. 2017). Expression differences between AMHs and Neanderthals have also been observed in genes related to skeletal, cardiovascular, and immune functions (Colbran et al. 2019). Several enhancer regions also show enrichment in Neanderthal alleles, such as adipose-related tissues and primary T cells, while others, including brain- and muscle-associated enhancers, demonstrate depletion (Telis et al. 2020). This is consistent with the differentiation between AMH and Neanderthal exomes in genes associated with pigmentation, skeletal morphology, and behavioral traits (Ahlquist et al. 2021). Moreover, 42 types of tissues in humans show significant enrichment of introgressed Neanderthal variants in enhancers. Among them, the highest rate of enrichment was identified in immune cells and adipose-related tissues (Silvert et al. 2019).

Finally, copy-number variations, fragments of the genome that vary in the number of times they are repeated in different people, also showed adaptive introgression from both Neanderthals and Denisovans near genes related to development, metabolism, and immunity (Hsieh et al. 2019). These results suggest that alleles, which helped admixed human populations to adapt to environmental pressures associated with high altitude (hypoxia tolerance), latitude and sun exposure (skin pigmentation), cold environments and dietary changes (lipid metabolism), and pathogens (immune response) increased in frequency after admixture, likely due to the important role they played in adaptation (Figure 6.4). Furthermore, most of these top candidate loci for adaptive introgression were not driven by the associative overdominance from the recessive deleterious mutations, and thus they may be the true adaptation signals (Zhang et al. 2020).

Functional Consequences of Adaptive Introgression from Neanderthal

There are many examples of local adaptation to diverse habitats driven by introgressed Neanderthal variants. Genes related to variation in skin pigmentation and hair morphology (*BNC2*, *MC1R*, *OCA2*) show the signals of positive selection acting on archaic variants as the result of adaptation to different degrees of insolation (Ahlquist et al. 2021). Local adaptation can be detected at the expression level as in the apelin receptor gene (*APLNR*). This gene plays a key role in early development such as gastrulation, blood vessel formation, and heart morphogenesis and regulates the cardiovascular and central nervous systems (CNSs) and oxygen levels. It shows a strong allele-specific expression of the Neanderthal allele in brain tissues, whereas in non-brain tissues it exhibits allele-specific expression favoring the modern human allele (McCoy et al. 2017). Similarly, several genes affecting keratin have been introgressed from Neanderthals into modern East Asians and Europeans, being driven by morphological adaptation in skin and hair to modern humans to deal with non-African environments (Sankararaman et al. 2014). The same is true for several genes involved in medically relevant phenotypes, such as those affecting primary biliary cirrhosis, systemic lupus erythematosus, optic disk size, Crohn's disease, interleukin levels, diabetes mellitus type 2, and smoking behavior (Sankararaman et al. 2014). More recently, McArthur et al. (2021) found that introgressed alleles were more associated with larger lung volumes, increased bone density, and increased likelihood of being a morning person. Some traits with heritability enrichment, like autoimmunity, have both genomic windows where Neanderthal alleles are associated with risk increase and other windows that associate with risk reduction. On the other hand, introgressed alleles have strong genome-wide unidirectional effects of protection from anorexia and schizophrenia. Even though Neanderthal variants contribute less to the heritability of these traits than expected, the introgressed alleles that remain are disproportionately risk-decreasing.

There are also a variety of examples of local adaptation driving regulatory variants, which result in population differences in immune responses (Dolgova and Lao 2018). Introgressed alleles that modify the expression of the *OAS1/OAS2/OAS3* genes are involved in innate immunity in different tissues (Abi-Rached et al. 2011). The archaic variants of *OAS* locus are related to diverse flavivirus resistance phenotypes. Further examples of local adaptation influencing levels of expression involve expression of the genes *CCR1*, limiting leukocyte recruitment and preventing inflammatory responses; *ERAP2*, affecting susceptibility to Crohn's disease; *HLA-DQA1*, associated with susceptibility to celiac disease; and *TLR1*, related to markedly lower levels of expression in inflammatory response genes (Nédélec et al. 2016). It was suggested that Neanderthal introgression also influenced the diversification of transcriptional responses to infection in human populations. European genomes carry archaic genetic segments that contain regulatory variants, which impact the steady-state expression and responses to influenza virus and *TLR7/8* stimulation (Quach and Quintana-Murci 2017). Moreover, the archaic variants of several expression quantitative trait loci (eQTLs) have been identified as potential candidates for adaptive introgression providing better adaptation through the regulation of gene expression. For example, the gene *DARS* is associated with neuroinflammatory and white matter disorders. It shows the strongest signatures of selection in archaic alleles, suggesting its possible protective nature. All these results indicate that archaic admixture and selection substantially impact modern interpopulation differences in immune responses, at least in terms of transcriptional variability.

Nevertheless, regardless of the potential benefits of archaic introgression in the past, alleles inherited from Neanderthals have been also related to several neurological, dermatological, and immunological phenotypes, indicating an association of ancient admixture with current disease risk in humans due to mismatches between past adaptations and modern environments (Dolgova and Lao 2018). Neanderthal SNPs explain a significant percentage of the risk in several traits: myocardial infarction, corns and callosities, actinic keratosis, depression, mood disorders, obesity, and seborrheic keratosis (Simonti et al. 2016); certain behavioral phenotypes including pain, chronotype/sleep, smoking, and alcohol consumption (Dannemann et al. 2022); and hair loss, younger age at menopause, and increased sunburn risk (McArthur et al. 2021). The enrichment for heritability of sunburn and tanning in introgressed alleles and the bias in the direction of effect in

AMH suggest that these introgressed alleles decreased hair and skin protection against sun exposure in ways that may have been beneficial, perhaps in response to decreased ultraviolet light at higher latitudes (McArthur et al. 2021).

Moreover, introgressed alleles associated with the immune system response can increase the risk of inflammation or autoimmunity under environmental factors changing over time (Corbett et al. 2018). There is evidence of Neanderthal introgression in the chemokine receptor (*CCR*) gene family (Taskent et al. 2017). Introgressed variants are the risk alleles for celiac disease, which were possibly maintained by selective forces in the early European population. Furthermore, population genetic analyses have shown that the high frequency of several risk alleles of genes such as *IL12A*, *IL18RAP*, and *SH2B3* associated with celiac disease in Europeans results from past positive selection events (Zhernakova et al. 2010). Another example comes from a nonsynonymous variant of the *ZNF365D* gene present in ~32% of Europeans, which was introgressed from Neanderthals, provoking a higher risk of Crohn's disease (Sankararaman et al. 2014). In the same way, introgressed alleles of gene cluster *TLR6-1-10*, which originated in Eurasians from interbreeding with Neanderthals and Denisovans, are related to greater susceptibility to allergies (Dannemann et al. 2016). Since the migration of modern human ancestors to Eurasia, sexual transmission from Neanderthals, Denisovans, or other hominin groups likely introduced certain human papillomavirus (HPV), herpesvirus, and ectoparasitic pathogens to the arriving group of modern human ancestors in Eurasia (reviewed in Pimenoff et al. 2018). These findings could explain, for example, an unusual aspect of HPV genetic diversity: the HPV16A variant virtually absent in sub-Saharan Africa and the most common outside Africa is likely introgressed from the Neanderthal population (Pimenoff et al. 2018). Finally, a more recent study identified Neanderthal inheritance in a gene cluster on chromosome 3, which is the major genetic risk factor for severe symptoms and hospitalization after severe acute respiratory syndrome coronavirus 2 (SARS-CoV-2) infection. The genomic segment of archaic ancestry is around 50 kilobases in size and is carried by around 50% of people in south Asia and around 16% of people in Europe (Zeberg and Pääbo 2020).

The results of these studies help to link genetic variants in a population to putative past beneficial effects, which likely only indirectly contribute to the pathology in present-day humans. A more detailed explanation of the consequences of Neanderthal introgression on brain and CNS functioning can be found below (see "Functional Consequences of Archaic Introgresssion in the Human Cranium and Brain-Related Traits").

Functional Consequences of Adaptive Introgression from Denisovan

The field of genomics is affected by a well-known Eurocentric bias, where European populations are more broadly sampled, and their genetic diversity is far better characterized, than other populations (Villanea and Witt 2022). Nevertheless, Europeans have some of the lowest proportions of both Neanderthal and Denisovan ancestry outside of Africa.

Therefore, a bigger potential for discovering novel adaptive archaic segments exists in non-European genomes (Villanea and Witt 2022). Thus, the majority of functional consequences of adaptive introgression from Denisovan ancestry are yet to be discovered.

The most compelling case of adaptive introgression of an archaic Denisovan allele to date is the high-altitude adaptation of Tibetans, achieved through selection for the archaic variants in the *EGLN1* and *EPAS1* genes. These genes are related to hemoglobin concentration and response to hypoxia (Zhang et al. 2021). The ancestral variant of *EPAS1* up-regulates hemoglobin levels to compensate for low oxygen levels—such as at high altitudes—but this is also associated with the trade-off of increasing blood viscosity (Zhang et al. 2021). On the other hand, the variant of Denisovan origin limits this increase in hemoglobin levels, thus resulting in a better altitude adaptation.

Another example of adaptive introgression detected in underrepresented populations is a Denisovan allele in *TBX15*, which is found at a high frequency in Greenlandic Inuit (Racimo et al. 2017). This Denisovan variant is also found in many Indigenous populations of the Americas at a high frequency (>0.8). *TBX15* is involved in many phenotypes, including lipid metabolism, especially at cold temperatures, which suggests that the Denisovan variant may have been advantageous for humans when they populated the Arctic (Racimo et al. 2017).

Additionally, Denisovan alleles may play a role in the immunity of western Asians. Two HLA-A (A*02 and A*11) and two HLA-C (C*15 and C*12:02) allotypes of Denisovan ancestry are common alleles in modern human populations, whereas one of the Denisovan's HLA-B allotypes constitutes a rare recombinant allele, while another is absent in modern humans (Abi-Rached et al. 2011). They were probably inherited by modern human ancestral populations from Denisovans as their independent preservation in both lineages is unlikely due to the high mutation rate of HLA alleles (Abi-Rached et al. 2011).

Functional Consequences of Archaic Introgression in the Human Cranium and Brain-Related Traits

Morphometric studies of hominin fossils have demonstrated substantial anatomical differences between the brains of Neanderthals and modern humans that may be consistent with divergent regulatory evolution targeting this organ. While overall brain size was similar, Neanderthal endocranial capacity was less than that of modern humans when adjusted for body size and the size of the visual system.

A recent study comparing European modern human high-coverage genomes with the Altai Neanderthal genome shows that archaic admixture is associated with several changes in the cranium and underlying brain morphology, suggesting changes in neurological function through Neanderthal-derived genetic variation (Gregory et al. 2017). Neanderthal variants are related to an extension of the posterolateral area of the modern human skull, expanding from the occipital and inferior parietal bones to bilateral temporal locales (Gregory et al. 2017). Regarding modern human brain morphology,

Neanderthal introgressed segments resulted in an increase in sulcal depth for the right intraparietal sulcus and high cortical complexity for the early visual cortex of the left hemisphere. Archaic admixture also produced an increase in the volume of white and gray matter localized in the right parietal region adjacent to the right intraparietal sulcus. In the primary visual cortex gyrification of the left hemisphere, the number of archaic alleles positively correlates with gray matter volume. The results also show evidence of a negative correlation between Neanderthal admixture and white matter volume in the orbitofrontal cortex, a structure that appears to have been laterally restricted in Neanderthals relative to modern *H. sapiens* (Gregory et al. 2017).

At the same time, brain regions had significantly lower expression of Neanderthal alleles than non-brain tissues, particularly in the neuron-rich cerebellum (BRNCHA) and basal ganglia regions (BRNCDT, BRNPTM, BRNNCC; McCoy et al. 2017). These brain regions have traditionally been associated with motor control and perception, but a broader role in cognitive function—including language processing—and behavior is now accepted. Interestingly, the cerebellum has undergone rapid expansion in the great ape lineage, and modern humans possess proportionally larger cerebella (greater cerebellum to total brain volume ratio) than did Neanderthals. This down-regulation suggests that modern humans and Neanderthals possibly experienced a relatively higher rate of divergence in these specific tissues (McCoy et al. 2017). One brain-specific gene that exemplifies this pattern of down-regulation is *NTRK2*, which encodes a neurotrophic tyrosine receptor kinase that regulates neuron survival and differentiation as well as synapse formation. This gene contains a pair of adjacent Neanderthal SNPs showing strong down-regulation signatures. Mutations and polymorphisms in this gene have been associated with a range of neuropsychiatric and neurological disorders including depression, suicide attempts, impaired speech and language development, severe obesity, autism, obsessive-compulsive disorder, Alzheimer's disease, anorexia nervosa, nicotine dependence, and pilocytic astrocytoma. Intriguingly, *NTRK2* is among a small set of brain-specific genes whose regulatory domains overlap signatures of modern human selective sweeps that occurred after divergence from Neanderthals. Furthermore, in Papuans assimilated Neanderthal alleles are found in the highest frequency in genes expressed in the brain, whereas Denisovan variants have the highest frequency in genes expressed in bones and other tissues (Akkuratov et al. 2018).

Neurological and psychiatric disorders showed the highest proportional Neanderthal DNA contribution (Simonti et al. 2016). Neanderthal alleles together explain a significant fraction of the variation in risk for depression, and individual Neanderthal alleles have been significantly associated with tobacco use and sleeping patterns (Dannemann and Kelso 2017). In some cases, such introgression-to-phenotype has been traced back to differences in tissue expression. For example, one of the identified introgressed SNPs (9.0% frequency in Europeans) is upstream of stromal interaction molecule 1 (*STIM1*). It was found that the Neanderthal allele was associated with a significantly decreased expression of *STIM1* in the caudate basal ganglia, a region of the brain related to bladder dysfunction, especially in connection with neurological impairments such as Parkinson's disease. Another archaic SNP was associated with tobacco use disorder with 0.5% frequency in Europeans in an intron of *SLC6A11*, a solute carrier family neurotransmitter transporter that is responsible for the reuptake of the neurotransmitter gamma-aminobutyric acid (GABA). Nicotine addiction reduces the expression of *SLC6A11*, disrupting GABAergic signaling in the brain. Furthermore, 29 Neanderthal SNPs were found to be significantly associated with brain *cis* eQTL in the cerebellum or temporal cortex. This represents significant enrichment for brain eQTL among Neanderthal SNPs as compared to the non-Neanderthal control SNPs (Simonti et al. 2016).

Noteworthy, when the Neanderthal associations with related but "non-direct disease" behavioral phenotypes were explored, signals became substantially stronger (Danneman et al. 2022). It was found that phenotypes defining the number of daily smoked cigarettes and smoking status were enriched for Neanderthal alleles (McArthur et al. 2021; Danneman et al. 2022). In addition, alcohol phenotypes characterizing regular intake frequencies for various alcoholic beverages, alcohol drinker status, and habit of consuming alcohol during meals were enriched for Neanderthal ancestry. Evolutionary origins have been postulated for addiction, suggesting a coevolution of the human brain, reward-seeking, and psychotropic substances (Saah 2005). An alternative—non-exclusive—hypothesis could be that this reflects self-medication for pain so that all of these associations (as well as the reported link with some pain medications) may be driven by the same selection processes (Dannemann et al. 2022).

Enrichment of Neanderthal alleles has also been identified in phenotypes related to chronotype and to sleep duration (McArthur et al. 2021; Danneman et al. 2022). A particularly extreme example was archaic SNPs in the region of chromosome 5 associated with chronotype and "getting up in the morning." This SNP reaches frequencies between 21.5% and 55.2% in present-day Europeans, South Asians, and native Americans, suggesting that they may have been positively selected at some point in the past (Dannemann et al. 2022). The archaic SNPs at this locus were associated with the modified expression of three genes (*GLRA1*, *LINC01933*, and *NMUR2*) in two brain regions and nerve tissue (Dannemann et al. 2022). Another case is regions near *NMUR2* linked to Neanderthal alleles that increase the propensity to be a morning person (McArthur et al. 2021).

Within the group of mental health phenotypes studied by genome-wide association studies (GWAS), the most notable archaic enrichment was linked to the length of a depressive episode, while the "longest period of unenthusiasm/disinterest" phenotype attained a substantially lower number of associations with archaic SNPs. On the individual GWAS level, only three tests for pain, including general pain, back pain, and knee pain, showed a larger number of significant archaic associations (Dannemann et al. 2022).

All in all, the enrichment of associations with traits such as chronotype, pain, alcohol use, and tobacco use rather than disease-related categories may thus reflect adaptations when Neanderthal variants could persist in modern humans due to their neutral or even potentially advantageous effects at some point during recent evolution (Dannemann et al. 2022). In contrast, traits related to eye structure, daily activities, and cognition were depleted. The depletion in cognitive traits suggests that the previously described strong reduction for Neanderthal alleles in regulatory regions expressed in the brain may be due to effects on brain-related complex traits such as cognition (McCoy et al. 2017; Rinker et al. 2020; Telis et al. 2020). Moreover, traits like anorexia and schizophrenia show depletion for heritability among introgressed variants (McArthur et al. 2021).

Gregory et al. (2021a), using a genome-wide measure of total "burden" of Neanderthal introgression, showed that patients with schizophrenia had less Neanderthal admixture than did controls in all ~9,500 individuals studied. Moreover, in 49 unmedicated patients with schizophrenia with more Neanderthal admixture, they predicted less severe positive symptoms, which included hallucinations and delusions. Exploring one potential mechanism for these effects, it was demonstrated that healthy individuals with a higher proportion of Neanderthal admixture had lower dopamine synthesis capacity in the striatum and pons, which is fundamentally important in the pathophysiology and treatment of psychosis. The symptoms in patients with higher scores of archaic admixtures responded less robustly to dopamine blockade with neuroleptics. At the same time, individuals with a greater degree of Neanderthal admixture had less severe psychotic symptoms when not receiving pharmacologic treatment (Gregory et al. 2021a). This suggests that dopaminergic mechanisms might be important in the relationship between schizophrenia and Neanderthal introgression. Moreover, it is interesting to speculate that the role of dopaminergic neurotransmission in the Neanderthal brain may have been different than in modern humans, particularly in relation to reward and pain processing. Overall, these results throw light on the evolutionary origin of schizophrenia as a condition specific to modern (as opposed to archaic) humans. This long-held theory was previously supported by the evidence that (1) the genetic risk for this disorder arose recently in hominid evolution after the divergence of Neanderthal and *H. sapiens*, (2) this genetic risk is enriched in human accelerated regions of our genome, and (3) there is a negative association between schizophrenia risk and metrics of Neanderthal selective sweep scores (Gregory et al. 2021a). Furthermore, another study examined whether specific SNPs with a high posterior probability of being derived from Neanderthals are associated with the heritability of schizophrenia risk. It did not find a significant relationship (Pardiñas et al. 2018).

Taken together, the results of all these studies suggest that Neanderthal variants are probably not directly linked to diagnostic entities of mental conditions but may have some indirect links, for example, via behavioral phenotypes such as smoking, pain, or sleep, which in turn are linked to diseases. Interestingly, genomic regions that differ most between Neanderthals and modern humans, including regions where no introgressed archaic DNA can be detected in people today, have previously been linked to brain-related genes (Dannemann et al. 2022). The consequences of Neanderthal admixture are of particular relevance for understanding uniquely human characteristics as such admixture biases the skull shape of living humans to more resemble Neanderthal fossil remains (Gregory et al. 2017), affects morphological characteristics of the modern human brain (Gregory et al. 2017, 2021b), is associated with neuronal gene expression (Simonti et al. 2016) and differential gene expression in brain tissues (McCoy et al. 2017), is overrepresented in genes known to be responsible for neurogenesis (Akkuratov et al. 2018), and is related to several behavioral phenotypes (McArthur et al. 2021; Dannemann et al. 2022). Considerable depletion of archaic heritability in brain-related variation, remarkable down-regulation of expression in genes affected by archaic introgression in brain tissues, along with the footprints of negative selection toward archaic-derived alleles in regions responsible for neurological and psychiatric traits indicate the extremely conservative nature of archaic allele inheritance in regions responsible for CNS development and brain functioning. The remains of archaic variation in the brain- and CNS-related traits in modern human populations are likely maintained mainly by the strong positive selection due to their protective nature with regard to behavioral and psychiatric impairments.

Unanswered Questions and Future Work

The sequencing data from thousands of individuals from different populations worldwide, including some archaic hominins and ancient AMH genomes, have offered novel insights into the evolutionary history of our species, demonstrating countless encounters between human populations. Genomic studies of introgression in early Eurasians from archaic human species, such as Neanderthals and Denisovans, provided a better understanding of the evolutionary and phenotypical consequences of such hybridization. The evidence for widespread selection against introgression across the genome is ubiquitous. Nevertheless, adaptive introgression can also be considered an important force driving the adaptation of modern humans to new environments.

However, the exact details of the number of admixture events, when and where they took place, between whom, and the related evolutionary and phenotypic consequences of such inbreeding—both for AMH and for the archaic population—remain contentious. Answering these questions requires a comprehensive and translational approach incorporating knowledge from different fields. Advances in archaeology on the understanding of the context of environmental and social changes during the time when AMH met archaic are becoming more accurate by combining classical studies of the biological composition of ancient biological remains from plants and animals that lived contemporaneously with humans with ancient DNA genomic analyses from that biological material (Ahlquist et al. 2021). Moreover, new paleontological techniques can explore questions of genetic inheritance from protein sequencing alone when DNA has already been degraded (Ahlquist et al. 2021). Such techniques would be useful to study archaic and modern population interactions in Southeast Asia, which has a long history of occupation by multiple archaic hominins, including *Homo erectus*, *H. floresiensis*, and *H. luzonensis* (Villanea and Witt 2022). Some of these hominins might have lived in the region simultaneously, and some likely coincided with humans, although admixture between these hominids and modern humans has yet to be identified (Teixeira et al. 2021).

Nevertheless, a key point to address these standing questions about hybridization is the need of retrieving the genetic variation of additional AMH and archaic genomes, particularly of Denisovans; the integration of different sources of information, such as studies of genetic variants with regulatory effects on gene expression (eQTL) across different types of tissues; and the development of new statistical and analytical methods. In particular, new data sets depicting a more comprehensive description of the human genetic variation, both present and past, are required to get an unbiased and fair picture of ancient times as current phenotypic, genomic, and gene expression analyses primarily use data sets of European individuals. This is particularly challenging in the case of African populations, where few modern individuals have been sampled and where archaic remains are shortened (Ahlquist et al. 2021). Nevertheless, African genetic variation represents a substantial amount of the whole genetic variability found in current humans, and sub-Saharan African populations are classically used as output groups for identifying introgressed regions in other populations. Similarly, if we want to increase our understanding of the evolutionary processes that shaped currently introgressed archaic fragments, a comprehensive temporal and spatial sampling of ancient samples is required (Ahlquist et al. 2021). With these longitudinal data, we expect to see an increase in methods that place ancient and archaic individuals in a relative temporal and geographic context and make better inferences about the structure of their populations, including gene flow from archaic and superarchaic "ghost" populations.

In addition to the generation of new data, the future challenge is to develop robust statistical models and computational methods for detecting selection, quantifying more widely the frequency of adaptive introgression, and understanding the circumstances where it played a predominant role in adaptation. Efficient inference from simulation techniques (Clemente et al. 2021) is opening up new lines of inquiry, integrating machine learning and natural computing with methods like approximate Bayesian computation, to perform inference without likelihood calculations, with or without predefined summary statistics, or based on raw data. Advances in existing inference and phylogenetic methods will be required to take full advantage of new large-scale and heterochronous samples.

Overall, while the last innovations and research into the legacy of archaic admixture have provided invaluable resources and increased our understanding of the molecular and cellular processes underlined by introgressed genetic variants, in the near future we expect even greater insight into the biomedical consequences of complex interactions between human species in the past.

Key Points to Remember

- The demographic history of AMH includes multiple migration events, population extinctions, genetic adaptations, and hybridization with archaic populations.
- Modern non-sub-Saharan African human genomes carry fragments of archaic origin associated with adaptive advantages as well as medical risks, both being mainly related to neurological, dermatological, and immunological phenotypes.
- The amount and nature of introgressed archaic fragments vary among the different human populations.
- As genome-wide data from complete genome sequencing become increasingly abundant and available even from extinct hominins, new insights into the consequences of archaic introgression on human health and disease are discovered.

References

Abi-Rached, L., Jobin, M. J., Kulkarni, S., McWhinnie, A., Dalva, K., Gragert, L., Babrzadeh, F., Gharizadeh, B., Luo, M., Plummer, F. A., Kumani, J., Carrington, M., Middleton, D., Rajalingam, R., Beksac, M., Marsh, S. G. E., Maiers, M., Guethlein, L. A., Tavoularis, S., . . . Parham, P. (2011). The shaping of modern human immune systems by multiregional admixture with archaic humans. *Science*, *334*(6052), 89–94. https://doi.org/10.1126/science.1209202

Ahlquist, K. D., Bañuelos, M. M., Funk, A., Lai, J., Rong, S., Villanea, F. A., & Witt, K. E. (2021). Our tangled family tree: New genomic methods offer insight into the legacy of archaic admixture. *Genome Biology and Evolution*, *13*(7), Article evab115. https://doi.org/10.1093/gbe/evab115

Akkuratov, E. E., Gelfand, M. S., & Khrameeva, E. E. (2018). Neanderthal and Denisovan ancestry in Papuans: A functional study. *Journal of Bioinformatics and Computational Biology*, *16*(2), Article 1840011. https://doi.org/10.1142/S0219720018400115

Browning, S. R., Browning, B. L., Zhou, Y., Tucci, S., & Akey, J. M. (2018). Analysis of human sequence data reveals two pulses of archaic Denisovan admixture. *Cell*, *178*(1), 53–61.e9. https://doi.org/10.1016/j.cell.2018.02.031

Clemente, F., Unterländer, M., Dolgova, O., Amorim, C. E. G., Coroado-Santos, F., Neuenschwander, S., Ganiatsou, E., Cruz Dávalos, D. I., Anchieri, L., Michaud, F., Winkelbach, L., Blöcher, J., Arizmendi Cárdenas, Y. O., Sousa da Mota, B., Kalliga, E., Souleles, A., Kontopoulos, I., Karamitrou-Mentessidi, G., Philaniotou, O., . . . Papageorgopoulou, C. (2021). The genomic history of the Aegean palatial civilizations. *Cell*, *184*(10), 2565–2586.e21. https://doi.org/10.1016/j.cell.2021.03.039

Colbran, L. L., Gamazon, E. R., Zhou, D., Evans, P., Cox, N. J., & Capra, J. A. (2019). Inferred divergent gene regulation in archaic hominins reveals potential phenotypic differences. *Nature: Ecology & Evolution*, *3*(11), 1598–1606. https://doi.org/10.1038/s41559-019-0996-x

Corbett, S., Courtiol, A., Lummaa, V., Moorad, J., & Stearns, S. (2018). The transition to modernity and chronic disease: Mismatch and natural selection. *Nature Reviews Genetics*, *19*(7), 419–430. https://doi.org/10.1038/s41576-018-0012-3

Dannemann, M., Andrés, A. M., & Kelso, J. (2016). Introgression of Neandertal- and Denisovan-like haplotypes contributes to adaptive variation in human Toll-like receptors. *American Journal of Human Genetics*, *98*(1), 22–33. https://doi.org/10.1016/j.ajhg.2015.11.015

Dannemann, M., & Kelso, J. (2017). The contribution of Neanderthals to phenotypic variation in modern humans. *American Journal of Human Genetics*, *101*(4), 578–589. https://doi.org/10.1016/j.ajhg.2017.09.010

Dannemann, M., Milaneschi, Y., Yermakovich, D., Stiglbauer, V., Kariis, H. M., Krebs, K., Friese, M. A., Otte, C., Estonian Biobank Research Team, Lehto, K., Penninx, B. W. J. H., Kelso, J., & Gold, S. M. (2022). Neandertal introgression partitions the genetic landscape of neuropsychiatric disorders and associated behavioral phenotypes. *Translational Psychiatry, 12*(1), Article 433. https://doi.org/10.1038/s41398-022-02196-2

Dolgova, O., & Lao, O. (2018). Evolutionary and medical consequences of archaic introgression into modern human genomes. *Genes, 9*(7), Article 358. https://doi.org/10.3390/genes9070358

Durvasula, A., & Sankararaman, S. (2020). Recovering signals of ghost archaic introgression in African populations. *Science Advances, 6*(7), Article eaax5097. https://doi.org/10.1126/sciadv.aax5097

Enard, D., & Petrov, D. A. (2018). Evidence that RNA viruses drove adaptive introgression between Neanderthals and modern humans. *Cell, 175*(2), 360–371.e13. https://doi.org/10.1016/j.cell.2018.08.034

Gouy, A., & Excoffier, L. (2020). Polygenic patterns of adaptive introgression in modern humans are mainly shaped by response to pathogens. *Molecular Biology and Evolution, 37*(5), 1420–1433. https://doi.org/10.1093/molbev/msz306

Green, R. E., Krause, J., Briggs, A. W., Maricic, T., Stenzel, U., Kircher, M., Patterson, N., Li, H., Zhai, W., Fritz, M. H., Hansen, N. F., Durand, E. Y., Malaspinas, A. S., Jensen, J. D., Marques-Bonet, T., Alkan, C., Prüfer, K., Meyer, M., Burbano, H. A., . . . Pääbo, S. A. (2010). A draft sequence of the Neandertal genome. *Science, 328*(5979), 710–722. https://doi.org/10.1126/science.1188021

Gregory, M. D., Eisenberg, D. P., Hamborg, M., Kippenhan, J. S., Kohn, P., Kolachana, B., Dickinson, D., & Berman, K. F. (2021a). Neanderthal-derived genetic variation in living humans relates to schizophrenia diagnosis, to psychotic symptom severity, and to dopamine synthesis. *American Journal of Medical Genetics. Part B, Neuropsychiatric Genetics, 186*(5), 329–338. https://doi.org/10.1002/ajmg.b.32872

Gregory, M. D., Kippenhan, J. S., Eisenberg, D. P., Kohn, P. D., Dickinson, D., Mattay, V. S., Chen, Q., Weinberger, D. R., Saad, Z. S., & Berman, K. F. (2017). Neanderthal-derived genetic variation shapes modern human cranium and brain. *Scientific Reports, 7*(1), Article 6308. https://doi.org/10.1038/s41598-017-06587-0

Gregory, M. D., Kippenhan, J. S., Kohn, P., Eisenberg, D. P., Callicott, J. H., Kolachana, B., & Berman, K. F. (2021b). Neanderthal-derived genetic variation is associated with functional connectivity in the brains of living humans. *Brain Connectivity, 11*(1), 38–44. https://doi.org/10.1089/brain.2020.0809

Hammer, M. F., Woerner, A. E., Mendez, F. L., Watkins, J. C., & Wall, J. D. (2011). Genetic evidence for archaic admixture in Africa. *Proceedings of the National Academy of Sciences of the United States of America, 108*(37), 15123–15128. https://doi.org/10.1073/pnas.1109300108

Hershkovitz, I., Weber, G. W., Quam, R., Duval, M., Grün, R., Kinsley, L., Ayalon, A., Bar-Matthews, M., Valladas, H., Mercier, N., Mercier, N., Arsuaga, J. L., Martinón-Torres, M., Bermúdez de Castro, J. M., Fornai, C., Martín-Francés, L., Sarig, R., May, H., Krenn, V. A., . . . Weinstein-Evron, M. (2018). The earliest modern humans outside Africa. *Science, 359*(6374), 456–459. https://doi.org/10.1126/science.aap8369

Hsieh, P., Vollger, M. R., Dang, V., Porubsky, D., Baker, C., Cantsilieris, S., Hoekzema, K., Lewis, A. P., Munson, K. M., Sorensen, M., Kronenberg, Z. N., Murali, S., Nelson, B. J., Chiatante, G., Maggiolini, F. A. M., Blanché, H., Underwood, J. G., Antonacci, F., Deleuze, J. F., & Eichler, E. E. (2019). Adaptive archaic introgression of copy number variants and the discovery of previously unknown human genes. *Science, 366*(6463), Article eaax2083. https://doi.org/10.1126/science.aax2083

Hublin, J. J., Ben-Ncer, A., Bailey, S. E., Freidline, S. E., Neubauer, S., Skinner, M. M., Bergmann, I., Le Cabec, A., Benazzi, S., Harvati, K., & Gunz, P. (2017). New fossils from Jebel Irhoud, Morocco and the pan-African origin of *Homo sapiens*. *Nature, 546*(7657), 289–292. https://doi.org/10.1038/nature22336

Huxley, T. (1890). The Aryan question and pre-historic man. In *Collected essays: Vol. 7. Man's place in nature*. Leopold Classic Library.

Kuhlwilm, M., Gronau, I., Hubisz, M. J., de Filippo, C., Prado-Martinez, J., Kircher, M., Fu, Q., Burbano, H. A., Lalueza-Fox, C., de la Rasilla, M., Rosas, A., Rudan, P., Brajkovic, D., Kucan, Ž., Gušic, I., Marques-Bonet, T., Andrés, A. M., Viola, B., Pääbo, S., . . . Castellano, S. (2016). Ancient gene flow from early modern humans into eastern Neanderthals. *Nature, 530*(7591), 429–433. https://doi.org/10.1038/nature16544

Lahr, M. M., Petraglia, M., Stokes, S., & Field, J. (2010). *Searching for traces of the Southern Dispersal: Environmental and historical research on the evolution of human diversity in southern Asia and Australo-Melanesia*. Archaeology Data Service. https://doi.org/10.5284/1000109

Lalueza-Fox, C. (2021). Neanderthal assimilation? *Nature: Ecology & Evolution*, 5(6), 711–712. https://doi.org/10.1038/s41559-021-01421-3

Larena, M., McKenna, J., & Sanchez-Quinto, F. (2021). Philippine Ayta possess the highest level of Denisovan ancestry in the world. *Current Biology*, 31(19), 4219–4230.e10. https://doi.org/10.1016/j.cub.2021.07.022

Lorente-Galdos, B., Lao, O., Serra-Vidal, G., Santpere, G., Kuderna, L. F. K., Arauna, L. R., Fadhlaoui-Zid, K., Pimenoff, V. N., Soodyall, H., Zalloua, P., Marques-Bonet, T., & Comas, D. (2019). Whole-genome sequence analysis of a pan African set of samples reveals archaic gene flow from an extinct basal population of modern humans into sub-Saharan populations. *Genome Biology*, 20(1), Article 77. https://doi.org/10.1186/s13059-019-1684-5

McArthur, E., Rinker, D. C., & Capra, J. A. (2021). Quantifying the contribution of Neanderthal introgression to the heritability of complex traits. *Nature Communications*, 12(1), Article 4481. https://doi.org/10.1038/s41467-021-24582-y

McCoy, R. C., Wakefield, J., & Akey, J. M. (2017). Impacts of Neanderthal-introgressed sequences on the landscape of human gene expression. *Cell*, 168(5), 916–927.e12. https://doi.org/10.1016/j.cell.2017.01.038

Mondal, M., Bertranpedt, J., & Lao, O. (2019). Approximate Bayesian computation with deep learning supports a third archaic introgression in Asia and Oceania. *Nature Communications*, 10(246), Article 246. https://doi.org/10.1038/s41467-018-08089-7

Nédélec, Y., Sanz, J., Baharian, G., Szpiech, Z. A., Pacis, A., Dumaine, A., Grenier, J. C., Freiman, A., Sams, A. J., Hebert, S., Pagé, S. A., Luca, F., Blekhman, R., Hernandez, R. D., Pique-Regi, R., Tung, J., Yotova, V., & Barreiro, L. B. (2016). Genetic ancestry and natural selection drive population differences in immune responses to pathogens. *Cell*, 167(3), 657–669.e21. https://doi.org/10.1016/j.cell.2016.09.025

Nielsen, R., Akey, J. M., Jakobsson, M., Pritchard, J. K., Tishkoff, S., & Willerslev, E. (2017). Tracing the peopling of the world through genomics. *Nature*, 541(7637), 302–310. https://doi.org/10.1038/nature21347

Pardiñas, A. F., Holmans, P., Pocklington, A. J., Escott-Price, V., Ripke, S., Carrera, N., Legge, S. E., Bishop, S., Cameron, D., Hamshere, M. L., Han, J., Hubbard, L., Lynham, A., Mantripragada, K., Rees, E., MacCabe, J. H., McCarroll, S. A., Baune, B. T., Breen, G., . . . Walters, J. T. R. (2018). Common schizophrenia alleles are enriched in mutation-intolerant genes and in regions under strong background selection. *Nature Genetics*, 50(3), 381–389. https://doi.org/10.1038/s41588-018-0059-2

Pimenoff, V. N., Houldcroft, C. J., Rifkin, R. F., & Underdown, S. (2018). The role of aDNA in understanding the coevolutionary patterns of human sexually transmitted infections. *Genes*, 9(7), Article 317. https://doi.org/10.3390/genes9070317

Prüfer, K., de Filippo, C., Grote, S., Mafessoni, F., Korlević, P., Hajdinjak, M., Vernot, B., Skov, L., Hsieh, P., Peyrégne, S., Reher, D., Hopfe, C., Nagel, S., Maricic, T., Fu, Q., Theunert, C., Rogers, R., Skoglund, P., Chintalapati, M., . . . Pääbo, S. (2017). A high-coverage Neandertal genome from Vindija Cave in Croatia. *Science*, 358(6363), 655–658. https://doi.org/10.1126/science.aao1887

Prüfer, K., Munch, K., Hellmann, I., Akagi, K., Miller, J. R., Walenz, B., Koren, S., Sutton, G., Kodira, C., Winer, R., Knight, J. R., Mullikin, J. C., Meader, S. J., Ponting, C. P., Lunter, G., Higashino, S., Hobolth, A., Dutheil, J., Karakoç, E., . . . Pääbo, S. (2012). The bonobo genome compared with the chimpanzee and human genomes. *Nature*, 486(7404), 527–531. https://doi.org/10.1038/nature11128

Quach, H., & Quintana-Murci, L. (2017). Living in an adaptive world: Genomic dissection of the genus *Homo* and its immune response. *Journal of Experimental Medicine*, 214(4), 877–894. https://doi.org/10.1084/jem.20161942

Racimo, F., Gokhman, D., Fumagalli, M., Ko, A., Hansen, T., Moltke, I., Albrechtsen, A., Carmel, L., Huerta-Sánchez, E., & Nielsen, R. (2017). Archaic adaptive introgression in TBX15/WARS2. *Molecular Biology and Evolution*, 34(3), 509–524. https://doi.org/10.1093/molbev/msw283

Reich, D., Green, R. E., Kircher, M., Krause, J., Patterson, N., Durand, E. Y., Viola, B., Briggs, A. W., Stenzel, U., Johnson, P. L., Maricic, T., Good, J. M., Marques-Bonet, T., Alkan, C., Fu, Q., Mallick, S., Li, H., Meyer, M., Eichler, E. E., . . . Pääbo, S. (2010). Genetic history of an archaic hominin group from Denisova Cave in Siberia. *Nature*, 46(7327), 1053–1060. https://doi.org/10.1038/nature09710

Rinker, D. C., Simonti, C. N., McArthur, E., Shaw, D., & Hodges, E. (2020). Neanderthal introgression reintroduced functional alleles lost in the human out of Africa bottleneck. *Nature: Ecology & Evolution*, 4(10), 1332–1341. https://doi.org/10.1038/s41559-020-1261-z

Rotival, M., Quach, H., & Quintana-Murci, L. (2019). Defining the genetic and evolutionary architecture of alternative splicing in response to infection. *Nature Communications, 10*(1), Article 1671. https://doi.org/10.1038/s41467-019-09689-7

Saah, T. (2005). The evolutionary origins and significance of drug addiction. *Harm Reduction Journal, 2*, Article 8. https://doi.org/10.1186/1477-7517-2-8

Sánchez-Quinto, F., Botigué, L. R., Civit, S., Arenas, C., Ávila-Arcos, M. C., Bustamante, C. D., Comas, D., & Lalueza-Fox, C. (2012). North African populations carry the signature of admixture with Neandertals. *PLOS ONE, 7*(10), Article e47765. https://doi.org/10.1371/journal.pone.0047765

Sankararaman, S., Mallick, S., Dannemann, M., Prüfer, K., Kelso, J., Pääbo, S., Patterson, N., & Reich, D. (2014). The genomic landscape of Neanderthal ancestry in present-day humans. *Nature, 507*(7492), 354–357. https://doi.org/10.1038/nature12961

Sankararaman, S., Mallick, S., Patterson, N., & Reich, D. (2016). The combined landscape of Denisovan and Neanderthal ancestry in present-day humans. *Current Biology, 26*(9), 1241–1247. https://doi.org/10.1016/j.cub.2016.03.037

Silvert, M., Quintana-Murci, L., & Rotival, M. (2019). Impact and evolutionary determinants of Neanderthal introgression on transcriptional and post-transcriptional regulation. *American Journal of Human Genetics,104*(6), 1241–1250. https://doi.org/10.1016/j.ajhg.2019.04.016

Simonti, C. N., Vernot, B., Bastarache, L., Bottinger, E., Carrell, D. S., Chisholm, R. L., Crosslin, D. R., Hebbring, S. J., Jarvik, G. P., Kullo, I. J., Li, R., Pathak, J., Ritchie, M. D., Roden, D. M., Verma, S. S., Tromp, G., Prato, J. D., Bush, W. S., Akey, J. M., . . . Capra, J. A. (2016). The phenotypic legacy of admixture between modern humans and Neandertals. *Science, 351*(6274), 737–741. https://doi.org/10.1126/science.aad2149

Skov, L., Coll Macià, M., Sveinbjörnsson, G., Mafessoni, F., Lucotte, E. A., Einarsdóttir, M. S., Jonsson, H., Halldorsson, B., Gudbjartsson, D. F., Helgason, A., Schierup, M. H., & Stefansson, K. (2020). The nature of Neanderthal introgression revealed by 27,566 Icelandic genomes. *Nature, 582*(7810), 78–83. https://doi.org/10.1038/s41586-020-2225-9

Slon, V., Mafessoni, F., Vernot, B., de Filippo, C., Grote, S., Viola, B., Hajdinjak, M., Peyrégne, S., Nagel, S., Brown, S., Douka, K., Higham, T., Kozlikin, M. B., Shunkov, M. V., Derevianko, A. P., Kelso, J., Meyer, M., Prüfer, K., & Pääbo, S. (2018). The genome of the offspring of a Neanderthal mother and a Denisovan father. *Nature, 561*(7721), 113–116. https://doi.org/10.1038/s41586-018-0455-x

Stringer, C. B., & Andrews, P. (1988). Genetic and fossil evidence for the origin of modern humans. *Science, 239*(4845), 1263–1268. https://doi.org/10.1126/science.3125610

Taskent, R. O., Alioglu, N. D., Fer, E., Melike Donertas, H., Somel, M., & Gokcumen, O. (2017). Variation and functional impact of Neanderthal ancestry in western Asia. *Genome Biology and Evolution, 9*(12), 3516–3524. https://doi.org/10.1093/gbe/evx216

Teixeira, J. C., Jacobs, G. S., Stringer, C., Tuke, J., Hudjashov, G., Purnomo, G. A., Sudoyo, H., Cox, M. P., Tobler, R., Turney, C. S. M., Cooper, A., & Helgen, K. M. (2021). Widespread Denisovan ancestry in Island Southeast Asia but no evidence of substantial super-archaic hominin admixture. *Nature: Ecology & Evolution, 5*(5), 616–624. https://doi.org/10.1038/s41559-021-01408-0

Telis, N., Aguilar, R., & Harris, K. (2020). Selection against archaic hominin genetic variation in regulatory regions. *Nature: Ecology & Evolution, 4*(11), 1558–1566. https://doi.org/10.1038/s41559-020-01284-0

Vernot, B., Tucci, S., Kelso, J., Schraiber, J. G., Wolf, A. B., Gittelman, R. M., Dannemann, M., Grote, S., McCoy, R. C., Norton, H., Scheinfeldt, L. B., Merriwether, D. A., Koki, G., Friedlaender, J. S., Wakefield, J., Pääbo, S., & Akey, J. M. (2016). Excavating Neandertal and Denisovan DNA from the genomes of Melanesian individuals. *Science, 352*(6282), 235–239. https://doi.org/10.1126/science.aad9416

Villanea, F. A., & Witt, K. E. (2022). Underrepresented populations at the archaic introgression frontier. *Frontiers in Genetics, 13*, Article 821170. https://doi.org/10.3389/fgene.2022.821170

Wohns, A. W., Wong, Y., Jeffery, B., Akbari, A., Mallick, S., Pinhasi, R., Patterson, N., Reich, D., Kelleher, J., & McVean, G. (2022). A unified genealogy of modern and ancient genomes. *Science, 375*(6583), Article eabi8264. https://doi.org/10.1126/science.abi8264

Zeberg, H., & Pääbo, S. (2020). The major genetic risk factor for severe COVID-19 is inherited from Neanderthals. *Nature, 587*(7835), 610–612. https://doi.org/10.1038/s41586-020-2818-3

Zhang, X., Kim, B., Lohmueller, K. E., & Huerta-Sanchez, E. (2020). The impact of recessive deleterious variation on signals of adaptive introgression in human populations. *Genetics*, *215*(3), 799–812. https://doi.org/10.1534/genetics.120.303081

Zhang, X., Witt, K. E., Bañuelos, M. M., Ko, A., Yuan, K., Xu, S., Nielsen, R., & Huerta-Sanchez, E. (2021). The history and evolution of the Denisovan-EPAS1 haplotype in Tibetans. *Proceedings of the National Academy of Sciences of the United States of America*, *118*(22), Article e2020803118. https://doi.org/10.1073/pnas.2020803118

Zhernakova, A., Elbers, C. C., Ferwerda, B., Romanos, J., Trynka, G., Dubois, P. C., de Kovel, C. G. F., Franke, L., Oosting, M., Barisani, D., Bardella, M. T., Finnish Celiac Disease Study Group, Joosten, L. A., Saavalainen, P., van Heel, D. A., Catassi, C., Netea, M. G., & Wijmenga, C. (2010). Evolutionary and functional analysis of celiac risk loci reveals SH2B3 as a protective factor against bacterial infection. *American Journal of Human Genetics*, *86*(6), 970–977. https://doi.org/10.1016/j.ajhg.2010.05.004

7

Goal-Directed and Habitual Behaviors
Anatomical and Functional Circuits in Health and Neurological Disease

Ledia F. Hernandez and Ignacio Obeso

> We are what we repeatedly do. Excellence, then, is not an act, but a habit.
> Aristotle cited by W. Durant, p. 87

> The more of the details of our daily life we can hand over to the effortless custody of automatism, the more our higher powers of mind will be set free for their own proper work.
> W. James (1890, p. 122)

Historical Background

Over a century ago, Thorndike's cat was confined in a cage until it pressed against a peddle which opened the cage door, giving the animal access to a piece of fish (Thorndike 1911). With repeated trials the animal gradually learned what it had to do so that, when placed in the cage, it was able to select the newly acquired act of peddle pressing and gain immediate access to the fish. This first formal demonstration of operant conditioning exemplifies reinforcement-driven action acquisition where an unexpected sensory reinforcer (the cage door opening) enabled relevant neural systems to select a relevant stimulus on which the cat will operate (the peddle) and allowed the animal to converge on the causal aspects of its behavior (the peddle press). As one of the vertebrate brain's basic computational units (Gerfen and Wilson 1996), the basal ganglia have been associated repeatedly (see the following section) with the fundamental processes of selection and reinforcement required for the type of behavioral modification that occurs in Thorndike's task.

Fundamental Behavioral Selection Mechanisms

Tying a shoelace, driving a car, handwriting, and so on are a few examples of instrumental behaviors acquired through two fundamental behavioral selection mechanisms. One is the goal-directed system (GDS), where actions (loops of the shoelace, fingers, etc.) are selected to obtain a predicted outcome or goal (tying a shoe). This kind of behavior is sensitive to the value of predicted outcome (making a loop at this juncture will likely be better than pulling the ends tight). If the action achieves the desired outcome, it will be more likely to be selected in future, and if not, the subject will look for another approach to reach the goal. As a form of exploratory learning, this kind of action acquisition is slow and requires conscious effort, attention, and repetition, leading to consolidation (Figure 7.1, Table 7.1).

When a particular task is repeated, a second system, the habitual system (HS), gradually takes over and the actions become automatized as a habit. When this occurs, the subject can automatically activate the individual task components without conscious voluntary intervention (McNamee et al. 2015). This process is less variable, is faster, and can

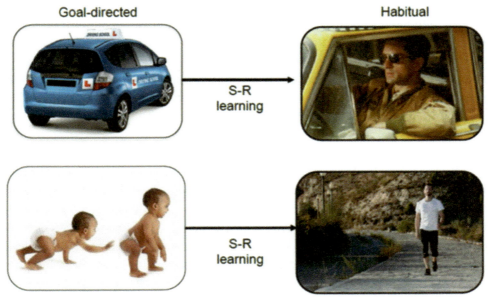

Figure 7.1 Depictions of goal-directed actions that start as stimulus–response (S-R) associations that become habitual with extended use and training.

TABLE 7.1 Characteristics of Goal-Directed and Habitual Behavior

Goal-directed	Habitual
Slow	Fast
Effortful	Automatic
Deliberate	Stereotypical
Flexible	Computationally efficient
Aware of outcome	Unaware of outcome
Sensitive to outcome value	Insensitive to motivational value
Enables planning	Slow to learn

be performed simultaneously with separate goal-directed behaviors, provided the same muscle groups are not used. The main difference between goal-directed and habitual control is that the former occurs predominantly at the early stages of learning and early in life (childhood; Figure 7.1, Table 7.1). The sequences of actions and individual movements required to achieve a particular goal early in learning are unclear, and conscious goal-directed effort is necessary.

However, "practice makes perfect," and even complicated motor sequences (e.g., balancing and walking) become automatic. Once a behavior becomes automatic, it is usually quite resistant to changes and is inflexible, even when the association between action and desired outcome no longer applies (e.g., pressing the elevator button to the floor of your old office, rather than the new one). Humans regularly combine automatic (habitual, HS-based) and goal-directed (voluntary, GDS-based) aspects of behavioral control simultaneously. Habitual routines allow us to perform well-practiced task components with minimal or no voluntary attention, which, from an evolutionary point of view, is very advantageous. The GDS can be freed to negotiate less predictable aspects of the world. However, habits can be hard to break. Reliance on the automatic HS is inflexible such that, in the presence of a triggering stimulus, the associated response will perseverate, even if it produces an outcome that is no longer desired. Perhaps it is this mechanism, operating throughout a lifetime, which makes it difficult for "old dogs to learn new tricks."

Why in brain evolution have the GDS and HS, as in integrated behavioral control systems, been developed? The simple answer is that as certain aspects of the world move from unpredictable and new to predictable and routine goals, they can be strategically achieved by stimulus–response routines that require no conscious voluntary intervention. To have such behaviors taken over by an automatic HS frees the conscious GDS to operate on higher-level behavioral planning, the kind that is required to set and achieve goals when situations are less predictable. This dual system allows overall behavior to be adaptive but also fast and automatic when needed, thereby preserving time, energy, and safety. Humans rely on automatic behavior to navigate daily routines, such as waking up, having coffee, and driving to work; and it is probably this ability that allows humans to

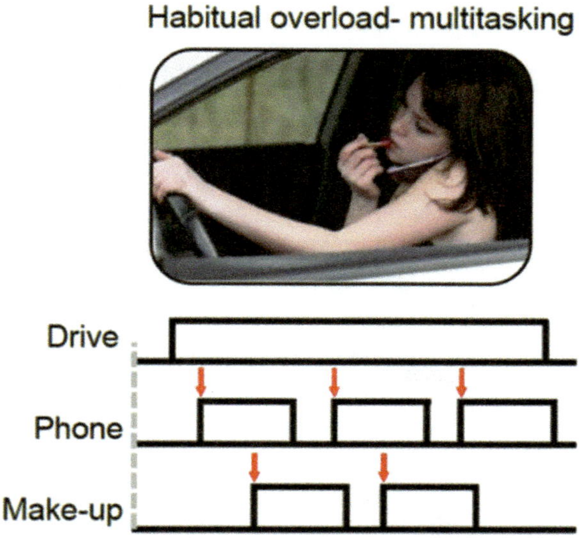

Figure 7.2 Habitual overload hypothesis. Diagram showing chunks of individualized actions simultaneously performed. Chunks require online goal-directed control momentarily (represented by the arrows) without altering automatic behavior.

simultaneously perform several behaviors and optimize time and resources (Figure 7.2). From the energy consumption point of view, it is also more efficient since fewer brain areas and neuronal networks are involved in the simple stimulus–response associations that characterize habitual behavior, thereby making goal achievement less demanding from a metabolic point of view. In fact, the transitional changes from the GDS to the HS occur throughout a human's lifetime, where experience allows more and more routine behaviors to be controlled by the HS (Figure 7.3). However, such transitions could be a double-edged sword: in situations where there may be an overreliance on automatic habits, components of the HS control system may suffer from overuse and start to malfunction. This maladaptation may be operative in the case of Parkinson's disease, discussed later (Hernandez et al. 2019). Hence, the evolutionary advantage of GDS and HS interactions is to have dual control mechanisms specialized to deal with aspects of the world that require conscious intervention to select adaptive responses and reject or override those that are not safe or efficient. Conversely, pathological conditions will be considered as the result of breakdown in the strategic balance between these two networks.

Goal-Directed and Habitual Behaviors from an Evolutionary Point of View

The GDS, where actions are selected to obtain a specific goal, is sensitive to outcome, slow and effortful, but flexible. If that behavior or particular task is repeated several times, it becomes automatized and becomes a habit. This transposition process can occur without

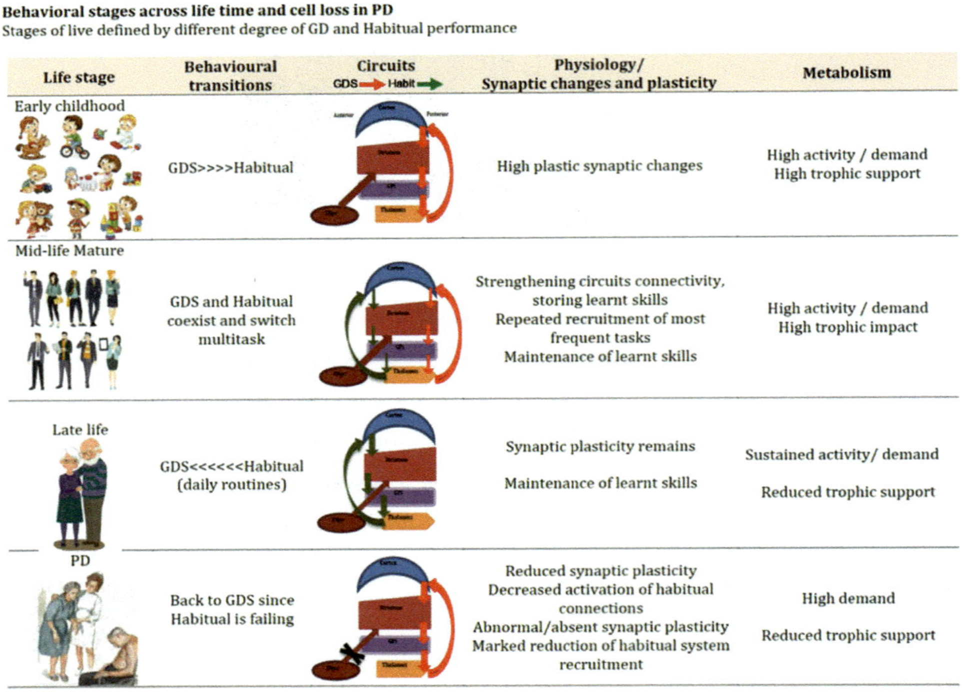

Behavioral stages across life time and cell loss in PD
Stages of live defined by different degree of GD and Habitual performance

Life stage	Behavioural transitions	Circuits GDS → Habit	Physiology/ Synaptic changes and plasticity	Metabolism
Early childhood	GDS>>>>Habitual		High plastic synaptic changes	High activity / demand High trophic support
Mid-life Mature	GDS and Habitual coexist and switch multitask		Strengthening circuits connectivity, storing learnt skills Repeated recruitment of most frequent tasks Maintenance of learnt skills	High activity / demand High trophic impact
Late life	GDS<<<<<<Habitual (daily routines)		Synaptic plasticity remains Maintenance of learnt skills	Sustained activity/ demand Reduced trophic support
PD	Back to GDS since Habitual is failing		Reduced synaptic plasticity Decreased activation of habitual connections Abnormal/absent synaptic plasticity Marked reduction of habitual system recruitment	High demand Reduced trophic support

Figure 7.3 Behavioral stages across lifetime and cell loss in Parkinson's disease. Stages of life defined by different degree of goal-directed and habitual performance. GD = goal-directed; GDS = goal-directed system; PD = Parkinson's disease.

voluntary concentration and effort, and ultimately the habit controlled by the HS is both less variable and faster. Obviously, from the evolutionary point of view, each kind of behavior has its advantages and drawbacks; but overall, the species that better adapt to the environment are the ones with more chances to survive, so relying on the HS may be efficient but must also carry the capacity to break and shift back to the GDS, in order to adapt to a changing environment. This plasticity or compensatory shift may be one of the reasons behind the observation that humans more easily break habits if those habits fail to bring the desired outcome or if a new option is presented, in which case, the HS shifts back to the GDS to adapt quickly to changes and establish new and optimized routines (new habits), shifting again to the HS when needed. Additionally, during the human life span, there is a gradual shift in the dominance of the GDS or HS with age. The different repertoires of needs and behaviors perpetually evolve with the explorative and learning (GDS-based) behaviors more predominant at the early stages of life (childhood and early youth). With normal aging, daily routine, lifestyles, and jobs that are habit-based, the HS dominates to navigate and perform over-learned routines in the most efficient manner, optimizing time and resources with less metabolic demand.

Theories for Goal-Directed and Habitual Behaviors

Since the origins of research on goal-directed and habitual behaviors, philosophers, psychologists, engineers, and neuroscientists have collaborated to develop the best tools for assessing whether an observed behavior is under GDS or HS control. A complication in the study of habits is the likely overlap with a wide range of other psychological concepts used to describe automatic habitual responses. These include skills, automatic cognition, routines and autonomous, inflexible and unconscious behavior (West and Brown 2013). Traditionally, and in line with the current view, the formal definition of habits relies on stimulus–response association that builds with repetition and is not modulated by outcome expectations (Dickinson 1985). The central element of this perspective is focused on the triggering stimulus that habitually evokes the response with which, through repetition, it has become associated. However, habitual behaviors are also consistent with other definitions (Gardner 2015) that place larger weight on statistical regularities of a contextual cue and the necessity of task repetition, thereby reducing the relevance of outcome expectations. More recent analyses give credit to goals and motivation as reinforcers of habitual performance (Wood et al. 2022). However, despite inconsistencies and discrepancies between the previous studies of habitual behavior, possibly due to the use of different descriptive terminologies (De Houwer 2019), there has been a high degree of conceptual overlap in the investigation of automatic habits that rely on stimulus–response associations established through repetition.

On the other hand, instrumental goal-directed behavior is oriented toward a particular objective. For an action to be considered goal-directed, selection has to be based on the relative values of predicted response outcomes. Thus, what is learned is the causal relationship between an action or response and its outcome, without being mediated by predictive relationships between environmental stimuli and the outcome (de Wit and Dickinson 2009). This is in contrast with Pavlovian conditioning procedures as they usually study predictive stimulus–outcome learning in animals. Here, a classically conditioned response is acquired to a stimulus that predicts the occurrence of a biologically relevant outcome. Typically, Pavlovian responses are adaptations to the environment in which they evolved, and such behavior can be maladaptive when the contingencies change.

The capacity to select actions instrumental in achieving desired goals (goal-directed behavior) is a fundamental evolutionary acquisition of animals' adaptive flexibility. Goal-directed behavior relies on two capabilities (Balleine and Dickinson 1998): first, the capacity to anticipate action outcomes and, second, the capacity to choose among different anticipated outcomes depending on their current value, based on the reward itself and the subject's motivational state.

Transition from Goal-Directed to Habitual Behavior: The Importance of Training

Although instrumental behavior where a particular outcome is obtained after a voluntary action, may start out as goal-directed, after repeated practice behavioral control gradually transitions into automatic habitual control. With this transition, the same action is performed (e.g., lever press), except it is now under automatic stimulus–response control, faster, and in most cases, more accurate. When automatic, the parameters of the movements continue to be refined, making them more efficient in terms of speed and energy expended. Given that simply observing an instrumental action may not be sufficient to determine whether a particular performance is being executed via goal-directed or habitual control, strategies must be devised to distinguish them. The most accepted test to differentiate the two behavioral control systems is to devaluate the outcome and examine whether the behavior continues (Dickinson 1985). Behavioral response is assessed during extinction (i.e., a non-reinforced probe test) to ensure the observed response is not influenced by new learning about outcome value (e.g., the reduced value of obtained food after having been pre-fed). This assessment demonstrates whether the outcome (goal) is part of the structure guiding behavior. Thus, if the behavior is goal-directed, responding will be reduced if the anticipated outcome is no longer desired or goal-linked. Conversely, if the behavior persists, it is deemed to be under stimulus control and independent from the goal representation, in which case it is classified as a habit. An additional approach to distinguish whether a behavior is under GDS or HS control relies on the strategy of contingency degradation. In this case, the association between the response and the outcome is degraded. Thus, the contingency is completely degraded when half the responses are rewarded and half the rewards are provided for free. In this situation the rate of reward is the same whether the subject responds or not. As with the outcome devaluation test, if responding is reduced, it is under GDS control; and if the behavior persists, then it is considered a stimulus–response habit. Behaviors such as learned motor responses, motor skills, procedural learning, and action-sequence learning have the overlapping feature of automaticity, share neural circuitry with habits (see following section), and sometimes may be defined as habits. Goal-directed behavior shares features with model-based reinforcement learning, where the behavior is driven by the prospective consideration of potential outcomes, while habits have been established to be akin to model-free reinforcement learning, where the behavior is driven by retrospective stimulus events (Daw et al. 2011).

Suppressing a habit is behaviorally demanding but not always physiologically challenging or energy-demanding. Paraphrasing Alexander Bain, William James stressed that in "the acquisition of a new habit, or the leaving off of an old one, we must take care to *launch ourselves with as strong and decided an initiative as possible*" (James, 1892, p. 145). Hence, the outcome devaluation or contingency degradation tests set up a situation where it is necessary to inhibit an acquired habit and revert to goal-directed efforts.

Yet, human experimental research shows that the cognitive capacity required to reverse acquired habits is relatively simple and thus constitutes a non-optimal measure of stimulus–response control. Possibly, humans are more flexible compared to other animals when overriding habits or require longer training protocols to reach habitual performance levels. Indeed, current views on the use of outcome devaluation procedures in humans have failed to replicate similar behavioral effects of task contingencies after training (de Wit et al. 2018). This may indicate that the methods used in humans are not appropriate to identify habitual operations.

To overcome the differences with animal behavioral protocols, research in humans needs to develop novel tools to distinguish goal-directed and habitual behavioral control. The imposition of time pressure in a visuomotor association task involving parallel cognitive processes (dual task; Hardwick et al. 2019) revealed a clear habitual dominance. Thus, speed costs seem to be more reliable as a habit marker using a reversal test where habits are no longer required (Luque et al. 2020). Recent accounts also investigate everyday-life routines in a controlled context, hence bringing established human habits into the experimental laboratory. Many behaviors experienced in modern life, including aspects of driving, eating, dancing, reading, talking, or walking, have significant stimulus–response components that can be performed automatically without thought while the person's conscious attention is directed elsewhere (Worringer et al. 2019). Such components have been acquired through frequent repetition throughout a lifetime of everyday trials. Such fully formed habitual associations can be investigated in the laboratory and can be shown to be independent of any new learning. To this, long established traffic-light color associations have been used to measure habitual initiation (green–go) and inhibition of actions (red–stop) in humans (Ceceli et al. 2020). Such behaviors with significant automatic stimulus–response load can be performed in the laboratory without further training. Examples of such tasks include reading, where comparisons are made between real words of different familiarity and emotional content, foreign words, and pseudo-words. Moreover, a large literature on brain mechanisms involved in walking and driving in humans is available. A recent meta-analysis on brain activation evoked by these behaviors proposed a consistent mapping of the habitual network in everyday-life activities (Guida et al. 2022). Experimental (two-stage reinforcement learning, probabilistic learning, or outcome devaluation tasks) and real-life activities (driving, walking, reading, and writing) with strong habitual components revealed a common mapping of activation in striatal subregions of the basal ganglia. However, the real-life tasks engaged larger posterior putamen volumes compared to habitual responding in experimental studies. These results suggest that (1) experimental acquired habits produce similar patterns of brain activation to those acquired in daily life and (2) sensorimotor regions of the posterior putamen are more engaged by real-life habitual tasks. Such research strategies can be developed further to investigate the neural mechanisms and circuits responsible for GDS and HS control in humans.

Goal-Directed and Habitual Neural Networks from an Evolutionary Perspective

The brain has developed parallel neural hubs dedicated for goal-directed and habitual behaviors. From the evolutionary perspective, the brain circuitry associated with these behaviors has been highly conserved across vertebrates (Suryanarayana et al. 2021).

In early learning stages, humans show prefrontal cortex (PFC) activities together with anterior sections of the striatum, including the anterior putamen and caudate, that are critical to reach expert levels in a novel task (Tricomi et al. 2009). Meanwhile, as behavior is consolidated and stabilized, the sensorimotor cortex and posterior striatum take control over habitual behavior (Tricomi et al. 2009). A similar transition occurs in animal studies (see Balleine and Odoherty 2010 for a comparison), where there is a different nomenclature for the equivalent areas in humans and other primates, so the PFC in rodents includes not only features of the medial and orbital areas in primates but also some features of the primate dorsolateral PFC (Uylings et al. 2003). Similarly, in the basal ganglia, the striatum in rodents is differentiated in the caudate and putamen in primates, separated by the internal capsule.

This section will cover brain-related areas associated with both goal-directed and habitual behaviors in human and non-human species.

The Basal Ganglia: From Goal-Directed to Habitual Behavior

A critical aspect of the neural basis for action is the evaluation of its outcome. There is a vast bibliography that points to the basal ganglia (and the habenula) within which the dopamine system plays a critical role as the brain areas critical for adaptive learning based on response outcome. From the evolutionary perspective, the brain circuitry, cell types, and neurotransmitters in the basal ganglia and associated structures have been conserved across vertebrates, from the evolutionary primitive lamprey to primates (Suryanarayana et al. 2021). In fact, the remarkable similarities between these conserved structures in the lamprey and mammals suggest that these aspects of the vertebrate brain go back to the dawn of vertebrate brain evolution (Grillner 2021). An important implication of the evolutionary conservation of brain structures like the basal ganglia is that the evolutionary problems they evolved to solve are unlikely to have changed materially over the course of vertebrate evolution. Modifying behavior on the basis of experience is likely to be one of those problems. Of course, there are major additions to the vertebrate system consequent on the appearance of limbs, the ability (capacity) to reach and grasp; however, these are most likely accommodated by the elaboration of structures external to conserved ones such as those comprising the basal ganglia and the cerebellum. These developments accommodate the behavioral requirements of the more advanced species (Grillner 2021). Also conserved are the limbic structures that in humans provide for complex emotions (love, lust, hate, fear, disgust, happiness, sadness, etc.). Typically, they are triggered by evolutionarily significant sensory stimuli that, in addition to overt behavioral responses, evoke these important emotions. In the absence of an affective system, sensory stimuli would not be associated with the positive and negative events to predispose the organism to make fast reactions. Ultimately, the social emotions including embarrassment, envy, and pride evolved and guided social interactions among humans living in communities and groups.

Early studies demonstrated that animals, under some experimental conditions (e.g., limited training), exhibited knowledge of anticipated outcomes, while under different conditions (e.g., extended training) they showed evidence of automatized habits (Tolman and Gleitman 1949). This observation led to the conclusion that these two types of learning could be mediated in parallel by different neural mechanisms. Packard and McGaugh (1996) reported that, after extended training on a cross-maze to reach for reward (food), the habitual strategy (i.e., body turn) eventually came to dominate the goal-oriented (place strategy, learned location within the room). They proceeded to show that dorsolateral striatum (DLS) inactivation prevented the expression of the habitual response with the goal-oriented, place-strategy response being reinstated in these animals. These findings showed first that habitual behavior is learned simultaneously with goal-directed behavioral control, that they are mediated by different neural systems, and that, when the overtrained habitual system is damaged, the conserved goal-directed system can be brought back into operation.

There is now a broad consensus demonstrating that the DLS is engaged in and necessary for the expression of habitual behavior, while the dorsomedial striatum (DMS) is engaged in and necessary for goal-directed behavioral control. Data supporting this dissociation come from a wide range of methodological approaches—lesion (Yin et al. 2004), inactivation (Corbit and Janak 2010) and neural recording studies (Barnes et al. 2005)—and using different species—mice (Hilario et al. 2012), rats (Corbit and Janak 2010), non-human primates (Miyachi et al. 2002) and humans (Tricomi et al. 2009). The dichotomy between the DLS control of habit and the DMS involvement in goal-directed control suggests that the two systems operate in competition to control behavioral output. Recent evidence suggests that these systems operate simultaneously during instrumental learning. The habitual system observes and learns the regularities in stimulus–response transitions during performance under goal-directed control. When regular stimulus–response associations become established, relevant stimuli are able to evoke habitual responses without reference to the value of predicted outcomes, linked to the posterior putamen activities in humans (Tricomi et al. 2009) and the DLS in rats (Smith and Graybiel 2013).

Cortical Mapping of Goal-Directed and Habitual Behavior

While the cortical representations of behavior in humans and other species are not directly comparable, the PFC is undoubtedly a leading hub to mediate novel behaviors and goal-directed control. Specifically, the orbitofrontal cortex (OFC) is key in early learning, but its activity fades as habits take control, transferring to posterior sensorimotor regions.

This subsection will highlight the cortical activity shifts in both human and non-human studies.

The orbitofrontal cortex

Experimental evidence in a range of species suggests that several "cortical" regions can influence behavior specifically related to goal-directed and habitual control systems. There are limbic and PFC regions that project to the DMS (McGeorge and Faull 1989) and have been shown to be engaged in goal-directed behavior (Ostlund and Balleine 2007). The OFC exhibits the greatest change in neuronal firing rate during the performance of goal-directed behavior but not when habitual control is operating (Gremel and Costa 2013). In humans, a transient deactivation has been reported as reward stimuli are repeated across time (O'Doherty et al. 2001). Additional findings show that updating goal expectations evokes plasticity in OFC dendritic spines, and this plasticity is required for goal-directed control (Whyte et al. 2019). Conversely, inhibiting OFC excitatory activity (Gremel and Costa 2013) disrupts goal-directed behavior. This suggests that the OFC plays a necessary role in the maintenance of goal-related performance. Using a within-subject task, it can be shown that training context promotes the use of either habitual or goal-directed behavioral control (Gremel and Costa 2013) and that OFC projections to the DMS are key for the performance of goal-directed responses. Thus, inhibition of OFC terminals in the DMS prevents goal-directed learning and forces reliance on the habitual system, even when the training context was designed to promote goal-directed control (Gremel et al. 2016). Following this, chronic ethanol exposure was shown to decrease the excitability of OFC projection neurons. The observed disruption of goal-directed behavior was interpreted as a failure of normal OFC output activity to the DMS (Renteria et al. 2018). Previous studies have shown that potential outcomes are represented in the OFC and that these are key for goal-directed behavior (Ostlund and Balleine 2007). Hence, in a system where both goal-directed and established habitual control operate in competition, effective communication must be maintained between the OFC and DMS to prevent habits taking control over behavior. In humans, a similar OFC link with goal-directed actions is reported. Using a two-step sequential learning task to explore whether choices were made under goal-directed control (predicting the outcome of each choice) or under stimulus–response habit (success of the preceding option), correlations were found between lower OFC gray matter volume and striatum with biased habitual performance (Voon et al. 2015). However, despite the dominance of the OFC, several additional brain areas have been implicated in the control of goal-directed behavior. For example, both the basolateral amygdala inputs from the DMS and thalamus (Bradfield et al. 2013) have been shown to be necessary for the acquisition and expression of goal-directed behavior.

What system-level changes occur following the transition from goal-directed to habitual performance? The most obvious is a rostrocaudal shift of involvement from associative to sensorimotor corticostriatal regions shown in both animals and humans (Yin and

Knowlton 2006; Figure 7.3). Thus, when well-learned sensory cues are detected, the habitual system primes motor (Killcross and Coutureau 2003) and premotor cortical areas to interact with the posterior putamen (de Wit et al. 2012; Figure 7.3).

The Infralimbic Cortex

The infralimbic cortex (ILC) is a critical region involved in acquisition and expression of habitual behavior in rats (Smith and Graybiel 2013). When habitual behavior, through training, has replaced goal-directed control of behavior, optogenetic inhibition of the IL restores the former goal-directed control (Coutureau and Killcross 2003).

Using a within-session dual reinforcement schedule (one under goal-directed behavior and another for promoting habitual responding), neural activity in the IL was suppressed during goal-directed actions but sustained while the animals responded on the reinforcement schedule that promoted habitual performance (Barker et al. 2018). Thus, it was shown that when the IL was inhibited immediately after each lever press on the habit-reinforcing schedule, animals were sensitive to contingency degradation, thereby demonstrating that their behavior had switched to goal-directed control (Barker et al. 2018). While these studies show that the IL is involved in habitual performance, the mechanism by which it promotes stimulus–response control at the expense of outcome, value-based goal-directed control, is unclear. One possibility is that the IL operates to suppress previously learned contingencies to promote habits (Coutureau and Killcross 2003). Given that the IL does not project directly to the DLS (Mailly et al. 2013), the basal ganglia region associated with habitual control, additional studies are needed to identify the system-level intricacies of habitual control circuitry. For example, what role the IL projections have to limbic regions related to behavioral inhibition play? Do they have the necessary selectivity to suppress goal-directed control circuits? Similarly, is it possible that IL projections to the ventral striatum compete with goal-directed control mechanisms in this region of the basal ganglia?

Thus, evidence from experimental animal studies, functional imaging in healthy humans, and behavioral studies in human patients with selective dysfunction in different subregions of the basal ganglia point to the conclusion that conscious goal-directed and automatic habitual control are exercised by segregated regions in the striatum that have been highly conserved from the evolutionary point of view.

Brain Connectivity in Both Behavioral Control Options

How cortical and subcortical operations interact to sustain goal-directed or habitual forms of behavioral control depends on the presence of contextual information. Repeatedly associated contextual cues will rapidly elicit bottom-up automatic stimulus processing that has been associated with and will evoke a predefined response. In contrast, uncertain or vague contextual information will engage more reflexive and explorative routes to achieve a specific goal. The cortex in humans detects well-known or unknown stimuli to then call out other connected regions, such as the striatum, in order to enact a rapid behavioral outcome. Depending on the contextual stimuli, whether the main properties are sensorimotor, motor planning, or reinforcement signals, the cortex will detect them and send information to the basal ganglia to expedite the last behavioral output. Indeed, functional neuroimaging, brain lesion and animal electrophysiological investigations all converge to suggest a dominant role of the striatum in the expression of associations between specific contextual cues and habitual movement kinematics. In this line, putaminal activity is related to specific contextual information that generates movement options, organizing action units into movement sequences even in individual finger movements (Andersen et al. 2020). Moreover, the role of the striatum is well established as participating in learning without motor plans, guided by reinforced choices, and during well-learned and contextually driven actions (Peters et al. 2021). Hence, the wide range of striatal functions in sensorimotor-related and high-order reinforcement activities supports a general hub for several habitual operations with varying qualitative context scenarios during processing stimulus–response associations.

The Corticostriatal Relationship during Habits: Top-Down or Bottom-Up?

How the cortex integrates and modulates striatal activity during habits is less clear. The input to the striatum comes from all cortical regions and many subcortical structures (Peters et al. 2021), thereby allowing the basal ganglia to receive an immense amount of potentially conflicting information (Haber and Knutson 2010). These inputs can represent potentially competing sensorimotor associations that can be expressed under habitual control. They include ventral premotor cortex (de Wit et al. 2012), extrastriate visual cortex (Luque et al. 2017), and bilateral insula and precentral gyrus (Eryilmaz et al. 2017). A connection relevant to the imposition of habitual control was described from the ventral premotor cortex to the posterior putamen during the acquisition of habits (de Wit et al. 2012). The dynamic association between corticobasal ganglia physiology and habitual behavior exhibits a similar transition from anterior to posterior cortical areas as habits build up. It is most unlikely that any single cortical area takes control over habitual behaviors. Stimuli from all sensory modalities, represented in different cortical regions, can be associated with responses that involve a complete range of body parts, the movements of which are also represented in different cortical regions; it is therefore unlikely that a single cortical region dedicated to habits will be found. However, while it seems that the PFC, including the OFC, the IL (in rats), or the dorsolateral PFC (in humans; Obeso et al. 2021), is a key player in initial associative learning to build habits, the exact nature of its role has yet to be determined.

Neurotransmitters Involved in Goal-Directed and Habitual Behaviors

All of the above systems are implemented by chemical neurotransmission. Among them, the dopamine system is most frequently implicated, but some studies also point to the involvement of the serotonin and opioid systems (Worbe et al. 2016). The short-latency phasic changes in dopamine activity appear to play an important reinforcement role in associative learning. Thus, in the past, dopamine has been linked to the model-free reinforcement model of habitual learning. Indeed, animal models investigating rodent behavior have shown that pharmacologically enhanced dopamine transmission boosts habit formation (Nelson and Killcross 2006). Although quantitatively different, selective dopaminergic nigrostriatal lesions restricted to input to the sensorimotor striatum impair habit formation in rats (Faure et al. 2005), again causing the behavior to revert to goal-directed control. Because of the difficulty of instituting selective regional changes in dopaminergic transmission, human evidence concerning the role of dopamine in habitual function is at best unclear. For example, dopamine depletion appears to increase habitual control (de Wit et al. 2012), while augmenting transmission via levodopa increases goal-directed control (Wunderlich et al. 2012) and reduces habitual control (Kroemer et al. 2019).

Other candidate neurotransmitters in goal-directed and habitual control are the opioid and serotonin systems. Diminished forebrain serotonin (5-HT) increased compulsive cocaine-seeking in rodents, also inducing larger dependence on habitual behaviors (Pelloux et al. 2012). In humans, a similar serotonin depletion also seems to boost habitual responding. An important distinction with serotonin depletion is the selective impairment of reward-related, goal-directed behavior at the expense of enhanced goal-directed control over possible punishments (Worbe et al. 2016). Alternatively, pharmacological blockade of opioid transmission has been shown to impair goal-directed learning together with greater habitual control of actions (Wassum et al. 2009).

Last, a further study explored the involvement of dopamine, serotonin and opioid transmission on goal-directed and habitual control. In healthy control subjects, three forms of positron emission tomography (PET) imaging—[^{18}F] F-DOPA, [^{11}C]MADAM, and [^{11}C]carfentanil—were used to explore presynaptic dopamine, serotonin transporter and mu-opioid receptor binding potential (Voon et al. 2020). A two-step task assessed how much decision-making was controlled by goal-directed or habitual behavioral strategies that provided reward and/or punishment outcomes. When working for rewards, greater serotonin transporter binding potential in prefrontal regions was associated with habitual control. In contrast, in loss outcomes the opioidergic system showed increased [^{11}C]carfentanil binding potential that was positively associated with goal-directed control and negatively associated with habit-directed performance. Given that different structures throughout the brain use the same neurotransmission system for widely differing functions, it is perhaps not surprising that generic systemic manipulation of neurotransmission systems yields complex results in the current context of behavioral control.

One can wonder then, what are the evolutionary advantages of having dual behavioral control systems that can almost certainly interfere with each other? The simple answer is to be able to relinquish repetitive tasks where the stimulus–response structure is highly predicted, from conscious goal-directed control, necessarily would free it to work on less predictable problems in the world. An inevitable consequence of off-loading overtrained tasks to the habitual system is that the more tasks that are overtrained, the greater number of routine tasks that will be performed under automatic habitual control. Small wonder that habitual responding is more frequently observed in senior citizens (Figure 7.4).

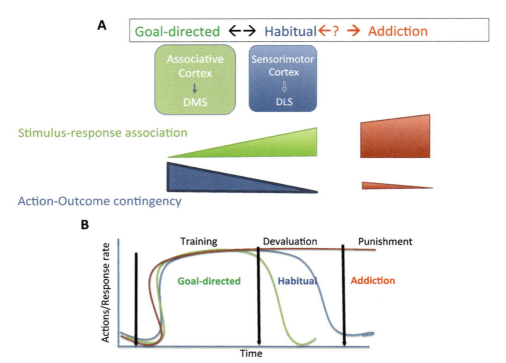

Figure 7.4 (A) The transition from goal-directed behavior to habitual behavior and then into compulsion, or addiction, is gradual. The shift from goal-directed to habitual behavior and then to compulsion/addiction corresponds to strengthened stimulus–response association and reduced action–outcome contingency. A behavior can shift from goal-directed to habitual performance and back again—but it is not clear if from addiction one can turn to habit/goal-directed behavior. (B) During instrumental training, the responding rates for a reward increase. Post-training reward devaluation reduces response rates more quickly for goal-directed behaviors than it does for habitual behaviors, which take many more extinction trials to disappear. Addiction is characterized by compulsive responding even after punishment. DLS = dorsolateral striatum; DMS = dorsomedial striatum.

Clinical and Therapeutic Implications

Corresponding brain regions controlling goal-directed and habitual control have been identified in humans. Neuroimaging studies have shown that initial goal-directed action outcome learning engages rostromedial (associative) striatum and that activation gradually shifts to caudolateral (sensorimotor) regions when habitual control takes over (Tricomi et al. 2009). Converging evidence for a critical involvement of the caudal striatum in automatic habitual behavior has been provided by several neuropsychological studies. For example, the loss of dopaminergic input from the midbrain to the striatum in Parkinson's disease is initially concentrated in the posterior putamen (Pineda-Pardo et al. 2022). This selective loss seems to alter automatic movements such as arm swinging during walking or facial expressions in social interactions. More recently, it has been shown that habitual components of typing (Bannard et al. 2019) and associative learning (Mi et al. 2021) are selectively reduced in patients with Parkinson's disease. In contrast, Huntington's disease induces neural atrophy specifically in the head of the caudate. This deficit seems to prevent the initial associations needed to integrate the repetition of stimulus–response associations into habits.

In contrast, Huntington's disease induces neural atrophy specifically in the head of the caudate. This deficit seems to prevent the initial associations needed to integrate the repetition of stimulus–response associations into habits. Recently, a novel hypothesis has been proposed where an overreliance on habitual control may lead to a selective functional overload of the dopaminergic system in sensorimotor regions of the basal ganglia. The resulting progressive degenerative cell loss could be an important contributory cause of Parkinson's disease (Hernandez et al. 2019; Figure 7.4).

An overreliance on habit is especially associated with the compulsivity that manifests in several neuropsychiatric conditions, including obsessive-compulsive disorder (Gillan et al. 2014), schizophrenia (Morris et al. 2015) and addiction (Sjoerds et al. 2013). Dopamine-related disorders such as schizophrenia are associated with motivational deficits and reduced instances of goal-directed behavior. Motivational deficits in schizophrenia are often measured using effort-based decision-making paradigms, a task whereby participants are asked to exert physical or cognitive efforts in order to get points in reward regimes. Relative to controls, alterations are found in effort expenditure in schizophrenia patients (Cooper et al. 2019). The abnormal neuronal activity and reduced gray matter volumes in frontostriatal regions seem to lead to an overreliance on habit (Fettes et al. 2017).

A dopamine-related "cascading loop" from ventral to dorsal striatum has been associated with habit formation (Yin et al. 2004). Reactivation of the hypoactivated PFC in cocaine-dependent rats restores control over cocaine-seeking for a period of time (Chen et al. 2013). Moreover, lesioning the dorsolateral striatum can reverse habitual drug-seeking in alcohol (Corbit et al. 2012) and cocaine (Zapata et al. 2010) studies.

In addicted humans, lost control over drug use contributes to an unbalanced system where an inability to abstain is driven by habitual control of high-level motivations (e.g., craving). The underlying mechanisms involve the debilitating impact of abused drugs on the lateral and ventromedial PFC responsible for controlling the brain's mesolimbic and striatal reward systems (Everitt and Robbins 2005). Hence, drug-related neural changes prioritize attention allocated to drug-related cues which, through repetition, become habitually associated with drug-related motivations (craving) and behaviors (Krebs et al. 2011). Together with prefrontal changes, functional abnormalities in the dorsolateral striatum and posterior putamen functional abnormalities contribute to disinhibiting drug-seeking and promoting relapse in many patients (Courtney et al. 2016). In alcohol-dependent subjects, enhanced activity in the posterior putamen was reported during an instrumental learning stimulus–response task (Sjoerds et al. 2013).

Compared with control subjects, when patients had to reverse and override their habit, larger activity in the posterior putamen was accompanied by reduced PFC activities. These results confirm that, in addition, there is a disturbed balance between goal-directed and habitual control in favor of an overreliance on inflexible habits. Overall, evidence indicates a transition in addictive behaviors from goal-directed ventral striatum toward increased control from the habitual dorsolateral striatum in both rodents and non-human primates.

Frontotemporal dementia is associated with enhanced perseveration and automaticity. Among others, patients with frontotemporal dementia exhibit variations in personality, apathy, or inertia, as well as behavioral disinhibition, loss of sympathy and dysexecutive behavior. Importantly, these patients show a range of enhanced stereotyped behavior expressed as simple repetitive movements through to rather more complex compulsive or ritualistic forms of behavior acts (Moheb et al. 2018). Such behavioral changes suggest that the deficits in normal conscious goal-directed behavior in demented patients have been supplanted by repetitive stimulus-response control.

Therapeutic Lessons and Opportunities

With the information so far reviewed, there are therapeutic implications that might prove useful in the understanding and treatment of various clinical conditions by modulation of the goal-directed or the habitual system. Overriding maladaptive habitual behaviors in, for example, drug addiction, would require enhanced PFC functioning related to cognitive control. This should promote the reorientation of attention, goal-planning and cognitive alert maintenance as well as greater flexibility (Yang et al. 2018), with the ultimate objective of overriding toxic habits. One possible intervention is cognitive reappraisal of intrusive repetitive and habitual thoughts. This is a cognitive behavioral option to tackle unwanted repetitive thoughts with the aim of transforming their emotional impact and thus control. By repetitively exposing the patient to their thoughts by daily inspection of the cues that automatically elicit them, the physiological responses, and feelings, the patient gains control of the motivational consequences of the sensory cues. Following this procedure, it seems to promote a PFC control role and improve the regulation of the attentional and emotional appraisal processes driven by the striatum, amygdala, and ventromedial PFC (Frank et al. 2014). The process of how cognitive reappraisal exerts control over volitional self-regulation is set upon increased awareness over the automatized psychophysiological reactivity to certain addictive cues, to change the neural response to cue reactivity in patients with substance use disorders. This technique indeed shows, after considerable training, increased activity in the dorsolateral PFC, inferior frontal gyrus, and dorsal anterior cingulate cortex as well as decreased limbic activity along the ventromedial PFC, ventral striatum and midbrain (Kober and Mell 2015). Thus, this treatment option represents a non-invasive way to recover behavioral and neural normal signals toward toxic cues.

Summary and Conclusions

There are two fundamental systems for controlling behavioral output: goal-directed and habitual. The goal-directed system selects actions that are instrumental in achieving valued outcomes. This system is sensitive to changes in outcome value and the contingency relationship between action and outcome. Typically, it is slow, requires conscious effort, but is flexible. When a particular task is repeated its stimulus–response structure is learned and becomes predictable. At this point it can be conducted under automatized habitual control. This simpler stimulus–response control is faster, is less variable, is often more accurate and can be conducted without voluntary intervention. Obviously, from the evolutionary point of view, each strategy of behavioral control has advantages and drawbacks depending on how predictable or unpredictable events in the world are. As the relationship between stimuli and actions becomes more predictable, it makes sense to have them controlled by a fast, reliable, automatic stimulus–response system. However, evolution of a parallel goal-directed system sensitive to the predicted outcome value of actions allows for more flexible and adaptive responding. During the human life span, there is a succession of different needs and behaviors. During the early stages of life, explorative (goal-directed) behavior will be predominant (childhood and early youth). As we grow old, the routines of our daily lifestyles and jobs enable the habitual system to take care of many tasks where the stimulus–response structure is reliable, is fully appreciated and can be conducted with little conscious mental effort.

There are different brain regions involved in each of these strategies for controlling behavior: the associative cortex and dorsomedial striatum (anterior putamen) are key brain areas for goal-directed behavior, while the sensorimotor cortex and dorsolateral striatum (posterior putamen) are the main brain regions controlling habitual behavior. Interestingly, these brain structures have been highly conserved during vertebrate brain evolution.

Alterations of goal-directed behaviors are key features of disorders such as schizophrenia, while overreliance on habitual behavior gives rise to various forms of compulsivity: obsessive-compulsive disorder, addiction, alcoholism, gambling, shopping, repetitive behaviors and compulsive overeating. Additionally, overreliance on habitual control and the consequent overactivation of the neuronal networks involved could lead to excessive energetic overload, high metabolic demand, and oxidative stress, which could be a leading cause of dopaminergic neurodegeneration. These factors could therefore be involved in the pathogenesis and progression of Parkinson's disease.

Unanswered Questions and Future Work

There is a major need in the field to implement and improve studies related to human habits in the lab. It is essential to be able to replicate and search for experiments that resemble more the natural expression of habits in everyday life. Therefore, there is an unfilled need to bring people's real-life habits into the lab for testing and study.

Additionally, although several putative habit substrates have been identified, the unique function of these substrates in behavioral control, as well as the mechanisms involved in arbitrating these regulatory systems, have not been clearly defined. It remains to be understood how habits are encoded in the ensemble activity of intrinsic basal ganglia microcircuitry. How does the cell type–specific ensemble activity change in each striatal subregion as habits develop? And, importantly, how do the habitual and goal-directed regions interact?

Finally, it would be relevant and revealing to determine if overreliance on habitual control is contributory to the selective degeneration of dopaminergic neurons in the substantia nigra and, as such, would represent an important etiological factor in Parkinson's disease.

Key Points to Remember

- Human behaviors operate under dual control mechanisms: goal-directed and habitual.
- The GDS operates when actions are selected based on the predicted outcome value. This kind of behavior is sensitive to changes in outcome value. It is slow, effortful, but flexible.
- When a task is repeated, its stimulus–response structure is learned to the point that it is performed automatically, without reference to the action outcome value. Such habitual responses can be performed without voluntary intervention. They are faster and more accurate than the volitional and conscious performance of the same task.
- The associative cortex and dorsomedial striatum (anterior putamen) are key brain areas for goal-directed behavior, while the sensorimotor cortex and dorsolateral striatum (posterior putamen) are the main brain regions exercising habitual control.

- The dual control systems have evolved to maximize adaptive flexibility (goal-directed), speed, and efficiency (habits).
- Deficits of goal-directed control have been associated with schizophrenic pathology where there is reduced motivation and less reliance on the value of action outcomes.
- Overreliance on habitual control leads to various forms of compulsivity including obsessive-compulsive disorder, addiction, alcoholism, gambling, shopping, repetitive behaviors, and compulsive overeating. Moreover, overreliance on habitual control can be considered a hypodopaminergic behavior and be associated potentially with selective dopaminergic neurodegeneration and Parkinson's disease.

References

Andersen, K. W., Madsen, K. H., & Siebner, H. R. (2020). Discrete finger sequences are widely represented in human striatum. *Scientific Reports*, *10*(1), Article 13189. https://doi.org/10.1038/s41598-020-69923-x

Balleine, B. W., & Dickinson, A. (1998). Goal-directed instrumental action: Contingency and incentive learning and their cortical substrates. *Neuropharmacology*, *37*(4–5), 407–419. https://doi.org/10.1016/s0028-3908(98)00033-1

Balleine, B. W., & O'Doherty, J. P. (2010). Human and rodent homologies in action control: Corticostriatal determinants of goal-directed and habitual action. *Neuropsychopharmacology*, *35*(1), 48–69. https://doi.org/10.1038/npp.2009.131

Bannard, C., Leriche, M., Bandmann, O., Brown, C. H., Ferracane, E., Sánchez-Ferro, A., Obeso, J., Redgrave, P., & Stafford, T. (2019). Reduced habit-driven errors in Parkinson's disease. *Scientific Reports*, *9*(1), Article 3423. https://doi.org/10.1038/s41598-019-39294-z

Barker, J. M., Glen, W. B., Linsenbardt, D. N., Lapish, C. C., & Chandler, L. J. (2018). Habitual behavior is mediated by a shift in response-outcome encoding by infralimbic cortex. *eNeuro*, *4*(6), Article ENEURO.0337-17.2017. https://doi.org/10.1523/ENEURO.0337-17.2017

Barnes, T. D., Kubota, Y., Hu, D., Jin, D. Z., & Graybiel, A. M. (2005). Activity of striatal neurons reflects dynamic encoding and recoding of procedural memories. *Nature*, *437*(7062), 1158–1161. https://doi.org/10.1038/nature04053

Bradfield, L. A., Bertran-Gonzalez, J., Chieng, B., & Balleine, B. W. (2013). The thalamostriatal pathway and cholinergic control of goal-directed action: Interlacing new with existing learning in the striatum. *Neuron*, *79*(1), 153–166. https://doi.org/10.1016/j.neuron.2013.04.039

Ceceli, A. O., Myers, C. E., & Tricomi, E. (2020). Demonstrating and disrupting well-learned habits. *PLOS ONE*, *15*(6), Article e0234424. https://doi.org/10.1371/journal.pone.0234424

Chen, B. T., Yau, H. J., Hatch, C., Kusumoto-Yoshida, I., Cho, S. L., Hopf, F. W., & Bonci, A. (2013). Rescuing cocaine-induced prefrontal cortex hypoactivity prevents compulsive cocaine seeking. *Nature*, *496*(7445), 359–362. https://doi.org/10.1038/nature12024

Cooper, J. A., Barch, D. M., Reddy, L. F., Horan, W. P., Green, M. F., & Treadway, M. T. (2019). Effortful goal-directed behavior in schizophrenia: Computational subtypes and associations with cognition. *Journal of Abnormal Psychology*, *128*, 710–722. https://doi.org/10.1037/abn0000443

Corbit, L. H., & Janak, P. H. (2010). Posterior dorsomedial striatum is critical for both selective instrumental and Pavlovian reward learning. *European Journal of Neuroscience*, *31*(7), 1312–1321. https://doi.org/10.1111/j.1460-9568.2010.07153.x

Corbit, L. H., Nie, H., & Janak, P. H. (2012). Habitual alcohol seeking: Time course and the contribution of subregions of the dorsal striatum. *Biological Psychiatry*, *72*(5), 389–395. https://doi.org/10.1016/j.biopsych.2012.02.024

Courtney, K. E., Schacht, J. P., Hutchison, K., Roche, D. J., & Ray, L. A. (2016). Neural substrates of cue reactivity: Association with treatment outcomes and relapse. *Addiction Biology*, *21*(1), 3–22. https://doi.org/10.1111/adb.12314

Coutureau, E., & Killcross, S. (2003). Inactivation of the infralimbic prefrontal cortex reinstates goal-directed responding in overtrained rats. *Behavioural Brain Research*, *146*(1–2), 167–174. https://doi.org/10.1016/j.bbr.2003.09.025

Daw, N. D., Gershman, S. J., Seymour, B., Dayan, P., & Dolan, R. J. (2011). Model-based influences on humans' choices and striatal prediction errors. *Neuron*, *69*(6), 1204–1215. https://doi.org/10.1016/j.neuron.2011.02.027

De Houwer, J. (2019). On how definitions of habits can complicate habit research. *Frontiers in Psychology*, *10*, Article 2642. https://doi.org/10.3389/fpsyg.2019.02642

de Wit, S., & Dickinson, A. (2009). Associative theories of goal-directed behaviour: A case for animal–human translational models. *Psychological Research*, *73*(4), 463–476. https://doi.org/10.1007/s00426-009-0230-6

de Wit, S., Kindt, M., Knot, S. L., Verhoeven, A. A. C., Robbins, T. W., Gasull-Camos, J., Evans, M., Mirza, H., & Gillan, C. M. (2018). Shifting the balance between goals and habits: Five failures in experimental habit induction. *Journal of Experimental Psychology: General*, *147*(7), 1043–1065. https://doi.org/10.1037/xge0000402

de Wit, S., Standing, H. R., Devito, E. E., Robinson, O. J., Ridderinkhof, K. R., Robbins, T. W., & Sahakian, B. J. (2012). Reliance on habits at the expense of goal-directed control following dopamine precursor depletion. *Psychopharmacology*, *219*(2), 621–631. https://doi.org/10.1007/s00213-011-2563-2

Dickinson, A. (1985). Actions and habits: The development of behavioral autonomy. *Philosophical Transactions of the Royal Society of London. Series B, Biological Sciences*, *308*(1135), 67–78. https://doi.org/10.1098/RSTB.1985.0010

Durant, W. (1926). *Story of Philosophy. The lives and opinions of the world's great philosophers*. Simon and Schuster, NY.

Eryilmaz, H., Rodriguez-Thompson, A., Tanner, A. S., Giegold, M., Huntington, F. C., & Roffman, J. L. (2017). Neural determinants of human goal-directed vs. habitual action control and their relation to trait motivation. *Scientific Reports*, *7*(1), Article 6002. https://doi.org/10.1038/s41598-017-06284-y

Everitt, B. J., & Robbins, T. W. (2005). Neural systems of reinforcement for drug addiction: From actions to habits to compulsion. *Nature Neuroscience*, *8*(11), 1481–1489. https://doi.org/10.1038/nn1579

Faure, A., Haberland, U., Condé, F., & El Massioui, N. (2005). Lesion to the nigrostriatal dopamine system disrupts stimulus-response habit formation. *Journal of Neuroscience*, *25*(11), 2771–2780. https://doi.org/10.1523/JNEUROSCI.3894-04.2005

Fettes, P., Schulze, L., & Downar, J. (2017). Cortico-striatal-thalamic loop circuits of the orbitofrontal cortex: Promising therapeutic targets in psychiatric illness. *Frontiers in Systems Neuroscience*, *11*, Article 25. https://doi.org/10.3389/fnsys.2017.00025

Frank, D. W., Dewitt, M., Hudgens-Haney, M., Schaeffer, D. J., Ball, B. H., Schwarz, N. F., Hussein, A. A., Smart, L. M., & Sabatinelli, D. (2014). Emotion regulation: Quantitative meta-analysis of functional activation and deactivation. *Neuroscience and Biobehavioral Reviews*, *45*, 202–211. https://doi.org/10.1016/j.neubiorev.2014.06.010

Gardner, B. (2015). A review and analysis of the use of "habit" in understanding, predicting and influencing health-related behavior. *Health Psychology Review*, *9*(3), 277–295. https://doi.org/10.1080/17437199.2013.876238

Gerfen, C. R., & Wilson, C. J. (1996). The basal ganglia. In A. Bjorklund, T. Hokfelt, & L. W. Swanson (Eds.), *Handbook of chemical neuroanatomy: Vol. 12. Integrated systems of the CNS* (Pt, 3, pp. 371–468). Elsevier.

Gillan, C. M., Morein-Zamir, S., Urcelay, G. P., Sule, A., Voon, V., Apergis-Schoute, A. M., Fineberg, N. A., Sahakian, B. J., & Robbins, T. W. (2014). Enhanced avoidance habits in obsessive-compulsive disorder. *Biological Psychiatry*, *75*(8), 631–638. https://doi.org/10.1016/j.biopsych.2013.02.002

Gremel, C. M., Chancey, J. H., Atwood, B. K., Luo, G., Neve, R., Ramakrishnan, C., Deisseroth, K., Lovinger, D. M., & Costa, R. M. (2016). Endocannabinoid modulation of orbitostriatal circuits gates habit formation. *Neuron*, *90*(6), 1312–1324. https://doi.org/10.1016/j.neuron.2016.04.043

Gremel, C. M., & Costa, R. M. (2013). Orbitofrontal and striatal circuits dynamically encode the shift between goal-directed and habitual actions. *Nature Communications*, *4*, Article 2264. https://doi.org/10.1038/ncomms3264

Grillner, S. (2021). Evolution of the vertebrate motor system—From forebrain to spinal cord. *Current Opinion in Neurobiology*, *71*, 11–18. https://doi.org/10.1016/j.conb.2021.07.016

Guida, P., Michiels, M., Redgrave, P., Luque, D., & Obeso, I. (2022). An fMRI meta-analysis of the role of the striatum in everyday-life vs laboratory-developed habits. *Neuroscience and Biobehavioral Review, 141*, Article 104826. https://doi.org/10.1016/j.neubiorev.2022.104826

Haber, S. N., & Knutson, B. (2010). The reward circuit: Linking primate anatomy and human imaging. *Neuropsychopharmacology, 35*(1), 4–26. https://doi.org/10.1038/npp.2009.129

Hardwick, R. M., Forrence, A. D., Krakauer, J. W., & Haith, A. M. (2019). Time-dependent competition between goal-directed and habitual response preparation. *Nature Human Behaviour, 3*(12), 1252–1262. https://doi.org/10.1038/s41562-019-0725-0

Hernandez, L. F., Obeso, I., Costa, R. M., Redgrave, P., & Obeso, J. A. (2019). Dopaminergic vulnerability in Parkinson disease: The cost of humans' habitual performance. *Trends in Neurosciences, 42*(6), 375–383. https://doi.org/10.1016/j.tins.2019.03.007

Hilario, M., Holloway, T., Jin, X., & Costa, R. M. (2012). Different dorsal striatum circuits mediate action discrimination and action generalization. *European Journal of Neuroscience, 35*(7), 1105–1114. https://doi.org/10.1111/j.1460-9568.2012.08073.x

James, W. (1890). *The principles of psychology* (Vol. 1). Henry Holt. https://doi.org/10.1037/10538-000

James, W. (1892). *Psychology. The Briefer Course*. Henry Holt.

Killcross, S., & Coutureau, E. (2003). Coordination of actions and habits in the medial prefrontal cortex of rats. *Cerebral Cortex, 13*(4), 400–408. https://doi.org/10.1093/cercor/13.4.400

Kober, H., & Mell, M. M. (2015). Neural mechanisms underlying craving and the regulation of craving. In S. J. Wilson (Ed.), *The Wiley handbook on the cognitive neuroscience of addiction* (pp. 195–218). Wiley Blackwell.

Krebs, R. M., Boehler, C. N., Egner, T., & Woldorff, M. G. (2011). The neural underpinnings of how reward associations can both guide and misguide attention. *Journal of Neuroscience, 31*(26), 9752–9759. https://doi.org/10.1523/JNEUROSCI.0732-11.2011

Kroemer, N. B., Lee, Y., Pooseh, S., Eppinger, B., Goschke, T., & Smolka, M. N. (2019). L-DOPA reduces model-free control of behavior by attenuating the transfer of value to action. *Neuroimage, 186*, 113–125. https://doi.org/10.1016/j.neuroimage.2018.10.075

Luque, D., Beesley, T., Morris, R. W., Jack, B. N., Griffiths, O., Whitford, T. J., & Le Pelley, M. E. (2017). Goal-directed and habit-like modulations of stimulus processing during reinforcement learning. *Journal of Neuroscience, 37*(11), 3009–3017. https://doi.org/10.1523/JNEUROSCI.3205-16.2017

Luque, D., Molinero, S., Watson, P., López, F. J., & Le Pelley, M. E. (2020). Measuring habit formation through goal-directed response switching. *Journal of Experimental Psychology: General, 149*(8), 1449–1459. https://doi.org/10.1037/xge0000722

Mailly, P., Aliane, V., Groenewegen, H. J., Haber, S. N., & Deniau, J. M. (2013). The rat prefrontostriatal system analyzed in 3D: Evidence for multiple interacting functional units. *Journal of Neuroscience, 33*(13), 5718–5727. https://doi.org/10.1523/JNEUROSCI.5248-12.2013

McGeorge, A. J., & Faull, R. L. M. (1989). The organization of the projection from the cerebral cortex to the striatum in the rat. *Neuroscience, 29*(3), 503–537. https://doi.org/10.1016/0306-4522(89)90128-0

McNamee, D., Liljeholm, M., Zika, O., & O'Doherty, J. P. (2015). Characterizing the associative content of brain structures involved in habitual and goal-directed actions in humans: A multivariate FMRI study. *Journal of Neuroscience, 35*(9), 3764–3771. https://doi.org/10.1523/JNEUROSCI.4677-14.2015

Mi, T. M., Zhang, W., Li, Y., Liu, A. P., Ren, Z. L., & Chan, P. (2021). Altered functional segregated sensorimotor, associative, and limbic cortical-striatal connections in Parkinson's disease: An fMRI investigation. *Frontiers in Neurology, 12*, Article 720293. https://doi.org/10.3389/fneur.2021.720293

Miyachi, S., Hikosaka, O., & Lu, X. (2002). Differential activation of monkey striatal neurons in the early and late stages of procedural learning. *Experimental Brain Research, 146*(1), 122–126. https://doi.org/10.1007/s00221-002-1213-7

Moheb, N., Charuworn, K., Ashla, M., Desarzant, R., Chavez, D., & Mendez, M. (2018). Repetitive behaviors in frontotemporal dementia: Compulsions or impulsions? *Journal of Neuropsychiatry and Clinical Neurosciences, 31*(2), 132–136. https://doi.org/10.1176/appi.neuropsych.18060148

Morris, R. W., Quail, S., Griffiths, K. R., Green, M. J., & Balleine, B. W. (2015). Corticostriatal control of goal-directed action is impaired in schizophrenia. *Biological Psychiatry, 77*(2), 187–195. https://doi.org/10.1016/j.biopsych.2014.06.005

Nelson, A., & Killcross, S. (2006). Amphetamine exposure enhances habit formation. *Journal of Neuroscience, 26*(14), 3805–3812. https://doi.org/10.1523/JNEUROSCI.4305-05.2006

Obeso, I., Herrero, M. T., Ligneul, R., Rothwell, J. C., & Jahanshahi, M. (2021). A causal role for the right dorsolateral prefrontal cortex in avoidance of risky choices and making advantageous selections. *Neuroscience, 458*, 166–179. https://doi.org/10.1016/j.neuroscience.2020.12.035

O'Doherty, J., Kringelbach, M. L., Rolls, E. T., Hornak, J., & Andrews, C. (2001). Abstract reward and punishment representations in the human orbitofrontal cortex. *Nature Neuroscience, 4*(1), 95–102. https://doi.org/10.1038/82959

Ostlund, S. B., & Balleine, B. W. (2007). The contribution of orbitofrontal cortex to action selection. *Annals of the New York Academy of Sciences, 1121*(1), 174–192. https://doi.org/10.1196/annals.1401.033

Packard, M. G., & McGaugh, J. L. (1996). Inactivation of hippocampus or caudate nucleus with lidocaine differentially affects expression of place and response learning. *Neurobiology of Learning and Memory, 65*(1), 65–72. https://doi.org/10.1006/nlme.1996.0007

Pelloux, Y., Dilleen, R., Economidou, D., Theobald, D., & Everitt, B. J. (2012). Reduced forebrain serotonin transmission is causally involved in the development of compulsive cocaine seeking in rats. *Neuropsychopharmacology, 37*(11), 2505–2514. https://doi.org/10.1038/npp.2012.111

Peters, A. J., Fabre, J. M. J., Steinmetz, N. A., Harris, K. D., & Carandini, M. (2021). Striatal activity topographically reflects cortical activity. *Nature, 591*(7850), 420–425. https://doi.org/10.1038/s41586-020-03166-8

Pineda-Pardo, J. A., Sánchez-Ferro, Á., Monje, M. H. G., Pavese, N., & Obeso, J. A. (2022). Onset pattern of nigrostriatal denervation in early Parkinson's disease. *Brain, 145*(3), 1018–1028. https://doi.org/10.1093/brain/awab378

Renteria, R., Baltz, E. T., & Gremel, C. M. (2018). Chronic alcohol exposure disrupts top-down control over basal ganglia action selection to produce habits. *Nature Communications, 9*(1), Article 211. https://doi.org/10.1038/s41467-017-02615-9

Sjoerds, Z., de Wit, S., van den Brink, W., Robbins, T. W., Beekman, A. T., Penninx, B. W., & Veltman, D. J. (2013). Behavioral and neuroimaging evidence for overreliance on habit learning in alcohol-dependent patients. *Translational Psychiatry, 3*(12), Article e337. https://doi.org/10.1038/tp.2013.107

Smith, K. S., & Graybiel, A. M. (2013). A dual operator view of habitual behavior reflecting cortical and striatal dynamics. *Neuron, 79*(2), 361–374. https://doi.org/10.1016/j.neuron.2013.05.038

Suryanarayana, S. M., Robertson, B., & Grillner, S. (2021). The neural bases of vertebrate motor behaviour through the lens of evolution. *Philosophical Transactions of the Royal Society of London B: Biological Sciences, 377*(1844), Article 20200521. https://doi.org/10.1098/rstb.2020.0521

Thorndike, E. L. (1911). *Animal intelligence: Experimental studies*. Macmillan.

Tolman, E. C., & Gleitman, H. (1949). Studies in spatial learning: VII. Place and response learning under different degrees of motivation. *Journal of Experimental Psychology, 39*(5), 653–659. https://doi.org/10.1037/h0059317

Tricomi, E., Balleine, B. W., & O'Doherty, J. P. (2009). A specific role for posterior dorsolateral striatum in human habit learning. *European Journal of Neuroscience, 29*(11), 2225–2232. https://doi.org/10.1111/j.1460-9568.2009.06796.x

Uylings, H. B., Groenewegen, H. J., & Kolb, B. (2003). Do rats have a prefrontal cortex? *Behavioural Brain Research, 146*(1–2), 3–17. https://doi.org/10.1016/j.bbr.2003.09.028

Voon, V., Derbyshire, K., Rück, C., Irvine, M. A., Worbe, Y., Enander, J., Schreiber, L. R., Gillan, C., Fineberg, N. A., Sahakian, B. J., Robbins, T. W., Harrison, N. A., Wood, J., Daw, N. D., Dayan, P., Grant, J. E., & Bullmore, E. T. (2015). Disorders of compulsivity: A common bias towards learning habits. *Molecular Psychiatry, 20*(3), 345–352. https://doi.org/10.1038/mp.2014.44

Voon, V., Joutsa, J., Majuri, J., Baek, K., Nord, C. L., Arponen, E., Forsback, S., & Kaasinen, V. (2020). The neurochemical substrates of habitual and goal-directed control. *Translational Psychiatry, 10*(1), Article 84. https://doi.org/10.1038/s41398-020-0762-5

Wassum, K., Cely, I., Maidment, N., & Balleine, B. (2009). Disruption of endogenous opioid activity during instrumental learning enhances habit acquisition. *Neuroscience, 163*(3), 770–780. https://doi.org/10.1016/j.neuroscience.2009.06.071

West, R., & Brown, J. (2013). *Theory of addiction*. Wiley Blackwell.

Whyte, A. J., Kietzman, H. W., Swanson, A. M., Butkovich, L. M., Barbee, B. R., Bassell, G. J., Gross, C., & Gourley, S. L. (2019). Reward-related expectations trigger dendritic spine plasticity in the mouse ventrolateral orbitofrontal cortex. *Journal of Neuroscience*, *39*(23), 4595–4605. https://doi.org/10.1523/JNEUROSCI.2031-18.2019

Wood, W., Mazar, A., & Neal, D. T. (2022). Habits and goals in human behavior: Separate but interacting systems. *Perspectives on Psychological Science*, *17*(2), 590–605. https://doi.org/10.1177/1745691621994226

Worbe, Y., Palminteri, S., Savulich, G., Daw, N. D., Fernandez-Egea, E., Robbins, T. W., & Voon, V. (2016). Valence-dependent influence of serotonin depletion on model-based choice strategy. *Molecular Psychiatry*, *21*(5), 624–629. https://doi.org/10.1038/mp.2015.46

Worringer, B., Langner, R., Koch, I., Eickhoff, S. B., Eickhoff, C. R., & Binkofski, F. C. (2019). Common and distinct neural correlates of dual-tasking and task-switching: A meta-analytic review and a neurocognitive processing model of human multitasking. *Brain Structure and Function*, *224*(5), 1845–1869. https://doi.org/10.1007/s00429-019-01870-4

Wunderlich, K., Smittenaar, P., & Dolan, R. J. (2012). Dopamine enhances model-based over model-free choice behavior. *Neuron*, *75*(3), 418–424. https://doi.org/10.1016/j.neuron.2012.03.042

Yang, Z., Oathes, D. J., Linn, K. A., Bruce, S. E., Satterthwaite, T. D., Cook, P. A., Satchell, E. K., Shou, H., & Sheline, Y. I. (2018). Cognitive behavioral therapy is associated with enhanced cognitive control network activity in major depression and posttraumatic stress disorder. *Biological Psychiatry: Cognitive Neuroscience and Neuroimaging*, *3*(4), 311–319. https://doi.org/10.1016/j.bpsc.2017.12.006

Yin, H. H., & Knowlton, B. J. (2006). The role of the basal ganglia in habit formation. *Nature Reviews: Neuroscience*, *7*(6), 464–476. https://doi.org/10.1038/nrn1919

Yin, H. H., Knowlton, B. J., & Balleine, B. W. (2004). Lesions of dorsolateral striatum preserve outcome expectancy but disrupt habit formation in instrumental learning. *European Journal of Neuroscience*, *19*(1), 181–189. https://doi.org/10.1111/j.1460-9568.2004.03095.x

Zapata, A., Minney. V. L., & Shippenberg, T. S. (2010). Shift from goal-directed to habitual cocaine seeking after prolonged experience in rats. *Journal of Neuroscience*, *30*(46), 15457–15463. https://doi.org/10.1523/JNEUROSCI.4072-10.2010

PART 2

How Human Brain Diseases Are Impacted by Human Evolution

8

Alzheimer's Disease, the Parietal Lobes, and the Evolution of the Human Genus

Emiliano Bruner and Heidi I. L. Jacobs

Introduction
Brain Evolution and the Human Natural History

One of the main features that characterizes the evolution of the human genus is the size and complexity of its brain, as easily evidenced when comparing our encephalic dimensions with any other animal taxa. Indeed, brain complexity and large size are general traits associated with the whole Primate order, although in humans the increase in both aspects is outstanding. There are still disagreements on whether dietary or social reasons have been the *primum movens* of such noticeable changes, and we should assume that both factors must have had their importance, probably in different evolutionary moments. The motivation behind the compelling (and sometimes compulsive) study of brain size, deeply rooted in the history of science, is not a matter of morphological interest or taxonomic enquiry but has a significant functional aspect: as a rule of thumb, in many mammals, and specifically in primates, we can observe a correlation between brain size and behavioral complexity (Herculano-Houzel 2017). This relationship is however not straight, and there are many disagreements on whether the absolute or relative size of the brain is the most interesting functional value, or whether brain size or brain internal organization is more crucial for cognitive evolution. Once more, we should consider that all these factors contribute because they are all relevant and, alone, each of them is not able to produce global predictions, namely reliable explanations valid for all species and able to explain the overall variability. Nonetheless, nobody would disagree when recognizing that brain evolution, in terms of size and complexity, is a key evolutionary feature for primates, and for all human species (i.e., belonging to the genus *Homo*) in particular.

Such evolutionary investment in brain complexity should have had, indeed, some advantages. It is worth remembering that, in evolution, "advantage" means, only and strictly, something that increases reproductive success. Features increasing the number of offspring of an individual, population, or species are considered advantages, while any trait that will decrease this parameter is a disadvantage and will be negatively selected. In the middle, there are many features that do not influence the evolutionary fitness, and hence will be transmitted or lost through secondary associations with other features or random effects. Nonetheless, biological traits rarely are inherited as independent and isolated features, being embedded in complex genetic, ontogenetic, and morphogenetic networks strongly entrenched in polygenic (one trait is influenced by many genes) and pleiotropic (one gene influences many traits) effects. Evolution does not support or reject single features but instead deals with "packages" of features, namely combinations of traits that can generate, at the same time, pros and cons. These packages of traits will be promoted or demoted, in terms of selective forces, depending on the overall effect on the evolutionary fitness, namely on the global reproductive success. This means that detrimental characters can be positively selected if they are associated with some very advantageous ones or if they do not influence the reproductive success. Indeed, features that are favorable for a species can be prejudicial for individuals. An example concerns those adverse features that affect the late life periods, when the reproductive ability is low or null (i.e., post-reproductive stages). This condition is particularly relevant for *Homo sapiens*, one of the few species with a conspicuous and extended post-reproductive life span (Ellis et al. 2018). With these premises in mind, we can wonder whether the evolution of our big brains may have involved some trade-off with the rest of the body functions, when considering its high costs and delicate management (Leonard et al. 2007). A big brain needs a lot of energy, and this can slow down the ontogenetic process of an organism, as well as all the stages of its life history (Aiello and Wheeler 1995; Leigh 2004). The advantage of a large brain is supposed to be a complex cognitive system, but what about the downsides? The balance between growth, development, energy, and life history is presumably a delicate one, which is likely to be affected by several metabolic and morphogenetic constraints. Indeed, this looks like a complicated package of biological and ecological features, where changes are channeled by an intricate assortment of pros and cons.

Species Diversity and Evolutionary Anthropology

In evolutionary biology, we need at least two sources of information to delineate properly the history and fate of those evolutionary packages of characters. In *neontological* study, comparative approaches are aimed at describing the diversity of living species. This allows us to understand the natural history of a zoological group and the patterns, rules, and strategies that have generated its diversification. All too often, living taxa are used to make inferences on ancestral-descendent changes, an approach that, in absence of a comprehensive phylogenetic evidence, may represent a superficial epistemological mistake. Chimps and macaques, for example, are frequently used as primitive models to propose hypotheses on human evolution, which is like the belief that our cousins are simultaneously our grandfathers. Living taxa represent independent and parallel lineages, with their own derived history and features. Hence, they represent the *product* of evolution, not the *process* (Bruner 2019). Indeed, to investigate the process, we must localize a true ancestral-descendent sequence, which means investigating fossils. The *paleontological* record can supply the information to investigate the actual changes that occurred along an evolutionary trajectory. In this case, however, we have a major limitation: we have information available only for some few species, only from few individuals, and only from their skeletal systems. Also in those few cases in which we have molecular data, these are limited to few specimens, incomplete sequencing, and a large set of methodological uncertainties. Namely, in paleontology, we have the direct evidence of the evolutionary changes but associated with poor samples and scarce biological and taxonomical representation. That's why evolutionary anthropology must rely on both neontological and paleontological information, taking advantage of both approaches but considering the respective limitations.

Paleoneurology and Endocranial Casts

The field in charge of investigating brain evolution in extinct species is called *paleoneurology*, and it is aimed at making anatomical inferences on the brain's gross morphology according to the evidence from the endocranial cavity (see Holloway et al. 2004; Bruner 2017). The mold of the endocranial cavity (called the "endocast"; Figure 8.1) can supply information on the brain volume, on the patterns of cortical folding, or on the meningeal vascular system. Working on an endocranial cast, the information we can obtain on the general brain anatomy is incomplete and partial, but it is the only direct evidence we can have on the evolution of the brain along an evolutionary lineage. In this case, we should take into account three considerations. First, there will be cerebral changes that cannot be detected according to the macroscopic anatomy of the endocranial cavity, for example, those associated with deeper brain regions, with connectivity, or with molecular and biochemical processes. The overall morphological correspondence between the brain and endocast is pretty good (Dumoncel et al. 2021), but it only concerns the external and superficial appearance. In this case, the absence of evidence is not the evidence of absence. For example, the paleoneurological record has not revealed, among the species of the genus *Homo*, major anatomical changes in the frontal lobe or in the pattern of hemispheric asymmetries (Bruner and Beaudet 2022). This does not mean that these features have not undergone changes in human evolution but just that possible variations of these characters have not concerned the macroanatomical appearance of the external

184 | PART 2 HOW HUMAN BRAIN DISEASES ARE IMPACTED BY HUMAN EVOLUTION

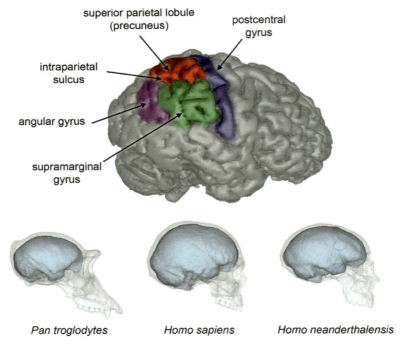

Figure 8.1 (Above) The main cortical elements of the human parietal lobe. (Below) Digital reconstruction of the skull and endocast of a chimpanzee, a modern human, and a Neanderthal (Saccopastore 1, Rome, dated to >120.000 years ago). In modern human, brain globularity is partially due to large and bulging parietal lobes. Images not at scale. Redrawn after Bruner (2017) and Bruner et al. (2018a).

brain. Second, there are brain (and endocranial) morphological changes that can be due to cranial constraints and not to primary changes of the brain anatomy. This is particularly frequent in those regions with spatial conflicts between the cerebral and cranial elements (e.g., the orbital and temporal cortices, housed in the anterior and middle cranial fossae respectively; Bruner 2017). Third, there are brain morphological changes that are indeed able to mold the cranial cavity, and hence can leave clues of actual cerebral variations on the endocranial form. In this latter group, an interesting example is the parietal lobe (Figure 8.1), which apparently underwent consistent changes in our own species (see Bruner 2018; Bruner et al. 2022).

Evolutionary Evidence Specifically Related to Human Health and Disease

Antagonistic Pleiotropy in Late-Onset Alzheimer's Disease

An important example of features that are disadvantageous to humans but are not negatively selected as they occur after reproductive age is sporadic or late-life Alzheimer's disease (AD), one of the most common forms of dementia, characterized by progressive cognitive and behavioral impairments evident from age 65. The advantageous features associated with the survival of this disease are most likely multifactorial including protection against infectious diseases, inflammation, or stress-related factors early in life. However, later in life the accumulation of AD-related proteinopathies, possessing specific genetic features related to AD risk, has been related to neuroinflammation, cognitive decline, and other clinical symptoms (Smith et al. 2019; Provenzano and Deleidi 2021). This represents a clear example of *antagonistic pleiotropy*. The neuropathologic hallmarks of AD are characterized by the extracellular aggregation of the β-amyloid protein–forming plaques and the intracellular aggregation of hyperphosphorylated tau protein–forming tangles (Braak and Braak 1991; Braak and Del Tredici 2011). Misfolding of these proteins into plaques and tangles starts approximately two to three decades before the onset of clinical symptoms, providing a long window where differential effects can take place (Jack et al. 2018). Furthermore, the primary genetic risk factor for late-onset AD is the apolipoprotein ε (APOE-ε) gene. Individuals carrying at least one ε4 isoform are at greater risk of AD later in life as compared to those carrying the common ε3 allele or the rare but protective ε2 allele. Presence of the APOE-ε4 allele has also been related to earlier and more abundant development of β-amyloid plaques (Yamazaki et al. 2019). Interestingly, the APOE-ε4 variant was the ancestral allele, while the ε3 allele emerged later during human evolution and the ε2 allele represents a recent variation, approximately 80,000 years of age (Fullerton et al. 2000). Furthermore, geographically, the distributions of the variants also differ, with enriched APOE-ε4 frequency in indigenous populations of Central Africa, Oceania, and Australia, suggesting that its role may have changed during the transition to the hunter–gatherer lifestyle or the origin of the human genus, when human longevity, including the post-reproductive life span, extended due to different lifestyles, nutritional changes, or other factors (Finch and Stanford 2004; Raichlen and Alexander 2014; Belloy et al. 2019). During midlife, the presence of the ε4 allele confers a protective effect on brain connectivity and has been associated with better cognitive performance, in particular reasoning and psychomotor speed (Gharbi-Meliani et al. 2021). African carriers of the ε4 allele develop AD less frequently than individuals with a European or Asian ancestry. Among populations with a high parasite and pathogen load in rural Ghana and the Amazonian forager–horticulturalists, carrying the APOE-ε4 allele has been associated with better cognitive performance, suggesting that the ε4 allele may provide protection against infections (Trumble et al. 2017; van Exel et al. 2017). The mechanism underlying this association remains unclear but might be related to stronger immune responses during reproductive life span, which can be detrimental later in life. Stronger immune responses have also been observed in young HIV-positive individuals with elevated β-amyloid and carrying an ε4 allele (Corder et al. 1998). The β-amyloid protein shares properties with viruses and can activate immune responses to fight off infections (Provenzano and Deleidi 2021). But chronic β-amyloid-related immune responses in response to infections can facilitate pathways toward toxic β-amyloid misfolding, neurodegeneration, and oxidative stress and increase the propensity to develop cognitive deficits (Hoeijmakers et al. 2016; Hashimoto et al. 2018). Alternatively, another potential mechanism explaining the opposing effects of APOE during the adult life span might stem from the elevated brain metabolism supporting cognitive functioning, but in the long term these metabolic demands may be disadvantageous (Smith et al. 2019).

In the following sections we will discuss how the evolutionary changes in specifically the parietal lobe are associated with advantageous consequences for the survival of humans but confer the disadvantage of developing AD later in life.

Parietal Lobes and Functional Morphology

The parietal cortex is particularly large, complex, and derived in primates, when compared with other mammals (Goldring and Krubitzer 2017). This expansion and complexity is supposedly associated with specializations in many distinct cognitive features, including, for example, eye–hand coordination, attention, and visuospatial cognition. Indeed, such parietal complexity is extreme in *H. sapiens*. Humans display a very large precuneus, even when compared with apes (Figure 8.2; Bruner et al. 2017). The precuneus and the superior parietal lobule (which should be regarded as a single cortical territory) are the meeting regions between somatic (body) and visual information and are divided into three main areas (Scheperjans et al. 2008). The anterior is more involved in body perception, the posterior is more involved in visual imaging, and the middle region integrates both kinds of signals. Traditionally, in terms of evolution, brain cortical changes

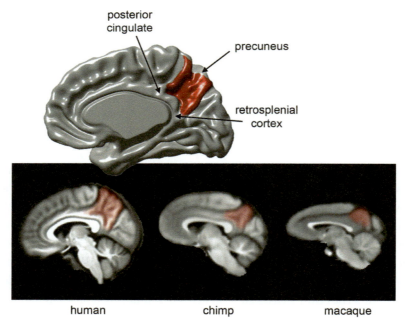

Figure 8.2 Magnetic resonance imaging scout views from the medial sections of the brains of a macaque, a modern human, and a chimpanzee, showing the general size and proportions of the precuneus in the three species. Images not at scale.

can be due to an increase/decrease of some specific areas or even to the formation of new ones. The large precuneus in modern humans can be due to the expansion of areas shared with other primates or else the evolution of new cortical elements. As an alternative explication, we should consider that the cerebral morphogenesis is characterized by gradients between the sensory regions, which can explain many aspects of the cortical organization (Huntenburg et al. 2017). In this case, therefore, the expansion of the precuneus should be intended as an increase in the amount of information exchanged and integrated between the somatic and visual systems. Further anatomical changes in humans can be associated with the white matter (i.e., connectivity), and in fact at least some connections between the precuneus and the posterior cingulate cortex are probably derived in our species (Goulas et al. 2014; Ardesch et al. 2019). The intraparietal sulcus is also particularly developed in humans, to a degree that makes difficult a proper recognition of the homologies between humans and nonhuman primates (Grefkes and Fink 2005). Finally, also the inferior parietal lobule shows many derived features in our species, and it is involved in human-only cognitive aspects that include tool use and language (Bzdok et al. 2016; Reyes et al. 2022). Because of the involvement in self-awareness and body–environment interaction, the parietal cortex is also crucial to social relationships, which adds a further key factor in terms of evolution.

If we take into account the spatial topology of the major brain regions, most parietal areas are not particularly constrained by the surrounding cortical elements, although they are part of an integrated posterior cortical block (Bruner et al. 2019), associated with a complex folding pattern (Zilles et al. 2013), and central to the functional and structural brain networks (Hagmann et al. 2008). However, a noticeable increase in anatomical complexity can be observed in the inferior part of the precuneus, below the subparietal sulcus, a region that displays an intricate spatial situation (Bruner 2022, 2023). The region formed by the inferior precuneus, posterior cingulate cortex, and retrosplenial cortex has an outstanding topological burden because it is in tight contact with many other neighboring elements and bridges the anterior and posterior cerebral modules (Figure 8.3). This region, furthermore, displays a remarkable individual variability, it is integrated with the rest of the precuneus, and it is also influenced by the biomechanical effects of the *tentorium cerebelli*, a connective tensor with a morphogenetic role in the brain–braincase ontogenetic balance. It is hence patent that such a position makes this region strongly constrained by structural influences associated with spatial and developmental factors and sensitive to any brain changes that can occur in the posterior cerebral district.

Brain Globularity and Parietal Development

Modern humans have large and bulging parietal lobes also when compared with extinct human species (Bruner 2018). Parietal lobes in *H. sapiens* are probably even larger when compared with Neanderthals, a species that shared with us a similar brain size (Pereira-Pedro et al. 2020). Therefore, it can be expected that brain globularity in modern humans is likely to be due, at least in part, to such large parietal volume. This feature was

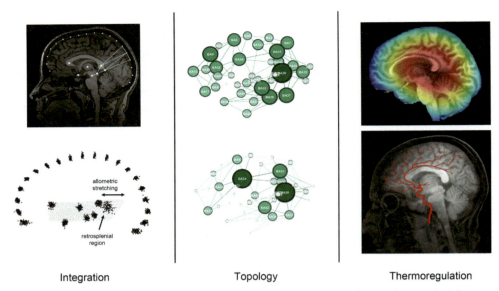

Figure 8.3 The region formed by the inferior precuneus, posterior cingulate, and retrosplenial cortex is influenced by several constraints. Integration: The retrosplenial areas are spatially integrated with the parietal surface (above, the wireframe shows a morphological module) and display a pronounced variation, influenced by allometric factors possibly associated with the developmental effect of the tentorium cerebelli (below). Topology: These regions (Brodmann areas 30 and 31) also display a high degree of spatial connection (above: degree centrality values) and are topological bridges between the anterior and posterior cerebral blocks (below: betweenness centrality values). Thermoregulation: According to a model of passive thermal distribution, these regions are included in the thermal core of the brain volume (above: higher thermal load in red, lower thermal load in blue); but, in terms of vessels, they only receive some minor posterior branches of the anterior cerebral artery (below). BA = Brodmann area. Images redrawn after Bruner (2023). Arterial scheme based on Lanfermann et al. (2019).

not present at the origin of our lineage but evolved lately, say in the last 100,000 years (Neubauer et al. 2018). During ontogeny, parietal bulging is apparent early after birth (Neubauer et al. 2009). In the morphogenesis of the braincase, the dorsal regions are generally shaped by the pressure of the growing brain (Moss and Young 1960) and, therefore, its posterior morphology is expected to be largely influenced, in its form and curvature, by the underlying parietal lobe. The parietal bone is integrated with the occipital bone, and the flattening of one of these elements is generally associated with the bulging of the other, even when comparing different human species (Gunz and Harvati 2007). Interestingly, parietal bulging in modern humans is also associated with an increased complexity of the parietal vascular networks, at least when dealing with the middle meningeal and diploic vessels. Our species is in fact characterized by a complex and reticulated vascular network, especially on the parietal surface, while extinct human species display vascular systems with few branches and no anastomosis (see Píšová et al. 2017). Other craniovascular features that can be relevant in this sense are the many microforamina scattered on the vault, mostly in association with the venous sinuses and bridging veins, connecting the marrow of the vault bones with the subarachnoid space, and probably involved in brain immunological response and inflammation (Bruner and Eisová 2024).

Because of the scarce metabolism of the meninges and vault bones, it has been proposed that the increase of the meningeal network may be associated with functions related to endocranial thermal regulation (Bruner et al. 2011). A recent study revealed that brain temperature is higher and more variable than previously speculated, fluctuating with circadian cycles and reaching the highest values around the thalamus (Rzechorzek et al. 2022). In extinct human species, the flat and smaller parietal surface was closer to the thermal core of the endocranial space, and it might eventually represent a useful region for heat exchange (in particular, brain/endocranial cooling). In modern humans, instead, heat distribution models suggest that brain globularization and parietal bulging may have generated higher heat loads in the geometric core of the endocranial space, especially in the inferior and deep parts of the parietal lobes (Bruner et al. 2014). Namely, at least for what concerns brain shape, a globular morphology does increase the heat burden in those central regions, when compared with a flatter (platycephalic) head form. Brain cooling is largely a matter of blood flow, and, in this sense, it is worth noting that the precuneus presents a peculiar vascular situation (Lanfermann et al. 2019). The superficial and lateral regions of the parietal cortex are vascularized by different networks, including the branches of the middle cerebral artery or the parietal veins stemming from the superior sagittal sinus. Instead, the medial parietal regions occupy a territory which is at the boundary between the anterior and posterior cerebral arteries (Mavridis et al. 2016; Kalamatianos et al. 2019). The precuneus is mainly irrigated by the terminal branches of the anterior cerebral artery (the superior and inferior precuneal arteries; see Figure 8.3) and, to a lesser extent, by the posterior cerebral artery (the parieto-occipital artery running into the parieto-occipital fissure). Current surgical practice suggests that the vascularization of the precuneus is neither exceptional nor poor (Mavridis, personal communication), although the existence of two systems (one anterior and one posterior) has led to the hypothesis of a sort of protection against stroke or other vascular impairments (Parvizi et al. 2021). We could speculate that depending on the individual variability in vascular architecture, certain vascularization patterns can provide vascular reserve for the supply and function of the medial parietal areas, as has been suggested for the hippocampus (Perosa et al. 2020). However, distal areas between arterial territories (also referred to as "watershed border zones") that are supplied by the most terminal arterioles, such as the medial parietal areas, may be at greater risk for (temporary) reductions in perfusion, ischemia, and small vessel disease (Torvik 1984; Mangla et al. 2011). The intermediate position might induce conflicts between the two arterial systems, or the small terminal arterioles may provide insufficient supply (mostly the anterior one, which represents the main supply) to these regions. The precuneus also displays a peculiar vascular feature: it is one of the few cortical regions that has interhemispheric anastomoses. This special combination of vascular traits in the precuneus (two alternative but marginal networks and interhemispheric flow) makes this topic an interesting one, mainly when considering that this region is characterized by high thermal loads. As mentioned above, this same region also displays important topological constraints because of its spatial position and contiguity with many other brain areas as well as with possible morphogenetic influences of the *tentorium cerebelli*. Indeed, a complicated anatomical situation, in which evolutionary changes must deal with a large set of structural and functional consequences.

AD and the Parietal Cortex

The neuropathologic hallmarks of AD are characterized by the extracellular aggregation of the β-amyloid protein–forming plaques and the intracellular aggregation of hyperphosphorylated tau protein–forming tangles (Braak and Braak 1991; Braak and Del Tredici 2011). Both of these proteins start to misfold and aggregate approximately two to three decades prior to the first clinical symptoms, highlighting the protracted course of the disease (Jack et al. 2018). The emergence and propagation of β-amyloid and tau follow a predictable spatiotemporal pattern in the majority of patients. The β-amyloid protein starts to accumulate in the neocortex, in which the initial vulnerable regions include the precuneus, retrosplenial and posterior cingulate cortex, anterior cingulate cortex, and sometimes the inferior temporal cortex (Bischof and Jacobs 2019). From the neocortex, β-amyloid progresses to the allocortex and striatum; when clinical AD dementia is evident, β-amyloid pathology can be detected in the cerebellum and brainstem. In contrast, tau pathology starts in the brainstem, including the locus coeruleus, around age 20 or 30 years (Jacobs et al. 2021). By age 50, half of the individuals exhibit elevated tau in the entorhinal cortex; and after age 65, when β-amyloid is also detectable, tau has engrossed the limbic system and ultimately will spread to the neocortex (Braak and Braak 1991; Braak and Del Tredici 2011).

Even though tau aggregates deposit prior to the accumulation of β-amyloid plaques, current research criteria for AD posit that the presence of elevated β-amyloid pathology is a prerequisite to establish underlying AD pathologic change (Jack et al. 2018). Key observations supporting a causally important role for β-amyloid are the fact that the proportion of clinically normal individuals with elevated β-amyloid PET by age mirrors the age-specific prevalence of individuals diagnosed with AD 20 years later (Rowe et al. 2010), along with the finding that β-amyloidosis is the first proteinopathy to occur in individuals with autosomal dominant AD (Bateman et al. 2012). Even though neocortical tau pathology correlates closely with the cognitive deficits and clinical progression, these relationships between tau and cognition are particularly observed in individuals with elevated β-amyloid pathology, indicating that rising β-amyloid pathology signals the onset of clinical impairment (Sperling et al. 2019).

The focus on the β-amyloid protein as a key criterion for AD and the predilection of the parietal lobe regions for the earliest β-amyloid protein accumulations emphasize the importance of the parietal lobe in AD's ontogeny and possibly phylogeny. Notably, the parietal lobe forms a convergence area of several critical disease-related changes in cognition, metabolism, and connectivity (Jacobs et al. 2012). Early β-amyloid accumulation is associated with disruptions in attentional processes, associated with the parietal lobe's role as a major operational hub for higher-order cognitive processes. The parietal lobe has this critical role because of its unique anatomical and functional properties, which also contribute to its vulnerability to AD-related pathology (Jacobs et al. 2012). As β-amyloid increases, deficits in encoding and retrieval of episodic information become more prominent, reflecting a disconnection between the medial temporal and medial parietal areas (Bischof and Jacobs 2019).

Structural and Functional Brain Changes in the Parietal Lobe in AD

Isolated medial temporal lobe changes are not specific for AD, and they can reflect other potential neurodegenerative changes, including vascular lesions, hippocampal sclerosis, TAR DNA-binding protein 43 (TDP-43), argyrophilic grain disease, or primary age-related tauopathy. However, the loss of connectivity between the initial regions vulnerable to β-amyloidosis, the medial parietal regions, and regions signaling the archetypical AD-related cognitive deficits and initial cortical tau accumulation, the medial temporal regions, conveys important and specific prognostic information about AD-related processes. Combining atrophy of the parietal lobe together with medial temporal lobe atrophy can increase the diagnostic accuracy of AD in the earliest stages by 22% (Jacobs et al. 2011b). The importance of a medial temporal–medial parietal loss of connection was recently shown in a multimodal neuroimaging study revealing that progression of tau from the medial temporal lobe to the retrosplenial cortex via the connecting pathways, the hippocampal cingulum bundle, was associated with memory decline over six years if individuals displayed elevated cortical β-amyloid (Jacobs et al. 2018).

Similar findings have been observed in resting-state functional magnetic resonance imaging (fMRI) studies. Resting-state fMRI measures spontaneous low-frequency fluctuations in the blood-oxygen-level-depending (BOLD) signal, an indirect measure of neuronal activity, providing a window into the functional architecture of the brain. One of the most often examined resting-state networks in the context of aging and AD is the default mode network, composed of the posterior cingulate cortex, precuneus, angular gyrus, medial temporal lobe areas, and medial prefrontal cortex. This network is active during passive rest and mind-wandering and has been ascribed a role in remembering the past, episodic memory, and autobiographical memory. Intriguingly, the topography of this resting-state network largely overlaps with the spatial layout of the initial accumulation of β-amyloid in the brain, with a strong preference for the precuneus (Buckner et al. 2005). Several resting-state fMRI studies reported loss of coactivation between the medial temporal lobe region and the precuneus and posterior cingulate cortex, starting from the early asymptomatic stages of AD (Pihlajamaki and Sperling 2009; Mevel et al. 2011). This loss of coactivation has important consequences for task-related neuronal activity as the brain will be less optimal in suppressing interference ("deactivation") from the default mode network during task performance (Jacobs et al. 2015). Interestingly, this increased interference in the precuneus and posterior cingulate cortex is exacerbated in individuals with cognitive impairment and comorbid small vessel disease (Papma et al. 2013). And indeed, in the earliest stages of the disease, the preclinical stages (before the appearance of clinical symptomatology), the precuneus exhibits a higher level of activation while performing a memory task. This higher level of activation was associated with lower behavioral performance, indicating that these increases are dysfunctional and interfere (Rami et al. 2014). Interestingly, cognitive performance and brain activity levels can temporarily be normalized by transcranial magnetic stimulation of the precuneus in prodromal AD patients (Koch et al. 2018). These medial parietal regions both exhibit prominent atrophy and lower cerebral blood flow in AD patients. Hypoperfusion has been reported in the earliest stages of the disease (Scott Miners et al. 2016). In particular, lower cerebral blood flow was localized to the distal branches of the posterior cerebral artery territories as well as the watershed areas (Huang et al. 2018), emphasizing the specific vulnerability of the medial parietal areas to AD-related pathology, neurodegeneration, as well as vascular changes.

This specific early vulnerability of the medial parietal lobe regions, posterior cingulate cortex, retrosplenial cortex, and precuneus for β-amyloid has also been attributed to the dense connectivity patterns and metabolic activity of this region. Similar to the fMRI insights, hypometabolic patterns in the medial temporal lobe and posterior cingulate cortex were detected in AD patients with typical memory impairments (Groot et al. 2021). Posterior cingulate cortex metabolism is lower in clinically normal older individuals with elevated β-amyloid and neocortical tau, which together predicted memory decline (Hanseeuw et al. 2017). Furthermore, hypometabolism in the medial parietal areas is more accurate in differentiating cognitively healthy older people from early AD patients than metabolic changes in other brain regions (Hunt et al. 2007). In summary, we hypothesize that the spatial colocalization between initial β-amyloid deposition in the parietal lobe with the architecture of the parietal lobe, as well as the high degree of connectivity and metabolic activity in medial parietal areas, in particular the retrosplenial and posterior cingulate cortices, contribute to the specific vulnerability of these regions to structural and functional damage, including β-amyloid-related tau spreading and β-amyloid-related neurodegeneration.

The Downside of a Complex Parietal Cortex

The special combination of features described in AD, as well as the high prevalence of this neurodegenerative disease, have predominantly been described for our species, *H. sapiens*. As mentioned, the medial parietal regions are specifically vulnerable to AD because they are a critical hub of early β-amyloidosis. Both neontological (living species) and paleontological (fossils) data suggest that these regions are particularly large and complex in modern humans only, so we may wonder whether there is an association between large and complex parietal lobes and AD factors.

The deposition of β-amyloid plaques has been described in the hippocampus and neocortex of nonhuman primate species and great apes (Perez et al. 2013, 2016). However, β-amyloid plaques were less fibrillar and more diffuse in nature, and concentrations did not reach the same level as in humans. Furthermore, neurofibrillary tangle pathology is rare in nonhuman primate brains, except in a caged, obese female chimpanzee with hypercholesterolemia (Rosen et al. 2008; Perez et al. 2016). In addition to nonhuman primates, β-amyloid has been detected in squirrel monkeys, vervets, lemurs, and dogs but at lower concentrations and with less fibrillization than in humans. Except for the presence of hyperphosphorylated tau in older dogs with disrupted sleep patterns, these animals are spared from the accumulation of tangles, neurodegeneration, and dementia (Walker and Jucker 2017; Freire-Cobo et al. 2021).

Thus, the AD phenotype is particularly associated with modern humans and concerns anatomical areas that are particularly developed only in modern humans. Therefore, we should consider the possibility that the evolutionary changes of those regions might be somehow related to the etiology of the disease (Bruner and Jacobs 2013). Indeed, brain regions that evolved recently have been shown to be more vulnerable to pathological events than regions characterized by higher evolutionary stability (Bartzokis 2009; Stricker et al. 2009; Jacobs et al. 2011a). In this case, we have described how the region formed by the inferior precuneus, posterior cingulate, and retrosplenial cortex may suffer biological constraints due to the spatial crowding with the neighboring elements, to a scarce vascularization, and to increased heat burden due to the deeper position in a globular braincase. We have also described how this same region is involved in metabolic impairments during the early stages of AD. There is hence the possibility that the change in shape and size of the parietal cortex in modern humans may have generated a sort of evolutionary conflict between the anatomical organization of this region and its functional balance. It is indeed striking that genetic (protective) changes of the apolipoprotein variants took place roughly at the same time of the parietal expansion, namely 80,000 years ago (Belloy et al. 2019).

The expansion of the superior parietal lobule has been hypothesized to be associated with enhanced and specialized visuospatial integration and body cognition, including crucial functions related to visual imaging, technological capacity, and social complexity (Bruner et al. 2018b). Visual imaging is a core function of the default mode network, enabling visual projections of past and future representations (Land 2014). Haptic perception and body cognition are key features of a "prosthetic capacity" that allows a structural and functional integration between brain, body, and tools (Bruner 2021). All these features are intimately entrenched with the cultural and social environments of modern humans (Bruner and Gleeson 2019). The evolutionary dynamics behind this parietal specialization are not clear, possibly including genetic, epigenetic, or environmental influences (Bruner and Iriki 2016). Nonetheless, the evolutionary advantages of this anatomical and cognitive package are clear and sufficient to hypothesize a direct effect on the evolutionary fitness and reproductive success of a population. The downside would be, simply speaking, an excessive anatomical and physiological burden of those areas subjected to metabolic or vascular unbalances, like heat stress or defects in blood clearance. In this sense, it can be interesting to consider whether such complex parietal morphology had influenced the glymphatic flow, associated with protein waste depletion and possibly related to distinct forms of dementia (Nedergaard and Goldman 2020). In evolutionary terms, it is also worth noting that all these mechanisms are related to sleep–wakefulness regulation, which has probably experienced major changes in the human genus as a consequence of bipedalism and open habitat (Coolidge and Wynn 2006).

As usually happens in evolutionary specialization, those who fly high can fall very far. The cognitive advantages of enhanced parietal functions are more useful for reproductive success than the drawbacks of dementia, which usually occurs in late life stages, rarely achieved in hunter–gatherers and, anyway, less involved in the reproductive function.

Counterarguments and Evidence against Evolutionary Influences

The hypothesis discussed here is, at present, largely speculative and based on the mere observation that, only in our species, there are large and complex parietal lobes and a very prevalent form of dementia that is associated with a specific parietal vulnerability. The association is, indeed, reasonable but far to be proven. Most evolutionary theories can be tentatively explored by studying fossil species, living primates, or the diversity of modern populations (especially those living according to an ecological and economical niche more affine with the origin of our species, namely hunter–gatherers). In this case, however, the hypothesis can be difficult to be conclusively verified, through any of those approaches.

In the case of fossil species, if this evolutionary association is true, we can, provocatively enough, make the prediction that extinct hominids were less prone to develop this kind of neurodegeneration, independently of life-span differences. Of course, such a prediction cannot be fully tested, at least directly. Concerning living species, a main limitation is associated with primate life history. In most primates, the post-reproductive stage is very short or absent and generally associated with an overall impairment of the biological functions. Humans are the only species in which the post-reproductive stage is not a short period associated with a general biological decadence but instead a large part of the life cycle, based on proper physiological mechanisms and ecological factors (Kaplan et al. 2000). We are also the only species with exceptionally complex parietal lobes. For both aspects, therefore, we are unique, a single and exceptional case. Consequently, being a singular exception, pathological conditions associated with post-reproductive stages (like AD) or the effect of large parietal lobes can hardly be investigated through an adequate comparative approach with other case studies. Finally, for what concerns comparisons with hunter–gatherers, a third limitation is associated with life expectancy that, in these populations, is, on average, lower than in modern ones. This hampers, at least in part, a proper statistical treatment of the issue because of the small samples and demographic biases.

Of course, these limitations point at the difficulties in testing the reliability of the hypothesis but are not against it. What is important, in this respect, is that the evolutionary perspective was able to suggest a factor (the complexity of the parietal lobe) that was not noticed before and that might deserve attention.

Clinical and Therapeutic Implications

The hypothesis of a relationship between parietal evolution and AD is not necessarily an all-or-nothing perspective, and it can be evaluated to a different degree. It can be hypothesized that AD might be (partially) a consequence of a complex and entangled parietal anatomy, so a price to pay for our powerful and specialized cerebral functions. Alternatively, it can be also proposed that such parietal organization might simply increase the sensitivity or the vulnerability to accumulation of AD-related processes, raising the incidence and progression of the disease. In the former case, the parietal architecture is the responsible variable for the onset of the pathology, while in the latter case the cause of AD is related to other factors (like genetic or biochemical factors) but the effects are amplified by the spatial, vascular, and thermal limitations of those regions. The theoretical and practical consequences of these two alternatives are very different, with different clinical implications. The first perspective would suggest that interventions need to target maintaining parietal health, possibly through lifestyle or pharmacological interventions to improve vascular clearance or heat loadings or non-invasive brain stimulation to entrain networks. However, the second position indicates that interventions focusing on the upstream determinants would be more promising and that parietal indicators can be meaningful for clinical prognosis. Nonetheless, in both cases, the evolution of a large and complex parietal anatomy is a key element of the AD etiology, and this should be carefully considered within a clinical perspective. If AD is somehow associated with the anatomical and physiological organization of the parietal cortex, then it should be possible to design proper medical approaches that take advantage of this information, at least on the above-mentioned two fronts.

Unanswered Questions and Future Work
Testing the Hypothesis

Many hypotheses have associated AD with some aspects of evolution, including lifespan extension, age-related selection bias, antagonistic pleiotropy, rapid brain evolution, and environmental mismatch (Fox 2018). Indeed, the biogeography of APOE-ε genetics suggest that natural selection is likely at work (Belloy et al. 2019). Nonetheless, hypotheses based on an evolutionary perspective can be difficult to test in a proper experimental way. Fossils are the only direct evidence of the evolutionary changes, and this is a major limitation when looking for an experimental approach to validate models and predictions. Also, evolutionary hypotheses are generally based on the concept of "adaptation": it is assumed that a specific adaptation can help in some way the success of a species or (like in this case) involve some negative drawbacks. However, an adaptation is easier to hypothesize than to test because, in the latter case, one should be able to check the reproductive success of a population over long time ranges—which is easier to do for flies than for primates. Therefore, in general, one should rely on a more practical ground, that is, by collecting multiple and independent types of evidence that apparently support or discard the hypothesis.

In the case of the possible relationship between parietal lobe evolutionary expansion and AD etiology, the first step should be looking for correlations between parietal anatomy and AD prevalence or markers. Such investigation should consider not only the size and proportions of the parietal cortex but also its shape, the sulcal patterns, and the associated vascular elements. At present, paradoxically, the information on these traits is rather scanty. Anatomy suffered a long stasis in the past century and the interest in this field was renewed recently, thanks to the advances in digital and computed morphology. Still today, much anatomical knowledge is based on schematic drawings or dissections of few subjects. For most features mentioned in this chapter, we still ignore the variability, the morphogenetic and ontogenetic patterns, the sexual or geographical differences, the homology in primates, or even the biological functions. Basically, what we need in this case is more research in basic anatomy.

Beyond this mandatory step, a second line of evidence can be supplied through modeling, that is, by the design of numerical models able to integrate anatomical and physiological information. Simulations can be used to investigate if, how, and to what extent vascular anatomy and cortical morphology can influence metabolic factors like heat production and distribution or blood clearance.

A third line of evidence, at first glance more speculative but congruent, can be the detection of osteological markers associated with AD. Many pathologies leave some sort of imprints in the bone as alterations of the bone structure (macroscopic or microscopic), metabolic modifications, or molecular traces. AD is associated with both metabolic and molecular changes, and we can wonder whether these changes can influence some aspects of bone biology which, although not necessarily relevant in terms of pathology, can be used as indirect clinical evidence. In some cases, the strict relationships between genes and pathological conditions can also be used as indirect markers, albeit with more caution. If such markers can be found for AD, then they can be used to investigate whether they are present in fossil remains, although with the sample limitations mentioned above.

Outlook: Dementia and the Social Brain

An additional and speculative note to discuss AD through a perspective in evolutionary anthropology concerns the social aspects associated with this disease. Neurodegenerative diseases have a profound social impact, in terms of social structure and emotions, and this must be considered when dealing with a zoological group that has evolutionarily invested so much in the social system. Many evolutionary features in primates are associated with their outstanding social complexity, and humans are the most social primates ever evolved. Brain size, in primates, is in part correlated with group size and with the complexity of the social organization (Dunbar 2018). The association cortex is particularly relevant in this relationship, and the parietal lobes are, in this sense, critical. The parietal cortex is particularly involved in somatosensorial management and hand coordination, and it is worth noting that, in primates, a large part of social bonding and group interactions are based on touch (Dunbar 2010). Furthermore, the integration between vision and body is essential to generate an autobiographical frame (the precuneus, because of its main role in this sense, has been sometimes called "the eye of the self"; Freton et al. 2014). Also, the body is a shared perceptual reference when managing space, time, and social relations (Peer et al. 2015; Hills et al. 2015; Maister et al. 2015). It is hence expected that key regions of our social capacity are also sensitive to impairments that profoundly affect the social structure. Beyond the consequence for the individual cognitive situation, the consequences of AD for the community also deserve a certain attention. Health care and the care for elderly people are null, scarce, or speculative for human species other than *H. sapiens*. Our species evolved, at the same time, a complex cognitive system, large and complex social groups, and self-awareness (Leary and Buttermore 2003). Therefore, AD likely appears in a species that was extremely social, conscious, and emotionally involved with its elders. Beyond the consequences on the individual health, we can expect that dementia should have triggered an admixture of astonishment, confusion, and suffering, in a society that, at the same time, was evolving some sort of spiritual or even mystical feelings. It can be speculated that, as soon as the prevalence of dementia increased because of an increase in life expectancy or reasons associated with brain biology, the social group experienced novel emotional conflicts and new questions dealing with the nature of the self. Such experiences should have represented, for the sapient hominid, a shocking revelation, questioning the integrity of the individual, its social role, and the role of the community and empirically revealing new conceptual challenges, like the impermanency of existence (even before death) or the mind–body dualism. Last but not least, the increasing prevalence of dementia in those populations of Upper Paleolithic hunter–gatherers should have represented, indeed, the very early roots of understanding the relationship between brain and behavior, namely the origin of neuropsychology.

Key Points to Remember

- The parietal lobes are particularly developed and specialized in primates, exceptionally in humans.
- The deeper regions of the parietal cortex (in particular the precuneus) are embedded in a complex and complicated structural and functional anatomical environment, influenced by spatial and metabolic constraints.
- These same regions are affected by β-amyloid proteinopathy, atrophy, and metabolic impairments during the early stages of AD.
- The expansion of the parietal cortex in humans may have increased the sensitivity to damages due to functional or structural imbalance between anatomical organization and physiological management.

Acknowledgment

E. B. is funded by the Spanish government (PID2021-122355NB-C33) and by the Italian Institute of Anthropology. H.I.L.J. is funded by the NIH-NIA (R01AG062559; R01AG068062; R01AG082006, R21AG074220).

References

Aiello, L. C., & Wheeler, P. (1995). The expensive-tissue hypothesis: The brain and the digestive system in human and primate evolution. *Current Anthropology, 36*(2), 199–221.

Ardesch, D. J., Scholtens, L. H., Li, L., Preuss, T. M., Rilling, J. K., & van den Heuvel, M. P. (2019). Evolutionary expansion of connectivity between multimodal association areas in the human brain compared with chimpanzees. *Proceedings of the National Academy of Sciences of the United States of America, 116*(14), 7101–7106. https://doi.org/10.1073/pnas.1818512116

Bartzokis, G. (2009). Alzheimer's disease as homeostatic responses to age-related myelin breakdown. *Neurobiology of Aging, 32*(8), 1341–1371. https://doi.org/10.1016/j.neurobiolaging.2009.08.007

Bateman, R. J., Xiong, C., Benzinger, T. L., Fagan, A. M., Goate, A., Fox, N. C., Marcus, D. S., Cairns, N. J., Xie, X., Blazey, T. M., Holtzman, D. M., Santacruz, A., Buckles, V., Oliver, A., Moulder, K., Aisen, P. S., Ghetti, B., Klunk, W. E., McDade, E., . . . Dominantly Inherited Alzheimer Network. (2012). Clinical and biomarker changes in dominantly inherited Alzheimer's disease. *New England Journal of Medicine, 367*(9), 795–804. https://doi.org/10.1056/NEJMoa1202753

Belloy, M. E., Napolioni, V., & Greicius, M. D. (2019). A quarter century of APOE and Alzheimer's disease: Progress to date and the path forward. *Neuron, 101*(5), 820–838. https://doi.org/10.1016/j.neuron.2019.01.056

Bischof, G. N., & Jacobs, H. I. L. (2019). Subthreshold amyloid and its biological and clinical meaning: Long way ahead. *Neurology, 93*(2), 72–79. https://doi.org/10.1212/WNL.0000000000007747

Braak, H., & Braak, E. (1991). Neuropathological staging of Alzheimer-related changes. *Acta Neuropathologica, 82*(4), 239–259. https://doi.org/10.1007/BF00308809

Braak, H., & Del Tredici, K. (2011). Alzheimer's pathogenesis: Is there neuron-to-neuron propagation? *Acta Neuropathologica, 121*(5), 589–595. https://doi.org/10.1007/s00401-011-0825-z

Bruner, E. (2017). The fossil evidence of human brain evolution. In J. Kaas (Ed.), *Evolution of nervous systems* (2nd ed., Vol. 4, pp. 63–92). Elsevier.

Bruner, E. (2018). Human paleoneurology and the evolution of the parietal cortex. *Brain, Behavior and Evolution, 91*(3), 136–147. https://doi.org/10.1159/000488889

Bruner, E. (2019). Human paleoneurology: Shaping cortical evolution in fossil hominids. *Journal of Comparative Neurology*, *527*(10), 1753–1765. https://doi.org/10.1002/cne.24591

Bruner, E. (2021). Evolving human brains: Paleoneurology and the fate of Middle Pleistocene. *Journal of Archaeological Method and Theory*, *28*(1), 76–94. https://doi.org/10.1007/s10816-020-09500-8

Bruner, E. (2022). A network approach to the topological organization of the Brodmann map. *Anatomical Record*, *305*(12), 3504–3515. https://doi.org/10.1002/ar.24941

Bruner, E. (2023). The evolution of the parietal lobes in the genus *Homo*: The fossil evidence. In E. Bruner (Ed.), *Cognitive archaeology, body cognition, and the evolution of visuospatial perception* (pp. 153–179). Elsevier.

Bruner, E., Amano, H., Pereira-Pedro, A. S., & Ogihara, N. (2018a). The evolution of the parietal lobes in the genus *Homo*. In E. Bruner, N. Ogihara, & H. C. Tanabe (Eds.), *Digital endocasts: From skulls to brains* (pp. 219–237). Springer.

Bruner, E., Battaglia-Mayer, A., & Caminiti, R. (2022). The parietal lobe evolution and the emergence of material culture in the human genus. *Brain Structure and Function*, *228*(1), 145–167. https://doi.org/10.1007/s00429-022-02487-w

Bruner, E., & Beaudet, A. (2022). The brain of *Homo habilis*: A tribute to the paleoneurology of Phillip Tobias. *Journal of Human Evolution*, *174*, Article 103281. https://doi.org/10.1016/j.jhevol.2022.103281

Bruner, E., de la Cuétara, J. M., Masters, M., Amano, H., & Ogihara, N. (2014). Functional craniology and brain evolution: From palaeontology to biomedicine. *Frontiers in Neuroanatomy*, *8*, Article 19. https://doi.org/10.3389/fnana.2014.00019

Bruner, E., & Eisová, S. (2024). Vascular microforamina and endocranial surface: normal variation and distribution in adult humans. *Anatomical Record*. https://doi.org/10.1002/ar.25426.

Bruner, E., Esteve-Altava, B., & Rasskin-Gutman, D. (2019). A network approach to brain form, cortical topology and human evolution. *Brain Structure and Function*, *224*(6), 2231–2245. https://doi.org/10.1007/s00429-019-01900-1

Bruner, E., & Gleeson, B. T. (2019). Body cognition and self-domestication in human evolution. *Frontiers in Psychology*, *10*, Article 1111. https://doi.org/10.3389/fpsyg.2019.01111

Bruner, E., & Iriki, A. (2016). Extending mind, visuospatial integration, and the evolution of the parietal lobes in the human genus. *Quaternary International*, *405*, 98–110. https://doi.org/10.1016/j.quaint.2015.05.019

Bruner, E., & Jacobs, H. I. L. (2013). Alzheimer's disease: The downside of a highly evolved parietal lobe? *Journal of Alzheimer's Disease*, *35*(2), 227–240. https://doi.org/10.3233/JAD-122299

Bruner, E., Mantini, S., Musso, F., de la Cuétara, J. M., Ripani, M., & Sherkat, S. (2011). The evolution of the meningeal vascular system in the human genus: From brain shape to thermoregulation. *American Journal of Human Biology*, *23*(1), 35–43. https://doi.org/10.1002/ajhb.21123

Bruner, E., Preuss, T., Chen, X., & Rilling, J. (2017). Evidence for expansion of the precuneus in human evolution. *Brain Structure and Function*, *222*(2), 1053–1060. https://doi.org/10.1007/s00429-015-1172-y

Bruner, E., Spinapolice, E., Burke, A., & Overmann, K. A. (2018b). Visuospatial integration: Paleoanthropological and archaeological perspectives. In L. D. Di Paolo, F. Di Vincenzo, & F. De Petrillo (Eds.), *Evolution of primate social cognition* (pp. 299–326). Springer.

Buckner, R. L., Snyder, A. Z., Shannon, B. J., LaRossa, G., Sachs, R., Fotenos, A. F., Sheline, Y. I., Klunk, W. E., Mathis, C. A., Morris, J. C., & Mintun, M. A. (2005). Molecular, structural, and functional characterization of Alzheimer's disease: Evidence for a relationship between default activity, amyloid, and memory. *Journal of Neuroscience*, *25*(34), 7709–7717. https://doi.org/10.1523/JNEUROSCI.2177-05.2005

Bzdok, D., Hartwigsen, G., Reid, A., Laird, A. R., Fox, P. T., & Eickhoff, S. B. (2016). Left inferior parietal lobe engagement in social cognition and language. *Neuroscience and Biobehavioral Reviews*, *68*, 319–334. https://doi.org/10.1016/j.neubiorev.2016.02.024

Coolidge, F., & Wynn, T. (2006). The effects of the tree-to-ground sleep transition in the evolution of cognition in early *Homo*. *Before Farming*, *2006*(4), 1–19. https://doi.org/10.3828/bfarm.2006.4.11

Corder, E. H., Robertson, K., Lannfelt, L., Bogdanovic, N., Eggertsen, G., Wilkins, J., & Hall, C. (1998). HIV-infected subjects with the E4 allele for APOE have excess dementia and peripheral neuropathy. *Nature Medicine*, *4*(10), 1182–1184. https://doi.org/10.1038/2677

Dumoncel, J., Subsol, G., Durrleman, S., Bertrand, A., de Jager, E., Oettlé, A. C., Lockhat, Z., Suleman, F. E., & Beaudet, A. (2021). Are endocasts reliable proxies for brains? A 3D quantitative comparison of the extant human brain and endocast. *Journal of Anatomy*, *238*(2), 480–488. https://doi.org/10.1111/joa.13318

Dunbar, R. I. M. (2010). The social role of touch in humans and primates: Behavioural function and neurobiological mechanisms. *Neuroscience and Biobehavioral Reviews*, *34*(2), 260–268. https://doi.org/10.1016/j.neubiorev.2008.07.001

Dunbar, R. I. M. (2018). The anatomy of friendship. *Trends in Cognitive Sciences*, *22*(1), 32–51. https://doi.org/10.1016/j.tics.2017.10.004

Ellis, S., Franks, D. W., Nattrass, S., Cant, M. A., Bradley, D. L., Giles, D., Balcomb, K. C., & Croft, D. P. (2018). Postreproductive lifespans are rare in mammals. *Ecology and Evolution*, *8*(5), 2482–2494. https://doi.org/10.1002/ece3.3856

Finch, C. E., & Stanford, C. B. (2004). Meat-adaptive genes and the evolution of slower aging in humans. *Quarterly Review of Biology*, *79*(1), 3–50. https://doi.org/10.1086/381662

Fox, M. (2018). Evolutionary medicine perspectives on Alzheimer's disease: Review and new directions. *Ageing Research Reviews*, *47*, 140–148. https://doi.org/10.1016/j.arr.2018.07.008

Freire-Cobo, C., Edler, M. K., Varghese, M., Munger, E., Laffey, J., Raia, S., In, S. S., Wicinski, B., Medalla, M., Perez, S. E., Mufson, E. J., Erwin, J. M., Guevara, E. E., Sherwood, C. C., Luebke, J. I., Lacreuse, A., Raghanti, M. A., & Hof, P. R. (2021). Comparative neuropathology in aging primates: A perspective. *American Journal of Primatology*, *83*(11), Article e23299. https://doi.org/10.1002/ajp.23299

Freton, M., Lemogne, C., Bergouignan, L., Delaveau, P., Lehéricy, S., & Fossati, P. (2014). The eye of the self: Precuneus volume and visual perspective during autobiographical memory retrieval. *Brain Structure and Function*, *219*(3), 959–968. https://doi.org/10.1007/s00429-013-0546-2

Fullerton, S. M., Clark, A. G., Weiss, K. M., Nickerson, D. A., Taylor, S. L., Stengård, J. H., Salomaa, V., Vartiainen, E., Perola, M., Boerwinkle, E., & Sing, C. F. (2000). Apolipoprotein E variation at the sequence haplotype level: Implications for the origin and maintenance of a major human polymorphism. *American Journal of Human Genetics*, *67*(4), 881–900. https://doi.org/10.1086/303070

Gharbi-Meliani, A., Dugravot, A., Sabia, S., Regy, M., Fayosse, A., Schnitzler, A., Kivimäki, M., Singh-Manoux, A., & Dumurgier, J. (2021). The association of APOE ε4 with cognitive function over the adult life course and incidence of dementia: 20 years follow-up of the Whitehall II study. *Alzheimer's Research & Therapy*, *13*(1), Article 5. https://doi.org/10.1186/s13195-020-00740-0

Goldring, A. B., & Krubitzer, L. A. (2017). Evolution of the parietal cortex in mammals: From manipulation to tool use. In J. Kass (Ed.), *Evolution of nervous systems* (2nd ed., Vol. 3, pp. 259–286). Elsevier.

Goulas, A., Bastiani, M., Bezgin, G., Uylings, H. B. M., Roebroeck, A., & Stiers, P. (2014). Comparative analysis of the macroscale structural connectivity in the macaque and human brain. *PLOS Computational Biology*, *10*(3), Article e1003529. https://doi.org/10.1371/journal.pcbi.1003529

Grefkes, C., & Fink, G. R. (2005). The functional organization of the intraparietal sulcus in humans and monkeys. *Journal of Anatomy*, *207*(1), 3–17. https://doi.org/10.1111/j.1469-7580.2005.00426.x

Groot, C., Risacher, S. L., Chen, J. Q. A., Dicks, E., Saykin, A. J., MacDonald, C. L., Mez, J., Trittschuh, E. H., Mukherjee, S., Barkhof, F., Scheltens, P., van der Flier, W. M., Ossenkoppele, R., Crane, P. K., & Alzheimer's Disease Neuroimaging Initiative (ADNI). (2021). Differential trajectories of hypometabolism across cognitively-defined Alzheimer's disease subgroups. *NeuroImage: Clinical*, *31*, Article 102725. https://doi.org/10.1016/j.nicl.2021.102725

Gunz, P., & Harvati, K. (2007). The Neanderthal "chignon": Variation, integration, and homology. *Journal of Human Evolution*, *52*(3), 262–274. https://doi.org/10.1016/j.jhevol.2006.08.010

Hagmann, P., Cammoun, L., Gigandet, X., Meuli, R., Honey, C. J., Wedeen, V. J., & Sporns, O. (2008). Mapping the structural core of human cerebral cortex. *PLOS Biology*, *6*(7), Article e159. https://doi.org/10.1371/journal.pbio.0060159

Hanseeuw, B. J., Betensky, R. A., Schultz, A. P., Papp, K. V., Mormino, E. C., Sepulcre, J., Bark, J. S., Cosio, D. M., LaPoint, M., Chhatwal, J. P., Rentz, D. M., Sperling, R. A., & Johnson, K. A. (2017). Fluorodeoxyglucose metabolism associated with tau-amyloid interaction predicts memory decline. *Annals of Neurology*, *81*(4), 583–596. https://doi.org/10.1002/ana.24910

Hashimoto, M., Ho, G., Sugama, S., Takamatsu Y., Shimizu, Y., Takenouchi, T., Waragai, M., & Masliah, E. (2018). Evolvability of amyloidogenic proteins in human brain. *Journal of Alzheimer's Disease, 62*(1), 73–83. https://doi.org/10.3233/JAD-170894

Herculano-Houzel, S. (2017). Numbers of neurons as biological correlates of cognitive capability. *Current Opinion in Behavioral Sciences, 16*, 1–7. https://doi.org/10.1016/j.cobeha.2017.02.004

Hills, T. T., Todd, P. M., Lazer, D., Redish, A. D., Couzin, I. D., & Cognitive Search Research Group. (2015). Exploration versus exploitation in space, mind, and society. *Trends in Cognitive Sciences, 19*, 46–54. https://doi.org/10.1016/j.tics.2014.10.004

Hoeijmakers, L., Heinen, Y., van Dam, A. M., Lucassen, P. J., & Korosi, A. (2016). Microglial priming and Alzheimer's disease: A possible role for (early) immune challenges and epigenetics. *Frontiers in Human Neuroscience, 10*, Article 398. https://doi.org/10.3389/fnhum.2016.00398

Holloway, R. L., Broadfield, D. C., & Yuan, M. S. (2004). *The human fossil record*: Vol. 3. *Brain endocasts: The paleoneurological evidence*. Wiley.

Huang, C. W., Hsu, S. W., Chang, Y. T., Huang, S. H., Huang, Y. C., Lee, C. C., Chang, W. N., Lui, C. C., Chen, N. C., & Chang, C. C. (2018). Cerebral perfusion insufficiency and relationships with cognitive deficits in Alzheimer's disease: A multi-parametric neuroimaging study. *Scientific Reports, 8*(1), Article 1541. https://doi.org/10.1038/s41598-018-19387-x

Hunt, A., Schönknecht, P., Henze, M., Seidl, U., Haberkorn, U., & Schröder, J. (2007). Reduced cerebral glucose metabolism in patients at risk for Alzheimer's disease. *Psychiatry Research: Neuroimaging, 155*(2), 147–154. https://doi.org/10.1016/j.pscychresns.2006.12.003

Huntenburg, J. M., Bazin, P. L., & Margulies, D. S. (2017). Large-scale gradients in human cortical organization. *Trends in Cognitive Sciences, 22*(1), 21–31. https://doi.org/10.1016/j.tics.2017.11.002

Jack, C. R., Jr., Bennett, D. A., Blennow, K., Carrillo, M. C., Dunn, B., Haeberlein, S. B., Holtzman, D. M., Jagust, W., Jessen, F., Karlawish, J., Liu, E., Molinuevo, J. L., Montine, T., Phelps, C., Rankin, K. P., Rowe, C. C., Scheltens, P., Siemers, E., Snyder, H. M., & Sperling, R. (2018). NIA-AA research framework: Toward a biological definition of Alzheimer's disease. *Alzheimer's and Dementia, 14*(4), 535–562. https://doi.org/10.1016/j.jalz.2018.02.018

Jacobs, H. I. L., Becker, J. A., Kwong, K., Engels-Dominguez, E., Prokopiou, P. C., Papp, K. V., Properzi, M., Hampton, O. L., d'Oleire Uquillas, F., Sanchez, J. S., Rentz, D. M., El Fakhri, G., Normandin, M. D., Price, J. C., Bennett, D. A., Sperling, R. A., & Johnson, K. (2021). In vivo and neuropathology data support locus coeruleus integrity as indicator of Alzheimer's disease pathology and cognitive decline. *Science Translational Medicine, 13*(612), Article eabj2511. https://doi.org/10.1126/scitranslmed.abj2511

Jacobs, H. I. L., Gronenschild, E. H. B. M., Evers, E. A. T., Ramakers, I. H. G. B., Hofman, P. A. M., Backes, W. H., Jolles, J., Verhey, F. R. J., & van Boxtel, M. P. J. (2015). Visuospatial processing in early Alzheimer's disease: A multimodal neuroimaging study. *Cortex, 64*, 394–406. https://doi.org/10.1016/j.cortex.2012.01.005

Jacobs, H. I. L., Hedden, T., Schultz, A. P., Sepulcre, J., Perea, R. D., Amariglio, R. E., Papp, K. V., Rentz, D. M., Sperling, R. A., & Johnson, K. A. (2018). Structural tract alterations predict downstream tau accumulation in amyloid-positive older individuals. *Nature Neuroscience, 21*(3), 424–431. https://doi.org/10.1038/s41593-018-0070-z

Jacobs, H. I. L., van Boxtel, M. P. J., Jolles, J., Verhey, F. R. J., & Uylings, H. B. M. (2012). Parietal cortex matters in Alzheimer's disease: An overview of structural, functional and metabolic findings. *Neuroscience and Biobehavioral Reviews, 36*(1), 297–309. https://doi.org/10.1016/j.neubiorev.2011.06.009

Jacobs, H. I. L., van Boxtel, M. P. J., Uylings, H. B. M., Gronenschild, E. H. B. M., Verhey, F. R. J., & Jolles, J. (2011a). Atrophy of the parietal lobe in preclinical dementia. *Brain and Cognition, 75*(2), 154–163. https://doi.org/10.1016/j.bandc.2010.11.003

Jacobs, H. I. L., van Boxtel, M. P. J., van der Elst, W., Burgmans, S., Smeets, F., Gronenschild, E. H. B. M., Verhey, F. R. J, Uylings, H. B. M., & Jolles, J. (2011b). Increasing the diagnostic accuracy of medial temporal lobe atrophy in Alzheimer's disease. *Journal of Alzheimer's Disease, 25*(3), 477–490. https://doi.org/10.3233/JAD-2011-102043

Kalamatianos, T., Mavridis, I. N., Karakosta, E., Drosos, E., Skandalakis, G. P., Kalyvas, A., Piagkou, M., Koutsarnakis, C., & Stranjalis, G. (2019). The parieto-occipital artery revisited: A microsurgical anatomic study. *World Neurosurgery, 126*, e1130-e1139. https://doi.org/10.1016/j.wneu.2019.02.215

Kaplan, H., Hill, K., Lancaster, J., & Hurtado, A. M. (2000). A theory of human life history evolution: Diet, intelligence, and longevity. *Evolutionary Anthropology, 9*(4), 156–185. https://doi.org/10.1002/1520-6505(2000)9:4<156::AID-EVAN5>3.0.CO;2-7

Koch, G., Bonnì, S., Pellicciari, M. C., Casula, E. P., Mancini, M., Esposito, R., . . . Bozzali, M. (2018). Transcranial magnetic stimulation of the precuneus enhances memory and neural activity in prodromal Alzheimer's disease. *Neuroimage, 169,* 302–311. https://doi.org/10.1016/j.neuroimage.2017.12.048

Land, M. F. (2014). Do we have an internal model of the outside world? *Philosophical Transactions of the Royal Society B: Biological Sciences, 369*(1636), Article 20130045. https://doi.org/10.1098/rstb.2013.0045

Lanfermann, H., Raab, P., Kretschmann, H. J., & Weinrich, W. (2019). *Cranial neuroimaging and clinical neuroanatomy.* Thieme.

Leary, M. R., & Buttermore, N. R. (2003). The evolution of the human self: Tracing the natural history of self-awareness. *Journal for the Theory of Social Behaviour, 33*(4), 365–404. https://doi.org/10.1046/j.1468-5914.2003.00223.x

Leigh, S. R. (2004). Brain growth, life history, and cognition in primate and human evolution. *American Journal of Primatology, 62*(3), 139–164. https://doi.org/10.1002/ajp.20012

Leonard, W. R., Snodgrass, J. J., & Robertson, M. L. (2007). Effects of brain evolution on human nutrition and metabolism. *Annual Review of Nutrition, 27,* 311–327. https://doi.org/10.1146/annurev.nutr.27.061406.093659

Maister, L., Slater, M., Sanchez-Vives, M. V., & Tsakiris, M. (2015). Changing bodies changes minds: Owning another body affects social cognition. *Trends in Cognitive Sciences, 19*(1), 6–12. https://doi.org/10.1016/j.tics.2014.11.001

Mangla, R., Kolar, B., Almast, J., & Ekholm, S. E. (2011). Border zone infarcts: Pathophysiologic and imaging characteristics. *RadioGraphics, 31*(5), 1201–1214. https://doi.org/10.1148/rg.315105014

Mavridis, I. N., Kalamatianos, T., Koutsarnakis, C., & Stranjalis, G. (2016). Microsurgical anatomy of the precuneal artery: Does it really exist? Clarifying an ambiguous vessel under the microscope. *Operative Neurosurgery, 12*(1), 68–76. https://doi.org/10.1227/NEU.0000000000001082

Mevel, K., Chételat, G., Eustache, F., & Desgranges, B. (2011). The default mode network in healthy aging and Alzheimer's disease. *International Journal of Alzheimer's Disease, 2011,* Article 535816. https://doi.org/10.4061/2011/535816

Moss, M. L., & Young, R. W. (1960). A functional approach to craniology. *American Journal of Physical Anthropology, 18*(4), 281–292. https://doi.org/10.1002/ajpa.1330180406

Nedergaard, M., & Goldman, S. A. (2020). Glymphatic failure as a final common pathway to dementia. *Science, 370*(6512), 50–56. https://doi.org/10.1126/science.abb8739

Neubauer, S., Gunz, P., & Hublin, J. J. (2009). The pattern of endocranial ontogenetic shape changes in humans. *Journal of Anatomy, 215*(3), 240–255. https://doi.org/10.1111/j.1469-7580.2009.01106.x

Neubauer, S., Hublin, J. J., & Gunz, P. (2018). The evolution of modern human brain shape. *Science Advances, 4*(1), Article eaao5961. https://doi.org/10.1126/sciadv.aao5961

Papma, J. M., den Heijer, T., de Koning, I., Mattace-Raso, F. U., van der Lugt, A., van der Lijn, F., van Swieten, J. C., Koudstaal, P. J., Smits, M., & Prins, N. D. (2013). The influence of cerebral small vessel disease on default mode network deactivation in mild cognitive impairment. *NeuroImage: Clinical, 2,* 33–42. https://doi.org/10.1016/j.nicl.2012.11.005

Parvizi, J., Braga, R. M., Kucyi, A., Veit, M. J., Pinheiro-Chagas, P., Perry, C., Sava-Segal, C., Zeineh, M., van Staalduinen, E. K., Henderson, J. M., & Markert, M. (2021). Altered sense of self during seizures in the posteromedial cortex. *Proceedings of the National Academy of Sciences of the United States of America, 118*(29), Article e2100522118. https://doi.org/10.1073/pnas.2100522118

Peer, M., Salomon, R., Goldberg, I., Blanke, O., & Arzy, S. (2015). Brain system for mental orientation in space, time, and person. *Proceedings of the National Academy of Sciences of the United States of America, 112*(35), 11072–11077. https://doi.org/10.1073/pnas.1504242112

Pereira-Pedro, A. S., Bruner, E., Gunz, P., & Neubauer, S. (2020). A morphometric comparison of the parietal lobe in modern humans and Neanderthals. *Journal of Human Evolution, 142,* Article 102770. https://doi.org/10.1016/j.jhevol.2020.102770

Perez, S. E., Raghanti, M. A., Hof, P. R., Kramer, L., Ikonomovic, M. D., Lacor, P. N., Erwin, J. M., Sherwood, C. C., & Mufson, E. J. (2013). Alzheimer's disease pathology in the neocortex and hippocampus of the western lowland gorilla (*Gorilla gorilla gorilla*). *Journal of Comparative Neurology*, *521*(18), 4318–4338. https://doi.org/10.1002/cne.23428

Perez, S. E., Sherwood, C. C., Cranfield, M. R., Erwin, J. M., Mudakikwa, A., Hof, P. R., & Mufson, E. J. (2016). Early Alzheimer's disease–type pathology in the frontal cortex of wild mountain gorillas (*Gorilla beringei beringei*). *Neurobiology of Aging*, *39*, 195–201. https://doi.org/10.1016/j.neurobiolaging.2015.12.017

Perosa, V., Priester, A., Ziegler, G., Cardenas-Blanco, A., Dobisch, L., Spallazzi, M., Assmann, A., Maass, A., Speck, O., Oltmer, J., Heinze, H. J., Schreiber, S., & Düzel, E. (2020). Hippocampal vascular reserve associated with cognitive performance and hippocampal volume. *Brain*, *143*(2), 622–634. https://doi.org/10.1093/brain/awz383

Pihlajamaki, M., & Sperling, R. A. (2009). Functional MRI assessment of task-induced deactivation of the default mode network in Alzheimer's disease and at-risk older individuals. *Behavioural neurology*, *21*(1), 77–91. https://doi.org/10.3233/BEN-2009-0231

Píšová, H., Rangel de Lázaro, G., Velemínský, P., & Bruner, E. (2017). Craniovascular traits in anthropology and evolution: From bones to vessels. *Journal of Anthropological Sciences*, *95*, 35–65. https://doi.org/10.4436/JASS.95003

Provenzano, F., & Deleidi, M. (2021). Reassessing neurodegenerative disease: Immune protection pathways and antagonistic pleiotropy. *Trends in Neurosciences*, *44*(10), 771–780. https://doi.org/10.1016/j.tins.2021.06.006

Raichlen, D. A., & Alexander, G. E. (2014). Exercise, APOE genotype and the evolution of the human lifespan. *Trends in Neurosciences*, *37*(5), 247–255. https://doi.org/10.1016/j.tins.2014.03.001

Rami, L., Sala-Llonch, R., Solé-Padullés, C., Fortea, J., Olives, J., Lladó, A., . . . Molinuevo, J. L. (2014). Distinct functional activity of the precuneus and posterior cingulate cortex during encoding in the preclinical stage of Alzheimer's disease. *Journal of Alzheimer's Disease*, *31*(3), 517–526. https://doi.org/10.3233/JAD-2012-120223

Reyes, L. D., Do Kim, Y., Issa, H., Hopkins, W. D., Mackey, S., & Sherwood, C. C. (2022). Cytoarchitecture, myeloarchitecture, and parcellation of the chimpanzee inferior parietal lobe. *Brain Structure and Function*, *228*(1), 63–82. https://doi.org/10.1007/s00429-022-02514-w

Rosen, R. F., Farberg, A. S., Gearing, M., Dooyema, J., Long, P. M., Anderson, D. C., Davis-Turak, J., Coppola, G., Geschwind, D. H., Pare, J. F., Duong, T. Q., Hopkins, W. D., Preuss, T. M., & Walker, L. C. (2008). Tauopathy with paired helical filaments in an aged chimpanzee. *Journal of Comparative Neurology*, *509*(3), 259–270. https://doi.org/10.1002/cne.21744

Rowe, C. C., Ellis, K. A., Rimajova, M., Bourgeat, P., Pike, K. E., Jones, G., Fripp, J., Tochon-Danguy, H., Morandeau, L., O'Keefe, G., Price, R., Raniga, P., Robins, P., Acosta, O., Lenzo, N., Szoeke, C., Salvado, O., Head, R., Martins, R., . . . Villemagne, V. L. (2010). Amyloid imaging results from the Australian Imaging, Biomarkers and Lifestyle (AIBL) study of aging. *Neurobiology of Aging*, *31*(8), 1275–1283. https://doi.org/10.1016/j.neurobiolaging.2010.04.007

Rzechorzek, N. M., Thrippleton, M. J., Chappell, F. M., Mair, G., Ercole, A., CENTER-TBI High Resolution ICU (HR ICU) Sub-Study Participants and Investigators, Rhodes, J., Marshall, I., & O'Neill, J. S. (2022). A daily temperature rhythm in the human brain predicts survival after brain injury. *Brain*, *145*(6), 2031–2048. https://doi.org/10.1093/brain/awab466

Scheperjans, F., Eickhoff, S. B., Hömke, L., Mohlberg, H., Hermann, K., Amunts, K., & Zilles, K. (2008). Probabilistic maps, morphometry, and variability of cytoarchitectonic areas in the human superior parietal cortex. *Cerebral Cortex*, *18*(9), 2141–2157. https://doi.org/10.1093/cercor/bhm241

Scott Miners, J. S., Palmer, J. C., & Love, S. (2016). Pathophysiology of hypoperfusion of the precuneus in early Alzheimer's disease. *Brain Pathology*, *26*(4), 533–541. https://doi.org/10.1111/bpa.12331

Smith, C. J., Ashford, J. W., & Perfetti, T. A. (2019). Putative survival advantages in young apolipoprotein ε4 carriers are associated with increased neural stress. *Journal of Alzheimer's Disease*, *68*(3), 885–923. https://doi.org/10.3233/JAD-181089

Sperling, R. A., Mormino, E. C., Schultz, A. P., Betensky, R. A., Papp, K. V., Amariglio, R. E., Hanseeuw, B. J., Buckley, R., Chhatwal, J., Hedden, T., Marshall, G. A., Quiroz, Y. T., Donovan, N. J., Jackson, J., Gatchel, J. R., Rabin, J. S., Jacobs, H., Yang, H. S., Properzi, M., . . . Johnson, K. A. (2019). The impact of amyloid-beta and tau on prospective cognitive decline in older individuals. *Annals of Neurology*, *85*(2), 181–193. https://doi.org/10.1002/ana.25395

Stricker, N. H., Schweinsburg, B. C., Delano-Wood, L., Wierenga, C. E., Bangen, K. J., Haaland, K. Y., Frank, L. R., Salmon, D. P., & Bondi, M. W. (2009). Decreased white matter integrity in late-myelinating fiber pathways in Alzheimer's disease supports retrogenesis. *NeuroImage, 45*(1), 10–16. https://doi.org/10.1016/j.neuroimage.2008.11.027

Torvik, A. (1984). The pathogenesis of watershed infarcts in the brain. *Stroke, 15*(2), 221–223. https://doi.org/10.1161/01.str.15.2.221

Trumble, B. C., Stieglitz, J., Blackwell, A. D., Allayee, H., Beheim, B., Finch, C. E., Gurven, M., & Kaplan, H. (2017). Apolipoprotein E4 is associated with improved cognitive function in Amazonian forager-horticulturalists with a high parasite burden. *FASEB Journal, 31*(4), 1508–1515. https://doi.org/10.1096/fj.201601084R

van Exel, E., Koopman, J. J. E., van Bodegom, D., Meij, J. J., de Knijff, P., Ziem, J. B., Finch, C. E., & Westendorp, R. G. J. (2017). Effect of APOE ε4 allele on survival and fertility in an adverse environment. *PLOS ONE, 12*(7), Article e0179497. https://doi.org/10.1371/journal.pone.0179497

Walker, L. C., & Jucker, M. (2017). The exceptional vulnerability of humans to Alzheimer's disease. *Trends in Molecular Medicine, 23*(6), 534–545. https://doi.org/10.1016/j.molmed.2017.04.001

Yamazaki, Y., Zhao, N., Caulfield, T. R., Liu, C. C., & Bu, G. (2019). Apolipoprotein E and Alzheimer disease: Pathobiology and targeting strategies. *Nature Reviews. Neurology, 15*(9), 501–518. https://doi.org/10.1038/s41582-019-0228-7

Zilles, K., Palomero-Gallagher, N., & Amunts, K. (2013). Development of cortical folding during evolution and ontogeny. *Trends in Neurosciences, 36*(5), 275–284. https://doi.org/10.1016/j.tins.2013.01.006

Parkinson's Disease

Overstrain Focused on Basal Ganglia and Brainstem Nuclei

Nico J. Diederich and Christopher G. Goetz

Introduction

Parkinson's disease (PD) is the second most common neurodegenerative disease. Men are more frequently affected than women, and increasing prevalence with age is robustly observed throughout the world. Due to increased longevity, the worldwide prevalence estimate in 2015 was 6 million subjects, and the 2040 projections predict 17 million PD patients worldwide (Dorsey et al. 2018). Traditionally the diagnosis is made when there is abnormal slowness of movements or bradykinesia, in conjunction with one or both of the following symptoms: muscle stiffness (rigidity) and rest or postural tremor. When the disease advances, the patients also show an instability of posture, and there is high risk for falls (postural instability). With best medical treatment, mean life expectancy is estimated to be at least 15 years from the beginning of first symptoms. Based on these symptoms, PD is primarily a dopaminergic disease with degeneration of the pars compacta of the substantia nigra (SN), and the subcortical anchor circuitry essentially involves the basal ganglia (BG). The neuropathological hallmarks are a substantial loss of neurons in the SN as well as local burden of cell inclusions, called Lewy bodies (LB) in the surviving cells. Protein aggregates, among them misfolded alpha-synuclein (aSyn), are the main components of LB (see Chapter 4). However, beyond the core motor syndrome, there is a panoply of non-motor symptoms, some encountered as forerunner symptoms, such as hyposmia or rapid eye movement (REM) sleep behavior disorder (see Chapter 12), and others such as dementia and severe dysphagia occurring late in the course of the disease. The concept of a PD complex has therefore been proposed, with the three core motor symptoms being only the "tip of the iceberg" (Langston 2006).

Why Search for Evolutionary Grounding of Human PD?

The pathophysiology of PD remains only partially elucidated despite tremendous progress. The most compelling hypothesis proposes a cell-to-cell propagation of misfolded aSyn (Prusiner 2001). Furthermore, aging, environmental toxins, and genetic predisposition are posited to play major roles. A single explanation embracing all these factors, whether triggering the disease or allowing progression, is still missing. While there has been success in animal modeling through toxin exposure or genetic manipulation and while antipsychotic-induced Parkinsonian syndrome can easily be provoked in animals, no naturally occurring model of PD exists, and therefore no perfect or near-perfect replication of the human PD phenotype has been developed in animals, and naturally occurring PD appears to be an exclusively human disease. Admittedly, we lack expert observational studies of mammals living in the wild, especially of large animal species; but the search has never revealed major candidates. A hasty answer could be that, in the wild, animals do not reach the age of PD susceptibility; but in the best-observed species living close to humans (dogs and cats), frailty and old age are well defined without any specific parkinsonism. Also, systematic in vivo or postmortem studies in aged mammals are rare and limited for addressing the specific PD issue, even in the closest natural nonhuman models, namely cage-living nonhuman primates. In a pioneer study, Emborg et al. (1998) identified significant motor impairments (slowed fine motor tasks, reduced home cage activity, and tremor) in aged rhesus monkeys. The tremor occurred during activity and, importantly, was not a rest tremor, the typical hallmark of human parkinsonism. At postmortem, the animals had age-dependent loss of tyrosine hydroxylase and dopamine transporter reactivity but no LB as the pathological hallmark of human PD. In another study (Hurley et al. 2011), aged female monkeys developed tremor and delayed initiation of movements but again had no LB and were refractory to dopaminergic treatment, the typical pharmacological drug class useful to abate human PD. Based on the present knowledge, it has been concluded that in these primates neuropathological data are compatible with aging and a possible "vulnerable pre-parkinsonian state," although the biological threshold for clinical and cellular pathological Parkinsonism has not been reached (Collier et al. 2017).

Unsettled Mystery on Phenotype Diversities

Beyond animal models, there are further unsolved questions. Despite the broad interindividual variability in PD expression, the four cardinal features of PD (bradykinesia, rigidity, tremor, and postural instability) cluster with remarkable consistency. Regardless of the comorbidities, exposure to external trigger factors, aging, or genetic predisposition, these seemingly disparate signs compose a neurological archetype of high distinction relative to other conditions. However, first symptoms and progression patterns vary widely, and unanswered questions focus on whether these features are explained by a single network or connectome defect or a multi-nuclear pathway that may intersect with non-motor functional determinants during the disease. Also, when the clinical syndrome enlarges (or is better differentiated), do we have to search for "fixed or predetermined" propagation routes, or can there be disease emergence almost similarly at different sites but with variable clinical expression?

By adopting an evolutionary point of view, this chapter revisits the PD clinical repertoire, emphasizing evidence for pathological involvement of evolutionarily old pathways. We posit that "deeply buried" and automatically functioning behaviors of evolutionarily old systems may no longer be clinically efficient in terms of energy costs (see Chapter 3) or functional accessibility to the human cortex or that, in the converse, there may be unchecked release of ancient reactions reminiscent of "fight or flight" reflexes. The discussion covers sensory deficits, numerous aspects of gait dysfunction, posture, facial expression, emotional dysregulation, dysautonomia, and visual functions. We will trace the phylogenetic history of the most affected areas ranging from the enteric nervous system (ENS) to brainstem nuclei, amygdala, etc. It will be important to point out that not all symptoms can be ranged in such a way. Tremor is difficult to insert in phylogenetic considerations. Also, in general, we lack a thorough understanding of the impact on disease expression by life history, epigenetic factors, environmental toxins, etc. In the outlook section we encourage research on potentially underlying evolutionarily rooted pathways by modern imaging techniques such as functional magnetic resonance imaging (fMRI), and we break a lance for a renaissance of comparative neurology and neuropathology with thorough observation of elderly subjects of other species.

Historical Background

Several neurological researchers have addressed evolutionary concepts when discussing the motor system. Among them, Jackson (1835–1911), surely one of the most influential, proposed an anatomical and physiological hierarchy of higher and lower centers of the nervous system: "The higher the centre, the more numerous, different, and more complex, and more special movements it represents, and the wider region it represents" (Jackson 1958; York and Steinberg 2011, p. 3109). Within the model, the prefrontal cortex represents the top level of the hierarchy. Jackson considered that the higher centers primarily suppress the function of the lower ones. Clinically "negative" symptoms manifest when loss of the controlling cortex occurs, and "positive" symptoms emerge from unsuppressed lower centers. The neuroscientist and psychiatrist MacLean (1913–2007) proposed a physiological model with evolutionary anchors in 1949, with refinements over the years, culminating in his legacy 1990 opus *The Triune Brain in Evolution* (MacLean 1990). In his three-layer brain model, the most archaic neural structures form the "reptilian complex" or "R-complex," which is evolutionarily traced back to reptiles and includes parts of the medulla, pons, cerebellum, midbrain (paleostriatum), and the olfactostriatum (part of the nucleus accumbens). Neural programs responsible for automatic gait patterns, balance, and autonomic functions are governed by this layer. The "paleomammalian complex," roughly corresponding to the limbic system, is the second layer and addresses emotions and instincts, fighting and fleeing, feeding and sexual behavior. Finally, the top evolutionary layer is the "neomammalian complex," the evolutionarily most recent brain development, responsible for more flexible, elaborate, and complex functions.

Catapulting to the late 20th century, historians highlight several neuroscientists closely interested in evolutionary studies of the motor system, some of them still actively involved in the field. These studies fall into two primary disciplines: neuropathology and neurophysiology. The renowned neuropathologist André Parent (b. 1944) compared the micro- and macro-anatomy of the BG in different species ranging from reptiles to primates. Among a vast array of contributions, he prophetically announced that "structures that appear early in evolution, are among the first to undergo involution in aging diseases" (Parent 1997). However, it has been already noted that the most elementary centers regulating breathing and feeding are not, or are only very lately, affected by the disease. Still more recently, Sten Grillner (b. 1941) and Karolinska colleagues first identified spinally located central pattern generators for locomotion in vertebrates and later published extensively on the complex circuitries in the BG, extant for more than 500 million years (Grillner and Robertson 2016). In neurophysiology, Redgrave and his team reported in 1989 that the tiny mammalian superior colliculus can distinguish events from emergencies (Dean et al. 1989) and dictates immediate, appropriate motor behaviors. His contributions also include the concept that the vertebrate BG handle a central switching device "specialized to resolve conflicts over access to limited motor and cognitive resources" (Redgrave et al. 1999, p. 1009). From the phylogenetic perspective, it has been concluded that old closed-loop connections between BG and brainstem nuclei, such as the superior colliculus, may be differentially sensitive to threats or appetitive stimuli ("emergencies or events") that lead to diseases involving motor systems of coordination and timing. While thoroughly tracing back the evolutionary history of the motor systems, these authors, except for Parent, did not directly allude to the clinical repercussions of such old toolboxes.

In contrast to these major scientific laboratory contributions, clinical researchers have been less productive in analyzing Parkinsonism from an evolutionary perspective. The renowned David Marsden (1938–1998), founder of modern clinical research in movement disorders and chair at the Institute of Neurology, Queen Square, provided a seminal Wartenberg Lecture in 1982 on the "Mysterious Motor Function of the Basal Ganglia" (Marsden 1982). He considered that the BG are responsible for the "automatic execution of learned motor programs" such as "fight/flight" and opened the field of movement disorders to a conceptual framework that included evolutionary considerations.

Evolutionary Evidence Specifically Related to Human Health and Disease

Evolutionary View on Pathological Lesions and Disruptive Pathways

In this section, *pathways* or *nuclei* involved in the PD process and as evidenced by neuroimaging or neuropathology will be presented. Both phylogenetic and ontogenetic considerations will be discussed as available. In the next section, we will present the resulting *clinical syndromes*, again emphasizing their postulated evolutionary rooting. This dissection/segregation may appear disruptive at first look as the usual Charcotian clinical–pathological paradigm is reversed. We justify our approach by newer connectomic concepts that will best anchor the clinical consideration and unify some signs that tend to occur in unison. For instance, we discuss the amygdala and the superior colliculi, which are involved in multiple pathways with varied clinical symptoms. If available, we first present *data from comparative neuroanatomy* because we will argue that the human size of a certain nucleus of the brainstem or the BG pertinent to PD may not be adequately proportional to the size of the human telencephalon. As such, our chapter theme is in part based on the premise that these evolutionarily old or archaic systems have been "left behind" during phylogenetic evolution (see also Chapter 22). However, we admit from the onset of our discussion that pure size measurements poorly reflect the extraordinary development of the human brain, which also shows cellular specialization and diversity and is more reactive to environmental stimuli during its development than in other mammals (Sherwood et al. 2012). We present our arguments in terms of *classical neuropathology* if in a certain area there is LB burden, cellular loss, atrophy of the nucleus. There is an ongoing debate on the rationale for the search for LB as possibly being a primary pathophysiological event (aggregation), a secondary protective event, or even an epiphenomenon (Espay et al. 2019). In contrast, the intricate interaction between aSyn and dopamine levels is without such debate. An in-depth discussion on this subject is beyond the topic of this essay. However, it is commonly accepted that the presence of LB at least documents local cellular interaction with the disease process. If available, we finally compare *fMRI* or other *imaging results* between healthy subjects and PD patients. Comparative fMRI studies between humans and other species are still very rare. Of note, we will not detail the cellular and subcellular commonalities between the areas affected, specifically the impact of high energy demands and cellular stress, because they are described in extensive detail in other chapters in this book (see especially Chapter 3).

BG: A 500-Million-Year-Old System at the Core of the Disease

Grillner and his team compared the mammalian lamprey BG, the lamprey having separated from other vertebrates 560 million years ago (Grillner and Robertson 2016). They found striking similarities in structure and function. Specifically, in terms of PD, across the vast epochs, a direct dopaminergic pathway responsible for execution of an automatic action and an indirect dopaminergic pathway inhibiting "competitive behaviors" can be identified. While in lampreys, the modules of action organized in this way are more simple and less numerous, with evolution, more modules have been added including identified centers for bipedal locomotion (in humans), posture, mastication, and ocular motility. Importantly, the newer modules retained the same design or functional structure as the older ones, a concept called "exaptation." Despite these new functions and the complexity of demands during the long human life, the cellular structures underlying the human BG have only evolved minimally. They currently function under continuous strain. In human PD, a tipping point is reached: the BG energy equilibrium between supply and demand collapses, and the system becomes deficient, as a function of the aging process. Several other disproportionalities come into play. In humans the quotient of neurons of the SN pars compacta in relation to a striatal volume unit is only a tenth of this quotient in rodents (Bolam and Pissadaki 2012). The striatal output to the neocortex is also disproportional: one striatal dopaminergic neuron in humans has over 10^6 synapses, in contrast to roughly over 10^5 synapses in rodents. The exceptional growth of the human telencephalon adds to this continuous strain, and it has been estimated that in humans the neocortical growth is 5 times that of the striatum (Stephan and Andy 1969). In PD the arguments for direct consequences in the BG are overwhelming. There is substantial neuronal cell loss in the SN, especially of the ventrolateral cells projecting to the posterolateral putamen. However, clinical PD signs are only seen when there is already a cell loss of 80%.

Olfactory and Visual Pathways

PD patients suffer from a substantial olfactory loss, called "hyposmia." It is often a forerunner syndrome, developing years before the onset of the motor signs that permit clinical diagnosis of PD. In comparison to other mammals, there is rudimentary development of the olfactory system in humans. The olfactory bulb is only 30% as large as it should be for a primate *brain* of human size but, remarkably, 1.6 times larger than expected for a primate *body* of our size. Neuropathological findings are also somehow contradictory: on one side there is heavy burden of LB but on the other side no volume reduction of the olfactory bulb.

For the visual system, PD patients often show deficits in light entrainment during their daily functions as well as subtle deficits in terms of missing color or contrast discrimination. Evolutionarily grounded dysfunctions are seen at various levels of the involved pathways (Lee et al. 2022). At the retinal level, melanopsin-containing ganglion cells have been identified as the phylogenetically old system responsible for nictemeral regulation. They innervate the suprachiasmatic nucleus and various other brain regions. Specifically, the number of these neurons essential for light entrainment during numerous body functions is small and even further reduced in PD. There is also reduction of retinal amacrine cells interacting with the melanopsin-containing cells (Ortuño-Lizarán et al. 2020). Finally, neuropathology studies reveal not precisely layered retinal aSyn immunoreactivity and LB deposits in the retina and suprachiasmatic nucleus.

The Amygdala as Orchestra Conductor of Numerous Musicians

The essential role of amygdalar dysfunction in numerous PD symptoms cannot be over-emphasized, ranging from hyposmia to decision-making to dysautonomia. Globally an amygdalar syndrome with both positive and negative signs can be proposed (Diederich et al. 2016; Figure 9.1). In the evolutionary perspective, the amygdala is primarily an alert system, although it also handles pleasure. Based on a crude information matrix,

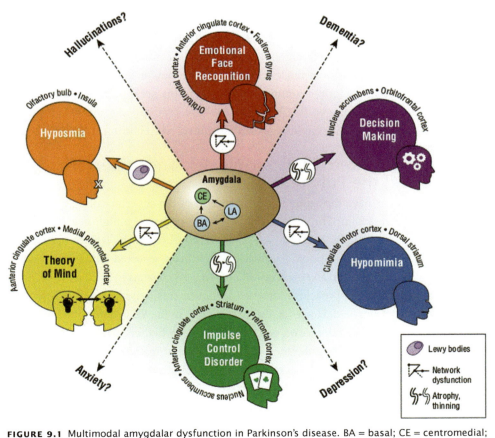

FIGURE 9.1 Multimodal amygdalar dysfunction in Parkinson's disease. BA = basal; CE = centromedial; LA = lateral.

fear-related cues are expediently detected and handled, and automatic or reflex responses are triggered in the domain of the autonomic nervous system (ANS). The amygdala can be divided into the evolutionarily older centromedial (CE) parts and the evolutionarily newer and larger laterobasal (LBA) parts. The central and corticomedial nuclei compose the CE. The laterobasal, lateral, basal, and mediobasal cell groups form the LBA, which operates as the recipient part of the amygdala. The phylogenetic evolution of both parts is highly discordant. In humans the central amygdaloid nucleus is only a third of the size expected for hominoids, and in consequence, a relative involution of CE must be considered. In contrast, there is relative size increase of the LBA in primates, even more in humans. The LBA has a 37% larger volume than predicted for an ape with a hemisphere volume of human size (Barger et al. 2014).

Neuroimaging of the amygdala in vivo is challenging, considering the tiny dimensions of this area, with even greater challenges to visually dissect its subcomponents. In PD, complex network dysfunctions and rearrangements must be presumed, although in anxious PD patients, *increased* functional connectivity between the amygdala and the orbitofrontal cortex has been consistently reported in several studies (Carey et al. 2021). Morphometry studied by MRI has shown a global amygdalar volume reduction in PD-related depression, a finding that markedly contrasts with the usual volumetric increase seen in non-Parkinsonian depressive patients. Does the pathological involvement of the amygdala by the PD process preferentially affect the older (central) parts of the amygdala, as we have shown for brainstem nuclei? While we have not yet resolved this question, Braak has reported severe lesions among PD patients in the areas of the CE that handle autonomic functions (Braak et al. 1994). Others have reported significant amygdalar volume reduction, without specifying the area most involved (Harding et al. 2002).

Mesencephalic and Thalamic Nuclei: The Superior Colliculi and the Pulvinar

Comparative studies have shown strong preservation of the superior colliculi (SC) as a coordinating alarm system. Throughout the vertebrate pedigree, a high degree of similarity and overlap has been identified. The SC are reminiscent of the tectum opticum of reptiles, and striking organizational parallelisms have been reported (King 2004). Their role in spatial orientation and alertness is well known, and neurons in the deeper SC layer register and respond to approaching or evasive movements, based on information received within the visual field. However, the alerting role of the SC is not limited to visual stimuli as the SC detect and decide on reaction to threats, whether certain or ambiguous, whether near or far, prompting the motor system with automatic escape movements as needed (Branco & Redgrave 2020). A provocative hypothesis has extended the SC role to initiate "fight or flight" also during REM sleep (see Chapter 12). The consequences of direct involvement of the SC in PD are far from being understood, missing blindsight being one of the most impressive syndromes. Indeed, the retino-colliculo-thalamo-amygdala

(RCTA) pathway targeting the amygdala runs through the SC. In healthy subjects, part of this pathway transmitting visual threat has been visualized, and the microstructure of the SC–amygdala pathway (as seen by fractional anisotropy) predicted the bias to orient saccades to the threat (Koller et al. 2019). In PD there is excessive constant inhibition of the SC by the overactive SN pars reticulata, thus inhibiting correct saccade initiation and disrupting the normal coupling of action and perception (Pretegiani et al. 2019). Therefore, the SC have been considered as the dysfunctional "bottleneck" of saccades (Terao et al. 2013). So far these SC dysfunctions have been assumed to be a collateral damage of the SN dysfunction. However, at least in patients with LB dementia and visual hallucinations, pronounced aSyn deposits in the deeper SC layers and reductions of neuronal density in the intermediate layers have been reported (Erskine et al. 2019). In contrast, the pulvinar, thalamic relay station of the RCTA, has only a low LB burden. It shows secondary changes to the neurodegeneration of connected nuclei, such as enriched expression of transcripts related to immune or synaptic functions (Erskine et al. 2019). The thalamus itself has a surprisingly sparse literature with an evolutionary focus useful for the study of PD. The key nuclei important to tremor regulation, one of the clinical hallmarks of PD, are the ventral lateral nuclei and the ventral anterior nuclei, often the targets for deep brain stimulation and lesions in PD. Both systems are intimately involved in movement and learning, with afferents from the BG and cerebellum and efferents directed cortically to the motor cortex, supplementary motor cortex, and prefrontal cortex. To the best of our knowledge, thalamic evolutionary changes relative to human disease have not been studied as such. However, in comparison to the dorsal thalamus, "exceptionally large cortices" due to expansion of higher-order regions of cortex have been reported in primates (Halley & Krubitzer 2019), thus indirectly alluding again to extended requirements of the thalamus in humans.

Brainstem Nuclei: The Pedunculopontine Nucleus and the Locus Coeruleus

Along the vertebrate pedigree the pedunculopontine nucleus (PPN) serves as relay between the BG and the spinal motor systems, and it is essential for gait pattern regulation. Together with the latero-dorsal tegmental nucleus, its cholinergic neurons "awaken" the brain and, more generally, update behaviors to environmental conditions (Mena-Segovia and Bolam 2017). PPN dysfunction in PD induces gait disturbances and lack of motivation and alertness, and it contributes to REM sleep behavior disorder (see Chapter 12). The wide cholinergic PPN innervation can be visualized by specific PET markers (Bohnen et al. 2022). Neuropathology studies reveal that there is early PPN involvement during the ascending disease propagation. The noradrenergic locus coeruleus (LC) and its surroundings are presented in detail in the chapter on REM sleep behavior disorder (see Chapter 12). aSyn immunoreactive inclusions have been described in the LC in incidental LB disease, presumed to advance to PD if the subjects would have lived longer.

From the Spinal Cord to the ENS

The involvement of the spinal cord in PD has so far only poorly been investigated. Central pattern generators generating evolutionarily old patterns of motion are located at this level. So far it is unknown if they are directly involved in the PD process, although LB have been described in PD patients downward of fourth thoracic segment (Del Tredici and Braak 2012).

The ANS is an early evolutionary acquisition present in all vertebrates. Conventionally, we divide it into the sympathetic, parasympathetic, and enteric nervous systems. While this separation is anatomically appropriate, there are numerous physiological intersections. Claude Bernard (1878) already assumed that the ANS assures stability of the interior environment (*milieu intérieur*), while Cannon (1929) proposed the ANS as the front line when there is risk of "fight or flight." The development of the ANS precedes the development of the central nervous system (CNS). Mammals show striking arrangement similarities with other vertebrates. In contrast to the spinal cord, there is overwhelming evidence of direct PD involvement of the ANS. At the cardiac level reduced noradrenergic uptake at the presynaptic level of postganglionic axons can be visualized by metaiodobenzylguanidine (MIBG)-SPECT, even at very early stages of the disease. There are no immediate consequences of this involvement, and it is thought that this potentially old network constitutes a last rescue pathway ready to accelerate heart rate in case of danger.

The ENS must also be considered in this discussion. Evolutionarily, it develops earlier than the CNS and has been therefore coined the "first brain" (Furness and Stebbing 2018). In humans, as the CNS has shown exponential development over time, the ENS has undergone relative shrinking. With the human dietary shift to more easily digestible food with higher calorie content, the human gastrointestinal system reduced its size and complexity at the same time, while the brain extended its development. In PD, gut mucosal barrier leakage of neurotoxins has been posited as the first port of entry for at least some cases of PD. Braak's conceptual framework of an ascending pathology from the gastrointestinal or olfactory system to the brainstem has been discussed for decades, and deposition of aSyn can appear in mesenteric neurites years prior to the clinical diagnosis of PD. Ganglionic atrophy in the ENS and local LB deposits have been reported in various autopsy series, but their specificity has been debated. Beyond the dopaminergic system, PET imaging by 5-[^{11}C]-methoxy-donepezil (^{11}C-donepezil) can visualize reduced acetylcholinesterase density of these cells (Horsager et al. 2020).

Commonalities between Nuclei and Pathways Affected in PD

The presented areas and nuclei are only a *selection* of presumed evolutionarily old and directly affected areas and nuclei in PD. The list could go on, with discussion of the habenula, claustrum, insula, or nucleus Meynert. As a crystallizing point, affected areas all play complex coordinating roles related primarily to automatized functions. Most could be considered as hubs, meaning centers integrating multiple secondary functions. However, it must be emphasized that evolutionary remodeling of hub circuitries may also occur as manifestations of adaptation or compensation. As such, at both the conceptual and clinical levels, the separation of primary disease involvement and secondary compensatory symptoms is often impossible to disentangle.

The most robust commonalities between these affected areas and pathways concern the cellular organization. As discussed in Chapter 3, most of these systems are continuously in action, inducing a high level of energy strain for organelles, especially for the mitochondria. The neurons in charge also have long axons when reaching out to the telencephalon or the periphery (see Chapter 4). Beyond these cellular characteristics, organizational common points must be emphasized. Some of the information provided—being it spinal general pattern generation for gait, crude olfactory information on disgusted odors, low contrast vision on approaching beings, perception of motion in the periphery, heart rate regulation, smooth gastrointestinal transit, information on light conditions—is a crucial prerequisite for automatized, subconscious, rapid patterns of action, thus "liberating" the telencephalon from laborious case-by-case decisions and executions. The clinical manifestations of these multiple themes in terms of PD will be discussed in the following section.

Evolutionary View of Clinical Syndromes of PD

While the previous section revisited the pathways and cerebral areas affected by PD, the following section addresses the resulting clinical signs of PD with emphasis on their evolutionary rooting. While the above presented primary motor triad retains its "gold standard" for diagnosing PD, other elements of the PD phenotype also are appropriate for an evolutionary discussion. Therefore, the discussion herein incorporates both the motor and non-motor symptoms as an evolutionary construct for analysis in a series of distinctive clinical PD hallmarks. As only one example, bradykinesia will be analyzed in the sections covering hypomimia (masked facies) but also within a larger discussion on emotional dysregulation and again in the section on dysfunctional gait. All selected salient clinical components of PD are anchored within the unifying concept that there is an evolutionary rooting of this symptom. Within such a framework, we propose three possibilities of evolutionary rooting or impregnation: first, there can be pathologically missing access to evolutionarily old pathways that otherwise function well, mostly involved in automaticity and reaction speed; second, primitive reactions pathologically lose their capacity to finely tune and adjust to external or internal cues; third, pathological unmasking or unleashing of formerly helpful, but now obsolete, reactions can impair the patient, the archetype being here sudden gait freezing or REM sleep behavior disorder (see Chapter 12).

Bipedal Gait—A Late Evolutionary Acquisition and Vulnerable Target in PD

Humans are the only hominoids practicing bipedal gait as adults. However, within the great apes including chimpanzees, gorillas, orangutans, and bonobos, the latter can also engage in bipedal ambulation for at least a distance of 20 m. It is reasonable to search for anatomical adaptations at the evolutionary level that distinguish quadrupedal from bipedal locomotor modes. Importantly, the distinction is not "either–or" since the bonobo example suggests that the anatomy engaged may remain polyvalent. The development of the specialized bipedalism gait and erect posture requires constraints of very specific types of locomotion (D'Août et al. 1989). The normal and healthy human adult posture is highly erect, and normal human gait is exclusively bipedal without falls. This locomotor pattern is 75% less energy-consumptive than quadrupedal gait, and humans execute repetitive low-cost locomotion. However, much higher organizational complexity is required for this mode of locomotion, and human muscle strength is lower in comparison to the chimpanzee (O'Neill et al. 2017).

Human ontogenetic gait development recapitulates phylogenetic evolution, starting with toddlers who crawl, then children who take small steps with a wide stance, low cadence, and short single-limb stance time, all maneuvers performed to reduce the risks of falls and injuries (Vaughan 2003). Later, a naturally erect posture develops, sometimes needing social training; and a strong, narrow-based gait with full arm swing and confident pivoting typifies the healthy adolescent and adult. With normal aging, the posture returns to mild flexion, step length naturally reduces, and a cautious, more irregular, protective navigation re-emerges. Speed, agility, flexibility, and most importantly automation of human gait are regulated by complex circuitries ranging from the basal ganglia to the PPN, the cuneiform and other nuclei in the mesencephalon and brainstem, and finally spinal central pattern generators (Kiehn 2016). Specifically, initiation of gait is mediated by mesencephalic locomotor nuclei and transmitted to the reticular formation in the hindbrain, the latter projecting to the spinal cord. Here, central pattern generators generate evolutionarily old motion patterns, thus assuring right/left or flexor/extensor alternating activation. Modulatory vestibular and proprioceptive signals give feedback information and assure stability of stance and gait. Importantly, the motion rhythm is completely automatized and without need of conscious cortical control.

Given the anatomical underpinnings of PD, it is reasonable to consider the gait of PD as a regression to earlier evolutionary modes. As one of the PD clinical hallmarks, gait automation is dramatically compromised and a significant source of impairment. The arm swinging, reminiscent of quadrupedal gait, becomes asymmetric or completely absent and thus no longer facilitates gait stabilization and acceleration. PD patients have problems of initiation gait and regularization of cadence and stride length. They may suddenly stop gait or "freeze," for instance, when entering through a door frame or encountering another walker. Internal rhythm or internal cueing normally assured by perfect functioning of the described complex pathways is missing, and PD patients must rely on external visual or auditory cues during ambulation. They must consciously focus on walking, and talk during walk is no more possible. They are unable to "rewire" automatic gait programs, for instance, by physiotherapy. In the final disease stages, they may show a "simian posture," with flexion of the trunk and elbows. In summary, PD patients lose the automaticity of gait; they may regress in part to former (ontogenetically transient) stepping patterns.

Emotional Dysregulation: From Poor Perception to Blunted Expression

Given that PD patients have overall normal day-to-day visual acuity, numerous authors have analyzed the perplexing lack of emotional expression and emotional recognition in PD patients. PD patients "selectively" show impairment in recognizing negative emotional facial expressions. At first glance, this observation may seem only of passing interest, but at the evolutionary survival level, correct identification of facial expressions with a negative valence is an essential prerequisite for recognition of potentially dangerous encounters and life threats. This failing emotion recognition, at least at the earliest phase, may be due to specific deficits in visual perception as distinct spatial frequencies are used in an instantaneous manner for processing faces and their emotional expressions (Vuilleumier et al. 2003). PD patients perform poorly when tested on low contrast sensitivity, and, indeed, a link between this visual impairment and poor recognition of human faces expressing "fear" has been reported in PD patients (Hipp et al. 2014). Moreover, at the central level, the temporal vision-associated cortices and the amygdala remain "silent" in PD patients exposed to fearful facial expressions Poor face recognition in PD also concerns the facial expressions of anger, sadness, and disgust. When confronted with these emotions in social communication, PD patients generally produce deficient responses in the BG (putamen and pallidum) and show trends for decreased responses in the mirror system, as evidenced by various neuroimaging studies (fMRI or PET) and in comparison to healthy controls (Arioli et al. 2022). In healthy subjects the amygdala is the master integrative hub for integrating pertinent afferents and assuring not only appropriate reciprocal face expression but also adequate autonomic reactions, such as increase of heart rate and respiration in preparation for fight or flight. In contrast, PD patients only develop poor startle reactions and blunted autonomic reactions when exposed in an experimental setting to aversive images of physical violence or aggressive animals (Bowers et al. 2006). Finally, the widely connected amygdala also informs prefrontal cortical areas on "first social impressions." In combination with memorized former encounters, it allows construction of a mental image of the emotional and cognitive insides of the counterpart, thus establishing the map of theory of mind. Again, PD patients show deficiencies in establishing such maps. They miss essential basic tools for efficient and fluent social communication with their surroundings. However, they retain the ability to laboriously establish the same matrices by cortical work, but they are put in pigeonholes as not interested in communication, lethargic, and inflexible. There is a fluent transition from the level of blunted facial expression to hypomimia, thus not only a consequence of the cardinal symptom bradykinesia.

Apathy or Missing Élan Vital

In PD, another transition is seen from slow social interaction speed to poor initiative in general, or apathy. This lack of drive must be distinguished from depression, for instance, by missing a feeling of culpability or inadequate sadness. Patients lose their interest in their hobbies, and they practice social withdrawal. PET studies have shown overlapping deficient networks responsible, for both apathy and impaired emotional facial recognition, in PD patients (Robert et al. 2014). However, beyond involvement of limbic areas, frontal, temporal, and cerebellar areas implicated in reward, emotion, and cognition are also involved, thus pointing to the multidimensionality of apathy in etiology and clinical expression (Pagonabarraga et al. 2015). In consequence, although apathy may have some rooting in evolutionarily old defective mechanisms, in its clinical complexity the syndrome goes far beyond such an origin.

Sensory Deficits: Essentially Preemptive Perceptions? The Examples of Hyposmia and Deficient Blindsight

PD patients preserve most sensory functions such as visual or auditory acuity or proprioception. Nevertheless, they may show difficulties in correctly apprehending various environmental stimuli and in appropriately reacting to them. In this perspective, it has been proposed that PD patients show "preemptive" visual deficits (Bodis-Wollner 2008), meaning that an initial visual input does not instantly assure preparedness for action. Normally a preemptive perception orders neuronal mechanisms in association with voluntary actions before action initiation. It implicates that a *conscious* perceptual identification of the goal is *not* forcefully needed as the first, very rapid, although incomplete and *unconscious* perception can already anticipate and prepare the appropriate answer by the organism. We propose here to apply this concept to olfactory deficits and to blindsight, both deficient and highly characteristic of PD patients. What are the commonalities behind both perceptual categories? It is thought that these sensory inputs assure a "first track" (emotional) appreciation/evaluation of incoming information, thus allowing immediate and possibly life-saving reactions.

In the olfactory system, there is only one synapse between the perception of an olfactory stimulus and the arrival of the nerve impulse in the olfactory bulb. From here the axons project to the olfactory cortex, including among others the lateral entorhinal cortex and the cortical nucleus of the amygdala. Second-range olfactory projections are remarkably diffuse, ranging from the hypothalamus, the nucleus basalis Meynert, and the hippocampus to the mesencephalic reticular system and the orbitofrontal cortex. The dense connections with the limbic system explain the emotional connation of olfactory stimuli, and connections with the cortical amygdaloid nuclei implicate autonomic reactions. However, the olfactory epithelium is poorly developed in humans, and there is "shrinking and simplification" of the secondary olfactory pathways. The olfactostriatum, widely developed in snakes, has only an abortive correspondence as the vomeronasal system in humans, although part of the accumbens shell/ventral pallidum has been considered a homologue of the reptilian olfactostriatum. Therefore, the olfactory system probably maintains some link with the ventral striatum, as a reward system.

In PD patients, *hyposmia* may be a forerunner syndrome of the core motor syndrome by years to decades. Odor detection thresholds are increased, odor discrimination and identification are reduced, and there is impairment of odor memorization. Additionally, secondary repercussions may be equally, if not more, relevant clinically. For example, given that unpleasant or pleasant odors strongly modulate facial expressiveness, does diminished olfactory information explain at least in part the blank stare or masked facies of the Parkinsonian patient and the poor recognition by PD patients of emotional facial content among people with whom they interact? Odor stimuli activate the ventral striatum, and thus the behavioral reward system. This system is deficient in PD. Although a direct link to hyposmia has not yet been proven, PD patients may search for immediate reward by impulsive choices of action. They inappropriately increase the intake of dopaminergic medication when suffering from the dopamine dysregulation syndrome. A significant correlation between hyposmia and executive dysfunction has been demonstrated. Finally, PD patients with the most expressed olfactory deficit are at higher risk for developing visual hallucinations (for review, see Doty 2012).

Blindsight, or unconscious vision, is the most typical preemptive sensory function as unconsciously it directly paves the way for action after having received a visual input. The concept was introduced by Weiskrantz in 1974 after noticing the retained visual capacities of a patient with bilateral occipital damage. The pathways involved reach the extrastriate visual cortex, bypassing the conscious V1 visual cortex (see section above). They represent phylogenetically old sensory systems, being fully developed, for instance, in fish and reptilians. They are also an ontogenetically early human acquisition, being active already in newborns. What is the purpose of blindsight? Without conscious vision, it allows one to immediately guess the location of a potentially threatening object; to initiate reflexive saccades, thus bringing this object to the central visual area; and to follow moving objects. The affective blindsight allows one to apprehend threatening facial expressions and to appropriately react to them by reciprocal facial expressions and by increasing autonomic preparedness. Taken together, both parts of blindsight allow the subject to instantly react to dangerous stimuli. Essential relay stations of the involved pathways are the amygdala, the pulvinar, and the SC. The latter are usually inhibited by the BG, but for initiating a saccade, instantaneous disinhibition is required. The SC also prepare the frontal eye fields and get feedback information. To achieve high efficiency, blindsight operates in an economic way: it uses only selective low-contrast wavelength bands and stimulates only red-sensitive receptors.

Deficient blindsight is a second example of sensory misalignment in PD. As described in the section above, this clinical problem is due to disease involvement of the main relay stations of blindsight, namely the amygdala, pulvinar, and SC (Diederich et al. 2014; Figure 9.2). First, PD patients show reduced perception of low-contrast visual signals as well as poor color discrimination. However, this deficit involves the whole color spectrum and is not limited to red color perception. Second, oculomotor deficits include deficient reflexive grasping saccades: they are incomplete ("hypometric"), fragmented, not updated with new visual information, and too slow to reach the area of interest "in time"; and there may be prolonged fixation time. Motion perception is impaired. Third, indirect proof of dysfunctional blindsight in PD is also given by the observation that PD patients do not react appropriately to visual stimuli with a negative valence (disgust, anger, sadness) popping up in their visual field. There can be delayed or deficient emotional face recognition. PD patients do not develop the immediate and appropriate autonomic reactions to threatening visual stimuli seen in healthy subjects ("blunted emotional reaction"). Finally, they may perceive signals of "false alarm" when they erroneously guess that there may be moving objects in their peripheral visual field, so-called *sensations de passage*. While not linked to an emotional state of anxiety or to heightened vigilance, remarkably such erroneous perceptions are seen already in de novo PD patients, meaning without any impact of the dopaminergic treatment. In consequence, PD patients must completely rely on the more time- and energy-consuming conscious visual perception when interacting with their environment.

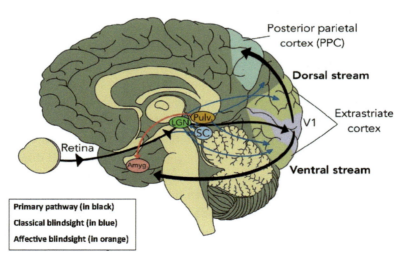

Figure 9.2 Affected areas and nuclei in deficient blindsight. The colored nuclei are responsible for functional blindsight. With the exception of the lateral geniculate nucleus (LGN), they are all directly or indirectly affected by the disease process in Parkinson's disease. Modified from Danckert et al. (2021) with permission. Amyg. = amygdala; LGN = lateral geniculate nucleus; Pulv = pulvinar; SC = superior colliculi.

Deficient Nictemeral Rhythm or Missing Light Entrainment

Humans naturally adopt a daytime rhythm for their routine activities. Imposed nighttime working hours remain challenging and not without long-term consequences on human health. Some schedules are given by so-called social *zeitgebers*, meaning by societal convenience. Most activities, however, are light entrained by a circadian clock in the suprachiasmatic nucleus. This clock gets its information by specific light-sensitive retinal receptors producing melanopsin. The information is further transmitted to the various organs, all obeying their internal clocks. Remarkably, we only dispose of a very small number of melanopsin-producing retinal receptors. However, the system is phylogenetically old, already fully in action in reptiles. There is also early ontogenetic maturation of the system. PD patients frequently suffer from severe sleep–wake dysregulation, producing fragmentation and disorganization of their daily activities, as there may be daytime sleepiness. In extreme situations PD patients completely lose the circadian regulation and show an inversion of the sleep–wake cycle: being awake during the night, being asleep during the day (Videnovic et al. 2014). Within the three proposed modes of evolutionary rooting, missing light entrainment can be considered another example of defective neuromodulation of human activities as PD patients lose the natural link to social *zeitgebers*.

Commonalities between Clinical Syndromes in PD

We have focused on the consequences of automation loss, specifically in the examples of gait and gastrointestinal functions. Also, in PD, we have cited the deficits of spontaneity in social interactions due to slow, aberrant, or insufficient information. Although almost all these actions and behaviors remain possible, unexpected problems in simple routine daily life actions or social interactions occur in PD. Fine-tuned neuromodulation impairments develop, for instance, the changes cited above in sensory input or autonomic reactions. More complicated, cortically driven, and controlled pathways take over but compete with the automatic functions, thereby compromising behaviors that used to be smooth and automatic. Compensatory networks involving, for instance, the cerebellum become active; and evolutionarily more recent "neural machinery" takes over, as reported for visual face processing (Tsao and Livingstone 2008). The analogy could be made that the PD brain is operating in the context of a broken-down computer system, where automated booking, terminal information, and boarding data malfunction, now leaving the system to return to older models of paperwork and printed boarding passes—systems that partially work but at a slower, more deliberate, and haltingly successful pace (Table 9.1).

TABLE 9.1 Comparison of Selected Deficient Behaviors and Affected Areas with Evolutionary Rooting in Parkinson's Disease

Description of deficient behavior	Areas involved	Evolutionary background	Proofs of involvement in PD
Deficient gait automatisms: slow gait initiation; irregular and small steps, freezing; sudden falls	Central pattern generators in spinal cord; pedunculopontine nucleus (PPN); basal ganglia (BG)	Central pattern generators in all vertebrates; BG over 500 millions years old	LB downward of fourth thoracic segment; LB in PPN and BG; substantial neuron loss in substantia nigra
Deficient colon motility: chronic constipation, even as forerunner syndrome	Enteric nervous system (ENS)	ENS develops ontogenetically and phylogenetically earlier than central nervous system	LB in ENS; shrinking of autonomic enteric ganglia. Leakage of colon epithelium as entry point of the PD process
Deficient blindsight: missing reflexive saccades; missing immediate appreciation of facial expression of counterparts; erroneous perception of motion and living beings in external visual field	Retina, amygdala, pulvinar, superior colliculi (SC)	Medial parts of the amygdala are evolutionarily old; SC are the analogue of the tectum opticum in reptiles	Cell loss and LB in amygdala; LB in SC
Deficient olfactory function: hyposmia may precede the core motor syndrome; missing disgust appreciation; potential impact on the reward system	Olfactory bulb and secondary olfactory areas	Oldest sensory system, involution in humans; rudimentary, reminiscent of olfactostriatum as part of nucleus accumbens	LB in olfactory bulb; could be another entry points of the PD process
Deficient nictemeral regulation: loss of natural day-/nighttime rhythmicity; somnolence during the day; insomnia during the night	Melanopsin-producing retinal cells (MPRC) and SCN	MPRC cells develop first; responsible for day-/nighttime rhythmicity even in animals with poor vision	LB at retinal level and in SCN

Counterarguments and Related Evidence against Evolutionary Influences

The concept of phylogenetically induced PD susceptibility can fit nicely into a holistic model of PD and does not need to be exclusionary of other proposed causative factors, genetic, toxic, inflammatory, or degenerative. It certainly does not exclude the crucial role of aging, which is the highest risk factor for PD (Reeve et al. 2014). Independently of the potentially different PD causes, we propose a logical approach to understand the vulnerability of pathways linked anatomically and functionally, emphasizing again that these evolutionarily old systems, whether subcortical or brainstem, whether nuclei, pathways, or circuits, are at specific risk for developing PD characteristic syndromes (Diederich et al. 2020).

At this point, admittedly several major primary and secondary PD syndromes "escape" any evolutionary explanation. As one example, resting or postural tremor, so characteristic of PD, does not fall into a clear evolutionary explanation. Although these tremor forms can be easily treated with deep brain stimulation or ablation or with surgical or ultrasound treatment of certain thalamic subnuclei, evolutionarily driven thalamic stress impacting the inhibitory thalamic oscillator has not been proven. The extent to which the thalamus keeps its multiple relay functions in PD remains highly speculative and defies a clear evolutionary model. The complex secondary syndrome of cognitive impairment starting as mild dysexecutive function and ending up in worst cases as dementia is probably multicausal and not immediately explicable from an evolutionary model. Here, enrolling disease propagation as progressive proteinopathy, Wallerian degeneration, and importantly, comorbidities must be considered. The same multidimensionality applies also for depression, often including reactive elements to a progressive loss of autonomy.

Clinical and Therapeutic Implications

Considering and reanalyzing PD symptoms from an evolutionary perspective kindles a ripe discussion of clinical pathophysiology in general and at the individual patient level since PD is characterized by a highly variable individual phenotype within a larger, stereotypic core. Braak's proposal that the pathology of PD progresses through an ascending anatomical propagation route from gut to brainstem to cortex has surely broadened our view of progressive PD pathology as it includes areas and nuclei responsible for both the motor and non-motor symptoms. The proposed cell-to-cell propagation of proteinopathy-based progression further facilitates our understanding of neurodegeneration and clinical decline. However, both mechanisms necessarily implicate a sequential disease progression and cannot directly explain early coexistence of symptoms ascribable to dysfunction in different pathways, not anatomically or physiologically related to one another (Diederich et al. 2019). In contrast, the consideration of evolutionarily induced disease susceptibility of various brain areas offers a different perspective. Here, we suggest that there may be isolated islands of disease vulnerability and initiation with, however, a strong common denominator. We suggest that this denominator is a gradual dysfunction of evolutionarily old neural structures as they no more deliver crude information on the potential hazardousness of the environment and no more trigger reflexive and automated behaviors. Depending on numerous local factors, these systems may "run out of energy," sequentially or simultaneously (see

Chapter 3). The exceptional human longevity may be the most important risk factor. A metaphor to facilitate the understanding of this concept of islands of susceptibility could be the risk of bush fires. Whole areas (the whole body) are exposed to this risk of fire, breaking out at exceptionally high temperatures (old age). However, local factors, such as hillside, local soil humidity, etc., may trigger the outbreak of the fire at some sites but not at others.

Proposals for Future Research

Future research can be directed to better disentangle primary disease involvement, as exposed above; secondary involvement through dysfunctional connectivity; Wallerian degeneration and concomitant disease; and, importantly, compensatory mechanisms already in place and misjudged as primary symptoms. We strongly suggest reinvestigating if there is some evolutionary model to explain the stress impact on the thalamus by the long human life. As composed of relay stations, the thalamus may be under constant cortical stress as well as receiving inputs from below to the point of collapse. In contrast, thalamic oscillators are easily treated with deep brain stimulation or ablation, surgical or ultrasound. This therapeutic efficiency shows that some suppressing influence has lost its power to unleash the devastation. Non-dopaminergic transmitters are involved in brainstem nuclei and autonomic ganglia. Primary recognition of these non-dopaminergic islands of disease onset should also propel pharmacological research in treatments beyond dopaminergic medications and *N*-methyl-D-aspartate antagonists. Newer neuroimaging techniques such as fMRI may help to identify evolutionarily old pathways. Better appreciation of local factors inhibiting or favoring disease initiation or progression in a certain pathway must be studied. Finally, we encourage in-depth comparative studies in elderly subjects of other species in order to disclose the mystery of how PD is (so far) an exclusively human disease.

Key Points to Remember

- PD occurs naturally only in humans.
- The affected circuitries and areas are evolutionarily old: some have stayed "behind" exponential neocortical expansion during hominization.
- These circuitries and areas are responsible for automatic behaviors and immediate reactions to the environment.
- These behaviors are deficient in PD, such as gait regulation, colon motility, fine autonomic tuning, and appropriate immediate reaction to environmental stimuli.
- Research in involved non-dopaminergic pathways and field studies in other long-living mammals are encouraged.

References

Arioli, M., Cattaneo, Z., Rusconi, M. L., Blandini, F., & Tettamanti, M. (2022). Action and emotion perception in Parkinson's disease: A neuroimaging meta-analysis. *NeuroImage: Clinical, 35*, Article 103031. https://doi.org/10.1016/j.nicl.2022.103031

Barger, N., Hanson, K. L., Teffer, K., Schenker-Ahmed, N. M., & Semendeferi, K. (2014). Evidence for evolutionary specialization in human limbic structures. *Frontiers in Human Neuroscience, 8*, Article 277. https://doi.org/10.3389/fnhum.2014.00277

Bernard, C. (1878). *Les phénomènes de la vie*. Éditions Baillère.

Bodis-Wollner, I. (2008). Pre-emptive perception. *Perception, 37*(3), 462–478. https://doi.org/10.1068/p5880

Bohnen, N. I., Yarnall, A. J., Weil, R. S., Moro, E., Moehle, M. S., Borghammer, P., Bedard, M. A., & Albin, R. L. (2022). Cholinergic system changes in Parkinson's disease: Emerging therapeutic approaches. *Lancet Neurology, 21*(4), 381–392. https://doi.org/10.1016/S1474-4422(21)00377-X

Bolam, J. P., & Pissadaki, E. K. (2012). Living on the edge with too many mouths to feed: Why dopamine neurons die. *Movement Disorders, 27*(12), 1478–1483. https://doi.org/10.1002/mds.25135

Bowers, D., Miller, K., Mikos, A., Kirsch-Darrow, L., Springer, U., Fernandez, H., Foote, K., & Okun, M. (2006). Startling facts about emotion in Parkinson's disease: Blunted reactivity to aversive stimuli. *Brain, 129*(12), 3356–3365. https://doi.org/10.1093/brain/awl301

Braak, H., Braak, E., Yilmazer, D., de Vos, R. A., Jansen, E. N., Bohl, J., & Jellinger, K. (1994). Amygdala pathology in Parkinson's disease. *Acta Neuropathologica, 88*(6), 493–500. https://doi.org/10.1007/BF00296485

Branco, T., & Redgrave, P. (2020). The neural basis of escape behavior in vertebrates. *Annual Review of Neuroscience, 43*, 417–439. https://doi.org/10.1146/annurev-neuro-100219-122527

Cannon, W. B. (1929). Organization for physiological homeostasis. *Physiological Reviews, 9*(3), 399–431. https://doi.org/10.1152/physrev.1929.9.3.399

Carey, G., Görmezoğlu, M., de Jong, J. J., Hofman, P. A., Backes, W. H., Dujardin, K., & Leentjens, A. F. (2021). Neuroimaging of anxiety in Parkinson's disease: A systematic review. *Movement Disorders, 36*(2), 327–339. https://doi.org/10.1002/mds.28404

Collier, T. J., Kanaan, N. M., & Kordower, J. H. (2017). Aging and Parkinson's disease: Different sides of the same coin? *Movement Disorders, 32*(7), 983–990. https://doi.org/10.1002/mds.27037

Danckert, J., Striemer, C., & Rossetti, Y. (2021). Blindsight. In J. J. S. Barton & A. Leff (Eds.), *Handbook of clinical neurology: Vol. 178. Neurology of vision and visual disorders* (pp. 297–310). Elsevier. https://doi.org/10.1016/B978-0-12-821377-3.00016-7

D'Août, K., Vereecke, E., Schoonaert, K., De Clercq, D., Van Elsacker, L., & Aerts, P. (2004). Locomotion in bonobos (*Pan paniscus*): Differences and similarities between bipedal and quadrupedal terrestrial walking, and a comparison with other locomotor modes. *Journal of Anatomy, 204*(5), 353–361. https://doi.org/10.1111/j.0021-8782.2004.00292.x

Dean, P., Redgrave, P., & Westby, G. M. (1989). Event or emergency? Two response systems in the mammalian superior colliculus. *Trends in Neurosciences, 12*(4), 137–147. https://doi.org/10.1016/0166-2236(89)90052-0

Del Tredici, K., & Braak, H. (2012). Spinal cord lesions in sporadic Parkinson's disease. *Acta Neuropathologica, 124*(5), 643–664. https://doi.org/10.1007/s00401-012-1028-y

Diederich, N. J., Goldman, J. G., Stebbins, G. T., & Goetz, C. G. (2016). Failing as doorman and disc jockey at the same time: Amygdalar dysfunction in Parkinson's disease. *Movement Disorders, 31*(1), 11–22. https://doi.org/10.1002/mds.26460

Diederich, N. J., Stebbins, G., Schiltz, C., & Goetz, C. G. (2014). Are patients with Parkinson's disease blind to blindsight? *Brain, 137*(6), 1838–1849. https://doi.org/10.1093/brain/awu094

Diederich, N. J., Surmeier, D. J., Uchihara, T., Grillner, S., & Goetz, C. G. (2019). Parkinson's disease: Is it a consequence of human brain evolution? *Movement Disorders, 34*(4), 453–459. https://doi.org/10.1002/mds.27628

Diederich, N. J., Uchihara, T., Grillner, S., & Goetz, C. G. (2020). The evolution-driven signature of Parkinson's disease. *Trends in Neurosciences, 43*(7), 475–492. https://doi.org/10.1016/j.tins.2020.05.001

Dorsey, E., Sherer, T., Okun, M. S., & Bloem, B. R. (2018). The emerging evidence of the Parkinson pandemic. *Journal of Parkinson's Disease*, *8*(S1), S3–S8. https://doi.org/10.3233/JPD-181474

Doty, R. L. (2012). Olfactory dysfunction in Parkinson disease. *Nature Reviews Neurology*, *8*(6), 329–339. https://doi.org/10.1038/nrneurol.2012.80

Emborg, M. E., Ma, S. Y., Mufson, E. J., Levey, A. I., Taylor, M. D., Brown, W. D., Holden, J. E., & Kordower, J. H. (1998). Age-related declines in nigral neuronal function correlate with motor impairments in rhesus monkeys. *Journal of Comparative Neurology*, *401*(2), 253–265. https://doi.org/10.1002/(SICI)1096-9861(19981116)401:2<253::AID-CNE7>3.0.CO;2-X

Erskine, D., Taylor, J. P., Thomas, A., Collerton, D., McKeith, I., Khundakar, A., Attems, J., & Morris, C. (2019). Pathological changes to the subcortical visual system and its relationship to visual hallucinations in dementia with Lewy bodies. *Neuroscience Bulletin*, *35*(2), 295–300. https://doi.org/10.1007/s12264-019-00341-4

Espay, A. J., Vizcarra, J. A., Marsili, L., Lang, A. E., Simon, D. K., Merola, A., Josephs, K. A., Fasano, A., Morgante, F., Savica, R., & Greenamyre, J. T. (2019). Revisiting protein aggregation as pathogenic in sporadic Parkinson and Alzheimer diseases. *Neurology*, *92*(7), 329–337. https://doi.org/10.1212/WNL.0000000000006926

Furness, J. B., & Stebbing, M. J. (2018). The first brain: Species comparisons and evolutionary implications for the enteric and central nervous systems. *Neurogastroenterology and Motility*, *30*(2), Article e13234. https://doi.org/10.1111/nmo.13234

Grillner, S., & Robertson, B. (2016). The basal ganglia over 500 million years. *Current Biology*, *26*(20), 1088–1100. https://doi.org/10.1016/j.cub.2016.06.041

Halley, A. C., & Krubitzer, L. (2019). Not all cortical expansions are the same: The coevolution of the neocortex and the dorsal thalamus in mammals. *Current Opinion in Neurobiology*, *56*, 78–86. https://doi.org/10.1016/j.conb.2018.12.003

Harding, A. J., Stimson, E., Henderson, J. M., & Halliday, G. M. (2002). Clinical correlates of selective pathology in the amygdala of patients with Parkinson's disease. *Brain*, *125*(11), 2431–2445. https://doi.org/10.1093/brain/awf251

Hipp, G., Diederich, N. J., Pieria, V., & Vaillant, M. (2014). Primary vision and facial emotion recognition in early Parkinson's disease. *Journal of the Neurological Sciences*, *338*(1–2), 178–182. https://doi.org/10.1016/j.jns.2013.12.047

Horsager, J., Andersen, K. B., Knudsen, K., Skjærbæk, C., Fedorova, T. D., Okkels, N., Schaeffer, E., Bonkat, S. K., Geday, J., Otto, M., Sommerauer, M., Danielsen, E. H., Bech, E., Kraft, J., Munk, O. L., Hansen, S. D., Pavese, N., Göder, R., Brooks, D. J., . . . Borghammer, P. (2020). Brain-first versus body-first Parkinson's disease: A multimodal imaging case-control study. *Brain*, *143*(10), 3077–3088. https://doi.org/10.1093/brain/awaa238

Hurley, P. J., Elsworth, J. D., Whittaker, M. C., Roth, R. H., & Redmond, D. E., Jr. (2011). Aged monkeys as a partial model for Parkinson's disease. *Pharmacology Biochemistry and Behavior*, *99*(3), 324–332. https://doi.org/10.1016/j.pbb.2011.05.007

Jackson, J. H. (1958). On some implications of dissolution of the nervous system. In J. Taylor (Ed.), *Selected writings of John Hughlings Jackson: Vol. 2. Evolution and dissolution of the nervous system, speech: Various papers, addresses and lectures* (pp. 22–44). Basic Books.

Kiehn, O. (2016). Decoding the organization of spinal circuits that control locomotion. *Nature Reviews Neuroscience*, *17*(4), 224–238. https://doi.org/10.1038/nrn.2016.9

King, A. J. (2004). The superior colliculus. *Current Biology*, *14*(9), R335–R338. https://doi.org/10.1016/j.cub.2004.04.018

Koller, K., Rafal, R. D., Platt, A., & Mitchell, N. D. (2019). Orienting toward threat: Contributions of a subcortical pathway transmitting retinal afferents to the amygdala via the superior colliculus and pulvinar. *Neuropsychologia*, *128*, 78–86. https://doi.org/10.1016/j.neuropsychologia.2018.01.027

Langston, J. W. (2006). The Parkinson's complex: Parkinsonism is just the tip of the iceberg. *Annals of Neurology*, *59*(4), 591–596. https://doi.org/10.1002/ana.20834

Lee, J. Y., Martin-Bastida, A., Murueta-Goyena, A., Gabilondo, I., Cuenca, N., Piccini, P., & Jeon, B. (2022). Multimodal brain and retinal imaging of dopaminergic degeneration in Parkinson disease. *Nature Reviews Neurology*, *18*(4), 203–220. https://doi.org/10.1038/s41582-022-00618-9

MacLean, P. D. (1990). *The triune brain in evolution: Role in paleocerebral functions*. Springer.
Marsden, C. D. (1982). The mysterious motor function of the basal ganglia: the Robert Wartenberg Lecture. *Neurology*, *32*(5), 514–539. https://doi.org/10.1212/WNL.32.5.514
Mena-Segovia, J., & Bolam, J. P. (2017). Rethinking the pedunculopontine nucleus: From cellular organization to function. *Neuron*, *94*(1), 7–18. https://doi.org/10.1016/j.neuron.2017.02.027
O'Neill, M. C., Umberger, B. R., Holowka, N. B., Larson, S. G., & Reiser, P. J. (2017). Chimpanzee super strength and human skeletal muscle evolution. *Proceedings of the National Academy of Sciences of the United States of America*, *114*(28), 7343–7348. https://doi.org/10.1073/pnas.1619071114
Ortuño-Lizarán, I., Sánchez-Sáez, X., Lax, P., Serrano, G. E., Beach, T. G., Adler, C. H., & Cuenca, N. (2020). Dopaminergic retinal cell loss and visual dysfunction in Parkinson disease. *Annals of Neurology*, *88*(5), 893–906. https://doi.org/10.1002/ana.25897
Pagonabarraga, J., Kulisevsky, J., Strafella, A. P., & Krack, P. (2015). Apathy in Parkinson's disease: Clinical features, neural substrates, diagnosis, and treatment. *Lancet Neurology*, *14*(5), 518–531. https://doi.org/10.1016/S1474-4422(15)00019-8
Parent, A. (1997). The brain in evolution and involution. *Biochemistry and Cell Biology*, *75*(6), 651–667. https://doi.org/10.1139/o97-094
Pretegiani, E., Vanegas-Arroyave, N., FitzGibbon, E. J., Hallett, M., & Optican, L. M. (2019). Evidence from Parkinson's disease that the superior colliculus couples action and perception. *Movement Disorders*, *34*(11), 1680–1689. https://doi.org/10.1002/mds.27861
Prusiner, S. B. (2001). Neurodegenerative diseases and prions. *New England Journal of Medicine*, *344*(20), 1516–1526. https://doi.org/10.1056/NEJM200105173442006
Redgrave, P., Prescott, T. J., & Gurney, K. (1999). The basal ganglia: A vertebrate solution to the selection problem? *Neuroscience*, *89*(4), 1009–1023. https://doi.org/10.1016/S0306-4522(98)00319-4
Reeve, A., Simcox, E., & Turnbull, D. (2014). Ageing and Parkinson's disease: Why is advancing age the biggest risk factor? *Ageing Research Reviews*, *14*(100), 19–30. https://doi.org/10.1016/j.arr.2014.01.004
Robert, G., Le Jeune, F., Dondaine, T., Drapier, S., Péron, J., Lozachmeur, C., Sauleau, P., Houvenaghel, J. F., Travers, D., Millet, B., Vérin, M., & Drapier, D. (2014). Apathy and impaired emotional facial recognition networks overlap in Parkinson's disease: A PET study with conjunction analyses. *Journal of Neurology, Neurosurgery, and Psychiatry*, *85*(10), 1153–1158. https://doi.org/10.1136/jnnp-2013-307025
Sherwood, C. C., Bauernfeind, A. L., Bianchi, S., Raghanti, M. A., & Hof, P. R. (2012). Human brain evolution writ large and small. In M. A. Hofman & D. Falk (Eds.), *Progress in brain research: Vol. 195. Evolution of the primate brain* (pp. 237–254). Elsevier. https://doi.org/10.1016/B978-0-444-53860-4.00011-8
Stephan, H., & Andy, O. J. (1969). Quantitative comparative neuroanatomy of primates: An attempt at a phylogenetic interpretation. *Annals of the New York Academy of Sciences*, *167*(1), 370–387. https://doi.org/10.1111/j.1749-6632.1969.tb20457.x
Terao, Y., Fukuda, H., Ugawa, Y., & Hikosaka, O. (2013). New perspectives on the pathophysiology of Parkinson's disease as assessed by saccade performance: A clinical review. *Clinical Neurophysiology*, *124*(8), 1491–1506. https://doi.org/10.1016/j.clinph.2013.01.021
Tsao, D. Y., & Livingstone, M. S. (2008). Mechanisms of face perception. *Annual Review of Neuroscience*, *31*, 411–437. https://doi.org/10.1146/annurev.neuro.30.051606.094238
Vaughan, C. L. (2003). Theories of bipedal walking: An odyssey. *Journal of Biomechanics*, *36*(4), 513–523. https://doi.org/10.1016/S0021-9290(02)00419-0
Videnovic, A., Lazar, A. S., Barker, R. A., & Overeem, S. (2014). "The clocks that time us"—Circadian rhythms in neurodegenerative disorders. *Nature Reviews Neurology*, *10*(12), 683–693. https://doi.org/10.1038/nrneurol.2014.206
Vuilleumier, P., Armony, J. L., Driver, J., & Dolan, R. J. (2003). Distinct spatial frequency sensitivities for processing faces and emotional expressions. *Nature Neuroscience*, *6*(6), 624–631. https://doi.org/10.1038/nn1057
Weiskrantz, L., Warrington, E. K., Sanders, M. D., & Marshall, J. (1974). Visual capacity in the hemianopic field following a restricted occipital ablation. *Brain*, *97*(1), 709–728. https://doi.org/10.1093/brain/97.1.709
York III, G. K., & Steinberg, D. A. (2011). Hughlings Jackson's neurological ideas. *Brain*, *134*(10), 3106–3113. https://doi.org/10.1093/brain/awr219

10

Brain Diseases Associated with Unstable Repeats

Katharine E. Shelly, Emily G. Allen, and Peng Jin

Historical Background

To appreciate the drivers of variation in brain function, it is necessary to look beyond single-nucleotide changes in the genetic code. Polymorphic tandem repeats (TRs) represent a class of variability that is as much as 100 times more common than single-nucleotide variation (Bhatia et al. 2015; Gymrek et al. 2016). However, the contribution of short TRs (STRs), with 2- to 6-bp repetitions, or variable number TRs (VNTRs), with 10–60 repeated base pairs, to pathological states was not appreciated until recently. Since the discovery of the dynamic trinucleotide CGG repeat tract in the fragile X messenger ribonucleoprotein (*FMR1*) genes in 1991, recognition of the influence of variable repeat tracts in the human genome has grown. Currently, the number of repeat expansion loci associated with pathology stands at almost 50 disorders. Many of these diseases primarily affect the central nervous system (CNS), even when the genes containing the repeat tracts are expressed ubiquitously (Albrecht and Mundlos 2005). This suggests critical neuronal or tissue-specific roles for these genes—or unstable repeat regions more generally—in modulating the nervous system. Further work demonstrates a role for the expanded repeats themselves in modulating activity-dependent gene expression in neurons (Rodriguez et al. 2020). In this chapter, we will endeavor to showcase the depth and breadth of repeat elements in genetic architecture and gene regulation. Of course, some parameters can be used to categorize dynamic repeat alleles. Features such as position within a genetic locus, sequence composition, and mechanism of pathogenesis create distinctions; but the picture is as complex as it is intriguing.

The Range of Repeat Expansion Disorders

Several features influence what the pathological mechanism and the functional consequences of repeat expansions may be. Among the close to 50 repeat-associated diseases identified so far, distinct patterns have emerged related to sequence composition, gene region, and inheritance pattern that enable general classification (Figure 10.1). Polymorphic repeats residing in protein coding regions are typically inherited in an autosomal dominant or X-linked pattern, while dynamic STRs within intronic regions tend to be autosomal dominant, and repeat tracts falling within 3' untranslated regions (UTRs) also appear to be inherited in an autosomal dominant mode (Depienne and Mandel 2021). Expansions within the promoter regions or 5' UTRs of host genes show more diversity in inheritance pattern, and pedigrees reveal some autosomal dominant diseases, such as neuronal intranuclear inclusion disease (NIID, *NOTCH2NLC*), oculopharyndistal myopathy type 1 (OPDM1, *LRP12*), spinocerebellar ataxia type 12 (SCA12, *PPP2R2B*), and folate-sensitive fragile site 12 (FRA12A, *DIP2B*) (Holmes et al. 1999; Winnepenninckx et al. 2007; Ishiura et al. 2019; Tian et al. 2019). There are also diseases within this group that follow autosomal recessive inheritance such as Baratela-Scott syndrome (BSS, *XYLT1*), progressive myoclonic epilepsy of Unverricht-Lundborg type 1A (EPM1, *CSTB*), folate-sensitive fragile site on chromosome 2 (FRA2A, *AFF3*), and global developmental delay, progressive ataxia and elevated glutamine (GDPAG, *GLS*) (Lalioti et al. 1997; Metsu et al. 2014; LaCroix et al. 2019; van Kuilenburg et al. 2019). Finally, a third group of diseases is X-linked. These include the folate-sensitive fragile site XE (FRAXE, *AFF2*) and fragile X syndrome (FXS, *FMR1*) (Verkerk et al. 1991; Knight et al. 1993). Inheritance pattern is one mode of classification, but other means of clustering are perhaps more informative. The location of repeat motifs within genes frequently offers general insights into disease mechanism.

Expansions in Promoters/5' UTRs—The Prototypic Fragile X–Associated Disorders

The polymorphic motifs residing in promoter regions or 5' UTRs commonly lead to changes in epigenetic states and pathologies driven by loss of gene function. Both BSS and GDPAG arise from loss of *XYLT1* and *GLS* transcription products due to epigenetic silencing, but the modifications are distinct between the two diseases. In BSS, DNA hypermethylation occurs, reducing transcription and translation of the locus. The loss of xylosyltransferase impairs the synthesis of glycosaminoglycans and ultimately chondroitin sulfate, leading to skeletal dysplasia, facial dysmorphia, and developmental delay (LaCroix et al. 2019). Similar increases in DNA methylation were associated with expanded repeat alleles in FRAXE and FXS (Verkerk et al. 1991; Knight et al. 1993).

Another mechanism that can lead to epigenetic silencing of a locus with a repeat expansion is observed with *GLS*. Using methylation-sensitive polymerase chain reaction (PCR) analyses, no increase in DNA methylation was detected within the GCA expansion

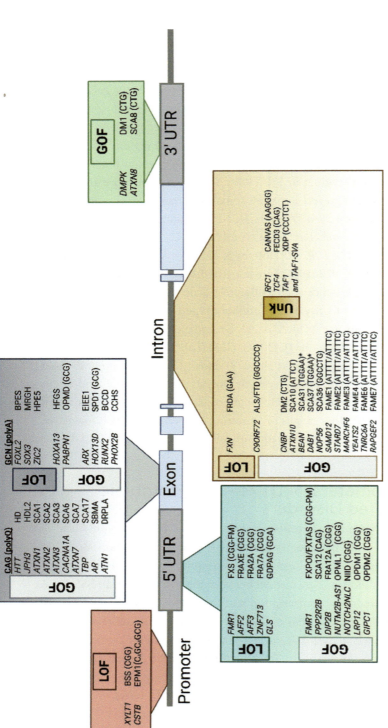

Figure 10.1 Repeat expansion disorders organized by gene region and mechanism of pathology. Each region shows the constellation of repeat motifs present and the general mechanisms through which the diseases operate. Loci where loss-of-function (LOF) and gain-of-function (GOF) boxes converge have evidence that both play a role in the genesis of diseases. *Indicates that disease is related to the insertion of an altered motif within the expansion. ALS/FTD = amyotrophic lateral sclerosis/frontotemporal dementia; BCCD = brachydactyly and cranial dysplasia; BPES = blepharophimosis, ptosis, and epicanthus inversus; BSS = Baratela-Scott syndrome; CANVAS = cerebellar ataxia, neuropathy, and vestibular areflexia syndrome; CCHS = congenital central hypoventilation syndrome; DM = myotonic dystrophy types 1 and 2; DRPLA = dentatorubral-pallidoluysian atrophy; EIEE1 = early infantile epileptic encephalopathy type 1; EPM1 = progressive myoclonic epilepsy type 1; FAME = familial adult myoclonic epilepsy types 1–4, 6, and 7; FECD3 = Fuchs endothelial corneal dystrophy type 3; FRA = folate-sensitive fragile site regions: on chromosome X site E (XE), chromosome 2 (2A), chromosome 7 (7A), chromosome 12 (12A); FRDA = Friedreich ataxia; FXPOI = fragile X-associated primary ovarian insufficiency; FXS = fragile X syndrome; FXTAS = fragile X-associated tremor/ataxia syndrome; GDPAG = global developmental delay, progressive ataxia, and elevated glutamine; HD = Huntington's disease; HDL2 = Huntington's-like 2; HFGS = hand-foot-genital syndrome; HPE5 = holoprosencephaly type 5; MRGH = mental retardation with isolated growth hormone deficiency; NIID = neuronal intranuclear inclusion disease; OPDM = ocularpharyngodistal myopathy types 1 and 2; OPMD = oculopharyngeal muscular dystrophy; OPML = ocular pharyngeal myopathy type 1; SBMA = spinal and bulbar muscular atrophy; SCA = spinocerebellar ataxia types 1, 2, 3, 6, 7, 8, 10, 12, 17, 31, 36, and 37; SPD1 = synpolydactyly type1; UTR = untranslated region. XDP = X-linked dystonia parkinsonism. (Created with BioRender)

in the *GLS* promoter. However, the repressive chromatin marker H3K9me3 was increased at the *GLS* promoter in cells from patients with expanded GCA alleles, and the transcription-permissive marker H3Kac was decreased in the same cells (van Kuilenburg et al. 2019). A similar epigenetic silencing mechanism was proposed for the dodecamer expansion underlying EPM1 since no DNA hypermethylation was observed in peripheral tissue samples from patients (Weinhaeusel et al. 2003). Testing for DNA or chromatin methylation status for multiple repeat expansion disorders is still dependent on detection of perturbations in blood samples, and the true epigenetic state of the relevant loci in neuronal tissues remains less clear. Emerging technologies that enable multiomic analyses of preserved postmortem brain tissues have the potential to provide some answers regarding tissue-specific markers in neurological diseases (Smajić et al. 2022).

The TR expansion in *FMR1* was one of the first identified loci and continues to demonstrate the complexity that expanded repeats within a single genetic locus can generate. CGG tracts that expand to >200 repeats lead to DNA hypermethylation of *FMR1* and a complete, or near complete, ablation of *FMR1* mRNA and FMRP. While the contribution of the expanded CGG repeat itself to pathology is an area of active investigation, recapitulation of almost all clinical features of FXS in individuals with point mutations in *FMR1* suggests that the loss of FMRP is the driver of pathology (Zang et al. 2009; Prieto et al. 2021). However, expansion of repeats to a smaller length of 55–200 CGGs is linked to at least two additional diseases: fragile X–associated primary ovarian insufficiency (FXPOI) and fragile X–associated tremor/ataxia syndrome (FXTAS) (Allingham-Hawkins et al. 1999; Hagerman et al. 2001). Notably, these expansions do not lead to robust DNA hypermethylation, and production of *FMR1* mRNA is, in contrast, increased in patient and animal model tissues expressing these alleles (Berman et al. 2014). Repeat-containing RNAs contribute to neuronal and ovarian cell dysfunction, but there is another component to the gain-of-function mechanism in disease. Aberrant translation mediated by the expanded repeats enables the ribosome machinery to utilize a near-AUG codon for initiation, called "repeat-associated non-AUG (RAN) translation" (Zu et al. 2011). Since the identification of this atypical translational mechanism and the detection of its products generated from repeat mRNAs, many of the diseases shown in Figure 10.1 have demonstrated that RAN translation plays at least some role in molecular pathogenesis. The mechanism of RAN translation, shown in Figure 10.2, will be discussed in further detail later in the text.

Repeats in Coding Regions—Exemplified by Huntington's Disease

Diseases associated with repetitive motifs located within the coding exons of genes are broadly broken into two clusters, the CAG or polyglutamine (polyQ) diseases and the GCN or polyalanine (polyA) diseases. These account for roughly 20 diseases and arise through a toxic gain-of-function mechanism. Pathology is driven by the aggregation of proteins driven by stretches of homopolymeric residues in the expanded alleles. The

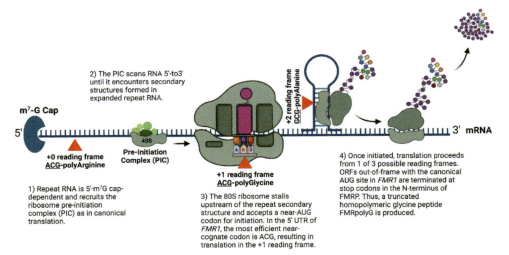

Figure 10.2 Repeat-associated non-AUG (RAN) translation describes the non-canonical mechanism used when secondary structure leads to initiation of translation from a near-AUG codon. RAN translation steps 1–4 are shown in the context of the 5' untranslated region of FMR1. RAN products have been detected from multiple pathogenic repeats, including those in ATXN8 (SCA8), DMPK (DM1), and C9ORF72 (ALS/FTD). Initiation can occur in all six (sense and antisense) reading frames, but efficiency in each reading frame is variable and generally leads to differences in accumulation of each RAN peptide. The three putative sense-strand reading frames for FMR1 are annotated with their positions relative to the AUG start of canonical FMRP, near-cognate start codon, and RAN product. Translational efficiency of RAN peptides is considerably lower than AUG-initiated translation and correlated with the length of repeats. Cellular stress increases RAN translation, and both stress response and increased RAN translation usage result in reduced production of the downstream AUG-initiated protein (Rodriguez et al., 2020). Requirements for initiation and elongation from each locus may vary, and active research to identify factors in each genomic context is ongoing. ORF = open reading frame. (Created with BioRender)

majority of dynamic repeats in exonic regions (illustrated in Figure 10.1) are inherited in a dominant fashion, with the exceptions being spinal and bulbar muscular atrophy (SBMA, *AR*) and mental retardation with isolated growth hormone deficiency (MRGH, *SOX3*), which are X-linked recessive disorders (La Spada et al. 1991; Laumonnier et al. 2002). Whether these polyQ or polyA tracts contribute to pathology via loss of function or by exerting a dominant negative effect on the product of the unexpanded allele varies by locus and disease.

Considerable work on the polyQ diseases has examined the production and stability of canonical proteins harboring expanded repeat tracts. While Huntington's disease (HD) is likely the first to come to mind, SCAs 1–6, 7, and 17 all bear hypermutable CAG repeats (Hannan 2018). Though each disease has its own range for age at onset, phenotypes are typically observed in adults (Albrecht and Mundlos 2005). In the Huntingtin gene (*HTT*) the normally 6–34 CAG span is increased beyond 40 repeats (Andrew et al. 1993). The identification of expanded CAG/polyQ tracts led to their association with increased protein aggregation, particularly involving the N-terminal portion of the HTT protein. Mutant HTT protein (mHTT) is also implicated in a broader range of cellular and molecular pathologies, including mislocalization of mHTT from cytoplasmic granules to the nucleus or perinuclear organelles. Since mHTT is commonly ubiquitinated, this shifts the localization of proteasome components to the nucleus (Trettel et al. 2000). PolyQ-containing proteins also perturb proteasome function and interact with chaperones. In HD, sequestered components like Hsp40 and Hsp70 are also limited in their abilities to degrade other elements necessary for removal from the cell. In HD models, overexpression of chaperones can ameliorate molecular pathologies (Jana et al. 2000). It is not certain whether protein aggregation is the proximal event driving excitotoxicity, altered synaptic plasticity, or other neuronal dysfunctions. Protein aggregation may occur in combination with other molecular mechanisms; but ultimately, key cellular processes are disrupted, and myriad downstream deficits lead to increased apoptosis. Other well-studied polyQ diseases appear to follow similar pathological trajectories, with increasing polyQ number correlating to more severe disease and/or earlier age at onset.

While some polyA diseases appear to share molecular phenotypes arising from the aggregation and impaired processing of proteins with expanded repeat tracts, this class of repeat disorders exhibits a few contrasting features. Unlike CAG repeats in *HTT*, CGG repeats in *FMR1*, or CTG repeats in *DMPK*, most GCN repeats are generally stable during meiosis and display comparatively low propensity for intergenerational expansion (Brais et al. 1998). Consequently, polyA stretches associated with disease also tend to be shorter than those found in polyQ disorders but still lead to robust cellular and neurological disruptions. Of the known genes harboring GCN repeats, all but *PABPN1* are developmental transcription factors (Figure 10.1; Albrecht and Mundlos 2005). Aggregates of the polyA-containing transcription factors appear in the cytoplasm and limit the transcription factors' ability to enter the nucleus and activate target gene expression (Albrecht et al. 2004). The resulting disease phenotypes result from both the loss of gene function and a toxic gain of function for the polyA-containing proteins. Oculopharyngeal muscular dystrophy (OPMD), arising from a GCG expansion in *PABPN1*, is exceptional among polyA diseases. Although this gene encodes a nuclear protein, it shuttles between the nucleus and cytoplasm, helping to regulate the length of the poly(A) tails on nascent mRNAs (Brais et al. 1998). Detection of inclusions with PABPN1-containing expanded polyA in neurons implicates protein aggregation/misfolding as the mechanism of disease in OPMD, but some neuronal populations can form PABPN1-containing inclusions even without the expanded polyA tract. This suggests that the small GCG repeat at the N terminus of the protein has a functional role that is not yet understood (Berciano et al. 2004).

Similarly, manipulation of CAG length in a mouse model of HD suggests that altered CNS function arises not only from expansion of CAG repeats since deletion of the polyQ tract is sufficient to produce contrasting phenotypes. Mice with expanded polyQ in the Htt protein have decreased life span, motor deficits, and protein aggregation; but full-length Htt missing the polyQ tract exhibited improvements in each phenotype as well as increased autophagosome capacity (Zheng et al. 2010). The findings related to autophagy are particularly interesting since overexpression of Htt protein with the normal CAG repeat length does not exhibit this increase in autophagosome production (Zheng et al. 2010). These results, with similar findings from other loci like *FMR1*, suggest that dynamic repeats themselves play a functional role in regulating the products of their resident genes. This indicates that similar therapeutic approaches to activate or silence *FMR1* or *HTT*, respectively, during development for temporal restriction of pathogenic alleles may be useful for diseases with pleiotropic effects throughout the life span.

Dynamic Repeats in Introns and a Spectrum of Pathologies—The Prototypic Friedreich Ataxia

Although the vast majority of polymorphic repeats found in the intronic regions of genes are inherited in an autosomal dominant fashion, their sequence content, repeat sizes, and pathological mechanisms are likely the most heterogeneous of the classes (Figure 10.1). The polymorphic repeats linked to Friedreich ataxia (FRDA, *FXN*) and cerebellar ataxia, neuropathy, and vestibular areflexia syndrome (CANVAS, *RFC1*) are autosomal recessive. GAA expansion is correlated with silencing of FXN, loss of frataxin production, and increasing repeat number linked to disease severity. The mechanism of disease is that loss of frataxin function in mitochondria causes impaired iron transport and protein accumulation and disrupted mitochondrial function, leading to cellular defects (Campuzano et al. 1997). Though FRDA is a neurological disease with the large neurons of the dorsal root ganglia being particularly affected, patient mortality is often the result of cardiomyopathy (Meyer et al. 2007). The other autosomal recessive condition is CANVAS. Interestingly, the disease-linked sequence of the AAGGG pentamer differs from the reference sequence (AAAAG) and shows increased repeat number (Cortese et al. 2019). Expansion of the mutant sequence in *RFC1* does not cause reduced gene expression in brain tissue, but it is yet unknown how altered or expanded alleles lead to dysfunction throughout the CNS and peripheral nervous system (Ishai et al. 2021). Notably for understanding the emergence of this type of repeat-associated allele is the intronic *Alu* element that is host to the pentamer repeat. Since these mobile retroelements appear within over 1 million loci in the human genome, it is unsurprising that another repeat-associated disease, X-linked dystonia parkinsonism (XDP, *TAF1*), has an expanded repeat within a SINE-VNTR-*Alu* feature (Deininger 2011; Reyes et al. 2022). In contrast to CANVAS, XDP phenotypes do result from the loss of TAF1 protein, but protein expression deficits do not explain the full spectrum of pathologies because the knockout mouse model for *Taf1* does not recapitulate the striatal degeneration observed in human patients (Reyes et al. 2022). As with fragile X premutation disorders, RAN translation is an additional mechanism that may contribute to pathology.

In recent years, the familial adult myoclonic epilepsy (FAME1–4, 6, 7) diseases have been identified (Figure 10.1). Each of these ATTTT expansions resides in a different gene but includes the insertion of an ATTCT motif that is associated with the presence of clinical phenotypes. (Ishiura and Tsuji 2020). The onset of disease is in adulthood with a myoclonic tremor and occasional seizures, but unlike many of the progressive neurological disorders mentioned in this chapter, there appears to be little change of severity in these diseases over long stretches of time (Ishiura and Tsuji 2020). FAMEs have also been referred to as "benign FAME" due to this lack of progressive degeneration.

The most prevalent repeat expansion disorder in humans, Fuchs' endothelial corneal dystrophy type 3 (FECD3), happens to reside in an intronic region of *TCF4*. FECD3, an age-related degeneration of the endothelial layer marked by corneal guttae, affects roughly 4% of White adults over 40 in the United States (Wieben et al. 2012; Mootha et al. 2014). Though individuals with mild disease may experience no symptoms, it remains the largest contributor to the corneal transplant need (Wieben et al. 2012). Although there is variable expressivity and incomplete penetrance of the allele carrying the CTG repeat expansion, with about 80% of individuals carrying the allele affected, there is currently little to no information about modifying genes or environmental factors. Although FECD3 is the most common repeat expansion disease, the expanded G_2C_4 hexanucleotide repeat in intron 1 of the *C9ORF72* locus is present in close to 40% of familial cases of amyotrophic lateral sclerosis and frontotemporal dementia (ALS/FTD; Majounie et al. 2012) and has understandably become the focus of intense investigation since its discovery in 2011. A tremendous amount of work has been required to dissect the pathogenesis of these expansions owing to the high GC content of the alleles. This sequence composition makes precise sizing difficult, and repeat tracts can range from just over 30 repeats into the thousands (Smeyers et al. 2021). This disorder is notable because of the complex pathology. Haploinsufficiency via reduction of C9ORF72 protein abundance drives a subset of defects; in addition, sequestration of RNA-binding proteins by repeat-containing RNAs leads to altered splicing events, and aggregation of the dipeptide produced via RAN translation leads to another subset of cell dysfunctions (Smeyers et al. 2021). The titration of RNA-binding proteins like TDP-43, in the case of ALS/FTD, into aggregates leads to additional perturbations like cryptic splicing events and further cell stress and phase separation through the accumulation of poly-PR and poly-GR peptides (Schmitz et al. 2021; Ma et al. 2022). Determining which molecular events are proximal to the repeat expansion and whether the RNA or protein gain of function predisposes neurons and other CNS cell types to stress or apoptosis remains a challenge that groups across the field are actively dissecting.

3' UTRs and Large Repeats

There is a distinct drop-off in the number of TRs identified in the 3' UTRs of genes, and only two are represented in Figure 10.1. The CTG tract is linked to myotonic dystrophy type 1 (DM1, *DMPK*), and the CTG expansion is linked to SCA8 (*ATXN8*; Mahadevan et al. 1992; Koob et al. 1999). SCA8 represents the first dominant form of SCA identified that is not associated with an expanded polyQ track, but individuals with expanded CTGs in ATXN8 do exhibit the impaired motor coordination and speech features in other SCAs (Koob et al. 1999). DM1, however, is a canonical example of genetic anticipation: growth in the number of repeats through generations and its link to increasing disease severity and earlier age at onset. DM1 also provides a clear example of RNA toxicity since the myotonia and progressive muscle weakness arise through RNA sequestration of RNA-binding proteins and subsequent cellular dysfunction (Ho et al. 2005). The CTG repeats in DM1 and SCA8 were the first in which RAN translation was demonstrated to occur, in a variety of reading frames (Zu et al. 2011).

Evolutionary Evidence Related to Human Health and Disease

Nearly 1,600 VNTRs with human-specific expansions can be detected, and 467, or slightly over one-fourth, have been mapped to a single genomic location (Sulovari et al. 2019; Course et al. 2021). This suggests they were derived independently from retrotransposon activity, typically associated with multiple integration events and loci within the genome. The overwhelming majority of the VNTRs fall within intergenic or intronic regions and, interestingly, are expanded in ancestral (i.e., Neanderthal/Denisovan) genomes as well (Course et al. 2021). The implementation of long-read sequencing technologies has been critical for the detection of these VNTRs by generating the long, overlapping reads needed to determine the mechanisms of each repeat expansion from ancestral genotypes. Since these longer reads reveal complex deletion and duplication events, they ultimately demonstrate how these human-specific VNTRs were generated. These longer-motif VNTRs not only reveal key features of repetitive genomic elements in an evolutionary context but also are associated with neurological diseases, such as the VNTR in WDR7 modulating risk for ALS and the repeat in *ABCA7* for Alzheimer's disease (De Roeck et al. 2018; Course et al. 2020). Despite these advances, the majority of repeats associated with diseases revolve around dynamic STRs.

In the initial human reference genome, over 1 million STRs were identified (Lander et al. 2001). That figure excludes the omitted sequences, which contain repetitive and GC-rich elements, that took another two decades to resolve (Nurk et al. 2022). Regulatory elements of genes are overrepresented as host regions for these STRs, suggesting that repeat expansions likely play a role in altering gene expression (Sawaya et al. 2013). Furthermore, when small STRs resided in these genetic elements, a more diverse gene expression profile in humans and a set of nonhuman apes was present than when STRs were absent. This finding was more robust when larger repeat elements were examined (Bilgin Sonay et al. 2015). Together, these findings suggest an evolutionary contribution of repeat elements to the regulation of gene expression. When examining the impact of these STRs in a clinical context, it is striking that most disorders with unstable repeat regions are associated with neurological diseases or disruptions of CNS function. Although this pattern may suggest homogeneity among STR-linked diseases, that is absolutely not the case.

From an evolutionary perspective, expansions within coding regions have the highest mutational constraints and generally have smaller expanded alleles, as may be expected (Gymrek et al. 2017). Polymorphic STRs within 5′ and 3′ UTRs exhibit less restriction, and intronic and intergenic regions harboring STRs show the least (Gymrek et al. 2017). Within these categories, further distinctions separate STRs. Among the TRs in coding regions, those producing expanded polyA stretches in proteins tend to be smaller than the repeats that encode expanded polyQ tracts (Albrecht and Mundlos 2005). In fact, compared with polyQ disorders, polyA diseases do not appear to have arisen from the same mechanism and are thought to be products of unequal homologous recombination during meiosis (Warren et al. 1997). This contrasts with the proposed mechanism of expansion of polyQ and other repeats via polymerase slippage. Interestingly, some data suggest that GCN repeats are under even tighter evolutionary constraints than other repeat alleles; but mutational analyses are highly dependent on the underlying assumptions of the model, and an updated review with recently developed computational tools and models might yield refined observations (Brais 2003). The number and heterogeneity of these TR-associated disorders demonstrate that pathologies can result in myriad ways.

Aside from mechanism of origin, it is important to consider the multiple roles that repeat-harboring genes or proteins play in brain development and maintenance. The role of *FMR1* in brain development has been known for three decades, and the loss of FMRP via the presence of CGG expansion leads to changes in developmental trajectory of neurons (Irwin et al. 2000). More recently, assessment of unmethylated full mutation expansions in *FMR1* has shown degenerative capacity, much like those premutation-sized repeats implicated in FXTAS and FXPOI (Berman et al. 2014). This connection between a protein required for development of the brain and gene products leading to adult-onset premutation disorders raises the question, is there a developmental role for expanded repeats that is not yet appreciated? There are some studies that suggest that there are indeed developmental roles for repeat-harboring proteins and that repeat expansion events do have developmental consequences. Research examining a mutant *HTT* variant has offered some insight into the potential role in neuronal generation in both humans and mice. Studying parallel periods of development, it was shown that mHTT was mislocalized in both developing human and mouse cortex. Due to increased colocalization of mHTT with junction complex proteins, HD cortical neurons showed subsequent deficits in cell cycle progression, altered cell polarity, and ultimately perturbed migration of progenitor cells in the cortex (Barnat et al. 2020). The subpopulations of cortical neurons investigated were those that are most vulnerable in adults with HD. This study and concordant studies performed in induced pluripotent stem cells and brain organoids suggest that expression of expanded repeats may play a critical role in generating vulnerabilities that are only clinically exposed in adulthood (Smajić et al. 2022). Findings from individuals with behavioral variant FTD (bvFTD) and repeat expansions in C9orf72 suggest that behavioral characteristics associated with neurodevelopmental disorders may be the first observable changes in bvFTD (Gossink et al. 2022).

One type of regulatory advantage STRs might offer these genes with dual roles in neuronal maintenance and development is control via RAN translation. This provided a new avenue to explore the pathogenesis of repeat expansion diseases in the context of neurological function. Since the first report regarding DM1 and SCA8 alleles in 2011, generation of peptide products using the non-standard translation mechanism has been observed/demonstrated in at least 10 of the diseases mentioned in this chapter. Most of these diseases, including FXPOI/FXTAS, ALS/FTD, HD, DM1, and SCA8 result from the presence and aggregation of RAN products—in addition to RNA toxicity that was initially linked to the deficits observed in each disorder (Banez-Coronel and Ranum 2019; Kearse et al. 2019; Schmitz et al. 2021). In the case of FXTAS where the CGG repeat is upstream of the canonical AUG start codon, RAN translation likely compounds cellular dysfunction and favors production through the RAN pathway through EIF2αα phosphorylation and reduced canonical translation (Kearse et al. 2019). This is not likely to be a generalizable mechanism, given the specific location of the CGG repeat; and the context of each repeat will be critical to understanding how RAN translation functions in each disease state.

For FMRP, a physiological role for RAN translation has been suggested for the CGG repeats as a control lever during neuronal activity (Rodriguez et al. 2020). Other diseases such as OPDM1, oculopharyngeal myopathy with leukoencephalopathy (OPML), and SCA12 do not have such a complex picture. RAN translation has been posited as the major driver of pathogenesis in these diseases, but the body of literature on this pathogenic contributor is still quite limited. Experimentally, it is quite difficult to separate the mRNA that enables the production of RAN peptides from the presence of the proteins themselves.

The presence of polymorphic repeats as a mechanism of physiological control suggests that there may be antagonistic pleiotropy that balances the need for elements in development that lead to late-onset degenerative processes. How is it then that the developmental and degenerative aspects of these loci are viewed as discrete from one another? The conservation of the variable CGG repeat in FMR1 is present in 44 mammalian species, suggesting survival through 150 million years of evolutionary forces and a likely functional role (Eichler et al. 1995). However, increasing length of the polymorphic repeats is restricted to primates, and the intergenerational expansions observed in human *FMR1* have not yet been modeled in animal systems, suggesting that the human-specific genomic elements may be key to repeat expansions. Regulatory control of FMRP translation appears to be a contributor for the developmental role of repeats in *FMR1*. That control is length-dependent, and increased repeat length leads to more dramatic phenotypes in adult-onset diseases (Rodriguez et al. 2020). This antagonistic relationship is also suggested by studies of the CAG repeat in the coding region of HTT. Assessments of mHTT in cortical development exhibited marked changes in cell polarization and connectivity (Barnat et al. 2020). In contrast, studies of children across the spectrum of mHTT revealed that those at the pathological threshold of 40 CAG repeats had the highest General Abilities Index, indicating that the possibility of mHTT production in development provided functional benefits (van der Plas et al. 2020). These studies are leading the way to determine the breadth of consequences for expanded repeats in development after decades of work described those resulting from degeneration. It is intriguing to consider the ability of these expanded repeats as regulatory elements that could exert control dependent on developmental, activation, or disease state. Perhaps this additional level of regulation in a higher-order neuronal network provides cells an evolutionary benefit but comes at a cost with age. However, many unanswered questions remain as to whether these findings suggest antagonistic pleiotropy or if there are underlying functions of expanded repeats that represent a spectrum of dysfunction from development to degeneration Studies of these mechanisms at the molecular level are still nascent. Their context in the evolution of the brain relies on a more complete understanding of functions, which is yet to come.

Counterarguments against Evolutionary Influences

Though the evidence for evolutionary influences on repetitive regions in the human genome is quite convincing, much of the focus on repeat expansions seems to focus on their roles in neurological function. To that end, it has become common to contextualize these events as evolutionary drivers exerted on brain function and disease. A recent paper showed that recurrent TR mutations were largely specific to cancer subtypes, but there were 160 recurrent expansions across seven cancer types (Erwin et al. 2023). The finding that these expansion events were represented uniformly across tissues differentiated from all primary germ layers suggests that there is not an influence on neuronal tissues or functions specifically (Erwin et al. 2023).

Perhaps these recurrent events do not select a particular tissue; it is important to remember that they are human-specific. This suggests that evolutionary pressures on the genome have enabled these events to occur uniquely within the human genome, despite the presence of retrotransposons and repetitive elements in nonhuman primates and lower-order species (Course et al. 2020). Regardless of whether the human brain is the primary target of selective pressures in the environment, identifying broader tandem repeat expansions in other pathological contexts has implications for clinical assessment and treatment of phenotypes resulting from repetitive elements. Further, the identification of a GAAA expansion similar to that in *FXN* led investigators to find a similar mechanism of pathology. By targeting the repeat expansion by the same means used in FRDA studies, Erwin et al. (2023) ablated the increased proliferation seen in liver and kidney cancer cell lines. As long-read sequencing comes online in the clinic to identify these expansion events, the likelihood that common mechanisms of pathogenesis and therapeutics can be applied across various human tissues increases.

Clinical and Therapeutic Implications

Modifiers of Expanded Alleles

An emerging consideration for the field of TR expansion diseases is the effect of *cis*- and *trans*-acting modifiers on penetrance and expressivity. Some of these *cis*-acting factors are interruptions in the TR tract itself. These modifiers have emerged as key considerations for identifying more precise genotype–phenotype correlations for repeat expansion. One striking example is the loss of the CAA interruption within the *HTT* CAG repeat. This simple change alters the age of HD onset by about a decade (Findlay Black et al. 2020). Both SCA8 and SCA10 have interruptions linked to functional changes, with CCG inclusions increasing RAN production in SCA8 and the interruption of an ATTCC in expanded ATTCT alleles of SCA10 associated with the presence of seizures (Perez et al. 2021; Morato Torres et al. 2022). Similarly to SCA10, FAME 1–7 shows that it is the inclusion of an ATTTC within the ATTTT expansions that leads to pathogenesis (Ishiura and Tsuji 2020). Beyond *cis*-acting elements, *trans*-acting factors also modify disease states, and DNA repair proteins were among the first evaluated. Loss of factors MSH2, MSH3, and MLH1 in the germline and another host of factors in somatic cells impacts transmission of disease alleles in HD, FXS, FRDA, and DM1 (Zhao et al. 2021). Oocytes, like neurons, are post-mitotic but are the cells through which many intergenerational expansions are transmitted. In individuals with hexanucleotide expansions, TDP-43 mutations are clear risk alleles for more severe ALS/FTD, leading to increased cryptic exon inclusion events (Ma et al. 2022). Other groups have looked for genetic modulators within disease cohorts using whole-genome sequencing (WGS) to identify variants/genes associated with severe disease. A modifier for FXTAS was found using this strategy, and *PSMB5* was identified as a suppressor of CGG-mediated toxicity in the disease (Kong et al. 2022). As WGS becomes even more widely utilized, it is expected that this approach, potentially coupled with other large-scale data sets, will be used to find modifiers for TR disorders.

Therapeutic Interventions and the Need for Temporal and Spatial Control

There are several therapeutic strategies to treat repeat expansion disorders including adenovirus-based gene replacement, gene editing or transcriptional silencing via clustered regularly interspersed short palindromic repeats (CRISPR)–CRISPR-associate 9 (Cas9), inhibition or removal of repeat-containing transcripts to aggregate, or treatment of the effects of repeat-containing gene products. The first two of these strategies have yielded particular focus since 2012 with the expanded study of antisense oligonucleotides

(ASOs) and CRISPR-Cas9. Particularly in the context of the RNA gain-of-function mechanism of pathology, the ASO paradigm is attractive as it does not alter DNA sequence but binds to mRNA transcripts and triggers cellular elimination. This approach has been examined extensively for correction of DM1, and work is ongoing to target toxic RNA present in FXTAS. While studies of DM1 have assessed multiple targets, ASOs directed to address CUG repeats themselves were successful in pre-clinical trials (Izzo et al. 2022). These ASOs have not shown success in human clinical trials, yet the dosage used has not been sufficient for measurable effects (Izzo et al. 2022). Parallel tactics are being employed in other diseases like FXTAS and HD, providing hope that this approach may be broadly applicable to several STR expansion disorders; but those studies have not yet been attempted in any clinical trials.

The other exciting prospective treatment paradigm is based on CRISPR-Cas9 technology. It is useful for gene editing, but modifications to Cas9 to eliminate its enzymatic activity have resulted in dead Cas9 (dCas9). This version of dCas9 enables transcriptional silencing of genes without DNA editing. It has been adapted further for transcriptional activation by fusing dCas9 to transcriptional activator VPR (a complex of VP64, p65, and RTA). Benefits of these molecules are that constructs are not dependent on repeat length in a given individual, and off-target effects appear to be low in available in vivo studies (Riedmayr et al. 2022). Key unknown factors are the window of therapeutic efficiency, whether delivery to the affected tissues is feasible, and whether control over timing of delivery or diminishing efficacy of activation/silencing over time will allow for temporal control. Ongoing work in DM1, HD, and FXTAS, among others, suggests that answers will become apparent shortly.

Unanswered Questions and Future Work

Sequencing Capabilities Are the Rate-Limiting Step in Identifying Pathological Repeat Expansions

Key considerations limiting the identification of de novo TR expansions revolve around the process of genome sequencing. These can be divided into three categories: pre-sequencing protocols, instrumentation, and computational pipelines used for analysis. Advancement in all three technological components have contributed to the markedly increased detection of disease-associated repeat expansions. Since 2017 at least 15 polymorphic STRs have been uncovered and linked to disease, via refining analytical tactics for widely available short-read next-generation sequencing data or linking information about known genomic STRs to gene expression profiles. Clear clinical utility can come from searching for epivariation in large cohorts and linking those variants back to putative disease loci, but exposing new loci with pathological functions must compensate for the bias of understanding the genomic architecture of analyzed regions or genomes at the outset (Garg et al. 2020; Mousavi et al. 2021).

PCR-Free Library Preparation and Instrumentation

Equally important to reducing bias in WGS is operationally reducing bias by increasing the breadth of reads generated through the input steps. Using PCR to amplify repeat tracts imposes size restrictions on pathological expansions, increases the likelihood that indels occur within the repeat-flanking sequences in amplicons, and ultimately skews detection of repetitive regions because the previous issues hamper the ability to map reads to a reference genome (Dolzhenko et al. 2017 Akimoto et al. 2014). The input requirements are similar even in clinical application of PCR-free methodologies, and the breadth and depth of coverage, particularly for copy number variations, is either comparable or improved (Zhou et al. 2022). While these technical improvements are practical, there is no substitute for the generation of long-read sequences. While short-read sequencing is widespread and affordable, the <500-bp reads do not allow for robust analysis of expanded repeat loci. Data generated by Pacific Biosciences' (PacBio) single-molecule real-time (SMRT) sequencing and Oxford Nanopore's minion (ONT) sequencing are being taken up readily and have achieved read lengths reaching up to 100 Kbp and up to 2.3 Mbp, respectively (Amarasinghe et al. 2020). Strikingly, both technologies perform well across a spectrum of GC-content samples. For large expansions with high GC richness like the CGG in *FMR1* and the G_2C_4 in *C9ORF72*, this capability to span an entire repeat region and produce reads is critical. Importantly both technologies could determine the methylation status of repeats. A caveat for these technologies is that both have a lower per-read accuracy than short-read methods, with error rates between 10% and 15% (Amarasinghe et al. 2020). Specific efforts have been made to improve accuracy with SMRTseq; extending the polymerase life span increased the number of reads for long molecules and improved accuracy in consensus base-calling. Current error rates may be as low as 1% with molecules having >4 reads (Amarasinghe et al. 2020). In the case of ONT, accuracy is highly dependent on the speed at which DNA molecules are moved through the pore. As this rate of movement slows, the quality of base calls is generally reduced. By reading the single strand and its complement with ONT's $1D^2$ protocol, the error rate for consensus sequences stands at about 2% (Amarasinghe et al. 2020).

Pipelines for Informatic Analyses

As adoption of long-read sequencing increases, efforts have been made to create bioinformatic tools to utilize extant short-read data in the search for expanded repeat alleles (Table 10.1). Early tools included lobSTR, which can use single- or paired-ends reads but lacks the ability to detect STRs longer than the read length (Mousavi et al. 2021). Groups adapt and release new tools as challenges and improvements become apparent, with the tools shown in Table 10.1 (Amarasinghe et al. 2020). These generally require some information about genetic loci for examination, and another set of algorithms were created to identify de novo polymorphic repeat tracts. These include ExpansionHunter and the newer iteration ExpansionHunter Denovo, exSTRa, TRhist, and TREDPARSE

(Amarasinghe et al. 2020; Depienne and Mandel 2021; Mousavi et al. 2021). Even these algorithms may necessitate user input for repeat size estimations.

The increasing availability of long-read sequencing data has spurred development of tools for analyses. A sampling of these algorithms includes Straglr, DeepRepeat, Repeat-HMM, tandem-genotypes, and NanoSatellite (Table 10.1). Though specific parameters vary for each program, these algorithms generally appear to accept long-read data generated by either SMRTseq or ONT. These programs already have demonstrated utility. NanoSatellite, which uses electric current data and is thus specific to the ONT platform, has demonstrated utility in the identification of a VNTR in *ABCA* associated with Alzheimer's disease (De Roeck et al. 2018). DeepRepeat and Straglr show strengths in call long STRs compared to older models Repeat-HMM and tandem genotypes using preselected candidate genes or on a genome-wide scale (Chiu et al. 2021; Fang et al. 2022). There are a remarkable number of programs with increasing functions and customization for each data set. As the availability of long-read data continues to expand and the limitations of long-read data sets become apparent, it is certain that these tools will evolve to meet investigators' demands.

TABLE 10.1 Computational Algorithms to Evaluate Repeat Expansions

Program	Data	Identified STRs >SR reads	Require user input for loci/ thresholds	Disease alleles identified	Year of method publication
lobSTR	SR	No	No		2016
gangster	SR	Yes	No		2019
ExpansionHunter	SR	Yes	Yes, repeat thresholds		2017
ExpansionHunter Denovo	SR	Yes	No	CANVAS	2020
exSTRa	SR	Yes	No		2017
TREDPARSE	SR	Yes	Yes, repeat thresholds		2017
TRhist	SR	Yes	Candidate STRs	FAME6/7, NIID, OPDM1, OPML	2014
DeepRepeat	LR-SMRT/ONT	Yes	No		2022
Straglr	LR-SMRT/ONT	Yes	No		2021
Repeat-HMM	LR-SMRT/ONT	Yes	No		2017
NanoSatellite	LR-ONT	Yes	No	VNTR in *ABCA*- AD	2019
tandem-genotypes	LR-SMRT/ONT	Yes	No		2019

Bioinformatic algorithms for the assessment of polymorphic short tandem repeats (STRs) in the genome have evolved from retrofitted pipelines to handle SR data to elegant scanners to identify putative STRs and output a possible pathogenic threshold based on outliers. LR = long read; ONT = Oxford Nanopore Technology; SMRT = single molecule, real-time; SR = short read.

Leveraging Polymorphic STR Discovery to Understand Neurological Function and Evolution

The ultimate goal of generating these algorithms is to probe genome-scale data for undiscovered polymorphic repeat loci. Identifying the full spectrum of expanded repeats will ultimately lead to a robust understanding of their functions and which evolutionary pressures influence their conservation in the human genome. The variety of genomic loci, be they protein coding or non-coding, in which these DNA repeats reside is likely to provide critical insight into developmental roles of normal or expanded repeat sizes in neuronal regulation. Comprehending how repeat expansions lead to dysfunction will ultimately provide the foundation to determine whether expanded STRs drive pathology, beginning in development but only becoming clinically apparent in adulthood. Alternatively, evidence may show that expanded repeat expression in development leads to functional neuronal networks and that aging is the key factor in the change to a pathological state. Evidence from a small number of loci that have been studied in both development and aging, HTT in Huntington's disease and FMRP in fragile X–related disorders, cannot yet answer these questions. Probing the range of known disease-associated repeat loci may provide key information since DM1, BSS, EPM1, and others already have well-described developmental phenotypes due to much earlier age at disease onset.

In summary, we have highlighted the impact and heterogeneity of repeat expansion alleles on human development, neuronal function, and technological advancement. The surge in identification of diseases associated with unstable repeat tracts is likely to continue, due to the ever-evolving field of WGS and the building of databases of long-read and ultra-long-read sequences. These data may also enable exploration for genomic architecture specific to humans, providing evidence for how and why these repeat elements are so prevalent and persistent in human DNA, as has been demonstrated for some VNTRs. What is most exciting about these possibilities is that they may offer even more insight into exactly how these dynamic repeats influence neurological function and, in some instances, drive pathology within the CNS.

Key Points to Remember

- The wide variation in tandem repeat length, sequence composition, and local context within the genome leads to a myriad of functional consequences in neurological disease.
- Repeat expansions may be expressed throughout development, leading to changes in early life. The developmental consequences of tandem repeats associated with neurodegeneration are only starting to be identified and appreciated.

- Control of gene expression by repeat elements, themselves, is an area of active exploration and may hold key information about evolutionary influences on neurological function.
- Evolving sequencing technology and computational tools are enabling the discovery of polymorphic repeats in the human genome, and there are likely to be many more disease-associated loci found.

References

Akimoto, C., Volk, A. E., van Blitterswijk, M., Van den Broeck, M., Leblond, C. S., Lumbroso, S., Camu, W., Neitzel, B., Onodera, O., van Rheenen, W., Pinto, S., Weber, M., Smith, B., Proven, M., Talbot, K., Keagle, P., Chesi, A., Ratti, A., van der Zee, J., . . . Kubisch, C. (2014). A blinded international study on the reliability of genetic testing for GGGGCC-repeat expansions in C9orf72 reveals marked differences in results among 14 laboratories. *Journal of Medical Genetics*, *51*(6), 419–424. https://doi.org/10.1136/jmedgenet-2014-102360

Albrecht, A., & Mundlos, S. (2005). The other trinucleotide repeat: Polyalanine expansion disorders. *Current Opinion in Genetics and Development*, *15*(3), 285–293. https://doi.org/10.1016/j.gde.2005.04.003

Albrecht, A. N., Kornak, U., Böddrich, A., Süring, K., Robinson, P. N., Stiege, A. C., Lurz, R., Stricker, S., Wanker, E. E., & Mundlos, S. (2004). A molecular pathogenesis for transcription factor associated polyalanine tract expansions. *Human Molecular Genetics*, *13*(20), 2351–2359. https://doi.org/10.1093/hmg/ddh277

Allingham-Hawkins, D. J., Babul-Hirji, R., Chitayat, D., Holden, J. J. A., Yang, K. T., Lee, C., Hudson, R., Gorwill, H., Nolin, S. L., Glicksman, A., Jenkins, E. C., Brown, W. T., Howard-Peebles, P. N., Becchi, C., Cummings, E., Fallon, L., Seitz, S., Black, S. H., . . . Vieri, F. (1999). Fragile X premutation is a significant risk factor for premature ovarian failure: The international collaborative POF in fragile X study—preliminary data. *American Journal of Medical Genetics*, *83*(4), 322–325. https://doi.org/10.1002/(SICI)1096-8628(19990402)83:4<322::AID-AJMG17>3.0.CO;2-B

Amarasinghe, S. L., Su, S., Dong, X., Zappia, L., Ritchie, M. E., & Gouil, Q. (2020). Opportunities and challenges in long-read sequencing data analysis. *Genome Biology*, *21*(1), Article 30. https://doi.org/10.1186/s13059-020-1935-5

Andrew, S. E., Goldberg, Y. P., Kremer, B., Telenius, H., Theilmann, J., Adam, S., Starr, E., Squitieri, F., Lin, B., Kalchman, M. A., Graham, R. K., & Hayden, M. R. (1993). The relationship between trinucleotide (CAG) repeat length and clinical features of Huntington's disease. *Nature Genetics*, *4*(4), 398–403. https://doi.org/10.1038/ng0893-398

Banez-Coronel, M., & Ranum, L. P. W. (2019). Repeat-associated non-AUG (RAN) translation: Insights from pathology. *Laboratory Investigation*, *99*(7), 929–942. https://doi.org/10.1038/s41374-019-0241-x

Barnat, M., Capizzi, M., Aparicio, E., Boluda, S., Wennagel, D., Kacher, R., Kassem, R., Lenoir, S., Agasse, F., Braz, B. Y., Liu, J. P., Ighil, J., Tessier, A., Zeitlin, S. O., Duyckaerts, C., Dommergues, M., Durr, A., & Humbert, S. (2020). Huntington's disease alters human neurodevelopment. *Science*, *369*(6505), 787–793. https://doi.org/10.1126/science.aax3338

Berciano, M. T., Villagra, N. T., Ojeda, J. L., Navascues, J., Gomes, A., Lafarga, M., & Carmo-Fonseca, M. (2004). Oculopharyngeal muscular dystrophy–like nuclear inclusions are present in normal magnocellular neurosecretory neurons of the hypothalamus. *Human Molecular Genetics*, *13*(8), 829–838. https://doi.org/10.1093/hmg/ddh101

Berman, R. F., Buijsen, R. A., Usdin, K., Pintado, E., Kooy, F., Pretto, D., Pessah, I. N., Nelson, D. L., Zalewski, Z., Charlet-Bergeurand, N., Willemsen, R., & Hukema, R. K. (2014). Mouse models of the fragile X premutation and fragile X–associated tremor/ataxia syndrome. *Journal of Neurodevelopmental Disorders*, *6*(1), Article 25. https://doi.org/10.1186/1866-1955-6-25

Bhatia, G., Gusev, A., Loh, P. R., Vilhjálmsson, B. J., Ripke, S., Purcell, S., Stahl, E., Daly, M., de Candia, T. R., Kendler, K. S., O'Donovan, M. C., Lee, S. H., Wray, N. R., Neale, B. M., Keller, M. C., Zaitlen, N. A., Pasaniuc, B., Yang, J., & Price, A. L. (2015). Haplotypes of common SNPs can explain missing heritability of complex diseases. bioRxiv, Article 022418. https://doi.org/10.1101/022418

Bilgin Sonay, T., Carvalho, T., Robinson, M. D., Greminger, M. P., Krützen, M., Comas, D., Highnam, G., Mittelman, D., Sharp, A., Marques-Bonet, T., & Wagner, A. (2015). Tandem repeat variation in human and great ape populations and its impact on gene expression divergence. *Genome Research*, 25(11), 1591–1599. https://doi.org/10.1101/gr.190868.115

Brais, B. (2003). Oculopharyngeal muscular dystrophy: A late-onset polyalanine disease. *Cytogenetic and Genome Research*, 100(1–4), 252–260. https://doi.org/10.1159/000072861

Brais, B., Bouchard, J. P., Xie, Y. G., Rochefort, D. L., Chrétien, N., Tomé, F. M. S., Lafrentére, R. G., Rommens, J. M., Uyama, E., Nohira, O., Blumen, S., Korcyn, A. D., Heutink, P., Mathieu, J., Duranceau, A., Codère, F., Fardeau, M., & Rouleau, G. A. (1998). Short GCG expansions in the *PABP2* gene cause oculopharyngeal muscular dystrophy. *Nature Genetics*, 18(2), 164–167. https://doi.org/10.1038/ng0298-164

Campuzano, V., Montermini, L., Lutz, Y., Cova, L., Hindelang, C., Jiralerspong, S., Trottier, Y., Kish, S. J., Faucheux, B., Trouillas, P., Authier, F. J., Dürr, A., Mandel, J. L., Vescovi, A., Pandolfo, M., & Koenig, M. (1997). Frataxin is reduced in Friedreich ataxia patients and is associated with mitochondrial membranes. *Human Molecular Genetics*, 6(11), 1771–1780. https://doi.org/10.1093/hmg/6.11.1771

Chiu, R., Rajan-Babu, I. S., Friedman, J. M., & Birol, I. (2021). Straglr: Discovering and genotyping tandem repeat expansions using whole genome long-read sequences. *Genome Biology*, 22(1), Article 224. https://doi.org/10.1186/s13059-021-02447-3

Cortese, A., Simone, R., Sullivan, R., Vandrovcova, J., Tariq, H., Yau, W. Y., Humphrey, J., Jaunmuktane, Z., Sivakumar, P., Polke, J., Ilyas, M., Tribollet, E., Tomaselli, P. J., Devigili, G., Callegari, I., Versino, M., Salpietro, V., Efthymiou, S., Kaski, D., . . . Houlden, H. (2019). Biallelic expansion of an intronic repeat in *RFC1* is a common cause of late-onset ataxia. *Nature Genetics*, 51(4), 649–658. https://doi.org/10.1038/s41588-019-0372-4

Course, M. M., Gudsnuk, K., Smukowski, S. N., Winston, K., Desai, N., Ross, J. P., Sulovari, A., Bourassa, C. V., Spiegelman, D., Couthouis, J., Yu, C. E., Tsuang, D. W., Jayadev, S., Kay, M. A., Gitler, A. D., Dupre, N., Eichler, E. E., Dion, P. A., Rouleau, G. A., & Valdmanis, P. N. (2020). Evolution of a human-specific tandem repeat associated with ALS. *American Journal of Human Genetics*, 107(3), 445–460. https://doi.org/10.1016/j.ajhg.2020.07.004

Course, M. M., Sulovari, A., Gudsnuk, K., Eichler, E. E., & Valdmanis, P. N. (2021). Characterizing nucleotide variation and expansion dynamics in human-specific variable number tandem repeats. *Genome Research*, 31(8), 1313–1324. https://doi.org/10.1101/gr.275560.121

Deininger, P. (2011). *Alu* elements: Know the SINEs. *Genome Biology*, 12(12), Article 236. https://doi.org/10.1186/gb-2011-12-12-236

Depienne, C., & Mandel, J. L. (2021). 30 years of repeat expansion disorders: What have we learned and what are the remaining challenges? *American Journal of Human Genetics*, 108(5), 764–785. https://doi.org/10.1016/j.ajhg.2021.03.011

De Roeck, A., Duchateau, L., Van Dongen, J., Cacace, R., Bjerke, M., Van den Bossche, T., Cras, P., Vandenberghe, R., De Deyn, P. P., Engelborghs, S., Van Broeckhoven, C., Sleegers, K., & BELNEU Consortium. (2018). An intronic VNTR affects splicing of *ABCA7* and increases risk of Alzheimer's disease. *Acta Neuropathologica*, 135(6), 827–837. https://doi.org/10.1007/s00401-018-1841-z

Dolzhenko, E., van Vugt, J. J. F. A., Shaw, R. J., Bekritsky, M. A., van Blitterswijk, M., Narzisi, G., Ajay, S. S., Rajan, V., Lajoie, B. R., Johnson, N. H., Kingsbury, Z., Humphray, S. J., Schellevis, R. D., Brands, W. J., Baker, M., Rademakers, R., Kooyman, M., Tazelaar, G. H. P., van Es, M. A., . . . Eberle, M. A. (2017). Detection of long repeat expansions from PCR-free whole-genome sequence data. *Genome Research*, 27(11), 1895–1903. https://doi.org/10.1101/gr.225672.117

Eichler, E. E., Kunst, C. B., Lugenbeel, K. A., Ryder, O. A., Davison, D., Warren, S. T., & Nelson, D. L. (1995). Evolution of the cryptic *FMR1* CGG repeat. *Nature Genetics*, 11(3), 301–308. https://doi.org/10.1038/ng1195-301

Erwin, G. S., Gürsoy, G., Al-Abri, R., Suriyaprakash, A., Dolzhenko, E., Zhu, K., Hoerner, C. R., White, S. M., Ramirez, L., Vadlakonda, A., Vadlakonda, A., von Kraut, K., Park, J., Brannon, C. M., Sumano, D. A., Kirtikar, R. A., Erwin, A .A., Metzner, T. J., Yuen, R. K. C., . . . Snyder, M. P. (2023). Recurrent repeat expansions in human cancer genomes. *Nature*, *613*(7942), 96–102. https://doi.org/10.1038/s41586-022-05515-1

Fang, L., Liu, Q., Monteys, A. M., Gonzalez-Alegre, P., Davidson, B. L., & Wang, K. (2022). DeepRepeat: Direct quantification of short tandem repeats on signal data from nanopore sequencing. *Genome Biology*, *23*(1), Article 108. https://doi.org/10.1186/s13059-022-02670-6

Findlay Black, H., Wright, G. E. B., Collins, J. A., Caron, N., Kay, C., Xia, Q., Arning, L., Bijlsma, E. K., Squitieri, F., Nguyen, H. P., & Hayden, M. R. (2020). Frequency of the loss of CAA interruption in the *HTT* CAG tract and implications for Huntington disease in the reduced penetrance range. *Genetics in Medicine*, *22*(12), 2108–2113. https://doi.org/10.1038/s41436-020-0917-z

Garg, P., Jadhav, B., Rodriguez, O. L., Patel, N., Martin-Trujillo, A., Jain, M., Metsu, S., Olsen, H., Paten, B., Ritz, B., Kooy, R. F., Gecz, J., & Sharp, A. J. (2020). A survey of rare epigenetic variation in 23,116 human genomes identifies disease-relevant epivariations and CGG expansions. *American Journal of Human Genetics*, *107*(4), 654–669. https://doi.org/10.1016/j.ajhg.2020.08.019

Gossink, F., Dols, A., Stek, M. L., Scheltens, P., Nijmeijer, B., Cohn Hokke, P., Dijkstra, A., Van Ruissen, F., Aalfs, C., & Pijnenburg, Y. A. L. (2022). Early life involvement in C9orf72 repeat expansion carriers. *Journal of Neurology, Neurosurgery and Psychiatry*, *93*(1), 93–100. https://doi.org/10.1136/jnnp-2020-325994

Gymrek, M., Willems, T., Guilmatre, A., Zeng, H., Markus, B., Georgiev, S., Daly, M. J., Price, A. L., Pritchard, J. K., Sharp, A. J., & Erlich, Y. (2016). Abundant contribution of short tandem repeats to gene expression variation in humans. *Nature Genetics*, *48*(1), 22–29. https://doi.org/10.1038/ng.3461

Gymrek, M., Willems, T., Reich, D., & Erlich, Y. (2017). Interpreting short tandem repeat variations in humans using mutational constraint. *Nature Genetics*, *49*(10), 1495–1501. https://doi.org/10.1038/ng.3952

Hagerman, R. J., Leehey, M., Heinrichs, W., Tassone, F., Wilson, R., Hills, J., Grigsby, J., Gage, B., & Hagerman, P. J. (2001). Intention tremor, parkinsonism, and generalized brain atrophy in male carriers of fragile X. *Neurology*, *57*(1), 127–130. https://doi.org/10.1212/wnl.57.1.127

Hannan, A. J. (2018). Tandem repeats mediating genetic plasticity in health and disease. *Nature Reviews Genetics*, *19*(5), 286–298. https://doi.org/10.1038/nrg.2017.115

Ho, T. H., Savkur, R. S., Poulos, M. G., Mancini, M. A., Swanson, M. S., & Cooper, T. A. (2005). Colocalization of muscleblind with RNA foci is separable from mis-regulation of alternative splicing in myotonic dystrophy. *Journal of Cell Science*, *118*(13), 2923–2933. https://doi.org/10.1242/jcs.02404

Holmes, S. E., O'Hearn, E. E., McInnis, M. G., Gorelick-Feldman, D. A., Kleiderlein, J. J., Callahan, C., Kwak, N. G., Ingersoll-Ashworth, R. G., Sherr, M., Sumner, A. J., Sharp, A. H., Ananth, U., Seltzer, W. K., Boss, M. A., Vieria-Saecker, A. M., Epplen, J. T., Riess, O., Ross, C. A., & Margolis, R. L. (1999). Expansion of a novel CAG trinucleotide repeat in the 5′ region of *PPP2R2B* is associated with SCA12. *Nature Genetics*, *23*(4), 391–392. https://doi.org/10.1038/70493

Irwin, S. A., Galvez, R., & Greenough, W. T. (2000). Dendritic spine structural anomalies in fragile-X mental retardation syndrome. *Cerebral Cortex*, *10*(10), 1038–1044. https://doi.org/10.1093/cercor/10.10.1038

Ishai, R., Seyyedi, M., Chancellor, A. M., McLean, C. A., Rodriguez, M. L., Halmagyi, G. M., Nadol, J. B., Jr., Szmulewicz, D. J., & Quesnel, A. M. (2021). The pathology of the vestibular system in CANVAS. *Otology and Neurotology*, *42*(3), e332–e340. https://doi.org/10.1097/MAO.0000000000002985

Ishiura, H., Shibata, S., Yoshimura, J., Suzuki, Y., Qu, W., Doi, K., Almansour, M. A., Kikuchi, J. K., Taira, M., Mitsui, J., Takahashi, Y., Ichikawa, Y., Mano, T., Iwata, A., Harigaya, Y., Matsukawa, M. K., Matsukawa, T., Tanaka, M., Shirota, Y., . . . Tsuji, S. (2019). Noncoding CGG repeat expansions in neuronal intranuclear inclusion disease, oculopharyngodistal myopathy and an overlapping disease. *Nature Genetics*, *51*(8), 1222–1232. https://doi.org/10.1038/s41588-019-0458-z

Ishiura, H., & Tsuji, S. (2020). Advances in repeat expansion diseases and a new concept of repeat motif–phenotype correlation. *Current Opinion in Genetics and Development*, *65*, 176–185. https://doi.org/10.1016/j.gde.2020.05.029

Izzo, M., Battistini, J., Provenzano, C., Martelli, F., Cardinali, B., & Falcone, G. (2022). Molecular therapies for myotonic dystrophy type 1: From small drugs to gene editing. *International Journal of Molecular Sciences, 23*(9), Article 4622. https://doi.org/10.3390/ijms23094622

Jana, N. R., Tanaka, M., Wang, G. H., & Nukina, N. (2000). Polyglutamine length-dependent interaction of Hsp40 and Hsp70 family chaperones with truncated N-terminal huntingtin: their role in suppression of aggregation and cellular toxicity. *Human Molecular Genetics, 9*(13), 2009–2018. https://doi.org/10.1093/hmg/9.13.2009

Kearse, M. G., Goldman, D. H., Choi, J., Nwaezeapu, C., Liang, D., Green, K. M., Goldstrohm, A. C., Todd, P. K., Green, R., & Wilusz, J. E. (2019). Ribosome queuing enables non-AUG translation to be resistant to multiple protein synthesis inhibitors. *Genes and Development, 33*(13–14), 871–885. https://doi.org/10.1101/gad.324715.119

Knight, S. J., Flannery, A. V., Hirst, M. C., Campbell, L., Christodoulou, Z., Phelps, S. R., Pointon, J., Middleton-Price, H. R., Barnicoat, A., & Pembrey, M. E. (1993). Trinucleotide repeat amplification and hypermethylation of a CpG island in FRAXE mental retardation. *Cell, 74*(1), 127–134. https://doi.org/10.1016/0092-8674(93)90300-f

Kong, H. E., Lim, J., Linsalata, A., Kang, Y., Malik, I., Allen, E. G., Cao, Y., Shubeck, L., Johnston, R., Huang, Y., Gu, Y., Guo, X., Zwick, M. E., Qin, Z., Wingo, T. S., Juncos, J., Nelson, D. L., Epstein, M. P., Cutler, D. J., . . . Jin, P. (2022). Identification of *PSMB5* as a genetic modifier of fragile X–associated tremor/ataxia syndrome. *Proceedings of the National Academy of Sciences of the United States of America, 119*(22), Article e2118124119. https://doi.org/10.1073/pnas.2118124119

Koob, M. D., Moseley, M. L., Schut, L. J., Benzow, K. A., Bird, T. D., Day, J. W., & Ranum, L. P. (1999). An untranslated CTG expansion causes a novel form of spinocerebellar ataxia (SCA8). *Nature Genetics, 21*(4), 379–384. https://doi.org/10.1038/7710

LaCroix, A. J., Stabley, D., Sahraoui, R., Adam, M. P., Mehaffey, M., Kernan, K., Myers, C. T., Fagerstrom, C., Anadiotis, G., Akkari, Y. M., Robbins, K. M., Gripp, K. W., Baratela, W. A. R., Bober, M. B., Duker, A. L., Doherty, D., Dempsey, J. C., Miller, D. G., Kircher, M., . . . Sol-Church, K. (2019). GGC repeat expansion and exon 1 methylation of *XYLT1* is a common pathogenic variant in Baratela-Scott syndrome. *American Journal of Human Genetics, 104*(1), 35–44. https://doi.org/10.1016/j.ajhg.2018.11.005

Lalioti, M. D., Mirotsou, M., Buresi, C., Peitsch, M. C., Rossier, C., Ouazzani, R., Baldy-Moulinier, M., Bottani, A., Malafosse, A., & Antonarakis, S. E. (1997). Identification of mutations in cystatin B, the gene responsible for the Unverricht-Lundborg type of progressive myoclonus epilepsy (EPM1). *American Journal of Human Genetics, 60*(2), 342–351.

Lander, E. S., Linton, L. M., Birren, B., Nusbaum, C., Zody, M. C., Baldwin, J., Devon, K., Dewar, K., Doyle, M., FitzHugh, W., Funke, R., Gage, D., Harris, K., Heaford, A., Howland, J., Kann, L., Lehoczky, J., LeVine, R., McEwan, P., . . . International Human Genome Sequencing Consortium. (2001). Initial sequencing and analysis of the human genome. *Nature, 409*(6822), 860–921. https://doi.org/10.1038/35057062

La Spada, A. R., Wilson, E. M., Lubahn, D. B., Harding, A. E., & Fischbeck, K. H. (1991). Androgen receptor gene mutations in X-linked spinal and bulbar muscular atrophy. *Nature, 352*(6330), 77–79. https://doi.org/10.1038/352077a0

Laumonnier, F., Ronce, N., Hamel, B. C. J., Thomas, P., Lespinasse, J., Raynaud, M., Paringaux, C., Van Bokhoven, H., Kalscheuer, V., Fryns, J. P., Chelly, J., Moraine, C., & Briault, S. (2002). Transcription factor *SOX3* is involved in X-linked mental retardation with growth hormone deficiency. *American Journal of Human Genetics, 71*(6), 1450–1455. https://doi.org/10.1086/344661

Ma, X. R., Prudencio, M., Koike, Y., Vatsavayai, S. C., Kim, G., Harbinski, F., Briner, A., Rodriguez, C. M., Guo, C., Akiyama, T., Schmidt, H. B., Cummings, B. B., Wyatt, D. W., Kurylo, K., Miller, G., Mekhoubad, S., Sallee, N., Mekonnen, G., Ganser, L., . . . Gitler, A. D. (2022). TDP-43 represses cryptic exon inclusion in the FTD-ALS gene *UNC13A*. *Nature, 603*(7899), 124–130. https://doi.org/10.1038/s41586-022-04424-7

Mahadevan, M., Tsilfidis, C., Sabourin, L., Shutler, G., Amemiya, C., Jansen, G., Neville, C., Narang, M., Barceló, J., & O'Hoy, K. (1992). Myotonic dystrophy mutation: An unstable CTG repeat in the 3′ untranslated region of the gene. *Science, 255*(5049), 1253–1255. https://doi.org/10.1126/science.1546325

Majounie, E., Renton, A. E., Mok, K., Dopper, E. G., Waite, A., Rollinson, S., Chiò, A., Restagno, G., Nicolaou, N., Simon-Sanchez, J., van Swieten, J. C., Abramzon, Y., Johnson, J. O., Sendtner, M., Pamphlett, R., Orrell, R. W., Mead, S., Sidle, K. C., Houlden, H., . . . Traynor, B. J. (2012). Frequency of the *C9orf72* hexanucleotide repeat expansion in patients with amyotrophic lateral sclerosis and frontotemporal dementia: A cross-sectional study. *Lancet Neurology*, *11*(4), 323–330. https://doi.org/10.1016/S1474-4422(12)70043-1

Metsu, S., Rooms, L., Rainger, J., Taylor, M. S., Bengani, H., Wilson, D. I., Chilamakuri, C. S. R., Morrison, H., Vandeweyer, G., Reyniers, E., Douglas, E., Thompson, G., Haan, E., Gecz, J., FitzPatrick, D. R., & Kooy, R. F. (2014). FRA2A is a CGG repeat expansion associated with silencing of *AFF3*. *PLOS Genetics*, *10*(4), Article e1004242. https://doi.org/10.1371/journal.pgen.1004242

Meyer, C., Schmid, G., Görlitz, S., Ernst, M., Wilkens, C., Wilhelms, I., Kraus, P. H., Bauer, P., Tomiuk, J., Przuntek, H., Mügge, A., & Schöls, L. (2007). Cardiomyopathy in Friedreich's ataxia—Assessment by cardiac MRI. *Movement Disorders*, *22*(11), 1615–1622. https://doi.org/10.1002/mds.21590

Mootha, V. V., Gong, X., Ku, H. C., & Xing, C. (2014). Association and familial segregation of CTG18.1 trinucleotide repeat expansion of *TCF4* gene in Fuchs' endothelial corneal dystrophy. *Investigative Ophthalmology and Visual Science*, *55*(1), 33–42. https://doi.org/10.1167/iovs.13-12611

Morato Torres, C. A., Zafar, F., Tsai, Y. C., Vazquez, J. P., Gallagher, M. D., McLaughlin, I., Hong, K., Lai, J., Lee, J., Chirino-Perez, A., Romero-Molina, A. O., Torres, F., Fernandez-Ruiz, J., Ashizawa, T., Ziegle, J., Jiménez Gil, F. J., & Schüle, B. (2022). ATTCT and ATTCC repeat expansions in the ATXN10 gene affect disease penetrance of spinocerebellar ataxia type 10. *Human Genetics and Genomics Advances*, *3*(4), Article 100137. https://doi.org/10.1016/j.xhgg.2022.100137

Mousavi, N., Margoliash, J., Pusarla, N., Saini, S., Yanicky, R., & Gymrek, M. (2021). TRTools: A toolkit for genome-wide analysis of tandem repeats. *Bioinformatics*, *37*(5), 731–733. https://doi.org/10.1093/bioinformatics/btaa736

Nurk, S., Koren, S., Rhie, A., Rautiainen, M., Bzikadze, A. V., Mikheenko, A., Vollger, M. R., Altemose, N., Uralsky, L., Gershman, A., & Aganezov, S. (2022). The complete sequence of a human genome. *Science*, *376*(6588), 44–53. https://doi.org/10.1126/science.abj6987

Perez, B. A., Shorrock, H. K., Banez-Coronel, M., Zu, T., Romano, L. E., Laboissonniere, L. A., Reid, T., Ikeda, Y., Reddy, K., Gomez, C. M., Bird, T., Ashizawa, T., Schut, L. J., Brusco, A., Berglund, J. A., Hasholt, L. F., Nielsen, J. E., Subramony, S. H., & Ranum, L. P. (2021). CCG•CGG interruptions in high-penetrance SCA8 families increase RAN translation and protein toxicity. *EMBO Molecular Medicine*, *13*(11), Article e14095. https://doi.org/10.15252/emmm.202114095

Prieto, M., Folci, A., Poupon, G., Schiavi, S., Buzzelli, V., Pronot, M., François, U., Pousinha, P., Lattuada, N., Abelanet, S., Castagnola, S., Chafai, M., Khayachi, A., Gwizdek, C., Brau, F., Deval, E., Francolini, M., Bardoni, B., Humeau, Y., . . . Martin, S. (2021). Missense mutation of *Fmr1* results in impaired AMPAR-mediated plasticity and socio-cognitive deficits in mice. *Nature Communications*, *12*(1), Article 1557. https://doi.org/10.1038/s41467-021-21820-1

Reyes, C. J., Asano, K., Todd, P. K., Klein, C., & Rakovic, A. (2022). Repeat-associated non-AUG translation of AGAGGG repeats that cause X-linked dystonia-parkinsonism. *Movement Disorders*, *37*(11), 2284–2289. https://doi.org/10.1002/mds.29183

Riedmayr, L. M., Hinrichsmeyer, K. S., Karguth, N., Böhm, S., Splith, V., Michalakis, S., & Becirovic, E. (2022). dCas9-VPR-mediated transcriptional activation of functionally equivalent genes for gene therapy. *Nature Protocols*, *17*(3), 781–818. https://doi.org/10.1038/s41596-021-00666-3

Rodriguez, C. M., Wright, S. E., Kearse, M. G., Haenfler, J. M., Flores, B. N., Liu, Y., Ifrim, M. F., Glineburg, M. R., Krans, A., Jafar-Nejad, P., Sutton, M. A., Bassell, G. J., Parent, J. M., Rigo, F., Barmada, S. J., & Todd, P. K. (2020). A native function for RAN translation and CGG repeats in regulating fragile X protein synthesis. *Nature Neuroscience*, *23*(3), 386–397. https://doi.org/10.1038/s41593-020-0590-1

Sawaya, S., Bagshaw, A., Buschiazzo, E., Kumar, P., Chowdhury, S., Black, M. A., & Gemmell, N. (2013). Microsatellite tandem repeats are abundant in human promoters and are associated with regulatory elements. *PLOS ONE*, *8*(2), Article e54710. https://doi.org/10.1371/journal.pone.0054710

Schmitz, A., Pinheiro Marques, J., Oertig, I., Maharjan, N., & Saxena, S. (2021). Emerging perspectives on dipeptide repeat proteins in C9ORF72 ALS/FTD. *Frontiers in Cellular Neuroscience*, *15*, Article 637548. https://doi.org/10.3389/fncel.2021.637548

Smajić, S., Prada-Medina, C. A., Landoulsi, Z., Ghelfi, J., Delcambre, S., Dietrich, C., Jarazo, J., Henck, J., Balachandran, S., Pachchek, S., Morris, C. M., Antony, P., Timmermann, B., Sauer, S., Pereira, S. L., Schwamborn, J. C., May, P., Grünewald, A., & Spielmann, M. (2022). Single-cell sequencing of human midbrain reveals glial activation and a Parkinson-specific neuronal state. *Brain*, *145*(3), 964–978. https://doi.org/10.1093/brain/awab446

Smeyers, J., Banchi, E. G., & Latouche, M. (2021). C9ORF72: What it is, what it does, and why it matters. *Frontiers in Cellular Neuroscience*, *15*, Article 661447. https://doi.org/10.3389/fncel.2021.661447

Sulovari, A., Li, R., Audano, P. A., Porubsky, D., Vollger, M. R., Logsdon, G. A., Human Genome Structural Variation Consortium, Warren, W. C., Pollen, A. A., Chaisson, M. J. P., & Eichler, E. E. (2019). Human-specific tandem repeat expansion and differential gene expression during primate evolution. *Proceedings of the National Academy of Sciences of the United States of America*, *116*(46), 23243–23253. https://doi.org/10.1073/pnas.1912175116

Tian, Y., Wang, J. L., Huang, W., Zeng, S., Jiao, B., Liu, Z., Chen, Z., Li, Y., Wang, Y., Min, H. X., Wang, X. J., You, Y., Zhang, R. X., Chen, X. Y., Yi, F., Zhou, Y. F., Long, H. Y., Zhou, C. J., Hou, X., . . . Shen, L. (2019). Expansion of human-specific GGC repeat in neuronal intranuclear inclusion disease–related disorders. *American Journal of Human Genetics*, *105*(1), 166–176. https://doi.org/10.1016/j.ajhg.2019.05.013

Trettel, F., Rigamonti, D., Hilditch-Maguire, P., Wheeler, V. C., Sharp, A. H., Persichetti, F., Cattaneo, E., & MacDonald, M. E. (2000). Dominant phenotypes produced by the *HD* mutation in ST*Hdh*^{Q111} striatal cells. *Human Molecular Genetics*, *9*(19), 2799–2809. https://doi.org/10.1093/hmg/9.19.2799

van der Plas, E., Schultz, J. L., & Nopoulos, P. C. (2020). The neurodevelopmental hypothesis of Huntington's disease. *Journal of Huntington's Disease*, *9*(3), 217–229. https://doi.org/10.3233/JHD-200394

van Kuilenburg, A. B. P., Tarailo-Graovac, M., Richmond, P. A., Drögemöller, B. I., Pouladi, M. A., Leen, R., Brand-Arzamendi, K., Dobritzsch, D., Dolzhenko, E., Eberle, M. A., Hayward, B., Jones, M. J., Karbassi, F., Kobor, M. S., Koster, J., Kumari, D., Li, M., MacIsaac, J., McDonald, C., . . . van Karnebeek, C. D. M. (2019). Glutaminase deficiency caused by short tandem repeat expansion in GLS. *New England Journal of Medicine*, *380*(15), 1433–1441. https://doi.org/10.1056/NEJMoa1806627

Verkerk, A. J., Pieretti, M., Sutcliffe, J. S., Fu, Y. H., Kuhl, D. P., Pizzuti, A., Reiner, O., Richards, S., Victoria, M. F., & Zhang, F. P. (1991). Identification of a gene (*FMR-1*) containing a CGG repeat coincident with a breakpoint cluster region exhibiting length variation in fragile X syndrome. *Cell*, *65*(5), 905–914. https://doi.org/10.1016/0092-8674(91)90397-h

Warren, S. T., Muragaki, Y., Mundlos, S., Upton, J., & Olsen, B. R. (1997). Polyalanine expansion in synpolydactyly might result from unequal crossing-over of HOXD13. *Science*, *275*(5298), 408–409. https://doi.org/10.1126/science.275.5298.408

Weinhaeusel, A., Morris, M. A., Antonarakis, S. E., & Haas, O. A. (2003). DNA deamination enables direct PCR amplification of the cystatin B (CSTB) gene–associated dodecamer repeat expansion in myoclonus epilepsy type Unverricht-Lundborg. *Human Mutation*, *22*(5), 404–408. https://doi.org/10.1002/humu.10276

Wieben, E. D., Aleff, R. A., Tosakulwong, N., Butz, M. L., Highsmith, W. E., Edwards, A. O., & Baratz, K. H. (2012). A common trinucleotide repeat expansion within the transcription factor 4 (*TCF4*, E2-2) gene predicts Fuchs corneal dystrophy. *PLOS ONE*, *7*(11), Article e49083. https://doi.org/10.1371/journal.pone.0049083

Winnepenninckx, B., Debacker, K., Ramsay, J., Smeets, D., Smits, A., FitzPatrick, D. R., & Kooy, R. F. (2007). CGG-repeat expansion in the *DIP2B* gene is associated with the fragile site FRA12A on chromosome 12q13.1. *American Journal of Human Genetics*, *80*(2), 221–231. https://doi.org/10.1086/510800

Zang, J. B., Nosyreva, E. D., Spencer, C. M., Volk, L. J., Musunuru, K., Zhong, R., Stone, E. F., Yuva-Paylor, L. A., Huber, K. M., Paylor, R., Darnell, J. C., & Darnell, R. B. (2009). A mouse model of the human fragile X syndrome I304N mutation. *PLOS Genetics*, *5*(12), Article e1000758. https://doi.org/10.1371/journal.pgen.1000758

Zhao, X., Kumari, D., Miller, C. J., Kim, G. Y., Hayward, B., Vitalo, A. G., Pinto, R. M., & Usdin, K. (2021). Modifiers of somatic repeat instability in mouse models of Friedreich ataxia and the fragile X–related disorders: Implications for the mechanism of somatic expansion in Huntington's disease. *Journal of Huntington's Disease*, *10*(1), 149–163. https://doi.org/10.3233/JHD-200423

Zheng, S., Clabough, E. B. D., Sarkar, S., Futter, M., Rubinsztein, D. C., & Zeitlin, S. O. (2010). Deletion of the Huntingtin polyglutamine stretch enhances neuronal autophagy and longevity in mice. *PLOS Genetics*, *6*(2), Article e1000838. https://doi.org/10.1371/journal.pgen.1000838

Zhou, G., Zhou, M., Zeng, F., Zhang, N., Sun, Y., Qiao, Z., Guo, X., Zhou, S., Yun, G., Xie, J., Wang, X., Liu, F., Fan, C., Wang, Y., Fang, Z., Tian, Z., Dai, W., Sun, J., Peng, Z., & Song, L. (2022). Performance characterization of PCR-free whole genome sequencing for clinical diagnosis. *Medicine*, *101*(10), Article e28972. https://doi.org/10.1097/MD.0000000000028972

Zu, T., Gibbens, B., Doty, N. S., Gomes-Pereira, M., Huguet, A., Stone, M. D., Margolis, J., Peterson, M., Markowski, T. W., Ingram, M. A. C., Nan, Z., Forster, C., Low, W. C., Schoser, B., Somia, N. V., Clark, H. B., Schmechel, S., Bitterman, P. B., Gourdon, G., . . . Ranum, L. P. W. (2011). Non-ATG-initiated translation directed by microsatellite expansions. *Proceedings of the National Academy of Sciences of the United States of America*, *108*(1), 260–265. https://doi.org/10.1073/pnas.1013343108

11

The Properties of Cortico-motoneuronal Connections and Their Evolutionary Significance for Amyotrophic Lateral Sclerosis

Roger N. Lemon

Historical Background

The involvement of the corticospinal tract (CST) was evident from Charcot's early definition of amyotrophic lateral sclerosis/motor neuron disease (ALS/MND). The evolutionary significance of ALS is apparent from the fact that it is a uniquely human disease. It is a disorder affecting both descending corticospinal projections and the motoneurons which they influence (Eisen 2021).

The key synaptic connections between the motor cortex and alpha motoneurons were first identified by Leyton and Sherrington (1917) in the chimpanzee. They made a localized lesion of the motor cortex arm area and several weeks later found degenerating corticospinal terminals close to motoneurons in the lower cervical cord. Definitive electrophysiological demonstration of the monosynaptic excitation of spinal motoneurons by cortico-motoneuronal (CM) connections had to wait for the studies of Bernhard et al. (1953).

The CM system is a subdivision of the corticospinal system, unique to primates: it is not yet known what proportion of the total corticospinal projection makes CM connections, but it is known that it varies across primate species, is best developed in humans, and represents a particularly strong and direct projection to motoneurons innervating muscles that are involved in the most voluntary actions characteristic of human movement, including speech and fine hand movements.

Andrew Eisen was the first to recognize the importance of the CM system in ALS (Eisen et al. 1992). He has pointed out that all neurodegenerative disorders, including ALS, are complex polygenic diseases, resulting in multisystem impairment of neocortical networks. The CM system and the motoneurons it innervates are the anatomical infrastructure of many early clinical features of ALS, which is a singularly human disorder. Early deficits target particular groups of bulbar or limb muscles and result in loss of complex vocalization, impaired capacity for true opposition of the digits and thumb required for precision grip and skilled manipulation, and difficulty with upright walking, especially when navigating challenging surfaces (Eisen et al. 2014). In addition is the association of frontotemporal dementia, causing language impairment, failing executive function, and deteriorating socialization (Eisen 2021).

An early demonstration of the evolutionary persistence of CM connections into the human nervous system came from postmortem evidence in a patient with an infarct in the motor cortex that showed degenerating corticospinal terminals in the ventral horn of the C8 segment, exactly in the location of motoneurons supplying hand and digit muscles (Schoen 1964). The introduction of non-invasive transcranial electrical and transcranial magnetic stimulation (TMS) in the 1980s revolutionized our understanding of the human CST, including the early demonstration of putative CM effects in a wide variety of upper and lower limb muscles (Day et al. 1989; Brouwer and Ashby 1990; Palmer and Ashby 1992; de Noordhout et al. 1999), with the largest effects seen in the intrinsic hand muscles (Palmer and Ashby 1992).

Eisen et al. (2014) argued that relatively recent evolutionary adaptations of the motor system, including speech, hand skill and bipedalism, and complex locomotion, are paralleled by the emergence of a number of neuroanatomical developments, including the CM system, which itself is a recently evolved feature. The growing evidence to support the hypothesis that ALS is a brain disease which specifically involves the CM system and its target muscles has been reviewed in detail elsewhere (Braak et al. 2013; Eisen et al. 2017; Ludolph et al. 2020). A particularly striking advance was made by the detailed study of the "split-hand syndrome" in ALS (Weber et al. 2000b; Eisen and Swash 2001). It was revealed that the well-known weakness and loss of motor units in hand muscles were more pronounced in thenar (thumb) than in hypothenar (little finger) muscles (Menon et al. 2014). This was consistent with the larger CM influence over motoneurons of thenar versus hypothenar muscles demonstrated in the nonhuman primate (Clough et al. 1968; Figure 11.1). For the intrinsic thumb motoneurons, it may be that the CM input represents such a large component of the finely structured excitatory drive that when this input is compromised by ALS, the result is both a weakness and a poverty of movement (Eisen and Lemon 2021). It is generally accepted that the earliest component of the motor evoked potential recorded from a muscle in response to TMS delivered over the motor

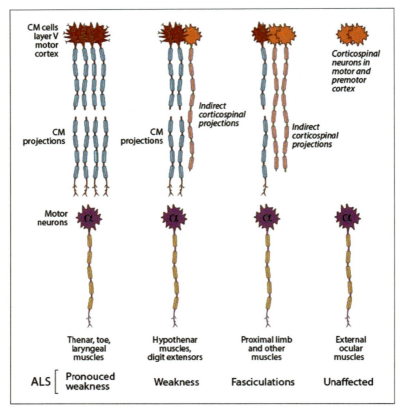

Figure 11.1 In humans it is probable that cortico-motoneuronal (CM) cells synapse with all motoneuron pools except those innervating the external ocular muscles and sphincter muscles; it is noteworthy that these muscles remain relatively spared in amyotrophic lateral sclerosis (ALS). CM synapses in humans may contribute a decisive proportion of the synaptic input to muscles involved in skilled hand function, pedal agility, complex vocalization and breathing, and other multineuronal bulbar functioning. Loss of CM input results in both weakness and poverty of movement (see Eisen and Lemon 2021). CM input is strongest to the low-threshold, early recruited motor units, the same ones in which ALS fasciculations can be initially recorded, and are thus appreciated early in a wider group of muscles than those mediating skilled motor activity. Reproduced, with permission, from Eisen et al. (2017).

cortex represents the action of fast, monosynaptic CM inputs to the motoneurons. Weber et al. (2000a) used TMS to show that the motor evoked potentials in the thenar muscles of ALS patients with split-hand syndrome were indeed smaller and more abnormal compared with those in the hypothenar muscles.

Interestingly, although the peripheral neuromuscular properties of both muscle groups showed pathological signs in ALS, these were rather similar in thenar and hypothenar muscles, so Menon et al. (2014) suggested that the "split-hand syndrome" reflects the particular vulnerability of the CM projections to the thenar motoneurons over and above a more general lower motoneuron disorder.

Further evidence for the particular vulnerability to ALS, with a possible evolutionary basis, of muscle groups with strong CM input has been accumulating. This includes

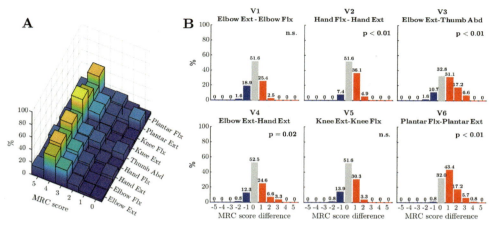

Figure 11.2 Pattern of weakness in upper and lower limb muscles in amyotrophic lateral sclerosis (ALS) patients. These data are from a prospective study of ALS patients in whom all four limbs were affected by the disease ($n = 61$ patients, $n = 122$ limbs). Color scale is autoscaled to the maximum strength score (Medical Research Council [MRC] scale 0–5) across all muscle groups. (A) Percentages of individual MRC scores for each muscle group. (B) Difference in MRC score between pairs of muscles for each muscle group. Bars indicate the percentage distribution of differences in MRC score. A positive difference (red bars) indicates that the muscle group receiving the less pronounced cortico-motoneuronal (CM) influence (first muscle group listed for each pair) was the stronger. Blue bars indicate a difference in the opposite direction. The p value derived from the sign test of these differences is given at the top of each panel. The data indicate that in each pair of muscles tested in these patients, the muscles whose motoneurons are known to receive the stronger CM projection are more severely weakened by ALS. Abd = abduction; Ext = extension; Flx = flexion; n.s. = not significant. Reproduced, with permission, from Ludolph et al. (2020).

studies of "split elbow," "split foot," and "split ankle" (Khalaf et al. 2019; Henderson and Eisen 2020). In all three syndromes, the more profound weakness is found in the muscle groups known to receive strong CM projections. A recent study by Ludolph et al. (2020) included both retrospective and prospective studies of a large cohort of ALS patients whose muscle strength was assessed in different pairs of upper and lower limb muscles, using the Medical Research Council scale. The study looked for any statistical sign of asymmetry, in each patient, between pairs of muscle groups, with one group known to receive stronger CM connections than the other (e.g., wrist extensors vs. wrist flexors). The results showed a greater relative weakness for the muscle groups which normally receive strong CM projections, consistent with the loss of CM influence during ALS. The results are presented in Figure 11.2.

Evolutionary Evidence Specifically Related to Human Health and Disease

In this section, the evolutionary status of the CM system is reviewed. Is its status consistent with its particular vulnerability to ALS? The starting point is the famous 1884 essay by Hughlings Jackson, "On Evolution and Dissolution of the Nervous System," where he pointed out that diseases of the human central nervous system (CNS) particularly target the "higher," complex systems of the CNS, which support the "most voluntary" functions, as exemplified by speech and by skilled hand movements. These "higher levels" are also the most recently evolved. During evolution the neocortical areas giving rise to the CST have expanded in size and number, while the number of fibers within the tract has increased and the range of fiber sizes widened. These changes have been accompanied by a relative diminution, in humans, of the importance of some brainstem pathways for motor control, such as the rubrospinal tract. As a consequence, the increasing reliance in humans upon the CST for these particular skilled movements means that damages and disorders of the CST have more devastating effects in humans than in most animals (Lemon 2008).

The CM system is important because it represents one of the key mechanisms used by the brain's motor network to generate and control complex movements involving multiple groups of muscles (Eisen et al. 2014; Eisen and Lemon 2021; Lemon 2021). The fractionation of activity across the 30 or more muscles controlling the hand and digits represents a good example of such control. The large number of possible combinations of these muscles underpins the many different types of grasping and manipulatory movements of which the human hand is capable and which are involved in myriad functions in technology, culture, medicine, music, and writing. There are, of course, also well-established links between gesture and language (Wilson 1998; Tallis 2004; Eisen et al. 2014). As Tallis (2004) points out, there are numerous theories which emphasize the importance of gesture as an intermediary activity between tool use/manufacture and the emergence of speech. The transfer to the mouth of a skill (communication) from the hand is made all the more plausible given an earlier transfer in which bipedal stance allowed the hand to replace the mouth as a gripping tool, manipulator, and tactile explorer, which is also used for collection and storage of food.

CM Connections and the Evolution of Dexterity

Does the evolutionary status of the CM system fit into this overall scheme? The comparative biology of the corticospinal system has been particularly well explored in relation to dexterity (Phillips 1971; Heffner and Masterton 1975, 1983; Kuypers 1981; Bortoff and Strick 1993; Porter and Lemon 1993; Nakajima et al. 2000). The key features revealed by this analysis are that CM connections are only well developed in humans, apes, some Old-World monkeys (macaque and baboon), and some New-World monkeys (e.g., capuchin [*Cebus apella*]). All of these primates have the capacity for tool use. In the New-World *Cebus*, tool use is very well developed, and it has copious CM connections, while the New-World squirrel monkey (*Saimiri sciureus*), which does not exhibit tool use, has very sparse CM connections (Bortoff and Strick 1993; Maier et al. 1997; Nakajima et al. 2000).

Heffner and Masterton (1975, 1983) attempted a correlation, across different species, between a number of anatomical parameters of the CST and an "index of dexterity." This measures the capacity of a particular species to make skilled and independent movements of its digits. They found a strong correlation between dexterity and the depth to which CST projections extend within the spinal gray matter, indicating that species with CM projections (those projecting deep into the motor nuclei within Rexed's lamina IX of the ventral horn) generally had a higher index of dexterity than those with projections to more dorsal regions of the spinal gray matter.

A quantitative study of the corticospinal projection from macaque motor cortex was made by Morecraft et al. (2013), who counted the number of boutons labeled by anterograde tracers injected into the arm/hand area of M1. They showed that a significant proportion (18%) of labeled terminals were located among spinal motoneurons in lamina IX, with a steady increase in the number of terminals from mid to low cervical segments (C5 to C8–T1) and with terminals concentrated among the intrinsic hand muscle motoneurons. A strong CM projection from the M1 hand area was also demonstrated using novel intersectional genetic approaches in the macaque monkey, which allow specific labeling of projections from the hand M1 cortex to motoneurons of particular muscles, such as digit or hand muscles (Sinopoulou et al. 2022).

Interestingly, recent work has identified regulatory plexin binding sites within layer V of the motor cortex, which appear to be necessary for corticospinal axon guidance and the establishment of CM connections during development and their maintenance in adulthood (Gu et al. 2017). These binding sites are found in the motor cortex of humans, chimpanzees, gorillas, orangutans, and baboons, all of which have CM connections, though they are lacking in other nonhuman primates such as marmosets and bushbabies and in nonprimates such as rabbits, rats, and mice, which have no CM connections (Lemon and Griffiths 2005; Lemon 2008).

Further details of the functions of the CM system were revealed many years ago in experimental studies using spike-triggered averaging methods. These allowed investigation of the influence over muscle activity of spikes in single neurons. These spikes were recorded from the macaque motor cortex during performance of precision grip and other skilled tasks (Fetz and Cheney 1980; Lemon et al. 1986). The results were consistent with earlier electrophysiological evidence for CM monosynaptic effects observed with stimulation techniques in anaesthetized animals (see Porter and Lemon 1993). They also revealed the focused selective action of many CM cells on a small group of functional synergist muscles. Given that CM connections have evolved most strongly in primates that use tools, it is interesting that corticospinal and CM neurons in the macaque monkey are strongly activated during active tool use (Quallo et al. 2012).

Evolution of Primate Motor Cortex and Primate Features of the CST

It is known that the mammalian primary motor cortex evolved from the primary somatosensory cortex approximately 100 million years ago, when placental mammals diverged from early mammals such as marsupials (Kaas et al. 2018). In the macaque the primary motor cortex (area 4, M1) has completely separated from the postcentral sensory cortex, and a number of important "secondary" motor areas have evolved in the lateral and medial premotor cortex (Strick et al. 2021). These areas now function as a closely interconnected network of areas during the preparation and execution of complex reach-to-grasp movements (Davare et al. 2011).

Within the arm/hand region of M1, an important further development is the separation of a caudal subdivision (M1c), which is characterized by the presence of numerous CM neurons with direct projections to the contralateral motor neuron pools supplying digit, hand, and arm muscles (Rathelot and Strick 2006, 2009). In the adjacent rostral division (M1r), CM neurons are present but in much smaller numbers, and the main descending projections are focused on spinal interneuronal mechanisms upstream from alpha motoneurons and on brainstem neurons that give rise to descending motor pathways, such as the reticulospinal tract. In addition, tactile afferent input from the glabrous (fingerprint, non-hairy) skin of the hand is mainly focused on the M1c division, while proprioceptive inputs dominate the afferent input to M1r (Strick and Preston 1982). In the human motor cortex these two subdivisions are known to be further specified by their contrasting cytoarchitecture and by different distributions of muscarinic neurotransmitters (Geyer et al. 1996).

The Range of Axon Sizes within the Human CST

In the great majority of mammals, the CST is made up of small-diameter fibers (<3 μm; see, e.g., Leenen et al. 1985). However, in carnivores and in large, dexterous primates, especially in humans, there is in addition a fast-conducting population of fibers with axons >8 μm in diameter—and up to 12 μm in the macaque and 20 μm in humans (Lassek and Rasmussen 1940; Firmin et al. 2014). This huge range in fiber sizes probably reflects the many different functions mediated by the CST (Lemon 2019). Although the thick axons, estimated to make up about 4% of the human CST (Lassek and Rasmussen 1940), are hugely outnumbered by fine fibers, it is likely that these fast-conducting fibers have a physiological significance for skilled dexterity out of all proportion to their numbers. Against this, Heffner and Masterton (1983) could not find any correlation between dexterity and the diameter of the *largest* fiber in the CST. But because there are so few published accounts of the complete distribution of fiber sizes across different species, they were unable to test whether there was any correlation with the population of *larger* fibers with diameters >8 μm. They did find a correlation between dexterity and the overall cross-sectional area of the pyramidal tract (PT) and showed that this was not simply a reflection of increases in body weight across species. However, reliable data from more species on the size distribution and total numbers of fibers within the CST are still needed to explain why the correlation with PT area is significant, but it could reflect the contribution of the larger fibers that are found in bigger, dexterous primates. The function of the finest fibers, with diameters of <1 μm, remains a mystery (Kraskov et al. 2019, 2020).

Betz Cells and the Fast-Conducting CM System

The motor cortex of primates contains very large or giant pyramidal cells in layer, called Betz cells (Rivara et al. 2003); and it is likely that they all give rise to large, fast-conducting axons and to CM connections, although neither conclusion has been definitively established in humans. Betz cells are certainly not the only source of CM connections since CM cell bodies, labeled by transsynaptic retrograde methods, come in a large range of sizes (Rathelot and Strick 2006).

One of the evolutionary advantages of the fast-conducting component of the CST is the capacity to reduce conduction time from the motor cortex to the spinal cord, and this might be a partial explanation of why larger CST axons (those with diameters >8 μm) have evolved in bigger primates (Kuypers 1981; Heffner and Masterton 1983). The reduction of conduction delays may be important in rapid transitions from posture to movement during manipulation (Venkadesan and Valero-Cuevas 2009).

A further property of neurons with these fast axons is the potential to support bursts of high-frequency impulses at movement onset, and it is interesting that the fastest PT neurons in the macaque M1 are characterized by brief or "thin" spikes (Vigneswaran et al. 2011; Lemon et al. 2021), which represent rapid repolarization of the axon after passage of an action potential spike, and therefore allow high-frequency discharges. The capacity for rapid repolarization may be related to the expression of the fast potassium channel Kv3.1b in the soma and dendrites of macaque M1 pyramidal neurons (Soares et al. 2017). By contrast, the rat motor cortex has pyramidal neurons with broad spikes (Barthó et al. 2004) which lack Kv3.1b expression (Soares et al. 2017). There may be also very important energy considerations for fast versus slow corticospinal axons (Perge et al. 2012).

Species Differences in the Corticospinal Connectome

The species-specific differences in the extent of CM connections are accompanied by some other striking changes in the pattern of descending projections from the CST. Thus, in the rodent and many other mammals, the major target of the CST is the dorsal horn of the spinal gray matter, which is somatosensory in function (Kuypers 1981). In all mammals a major function of the CST is the control of somatosensory input (Lemon 2008; Liu et al. 2018; Moreno-Lopez et al. 2021). Motor control theories suggest that the prediction of somatosensory inputs generated by a voluntary movement must be a key feature in planning and executing a skilled movement (Franklin and Wolpert 2011; Adams et al. 2013). It is suggested that the corticospinal system carries information about the upcoming movement and that this information is delivered to dorsal horn mechanisms receiving the proprioceptive and cutaneous feedback from the limb, allowing a close interrogation of the inputs to determine if the predicted sensory input matches the actual input. In primates, there is a separate and specific CST projection to the dorsal horn, which comes from the parietal cortex and S1 in particular (Ralston and Ralston 1985) but not from M1 (Morecraft et al. 2013; Sinopoulou et al. 2022).

The corticospinal axon originating from pyramidal neurons in layer V of the sensory and motor cortices is, of course, just one of many belonging to the "connectome" of the neuron, that is, the entire set of connections established by the corticospinal neuron, through its axon to the spinal cord as well as through the collaterals given off from this axon as it passes through the cerebrum, midbrain, and brainstem. For example, most neurons projecting to the spinal cord (i.e., CST neurons) also give a collateral to the pontine nuclei, a major source of mossy fiber input to the cerebellum (Keizer and Kuypers 1984). Recent intersectional genetic approaches suggest that there are again major species differences in the structure of the connectome of CST neurons in the M1 hand region. The connectome is more widespread in the rat and mostly targeting somatosensory structures, while in the macaque it is more focused and directed mainly to motor structures, such as the pontine nuclei and red nucleus (Sinopoulou et al. 2022).

Evolutionary Arguments: Impact on Selecting Animal Models to Study ALS

Current understanding of ALS would suggest that there are strong evolutionary pressures rendering the human motor system vulnerable to the disease. In particular, there is the special vulnerability of long, fast-conducting corticofugal systems which directly innervate alpha motoneurons and which are involved in the generation of complex movements, including speech and manipulation. This is a strong indication that in selecting animal models for the study of ALS, a primate model should be included (Braak et al. 2013; Lemon 2021; Li et al. 2022). The motor system of the macaque monkey has most of the properties just listed and is the best available model of the human motor system (Kuypers 1981; Box 11.1).

That said, there are no reports of the distinctive pathology of ALS occurring spontaneously in any animal model and crucially in one that exhibits skilled hand control (Del Tredici and Braak 2022). An emerging concept is that this pathology involves a prion-like propagation of the abnormal form of the protein TDP43 (Braak et al. 2013). There is, however, a report that overexpression of TDP43 in the macaque spinal cord, via an adenovirus vector, can result in ALS-like symptoms, including muscle weakness, atrophy, and fasciculation in hand muscles. These signs were not observed after similar expression in rats (Uchida et al. 2012; Li et al. 2022).

Among the reasons there appear to be no reports of ALS-like disease in naturally aging nonhuman primates might be that these animals have considerably shorter life spans than humans and that there have been few specific studies of older monkeys—and, of course, they have not been exposed to the same environment as humans, particularly not to the same environmental toxins.

Bulbar involvement and effects on speech are the initial signs in some ALS patients, and it has been noted that in humans there is good evidence for direct cortical projections

BOX 11.1 An Animal Model for ALS

Animal models have always been and continue to be important for improved understanding of basic disease mechanisms and for translational research. Amyotrophic lateral sclerosis (ALS) is a human disease which poses some particular challenges in finding an appropriate animal model: it is a disease that develops over a long period of time, and it is a disease of aging. Most animal models may not have a natural life span long enough for robust expression of the pathology that leads to ALS. Mice and rats also lack the cortico-motoneuronal (CM) projection whose particular vulnerability seems to trigger some forms of ALS in humans. However, genetic, molecular, cell biological, and neurobiological aspects of ALS can still be studied in these models.

One important consideration in selecting an animal model for studying ALS is the need to understand and eventually treat the effects of ALS on the axons of long corticofugal neurons. It will be particularly critical to study the long, fast-conducting axons that belong to CM neurons, with their direct synaptic connections with key groups of muscles in the larynx, hand, and foot. These effects could cover a range of axonal functions, including axonal transport, metabolism and support, and the successful conduction of the high-frequency impulses generated by fast corticospinal neurons.

Neuropathological studies indicate loss of myelin from corticospinal axons and selective loss of fast-conducting axons; these would lead to a slowing of responses in limb muscles to cortical stimulation and degradation of skilled actions such as speech and manipulation.

Therefore, in addition to rodent models, it will be essential to develop nonhuman primate models for study: a macaque monkey model offers three key features that are lacking in rodents. First, macaques have the same fast-conducting axons as are found in the human corticospinal tract; second, these axons make CM connections with the same muscle groups that are vulnerable to ALS in humans; and finally, they exhibit skilled manipulation similar to that in humans.

There are, of course, serious ethical and welfare issues involved in the use of animal models, and these are particularly important for the use of sensitive and valuable animals such as macaque monkeys. It will be important to model ALS in these animals in a controlled and restricted way, which produces limited pathology that the animals can tolerate, while still allowing insights into the wider disease process. The numbers of monkeys involved will be relatively small but could produce important advances not yet delivered by rodent models.

to the nucleus ambiguus, which provides the motor innervation of the laryngeal muscles (Kuypers 1981; Iwatsubo et al. 1990; Figure 11.1), while cortical innervation is indirect in nonhuman primates (Simonyan 2014).

Because rodents lack the CM projection and because the motor consequences of either stimulating or lesioning the rodent corticospinal system are generally rather modest, use of rodents as a model for ALS can be criticized (Braak et al. 2013; Izpisua Belmonte et al. 2015; Lemon 2021). However, it is clear that studies in the mouse, in particular, can make important contributions to our understanding of the basic neurobiology of the disease (Fogarty et al. 2016; Gautam et al. 2019). For example, knockout studies in the mouse have shown that genetic deletion of corticofugal projections has important consequences for the development of spinal circuitry (Han et al. 2013) and can ameliorate symptoms in a mouse model of ALS (Burg et al. 2020).

The Evolutionary Involvement of Skilled Hand Movements and ALS

Although the CST has been long been implicated in the pathology of ALS, the nature of that involvement is now increasingly being understood from both neurophysiological and neuropathological standpoints (Eisen et al. 1992, 2017; Henderson and Eisen 2020; Eisen 2021). There is now neuropathological evidence for selective impact of ALS on corticofugal neurons in the human motor cortex with long axons (Braak et al. 2013; Figure 11.1). This results in transmission of TDP43 through the CM system to target motoneurons in the brainstem and spinal cord and through other collaterals to involve other connected cortical regions at later stages of the disease, possibly explaining the link between ALS and frontotemporal dementia (Braak et al. 2013; Del Tredici and Braak 2022; Ma et al. 2022). In some autosomal recessive forms of the disease, there is evidence for preservation of a slowly conducting CM system (Weber et al. 2000a). This may have a different cortical origin: while the caudal division of M1 ("new" M1, or BA 4p) is characterized by many CM neurons including the fastest-conducting axons, most of the rostral division of M1 ("old" M1, or BA 4a) is characterized by smaller pyramidal neurons with more slowly conducting outputs (Witham et al. 2016).

It is also known that the larger CS fibers are particularly vulnerable to disease and to trauma, for example, during spinal cord injury (Quencer et al. 1992). There have been some important, but rarely cited, pathological studies of the human CST in ALS and related disorders. For example, Sobue et al. (1987) looked at the distribution of axonal diameters in the CST at the T7 level and found a disproportionate loss of large axons. A postmortem study of the spinal cords from 13 ALS patients by Oyanagi et al. (1995) compared the distribution of fiber diameters in the lateral CST (LCST) of the dorsolateral funiculus and of fibers in the anterolateral fasciculus, both sampled at the cervical level. They found a significant loss of larger fibers from the LCST, but loss of these fibers did not always correlate with clinical "upper motoneuron" signs. A more recent study (Riku et al. 2014) looked at the distribution of fibers in the LCST at the C5–C6 spinal level in 93 patients with ALS who initially presented with a variety of symptoms (upper limb, lower limb, and bulbar) and again found evidence for a disproportionate loss of larger axon diameters. They determined the percentage of axons in the LCST with a diameter >1 µm; this was only 18.5% in the ALS patient group compared with 32.3% in healthy controls.

The contribution to pathology of the majority of fine axons in the CST demands further attention; it would also be useful to look at the distribution of CST fibers within the PT since there are normative data available for comparison with patient material, and one can be sure that, at least in the caudal PT, all of the fibers are corticospinal. This is uncertain within the LCST because of some intermingling with other ascending and descending fiber tracts.

Counterarguments and Evidence against Evolutionary Influences

ALS is characterized by degenerative changes in both spinal/cranial motoneurons and corticospinal neurons, including those with CM connections. CM connections evolved relatively late, whereas motoneurons are highly conserved across many species (Alaynick et al. 2011). One might speculate that this major difference in evolutionary timetable may be reflected in the very different ways in which pyramidal cells and alpha motoneurons process TDP43 (Braak et al. 2017; Del Tredici and Braak 2022). This observation supports the emerging concept of prion-like propagation of abnormal proteins in noninfectious neurodegenerative diseases.

Clinical and Therapeutic Implications

Many investigators have emphasized ALS as a motoneuron disease, and the major diagnostic tests for confirming ALS/MND involve tests of muscle function and strength. However, another view is that ALS is a brain disease whose major impact is on the generation and control of complex movements (Eisen and Lemon 2021). Recognition of the evolutionary significance of the CM system's vulnerability to ALS should result in more attention being directed to the control of skilled movement, in addition to measures of muscle strength. A given muscle can be used in many different ways, and one of the key features of the CM system is the task-specific recruitment of particular groups of muscles. There is thus a task-specific flexibility between activity in a CM neuron (the "upper motoneuron") and its target ("lower") motoneurons (Lemon 2021). Tests for the execution of complex movement could be carried out to explore these ideas. For example, Czell et al. (2019) found that the time to complete the standard Nine Hole Peg Test was significantly longer in 20 ALS patients than in age-matched controls, and this measure increased with disease progression.

Unanswered Questions and Future Work

There are, of course, a great many unanswered questions; but the neuropathological, neurophysiological, and behavioral research into ALS referred to in this review underlines the increasing realization that dysfunction of corticofugal neurons is central to many forms of ALS. Despite the huge challenges of undertaking neurophysiological studies in patients suffering from ALS, these studies are needed to understand the nature and time course of CM dysfunction. It is evident that corticospinal axons play a crucial role in both the progression of the disease and influence over key muscle groups, but there is a need to know a lot more about how the disease affects human CST axons, and postmortem studies of axons in the PT would be essential to progress.

Key Points to Remember

- There is ample evidence that primate evolution included the development of fast, direct projections from the motor cortex to motoneurons, allowing direct control of muscles from the neocortex.
- There is also evidence that these direct, CM projections are best developed in primates that are tool users and neurophysiological evidence that these projections are activated during skilled hand function and during tool use.
- Long corticofugal pathways, including the CM projections, are particularly vulnerable to ALS. This evidence fits with the observation that skilled actions, including those involving the hand, are affected early in disease progression.
- In terms of understanding the disease, and possibly finding better treatments for it, it will be necessary to use nonhuman primate models, which share with humans a well-developed CM system.

Acknowledgments

I would like to thank all my collaborators and Andrew Eisen for reviewing a draft of this chapter.

References

Adams, R. A., Shipp, S., & Friston, K. J. (2013). Predictions not commands: Active inference in the motor system. *Brain Structure and Function*, *218*(3), 611–643. https://doi.org/10.1007/s00429-012-0475-5

Alaynick, W. A., Jessell, T. M., & Pfaff, S. L. (2011). SnapShot: Spinal cord development. *Cell*, *146*(1), 178–178.e1. https://doi.org/10.1016/j.cell.2011.06.038

Barthó, P., Hirase, H., Monconduit, L., Zugaro, M., Harris, K. D., & Buzsáki, G. (2004). Characterization of neocortical principal cells and interneurons by network interactions and extracellular features. *Journal of Neurophysiology*, *92*(1), 600–608. https://doi.org/10.1152/jn.01170.2003

Bernhard, C. G., Bohm, E., & Petersén, I. (1953). Investigations on the organization of the corticospinal system in monkeys (*Macaca mulatta*). *Acta Physiologica Scandinavica*, *29*(S106), 79–105. https://doi.org/10.1111/j.1365-201X.1953.tb10772.x

Bortoff, G. A., & Strick, P. L. (1993). Corticospinal terminations in two New-World primates: Further evidence that corticomotoneuronal connections provide part of the neural substrate for manual dexterity. *Journal of Neuroscience*, *13*(12), 5105–5118. https://doi.org/10.1523/JNEUROSCI.13-12-05105.1993

Braak, H., Brettschneider, J., Ludolph, A. C., Lee, V. M., Trojanowski, J. Q., & Del Tredici, K. (2013). Amyotrophic lateral sclerosis—A model of corticofugal axonal spread. *Nature Reviews: Neurology, 9*(12), 708–714. https://doi.org/10.1038/nrneurol.2013.221

Braak, H., Ludolph, A. C., Neumann, M., Ravits, J., & Del Tredici, K. (2017). Pathological TDP-43 changes in Betz cells differ from those in bulbar and spinal α-motoneurons in sporadic amyotrophic lateral sclerosis. *Acta Neuropathologica, 133*, 79–90. https://doi.org/10.1007/s00401-016-1633-2

Brouwer, B., & Ashby, P. (1990). Corticospinal projections to upper and lower limb spinal motoneurons in man. *Electroencephalography and Clinical Neurophysiology, 76*(6), 509–519. https://doi.org/10.1016/0013-4694(90)90002-2

Burg, T., Bichara, C., Scekic-Zahirovic, J., Fischer, M., Stuart-Lopez, G., Brunet, A., Lefebvre, F., Cordero-Erausquin, M., & Rouaux, C. (2020). Absence of subcerebral projection neurons is beneficial in a mouse model of amyotrophic lateral sclerosis. *Annals of Neurology, 88*(4), 688–702. https://doi.org/10.1002/ana.25833

Clough, J. F. M., Kernell, D., & Phillips, C. G. (1968). The distribution of monosynaptic excitation from the pyramidal tract and from primary spindle afferents to motoneurones of the baboon's hand and forearm. *Journal of Physiology, 198*(1), 145–166. https://doi.org/10.1113/jphysiol.1968.sp008598

Czell, D., Neuwirth, C., Weber, M., Sartoretti-Schefer, S., Gutzeit, A., & Reischauer, C. (2019). Nine-hole peg test and transcranial magnetic stimulation: Useful to evaluate dexterity of the hand and disease progression in amyotrophic lateral sclerosis. *Neurology Research International, 2019*, Article 7397491. https://doi.org/10.1155/2019/7397491

Davare, M., Kraskov, A., Rothwell, J. C., & Lemon, R. N. (2011). Interactions between areas of the cortical grasping network. *Current Opinion in Neurobiology, 21*(4), 565–570. https://doi.org/10.1016/j.conb.2011.05.021

Day, B. L., Dressler, D., Maertens de Noordhout, A., Marsden, C. D., Nakashima, K., Rothwell, J. C., & Thompson, P. D. (1989). Electric and magnetic stimulation of human motor cortex: Surface EMG and single motor unit responses. *Journal of Physiology, 412*, 449–473. https://doi.org/10.1113/jphysiol.1989.sp017626

Del Tredici, K., & Braak, H. (2022). Neuropathology and neuroanatomy of TDP-43 amyotrophic lateral sclerosis. *Current Opinion in Neurology, 35*(5), 660–671. https://doi.org/10.1097/WCO.0000000000001098

de Noordhout, A. M., Rapisarda, G., Bogacz, D., Gérard, P., De Pasqua, V., Pennisi, G., & Delwaide, P. J. (1999). Corticomotoneuronal synaptic connections in normal man: An electrophysiological study. *Brain, 122*(7), 1327–1340. https://doi.org/10.1093/brain/122.7.1327

Eisen, A. (2021). The dying forward hypothesis of ALS: Tracing its history. *Brain Sciences, 11*(3), Article 300. https://doi.org/10.3390/brainsci11030300

Eisen, A., Braak, H., Del Tredici, K., Lemon, R., Ludolph, A. C., & Kiernan, M. C. (2017). Cortical influences drive amyotrophic lateral sclerosis. *Journal of Neurology, Neurosurgery and Psychiatry, 88*(11), 917–924. https://doi.org/10.1136/jnnp-2017-315573

Eisen, A., Kim, S., & Pant, B. (1992). Amyotrophic lateral sclerosis (ALS): A phylogenetic disease of the corticomotoneuron? *Muscle & Nerve, 15*(2), 219–224. https://doi.org/10.1002/mus.880150215

Eisen, A., & Lemon, R. (2021). The motor deficit of ALS reflects failure to generate muscle synergies for complex motor tasks, not just muscle strength. *Neuroscience Letters, 762*, Article 136171. https://doi.org/10.1016/j.neulet.2021.136171

Eisen, A., & Swash, M. (2001). Clinical neurophysiology of ALS. *Clinical Neurophysiology, 112*(12), 2190–2201. https://doi.org/10.1016/S1388-2457(01)00692-7

Eisen, A., Turner, M. R., & Lemon, R. (2014). Tools and talk: An evolutionary perspective on the functional deficits associated with amyotrophic lateral sclerosis. *Muscle & Nerve, 49*(4), 469–477. https://doi.org/10.1002/mus.24132

Fetz, E. E., & Cheney, P. D. (1980). Postspike facilitation of forelimb muscle activity by primate corticomotoneuronal cells. *Journal of Neurophysiology, 44*(4), 751–772. https://doi.org/10.1152/jn.1980.44.4.751

Firmin, L., Field, P., Maier, M. A., Kraskov, A., Kirkwood, P. A., Nakajima, K., Lemon, R. N., & Glickstein, M. (2014). Axon diameters and conduction velocities in the macaque pyramidal tract. *Journal of Neurophysiology, 112*(6), 1229–1240. https://doi.org/10.1152/jn.00720.2013

Fogarty, M. J., Klenowski, P. M., Lee, J. D., Drieberg-Thompson, J. R., Bartlett, S. E., Ngo, S. T., Hilliard, M. A., Bellingham, M. C., & Noakes, P. G. (2016). Cortical synaptic and dendritic spine abnormalities in a presymptomatic TDP-43 model of amyotrophic lateral sclerosis. *Scientific Reports, 6*, Article 37968. https://doi.org/10.1038/srep37968

Franklin, D. W., & Wolpert, D. M. (2011). Computational mechanisms of sensorimotor control. *Neuron, 72*(3), 425–442. https://doi.org/10.1016/j.neuron.2011.10.006

Gautam, M., Jara, J. H., Kocak, N., Rylaarsdam, L. E., Kim, K. D., Bigio, E. H., & Hande Özdinler, P. (2019). Mitochondria, ER, and nuclear membrane defects reveal early mechanisms for upper motor neuron vulnerability with respect to TDP-43 pathology. *Acta Neuropathologica, 137*(1), 47–69. https://doi.org/10.1007/s00401-018-1934-8

Geyer, S., Ledberg, A., Schleicher, A., Kinomura, S., Schormann, T., Bürgel, U., Klingberg, T., Larsson, J., Zilles, K., & Roland, P. E. (1996). Two different areas within the primary motor cortex of man. *Nature, 382*(6594), 805–807. https://doi.org/10.1038/382805a0

Gu, Z., Kalambogias, J., Yoshioka, S., Han, W., Li, Z., Kawasawa, Y. I., Pochareddy, S., Li, Z., Liu, F., Xu, X., Wijeratne, H. R. S., Ueno, M., Blatz, E., Salomone, J., Kumanogoh, A., Rasin, M. R., Gebelein, B., Weirauch, M. T., Sestan, N., . . . Yoshida, Y. (2017). Control of species-dependent cortico-motoneuronal connections underlying manual dexterity. *Science, 357*(6349), 400–404. https://doi.org/10.1126/science.aan3721

Han, Q., Feng, J., Qu, Y., Ding, Y., Wang, M., So, K. F., Wu, W., & Zhou, L. (2013). Spinal cord maturation and locomotion in mice with an isolated cortex. *Neuroscience, 253*, 235–244. https://doi.org/10.1016/j.neuroscience.2013.08.057

Heffner, R. S., & Masterton, R. B. (1975). Variation in form of the pyramidal tract and its relationship to digital dexterity. *Brain, Behavior and Evolution, 12*(3), 161–200. https://doi.org/10.1159/000124401

Heffner, R. S., & Masterton, R. B. (1983). The role of the corticospinal tract in the evolution of human digital dexterity. *Brain, Behavior and Evolution, 23*(3–4), 165–183. https://doi.org/10.1159/000121494

Henderson, R. D., & Eisen, A. (2020). ALS split phenotypes—To what extent do they exist? *Clinical Neurophysiology, 131*(4), 847–849. https://doi.org/10.1016/j.clinph.2019.12.417

Hughlings Jackson, J. (1884). Croonian lectures on the evolution and dissolution of the nervous system. *British Medical Journal, 1*(1213), 593–595. https://doi.org/10.1136/bmj.1.1213.591

Iwatsubo, T., Kuzuhara, S., Kanemitsu, A., Shimada, H., & Toyokura, Y. (1990). Corticofugal projections to the motor nuclei of the brainstem and spinal cord in humans. *Neurology, 40*(2), 309–312. https://doi.org/10.1212/wnl.40.2.309

Izpisua Belmonte, J. C., Callaway, E. M., Caddick, S. J., Churchland, P., Feng, G., Homanics, G. E., Lee, K. F., Leopold, D. A., Miller, C. T., Mitchell, J. F., Mitalipov, S., Moutri, A. R., Movshon, J. A., Okano, H., Reynolds, J. H., Ringach, D., Sejnowski, T. J., Silva, A. C., Strick, P. L., . . . Zhang, F. (2015). Brains, genes, and primates. *Neuron, 86*(3), 617–631. https://doi.org/10.1016/j.neuron.2015.03.021

Kaas, J. H., Qi, H. X., & Stepniewska, I. (2018). The evolution of parietal cortex in primates. In G. Vallar & H. B. Coslett (Eds.), *Handbook of clinical neurology: Vol. 151. The parietal lobe* (pp. 31–52). Elsevier. https://doi.org/10.1016/B978-0-444-63622-5.00002-4

Keizer, K., & Kuypers, H. G. (1984). Distribution of corticospinal neurons with collaterals to lower brain stem reticular formation in cat. *Experimental Brain Research, 54*(1), 107–120. https://doi.org/10.1007/BF00235823

Khalaf, R., Martin, S., Ellis, C., Burman, R., Sreedharan, J., Shaw, C., Leigh, P. N., Turner, M. R., & Al-Chalabi, A. (2019). Relative preservation of triceps over biceps strength in upper limb-onset ALS: The "split elbow." *Journal of Neurology, Neurosurgery and Psychiatry, 90*(7), 730–733. https://doi.org/10.1136/jnnp-2018-319894

Kraskov, A., Baker, S. N., Soteropoulos, D., Kirkwood, P., & Lemon, R. (2019). The corticospinal discrepancy: Where are all the slow pyramidal tract neurons? *Cerebral Cortex, 29*(9), 3977–3981. https://doi.org/10.1093/cercor/bhy278

Kraskov, A., Soteropoulos, D. S., Glover, I. S., Lemon, R. N., & Baker, S. N. (2020). Slowly conducting pyramidal tract neurons in macaque and rat. *Cerebral Cortex, 30*(5), 3403–3418. https://doi.org/10.1093/cercor/bhz318

Kuypers, H. G. J. M. (1981). Anatomy of the descending pathways. In V. Brooks (Ed.), *Handbook of physiology: Sect. 1. The nervous system: Vol. 2. Motor control* (pp. 597–666). Williams and Wilkins. https://doi.org/10.1002/cphy.cp010213

Lassek, A. M., & Rasmussen, G. L. (1940). A comparative fiber and numerical analysis of the pyramidal tract. *Journal of Comparative Neurology, 72*(2), 417–428. https://doi.org/10.1002/cne.900720209

Leenen, L. P., Meek, J., Posthuma, P. R., & Nieuwenhuys, R. (1985). A detailed morphometrical analysis of the pyramidal tract of the rat. *Brain Research, 359*(1–2), 65–80. https://doi.org/10.1016/0006-8993(85)91413-1

Lemon, R. (2019). Recent advances in our understanding of the primate corticospinal system. *F1000Research, 8*, Article 274. https://doi.org/10.12688/f1000research.17445.1

Lemon, R. N. (2008). Descending pathways in motor control. *Annual Review of Neuroscience, 31*, 195–218. https://doi.org/10.1146/annurev.neuro.31.060407.125547

Lemon, R. N. (2021). The cortical "upper motoneuron" in health and disease. *Brain Sciences, 11*(5), Article 619. https://doi.org/10.3390/brainsci11050619

Lemon, R. N., Baker, S. N., & Kraskov, A. (2021). Classification of cortical neurons by spike shape and the identification of pyramidal neurons. *Cerebral Cortex, 31*(11), 5131–5138. https://doi.org/10.1093/cercor/bhab147

Lemon, R. N., & Griffiths, J. (2005). Comparing the function of the corticospinal system in different species: Organizational differences for motor specialization? *Muscle & Nerve, 32*(3), 261–279. https://doi.org/10.1002/mus.20333

Lemon, R. N., Mantel, G. W., & Muir, R. B. (1986). Corticospinal facilitation of hand muscles during voluntary movement in the conscious monkey. *Journal of Physiology, 381*, 497–527. https://doi.org/10.1113/jphysiol.1986.sp016341

Leyton, A. S. F., & Sherrington, C. S. (1917). Observations on the excitable cortex of the chimpanzee, orangutan and gorilla. *Quarterly Journal of Experimental Physiology, 11*(2), 135–222. https://doi.org/10.1113/expphysiol.1917.sp000240

Li, B., He, D. J., Li, X. J., & Guo, X. Y. (2022). Modeling neurodegenerative diseases using non-human primates: Advances and challenges. *Ageing and Neurodegenerative Diseases, 2*(3), Article 12. https://doi.org/10.20517/and.2022.14

Liu, Y., Latremoliere, A., Li, X., Zhang, Z., Chen, M., Wang, X., Fang, C., Zhu, J., Alexandre, C., Gao, Z., Chen, B., Ding, X., Zhou, J. Y., Zhang, Y., Chen, C., Wang, K. H., Woolf, C. J., & He, Z. (2018). Touch and tactile neuropathic pain sensitivity are set by corticospinal projections. *Nature, 561*(7724), 547–550. https://doi.org/10.1038/s41586-018-0515-2

Ludolph, A. C., Emilian, S., Dreyhaupt, J., Rosenbohm, A., Kraskov, A., Lemon, R. N., Del Tredici, K., & Braak, H. (2020). Pattern of paresis in ALS is consistent with the physiology of the corticomotoneuronal projections to different muscle groups. *Journal of Neurology, Neurosurgery and Psychiatry, 91*(9), 991–998. https://doi.org/10.1136/jnnp-2020-323331

Ma, X. R., Prudencio, M., Koike, Y., Vatsavayai, S. C., Kim, G., Harbinski, F., Briner, A., Rodriguez, C. M., Guo, C., Akiyama, T., Schmidt, H. B., Cummings, B. B., Wyatt, D. W., Kurylo, K., Miller, G., Mekhoubad, S., Sallee, N., Mekonnen, G., Ganser, L., . . . Gitler, A. D. (2022). TDP-43 represses cryptic exon inclusion in the FTD-ALS gene UNC13A. *Nature, 603*(7899), 124–130. https://doi.org/10.1038/s41586-022-04424-7

Maier, M. A., Olivier, E., Baker, S. N., Kirkwood, P. A., Morris, T., & Lemon, R. N. (1997). Direct and indirect corticospinal control of arm and hand motoneurons in the squirrel monkey (*Saimiri sciureus*). *Journal of Neurophysiology, 78*(2), 721–733. https://doi.org/10.1152/jn.1997.78.2.721

Menon, P., Kiernan, M. C., & Vucic, S. (2014). ALS pathophysiology: Insights from the split-hand phenomenon. *Clinical Neurophysiology, 125*(1), 186–193. https://doi.org/10.1016/j.clinph.2013.07.022

Morecraft, R. J., Ge, J., Stilwell-Morecraft, K. S., McNeal, D. W., Pizzimenti, M. A., & Darling, W. G. (2013). Terminal distribution of the corticospinal projection from the hand/arm region of the primary motor cortex to the cervical enlargement in rhesus monkey. *Journal of Comparative Neurology, 521*(18), 4205–4235. https://doi.org/10.1002/cne.23410

Moreno-Lopez, Y., Bichara, C., Delbecq, G., Isope, P., & Cordero-Erausquin, M. (2021). The corticospinal tract primarily modulates sensory inputs in the mouse lumbar cord. *eLife, 10*, Article e65304. https://doi.org/10.7554/eLife.65304

Nakajima, K., Maier, M. A., Kirkwood, P. A., & Lemon, R. N. (2000). Striking differences in transmission of corticospinal excitation to upper limb motoneurons in two primate species. *Journal of Neurophysiology, 84*(2), 698–709. https://doi.org/10.1152/jn.2000.84.2.698

Oyanagi, K., Makifuchi, T., & Ikuta, F. (1995). The anterolateral funiculus in the spinal cord in amyotrophic lateral sclerosis. *Acta Neuropathologica, 90*(3), 221–227. https://doi.org/10.1007/BF00296504

Palmer, E., & Ashby, P. (1992). Corticospinal projections to upper limb motoneurones in humans. *Journal of Physiology, 448*(1), 397–412. https://doi.org/10.1113/jphysiol.1992.sp019048

Perge, J. A., Niven, J. E., Mugnaini, E., Balasubramanian, V., & Sterling, P. (2012). Why do axons differ in caliber? *Journal of Neuroscience, 32*(2), 626–638. https://doi.org/10.1523/JNEUROSCI.4254-11.2012

Phillips, C. G. (1971). Evolution of the corticospinal tract in primates with special reference to the hand. In W. Leutenegger (Ed.), *Proceedings of the 3rd International Congress of Primatology* (Vol. 2, pp. 2–23). Karger.

Porter, R., & Lemon, R. N. (1993). *Corticospinal function and voluntary movement*. Oxford University Press. https://doi.org/10.1093/acprof:oso/9780198523758.001.0001

Quallo, M. M., Kraskov, A., & Lemon, R. N. (2012). The activity of primary motor cortex corticospinal neurons during tool use by macaque monkeys. *Journal of Neuroscience, 32*(48), 17351–17364. https://doi.org/10.1523/JNEUROSCI.1009-12.2012

Quencer, R. M., Bunge, R. P., Egnor, M., Green, B. A., Puckett, W., Naidich, T. P., Post, M. J., & Norenberg, M. (1992). Acute traumatic central cord syndrome: MRI-pathological correlations. *Neuroradiology, 34*(2), 85–94. https://doi.org/10.1007/BF00588148

Ralston, D. D., & Ralston, H. J., III. (1985). The terminations of corticospinal tract axons in the macaque monkey. *Journal of Comparative Neurology, 242*(3), 325–337. https://doi.org/10.1002/cne.902420303

Rathelot, J. A., & Strick, P. L. (2006). Muscle representation in the macaque motor cortex: An anatomical perspective. *Proceedings of the National Academy of Sciences of the United States of America, 103*(21), 8257–8262. https://doi.org/10.1073/pnas.0602933103

Rathelot, J. A., & Strick, P. L. (2009). Subdivisions of primary motor cortex based on cortico-motoneuronal cells. *Proceedings of the National Academy of Sciences of the United States of America, 106*(3), 918–923. https://doi.org/10.1073/pnas.0808362106

Riku, Y., Atsuta, N., Yoshida, M., Tatsumi, S., Iwasaki, Y., Mimuro, M., Watanabe, H., Ito, M., Senda, J., Nakamura, R., Koike, H., & Sobue, G. (2014). Differential motor neuron involvement in progressive muscular atrophy: A comparative study with amyotrophic lateral sclerosis. *BMJ Open, 4*(5), Article e005213. https://doi.org/10.1136/bmjopen-2014-005213

Rivara, C. B., Sherwood, C. C., Bouras, C., & Hof, P. R. (2003). Stereologic characterization and spatial distribution patterns of Betz cells in the human primary motor cortex. *Anatomical Record, 270A*(2), 137–151. https://doi.org/10.1002/ar.a.10015

Schoen, J. H. R. (1964). Comparative aspects of the descending fibre systems in the spinal cord. *Progress in Brain Research, 11*, 203–222. https://doi.org/10.1016/s0079-6123(08)64049-2

Simonyan, K. (2014). The laryngeal motor cortex: Its organization and connectivity. *Current Opinion in Neurobiology, 28*, 15–21. https://doi.org/10.1016/j.conb.2014.05.006

Sinopoulou, E., Rosenzweig, E. S., Conner, J. M., Gibbs, D., Weinholtz, C. A., Weber, J. L., Brock, J. H., Nout-Lomas, Y. S., Ovruchesky, E., Takashima, Y., Biane, J. S., Kumamaru, H., Havton, L. A., Beattie, M. S., Bresnahan, J. C., & Tuszynski, M. H. (2022). Rhesus macaque versus rat divergence in the corticospinal projectome. *Neuron, 110*(18), 2970–2983.e4. https://doi.org/10.1016/j.neuron.2022.07.002

Soares, D., Goldrick, I., Lemon, R. N., Kraskov, A., Greensmith, L., & Kalmar, B. (2017). Expression of Kv3.1b potassium channel is widespread in macaque motor cortex pyramidal cells: A histological comparison between rat and macaque. *Journal of Comparative Neurology, 525*(9), 2164–2174. https://doi.org/10.1002/cne.24192

Sobue, G., Hashizume, Y., Mitsuma, T., & Takahashi, A. (1987). Size-dependent myelinated fiber loss in the corticospinal tract in Shy-Drager syndrome and amyotrophic lateral sclerosis. *Neurology, 37*(3), 529–532. https://doi.org/10.1212/wnl.37.3.529

Strick, P. L., Dum, R. P., & Rathelot, J. A. (2021). The cortical motor areas and the emergence of motor skills: A neuroanatomical perspective. *Annual Review of Neuroscience, 44*, 425–447. https://doi.org/10.1146/annurev-neuro-070918-050216

Strick, P. L., & Preston, J. B. (1982). Two representations of the hand in area 4 of a primate. II. Somatosensory input organization. *Journal of Neurophysiology*, *48*(1), 150–159. https://doi.org/10.1152/jn.1982.48.1.150

Tallis, R. (2004). *The hand: A philosophical enquiry into human being*. Edinburgh University Press.

Uchida, A., Sasaguri, H., Kimura, N., Tajiri, M., Ohkubo, T., Ono, F., Sakaue, F., Kanai, K., Hirai, T., Sano, T., Shibuya, K., Kobayashi, M., Yamamoto, M., Yokota, S., Kubodera, T., Tomori, M., Sakaki, K., Enomoto, M., Hirai, Y., . . . Yokota, T. (2012). Non-human primate model of amyotrophic lateral sclerosis with cytoplasmic mislocalization of TDP-43. *Brain*, *135*(3), 833–846. https://doi.org/10.1093/brain/awr348

Venkadesan, M., & Valero-Cuevas, F. J. (2009). Effects of neuromuscular lags on controlling contact transitions. *Philosophical Transactions. Series A: Mathematical, Physical, and Engineering Sciences*, *367*(1891), 1163–1179. https://doi.org/10.1098/rsta.2008.0261

Vigneswaran, G., Kraskov, A., & Lemon, R. N. (2011). Large identified pyramidal cells in macaque motor and premotor cortex exhibit "thin spikes": Implications for cell type classification. *Journal of Neuroscience*, *31*(40), 14235–14242. https://doi.org/10.1523/JNEUROSCI.3142-11.2011

Weber, M., Eisen, A., Stewart, H. G., & Andersen, P. M. (2000a). Preserved slow conducting corticomotoneuronal projections in amyotrophic lateral sclerosis with autosomal recessive D90A CuZn-superoxide dismutase mutation. *Brain*, *123*(7), 1505–1515. https://doi.org/10.1093/brain/123.7.1505

Weber, M., Eisen, A., Stewart, H., & Hirota, N. (2000b). The split hand in ALS has a cortical basis. *Journal of the Neurological Sciences*, *180*(1–2), 66–70. https://doi.org/10.1016/s0022-510x(00)00430-5

Wilson, F. R. (1998). *The hand: How its use shapes the brain, language, and human culture*. Pantheon.

Witham, C. L., Fisher, K. M., Edgley, S. A., & Baker, S. N. (2016). Corticospinal inputs to primate motoneurons innervating the forelimb from two divisions of primary motor cortex and area 3a. *Journal of Neuroscience*, *36*(9), 2605–2616. https://doi.org/10.1523/JNEUROSCI.4055-15.2016

12

REM Sleep Behavior Disorder
Nocturnal Replay of "Fight or Flight"?

Nico J. Diederich and Isabelle Arnulf

Introduction
What Is REM Sleep Behavior Disorder?
Rapid eye movement (REM) sleep behavior disorder (RBD) is a parasomnia (abnormal behavior during sleep) of the REM sleep. Its prevalence has been estimated at 1% in middle-aged to older adults. RBD is a clinical diagnosis supported by typical video-polysomnographic (vPSG) findings. It is characterized by various motor behaviors and vocalizations during REM sleep. While violent movements suggesting defeating an aggressor or fighting with a predator are the most frequently reported features of RBD, slow, quiet behaviors and soft or subtle movements may also be observed (Oudiette et al. 2009). Often there is concordant dream content, which is why "acting out of dreams" has been postulated. Injuries to the patient and the bed partner may be the worst consequence. PSG shows REM sleep *without* the usual muscle atonia, which by itself can be a precursor syndrome of the full-blown syndrome. RBD is a harbinger of several neurodegenerative disorders, most prominently, but not exclusively, of synucleinopathies, such as Parkinson's disease (PD), multiple system atrophy, and Lewy body dementia. In terms of the two most associated genetic mutations linked with PD, there is evidence of a higher risk for RBD in glucocerebrosidase (GBA), but not in leucine rich repeat kinase 2 (LRRK2) mutation carriers. Higher risk for RBD has been reported with a variant in the 5' region of the SNCA gene, although the underlying pathomechanism is unknown (Krohn et al. 2020). It is not elucidated why RBD can also manifest later in the course of these diseases. RBD has been observed in narcolepsy, and it can be triggered by psychotropic medication or alcohol intake. Less convincingly, there are reports of RBD in

patients suffering from post-traumatic stress disorder (PTSD; Ross et al. 1994). In neurodegenerative disorders with RBD, dysfunction of the pontine tegmentum (which contains the REM atonia system) has been shown by neurophysiological and neuroimaging techniques. Autopsies confirm the causal role of brainstem nuclei. Indeed, high burden of Lewy bodies and substantial cell loss have been seen in REM sleep regulating nuclei, such as the cholinergic pedunculopontine nucleus (PPN) or the glutamatergic locus subcoeruleus (LSC; Iranzo et al. 2013).

Why Consider an Evolutionary Basis in RBD?

While the underlying pathophysiology of the lack of REM muscle atonia is sufficiently understood, there has been no detailed mechanistic explanation on how nuclei responsible for REM sleep muscle atonia are affected by a neurodegenerative process. Even more, there is no all-encompassing explanation on why "the acting out of violent dreams" should be pathognomonic for RBD and, more generally, why dreams often have a negative content. In the following essay it will be suggested that these behaviors are largely supported by the *threat simulation theory* (TST), an evolutionary theory of dreams, first conceptualized in 2000 (Revonsuo 2000). Phylogenetic and ontogenetic aspects of REM sleep will be revisited. Detailed animal experiments showing RBD-like movements with specific brainstem lesions or genetic manipulations with impact on REM sleep regulatory systems will be described. It will also be debated if ontogenetically early motor patterns as seen during the "active sleep" of newborns may resurface in late life as RBD features. Other evolutionarily based explanatory hypotheses will be critically reviewed as well. The nuclei responsible for RBD may be "left behind" during human evolution marked by exponential telencephalization, and thus they may be at higher risk for neurodegeneration. Finally, the influence of life history, psychodynamic considerations, and the transient RBD manifestation after various medications must be considered.

Historical Background

In Latin times, Pliny the Elder observed that horse, dog, beef, goat, and sheep were having movements when dreaming; and in the late 18th century the French naturalist Buffon observed birds singing while asleep. The philosopher Montaigne insightfully observed twitches and yelps of sleeping dogs and contributed them to canine dreams. He also formulated the adage "We wake sleeping and sleep waking." James Parkinson reported tumultuous sleep in one of the patients he had observed in the streets of London. More recently, Japanese scientists reported agitated REM sleep in patients with delirium, although not considering this syndrome as a new parasomnia entity. A concise report of REM sleep without muscular atonia in PD was first given in 1969 (Traczynska-Kubin et al. 1969). Finally, in 1986 Schenck, Mahowald, and others defined in a landmark study RBD as a new entity of REM sleep parasomnia (Schenck et al. 1986). They identified RBD as a potential forerunner syndrome of neurodegenerative syndromes: 38% of the patients in their pilot cohort developed a Parkinsonian syndrome within 3.7 years on average. Today, we know that 12 years after initial RBD manifestation, 74% of patients with RBD will have developed a synucleinopathy (Postuma et al. 2019).

Evolutionary Evidence Specifically Related to Human Health

In order to understand RBD as a neurological disorder, we must first revisit phylogenetic and ontogenetic aspects of REM sleep in general and, more specifically, body movements during REM sleep.

Is REM Sleep a "Late" Phylogenetic Acquisition?

Although it has been assumed since the early 1970s that REM sleep (defined by rapid electroencephalographic [EEG] rhythm, rapid eye movements, and muscle atonia, often interrupted by muscle twitches as in humans) would be exclusive to birds and mammals (which are homeothermic animals) and not to be found in "cold-blooded" animals, several recent findings challenge this belief. Sleep stages resembling REM sleep are seen in bearded dragons, tegu lizards, octopuses, and even spiders. Some characteristics and functions of human REM sleep may be "missing" in other species; others may be very prominent (see Figure 12.1). For instance, REM sleep in other species may only be characterized by the presence of muscle atonia, often interrupted by muscle twitches. There may also be bursts of REMs but absence of desynchronized EEG. Periods of rest and activity and autonomic fluctuations are seen in fish, amphibians, and non-avian reptiles; but "typical" REM sleep as seen in mammals is difficult to identify in these animals (Blumberg

Figure 12.1 Different characteristics of REM sleep in different species. Reprinted from Blumberg et al. (2020), with permission.

et al. 2020). Of interest, the invertebrate octopus has an active state of sleep, including dynamic changes in skin color and texture as well as movements. Jumping spiders, as terrestrial invertebrates, have periodic bouts of retinal movements coupled with limb twitching and stereotyped leg curling behaviors (Rößler et al. 2022). Frequency, duration, and features of REM sleep may vary across species, size, maturation, and environmental pressure (for a detailed review, see Siegel 2022). The quantity of REM sleep is increased at birth in altricial species. Their newborns are partially immobile and immature, and thus in need of further motor and cognitive development. For instance, immature rats show 70% of REM sleep in the perinatal period, in contrast to relatively mature guinea pigs with only 7% in the perinatal period (Jouvet-Mounier et al. 1969). Also, in mammals with larger amounts of REM sleep there is higher degree of encephalization and intracerebral interconnectivity as well as development of more complex behavior (Peever and Fuller 2017). Finally, the life conditions per se may have an impact on the amount of REM sleep. Species living in open locations and herbivorous species have less REM sleep, and it has been proposed that the amount is reduced when there is higher risk for a predator attack (Lesku et al. 2008), possibly because the threshold for arousal is higher in REM than in non-REM sleep (NREM).

Ontogenetic "Maturation" of REM Sleep

Frank has succinctly summarized that "first oscillations in autonomic function and brain activity later coalesce into . . . immature forms of NREM and REM sleep" (2011, p. 49). Indeed, "active sleep" in the prenatal and postnatal phases can be considered as a precursor phase of REM sleep. It is most prominent in animals less developed at birth, and it contains both rapid *body* movements and rapid *eye* movements. Newborns spend more than 50% of the time in this sleep stage. At first look the rapid body movements seem to be random, spontaneous, senseless, but often stereotyped. In the prenatal period the movements of "active sleep" are triggered by neuronal excitations within the spinal cord and the brainstem (Corner and Schenck 2015). Later myelination of pathways running from the brainstem to the cortex facilitates the progressive transition from "active" sleep to REM sleep, with also reduction of body movements by GABAergic inhibition.

"Active sleep" and early REM sleep have been viewed as major contributors to various crucial maturation processes (Roffwarg et al. 1966), although more recent experiences found that NREM sleep is crucial for these maturation processes as well. Thus, twitches are *not* random but serve a developmental purpose in shaping the sensory-motor cortex. During wakefulness, the nervous system drives movements and simultaneously generates a copy of the motor command (the corollary discharge) to inform the sensory cortex on the changes to be expected. This process helps to distinguish sensations that are self-generated from those that are external. In contrast, the twitches during REM sleep are not accompanied by corollary discharges and are not generated by the motor cortex. These brief muscle contractions "surprise" the sensory cortex and trigger strong activity in the primary motor cortex, not seen in response to passive movements of the limb at the awake state (Tiriac et al. 2014). The twitches might contribute to an activity-dependent development of the spinal cord, cerebellum, and forebrain and to the construction of internal models. Early REM sleep may also contribute to visual development. When a kitten is raised with one eye closed, a rapid remodeling of synaptic weights and morphology in thalamocortical circuits occurs and reduces the cortical territories innervated by the deprived eye (Antonini and Stryker 1993). This change is viewed as a plastic brain adaptation to a deficient organ, possibly via ponto-geniculo-occipital (PGO) waves. The lateral geniculate nucleus remains smaller when there is REM sleep deprivation during early life (Lopez et al. 2008). Most importantly, early REM sleep allows hippocampal long-term potentiation, and REM sleep deprivation during the third postnatal week destabilizes such long-term potentiation (Lopez et al. 2008). Finally, in adult humans REM sleep consolidates procedural learning of a visual discrimination task (Stickgold et al. 2000).

Functions of REM Sleep and Dreaming during Adult Life

Generally speaking, there is fine-tuning of up- and down-regulation of gene expression during the different sleep–wake states (Blumberg and Plumeau 2016). Detailed description of the role of REM sleep during adult life is beyond the topic of this essay on RBD. Importantly, dreams also occur in NREM sleep. Consequently, the postulated functions of dreams, as presented here, may also apply to NREM sleep dreaming. In the following, we are only focusing on hypotheses of interest from an evolutionary perspective (for a succinct review on other hypotheses, see Siclari et al. 2020).

Thermoregulation

During REM sleep the body temperature increases or decreases, depending on the environmental conditions. In contrast, the brain temperature is kept at an increased level, and it has been postulated that REM sleep is "like shivering for the brain" (Wehr 1992; Siegel, 2022). In this way functionality of the brain is maintained. Furthermore, the reduction or even the complete arrest of muscle activity saves energy, and it prevents dream-enacting behaviors (Jouvet 1967; Siegel 2022).

Regulation of Emotions

REM sleep more than other sleep stages facilitates the regulation of emotions experienced by humans the day before (van der Helm et al. 2011). Adults exposed to scary images strongly activate their amygdala (and heart rate) but do not do so when re-exposed after a night's sleep. Indeed, the daytime negative emotions may be replayed during REM sleep dreams (when the amygdala is strongly activated) but without their daytime physical consequences (increased heart rate, vasoconstriction). During the dreams the events may be associated with another scenario. By cortico-amygdala-hippocampus dialogue there is desensitization or down-regulation of the memory-associated negative emotions but up-regulation or consolidation of the memories.

Rehearsing Instinctual Behaviors

In 1980 Jouvet proposed that during REM sleep there is rehearsal of innate motor programs ("instinctive behaviors"), enabling animals to act and react appropriately, when the releasing stimulus is presented for the first time. The internal programmer of these behaviors could be the PGO waves, as strongly associated with such behaviors in Jouvet's cat experiments. Similarly, Hobson postulated that REM sleep provides a "virtual reality model of the world that is of functional use . . . and prepares for instinctual responsiveness" (2009, p. 803). In both theories REM sleep may trigger "offline" learning. However, the exact mechanisms remain unknown (Peever and Fuller 2017).

Simulating Threats

Based on evolutionary concepts, anthropological observations, and dream recollections, the TST proposes that, first, dreams simulate threatening life events experienced during wakefulness over and over and in rich variations. Second, dreams rehearse threat perception; and third, dreams entrain threat escaping strategies. Such replay may have been crucial for mastering survival maneuvers in early epochs of humankind. Today, hunting Mehinaku Indians in Brazil still report more physically aggressive dream content than a normative American cohort, and only 20% of their dream content is without physical aggressiveness or threatening (Gregor 1981). However, also in subjects living in Western societies aggression keeps being the most frequent manifestation of social interaction in dreams (Hall and Van de Castle 1966), and two-thirds of emotions mentioned in dreams are negative, with fear and anger ranking first (Snyder et al. 1970). Animals and male strangers are cited as the adversaries of the dreamer. All these observations have been interpreted as reminiscent of the risks experienced by ancestral human societies, with presumably dangerous intergroup interactions or direct exposure to predation by large carnivores. The nocturnal rehearsal of efficiently defeating such events may have assured daytime mastering of life-threatening encounters, with direct impact on physical survival and reproductive success. As modern humans and their ancestors have experienced dangerous events during 99% of their evolutionary history, the nocturnal threat simulation of such threats remains robustly anchored, even in modern times. Also supportive of this theory is the observation that postpartum women and pregnant women experience a high frequency of dreams about their baby in danger, when compared to nulliparous women (Nielsen and Paquette 2007). In prehistoric times social exclusion was synonymous with death, and in present societies it still reflects the most common threat concern (Revonsuo 2000).

Evolutionary Evidence Specifically Related to Human Disease
Observations on RBD-Like Syndromes in Animals

Jouvet's Pioneer Studies in Cats

While RBD in humans was only described in 1986, Michel Jouvet reported RBD-like symptoms in cats already in 1967 (Jouvet 1967). Using selective sections of the brainstem, they found that the centers responsible for the observed muscle atonia during REM sleep in these cats were located in the pontine tegmentum, more precisely in the peri-locus coeruleus (LC) alpha corresponding to the LSC. Moreover, these cats, friendly during wakefulness, produced various stereotypical behaviors during REM sleep in conjunction with loss of muscle atonia. They lifted the head, produced intermittent general body jerking, and exhibited complex behaviors, including grooming and leaking. Most remarkably, they showed movements compatible with rage and fight, a predatory approach, or an attack (21%–41%; Table 12.1; Sastre and Jouvet 1979). Indeed, attack was the most frequent behavior occurring during REM sleep. Amazingly, sexual behavior was not noticed. Jouvet named these behaviors "oneiric behaviors" or "dreamlike behaviors" and hypothesized that they reflect the unmasking of instinctual behaviors (grooming, fighting; Jouvet, 1980). He demonstrated that the pontine lesion induced by sectioning at this level damaged the descending pathway which relays the pontine area to the medulla. When normally active, this pathway releases glycine as potent inhibitor of *lower* motor neurons in the spinal cord, thereby paralyzing these neurons and leading to postural and limb muscle atonia. In contrast, the *upper* motor neurons remain highly active during REM sleep (McCarley and Hobson 1970).

RBD-Like Behaviors in Dogs

While RBD-like behaviors in cats were experimentally induced, naturally occurring RBD has so far only been reported in dogs. In an observational study (Shea et al. 2018), almost half of the dogs surviving canine tetanus showed twitching, running, vocalization, and jaw chomping during REM sleep, suggesting abnormal "dream enactment," clinically consistent with RBD. There was spontaneous resolution within 6 months in half of the affected animals. A smaller study similarly reported RBD even in dogs without a specific pathological syndrome (Schubert et al. 2011). Again, sleep episodes characterized by violent limb movements, howling, barking, biting, etc. were reported. These observations paved the way for the description of a novel neurodegenerative disorder in retrievers (Barker et al. 2016), starting again with marked movements during sleep and later gait abnormalities and anxiety.

TABLE 12.1 Exemplary Comparison of Movement Phenomenology in Different Species and Different Underlying Diseases or Lesions

Species	Underlying disease/ lesion	Observed movements (violent and nonviolent)	Reference
Human	Parkinson's disease	"Fighting with an invisible foil, with great agility." "Sitting on the bed rowing without paddles, shouting 'Help, caimans!,' getting hold of a heavy oak bedside table and throwing it across the room."	Cochen De Cock et al. (2011)
		He sang "C'est le plus beau tango du monde." (a classic French love song)	Oudiette et al. (2009)
Human	Narcolepsy, type 1	"Patient 27 talks agitatedly; he lifts and shakes his head; he makes a fist in the air."	Cipolli et al. (2011)
		"The patient, a former smoker, seemed to smoke an invisible cigarette and then to throw it in an ashtray."	Oudiette et al. (2009)
Cat	Selective bilateral pontine tegmental lesions in the region peri-locus coeruleus alpha	"Upside on its paws, the animal seems to fight phantom enemies. Its behavior is suggestive of rage, and it seems to be completely alert, although it does not react to various sensory stimuli."	Sastre and Jouvet (1979)
		"The cat groomed itself during REM sleep."	Sastre and Jouvet (1979)
Dog	Canine tetanus	"twitching, running, vocalization and jaw chomping"	Shea et al. (2018)
Mouse		"limb jerks, tail sweepings, forward leaping"	Shen et al. (2020)
Rat	Genetic inactivation of glutamate neurons in sublaterodorsal tegmental nucleus	"non-elaborated behaviors . . . loosely arranged at the level of the head, fore and hind limbs, tail or nose. . . . Occasional . . . seeking for food with the snout in woodchips, eating a virtual pellet, trying to run or jump."	Valencia Garcia et al. (2017)

Quotations from examiners' or patients' reports. Of note, both violent and nonviolent behaviors have been seen in all species.
REM = rapid eye movement.

Rodent Models of RBD

Although models of muscle atonia during REM sleep differ between cats and rodents (Arnulf 2012), Jouvet's observations have indirectly been reproduced in various rodents by anatomical or biochemical lesions of the ventromedial medulla and the pontine tegmentum (Holmes and Jones 1994). Glutamatergic neurons located within the sublaterodorsal tegmental nucleus have a causal role in RBD in rodents, as demonstrated by genetic inactivation of these neurons in a rat model (Valencia Garcia et al. 2017). Recapitulation of a progressive synucleinopathy in wild-type mice by injection of preformed α-synuclein fibrils in the sublaterodorsal tegmental nucleus has been successful

(Shen et al. 2020). As predicted, these animals first showed an RBD-like behavior disorder and later developed Parkinsonian features. Interestingly, in rats pharmacological suppression of REM sleep after birth elucidates exaggerated phasic motility during REM sleep at middle age (Mirmiran et al. 1981). Of note, in monkeys RBD has so far only been pharmacologically triggered in marmosets in a 1-methyl-4-phenyl-1,2,3,6-tetrahydropyridine model.

RBD Pathophysiology in Humans

Despite the progress in animal modeling of RBD, the mechanisms of human RBD are not fully understood. Results of animal studies can only conditionally be applied to humans for various reasons (Arnulf 2012). Briefly, there is need for explanation of three overlapping phenomena:

- loss of REM sleep muscle atonia
- appearance of complex movements during RBD
- suitable dream content (isomorphism)

Loss of Muscle Atonia in REM Sleep

In order to understand the pathophysiology behind loss of muscle atonia in RBD, we must first revisit the mechanisms inducing and maintaining muscle atonia in normal REM sleep. It is not surprising that phylogenetically old brainstem nuclei play a predominant role. As a reminder, there is usually subtle interplay of REM-on and REM-off neuronal networks ("flip-flop twitching systems"). Normal REM sleep is promoted by activating projections (mostly cholinergic) from the PPN area to the pons. The physiological loss of muscle tone during REM sleep results from two mechanisms, one passive and one active (Siegel 2006). Serotonergic neurons descending to the spinal cord reduce their firing during sleep. As a consequence, the muscle tone diminishes progressively from sleep onset to slow-wave NREM sleep, leading first to hypotonia (postural muscle tone is reduced but still present) and illustrated by the drop of eyelids and head at sleep onset, as well as dropping objects held tightly in hand (Lacaux et al. 2021). When REM sleep starts, these serotonergic neurons descending to motor neurons cease firing completely. Based on animal data, we assume that in addition to this *passive* mechanism, an *active* paralysis of postural muscle tone (named "atonia") occurs during REM sleep. Here, while the LSC is continuously blocked by GABA input, this blockade is stopped in REM sleep during atonia; and the cholinergic PPN activates the LSC, which stimulates the magnocellular nucleus, which in turn blocks spinal lower motor neurons by glycinergic inhibition and thus produces muscle atonia (Figure 12.2A).

If there are lesions of the pathway running from the LSC to the medulla oblongata, as seen in synucleinopathies like PD (see Chapter 9), the motor output from the motor cortex traveling via the upper motor neuron can directly be transmitted to lower motor neurons (Figure 12.2B). Indeed, the primary motor pathway is acting in parallel to the

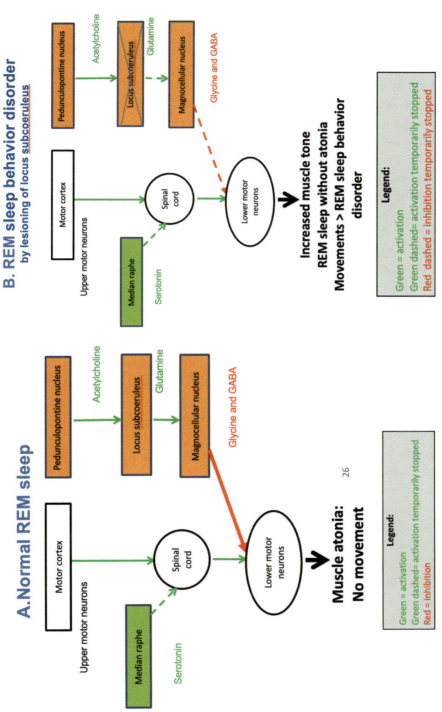

Figure 12.2 Proposed functional pathways in human rapid eye movement (REM) sleep behavior disorder, as seen in synucleinopathies. (A) Normal interplay of brainstem nuclei regulating REM sleep atonia. (B) Dysfunctional interplay of brainstem nuclei in REM sleep behavior disorder.

REM sleep–atonia system, and the upper motor neurons remain normally active during REM sleep (McCarley and Hobson 1970), although their effect on peripheral muscles is blocked by the muscle atonia. Animal modeling predicts that lesions in the pontine dorsal tegmentum and in the ventral part of the medulla lead to symptomatic RBD, when stroke, inflammatory lesions, or Lewy body burden occur in these areas (Iranzo et al. 2013; McCarter et al. 2015). Also, in humans, drugs facilitating serotonin activity (e.g., fluoxetine, venlafaxine, and paroxetine) and drugs blocking acetylcholine transmission (tricyclics such as clomipramine) can induce RBD and REM sleep without atonia, possibly because they prevent the normal sleep-related hypotonia (serotoninergic drugs) or the normal REM sleep–related atonia (anticholinergics).

The neuropathological findings in deceased patients with RBD have been substantiated in living patients with RBD by advanced neuroimaging including neuromelanin-sensitive magnetic resonance tomographic imaging. There is reduced signal intensity in the LC/LSC in these patients (Ehrminger et al. 2016). Higher amounts of REM sleep without atonia also correlate with lower signal intensity in this area in PD patients (García-Lorenzo et al. 2013), further supporting a direct relationship between LSC damage and loss of atonia during REM sleep.

Appearance of Complex Movements and Behaviors during RBD

There are two perspective levels here to discuss. First, at the clinical level, RBD movements are various, not stereotyped, and purposeful, including learned behaviors (e.g., smoking a fictive cigarette, scoring a goal, giving a lecture, eating with a fictive fork). There are also behaviors dependent on maintenance of intact eye–limb coordination (Leclair-Visonneau et al. 2010). Patients speak during RBD, using a semantically correct language. This suggests that the Wernicke and Broca areas at play during speech in the waking state are also involved in RBD speeches (Arnulf et al. 2017). At the neuroanatomical level, neuroimaging studies confirm that there is motor cortex activation during the body movements observed in RBD. Single photon emission tomography (SPECT) showed increased metabolic rates in the motor cortex in four patients during episodes of RBD, one with idiopathic RBD, one with PD and RBD, and two with narcolepsy and RBD (Mayer et al. 2015). Similarly, increased perfusion in supplementary motor area has been reported during RBD (Dauvilliers et al. 2011). The patients concordantly displayed activation in the bilateral premotor areas, the interhemispheric cleft, the periaqueductal area, the dorsal and ventral pons, and the anterior lobe of the cerebellum but not in the basal ganglia. Measurements of the basal ganglia and cortex activity during RBD using internal electrodes in sleeping PD patients confirm this hypothesis by showing that the subthalamus in the basal ganglia is "silent" during RBD movements (Hackius et al. 2016). In conclusion, the movements that we see during REM sleep result from the premotor/motor activation (pyramidal pathway), devoid of the influence of the extrapyramidal pathway.

TABLE 12.2 Most Frequent Causes of REM Sleep Behavior Disorder in Humans

Disease	Comment
Synucleinopathies	
Parkinson's disease	Common, both as a forerunner syndrome and during the course of the disease. Confirmed by polysomnography.
Lewy body dementia	
Multiple system atrophy	
Tauopathies	
Alzheimer' disease	Case reports, may be based only on questionnaire.
Progressive supranuclear palsy	
Corticobasal degeneration	
Narcolepsy	
Narcolepsy type 1	Not as frequent as in synucleinopathies, violent acting out rather rare.
Antibody-mediated diseases	
IG-Lon5 parasomnia	Part of the pathognomonic syndrome, presenting as NREM and/or REM parasomnia behaviors
Voltage-gated K-antibody encephalitis	Rare descriptions
Brainstem lesions of various origins	
Vascular lesions	When available, neuroimaging or neuropathology evidences lesions involving locus subcoeruleus area.
Inflammatory lesions (multiple sclerosis, etc.)	
Neoplasms	
Medication- or drug-induced	
Serotonin and noradrenaline reuptake inhibitors	May be suggestive of risk for later neurodegeneration.
Tricyclic and tetracyclic antidepressants	
Alcohol or cocaine withdrawal	

NREM = non-REM sleep; REM = rapid eye movement.

The proposed cascade of events applies to patients with RBD in the context of a synucleinopathy and, most probably, to patients with RBD due to a brainstem lesion or inflammatory disease. In narcolepsy there are unconsolidated REM sleep episodes, often containing enhanced muscle tone, and the RBD phenomenology is less expressed (Table 12.2).

Dream Content Corresponding to RBD Behaviors

The dreaming features in RBD are less studied than the motor aspects of RBD. The behaviors usually manifest as attempted enactments of unpleasant, action-filled, and violent dreams or nightmares in which the individual is being confronted, attacked, or chased by unfamiliar people or animals. Typically, at the end of an episode, the individual awakens quickly, becomes rapidly alert, and reports a dream with a coherent story. However, dream content during RBD is often not recalled afterward, exactly like dream contents which are also often not recalled after awakening in normal subjects. Despite these limitations, we assume that the dream action corresponds closely to the observed sleep behaviors, a condition named dream–action "isomorphism," confirmed by numerous incidental reports and a controlled study (Valli et al. 2012). The main hypothesis here is that the dream scenario is elaborated in the cortex, with activation of the upper motor neurons according to the dream content, and that the descending output is transmitted to the lower motor neurons (resulting in RBD movements) because of missing blockage by the dysfunctional LSC. An alternative hypothesis (Blumberg and Plumeau 2016) proposes that the RBD movements are primarily generated by the red nucleus (like muscle twitches in babies) and other brainstem nuclei. They are transmitted to the lower motor neurons because the dysfunctional LSC does not block them. These RBD movements are then perceived by the sensory cortex, which tries to generate suitable dream content. However, this model cannot explain several complex aspects of RBD movements such as well-organized speeches or learned behaviors (e.g., smoking a cigarette).

The crucial question remains whether RBD could also be a *dreaming* disorder, with a change in dreams toward more negative and violent or at least more physically active content. Indeed, many RBD-associated dreams differ from "normal" dreaming as they have a more active/violent content. When dream reports are retrospectively collected over the last month or year during a single clinical interview, a higher frequency of aggressive content (being chased by people or needing to defend themselves against others) is found in patients with RBD than in healthy controls (Fantini et al. 2005) and in PD patients *with* than *without* RBD (Borek et al. 2007). However, a bias toward selective remembering of violent dreams has been proposed. Indeed, dream recall in subjects with or without RBD is highly dependent on immediate awakening in order to provide temporally concordant encoding of the dreams, and violent dreams lead to more frequent awakening and recall. On vPSG, RBD patients also show common quiet dreaming behaviors including smoking, eating, drinking, laughing, scoring a goal, bicycling, singing, kissing, laughing, performing one's job, lecturing, playing, picking apples, fishing, and repeating the finger-tapping task of the Unified Parkinson's Disease Rating Scale (Oudiette et al. 2009; Uguccioni et al. 2013). Surprisingly, smiles and laughs are more frequent than anger, fear, and disgust (Maranci et al. 2022). Evidently, such "neutral" behaviors are less disturbing and may therefore be missed by casual reports. When reports are collected just upon REM sleep awakening, dreams of PD patients with and without RBD are similarly action-filled and vivid, and both contain similar threat (Valli et al. 2012). PD dreams prospectively collected over 2 weeks contain almost as frequent aggression (22% vs. 15%) than those of healthy controls, but patients more frequently suffer from physical aggression or have a passive role during the aggressive dream episode (Bugalho and Viana-Baptista 2013). Of interest, the aggressive content in dreams was not higher in PD with than without RBD, but it was increased with frontal cognitive impairment. Taken together, these results suggest that the frequent threats in RBD dreams are due not to RBD (and damaged LSC) itself but to a concomitant loss of the prefrontal inhibition of the limbic system during REM sleep, resulting in intensification of the limbic stimulation of dreams (van der Helm et al. 2011).

Is there a general mechanism responsible for the RBD dream content, independently of its negative or positive content? According to the TST, this may be the case. As detailed below, central pattern generators (CPGs) may be released during RBD, similarly to normal dreams (Fantini et al. 2005; Tassinari et al. 2012). They may be located, at least in part, in the superior colliculus (SC), which is also engaged in the modulation of threat responses and handling primitive "fight or flight" reactions (Maior et al. 2012). Observations in rodents and macaques also support this function of the SC, beyond its role in spatial orientation. Indeed, after pharmacological disinhibition of the SC, macaques show not only defense-like behaviors including vocalizations but also active behaviors including attack- or escape-like movements (DesJardin et al. 2013). Therefore, it has been proposed that not only the motor patterns but also the suitable dream content during REM sleep may be generated by the SC "having access to a large family of defense related 'fight or flight reactions'" (Dean et al. 1989, p. 141). "Sensations arising from ... violent limb movements" (Blumberg and Plumeau 2016, p. 39) could be inserted into dreams, and there may be concordant SC activation of the cortex. Across species and including humans it is thought that the SC has kept a crucial role in rolling out motor patterns responding to threat, and thus was essential for survival in ancient human times or in phylogenetic ancestors (DesJardin et al. 2013). Although this has not yet been confirmed in humans, it should be noted that in patients with Parkinsonian syndromes—the prototypic population at risk for RBD—there is substantial dysfunction of the SC, as seen by missing blindsight (Diederich et al. 2014). Neuronal loss and α-synuclein deposits in the SC have been reported in patients with Lewy body dementia and visual hallucinations, although not yet in patients with RBD (Erskine et al. 2017). Beyond direct involvement of the SC by the degenerative disease process, there may also be missing modulation of the SC by other brainstem nuclei, if affected themselves by the degenerative process (Dugger et al. 2012). So far we lack direct neuroimaging or neurophysiological proof of SC dysfunction during human RBD episodes. Despite these restrictions, a model focusing on concomitant dysfunction/degeneration of several different brainstem, hypothalamic, limbic, and prefrontal nuclei beyond the LSC may hypothetically best explain all components of RBD.

Evolutionary Constraints in Humans

Having a crude idea how there is dysfunction of nuclei involved in RBD, it remains open why there is initiation of the degenerative process in these nuclei. Again, the evolutionary perspective could be helpful. The SC with its phylogenetic correspondent of the tectum opticum and the PPN with its homolog in lamprey are evolutionarily conserved among all vertebrates, thus reflecting phylogenetically old structures (Isa et al. 2021). We must consider the relative size of these nuclei in relation to the telencephalon (Diederich et al. 2019). While it has been demonstrated that the exponential telencephalization of humans is directly linked to human-specific acquisitions such as language, reasoning, and theory of mind, the brainstem nuclei seem to have been outpaced in growth by neocortical structures, thus becoming more susceptible to cellular and subcellular overstraining and, consequently, to neurodegenerative processes during the exceptionally long human life (see Chapter 4). For instance, the LC shows relative involution in humans (Sharma et al. 2010). The number of tyrosine hydroxylase–immunoreactive neurons in the human LC is significantly lower than expected, based on the cerebellum volume and the neocortical gray matter volume.

The neurotransmitters responsible for muscle atonia during REM sleep may be different across species. While well characterized in rodents and cats, the responsible neurotransmitters in humans have not been characterized, although the effects of pro- and anticholinergic drugs on REM sleep and REM sleep atonia, as mentioned above, suggest that acetylcholine plays the same role in human as in cats.

Specificities of Human RBD Phenomenology

RBD Movements Are Jerky and Faster

RBD is a frequent syndrome in patients suffering from PD. During daytime these patients characteristically show slowness of voluntary movements (bradykinesia) and stiff muscle tone (rigidity). However, surprisingly PD patients show a restoration of normal motor control during REM sleep (Cochen De Cock et al. 2007). In a large study, 59 out of 100 PD patients had clinical RBD, and there was a testifying bed partner available in 53 of them. All these partners reported improvement of at least one type of movement during RBD. They also testified on faster, stronger, and smoother movements; and they noticed an amelioration of speech (better articulated, louder) in three-quarters of the movements. Nevertheless, the movements were not completely normal as they were repetitive, violent, faster, and jerky, meaning without the fine-tuning of daytime movements. They were comparable to movements shown by Charlie Chaplin in his movies (Arnulf 2012). In patients with multiple system atrophy a similar restoration of peripheral or facial movements was observed by the spouses (Cochen De Cock et al. 2011). The authors speculated on a "transient, levodopa-like reestablishment of the movements," possibly due to a direct pathway from the motor cortex to lower motor neurons, thus bypassing the dysfunctional extrapyramidal system. Direct cortical activation of the brainstem with circumvention of the basal ganglia suggests that there may be a second, "simpler," faster, and possibly phylogenetically older motor pathway for rescue use in case of an emergency.

RBD-Associated Symptoms in Humans

Numerous associated abnormalities have been seen in patients with idiopathic RBD. Their detailed description is beyond the topic of this review. They recapitulate and herald the non-motor and subtle motor abnormalities described in PD (for a succinct description, see Miglis et al. 2021). Briefly, sensory abnormalities, subtle motor impairment (not yet fulfilling the criteria of idiopathic PD), dysautonomia, and cognitive deficits have been linked to human RBD. Furthermore, hallucinations in PD are frequently embedded in bizarre, flickering, although vivid scenarios, reminiscent of dreaming imagery. Although the patients may experience these scenes as "neutral bystanders," a relationship with RBD has been postulated, and such visual hallucinations could be considered as fragments of REM sleep, exported into wakefulness (Arnulf et al. 2000). It remains obvious that these deficits cannot be explained by the nuclei primarily involved in RBD. Furthermore, Braak's ascending propagation theory cannot explain the simultaneous manifestation of these symptoms. Possibly, these clinical findings suggest parallel initiation of the disease process at anatomically distinct sites (Diederich et al. 2019).

Why Human RBD May Be Evolutionarily Impregnated

Ontogenetic Tracking Back of Network Dysregulation?

As body movements during sleep (active sleep) are a natural phenomenon during phylogenetic evolution and our own ontogenetic evolution, it has been speculated that an ontogenetically early dysregulation of neuromotor systems could later cause RBD (Corner and Schenck 2015). However, such an argumentation just advances the time to prove why there must be a specific vulnerability of the nuclei regulating muscle atonia during REM sleep. It does not explain the vulnerability of the LSC itself or the degeneration of other nuclei beyond REM sleep regulation.

Disinhibition of Phylogenetically Old Central Pattern Generators?

Some rare RBD movements have similarities with movements seen during limbic seizures (Tassinari et al. 2012). In both syndromes the nocturnal movements could be caused by disinhibited central pattern generators (CPGs), located in the mesencephalon, pons, and spinal cord. CPGs are phylogenetically well-preserved cellular networks seen in invertebrates and vertebrates. These neuronal networks allow automatic, although complex, behaviors such as deglutition, mastication, respiration, etc. (Steuer and Guertin 2019). Such automatic behaviors are needed for feeding, locomotion, and reproduction and are crucial for survival. CPGs have specifically been analyzed as paving networks of locomotion, including human bipedal gait. Usually CPG-mediated movements, although automatically executed, remain under control of the neocortex. This may no more be the case in RBD, and the archaic appeal of these movements re-emerges (Tassinari et al. 2012). For instance, in RBD the expansive gestures of all four extremities could reflect evolutionarily old CPG movements—now out of control of the neocortex. Such disinhibited CPGs may explain the violent or aggressive momentum of RBD movements. However, they cannot explain the majority of most RBD dreams showing a "neutral" content. Also, why are only some of the CPG-triggered movements seen and not others, such as pelvic thrusting, biting, and teeth chattering? All in all, the concept that RBD behaviors are mainly the expression of dysfunctional or disinhibited CPGs is not sufficiently supported by clinical observations.

Purposeful as Essential for Survival?

According to the TST, humans and animals are virtually exposed to stressful situations during dreams, in order to be trained and face them adequately during daytime (Revonsuo 2000). Rehearsals of the "fight or flight" survival behavior may have conferred an advantage to humans in prehistoric times, explaining its conservation in modern times. Such a hypothesis is well in line with the purpose of sleep in general and supported by phylogenetic evolution and ontogenetic development. Survival may be enhanced by constant repetition and execution of gestures essential for life. Probing them can be done mentally or physically. The *mental* rehearsal is feasible in all subjects: dream imagination of an attack, strong emotions during such situations, total belief that they are real while they are not. We propose that the *motor* rehearsal takes place only in patients with RBD. Notably, young adults with NREM parasomnia also display behaviors and associated sleep mentation containing major life threats (Oudiette et al. 2009). However, when comparing patients with RBD to sleepwalkers, the response to these threats is different: sleepwalkers mostly flee from a disaster, while most patients with RBD counterattack when assaulted (Uguccioni et al. 2013). These major differences in the type of dreamer's response are again reminiscent of the fight or flight response to threats. During RBD dreams the patients defend themselves or their family from attackers (mostly human strangers and animals), and they are rarely the first attacker. In conclusion, RBD-associated aggression may be a disorder of enacting dreams (aggression dreams because aggression is frequent in REM sleep) rather than a disorder of dreaming. The threats in RBD may be the exacerbation of systems that train humans to appropriately react during the daytime to a wide spectrum of dangers.

Counterarguments and Related Evidence against Evolutionary Influences

As shown, a concise and all-encompassing evolutionary concept of REM sleep and even more of RBD remains challenging. First, only rhythmic fluctuations of autonomic activity may be seen. Later on, wakefulness can be clearly distinguished from sleeping states. However, REM sleep may for long epochs of evolution and even during early ontogenetic development be intermingled with NREM sleep. Remarkably, in patients with *advanced* stages of neurodegeneration, dissociated states (meaning mixed states of REM, NREM, and wakefulness) have been described. In those individuals it can become hard to assume that there is "acting out of dreams" as the observed movements contributing to dysfunctional REM sleep may be fragmentary awake phenomena. However, the progressive degeneration of brainstem nuclei, in particular those already "left behind" human exponential telencephalization, may further enhance such intermingling of states of consciousness.

Another counterargument against an all-encompassing explanation of RBD as an evolutionary remnant concerns minor, slow, and soft REM movements (Oudiette et al. 2009). Also, only loss of muscle atonia may be seen, without enactment of dreams. Slow evolution from pure REM sleep atonia to associative REM sleep events and finally dream enactment at the advanced stage has been proposed (Högl et al. 2018). Is full disclosure of phylogenetically old pathways only operational in advanced stages of RBD but not in its earliest stages? However, patients with RBD may display within the same REM sleep different patterns ranging from normal atonia over phasic soft movements to fighting movements. If one compares REM sleep to a mental theater, the curtain would be muscle atonia and the phasic movements and behaviors, the theatrical play. The curtain might be closed (atonia) or raised (enhanced chin muscle tone) over an empty theatrical play (no movement or no participation of the dreamer) or over a fully developed scenario.

We have ignored so far the impact of specific life histories on the occurrence of RBD. Recent life events are frequently incorporated into normal dreams, as supported by the continuity hypothesis, which stipulates that dreams are embodied simulations that dramatize our conceptions and concerns (Schredl and Hofmann 2003). The experience of a recently learned stressful text can be integrated into an RBD episode (Uguccioni et al. 2013). Is the experience of traumatic stress sufficient to elucidate the manifestation of RBD, or is there requirement of a neurodegenerative predisposition? Alternatively, is RBD reported in patients with PTSD only fragmentary, reflecting primarily anxiety-linked REM episodes? So far, vPSG data in PTSD are rare or incomplete. Other epigenetically acting possible inducers have not been explored. But the observation that RBD patients do not show increased daytime aggressiveness easily excludes a purely psychodynamic explanation.

Clinical and Therapeutic Implications

The medical treatment of RBD remains symptomatic. Beside physical protection measures, clonazepam and melatonin are commonly used. In case of a medication induced RBD, the withdrawal of the psychotropic medication in question may be sufficient. When there is suspicion of an ongoing neurodegenerative process, there is reasonable hope that in the future early detection of RBD may help to offer better targeted treatment, slowing down the disease trajectory (Miglis et al. 2021). Notably, it may be comforting for both the patient and the partner to hear that violent dream content and even enactment of violent dream content could be an evolutionarily old, formerly purposeful mechanism for self-protection. We have also to explore if cognitive behavioral treatment (CBT) may induce some reprogramming of altered pathways, leading to recurrent negative dreams.

CBT uses image rehearsing therapy, and it is efficiently applied in nightmares disorder for changing a nightmare into a more acceptable scenario.

Unanswered Questions and Future Work

The occurrence of RBD in ongoing neurodegenerative processes, narcolepsy, or by psychotropic treatment suggests that *different* triggering events may be possible, although the final common way seems to consistently be the REM atonia system. It is unknown why in some patients the syndrome can be transiently triggered by anticholinergic or serotonergic medications. Are these patients at higher risk for later development of a stable RBD syndrome announcing neurodegeneration? What is the specific susceptibility of the LSC for the neurodegenerative process? Why is the aimed shot "so precise," at least in the beginning, without neighboring collateral damage? Is it due to the evolutionary rooting and to the lifelong (over)use of the LSC during REM sleep? We do not know if early ontogenetic damage or later epigenetic influences have any impact on human RBD. Is the prognosis different in patients with more aggressive RBD content in comparison to those with less aggressive content? This could help to resolve the paradox of RBD in otherwise non-aggressive subjects. Do daytime events influence enacting of dreams?

The fight behaviors observed during RBD impressively confirm the TST: dreaming is a place where we are virtually trained to face dangers. This is also the case when non-violent, learned behaviors are observed. Phylogenetic comparison evidences striking similarities between human RBD that is naturally occurring and experimentally induced RBD in other mammals, showing also nocturnal replay of a "fight or flight" maneuver or more ordinary species-specific behaviors. In humans the movements (and especially the defense/attack movements) during RBD may be driven by dysfunctional brainstem (rather than cortical) nuclei underlying archaic defense behaviors. However, RBD phenomenology and RBD-associated dream content can only partially be explained by the TST-based "fight or flight" hypothesis as complex but entirely peaceful daily-life movements are observed as well. With this thought in mind, RBD provides a direct access to dream-associated movements, facial emotional expressions, and speeches. These "solid mental images" will allow us to progress in understanding the functions of dreaming in general.

Key Points to Remember

- RBD is characterized by loss of muscle atonia in REM sleep, acting out of dreams, and concordant dream content.
- These dreams are often, but not always, violent; soft and purposeful movements of daily life may also be observed.
- RBD can be a harbinger of neurodegenerative diseases, in particular PD.

- The underlying brainstem nuclei are evolutionarily old and seem to have been outpaced in growth by neocortical structures.
- There are several evolutionarily based theories, all proposing motor rehearsal of threatening life events; mental rehearsal of such life events is seen as a purpose of REM sleep itself.

References

Antonini, A., & Stryker, M. (1993). Rapid remodeling of axonal arbors in the visual cortex. *Science*, *260*(5115), 1819–1821. https://doi.org/10.1126/science.8511592

Arnulf, I. (2012). REM sleep behavior disorder: Motor manifestations and pathophysiology. *Movement Disorders*, *27*(6), 677–689. https://doi.org/10.1002/mds.24957

Arnulf, I., Bonnet, A. M., Damier, P., Bejjani, B. P., Seilhean, D., Derenne, J. P., & Agid, Y. (2000). Hallucinations, REM sleep, and Parkinson's disease: A medical hypothesis. *Neurology*, *55*(2), 281–288. https://doi.org/10.1212/wnl.55.2.281 Arnulf, I., Uguccioni, G., Gay, F., Baldayrou, E., Golmard, J. L., Gayraud, F., & Devevey, A. (2017). What does the sleeping brain say? Syntax and semantics of sleep talking in healthy subjects and in parasomnia patients. *Sleep*, *40*(11), Article zsx159. https://doi.org/10.1093/sleep/zsx159

Barker, E. N., Dawson, L. J., Rose, J. H., Van Meervenne, S., Frykman, O., Rohdin, C., Leijon, A., Soerensen, K. E., Järnegren, J., Johnson, G. C., O'Brien, D. P., & Granger, N. (2016). Degenerative encephalopathy in Nova Scotia duck tolling retrievers presenting with a rapid eye movement sleep behavior disorder. *Journal of Veterinary Internal Medicine*, *30*(5), 1681–1689. https://doi.org/10.1111/jvim.14575

Blumberg, M. S., Lesku, J. A., Libourel, P. A., Schmidt, M. H., & Rattenborg, N. C. (2020). What is REM sleep? *Current Biology*, *30*(1), R38–R49. https://doi.org/10.1016/j.cub.2019.11.045

Blumberg, M. S., & Plumeau, A. M. (2016). A new view of "dream enactment" in REM sleep behavior disorder. *Sleep Medicine Reviews*, *30*, 34–42. https://doi.org/10.1016/j.smrv.2015.12.002

Borek, L. L., Kohn, R., & Friedman, J. H. (2007). Phenomenology of dreams in Parkinson's disease. *Movement Disorders*, *22*(2), 198–202. https://doi.org/10.1002/mds.21255

Bugalho, P., & Viana-Baptista, M. (2013). REM sleep behavior disorder and motor dysfunction in Parkinson's disease—A longitudinal study. *Parkinsonism & Related Disorders*, *19*(12), 1084–1087. https://doi.org/10.1016/j.parkreldis.2013.07.017

Cipolli, C., Franceschini, C., Mattarozzi, K., Mazzetti, M., & Plazzi, G. (2011). Overnight distribution and motor characteristics of REM sleep behaviour disorder episodes in patients with narcolepsy–cataplexy. *Sleep Medicine*, *12*(7), 635–640. https://doi.org/10.1016/j.sleep.2010.12.016

Cochen De Cock, V., Debs, R., Oudiette, D., Leu, S., Radji, F., Tiberge, M., Yu, H., Bayard, S., Roze, E., Vidailhet, M., Dauvilliers, Y., Rascol, O., & Arnulf, I. (2011). The improvement of movement and speech during rapid eye movement sleep behaviour disorder in multiple system atrophy. *Brain*, *134*(3), 856–862. https://doi.org/10.1093/brain/awq379

Cochen De Cock, V., Vidailhet, M., Leu, S., Texeira, A., Apartis, E., Elbaz, A., Roze, E., Willer, J. C., Derenne, J. P., Agid, Y., & Arnulf, I. (2007). Restoration of normal motor control in Parkinson's disease during REM sleep. *Brain*, *130*(2), 450–456. https://doi.org/10.1093/brain/awl363

Corner, M. A., & Schenck, C. H. (2015). Perchance to dream? Primordial motor activity patterns in vertebrates from fish to mammals: Their prenatal origin, postnatal persistence during sleep, and pathological reemergence during REM sleep behavior disorder. *Neuroscience Bulletin*, *31*(6), 649–662. https://doi.org/10.1007/s12264-015-1557-1

Dauvilliers, Y., Boudousq, V., Lopez, R., Gabelle, A., Cochen De Cock, V., Bayard, S., & Peigneux, P. (2011). Increased perfusion in supplementary motor area during a REM sleep behaviour episode. *Sleep Medicine*, *12*(5), 531–532. https://doi.org/10.1016/j.sleep.2011.02.003

Dean, P., Redgrave, P., & Westby, G. M. (1989). Event or emergency? Two response systems in the mammalian superior colliculus. *Trends in Neurosciences*, *12*(4), 137–147. https://doi.org/10.1016/0166-2236(89)90052-0

DesJardin, J. T., Holmes, A. L., Forcelli, P. A., Cole, C. E., Gale, J. T., Wellman, L. L., Gale, K., & Malkova, L. (2013). Defense-like behaviors evoked by pharmacological disinhibition of the superior colliculus in the primate. *Journal of Neuroscience*, *33*(1), 150–155. https://doi.org/10.1523/JNEUROSCI.2924-12.2013

Diederich, N. J., Stebbins, G., Schiltz, C., & Goetz, C. G. (2014). Are patients with Parkinson's disease blind to blindsight? *Brain*, *137*(6), 1838–1849. https://doi.org/10.1093/brain/awu094

Diederich, N. J., Surmeier, D. J., Uchihara, T., Grillner, S., & Goetz, C. G. (2019). Parkinson's disease: Is it a consequence of human brain evolution? *Movement Disorders*, *34*(4), 453–459. https://doi.org/10.1002/mds.27628

Dugger, B. N., Murray, M. E., Boeve, B. F., Parisi, J. E., Benarroch, E. E., Ferman, T. J., & Dickson, D. W. (2012). Neuropathological analysis of brainstem cholinergic and catecholaminergic nuclei in relation to rapid eye movement (REM) sleep behaviour disorder. *Neuropathology and Applied Neurobiology*, *38*(2), 142–152. https://doi.org/10.1111/j.1365-2990.2011.01203.x

Ehrminger, M., Latimier, A., Pyatigorskaya, N., García-Lorenzo, D., Leu-Semenescu, S., Vidailhet, M., Lehericy, S., & Arnulf, I. (2016). The coeruleus/subcoeruleus complex in idiopathic rapid eye movement sleep behaviour disorder. *Brain*, *139*(4), 1180–1188. https://doi.org/10.1093/brain/aww006

Erskine, D., Thomas, A. J., Taylor, J. P., Savage, M. A., Attems, J., McKeith, I. G., Morris, C. M., & Khundakar, A. A. (2017). Neuronal loss and α-synuclein pathology in the superior colliculus and its relationship to visual hallucinations in dementia with Lewy bodies. *American Journal of Geriatric Psychiatry*, *25*(6), 595–604. https://doi.org/10.1016/j.jagp.2017.01.005

Fantini, M. L., Corona, A., Clerici, S., & Ferini-Strambi, L. (2005). Aggressive dream content without daytime aggressiveness in REM sleep behavior disorder. *Neurology*, *65*(7), 1010–1015. https://doi.org/10.1212/01.wnl.0000179346.39655.e0

Frank, M. G. (2011). The ontogeny and function (s) of REM sleep. In B. Mallick, S. Pandi-Perumal, S. R. McCarley, & A. Morrison (Eds.), *Rapid eye movement sleep: Regulation and function* (pp. 49–57). Cambridge University Press.

García-Lorenzo, D., Longo-Dos Santos, C., Ewenczyk, C., Leu-Semenescu, S., Gallea, C., Quattrocchi, G., Pita Lobo, P., Poupon, C., Benali, H., Arnulf, I., Vidailhet, M., & Lehericy, S. (2013). The coeruleus/subcoeruleus complex in rapid eye movement sleep behaviour disorders in Parkinson's disease. *Brain*, *136*(7), 2120–2129. https://doi.org/10.1093/brain/awt152

Gregor, T. (1981). A content analysis of Mehinaku dreams. *Ethos*, *9*(4), 353–390. http://www.jstor.org/stable/639915

Hackius, M., Werth, E., Sürücü, O., Baumann, C. R., & Imbach, L. L. (2016). Electrophysiological evidence for alternative motor networks in REM sleep behavior disorder. *Journal of Neuroscience*, *36*(46), 11795–11800. https://doi.org/10.1523/JNEUROSCI.2546-16.2016

Hall, C. S., & Van de Castle, R. L. (1966). *The content analysis of dreams*. Appleton-Century-Crofts.

Hobson, J. A. (2009). REM sleep and dreaming: Towards a theory of protoconsciousness. *Nature Reviews Neuroscience*, *10*(11), 803–813. https://doi.org/10.1038/nrn2716

Högl, B., Stefani, A., & Videnovic, A. (2018). Idiopathic REM sleep behaviour disorder and neurodegeneration—An update. *Nature Reviews Neurology*, *14*(1), 40–55. https://doi.org/10.1038/nrneurol.2017.157

Holmes, C. J., & Jones, B. E. (1994). Importance of cholinergic, GABAergic, serotonergic and other neurons in the medial medullary reticular formation for sleep–wake states studied by cytotoxic lesions in the cat. *Neuroscience*, *62*(4), 1179–1200. https://doi.org/10.1016/0306-4522(94)90352-2

Iranzo, A., Tolosa, E., Gelpi, E., Molinuevo, J. L., Valldeoriola, F., Serradell, M., Sanchez-Valle, R., Vilaseca, I., Lomeña, F., Vilas, D., Lladó, A., Gaig, C., & Santamaria, J. (2013). Neurodegenerative disease status and post-mortem pathology in idiopathic rapid-eye-movement sleep behaviour disorder: An observational cohort study. *Lancet Neurology*, *12*(5), 443–453. https://doi.org/10.1016/S1474-4422(13)70056-5

Isa, T., Marquez-Legorreta, E., Grillner, S., & Scott, E. K. (2021). The tectum/superior colliculus as the vertebrate solution for spatial sensory integration and action. *Current Biology*, *31*(11), R741–R762. https://doi.org/10.1016/j.cub.2021.04.001

Jouvet, M. (1980). Paradoxical sleep and the nature–nurture controversy. *Progress in Brain Research*, *53*, 331–346. https://doi.org/10.1016/S0079-6123(08)60073-4

Jouvet M. (1967). Neurophysiology of the states of sleep. *Physiological Reviews*, *47*(2), 117–177. https://doi.org/10.1152/physrev.1967.47.2.117

Jouvet-Mounier, D., Astic, L., & Lacote, D. (1969). Ontogenesis of the states of sleep in rat, cat, and guinea pig during the first postnatal month. *Developmental Psychobiology*, *2*(4), 216–239. https://doi.org/10.1002/dev.420020407

Krohn, L., Wu, R. Y., Heilbron, K., Ruskey, J. A., Laurent, S. B., Blauwendraat, C., Alam, A., Arnulf, I., Hu, M. T. M., Dauvilliers, Y., Högl, B., Toft, M., Bjørnarå, K. A., Stefani, A., Holzknecht, E., Monaca, C. C., Abril, B., Plazzi, G., Antelmi, E., . . . Gan-Or, Z. (2020). Fine mapping of *SNCA* in rapid eye movement sleep behavior disorder and overt synucleinopathies. *Annals of Neurology*, *87*(4), 584–598. https://doi.org/10.1002/ana.25687

Lacaux, C., Andrillon, T., Bastoul, C., Idir, Y., Fonteix-Galet, A., Arnulf, I., & Oudiette, D. (2021). Sleep onset is a creative sweet spot. *Science Advances*, *7*(50), Article eabj5866. https://doi.org/10.1126/sciadv.abj5866

Leclair-Visonneau, L., Oudiette, D., Gaymard, B., Leu-Semenescu, S., & Arnulf, I. (2010). Do the eyes scan dream images during rapid eye movement sleep? Evidence from the rapid eye movement sleep behaviour disorder model. *Brain*, *133*(6), 1737–1746. https://doi.org/10.1093/brain/awq110

Lesku, J. A., Bark, R. J., Martinez-Gonzalez, D., Rattenborg, N. C., Amlaner, C. J., & Lima, S. L. (2008). Predator-induced plasticity in sleep architecture in wild-caught Norway rats (*Rattus norvegicus*). *Behavioural Brain Research*, *189*(2), 298–305. https://doi.org/10.1016/j.bbr.2008.01.006

Lopez, J., Roffwarg, H. P., Dreher, A., Bissette, G., Karolewicz, B., & Shaffery, J. P. (2008). Rapid eye movement sleep deprivation decreases long-term potentiation stability and affects some glutamatergic signaling proteins during hippocampal development. *Neuroscience*, *153*(1), 44–53. https://doi.org/10.1016/j.neuroscience.2008.01.072

Maior, R. S., Hori, E., Uribe, C. E., Saletti, P. G., Ono, T., Nishijo, H., & Tomaz, C. (2012). A role for the superior colliculus in the modulation of threat responsiveness in primates: Toward the ontogenesis of the social brain. *Reviews in the Neurosciences*, *23*(5–6), 697–706. https://doi.org/10.1515/revneuro-2012-0055

Maranci, J. B., Nigam, M., Masset, L., Msika, E. F., Vionnet, M. C., Chaumereil, C., Vidailhet, M., Leu-Semenescu, S., & Arnulf, I. (2022). Eye movement patterns correlate with overt emotional behaviours in rapid eye movement sleep. *Scientific Reports*, *12*(1), Article 1700. https://doi.org/10.1038/s41598-022-05905-5

Mayer, G., Bitterlich, M., Kuwert, T., Ritt, P., & Stefan, H. (2015). Ictal SPECT in patients with rapid eye movement sleep behaviour disorder. *Brain*, *138*(5), 1263–1270. https://doi.org/10.1093/brain/awv042

McCarley, R. W., & Hobson, J. A. (1970). Cortical unit activity in desynchronized sleep. *Science*, *167*(3919), 901–903. https://doi.org/10.1126/science.167.3919.901

McCarter, S. J., Tippmann-Peikert, M., Sandness, D. J., Flanagan, E. P., Kantarci, K., Boeve, B. F., Silber, M. H., & St. Louis, E. K. (2015). Neuroimaging-evident lesional pathology associated with REM sleep behavior disorder. *Sleep Medicine*, *16*(12), 1502–1510. https://doi.org/10.1016/j.sleep.2015.07.018

Miglis, M. G., Adler, C. H., Antelmi, E., Arnaldi, D., Baldelli, L., Boeve, B. F., Cesari, M., Dall'Antonia, I., Diederich, N. J., Doppler, K., Dušek, P., Ferri, R., Gagnon, J. F., Gan-Or, Z., Hermann, W., Högl, B., Hu, M. T., Iranzo, A., Janzen, A., . . . Oertel, W. H. (2021). Biomarkers of conversion to α-synucleinopathy in isolated rapid-eye-movement sleep behaviour disorder. *Lancet Neurology*, *20*(8), 671–684. https://doi.org/10.1016/S1474-4422(21)00176-9

Mirmiran, M., Van de Poll, N. E., Corner, M. A., Van Oyen, H. G., & Bour, H. L. (1981). Suppression of active sleep by chronic treatment with chlorimipramine during early postnatal development: Effects upon adult sleep and behavior in the rat. *Brain Research*, *204*(1), 129–146. https://doi.org/10.1016/0006-8993(81)90657-0

Nielsen, T., & Paquette, T. (2007). Dream-associated behaviors affecting pregnant and postpartum women. *Sleep*, *30*(9), 1162–1169. https://doi.org/10.1093/sleep/30.9.1162

Oudiette, D., Cochen De Cock, V., Lavault, S., Leu, S., Vidailhet, M., & Arnulf, I. (2009). Nonviolent elaborate behaviors may also occur in REM sleep behavior disorder. *Neurology*, *72*(6), 551–557. https://doi.org/10.1212/01.wnl.0000341936.78678.3a

Peever, J., & Fuller, P. M. (2017). The biology of REM sleep. *Current Biology*, *27*(22), R1237–R1248. https://doi.org/10.1016/j.cub.2017.10.026

Postuma, R. B., Iranzo A., Hu, M., Högl, B., Boeve, B. F., Manni, R., Oertel, W. H., Arnulf, I., Ferini-Strambi, L., Puligheddu, M., Antelmi, E., Cochen De Cock, V., Arnaldi, D., Mollenhauer, B., Videnovic, A., Sonka, K., Jung, K. Y., Kunz, D., Dauvilliers, Y., . . . Pelletier, A. (2019). Risk and predictors of dementia and parkinsonism in idiopathic REM sleep behaviour disorder: A multicentre study. *Brain*, *142*(3), 744–759. https://doi.org/10.1093/brain/awz030

Revonsuo, A. (2000). The reinterpretation of dreams: An evolutionary hypothesis of the function of dreaming. *Behavioral and Brain Sciences*, *23*(6), 877–901. https://doi.org/10.1017/s0140525x00004015

Roffwarg, H. P., Muzio, J. N., & Dement, W. C. (1966). Ontogenetic development of the human sleep–dream cycle: The prime role of "dreaming sleep" in early life may be in the development of the central nervous system. *Science*, *152*(3722), 604–619. https://doi.org/10.1126/science.152.3722.604

Ross, R. J., Ball, W. A., Dinges, D. F., Kribbs, N. B., Morrison, A. R., Silver, S. M., & Mulvaney, F. D. (1994). Motor dysfunction during sleep in posttraumatic stress disorder. *Sleep*, *17*(8), 723–732. https://doi.org/10.1093/sleep/17.8.723

Rößler, D. C., Kim, K., De Agrò, M., Jordan, A., Galizia, C. G., & Shamble, P. S. (2022). Regularly occurring bouts of retinal movements suggest an REM sleep–like state in jumping spiders. *Proceedings of the National Academy of Sciences of the United States of America*, *119*(33), Article e2204754119. https://doi.org/10.1073/pnas.2204754119

Sastre, J. P., & Jouvet, M. (1979). Le comportement onirique du chat. *Physiology & Behavior*, *22*(5), 979–989. https://doi.org/10.1016/0031-9384(79)90344-5

Schenck, C. H., Bundlie, S. R., Ettinger, M. G., & Mahowald, M. W. (1986). Chronic behavioral disorders of human REM sleep: A new category of parasomnia. *Sleep*, *9*(2), 293–308. https://doi.org/10.1093/sleep/9.2.293

Schredl, M., & Hofmann, F. (2003). Continuity between waking activities and dream activities. *Consciousness and Cognition*, *12*(2), 298–308. https://doi.org/10.1016/s1053-8100(02)00072-7

Schubert, T. A., Chidester, R. M., & Chrisman, C. L. (2011). Clinical characteristics, management, and long-term outcome of suspected rapid eye movement sleep behavior disorder in 14 dogs. *Journal of Small Animal Practice*, *52*(2), 93–100. https://doi.org/10.1111/j.1748-5827.2010.01026.x

Sharma, Y., Xu, T., Graf, W. M., Fobbs, A., Sherwood, C. C., Hof, P. R., Allman, J. M., & Manaye, K. F. (2010). Comparative anatomy of the locus coeruleus in humans and nonhuman primates. *Journal of Comparative Neurology*, *518*(7), 963–971. https://doi.org/10.1002/cne.22249

Shea, A., Hatch, A., De Risio, L., & Beltran, E. (2018). Association between clinically probable REM sleep behavior disorder and tetanus in dogs. *Journal of Veterinary Internal Medicine*, *32*(6), 2029–2036. https://doi.org/10.1111/jvim.15320

Shen, Y., Yu, W. B., Shen, B., Dong, H., Zhao, J., Tang, Y. L., Fan, Y., Yang, Y. F., Sun, Y. M., Luo, S. S., Chen, C., Liu, F. T., Wu, J. J., Xiao, B. G., Yu, H., Koprich, J. B., Huang, Z. L., & Wang, J. (2020). Propagated α-synucleinopathy recapitulates REM sleep behavior disorder followed by Parkinsonian phenotypes in mice. *Brain*, *143*(11), 3374–3392. https://doi.org/10.1093/brain/awaa283

Siclari, F., Valli, K., & Arnulf, I. (2020). Dreams and nightmares in healthy adults and in patients with sleep and neurological disorders. *Lancet Neurology*, *19*(10), 849–859. https://doi.org/10.1016/S1474-4422(20)30275-1

Siegel, J. M. (2006). The stuff dreams are made of: Anatomical substrates of REM sleep. *Nature Neuroscience*, *9*(6), 721–722. https://doi.org/10.1038/nn0606-721

Siegel, J. M. (2022). Sleep function: An evolutionary perspective. *Lancet Neurology*, *21*(10), 937–946. https://doi.org/10.1016/S1474-4422(22)00210-1

Snyder, F. (1970). The phenomenology of REM dreaming. In H. Madow & L. H. Snow (Eds.), *The psychodynamic implications of physiological studies on dreams* (pp. 124–151). Charles C. Thomas.

Steuer, I., & Guertin, P. A. (2019). Central pattern generators in the brainstem and spinal cord: An overview of basic principles, similarities and differences. *Reviews in the Neurosciences*, *30*(2), 107–164. https://doi.org/10.1515/revneuro-2017-0102

Stickgold, R., James, L., & Hobson, J. A. (2000). Visual discrimination learning requires sleep after training. *Nature Neuroscience*, *3*(12), 1237–1238. https://doi.org/10.1038/81756

Tassinari, C. A., Gardella, E., Cantalupo, G., & Rubboli, G. (2012). Relationship of central pattern generators with parasomnias and sleep-related epileptic seizures. *Sleep Medicine Clinics, 7*(1), 125–134. https://doi.org/10.1016/j.jsmc.2012.01.003

Tiriac, A., Del Rio-Bermudez, C., & Blumberg, M. S. (2014). Self-generated movements with "unexpected" sensory consequences. *Current Biology, 24*(18), 2136–2141. https://doi.org/10.1016/j.cub.2014.07.053

Traczynska-Kubin, D., Atzef, E., & Petre-Quadens, O. (1969). Sleep in parkinsonism. *Acta Neurologica et Psychiatrica Belgica, 69*(9), 727–733.

Uguccioni, G., Golmard, J. L., de Fontréaux, A. N., Leu-Semenescu, S., Brion, A., & Arnulf, I. (2013). Fight or flight? Dream content during sleepwalking/sleep terrors vs. rapid eye movement sleep behavior disorder. *Sleep Medicine, 14*(5), 391–398. https://doi.org/10.1016/j.sleep.2013.01.014

Valencia Garcia, S., Libourel, P. A., Lazarus, M., Grassi, D., Luppi, P. H., & Fort, P. (2017). Genetic inactivation of glutamate neurons in the rat sublaterodorsal tegmental nucleus recapitulates REM sleep behaviour disorder. *Brain, 140*(2), 414–428. https://doi.org/10.1093/brain/aww310

Valli, K., Frauscher, B., Gschliesser, V., Wolf, E., Falkenstetter, T., Schoenwald, S. V., Ehrmann, L., Zangerl, A., Marti, I., Boesch, S. M., Revonsuo, A., Poewe, W., & Högl, B. (2012). Can observers link dream content to behaviours in rapid eye movement sleep behaviour disorder? A cross-sectional experimental pilot study. *Journal of Sleep Research, 21*(1), 21–29. https://doi.org/10.1111/j.1365-2869.2011.00938.x

van der Helm, E., Yao, J., Dutt, S., Rao, V., Saletin, J. M., & Walker, M. P. (2011). REM sleep depotentiates amygdala activity to previous emotional experiences. *Current Biology, 21*(23), 2029–2032. https://doi.org/10.1016/j.cub.2011.10.052

Wehr, T. A. (1992). A brain-warming function for REM sleep. *Neuroscience and Biobehavioral Reviews, 16*(3), 379–397. https://doi.org/10.1016/S0149-7634(05)80208-8

13

An Evolutionary Psychoneuroimmunological Approach to Major Depressive Disorder

Markus J. Rantala and Javier I. Borráz-León

Introduction

According to the *Diagnostic and Statistical Manual of Mental Disorders*, Fifth Edition (DSM-5), major depressive disorder (MDD) diagnosis requires the presence of five or more of the following symptoms lasting for 2 or more weeks causing significant emotional distress and/or functioning impairment. The symptoms include being sad or depressed, anhedonia (loss of interest or enjoyment in activities that were previously enjoyed), feelings of worthlessness or guilt, thoughts of death or suicide, concentration problems, fatigue or loss of energy, psychomotor retardation or agitation, weight loss or gain, and difficulty sleeping or sleeping too much (American Psychiatric Association 2013).

MDD has become one of the leading sources of disability worldwide and is believed to be a major contributor to the overall global disease burden. It has been estimated that over 300 million people suffer from MDD, equating to approximately 4.4% of the world's population (World Health Organization 2017). Surprisingly, its etiology is still poorly known, and there has been no significant improvement in its medical treatment for decades.

In this chapter, we present an evolutionary psychoneuroimmunological approach to MDD (Rantala et al. 2018; Rantala and Luoto 2022). This approach has the strength of connecting evolutionary psychological (ultimate-level) explanations, with neurobiological and immunological (proximate-level) explanations.

Major Depressive Disorder

MDD as a Disease of Modern Lifestyles

There is substantial variance in the prevalence of MDD among countries. For example, a study in 11 countries organized by the World Health Organization showed that the lifetime prevalence of MDD varied from 3.3% in Romania to 19.2% in the United States (Merikangas et al. 2007). Interestingly, cross-cultural analyses of women living in developed and developing countries have reported that the degree of modernization positively correlates with a higher prevalence of MDD (Colla et al. 2006). It has also been calculated that Chinese people born after 1966 are 22.4 times more likely to have at least one episode of MDD during their lifetime than Chinese people born before 1973 (Lee et al. 2007), which further supports the link between "Western" or modern lifestyles and the development of mood disorders. Likewise, a meta-analysis from studies on American college and high school students found that in 2007 they were six to eight times more likely to meet the diagnostic criteria of MDD compared to their peers in 1938 (discussed in Rantala and Luoto 2022). In Lundby, Sweden, the point prevalence of MDD increased from 0.8% in 1957 to 2.6% in 1972, whereas the overall MDD prevalence in Sweden was as high as 10.8% in 2009 (see Rantala and Luoto 2022). In addition, when comparing different countries, MDD becomes more common in those that have fully adopted modern "Western" lifestyles (Colla et al. 2006). A study on Indigenous populations living in arctic areas in Canada reported that transitioning from traditional to modern "Western" lifestyles was associated with a threefold increase in suicide rates in just a decade (Shephard and Rode 1996).

Rantala et al. (2018) suggested that the explanation for the current MDD "epidemic" around the world is that our mind and body have simply not adapted to the modern "Western" lifestyle, and therefore, previously adaptive behaviors associated with mood changes may have become maladaptive (see also Gilbert 2006). This evolutionary mismatch hypothesis is supported by anthropological reports that have documented the rarity of depression episodes that fulfill the diagnostic criteria of MDD in hunter–gatherer societies that have a more traditional lifestyle (discussed in Rantala et al. 2021; Rantala and Luoto 2022). For instance, there is evidence showing that depression episodes are more common in hunter–gatherer societies that have adopted agricultural practices such as in Ik people from Uganda (Stevens and Price 2000).

There is another line of evidence suggesting that hunter–gatherers have higher psychological well-being than people with "Western" lifestyles. A study on the subjective happiness of Hadza people in Tanzania found that they had significantly higher scores of happiness than all 12 industrialized societies where the same scale has been applied (Frackowiak et al. 2020). Likewise, previous studies have shown that Hadza people do not suffer from physiological chronic stress, and they often describe their life as "relaxed" when asked whether they experience any worries or anxieties (Fedurek et al. 2023).

There are also some counterarguments in the literature suggesting that depression might not be as rare among hunter–gatherers as anthropologists living with them have reported (Chaudhary and Salali 2022). However, these counterarguments do not stand closer scrutiny. For example, a study on postpartum depression prevalence among the Hadza people in Tanzania using the Edinburgh Postnatal Depression Scale reported that 52% of women with infants under the age of 12 months had scores that are commonly used as a threshold for postpartum depression (Herlosky et al. 2020). However, the problem in the study was, as the authors themselves reported, that interviewed participants "never used any words that described 'depression', or any other comparably translated term." Instead, "they tended to associate labor pain with unhappiness." Since one question on the Edinburgh Postnatal Depression Scale asks whether the mother has felt unhappy that she has been crying, whereas another question asks whether she has been so unhappy that she has had difficulty sleeping, the apparently high prevalence of postpartum depression seems to be the result of misunderstanding the terms used in the test. Thus, this study should not be used as evidence for a high prevalence of postpartum depression among hunter–gatherers.

Comparing the prevalence of mood disorders between people who still live traditional lifestyles and people living modern "Western" lifestyles can help to assess the extent to which modern Western lifestyles may contribute to the development of mood disorders. For example, the Old Order Amish are known to live a lifestyle that was more typical in the 18th century. They still live without electricity and plow their fields with horses. A 5-year-long mental health study on a population of 12,500 Old Order Amish, of which 8,186 were adults, found that the 5-year prevalence of MDD was only 0.5% among Old Order Amish people in the United States (Egeland and Hostetter 1983). A more recent study on Old Order Amish and other old order groups found that the point prevalence of depression was 1% in Lancaster Amish, 1% in Groffdale Mennonite, 0% in Weaverland Mennonite, 1% in Mifflin County Amish, and 4% in Somerset County Amish (Yost et al. 2016). The point prevalence of depression in other people living in North America has been reported to be as high as 13.4% (Lim et al. 2018). Thus, the prevalence of MDD among Old Order Amish and other old order groups is substantially lower than that observed among Americans living "Western" lifestyles.

Why Does Modern Lifestyle Increase the Risk of MDD?

Several potential factors associated with modern "Western" lifestyles may contribute to the "epidemic" of MDD (Rantala et al. 2018, Rantala and Luoto 2022). These factors of modern lifestyles include reduced food nutritional quality, obesity, reduced contact with nature, sedentary lifestyle, sleep deprivation, and social isolation. In general, these lifestyle factors can lead to poor physical health, chronic inflammation, and increased susceptibility to stress (Miller and Raison 2016). Thus, in societies with modern lifestyles, adverse life events hypothetically lead to a state of chronic stress or "a runaway stress response" more often than in traditional cultures (Rantala and Luoto 2022).

Modern lifestyles are also associated with the so-called diseases of modernity, which include atherosclerosis, obesity, hormone-related and gastrointestinal cancers, type 2 diabetes, and osteoporosis (Lindeberg 2010). Common to all these conditions is systemic low-grade inflammation (Shoelson et al. 2007) and that they are practically absent in traditional hunter–gatherer societies (Lindeberg 2010). Rantala et al. (2018) suggested that neuroinflammation caused by systematic low-grade inflammation and chronic stress may turn the short-term adaptive mood change into chronic abnormal or maladaptive depressive states by preventing the normalization of mood after adverse life events. Neuroinflammation also causes symptoms of "sickness behavior," many of which are the same as those observed in MDD. Sickness behavior symptoms include low mood, lethargy, anxiety, malaise, loss of appetite, hyperalgesia, sleepiness, and problems concentrating (Rantala and Luoto 2022).

Inflammation and MDD

The association between MDD and systemic low-grade inflammation has been known for decades. A meta-analysis of studies testing the association between MDD and markers of systemic low-grade inflammation in blood found that patients with MDD had increased levels of C-reactive protein (CRP) and pro-inflammatory cytokine interleukin-6 (IL-6) compared to controls (Haapakoski et al. 2015). Another meta-analysis found that about a quarter of patients with MDD show evidence of systemic low-grade inflammation measured by high CRP levels, while over half of patients show mildly elevated CRP levels (Osimo et al. 2019). Moreover, patients with coronary heart disease—a condition where inflammation is one of the major contributing factors—have a threefold risk of suffering from MDD compared with the general population (Frasure-Smith and Lespérance 2006). Likewise, the prevalence of MDD is substantially higher in patients with autoimmune diseases where pro-inflammatory cytokine levels are continuously elevated. For example, of patients with rheumatoid arthritis about 70% will suffer from MDD during their lifetime (see Rantala and Luoto 2022).

The strongest evidence for the role of systemic low-grade inflammation in the etiology of MDD comes from experimental studies that have shown that activation of the immune system by vaccination against typhoid (Harrison et al. 2009) or by injecting pro-inflammatory cytokines induces symptoms of MDD in previously healthy people (Afridi and Suk 2021). Likewise, it has been found that up to 58% of patients without previous mental health problems who received therapy with interferons (a cytokine released in response to a viral infection) developed MDD (Raison et al. 2005).

Systemic low-grade inflammation is able to trigger inflammation in the brain, known as "neuroinflammation" (Troubat et al. 2021). In addition to systemic low-grade inflammation, chronic stress is known to be able to trigger neuroinflammation (Kim and Won 2017). During neuroinflammation, microglia and/or astrocytes are activated (DiSabato et al. 2016). Microglia are innate immune cells, functionally similar to peripheral macrophages, which are able to produce proinflammatory cytokines. Astrocytes are glial cells with many functions, including control of the blood–brain barrier (Dai et al. 2022).

There is robust evidence showing that MDD is associated with neuroinflammation (reviewed in Troubat et al. 2021). For example, previous neuroimaging studies have shown that patients with MDD have neuroinflammation in their brains. In addition, a meta-analysis of studies testing the presence of pro-inflammatory cytokine levels in cerebrospinal fluid found increased levels of IL-6, tumor necrosis factor-α, and IL-8 compared to controls, which also suggests the presence of neuroinflammation in patients with MDD (Enache et al. 2019).

Neuroinflammation is known to influence serotonin and dopamine levels, which may reduce mood but also prevent the normalization of mood after adverse life events. Likewise, other neurotransmitters that are known to play a role in MDD are influenced by neuroinflammation, like gamma-aminobutyric acid (GABA), which is the main inhibitory neurotransmitter, and glutamate, which is the most abundant excitatory neurotransmitter in the brain (Troubat et al. 2021). There is also evidence suggesting that severe neuroinflammation can lower mood even in the absence of any adverse life events (Sandiego et al. 2015). It is important to note that systemic low-grade inflammation and neuroinflammation are not specific to mood disorders but also occur in many other mental disorders such as anorexia and schizophrenia, where symptoms of depression are also common (Rantala et al. 2019, 2022).

Another line of evidence that links inflammation to MDD comes from a meta-analysis of placebo-controlled double-blind studies that have found that drugs with anti-inflammatory properties have moderate antidepressant effects (Köhler-Forsberg et al. 2019; Rantala and Luoto 2022). Furthermore, candidate gene and genome-wide association studies that have tried to identify risk alleles for MDD have found that alleles that increase the risk of MDD are associated with the function of the immune system (Raison and Miller 2012). Thus, in the light of current evidence, systemic inflammation and neuroinflammation caused by "Western" lifestyles seem to play major roles in the etiology of MDD (Figure 13.1).

The crucial question is why natural selection has made the human nervous system so vulnerable that neuroinflammation is able to cause MDD. It is also important to understand that MDD is not a uniform disease and that not all patients with low mood have systemic low-grade inflammation or neuroinflammation. Rantala et al. (2018) proposed that MDD is not a single disorder but at least 12 different subtypes, which can explain the presence of hundreds of unique and specific symptom profiles in MDD patients.

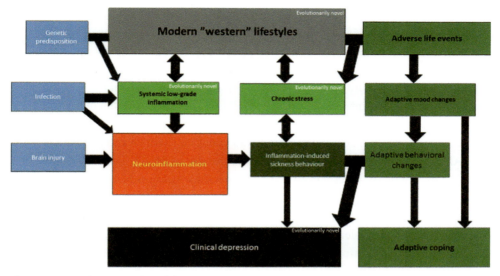

Figure 13.1 Pathways to clinical depression (major depressive disorder) and adaptive coping.

MDD Subtypes

The hypothesis that MDD is not a single disorder is supported by the fact that MDD patients can express a wide variety of, and sometimes opposite, patterns of symptoms. For example, whereas some depressed individuals gain weight, others can lose it; likewise, whereas some depressive patients can suffer from excessive sleepiness, others can experience poor sleep quality or insomnia. Empirical evidence suggests that depressive episodes may have different triggering factors and that they have different symptom profiles, even for the same person at different times. Thus, from an evolutionary psychological point of view, it is feasible to classify MDD episodes according to the proximate mechanisms and ultimate functions that trigger the mood change that leads to depression. Here, we briefly summarize the 12 subtypes of depression that have been previously described (see Rantala et al. 2018 for detailed descriptions). Some of these subtypes are adaptations, some of them are byproducts of other adaptations, and some of them are just pathological states without any benefit for patients. Even adaptive ones may turn into a pathological state of MDD as a result of low-grade inflammation or chronic stress caused by "Western" lifestyles. Some depression subtypes may occur simultaneously in the same patient, but subtyping them based on triggering factors may help to find more effective treatments focused on specific underlying factors for each patient (Rantala and Luoto 2022).

Infection-Induced Depression (Sickness Behavior)

To deal with pathogens and other parasites, natural selection has sculpted our immune system along with other adaptations such as the behavioral immune system. Within the behavioral immune system, sickness behavior can help individuals overcome microbial infections by reducing metabolic resources for the use of the immune system through physiological and behavioral changes which include the expression of anorexia, low mood, anergia, sleep disturbances, psychomotor retardation, listlessness, weakness, hyperalgesia, malaise, and weakened concentration (Dantzer 2001; Anders et al. 2013). In addition, it causes a desire for social isolation, which reduces the likelihood of infecting other individuals if the infection is contagious. It is noteworthy to highlight that systemic low-grade inflammation and neuroinflammation caused by an unhealthy lifestyle can lead to a similar psychopathological state as infections do (Rantala and Luoto 2022).

Although traditional lifestyles may reduce the risk of MDD, it does not mean that people living traditional lifestyles would not experience sickness behavior as a result of infection. For example, Stieglitz et al. (2015) studied the prevalence of symptoms of depression among the Tsimane of Bolivia, who are forager–horticulturalists with a high pathogen infection prevalence—19.4% having a CRP value >10 mg/dl, a level that indicates acute or chronic infections (Vasunilashorn 2010). Stieglitz and colleagues (2015) found that markers of the activation of the immune system (e.g., high CRP and some pro-inflammatory cytokine levels) were strongly associated with symptoms of depression. The point prevalence of depression among the Tsimane was 10%. Although the point prevalence of depression might sound high, it is low compared to the prevalence of MDD in most South American countries (mean = 20.6%; Lim et al. 2018). Unfortunately, the authors did not screen participants for acute infections, which would have allowed them to exclude the sick ones. Due to the high prevalence of acute infections, many depressed members of the Tsimane people would not have filled the diagnostic criteria of the DSM-5 for MDD since it excludes cases with physical illness (American Psychiatric Association 2013). Interestingly, despite the exceptionally high prevalence of acute infections, Tsimane people do not have chronic low-grade inflammation and do not suffer from other diseases associated with chronic low-grade inflammation (Kaplan et al. 2017). Thus, their lifestyle seems to protect them against the diseases of the modern lifestyle, including MDD; but unfortunately, it does not protect them against infections since they live in a habitat with high infection risk.

Depression Induced by Long-Term Stress (Burnout)

It is well known that long-term psychological stress negatively impacts physical and mental health by, for example, affecting the correct functioning of the endocrine and immune systems, promoting low-grade inflammation and neuroinflammation, which affects neurotransmitter and mood regulation (Berk et al. 2013). Among the symptoms of depression that can be associated with long-term stress are fatigue, reduced mood, appetite problems, self-blame, concentration problems, trouble sleeping, suicidal ideation, and anhedonia. Likewise, long-term stress up-regulates the immune system by causing leaky gut syndrome, which causes an increase of pro-inflammatory cytokines (Miller and Raison 2016, for a review), promoting sickness behavior that may lead to a state of depression in predisposed individuals. The trigger of chronic stress does not need to be a life-threatening danger; even social and financial problems are often stressful enough to cause severe chronic stress, leading to a state of depression known as "burnout." Burnout is a maladaptive byproduct of chronic stress that plagues people living modern "Western" lifestyles.

Depression Induced by Loneliness

Since humans are highly social animals, loneliness can be seen as an adaptation produced by natural selection to promote the search for the company of conspecifics in an individual (Rantala et al. 2018). During our evolutionary history, separation from a social group could have been a life-threatening danger compromising survival and reproduction. Thus, a person unconsciously perceives loneliness as a threat, causing activation of the amygdala and an increase in cortisol levels. If loneliness continues for a long time, it may lead to chronic stress that can trigger neuroinflammation causing sickness behavior, which leads to loneliness-induced depression (Rantala et al. 2018). Depression induced by loneliness is a maladaptive state because it might even reduce the odds that a person finds the company of conspecifics.

Depression Induced by Traumatic Events

Some people develop post-traumatic stress disorder (PTSD) after traumatic events in their life. The symptoms include anxiety, fear, frightfulness, emotional numbness, constant alertness, nightmares, and traumatic memories of the traumatic events that may come back as flashbacks. These symptoms may cause individuals to avoid places, situations, objects, and people that remind them of the traumatic event. Thus, these symptoms can be seen as adaptations that help individuals to decrease the likelihood of ending up in a similar situation again. However, Rantala and Luoto (2022) suggested that contemporary "Western" lifestyles may promote chronic low-grade inflammation, changing the previously adaptive function of PTSD symptoms that, along with symptoms of sickness behavior, may lead to a maladaptive state of MDD induced by traumatic events.

Depression Induced by Hierarchy Conflict

Like nonhuman animals, humans of both sexes have an urge to climb within the social hierarchy of the group to access valuable resources and high-quality mates (Miller 2009). Depressive symptoms may appear after a social defeat to signal that an individual has given up after hierarchy conflict. In this sense, depressive symptoms may function as an honest signal that the defeated individual is not a threat to other members of the group, which can facilitate the permanence of the depressed individual within the social group (Price et al. 1994; Rohde 2001). Decreased self-esteem may prevent a defeated individual from rechallenging the winner, which helps them to conserve their energy and sometimes even save their life.

Although direct hierarchy conflicts are rare in contemporary modern societies, living experiences such as being bullied at school or being fired from work may trigger depression induced by hierarchy conflict. Although depression triggered by hierarchy might have been adaptive in the evolutionary history of the human species, it might not always be adaptive in people living modern "Western" lifestyles since low-grade inflammation may exacerbate the symptoms and mix them with symptoms of sickness behavior (Rantala and Luoto 2022).

Depression Induced by Grief

Humans, like many other animals, grieve after the death of individuals with whom they have formed long-term attachment relationships (e.g., family, friends, pets; Darwin 1871). Grief seems to be a byproduct of our ability to form long-term attachments without any adaptive benefits because it exists also in non-social species when their pups die (Rantala and Luoto 2022). On the other hand, the knowledge that we will experience grief and pain after the death of our loved ones makes us take better care of them (Rantala et al. 2018). In some predisposed individuals, grief does not go away with time but instead changes into MDD induced by grief.

Losing a loved one increases cortisol levels sharply and may produce a state of chronic stress in vulnerable individuals, which may explain why grief may turn into MDD in some individuals. Fagundes et al. (2019) studied individuals who were recently widowed and found that those individuals with the highest levels of pro-inflammatory cytokines had the strongest symptoms of grief and that they had higher odds that the grief turned into an episode of depression. It seems that "Western" lifestyles may also exacerbate the expression of grief and symptoms of depression by reducing stress sensitivity and increasing low-grade inflammation.

Depression Induced by Romantic Rejection

There is evidence reporting that romantic rejection may lead to MDD and, in extreme cases, suicide or homicide (Rantala et al. 2018). Under an adaptive lens, some depression symptoms associated with romantic rejection, such as rumination, can help an individual avoid repeating the same mistakes in future relationships (Andrews and Thomson 2009). As Rantala et al. (2018) suggested, it is also possible that the expression of depression symptoms in the rejected partners could be interpreted as an honest signal of true love by the rejecting partner, which could help to win back the lost love. What is more, lowered self-esteem linked to romantic rejection can make an individual lower their self-perceived mate value, which would help the dumped person to find a new mate who matches better with their marketing value (Rantala et al. 2018). Although the mood change might have been an adaptation, it may become maladaptive if low-grade inflammation makes the symptoms stronger than what is beneficial for a person due to "Western" lifestyles or if the low-grade inflammation causes neuroinflammation and produces symptoms of sickness behavior that mix with the adaptive symptoms (Rantala et al. 2018).

Postpartum Depression

Previous studies suggest that between 10% and 15% of women will develop postpartum depression within 6 months after giving birth. These women may experience low mood, bouts of anger, crying, suicidal ideation, anxiety, loss of interest in their baby, and even harmful intentions toward them (Brummelte and Galea 2016). Postpartum depression may arise from the mother's perception that she is not receiving proper childcare support from the father or kin, which leads to a state of hopelessness (Hagen 1999). In the human evolutionary history, it would not have been possible for a woman to raise her baby alone without the proper help of her relatives and partner. Thus, a woman who ceased her investment in the baby and postponed her reproduction, in a situation where the future of the baby looked bad, would have higher reproductive success during her lifetime than a woman who would have continued investing in a baby that would have died anyway. Since child health influences the child's odds of surviving, it is not surprising that health problems in the child increase the risk of postpartum depression in the mother (Rantala et al. 2018).

It seems that the primary function of postpartum low mood (depression) is not to desert the baby. Instead, symptoms of depression may work as a signal for kin and partner that the mother needs more help with the baby (Hagen 1999; Rantala et al. 2018). However, if the help is not received, with time the postpartum depression may lead to MDD, a state of mind in which desertion of the baby is easier for the mother to do. Thus, postpartum low mood (depression) might have been adaptive during our evolutionary history, but due to the environmental mismatch, it might not always be adaptive in modern conditions because it may lead to MDD (Rantala and Luoto 2022).

Since postpartum depression also occurs in fathers and adoptive mothers when they feel hopelessness due to poor childcare support, it is clear that the hormonal changes due to childbirth are not the causative factors behind postpartum depression, as has been usually explained in psychiatry (Hagen 1999).

Season-Related Depression

Seasonal affective disorder (SAD) is a subtype of MDD that affects an individual at the same time each year (Rosenthal et al. 1984). Two forms of SAD have been described in the literature:

1. *Winter depression*: A SAD common in temperate and northern latitudes characterized by an increased need to sleep, decreased libido, general fatigue, and increased appetite, especially for carbohydrates (Rohan et al. 2003). These symptoms typically begin during autumn and end in spring and early summer. Although it has previously been suggested that SAD may have evolved as an adaptation to save energy during wintertime in northern latitudes during periods of food shortage (Davis and Levitan 2005), the evidence suggests that the most plausible explanation for winter depression

is that it is a maladaptive byproduct of the failure of an individual's circadian rhythm to match the reduced amount of light (Rantala et al. 2018).
2. *Spring depression*: This form of SAD occurs during spring and early summer, coincident with the higher prevalence of suicides in a temperate climate (Reutfors et al. 2009). Spring depression is associated with symptoms that include low mood, decreased appetite, weight loss, and waking up earlier (Boyce and Parker 1988). It is possible that allergenic reactions to pollen that increase low-grade inflammation together with increased sunlight exposure that excessively elevates the levels of serotonin in the brain, leading to symptoms of depression and anxiety in persons with chronic stress, can be the proximate mechanisms underlying this form of SAD (Rantala et al. 2018). Thus, spring depression may be interpreted as a maladaptive byproduct of seasonal changes in the amount of daylight and/or allergens (Rantala et al. 2018).

Chemically Induced Depression

According to the American Psychiatric Association (2013), depression symptoms such as insomnia, irritability, and concentration problems can be derived from substance abuse (e.g., alcohol, cocaine, opiates) or be a side effect of some medicines (e.g., corticosteroids, benzodiazepine tranquilizers). Unlike other subtypes of depression, symptoms of chemically induced depression do not last all day or occur every day, as observed in alcoholics. In fact, the symptoms associated with chemically induced depression usually decrease or disappear rapidly with abstinence, making the identification of this depression subtype easier (Rantala et al. 2018).

Depression Induced by Somatic Diseases

It has been observed that some neurological conditions such as Parkinson's disease, Alzheimer's disease, stroke, epilepsy, migraine, and traumatic brain injury are associated with an increased risk of MDD (Bulloch et al. 2015). Similar associations between MDD and endocrine and organic diseases such as Cushing's disease, hypothyroidism, and different kinds of cancers have also been observed. Three main proximate causes have been suggested to explain why cancer may induce depression: (1) cancer diagnosis may cause chronic stress and anxiety that may lead to MDD; (2) chemotherapy, radiotherapy, surgery, and other medical interventions may cause depression induced by sickness behavior; and (3) cancer tissue itself is known to increase pro-inflammatory cytokine levels which may lead to sickness behavior (Rantala et al. 2018). In any case, depression induced by somatic diseases can be also interpreted as a maladaptive state if it prevents an optimal recovery, increasing the possibility of death (Rantala et al. 2018).

Starvation-Induced Depression

Famines were common during human evolutionary history. Starvation-induced low mood (depression) may be interpreted as a psychological adaptation that helps an individual to overcome famine (Rantala et al. 2018). During starvation, the body inhibits functions that are not essential for immediate survival such as growth, reproduction, and immune function, whereas it increases physical activity, increasing the chance of finding food (Exner et al. 2000). Thus, starvation-induced depression helps to save energy in order to increase the chance of surviving a famine (Rantala et al. 2018).

The Adaptive Function of MDD Symptoms

Understanding the proximate mechanisms and ultimate functions behind the symptoms of MDD will allow the development of optimal treatment and recovery options (Rantala et al. 2018). For example, since many patients with MDD have increased levels of pro-inflammatory cytokines when compared to individuals without mood disorders, some of the symptoms of these patients may be symptoms of sickness behavior that are caused by neuroinflammation (Rantala et al. 2018, 2021). These include changes in appetite, anhedonia, social withdrawal, and suicidal ideation (Miller and Raison 2016). Thus, while some symptoms associated with mood disorders can be maladaptive byproducts of normally functioning psychological adaptations, others can be adaptations aimed at solving specific adaptive problems (Rantala et al. 2018, 2021).

MDD Symptoms

Emotional Pain

It has been proposed that emotional pain may teach an individual to avoid behaviors and situations that have caused the pain, in the same way that physical pain helps an individual to avoid behaviors or activities that harm their bodies (Keller and Nesse 2006). Thus, emotional pain, like physical pain, can be seen as an adaptation that helps people avoid behaviors that have caused fitness costs throughout evolutionary history (Rantala et al. 2018).

Suicide Proneness

It is common that depressed subjects experience suicidal thoughts and suicidal proneness. Although suicide could be maladaptive in current modern human environments, during our evolutionary history it may have increased the inclusive fitness of our ancestors under certain circumstances. For example, suicide may increase the success of a person's genes in a situation where the reproductive potential of an individual is weak or non-existent and the individual is just a burden for their relatives. By committing suicide, an individual may contribute to the reproductive success of their close relatives and thus the proliferation of their own genes while diminishing the relatives' burden (e.g., the absence of a sick individual to care for and one mouth less to feed; de Cantanzaro 1986). In support of this hypothesis, several studies have shown that suicidal thoughts and suicidal proneness are recurrent in individuals who have poor chances of reproduction and who feel they are a burden to their families (see Rantala et al. 2018). Consistent with this hypothesis, suicide incidence is nine times higher among patients with Huntington's disease than in the control population, with a median age at suicide of 55 years (van Duijn et al. 2021), suggesting that suicide is committed after the reproductive years.

Since it has been previously observed that injecting pro-inflammatory cytokines into the bloodstream can cause suicidal thoughts in some healthy subjects (Capuron et al. 2002), it is feasible to suggest that increases in the levels of pro-inflammatory cytokines associated with mood disorders, which are exacerbated by modern "Western" lifestyles, may produce maladaptive suicidal ideation. It could also be possible that the threat of suicide may serve as an effective way to get attention and support from relatives, friends, and community members who benefit from the existence of the individual (Rantala et al. 2018).

Rumination

The obsessive replaying of negative events (i.e., rumination) is commonly seen as a harmful symptom by mental health clinicians. However, continuous rumination about behaviors or events that led to the development of depression symptoms may help the individual to avoid similar situations in the future (Andrews and Thompson 2009). In fact, rumination is more common in contexts where the same situation can happen again (Keller and Nesse 2006). By changing neurotransmitter levels, neuroinflammation may cause a person to ruminate too much or chronically on things that are not associated with the triggering factors of mood changes. Thus, neuroinflammation may turn previously adaptive rumination into pathological rumination (Rantala et al. 2018).

Appetite Changes

It is common that mood disorder patients experience either an increase or a decrease in their appetite, contributing to weight changes. On the one hand, an increase in appetite, especially for sweet and fatty foods, may be related to comfort eating, in which a depressed person consumes and self-medicates themselves with food that momentarily raises mood. Comfort eating elevates mood by increasing serotonin and dopamine levels momentarily (Rantala et al. 2019). However, this comfort eating may lead to a vicious cycle that will contribute to the development of severer depression by increasing peripheral low-grade inflammation and gut dysbiosis. In some cases, increased appetite may be a byproduct of sleeping problems that are common in depression (Rantala and Luoto 2022).

On the other hand, a decrease in appetite may cause an increase in pro-inflammatory cytokine levels in MDD patients with immune activation by either infection or chronic stress. Although one could think that the decrease in appetite might be harmful, it is possible that in the case of acute infection it is adaptive because fasting can enhance the function of the immune system in many ways, and it is a common symptom of sickness behavior (Rantala and Luoto 2022). However, in the case of chronic stress (which can trigger neuroinflammation and sickness behavior), reduction of appetite might not be adaptive for the individual.

Anhedonia

Feelings of pleasure and joy are psychological adaptations that motivate organisms to behave in ways that enhance the fitness of their ancestors (Barron et al. 2010). In contrast, the loss of interest in activities that used to give pleasure and joy to the person who suffers from anhedonia can be adaptive for them because it reduces activity and can help a person to save energy for the use of the immune system. However, anhedonia caused by low-grade inflammation or neuroinflammation triggered by "Western" lifestyles rather than pathogenic infection is maladaptive (Rantala et al. 2018).

Sleep Problems

Most individuals suffering from mood disorders have sleeping problems, which often precede MDD and bipolar disorder. Depressed people may experience an increased need to sleep, problems falling asleep, waking up at night, or waking up too early. An increased need to sleep may help to conserve energy for the immune system, as occurs in sickness behavior (Rantala et al. 2018). However, an individual suffering from chronic stress may also have a lighter sleep, waking up to the slightest sound. Chronic stress may cause sleep problems because people are in their most vulnerable and defenseless state during sleep. While in modern societies sources of sound during the night are mostly harmless, they could have indicated real danger during human evolutionary history by indicating the presence of a predator or enemies (Rantala et al. 2018). Sleeping problems might also be byproducts of excessive rumination (Rantala and Luoto 2022). In the case of SAD, sleeping problems are caused by malfunction of the internal clock, without any adaptive function (Rantala et al. 2018).

Fatigue

Fatigue is a symptom of sickness behavior that may also help the organism to save energy for the use of the immune system to fight against infection (Rantala et al. 2018). It can also be a byproduct of sleeping problems without any adaptive benefits for the individual. However, fatigue can also be adaptive under certain contexts, such as in depression induced by hierarchy conflict, since it helps to save energy while reducing the probability of challenging individuals that have a higher rank within the hierarchy (Rantala et al. 2018).

Psychomotor Agitation or Retardation

In the case of MDD, psychomotor agitation can be interpreted as a byproduct of anxiety without any adaptive component (Rantala et al. 2018). Instead, psychomotor retardation (slowing down of thoughts and movements) is one symptom of sickness behavior that helps save energy for the immune system. It can be caused by acute infection, or it can be a pathological consequence of low-grade inflammation and neuroinflammation caused or exacerbated by "Western" lifestyles. In some individuals, psychomotor retardation can also be a byproduct of an increased allocation of cognitive capacity to rumination. Psychomotor retardation can also be a side effect of sleeping problems without any adaptive function (Rantala et al. 2018).

Pessimism
Pessimism is one of the most common symptoms experienced by depressed individuals. It has been shown that people are normally optimistic about their abilities and skills while having a positive view of the future; however, depression usually dissolves this optimism bias. Thus, it could be possible that under situations in which failure is more likely than success, pessimism can be seen as adaptive, especially in contexts in which previous failures predict future ones. However, neuroinflammation may make a person more pessimistic than what would be adaptive for the individual (Rantala et al. 2018).

Self-Accusations and Guilt
Thoughts of guilt and self-accusations about negative or traumatic events may trigger rumination in some individuals. In general, the greater the person's involvement in negative events, the greater the feelings of guilt (Keller and Nesse 2006). Thus, self-accusations and guilt can show the individual how their own behavior led to the negative outcome, helping to avoid similar behaviors and situations in the future (Rantala et al. 2018).

Loss of Self-Confidence
Since high self-confidence regulates a person's status in the social hierarchy, promoting positive outcomes (Borráz-León et al. 2018; Rantala et al. 2018), a decreased self-esteem resulting from, for example, a defeat in hierarchy conflict may prevent a person from challenging individuals who have a higher-rank position, which protects the individual from new problems (Rantala et al. 2018). However, neuroinflammation may reduce self-esteem to a maladaptive level.

Gut Microbiota and Mood Disorders
In addition to genetic differences, one plausible explanation for why chronic stress can cause MDD in some persons while causing bipolar disorder or other psychiatric disorders such as eating disorders or schizophrenia in other individuals may be related to between-individual differences in gut microbiota (Rantala et al. 2021, 2022). Dysbiosis, a state of imbalanced microbiota characterized by a decrease in microbial diversity and an increase in pro-inflammatory species, causes inflammation in the gut (Lobionda et al. 2019), which may be associated with increases in the production of many pro-inflammatory cytokines

that are able to cross the blood–brain barrier, causing neuroinflammation. There is also evidence linking gut microbiota to stress sensitivity, which provides a mechanism for how gut microbiota influences the risk of developing mood disorders (Rantala et al. 2021; Rantala and Borráz-León 2024).

Although gut dysbiosis seems to play a role in the pathophysiology of mood disorders and other mental health problems, it is not a counterargument against evolutionary explanations for them. It is well known that "Western" lifestyles cause a decrease in microbiota diversity (e.g., Barone et al. 2019). Thus, the role of the gut microbiome in the pathophysiology of mood disorders supports the environmental mismatch hypothesis of mood disorders (Rantala et al. 2018, 2021; Gondalia et al. 2019).

Evolution-Based Treatment for MDD

In the light of this evolutionary psychoneuroimmunological approach, the treatment of MDD should be made more personalized by identifying which subtype of depression is experienced, facilitating the development of a better treatment (Rantala et al. 2018). For example, if a depression episode appears to be a response to an adverse life event, it should be evaluated whether the symptoms are adaptive for the patient or whether the depressive episode has exacerbated into a harmful pathological depression due to neuroinflammation (Rantala and Luoto 2022; Rantala and Borráz-León 2024). For this purpose, it should also be a standard routine in psychiatry and clinical psychology to conduct a blood test that can reveal systemic low-grade inflammation. In addition, measurements of cortisol levels could help to identify the subtype of MDD experienced. Together, these measurements would *help* to assess whether the depressive episode has maladaptive features of sickness behavior, which would help to track the treatment effectiveness (Rantala et al. 2018).

For example, if a symptom is adaptive (e.g., rumination, pessimism), the most effective intervention might be to help the depressed person to solve the adaptive problem that triggered the depressive episode. However, if the depressive episode has exacerbated into a maladaptive level, including symptoms of sickness behavior, chronic stress, and neuroinflammation, then the most effective treatment for it would be to reduce neuroinflammation, stress, and stress sensitivity (Rantala et al. 2018). Since neuroinflammation plays an important role in MDD, the treatments should focus on lowering neuroinflammation and peripheral low-grade inflammation, not just alleviating the symptoms pharmacologically. Low-grade inflammation and stress are known to be reduced effectively by lifestyle interventions including adopting a healthy diet, engaging in regular exercise, mindfulness, yoga, increasing contact with nature, and avoiding alcohol consumption and smoking (discussed in Rantala et al. 2021). In addition to lifestyle changes, psychotherapy and cognitive behavioral therapy may be used to decrease chronic stress and inflammation (see Rantala et al. 2021).

Key Points to Remember

- Mood disorders seem to be the result of an evolutionary mismatch.
- "Western" lifestyles cause systemic low-grade inflammation and chronic stress, leading to neuroinflammation.
- Neuroinflammation triggers sickness behavior, which may change previously adaptive mood swings to a maladaptive state of MDD.
- The treatment of mood disorders should target the underlying root reasons rather than just alleviating symptoms with drugs.

References

Afridi, R., & Suk, K. (2021). Neuroinflammatory basis of depression: Learning from experimental models. *Frontiers in Cellular Neuroscience, 15*, Article 691067. https://doi.org/10.3389/fncel.2021.691067

American Psychiatric Association. (2013). *Diagnostic and statistical manual of mental disorders: DSM-5* (5th ed.). American Psychiatric Publishing.

Anders, S., Tanaka, M., & Kinney, D. K. (2013). Depression as an evolutionary strategy for defense against infection. *Brain, Behavior, and Immunity, 31*, 9–22. https://doi.org/10.1016/j.bbi.2012.12.002

Andrews, P. W., & Thomson, J. A., Jr. (2009). The bright side of being blue: Depression as an adaptation for analyzing complex problems. *Psychological Review, 116*(3), 620–654. https://doi.org/10.1037/a0016242

Barone, M., Turroni, S., Rampelli, S., Soverini, M., D'Amico, F., Biagi, E., Brigidi, P., Troiani, E., & Candela, M. (2019). Gut microbiome response to a modern Paleolithic diet in a Western lifestyle context. *PLOS ONE, 14*(8), Article e0220619. https://doi.org/10.1371/journal.pone.0220619

Barron, A. B., Søvik, E., & Cornish, J. L. (2010). The roles of dopamine and related compounds in reward-seeking behavior across animal phyla. *Frontiers in Behavioral Neuroscience, 4*, Article 163. https://doi.org/10.3389/fnbeh.2010.00163

Berk, M., Williams, L. J., Jacka, F. N., O'Neil, A., Pasco, J. A., Moylan, S., Allen, N. B., Stuart, A. L., Hayley, A. C., Byrne, M. L., & Maes, M. (2013). So depression is an inflammatory disease, but where does the inflammation come from? *BMC Medicine, 11*, Article 200. https://doi.org/10.1186/1741-7015-11-200

Borráz-León, J. I., Cerda-Molina, A. L., Rantala, M. J., & Mayagoitia-Novales, L. (2018). Choosing fighting competitors among men: Testosterone, personality, and motivations. *Evolutionary Psychology, 16*(1), Article 1474704918757243. https://doi.org/10.1177/1474704918757243

Boyce, P., & Parker, G. (1988). Seasonal affective disorder in the southern hemisphere. *American Journal of Psychiatry, 145*(1), 96–99. https://doi.org/10.1176/ajp.145.1.96

Brummelte, S., & Galea, L. A. M. (2016). Postpartum depression: Etiology, treatment and consequences for maternal care. *Hormones and Behavior, 77*, 153–166. https://doi.org/10.1016/j.yhbeh.2015.08.008

Bulloch, A. G. M., Fiest, K. M., Williams, J. V. A., Lavorato, D. H., Berzins, S. A., Jetté, N., Pringsheim, T. M., & Patten, S. B. (2015). Depression—A common disorder across a broad spectrum of neurological conditions: A cross-sectional nationally representative survey. *General Hospital Psychiatry, 37*(6), 507–512. https://doi.org/10.1016/j.genhosppsych.2015.06.007

Capuron, L., Gumnick, J. F., Musselman, D. L., Lawson, D. H., Reemsnyder, A., Nemeroff, C. B., & Miller, A. H. (2002). Neurobehavioral effects of interferon-α in cancer patients: Phenomenology and paroxetine responsiveness of symptom dimensions. *Neuropsychopharmacology, 26*(5), 643–652. https://doi.org/10.1016/S0893-133X(01)00407-9

Chaudhary, N., & Salali, G. D. (2022). Hunter–gatherers, mismatch and mental disorders. In R. Abed & P. St. John-Smith (Eds.), *Evolutionary psychiatry: Current perspectives on evolution and mental health* (pp. 64–83). Cambridge University Press. https://doi.org/10.1017/9781009030564.007

Colla, J., Buka, S., Harrington, D., & Murphy, J. M. (2006). Depression and modernization: A cross-cultural study of women. *Social Psychiatry and Psychiatric Epidemiology, 41*(4), 271–279. https://doi.org/10.1007/s00127-006-0032-8

Dai, N., Jones, B. D. M., & Husain, M. I. (2022). Astrocytes in the neuropathology of bipolar disorder: Review of current evidence. *Brain Sciences, 12*(11), Article 1513. https://doi.org/10.3390/brainsci12111513

Dantzer, R. (2001). Cytokine-induced sickness behavior: Mechanisms and implications. *Annals of the New York Academy of Sciences, 933*, 222–234. https://doi.org/10.1111/j.1749-6632.2001.tb05827.x

Darwin, C. (1871). *The descent of man*. D. Appleton.

Davis, C., & Levitan, R. D. (2005). Seasonality and seasonal affective disorders (SAD): An evolutionary viewpoint tied to energy conservation and reproductive cycles. *Journal of Affective Disorders, 87*(1), 3–10. https://doi.org/10.1016/j.jad.2005.03.006

de Cantanzaro, D. (1986). A mathematical model of evolutionary pressures regulating self-preservation and self-destruction. *Suicide and Life-Threatening Behavior, 16*(2), 166–181. https://doi.org/10.1111/j.1943-278X.1986.tb00350.x

DiSabato, D. J., Quan, N., & Godbout, J. P. (2016). Neuroinflammation: The devil is in the details. *Journal of Neurochemistry, 139*(S2), 136–153. https://doi.org/10.1111/jnc.13607

Egeland, J. A., & Hostetter, A. M. (1983). Amish study, I: Affective disorders among the Amish, 1976–1980. *American Journal of Psychiatry, 140*(1), 56–61. https://doi.org/10.1176/ajp.140.1.56

Enache, D., Pariante, C. M., & Mondelli, V. (2019). Markers of central inflammation in major depressive disorder: A systematic review and meta-analysis of studies examining cerebrospinal fluid, positron emission tomography and post-mortem brain tissue. *Brain, Behavior, and Immunity, 81*, 24–40. https://doi.org/10.1016/j.bbi.2019.06.015

Exner, C., Hebebrand, J., Remschmidt, H., Wewetzer, C., Ziegler, A., Herpertz, S., Shcweiger, U., Blum, W. F., Preibisch, G., Heldmaier, G., & Klingenspor, M. (2000). Leptin suppresses semi-starvation induced hyperactivity in rats: Implications for anorexia nervosa. *Molecular Psychiatry, 5*(5), 476–481. https://doi.org/10.1038/sj.mp.4000771

Fagundes, C. P., Brown, R. L., Chen, M. A., Murdock, K. W., Saucedo, L., LeRoy, A., Wu, E. L., Garcini, L. M., Shahane, A. D., Baameur, F., & Heijnen, C. (2019). Grief, depressive symptoms, and inflammation in the spousally bereaved. *Psychoneuroendocrinology, 100*, 190–197. https://doi.org/10.1016/j.psyneuen.2018.10.006

Fedurek, P., Lehmann, J., Lacroix, L., Aktipis, A., Cronk, L., Makambi, E. J., Mabulla, I., & Berbesque, J. C. (2023). Status does not predict stress among Hadza hunter–gatherer men. *Scientific Reports, 13*(1), Article 1327. https://doi.org/10.1038/s41598-023-28119-9

Frackowiak, T., Oleszkiewicz, A., Butovskaya, M., Groyecka, A., Karwowski, M., Kowal, M., & Sorokowski, P. (2020). Subjective happiness among Polish and Hadza people. *Frontiers in Psychology, 11*, Article 1173. https://doi.org/10.3389/fpsyg.2020.01173

Frasure-Smith, N., & Lespérance, F. (2006). Depression and coronary artery disease. *Herz, 31*(S3), 64–68.

Gilbert, P. (2006). Evolution and depression: Issues and implications. *Psychological Medicine, 36*(3), 287–297. https://doi.org/10.1017/S0033291705006112

Gondalia, S., Parkinson, L., Stough, C., & Scholey, A. (2019). Gut microbiota and bipolar disorder: A review of mechanisms and potential targets for adjunctive therapy. *Psychopharmacology, 236*(5), 1433–1443. https://doi.org/10.1007/s00213-019-05248-6

Haapakoski, R., Mathieu, J., Ebmeier, K. P., Alenius, H., & Kivimäki, M. (2015). Cumulative meta-analysis of interleukins 6 and 1β, tumor necrosis factor α and C-reactive protein in patients with major depressive disorder. *Brain, Behavior, and Immunity, 49*, 206–215. https://doi.org/10.1016/j.bbi.2015.06.001

Hagen, E. H. (1999). The functions of postpartum depression. *Evolution and Human Behavior, 20*(5), 325–359. https://doi.org/10.1016/S1090-5138(99)00016-1

Harrison, N. A., Brydon, L., Walker, C., Gray, M. A., Steptoe, A., & Critchley, H. D. (2009). Inflammation causes mood changes through alterations in subgenual cingulate activity and mesolimbic connectivity. *Biological Psychiatry, 66*(5), 407–414. https://doi.org/10.1016/j.biopsych.2009.03.015

Herlosky, K. N., Benyshek, D. C., Mabulla, I. A., Pollom, T. R., & Crittenden, A. N. (2020). Postpartum maternal mood among Hadza Foragers of Tanzania: A mixed methods. *Approach. Culture Medicine and Psychiatry, 44*, 305–332. https://doi.org/10.1007/s11013-019-09655-4

Kaplan, H., Thompson, R. C., Trumble, B. C., Wann, L. S., Allam, A. H., Beheim, B., Frohlich, B., Sutherland, M. L., Sutherland, J. D., Stieglitz, J., Rodriguez, D. E., Michalik, D. E., Rowan, C. J., Lombardi, G. P., Bedi, R., Garcia, A. R., Min, J. K., Narula, J., Finch, C. E., . . . Thomas, G. S. (2017). Coronary atherosclerosis in Indigenous South American Tsimane: A cross-sectional cohort study. *Lancet, 389*(10080), 1730–1739. https://doi.org/10.1016/S0140-6736(17)30752-3

Keller, M. C., & Nesse, R. M. (2006). The evolutionary significance of depressive symptoms: Different adverse situations lead to different depressive symptoms patterns. *Journal of Personality and Social Psychology, 91*(2), 316–330. https://doi.org/10.1037/0022-3514.91.2.316

Kim, Y. K., & Won, E. (2017). The influence of stress on neuroinflammation and alterations in brain structure and function in major depressive disorder. *Behavioural Brain Research, 329*, 6–11. https://doi.org/10.1016/j.bbr.2017.04.020

Köhler-Forsberg, O., Lydholm, C. N., Hjorthøj, C., Nordentoft, M., Mors, O., & Benros, M. E. (2019). Efficacy of anti-inflammatory treatment on major depressive disorder or depressive symptoms: Meta-analysis of clinical trials. *Acta Psychiatrica Scandinavica, 139*(5), 404–419. https://doi.org/10.1111/acps.13016

Lee, S., Tsamg, A., Zhang, M. Y., Huang, Y. Q., He, Y. L., Liu, Z. R., Shen, Y. C., & Kessler, R. C. (2007). Lifetime prevalence and inter-cohort variation in DSM-IV disorders in metropolitan China. *Psychological Medicine, 37*(1), 61–71. https://doi.org/10.1017/S0033291706008993

Lim, G. Y., Tam, W. W., Lu, Y., Ho, C. S., Zhang, M. W., & Ho, R. C. (2018). Prevalence of depression in the community from 30 countries between 1994 and 2014. *Scientific Reports, 8*(1), Article 2861. https://doi.org/10.1038/s41598-018-21243-x

Lindeberg, S. (2010). *Food and Western disease: Health and nutrition from an evolutionary perspective.* Wiley-Blackwell. https://doi.org/10.1002/9781444317176

Lobionda, S., Sittipo, P., Kwon, H. Y., & Lee, Y. K. (2019). The role of gut microbiota in intestinal inflammation with respect to diet and extrinsic stressors. *Microorganisms, 7*(8), Article 271. https://doi.org/10.3390/microorganisms7080271

Merikangas, K. R., Akiskal, H. S., Angst, J., Greenberg, P. E., Hirschfeld, R. M. A., Petukhova, M., & Kessler, R. C. (2007). Lifetime and 12-month prevalence of bipolar spectrum disorder in the National Comorbidity Survey replication. *Archives of General Psychiatry, 64*(5), 543–552. https://doi.org/10.1001/archpsyc.64.5.543

Miller, A. H., & Raison, C. L. (2016). The role of inflammation in depression: From evolutionary imperative to modern treatment target. *Nature Reviews Immunology, 16*(1), 22–34. https://doi.org/10.1038/nri.2015.5

Miller, G. (2009). *Spent: Sex, evolution, and consumer behavior.* Viking.

Osimo, E. F., Baxter, L. J., Lewis, G., Jones, P. B., & Khandaker, G. M. (2019). Prevalence of low-grade inflammation in depression: A systematic review and meta-analysis of CRP levels. *Psychological Medicine, 49*(12), 1958–1970. https://doi.org/10.1017/S0033291719001454

Price, J., Sloman, L., Gardner, R., Jr., Gilbert, P., & Rohde, P. (1994). The social competition hypothesis of depression. *British Journal of Psychiatry, 164*(3), 309–315. https://doi.org/10.1192/bjp.164.3.309

Raison, C. L., Demetrashvili, M., Capuron, L., & Miller, A. H. (2005). Neuropsychiatric adverse effects of interferon-alpha: Recognition and management. *CNS Drugs, 19*(2), 105–123. https://doi.org/10.2165/00023210-200519020-00002

Raison, C. L., & Miller, A. H. (2012). The evolutionary significance of depression in pathogen host defense (PATHOS-D). *Molecular Psychiatry, 18*(1), 15–37. https://doi.org/10.1038/mp.2012.2

Rantala, M. J., & Borráz-León, J. I. (2024). The environmental mismatch model of bipolar disorder: The role of stress, gut microbiota, lifestyle factors, and neuroinflammation. In G. Fink (Ed.), *Handbook of stress series: Vol. 5. Stress, immunology, and inflammation* (pp. 215–222). Elsevier.

Rantala, M. J., & Luoto, S. (2022). Evolutionary perspectives on depression. In R. Abed & P. St. John-Smit (Eds.), *Evolutionary psychiatry: Current perspectives on evolution and mental health* (pp. 117–133). Cambridge University Press. https://doi.org/10.1017/9781009030564.010

Rantala, M. J., Luoto, S., Borráz-León, J. I., & Krams, I. (2021). Bipolar disorder: An evolutionary psychoneuroimmunological approach. *Neuroscience and Biobehavioral Reviews, 122*, 28–37. https://doi.org/10.1016/j.neubiorev.2020.12.031

Rantala, M. J., Luoto, S., Borráz-León, J. I., & Krams, I. (2022). The environmental mismatch model of bipolar disorder is supported by evidence: A response to Partonen et al. *Neuroscience and Biobehavioral Reviews, 136*, Article 104631. https://doi.org/10.1016/j.neubiorev.2022.104631

Rantala, M. J., Luoto, S., Borráz-León, J. I., & Krams, I. (2022). Schizophrenia: The new etiological synthesis. *Neuroscience and Biobehavioral Reviews, 142*, Article 104894. https://doi.org/10.1016/j.neubiorev.2022.104894

Rantala, M. J., Luoto, S., Krams, I., & Karlsson, H. (2018). Depression subtyping based on evolutionary psychiatry: Proximate mechanisms and ultimate functions. *Brain, Behavior, and Immunity, 69*, 603–617. https://doi.org/10.1016/j.bbi.2017.10.012

Rantala, M. J., Luoto, S., Krama, T., & Krams, I. (2019). Eating disorders: An evolutionary psychoneuroimmunological approach. *Frontiers in Psychology, 10*, Article 2200. https://doi.org/10.3389/fpsyg.2019.02200

Reutfors, J., Osby, U., Ekbom, A., Nordstrom, P., Jokinen, J., & Papadopoulos, F. C. (2009). Seasonality of suicide in Sweden: Relationship with psychiatric disorder. *Journal of Affective Disorders, 199*(1–3), 59–65. https://doi.org/10.1016/j.jad.2009.02.020

Rohan, K. J., Sigmon, S. T., & Dorhofer, D. M. (2003). Cognitive-behavioral factors in seasonal affective disorder. *Journal of Consulting and Clinical Psychology, 71*(1), 22–30. https://doi.org/10.1037/0022-006X.71.1.22

Rohde, P. (2001). The relevance of hierarchies, territories, defeat for depression in humans: Hypotheses and clinical predictions. *Journal of Affective Disorders, 65*(3), 221–230. https://doi.org/10.1016/S0165-0327(00)00219-6

Rosenthal, N. E., Sack, D. A., Gillin, J. C., Lewy, A. J., Goodwin, F. K., Davenport, Y., Mueller, P. S., Newsome, D. A., & Wehr, T. A. (1984). Seasonal affective disorder: A description of the syndrome and preliminary findings with light therapy. *Archives of General Psychiatry, 41*(1), 72–80. https://doi.org/10.1001/archpsyc.1984.01790120076010

Sandiego, C. M., Gallezot, J. D., Pittman, B., Nabulsi, N., Lim, K., Lin, S. F., Matuskey, D., Lee, J. Y., O'Connor, K. C., Huang, Y., Carson, R. E., Hannestad, J., & Cosgrove, K. P. (2015). Imaging robust microglial activation after lipopolysaccharide administration in humans with PET. *Proceedings of the National Academy of Sciences of the United States of America, 112*(40), 12468–12473. https://doi.org/10.1073/pnas.1511003112

Shephard, R. J., & Rode, A. (1996). *The health consequences of "modernization": Evidence from circumpolar peoples*. Cambridge University Press.

Shoelson, S. E., Herrero, L., & Naaz, A. (2007). Obesity, inflammation, and insulin resistance. *Gastroenterology, 132*(6), 2169–2180. https://doi.org/10.1053/j.gastro.2007.03.059

Stevens, A., & Price, J. (2000). *Evolutionary psychiatry: A new beginning*. Routledge. https://doi.org/10.4324/9781315740577

Stieglitz, J., Trumble, B. C., Thompson, M. E., Blackwell, A. D., Kaplan, H., & Gurven, M. (2015). Depression as sickness behavior? A test of the host defense hypothesis in a high pathogen population. *Brain, Behavior, and Immunity, 49*, 130–139. https://doi.org/10.1016/j.bbi.2015.05.008

Troubat, R., Barone, P., Leman, S., Desmidt, T., Cressant, A., Atanasova, B., Brizard, B., El Hage, W., Surget, A., Belzung, C., & Camus, V. (2021). Neuroinflammation and depression: A review. *European Journal of Neuroscience, 53*(1), 151–171. https://doi.org/10.1111/ejn.14720

van Duijn, E., Fernandes, A. R., Abreu, D., Ware, J. J., Neacy, E., & Sampaio, C. (2021). Incidence of completed suicide and suicide attempts in a global prospective study of Huntington's disease. *British Journal Open, 7*(5), Article e158. https://doi.org/10.1192/bjo.2021.969

Vasunilashorn, S., Crimmins, E. M., Kim, J. K., Winking, J., Gurven, M., Kaplan, H., & Finch, C. E. (2010). Blood lipids, infection, and inflammatory markers in the Tsimane of Bolivia. *American Journal of Human Biology, 22*(6), 731–740. https://doi.org/10.1002/ajhb.21074

World Health Organization. (2017). *Depression and other common mental disorders: Global health estimates*. https://www.who.int/publications/i/item/depression-global-health-estimates

Yost, B., Thompson, S., Miller, K., & Abbott, C. (2016). *Physical and mental health conditions in five Plain communities*. Young Center International Conference. https://www.fandm.edu/uploads/files/953350868745554699-physical-and-mental-health-conditions-in-five-plain-communities.pdf

14

Schizophrenia

Embracing the Spectrum

John S. Allen

Schizophrenia—A Human Condition

Schizophrenia is an illness that takes a heavy toll on those who suffer from it, their families and caregivers, and society as a whole. While nearly all forms of mental illness confer some level of social stigma, schizophrenia may be the most stigmatized psychiatric condition of all (Thornicroft et al. 2022). A diagnosis of schizophrenia is akin to a diagnosis of "otherness." This chapter is about the evolution of schizophrenia, and it may seem odd to start it by talking about stigma. However, all evolutionary approaches to schizophrenia share a perspective that places the condition in the context of the wider spectrum of human behavior, in which schizophrenia can be identified as a discrete entity within a continuous range of variation. While this perspective on its own cannot remove the stigma attached to schizophrenia, it makes clear that the "otherness" of schizophrenia is arbitrary and artificial.

Schizophrenia has complex and varied clinical, cognitive, anatomical, and biochemical presentations (Jauhar et al. 2022). The most common symptoms for schizophrenia cluster into three reasonably discrete groups: reality distortion (delusions and hallucinations), disorganization (thought disorder and disorganized behavior), and negative symptoms (e.g., flat affect or poverty of speech; Liddle 1987). Clinical representations of these symptom groups include, for example, persecutory or grandiose delusions, auditory delusions that are perceived to have a source other than the individual's own thoughts, incoherent speech or "word salad," or grossly disorganized behaviors, such as childlike "silliness" or extreme agitation (American Psychiatric Association 2013).

Catatonia has historically been considered a subtype of schizophrenia but is now seen to be a more generalized signifier of schizophrenia and other psychiatric and neurological conditions (Tandon et al. 2013).

In addition to clinical symptoms, schizophrenia is associated with a range of cognitive or intellectual impairments (in working memory, attention, executive functions, language, and several other areas) that are independent of the course of psychiatric illness (Reichenberg 2010). Birth cohort studies show that people (or children) who will develop schizophrenia have a lifelong 7–8-point deficit in IQ test score performance (equivalent to half a standard deviation) compared to those who will not develop schizophrenia (Welham et al. 2009). Although the diagnosis of schizophrenia separates people with the condition from those not diagnosed, it is clear from studies of unaffected siblings of patients that they are at higher risk for some schizophrenia-associated cognitive impairments, including those related to social cognition, than others in the general population (Egan et al. 2001; Andric et al. 2016). This is consistent with the endophenotype or intermediate phenotype perspective in psychiatry, which sees schizophrenia and other diseases as genetically complex conditions expressed in the context of a phenotypic spectrum (Meyer-Lindenberg and Weinberger 2006).

It is often stated that schizophrenia affects about 1% of the population. This figure reflects an average global incidence rate, but incidence and prevalence rates for the condition can vary among populations (McGrath et al. 2008; Jauhar et al. 2022). Schizophrenia is more common in men than woman (incidence ratio of about 1.7), with men having a distinct age-at-onset peak in their early twenties. Women also have a peak in their twenties; but it is much less pronounced, and the incidence rate is steadier across the life span. Schizophrenia is a long-term condition, and only one in seven of those diagnosed with it have an adaptive clinical outcome (Jääskeläinen et al. 2013). Prognosis does not appear to vary systematically across populations, although there have been claims that prognosis is worse in developed, industrialized societies compared to others (Cohen et al. 2008; but see Allen 1997). The economic impact of schizophrenia on society is significant; in 2019, the total direct and indirect costs of schizophrenia in the United States were estimated to be $343.2 billion (Kadakia et al. 2022).

Schizophrenia is a brain disease. At the neurochemical level, a primary role for the dopamine system in the genesis of schizophrenia remains the strongest hypothesis; a role for the glutamate system has also been suggested—like the dopamine hypothesis, based primarily on pharmacological insights—but the evidence for it is less strong (Jauhar et al. 2022). Structural neuroimaging studies have shown modest but consistent differences in brain volumes between schizophrenia patients and healthy controls. Ventricle size is increased, while total brain volume is decreased, in schizophrenia patients, in both the white and gray volumes; gray matter (cortical and subcortical) showed greater declines than white matter; gray matter differences were present in medication-naive, early-stage patients but became more pronounced as the illness progressed (Haijma et al. 2013). Functional neuroimaging studies have shown that frontal activation is abnormal in schizophrenia, although depending on task, this can result in hypo- or hyperactivity in frontal areas (Hill et al. 2004; Minzenberg et al. 2009).

Some of the most exciting developments in schizophrenia brain imaging are coming from studies of the connectome—the network of white matter pathways that functionally connect various regions of the brain. For more than a century, researchers have speculated that psychosis and disorganized thinking and behavior might be the result of impaired connectivity among brain regions; but until the development of newer imaging technologies in the early 21st century, especially diffusion tensor imaging (DTI), the ability to identify impaired networks has been limited (Collin et al. 2016). With DTI, researchers have found that white matter tract alterations are widespread in patients with chronic schizophrenia and in first-episode and drug-naive patients. These regions include the cingulum bundles, uncinate fasciculi, internal capsules, and corpus callosum; and there is some variation in affected tracts depending on primary symptoms (Wheeler and Voinekos 2014). Unimpaired first-degree relatives of patients also show tract alterations, often at an intermediate level between patients and controls (Wheeler and Voinekos 2014; Collin et al. 2014).

As will be discussed below, recent genomic research strongly suggests that schizophrenia is an emergent phenotype resulting from the myriad genetic changes that underlie the evolution of human brain and behavior over the past 6–7 million years. Van den Heuvel and colleagues (2019) have provided anatomical evidence for this emergence by comparing the connectomes of humans and our closest relatives, chimpanzees. They found that while the vast majority (94%) of white matter tract connections are shared by the two species, 27 connections are unique to humans. Adding a schizophrenia data set to this comparison, they then discovered that connections that were unique to humans were significantly more likely to be disrupted in schizophrenia compared to connections shared by humans and chimpanzees. Van den Heuvel et al. concluded (2019, p. 4000), "Our cross-species connectome comparison suggests that human specializations in brain connectivity may potentially be enriched for domains affected in schizophrenia . . . the evolution of the human connectome in service of developing more complex brain function may be paralleled with higher risk for brain dysfunction" (Figure 14.1).

This thumbnail sketch of schizophrenia hints at the complexities that underlie the evolution of this particular phenotype or spectrum of phenotypes. As I will discuss below, the genetics of schizophrenia are undoubtedly more complex than anyone would have thought 30 years ago. In this sense, schizophrenia is similar to any of the other innumerable, polygenic, complex phenotypes, morphological or behavioral, that have evolved in the animal world (Barton 2022).

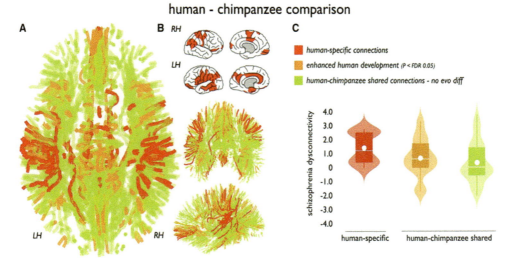

Figure 14.1 Cortico-cortical connections of the human cerebral connectome. (A) Red connections identify human-specific connections, orange connections show human–chimpanzee shared connections but with human enhancement, and green connections indicate human–chimpanzee shared connections without any species difference. (B) Subset of cortical regions that displayed human-specific cortico-cortical connections. (C) Schizophrenia effects across human-specific (red) and connections shared between humans and chimpanzees. FDR = false discovery rate; LH = left hemisphere; no evo diff = no evolutionary difference; RH = right hemisphere. Modified from van den Heuvel et al. (2019) with permission.

Historical Background

As anthropology and ethnography were developing as academic disciplines in the 19th and early 20th centuries, there was much discussion about the apparent absence of insanity or other forms of mental illness in so-called primitive societies, at least relative to what was seen in "civilized" populations (Ackerknecht 1943; Allen 1997). This observation was slotted into the dominant cultural evolutionary paradigm of the time, in which complex, civilized populations were seen to have stressors that were absent from the lives led by "primitive" peoples. Often accompanying this cultural evolutionary perspective was an implicit biological/racialist viewpoint that saw madness as a "cost" of the evolution of advanced civilization: "Opinion is widespread among students of society that the conditions of modern life are conducive to the development of mental disorders... that nervous diseases are an accompaniment of complex civilization" (Winston 1934, p. 234). In a seminal paper published in 1934, sociologist Ellen Winston, influenced by the critique of racialist ethnography led by Franz Boas and his students, directly addressed the issue of "the alleged lack of mental diseases among primitive groups." In addition to collating descriptive accounts from various ethnographers actually describing individual cases of mental illness in various traditional cultures, Winston used data that Margaret Mead had collected during her fieldwork in Samoa to estimate age-structured rates of mental illness in that society. She then compared the Samoan rates to those she determined for the American population as a whole. This unusual-for-its-time quantitative analysis showed that rates of mental illness in the United States and Samoa were broadly similar, leading Winston (1934, p. 238) to conclude, "The data are admittedly slight. Nevertheless, they are sufficient to raise the issue as to whether the simpler peoples are actually free from mental disorders and whether there is an excessive rate of mental disease in complex cultures."

Although Winston's work said nothing substantive about the relative rates of schizophrenia among different populations, it did signal a departure from the idea that it was a unique sign of civilized life. Still, there was a pervasive view that schizophrenia was rarer in small-scale, traditional societies than in large-scale, developed, industrial societies (Torrey 1980). The 1930s and 1940s saw the beginnings of social labeling theory, which in the 1960s and 1970s exploded into a full-on critique of all psychiatric diagnoses (Waxler 1974). The idea that in traditional societies people exhibiting schizophrenia-like behaviors could find meaningful roles as shamans or other kinds of ritual leaders, and thereby avoid being labeled deviant, was an early (and persistently popular) example of labeling theory being applied.

Throughout the second half of the 20th century, the schizophrenia–shaman connection was raised and debunked in turn by a series of investigators (see Allen 1997). Experiential analyses of the highly prescribed roles of shamans and similar religious leaders in traditional societies make clear that people with schizophrenia-like tendencies would probably have difficulty filling them (Steadman and Palmer 1994). Outside observers have not always been clear on what they were seeing when watching ritualistic behavior. As an Inuit individual from Baffin Island pointed out, "When the shaman is healing, he is out of his mind, *but he is not crazy*" (in Murphy 1976, p. 1022, emphasis in original).

The shaman hypothesis was not directly tied by its proponents to an evolutionary view of schizophrenia (but see Polimeni 2012). However, in positing a potential compensatory role for people with a tendency toward developing schizophrenia, it anticipated later evolutionary models. In addition, the shaman hypothesis carries an implicit spectrum perspective of schizophrenia since it does not rule out that there would be individuals too impaired to carry out the role, even while others, less impaired, might be able to.

In the late 1950s and 1960s, the introduction of reasonably effective neuroleptic drugs and the results of pedigree studies, especially those derived from comprehensive birth registries in Scandinavia, convinced most scientists and clinicians that schizophrenia was a biological brain disease with high heritability (ironically, at the same time the anti-psychiatry movement—"schizophrenia is a sane reaction to an insane world"—was at its peak). The first small batch of papers on the evolution of schizophrenia soon followed.

In 1964, Julian Huxley, Ernst Mayr, Humphry Osmond, and Abram Hoffer (an extraordinary author list) published a paper considering "schizophrenia as a genetic morphism." Using the logic that would inform all subsequent efforts before the modern genetic era to explain the evolution of schizophrenia, they argued that if schizophrenia is the product of a single major allele with incomplete penetrance, it is too common (approximately 1% prevalence) to be maintained by mutation rates alone, and therefore positive selection acting on non-affected carriers of the gene ("crypto-schizophrenic carriers") must be present to compensate for the reduced fitness seen in individuals with overt schizophrenia. They suggested that physiological advantages in the carriers allowed the schizophrenia genotype to persist. These advantages might include resistance to wound shock; exposure to dangerously high levels of hormones, such as insulin; and resistance to infectious disease. L. Erlenmeyer-Kimling initially pointed out that there were no data to support a psychosocial or physiological advantage for the relatives (Erlenmeyer-Kimling and Paradowski 1966), but in attempting to test the Huxley et al. hypothesis, she examined a longitudinal demographic data set of schizophrenia patients and found that the offspring of parents with schizophrenia had lower mortality in the first year of life compared to non-affected controls (Erlenmeyer-Kimling 1968). In a study focused on two general practices, Carter and Watts (1971) found that the relatives of patients with schizophrenia seemed to have lower rates of viral infections, as well as accidents, compared to matched controls, again lending some preliminary weight to the physiological advantage hypothesis.

As it has turned out, these preliminary hints of a physiological advantage for the relatives of patients with schizophrenia did not lead to definitive studies showing any kind of compensatory physiological advantage. In fact, subsequent hypotheses offered to explain the evolution of schizophrenia genotypes have focused on behavioral, psychological, and cognitive factors that might confer an advantage. Nonetheless, these early studies laid the groundwork for considering schizophrenia in an evolutionary and demographic context.

Evolutionary Evidence Specifically Related to Human Health and Disease

There are several prerequisites that should be fulfilled before considering schizophrenia in an evolutionary perspective. Huxley et al. (1964) identified two of the primary ones: heritability and reduced fertility. Decades of research have demonstrated that genetics plays an essential role in the etiology of schizophrenia. In a definitive study, using Swedish hospital discharge records for over 9 million individuals in over 2 million nuclear families (biological and adoptive), Lichtenstein et al. (2009) calculated a heritability for schizophrenia of 64%, while environmental effects were calculated at 4.5%. This heritability value was a bit lower than the 80% determined in a meta-analysis of twin studies (Sullivan et al. 2003), but Lichtenstein et al. (2009) caution that twin studies lack the statistical power of their large pedigree analysis. A heritability of 70% for schizophrenia is generally accepted as a reasonable estimate in the current scientific literature.

Several studies over the years have shown reduced fertility in patients with schizophrenia, especially among men. A meta-analysis by Bundy et al. (2010) found that schizophrenia patients had a fertility ratio of 0.39 compared to non-affected controls and that the fertility ratio of men with schizophrenia versus women was 0.54 (it is interesting to note that of 1,332 studies that could potentially be included in the meta-analysis, only six were deemed acceptable). A subsequent large-scale study, based again on Swedish records (2.3 million individuals), found that men with schizophrenia had a fertility ratio of 0.23, while for women, it was 0.47 (Power et al. 2013). The lower rate among men is generally linked to the earlier age at onset, which peaks around the time when young people are forming reproductive relationships. Neither of these studies found evidence that the fertility of unaffected parents or siblings compensated for the decreased fertility of individuals with schizophrenia.

In addition to heritability and reduced fertility, Timothy Crow (2008) has suggested three other "principles" that must necessarily hold to form the basis of an evolutionary theory of schizophrenia: universality, constitutionality, and continuity. Evidence for *universality* is clear: schizophrenia has been identified in biologically and culturally diverse populations and societies across the globe; no population of reasonable size has been found to not have individuals with schizophrenia. More contentious is the claim that all populations have essentially the same rate of schizophrenia. Crow and many others look to the landmark World Health Organization (WHO) Ten Country Study (Jablensky et al. 1992) as evidence that schizophrenia appears with approximately the same frequency in all human populations. Others have disputed this, and in a comprehensive epidemiological analysis, John McGrath and colleagues (2008) argue that there is strong evidence for considerable variation among populations in schizophrenia rates. For example, they found that the central 80% of estimates for schizophrenia incidence varied fivefold (7.7–43.0/1,000) across populations. McGrath et al. (2008) also determined that the lifetime individual risk for schizophrenia, usually quoted as being on the order of 1%, should be more accurately characterized as 7/1,000. Crow (2008) and others who support the common interpretation of the WHO Ten Country Study argue that population size needs to be quite large (100,000) before an accurate assessment of schizophrenia rates can be obtained; in addition, Crow (2008) claims that narrower diagnostic criteria focusing on more "nuclear" symptoms of schizophrenia tend to reduce variation between populations in epidemiological studies.

It is safe to say that schizophrenia is universally represented in all human populations. It is also clearly unique to humans as there has never been a report of a homologous condition among the great apes, our closest relatives, in which there is (for example) an onset of disorganized or psychotic behavior in early adulthood. The issue of uniform frequency in human populations is less settled—and there is strong evidence for varied rates—but critically important in how evolutionary models of schizophrenia are constructed. Crow's (2008) model directly linking the emergence of schizophrenia to the evolution of language depends on uniform frequency since it posits a species-wide evolutionary event. In contrast, some models, such as the frequency-dependent balanced polymorphism model proposed by Allen and Sarich (1988), recognize schizophrenia as being universal but not uniform in its distribution in human populations. A better understanding of underlying population variation for schizophrenia-risk loci (see the following section) will help clarify our understanding of schizophrenia universality and population uniformity.

Crow's second principle is *constitutionality*, meaning in this context that schizophrenia is a neurodevelopmental condition that does not depend on an exogenous or environmental factor to trigger its expression. Viral triggers for schizophrenia have been enthusiastically investigated, with no conclusive evidence that they are involved; obstetric complications have also been examined, again with no good evidence that they play a role (Crow 2008). A somewhat controversial, potentially triggering environmental factor is cannabis use, which has been linked to an increased risk for developing schizophrenia. In their comprehensive overview of schizophrenia, Sameer Jauhar and colleagues (2022, p. 480) write that the evidence for this link is "somewhere between strong and overwhelming" and that daily users of high-potency cannabis in the United Kingdom have a four-times greater risk of developing schizophrenia than non-users.

Presumably, cannabis use does not cause schizophrenia without an underlying genetic predisposition, which suggests that there is a spectrum of risk for developing schizophrenia connected to cannabis effects. The idea of a spectrum of risk and expression in relation to schizophrenia is embodied in Crow's third principle of *continuity*, or "that schizophrenia is not a disease entity distinct from other conditions, but that states described as schizophrenic merge imperceptibly into affective states and non-psychotic conditions" (Crow 2008, p. 32). Since the turn of the 21st century, the extent of the clinical spectrum in which schizophrenia is expressed has become increasingly apparent. For example, in the Lichtenstein et al. (2009) Swedish pedigree study discussed above, a high level of comorbidity was found for schizophrenia and bipolar illness, such that all classes of biological relatives of probands with bipolar illness had a substantially enhanced risk of having schizophrenia. These results suggest a substantial genetic overlap between the two conditions, with non-shared genetic factors influencing the separate expressions of the two diseases. I will look more at the emerging evolutionary landscape of psychiatric genetics in the next section.

Evolutionary modelers have long paid attention to the possible non-clinical spectrum of schizophrenia. This was especially true for advocates of a balanced polymorphism approach, who identified factors such as creativity (Karlsson 1970; Nettle and Clegg 2006), functional asociality (Allen and Sarich 1988), and religious and shamanistic expression (Polimeni and Reiss 2002), as the potential fitness-enhancing features present in relatives that could compensate for the reduced fitness associated with overt schizophrenia. Even models that characterize schizophrenia as an evolutionary byproduct, related to the evolution of language (Crow 2008) or social behavior (Burns 2004), or as an emergent phenotype in complex neurogenetic systems (Keller and Miller 2006) place schizophrenia along a clinical–non-clinical spectrum of behavior. These byproduct models do not require that there be direct fitness compensation for the patients among relatives since schizophrenia is seen as a low-frequency expression of genes and alleles essential for the human species as a whole.

It is beyond the scope of this chapter to review and assess the merits of the various models for the evolution of schizophrenia that have been proposed (see excellent reviews by Brüne 2004; van Dongen and Boomsa 2013; Nesic et al. 2019). The paradox of the persistence of schizophrenia as a genetic condition with reduced fitness among its sufferers remains to be fully explained. As I will discuss in the next section, the genomic revolution points us toward new solutions, even as the diagnostic category of "schizophrenia" is seen as less bounded than ever.

Counterarguments and Evidence against Evolutionary Influences

There is no plausible counterargument to the hypothesis that schizophrenia has an evolutionary history. As discussed above, heritability combined with universality, constitutionality, and continuity provides a firm foundation for examining the evolutionary influences that have shaped schizophrenia phenotypes. However, new developments in genomic science have fundamentally changed our perspective on what those influences might be. More recent evolutionary models of schizophrenia almost all emphasize "byproduct" or "genetic load" reasoning rather than hypothesizing a fitness advantage among close relatives of patients. A non-adaptationist perspective is sometimes seen as non-evolutionary, but of course, that is not true. For example, Crow's (2008) byproduct model embedded schizophrenia in a strong phylogenetic framework, making its emergence a discrete evolutionary event. The current genetic picture we have of schizophrenia suggests that things were not as simple or straightforward as any single-gene model might suggest. However, no matter how complex the underlying genetics, the schizophrenia phenotype in human populations has a single evolutionary history. That history may be extraordinarily complicated, reflecting the interacting effects of pleiotropy and population history among hundreds of genes; but it is still a history that theoretically can be reconstructed.

In the late 20th century, research on heredity moved beyond classical genetics and into the modern genomic era. Very quickly, via genome-wide linkage studies, a host of candidate genes were identified that appeared to be critical for developing schizophrenia (and many other illnesses); however, there were problems replicating some of these results, and even when they were statistically confirmed, they did not enhance risk assessment (Pearlson and Folley 2008; van de Leemput et al. 2016). Genome-wide association studies (GWAS) subsequently provided more methodological and statistical power to the search for schizophrenia-related alleles (and for a multitude of other diseases and complex traits; Abdellaoui et al. 2023). In a landmark GWAS study involving 36,989 schizophrenia cases and 113,075 controls, the Schizophrenia Working Group of the Psychiatric Genomics Consortium (2014) identified 108 schizophrenia-associated loci. Further research has now raised that total to over 270 (Legge et al. 2021). Some of these alleles are quite rare and may provide particular insights into molecular processes underlying the disease. Godfrey Pearlson and Bradley Folley (2008, p. 725) provide a succinct summary of our current understanding of this genetic situation:

> Schizophrenia is a disorder caused, in part, by multiple susceptibility alleles, each of small effect. The cumulative effect of evolutionary selection, mutations, and by-products that deposit liability on the population may result in observable traits, behaviors, and cognitive abilities that are distributed across and within individuals. Multiple genetic susceptibilities . . . interact with the environment . . . leading to positive and negative symptoms, cognitive distortions, or in combination, the overt clinical manifestation of the disorder.

The GWAS results make clear the extent to which psychiatric illnesses correlate genetically with one another, giving strong support to the idea that schizophrenia sits within a spectrum of clinically recognized illnesses. The overlap of susceptibility loci with bipolar illness is most pronounced, but significant associations are also found (in decreasing levels of correlation) for major depressive disorder, obsessive-compulsive disorder, anorexia nervosa, and autism spectrum disorder (ASD; Legge et al. 2021). These genetic overlaps open up multiple opportunities for understanding the underlying neuroprocesses involved in schizophrenia. For example, de novo variants associated with intellectual disability and ASD are also enriched in schizophrenia, suggesting that some expressions of schizophrenia—which has a co-morbidity with intellectual disability of 3%–5%—lie on a shared neurodevelopmental disorders continuum (Rees et al. 2021). One genetic connection between schizophrenia and ASD is possibly mediated by the DUF1220 protein domain, which is encoded by genes of variable copy number that are important in neurodevelopment (Searles Quick et al. 2015). Compared to chimpanzees, our closest extant relatives, copy number for these genes has increased dramatically in the human lineage, suggesting an important role for this system in hominid brain evolution. Decreased copy number variation in this region (within the human range) is associated with positive symptoms in schizophrenia, while increased copy number is correlated with autism and negative symptoms in schizophrenia (Searles Quick et al. 2015). Copy number variants in DUF1220, along with variation in other domains, alleles, and pathways, indicate that schizophrenia and autism reflect "diametric conditions with regard to their genomic underpinnings" (Crespi et al. 2010, p. 1736).

The exploration of the likely genetic connection between schizophrenia and ASD is one example of how an evolutionary perspective on schizophrenia is evolving in the post-genomic era. As the DUF1220 results suggest, schizophrenia may to some extent be an emergent property of evolved changes in this particular genetic domain; it shares this quality with ASD. The GWAS results indicate that there are many other genetic systems that influence the expression of schizophrenia, which are also important in the etiologies of other psychopathological conditions. From an evolutionary perspective, rather than being the central character in a genetically circumscribed narrative, schizophrenia has become a supporting player in much broader narratives (i.e., incorporating multiple genetic systems, neurocognitive processes, and psychopathologies) concerning the genetics underlying the evolution of the human brain and behavior. This is consistent with a model of schizophrenia as a condition that emerges from continuous variation across a phenotypic spectrum.

The genomic basis of the emergent nature of schizophrenia has been supported by association studies in which schizophrenia-risk alleles are examined in reference to human accelerated regions (HARs; Xu et al. 2015; Bhattacharyya et al. 2021). These HARs are genomic markers that are conserved in nonhuman primates but which have had accelerated substitution rates in the human lineage since our divergence from chimpanzees. Xu and colleagues found that HARs are relatively enriched in schizophrenia-associated loci, especially those associated with the gamma-aminobutyric acid-ergic (GABAergic) neurons and synaptic formation. Bhattacharyya and colleagues identified a group of 49 single-nucleotide polymorphisms (among the 2,737 known HARs) that are likely to have a role in neurodevelopment and found that four of these were schizophrenia-associated loci (based on a cohort of over 2,000 South Asian schizophrenia patients). These HAR studies indicate that over the approximately 6 million years of evolution since humans and chimpanzees split, genomic regions that have been important for the evolution of the human brain and behavior are uniquely associated with schizophrenia.

Other investigators have looked at the more recent evolutionary past (hundreds of thousands of years) for clues about the genomic uniqueness of schizophrenia in humans, by making use of the complete Neanderthal genomes that are now available for comparative study. Srinivasan and colleagues (2016) examined schizophrenia-risk loci with reference to genetic regions known to have been influenced by positive selection as determined by comparing Neanderthal and modern human genomes. Although the exact nature of differences in the cognitive capacities of ourselves and Neanderthals remains highly speculative, it is reasonable to postulate that some of them may have touched on language and creative thinking domains. Srinivasan et al. found that schizophrenia loci were much more likely than those for other psychiatric and neurological conditions to be found in regions that have diverged the most since humans diverged from Neanderthals. The effect size for these schizophrenia loci was at about the magnitude of those for height and body mass index (there are substantial differences in body morphology between modern humans and Neanderthals). Several schizophrenia loci that are included in this positive selection sweep are (unsurprisingly) related to cognitive function. Srinivasan et al. conclude that these data support the contention that schizophrenia emerged as a "byproduct" or "side effect" of strong positive selection for these cognition-related loci. It is important to note that while the results of Srinivasan et al. are intriguing, other studies of selection signatures related to schizophrenia-related alleles have found that they are essentially neutral (Yao et al. 2020) or that negative selection is acting on these alleles (Liu et al. 2019).

Most genomic studies have focused on comparing ourselves to our closest phylogenetic relatives, but Kasap and colleagues (2018) have taken a very different approach, by looking at the conservation of risk genes for schizophrenia in the nematode *Caenorhabditis elegans*, *Drosophila*, and zebrafish. Although these species are all very distantly related to us (especially compared to Neanderthals and other primates), Kasap et al. (2018) found that schizophrenia-related genes had conservation rates of over 80% for both *C. elegans* and zebrafish, compared to 40%–70% for genes in general. The high level of conservation of the schizophrenia loci suggests that these genes are related to essential functions, which are shared across disparate species. The exact specific functional significance for schizophrenia of these highly conserved loci has yet to be determined, just as the significance of the loci linked to positive selection since the Neanderthal divergence can only be speculated upon. Kasap et al. (2018) come to a very similar conclusion as Srinivasan et al. in that they see schizophrenia emerging as a "bad luck" combination of alleles, whose independent expressions are too minor to be subject to strong purifying selection.

We are just at the beginning of understanding schizophrenia genetics at a fully genomic level and how the loci involved translate into neurophysiology and then behavior. It is clear that these new data will allow us to place schizophrenia in the wider context of human brain evolution. Although schizophrenia seems to definitively emerge as an evolutionary "byproduct" due to bad luck or the "cost" of evolving something profoundly beneficial, these conceptualizations are inadequate and misleading. A true spectrum-based perspective on psychiatric illness and human cognition should avoid such essentialist concepts and terminology. We may be at a point where concepts and terminology have not caught up with technology, but genomics combined with a broader perspective on schizophrenia (and other psychiatric illness) phenotypes will provide insights into the evolution of human behavior. This will happen more quickly if these clinical conditions are not simply seen as part of the probabilistic genetic load our species carries.

Clinical and Therapeutic Implications

Although there is little evidence that an evolutionary approach has ever had much of an influence on the treatment of schizophrenia at any interventional level, it could certainly be relevant for both psychosocial therapies and public health education. Psychosocial interventions for schizophrenia patients can be very effective when used in conjunction with pharmacological treatment; unfortunately, in many care settings, access to them lags behind recommended levels (McDonagh et al. 2022). Many of these interventions focus on the improvement of cognitive, social, and interpersonal skills, in both community and family contexts. There is a wealth of information available on the evolutionary basis of human social behavior and cognition, which could be used to enhance psychosocial therapy and communication with patients about why such therapies have a chance of being effective in a community context.

In an interesting pilot study, Dodgson and Gordon (2009) describe a therapeutic approach built around the idea that certain auditory hallucinations reflect "hypervigilance" errors. Dodgson and Gordon argue that such errors arise out of the perceptual vigilance system, where false positives are a byproduct of a system that has evolved to minimize (potentially dangerous) false negatives. They provide a case history of an individual with schizophrenia ("Michael") who was highly distressed by auditory hallucinations (falsely) accusing him of being a pedophile. When Michael had these hallucinations, he became anxious, slept poorly, increased his drug use, and limited his social interactions. He became caught in a hypervigilance cycle that eventually led to a psychotic break and hospitalization. Dodgson and Gordon (2009, p. 332) found that their evolutionary model was helpful to those they were treating: "service users found the model easy to understand and powerfully normalizing. Insight into the cognitive processes leading to false positives enabled service users to question their response to sensory stimuli and reduced arousal and preoccupation . . . the service user no longer has to develop delusional explanations for their anomalous experience." In other words, the patient's insight was enhanced by understanding the possible evolutionary basis of a mental phenomenon being experienced.

Dodgson and Gordon mention "normalizing," which is usually the hoped-for outcome of schizophrenia therapy (or any therapy). To be cured of a disease is to be returned to "normal." As we increasingly understand the spectrum of genes and their behavioral phenotypes associated with the clinical entity schizophrenia, we need to ask, what is normal and how should it be defined? Anthropologist Andrea Wiley (2021) argues that defining normal is no simple matter. She writes that "biological normalcy" arises from "statistical norms (averages and variances) and normative assessments (culturally informed ideas about what bodies [minds] should be like)" expressed in terms of "normative or evaluative language, which stems from, reflects, and reifies cultural assumptions about what constitutes 'normal' human biology" (Wiley 2021, p. 2).

Biological normalcy reflects both the product and process of organic and cultural evolution. By this I mean that what we think of as normal emerges statistically from heritable variation shaped by a range of evolutionary forces acting on populations. The schizophrenia phenotypes that emerge from the complex evolutionary processes underlying the condition are culturally defined as not normal, and the therapeutic systems in which people with schizophrenia are treated typically have normalization as a goal. This puts people with schizophrenia in a therapeutic bind, in that the goal of normalcy is both practically unattainable and, over cultural time and space, unstable.

A fascinating story of community-based treatment, normalcy, and schizophrenia has been told by anthropologist Karen Nakamura (2013), about the therapeutic Bethel community that arose within a small, economically depressed town in northern Japan.

Under the guidance of a local psychiatrist who emphasized living with schizophrenia rather than curing or controlling it, schizophrenic members of this community have a strong say in their own governance. Signs of illness, such as auditory hallucinations or not being able to focus on work or a productive activity, are acknowledged as part of life (i.e., normalized). The community has gained a certain amount of notoriety in Japan, and its annual festival is attended by large numbers of people from outside the local town. One of the highlights of the festival is the annual awarding of a prize for best hallucination or delusion.

It is impossible to summarize the Bethel story in one paragraph, and it is worth reading Nakamura's ethnography to gain a fuller picture of what has happened there. At first glance, Bethel does not appear to have much to do with the evolutionary basis of schizophrenia. But this story clearly and poignantly demonstrates that what is described as schizophrenia is one way, among many, of being human.

The evolutionary perspective shows us that schizophrenia emerges in all populations and cultures, as a result of genetic and historical forces that we are just beginning to understand. It is a relatively common manifestation of human biological variation. This does not make schizophrenia normal or legitimate or pathological—those are judgments that cultures and medical systems make. The evolutionary perspective offers an insight into the place of schizophrenia in our species' story. Service providers may be able to translate that knowledge into therapies that allow people with schizophrenia to form insights about their own places in their societies.

Unanswered Questions and Future Work

The spectrum perspective on schizophrenia, fundamental to evolutionary inquiry in schizophrenia, is now becoming more broadly appreciated. Although it has always been outside the psychiatric mainstream, an evolutionary perspective can help inform a range of issues related to schizophrenia. The opposite is also true; for example, GWAS research on schizophrenia may someday provide us with insights into how the human brain and behavior evolved. Here are some areas of inquiry about schizophrenia which can be directly or indirectly influenced by an evolutionary approach.

> *Reducing the cost of stigma.* Schizophrenia is found in all human cultures, reflecting its universal nature. This universality is probably related in some way—as yet undetermined—to how the human brain and behavior evolved. Seeing schizophrenia as fundamentally human can help reduce the widespread stigma against the condition. In contrast, referring to schizophrenia as a "cost of x" or a "byproduct" of our evolved genetics is probably not all that helpful. If we hope that evolutionary insights can encourage individuals to develop personal insights, we should probably not reduce their existence to an economic metaphor.

Cross-cultural and biocultural research. Although there has been a long history of cross-cultural research on schizophrenia and other psychiatric illnesses, more can now be done to combine that approach with contemporary knowledge of genetics, the psychopathological spectrum, and the spectrum of social cognitive deficits that people diagnosed with schizophrenia share with their relatives and the general population. Understanding the variation across the various schizophrenia spectrums will be enhanced by systematic, evolution-informed, comparative research across cultures (i.e., biocultural studies; Shattuck 2019). For example, schizophrenia-risk alleles undoubtedly vary to some extent across human populations; does this phylogenetic variation coincide with variation in incidence rates or prognosis? And if schizophrenia is an emergent property of the evolution of the human brain and behavior, characterizing cross-cultural variation in symptoms may inform selection forces that shaped that evolution.

Evolving GWAS research on schizophrenia and evolution. The effort to put the GWAS findings on schizophrenia-risk loci into some kind of evolutionary context has had a good, albeit somewhat inconsistent, beginning. I see two issues with this research: (1) they are based on a somewhat impoverished view of the last 7 million years of human evolution (i.e., there is more to our story than language and creativity) and (2) the availability of the Neanderthal genome for comparative purposes, while exciting, can also lead to a narrowing of perspective on the evolutionary issues that are addressed (see Allen 2009). The better we understand the expression of schizophrenia-risk loci in contemporary populations, the more useful they will be to help us understand not only the evolution of schizophrenia but also the evolution of the human brain and behavior in the full phylogenetic context of hominid evolution. Further examination of the selective forces acting on schizophrenia-risk alleles (current studies provide ambiguous, contradictory, or minimally contextualized results) is warranted and could help answer the basic question: should we expect schizophrenia to become more or less common in the future? That is, are schizophrenia-risk alleles being slowly selected out of the population, or will they be maintained at more or less the current frequency due to the relatively small effect each allele has on reducing or enhancing fitness?

Evolution-informed, bottom-up approaches to schizophrenia genomics. The large number of identified schizophrenia-risk loci (some common, some rare; some regulatory, some structural, etc.) will require strategies to ultimately link the genes to brain anatomy and function. Clara Moreau and colleagues (2021) advocate a bottom-up approach, in which variants are identified and then explored in their imaging and cognitive dimensions, not just in psychiatric populations but in the general population as well. Using evolutionary insights to guide investigators through the large number of risk loci would help in selecting those that could yield functional insights with a bottom-up approach.

Key Points to Remember

- **Despite recent changes in how researchers perceive the evolutionary landscape surrounding schizophrenia, the basic factors outlining the problem remain the same: high heritability combined with low fertility, universal distribution (expressed across all human populations), no exogenous triggering agent, phenotypic continuity across pathological and non-pathological states.**
- **The GWAS-based discovery that schizophrenia phenotypes emerge from various combinations of hundreds of low-effect alleles provides ample opportunities for reconstructing the evolutionary narrative of schizophrenia in the context of human brain evolution.**
- **Schizophrenia sits on psychopathological, neurodevelopmental, and sociocognitive spectrums. Understanding the evolution of schizophrenia will require characterizing the variable expression of the condition along each of these spectrums.**
- **Educating people about how schizophrenia phenotypes emerge as a direct result of forces that have shaped the evolution of the human brain and behavior could help reduce stigma toward the condition among the general public and provide personal insights for those who suffer from it.**

References

Abdellaoui, A., Yengo, L., Verweij, K. J. H., & Visscher, P. M. (2023). 15 years of GWAS discovery: Realizing the promise. *American Journal of Human Genetics, 110*(2), 179–194. https://doi.org/10.1016/j.ajhg.2022.12.011

Ackerknecht, E. (1943). Psychopathology, primitive medicine, and primitive culture. *Bulletin of the History of Medicine, 14*, 30–67.

Allen, J. S. (1997). Are traditional societies schizophrenogenic? *Schizophrenia Bulletin, 23*(3), 357–364. https://doi.org/10.1093/schbul/23.3.357

Allen, J. S. (2009). *The lives of the brain: Human evolution and the organ of mind.* Harvard University Press.

Allen, J. S., & Sarich, V. M. (1988). Schizophrenia in an evolutionary perspective. *Perspectives in Biology and Medicine, 32*(1), 132–153. https://doi.org/10.1353/pbm.1988.0039

American Psychiatric Association. (2013). *Diagnostic and statistical manual of mental disorders* (5th ed.). American Psychiatric Publishing.

Andric, S., Maric, N. P., Mihaljevic, T., & van Os, J. (2016). Familial covariation of facial recognition and IQ in schizophrenia. *Psychiatry Research, 246*, 52–57. https://doi.org/10.1016/j.psychres.2016.09.022

Barton, N. H. (2022). The "new synthesis." *Proceedings of the National Academy of Sciences of the United States of America, 119*(3), Article e2122147119. https://doi.org/10.1073/pnas.2122147119

Bhattacharyya, U., Deshpande, S. N., Bhatia, T., & Thelma, B. K. (2021). Revisiting schizophrenia from an evolutionary perspective: An association study of recent evolutionary markers and schizophrenia. *Schizophrenia Bulletin, 47*(3), 827–836. https://doi.org/10.1093/schbul/sbaa179

Brüne, M. (2004). Schizophrenia—An evolutionary enigma? *Neuroscience and Biobehavioral Reviews, 28*(1), 41–53. https://doi.org/10.1016/j.neubiorev.2003.10.002

Bundy, H., Stahl, D., & MacCabe, J. H. (2010). A systematic review and meta-analysis of the fertility of patients with schizophrenia and their unaffected relatives. *Acta Psychiatrica Scandinavica, 123*, 98–106. https://doi.org/10.1111/j.1600-0447.2010.01623.x

Burns, J. K. (2004). An evolutionary theory of schizophrenia: Cortical connectivity, metarepresentation, and the social brain. *Behavioral and Brain Sciences*, *27*(6), 831–885. https://doi.org/10.1017/s0140525x04000196

Carter, M., & Watts, C. A. H. (1971). Possible biological advantages among schizophrenics' relatives. *British Journal of Psychiatry*, *118*(545), 453–460. https://doi.org/10.1192/bjp.118.545.453

Cohen, A., Patel, V., Thara, R., & Gureje, O. (2008). Questioning an axiom: Better prognosis for schizophrenia in the developing world? *Schizophrenia Bulletin*, *34*(2), 229–244. https://doi.org/10.1093/schbul/sbm105

Collin, G., Kahn, R. S., de Reus, M. A., Cahn, W., & van den Heuvel, M. P. (2014). Impaired rich club connectivity in unaffected siblings of schizophrenia patients. *Schizophrenia Bulletin*, *40*(2), 438–448. https://doi.org/10.1093/schbul/sbt162

Collin, G., Turk, E., & van den Heuvel, M. P. (2016). Connectomics in schizophrenia: From early pioneers to recent brain network findings. *Biological Psychiatry: Cognitive Neuroscience and Neuroimaging*, *1*(3), 199–208. https://doi.org/10.1016/j.bpsc.2016.01.002

Crespi, B., Stead, P., & Elliot, M. (2010). Comparative genomics of autism and schizophrenia. *Proceedings of the National Academy of Sciences of the United States of America*, *107*(S1), 1736–1741. https://doi.org/10.1073/pnas.0906080106

Crow, T. J. (2008). The "big bang" theory of the origin of psychosis and faculty of language. *Schizophrenia Research*, *102*(1–3), 31–52. https://doi.org/10.1016/j.schres.2008.03.010

Dodgson, G., & Gordon, S. (2009). Avoiding false negatives: Are some auditory hallucinations an evolved design flaw? *Behavioural and Cognitive Psychotherapy*, *37*(3), 325–334. https://doi.org/10.1017/S1352465809005244

Egan, M. F., Goldberg, T. E., Gscheidle, T., Weirich, M., Rawlings, R., Hyde, T. M., Bigelow, L., & Weinberger, D. R. (2001). Relative risk for cognitive impairments in siblings of patients with schizophrenia. *Biological Psychiatry*, *50*(2), 98–107. https://doi.org/10.1016/S0006-3223(01)01133-7

Erlenmeyer-Kimling, L. (1968). Mortality rates in the offspring of schizophrenic patients and a physiological advantage hypothesis. *Nature*, *220*(5169), 798–800. https://doi.org/10.1038/220798a0

Erlenmeyer-Kimling, L., & Paradowski, W. (1966). Selection and schizophrenia. *American Naturalist*, *100*(916), 651–665.

Haijma, S. V., Van Haren, N. E. M., Cahn, W., Koolschijn, P. C. M. P., Hulshoff Pol, H. E., & Kahn, R. S. (2013). Brain volumes in schizophrenia: A meta-analysis in over 18000 subjects. *Schizophrenia Bulletin*, *39*(5), 1129–1138. https://doi.org/10.1093/schbul/sbs118

Hill, K., Mann, L., Laws, K. R., Stephenson, C. M. E., Nimmo-Smith, I., & McKenna, P. J. (2004). Hypofrontality in schizophrenia: A meta-analysis of functional imaging studies. *Acta Psychiatrica Scandinavica*, *110*(4), 243–256. https://doi.org/10.1111/j.1600-0447.2004.00376.x

Huxley, J., Mayr, E., Osmond, H., & Hoffer, A. (1964). Schizophrenia as a genetic morphism. *Nature*, *204*, 220–221. https://doi.org/10.1038/204220a0

Jääskeläinen, E., Juola, P., Hirvonen, N., McGrath, J. J., Saha, S., Isohanni, M., Veijola, J., & Miettunen, J. (2013). A systematic review and meta-analysis of recovery in schizophrenia. *Schizophrenia Bulletin*, *39*(6), 1296–1306. https://doi.org/10.1093/schbul/sbs130

Jablensky, A., Sartorius, N., Emberg, G., Anker, M., Korten, A., Cooper, J. E., Day, R., & Bertelsen, A. (1992). Schizophrenia: Manifestations, incidence, and course in different cultures. A World Health Organization ten country study. *Psychological Medicine Monograph Supplement*, *20*, 1–97. https://doi.org/10.1017/s0264180100000904

Jauhar, S., Johnstone, M., & McKenna, P. J. (2022). Schizophrenia. *Lancet*, *399*(10323), 473–486. https://doi.org/10.1016/S0140-6736(21)01730-X

Kadakia, A., Catillon, M., Fan, Q., Williams, G. R., Marden, J. R., Anderson, A., Kirson, N., & Dembek, C. (2022). The economic burden of schizophrenia in the United States. *Journal of Clinical Psychiatry*, *83*(6), Article 22m14458. https://doi.org/10.4088/JCP.22m14458

Karlsson, J. L. (1970). Genetic association of giftedness and creativity with schizophrenia. *Hereditas*, *66*(2), 177–182. https://doi.org/10.1111/j.1601-5223.1970.tb02343.x

Kasap, M., Rajani, V., Rajani, J., & Dwyer, D. S. (2018). Surprising conservation of schizophrenia risk genes in lower organisms reflects their essential function and the evolution of genetic liability. *Schizophrenia Research, 202*, 120–128. https://doi.org/10.1016/j.schres.2018.07.017

Keller, M. C., & Miller, G. (2006). Resolving the paradox of common, harmful, heritable mental disorders: Which evolutionary genetic models work best? *Behavioral and Brain Sciences, 29*(4), 385–452. https://doi.org/10.1017/S0140525X06009095

Legge, S. E., Santoro, M. L., Periyasamy, S., Okewole, A., Arsalan, A., & Kowalec, K. (2021). Genetic architecture of schizophrenia: A review of major advancements. *Psychological Medicine, 51*(13), 2168–2177. https://doi.org/10.1017/S0033291720005334

Lichtenstein, P., Yip, B. H., Björk, C., Pawitan, Y., Cannon, T. D., Sullivan, P. F., & Hultman, C. M. (2009). Common genetic determinants of schizophrenia and bipolar disorder in Swedish families: A population-based study. *Lancet, 373*(9659), 234–239. https://doi.org/10.1016/S0140-6736(09)60072-6

Liddle, P. F. (1987). Schizophrenic syndromes, cognitive performance, and neurological dysfunction. *Psychological Medicine, 17*(1), 49–57. https://doi.org/10.1017/S0033291700012976

Liu, C., Everall, I., Pantelis, C., & Bousman, C. (2019). Interrogating the evolutionary paradox of schizophrenia: A novel framework and evidence supporting recent negative selection of schizophrenia risk alleles. *Frontiers in Genetics, 10*, Article 389. https://doi.org/10.3389/fgene.2019.00389

McDonagh, M. S., Dana, T., Kopelovich, S. L., Monroe-DeVita, M., Blazina, I., Bougatsos, C., Grusing, S., & Selph, S. S. (2022). Psychosocial interventions for adults with schizophrenia: An overview and update of systematic reviews. *Psychiatric Services, 73*(3), 299–312. https://doi.org/10.1176/appi.ps.202000649

McGrath, J., Saha, S., Chant, D., & Welham, J. (2008). Schizophrenia: A concise overview of incidence, prevalence, and mortality. *Epidemiologic Reviews, 30*, 67–76. https://doi.org/10.1093/epirev/mxn001

Meyer-Lindenberg, A., & Weinberger, D. R. (2006). Intermediate phenotypes and genetic mechanisms of psychiatric disorders. *Nature Reviews Neuroscience, 7*(10), 818–827. https://doi.org/10.1038/nrn1993

Minzenberg, M. J., Laird, A. R., Thelen, S., Carter, C. S., & Glahn, D. C. (2009). Meta-analysis of 41 functional neuroimaging studies of executive function in schizophrenia. *Archives of General Psychiatry, 66*(8), 811–822. https://doi.org/10.1001/archgenpsychiatry.2009.91

Moreau, C. A., Raznahan, A., Bellec, P., Chakravarty, M., Thompson, P. M., & Jacquemont, S. (2021). Dissecting autism and schizophrenia through neuroimaging genomics. *Brain, 144*(7), 1943–1957. https://doi.org/10.1093/brain/awab096

Murphy, J. (1976). Psychiatric labeling in cross-cultural perspective. *Science, 191*(4231), 1019–1028. https://doi.org/10.1126/science.1251213

Nakamura, K. (2013). *A disability of the soul: An ethnography of schizophrenia and mental illness in contemporary Japan*. Cornell University Press.

Nesic, M., Stojkovic, B., & Maric, N. P. (2019). On the origin of schizophrenia: Testing evolutionary theories of the post-genomic era. *Psychiatry and Clinical Neurosciences, 73*(12), 723–730. https://doi.org/10.1111/pcn.12933

Nettle, D., & Clegg, H. (2006). Schizotypy, creativity, and mating success in humans. *Proceedings of the Royal Society B: Biological Sciences, 273*(1586), 611–615. https://doi.org/10.1098/rspb.2005.3349

Pearlson, G. D., & Folley, B. S. (2008). Schizophrenia, psychiatric genetics, and Darwinian psychiatry: An evolutionary framework. *Schizophrenia Bulletin, 34*(4), 722–733. https://doi.org/10.1093/schbul/sbm130

Polimeni, J. (2012). *Shamans among us: Schizophrenia, shamanism, and the evolutionary origins of schizophrenia*. Lulu Press.

Polimeni, J., & Reiss, J. P. (2002). How shamanism and group selection many reveal the origins of schizophrenia. *Medical Hypotheses, 58*(3), 244–248. https://doi.org/10.1054/mehy.2001.1504

Power, R. A., Kyaga, S., Uher, R., MacCabe, J. H., Långström, N., Landen, M., McGuffin, P., Lewis, C. M., Lichtenstein, P., & Svensson, A. C. (2013). Fecundity of patients with schizophrenia, autism, bipolar disorder, depression, anorexia nervosa, or substance abuse vs their unaffected siblings. *JAMA Psychiatry, 70*(1), 22–30. https://doi.org/10.1001/jamapsychiatry.2013.268

Rees, E., Creeth, H. D. J., Hwu, H. G., Chen, W. J., Tsuang, M., Glatt, S. J., Rey, R., Kirov, G., Walters, J. T. R., Holmans, P., Owen, M. J., & O'Donovan, M. C. (2021). Schizophrenia, autism spectrum disorders and developmental disorders share specific disruptive coding mutations. *Nature Communications, 12*(1), Article 5353. https://doi.org/10.1038/s41467-021-25532-4

Reichenberg, A. (2010). The assessment of neuropsychological functioning in schizophrenia. *Dialogues in Clinical Neuroscience*, *12*(3), 383–392. https://doi.org/10.31887/DCNS.2010.12.3/areichenberg

Schizophrenia Working Group of the Psychiatric Genomics Consortium. (2014). Biological insights from 108 schizophrenia-associated genetic loci. *Nature*, *511*(7510), 421–427. https://doi.org/10.1038/nature13595

Searles Quick, V. B., Davis, J. M., Olincy, A., & Sikela, J. M. (2015). DUF1220 copy number is associated with schizophrenia risk and severity: Implications for understanding autism and schizophrenia as related diseases. *Translational Psychiatry*, *5*(12), Article e697. https://doi.org/10.1038/tp.2015.192

Shattuck, E. C. (2019). A biocultural approach to psychiatric illness. *Psychopharmacology*, *236*(10), 2923–2936. https://doi.org/10.1007/s00213-019-5178-7

Srinivasan, S., Bettella, F., Mattingsdal, M., Wang, Y., Witoelar, A., Schork, A. J., Thompson, W. K., Zuber, V., The Schizophrenia Working Group of the Psychiatric Genetics Consortium, The International Headache Genetics Consortium, Winsvold, B. S., Zwart, J. A., Collier, D. A., Desikan, R. S., Melle, I., Werge, T., Dale, A. M., Djurovic, S., & Andreassen, O. A. (2016). Genetic markers of human evolution are enriched in schizophrenia. *Biological Psychiatry*, *80*(4), 284–292. https://doi.org/10.1016/j.biopsych.2015.10.009

Steadman, L. B., & Palmer, C. T. (1994). Visiting dead ancestors: Shamans as interpreters of religious traditions. *Zygon*, *29*(2), 173–189. https://doi.org/10.1111/j.1467-9744.1994.tb00659.x

Sullivan, P. F., Kendler, K. S., & Neale, M. C. (2003). Schizophrenia as a complex trait: Evidence from a meta-analysis of twin studies. *Archives of General Psychiatry*, *60*(12), 1187–1192. https://doi.org/10.1001/archpsyc.60.12.1187

Tandon, R., Heckers, S., Bustillo, J., Farch, D. M., Gaebel, W., Gur, R. E., Malaspina, D., Owen, M. J., Schultz, S., Tsuang, M., van Os, J., & Carpenter, W. (2013). Catatonia in DSM-5. *Schizophrenia Research*, *150*(1), 26–30. https://doi.org/10.1016/j.schres.2013.04.034

Thornicroft, G., Sunkel, C., Alikhon Aliev, A., Baker, S., Brohan, E., El Chammay, R., Davies, K., Demissie, M., Duncan, J., Fekadu, W., Gronholm, P. C., Guerrero, Z., Gurung, D., Habtamu, K., Hanlon, C., Heim, E., Henderson, C., Hijazi, Z., Hoffman, C., . . . Winkler, P. (2022). The *Lancet* Commission on ending stigma and discrimination in mental health. *Lancet*, *400*(10361), 1438–1480. https://doi.org/10.1016/S0140-6736(22)01470-2

Torrey, E. H. (1980). *Schizophrenia and civilization*. Jason Aronson.

van de Leemput, J., Hess, J. L., Glatt, S. J., & Tsuang, M. T. (2016). Genetics of schizophrenia: Historical insights and prevailing evidence. In T. Friedmann, J. C. Dunlap, & S. F. Goodwin (Eds.), *Advances in genetics* (Vol. 96, pp. 99–141). Elsevier. https://doi.org/10.1016/bs.adgen.2016.08.001

van den Heuvel, M. P., Scholtens, L. H., de Lange, S. C., Pijnenburg, R., Cahn, W., Van Haren N. E. M., Sommer, I. E., Bozzali, M., Koch, K., Boks, M. P., Repple, J., Pievani, M., Li, L., Preuss, T. M., & Rilling, J. K. (2019). Evolutionary modifications in human brain connectivity associated with schizophrenia. *Brain*, *142*(12), 3991–4002. https://doi.org/10.1093/brain/awz330

van Dongen, J., & Boomsma, D. I. (2013). The evolutionary paradox and the missing heritability of schizophrenia. *American Journal of Medical Genetics Part B*, *162*(2), 122–136. https://doi.org/10.1002/ajmg.b.32135

Waxler, N. (1974). Culture and mental illness: A social labeling perspective. *Journal of Nervous and Mental Disease*, *159*(6), 379–395. https://doi.org/10.1097/00005053-197412000-00001

Welham, J., Isohanni, M., Jones, P., & McGrath, J. (2009). The antecedents of schizophrenia: A review of birth cohort studies. *Schizophrenia Bulletin*, *35*(3), 603–623. https://doi.org/10.1093/schbul/sbn084

Wheeler, A. L., & Voineskos, A. N. (2014). A review of structural neuroimaging in schizophrenia: From connectivity to connectomics. *Frontiers in Human Neuroscience*, *8*, Article 653. https://doi.org/10.3389/fnhum.2014.00653

Wiley, A. S. (2021). Pearl lecture: Biological normalcy: A new framework for biocultural analysis of human population variation. *American Journal of Human Biology*, *33*(5), Article e23563. https://doi.org/10.1002/ajhb.23563

Winston, E. (1934). The alleged lack of mental diseases among primitive groups. *American Anthropologist*, *36*, 234–238.

Xu, K., Schadt, E. E., Pollard, K. S., Roussos, P., & Dudley, J. T. (2015). Genomic and network patterns of schizophrenia genetic variation in human evolutionary accelerated regions. *Molecular Biology and Evolution*, *32*(5), 1148–1160. https://doi.org/10.1093/molbev/msv031

Yao, Y., Yang, J., Xie, Y., Liao, H., Yang, B., Xu, Q., & Rao, S. (2020). No evidence for widespread positive selection in common risk alleles associated with schizophrenia. *Schizophrenia Bulletin*, *46*(3), 603–611. https://doi.org/10.1093/schbul/sbz048

15

Williams Syndrome and Autism
Dysfunction of Frontal Networks

Isabel August and Katerina Semendeferi

Introduction

Humans and other primates are particularly social mammals who inhabit complex social environments. This complexity has been proposed as a principal driving force in brain evolution. Certain human disorders, such as William's syndrome (WS) and autism spectrum disorders (ASD), are known to affect social behaviors in opposing ways to some extent. WS is characterized by hypersocial behaviors, while ASD is, in general, characterized by hyposocial behaviors. Understanding the brain changes characteristic of these disorders, particularly in the context of human-specific brain changes, might provide some insight into the neuroanatomical underpinnings of certain social behaviors in the context of the evolution of primate sociality. Here we review findings from research on the brain in WS and ASD before discussing in more detail the neuroanatomy of social brain regions which have undergone recent evolutionary change (changes which have occurred after the last common ancestor with chimpanzees) and which are significantly impacted in both WS and ASD. It should, however, be noted that ASD has been much more extensively studied than WS and that for several of the topics discussed below data on WS are not available.

WS is a rare neurodevelopmental disorder (<1 in 7,500) caused by the hemizygous deletion of approximately 25 genes on chromosome band 7q11.23. Individuals with WS exhibit consistent somatic, cognitive, and behavioral effects. Somatic effects of the disorder include connective tissue abnormalities, high incidence of cardiovascular disease, and a distinctive facial morphology. Individuals with WS also exhibit cognitive and motor impairments, increased interest in and emotional reactivity to music, and weak visuospatial skills. WS is further characterized by changes in behavior in a social environment. In particular, individuals with WS are described as being hypersocial with a particular willingness to approach strangers. Additionally, individuals with WS tend to have increased levels of non-social anxiety (Bellugi et al. 2000).

Unlike WS, ASD is a heterogenous group of disorders which are much more common (1 in 68) than WS. The exact etiology of ASD is unknown, but given the heritability of the disorders, it seems that genetics are central to their etiology. ASD is broadly characterized by differences in communication, social interaction, and restricted and repetitive behaviors. Differences in communication affect both verbal and nonverbal communication. In the verbal domain, these differences exist along a continuum from complete absence of verbal communication to hyperlexia. Moreover, autistic individuals may have monotone speech and a restricted range of facial expressions. Differences in social interaction are interrelated with differences in social communication and exist along a continuum. Finally, the restricted and repetitive behaviors of autistic individuals can result in highly focused interests and a lack of behavioral flexibility, although differences in this area are also represented by a broad spectrum of behaviors (Cashin and Barker 2009). Given the significance of social behavior in humans and other primates, these disorders have attracted significant interest in the context of evolutionary studies, in addition to the medical literature.

Historical Background

Behavioral observations on WS date back to its first identification in 1961, and behavioral observations of ASD date back even further to its first identification in 1943. In 1972, Niko and Elizabeth Tinbergen attempted to understand autistic behaviors from an ethological perspective and developed a hypothesis of autism as a functional behavior. They noted that certain behaviors characteristic of ASD, such as gaze avoidance and withdrawal, can also be observed in typically developing (TD) children. They hypothesized that children experience a "motivational conflict" between the desire to approach others and the fear of harm from strangers, and behaviors characteristic of ASD, rather than being fundamentally different, are simply an extreme form of behaviors exhibited by TD children (Tinbergen and Tinbergen 1972).

Studies of the neuroanatomical underpinnings of WS and ASD did not begin until the 1980s and 1990s, respectively. Early investigations of the brain in WS addressed total and regional brain volumes, cell-packing density, and neuronal organization in post-mortem tissue. These early studies revealed that total brain volume is reduced in WS compared to TD controls, particularly in posterior regions. Qualitative architectonic observations indicated increased cell-packing density, abnormally clustered and oriented neurons, and exaggerated horizontal, immature organization of neurons, which was most striking in area 17 (primary visual cortex; Galaburda et al. 1994). Quantitative follow-ups to this early work investigated cell size and cell-packing densities and found an increase in cell size and a decrease in cell-packing density in WS (Galaburda and Bellugi 2000). While it may at first seem odd to find such differences in the primary visual cortex, as opposed to areas involved in socioemotional processing, it should be noted that individuals with WS exhibit weak visuospatial skills, which may be linked to these differences in the visual cortex. Early investigations of the brain of individuals on the autism spectrum focused on similar parameters within limbic brain regions. Results of these studies revealed increased cell-packing density and reduced cell size in the anterior cingulate cortex (ACC), hippocampus, subiculum, entorhinal cortex, amygdala, mamillary body, and septal nucleus in ASD. Additionally, studies revealed age-related abnormalities in the number and size of Purkinje cells in the cerebellum and cerebellar circuits in ASD (Bauman 1991).

Evolutionary Evidence Related to Human Neurodevelopmental Disorders

Since the work of Tinbergen and Tinbergen (1972), a number of other evolutionary hypotheses for human neurodevelopmental disorders, including WS and ASD, have been put forward. These hypotheses include genomic imprinting for both WS and ASD, the extreme male brain hypothesis for ASD, and the solitary forager model of ASD (Badcock and Crespi 2006; Reser 2011; Crespi and Procyshyn 2017). Genomic imprinting is said to evolve out of intragenomic conflicts between maternally and paternally expressed genes (maternally expressed genes should decrease costs to the mother, and paternally expressed genes should increase costs to the mother and benefits to the offspring) and result in the differential expression of particular genes based on their parent of origin and their effects on the fitness of both mothers and offspring (Badcock and Crespi 2006). In the case of WS, *GTF2I* (one of the genes deleted in WS) is reported to have a maternal expression bias. Therefore, deletion of this gene leads to a paternal expression bias in WS (Crespi and Procyshyn 2017). Interestingly, despite their opposing sociobehavioral phenotypes, it has also been proposed that ASD is related to a paternal expression bias due, in part, to the significant role of limbic system structures in ASD. Studies conducted in mice show that paternally expressed genes differentially affect the limbic system, whereas maternally expressed genes differentially affect the neocortex (Badcock and Crespi 2006). The extreme male brain hypothesis provides an evolutionary framework for understanding ASD in terms of male–female differences. This hypothesis suggests that males tend to be more systematizing, while females tend to be more empathetic, and that ASD is an extreme example of male-typical behavior (Badcock and Crespi 2006). Finally, the solitary forager model of ASD suggests that many traits characteristic of ASD might have made individuals with these traits particularly suited to solitary foraging and been adaptive in the context of sparsely distributed food resources (Reser 2011). While comprehensive hypotheses such as those discussed above are attractive, they can be difficult to test; and the heterogeneity and multifactorial nature of ASD in particular can act as confounds to the explanatory power of such models.

However, considering the neuroanatomical profiles of WS and ASD in the context of comparative evolutionary studies does raise the possibility of evolutionary influences on the development of these disorders. The human brain has not only increased in size over the course of human evolution but also undergone a significant amount of reorganization in both cortical and subcortical regions. In particular, there are regions in the frontal and temporal lobes for which there is accumulating evidence of human-specific changes. A number of these regions are also significantly impacted in WS and ASD, which suggests a link between recent evolutionary change and vulnerability to typical development. In the remainder of this chapter, we will discuss evidence for recent evolutionary changes in regions of the frontal and temporal lobes as well as evidence for differences in these regions in WS and ASD. In particular, we will focus on the prefrontal cortex (PFC) in the frontal lobe and the amygdala in the temporal lobe as these regions have been most extensively studied.

Frontal Lobe

Neuroimaging studies of the frontal lobe in humans, apes, and other primates have compared total frontal lobe volume, cortical volume, and white matter volume. Results of these studies show that total frontal lobe volume is as large as expected for an ape of human brain size. Total white matter volume in the frontal lobe is also as large as expected for an ape of human brain size. However, volumetric estimates of gyral and core white matter reveal that gyral white matter is enlarged in the human frontal lobe. Gyral white matter lies just under the cortex and approximates the size of connections between neighboring cortical regions. Enlarged gyral white matter suggests increased emphasis on short-range connections (Semendeferi et al. 2010).

Neuroimaging studies of WS and ASD have compared total and regional volumes, cortical thickness, cortical surface area, and gyrification in various frontal regions. Increased volume has been reported in the ventral anterior PFC (Gothelf et al. 2008), and increased gyrification has been reported in the mesial PFC (Gaser et al. 2006) in WS compared to TD controls. There is also an increase in cortical thickness in the medial and lateral orbitofrontal cortex (OFC) and a decrease in cortical surface area in the lateral OFC in WS subjects compared to TD subjects when controlling for total intracranial volume (Meda et al. 2012). In WS children frontal lobe gray matter volume is increased, while gray matter volume in posterior regions is decreased (Campbell et al. 2009). This is consistent with findings in adults with WS (Chiang et al. 2007). In ASD, on the other hand, there is an early increase in brain volume, primarily in frontal and temporal regions, followed by an accelerated rate of volumetric decline linked to an accelerated rate of synaptic pruning during adolescence (reviewed in Rafiee et al. 2022). Findings from neuroimaging studies of adults and children with ASD have also revealed regionally specific increases in gray matter volume in the PFC, superior and middle frontal gyri, and ACC (reviewed in Rafiee et al. 2022; Figure 15.1).

PFC

There is evidence of human-specific microstructural changes in relative size of cortical areas, neuron density, minicolumnar organization, and dendritic morphology. The human OFC is 11% larger than expected for an ape of human brain size, but some cortical areas within this part of the brain, like Brodmann area (BA) 13, are smaller compared to apes. In contrast, BA 10, which forms the frontal pole in humans, is twice as large as expected. There is also a decrease in neuron density and an increase in neuropil space in BA 10 in humans (Lew and Semendeferi 2017). Minicolumns represent vertically arranged neurons that are primarily identifiable in cortical layers III, V, and VI and represent a basic unit of cortical organization. Size and morphology of minicolumns differ across regions and species, reflecting functional differences as well as differences in cortical organization (Buxhoeveden and Casanova 2002). Minicolumns are identifiable in Nissl-stained histological sections that reveal the cell bodies and their spacing. Identification and analysis of minicolumns are also used as a proxy to the more methodologically challenging visualization of the complete dendritic trees of neurons in the cortex of humans and other primates (e.g., Golgi stain). Studies targeting dendritic morphology have shown it to be altered in pyramidal neurons in BA 10, with more branched neurons and increased number of dendritic spines in humans compared to chimpanzees (Bianchi et al. 2013). (For a map of Brodmann areas, see Figure 15.2.)

A number of studies demonstrate alterations in these same parameters in both WS and ASD. In WS there is a pattern of decreased neuron densities in PFC areas involved in executive functioning and higher-order socioemotional processing, namely BA 10 and BA 11, which is not seen in other, unimodal areas such as sensory (BA 3), motor (BA 4), and visual (BA 18) areas. This decrease is most evident in the infragranular layers, V and VI. However, results only reached statistical significance in BA 10 (Lew et al. 2017). Neuron density is also decreased in the ventromedial PFC, BA 25, in WS, particularly in the supragranular layers, II and III (Wilder et al. 2018). In BA 10 this decrease may be related to a decrease in SMI-32 immunoreactive (SMI-32ir) neurons. SMI-32ir neuron density and soma volume are decreased in BA 10 in WS subjects (Hrvoj-Mihic and Semendeferi 2019). (For an overview of microstructural findings in WS, see Table 15.1.)

In ASD there is an increase in neuron number and density in prefrontal regions. Increased neuron number has been reported in the dorsolateral (dl) and mesial subdivisions of the PFC (Courchesne et al. 2011), and increased density has been reported in BA 9 (Casanova et al. 2006). Reports on cell size in prefrontal regions in ASD have been conflicting, with some reporting no difference in cell size (Courchesne et al. 2011) and others reporting a decrease in cell size (Karst and Hutsler 2016). (For an overview of microstructural findings in ASD, see Table 15.2.)

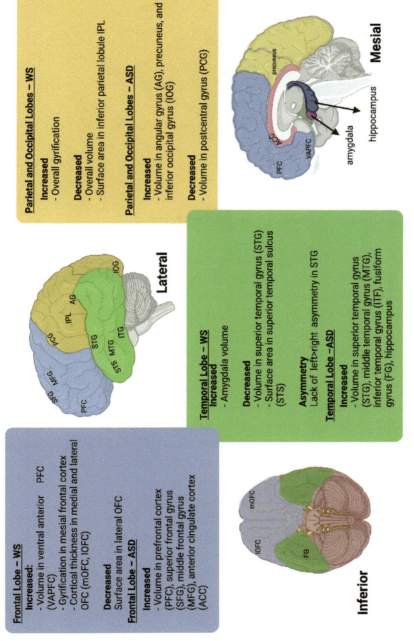

Global Trends in WS Cerebrum: Decreased cerebral volume and cortical surface area, Increased gyrification and cortical thickness.

Frontal and temporal lobes in ASD: Increased volume early in development, followed by accelerated rate of decline; increased amygdala volume during childhood – disappears during adolescence

Figure 15.1 Visual summary of structural neuroimaging findings in Williams syndrome and autism spectrum disorder (created with BioRender). ASD = autism spectrum disorder; OFC = orbitofrontal cortex; WS = Williams syndrome.

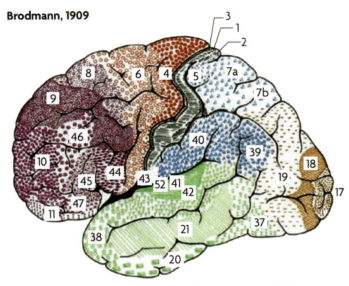

Figure 15.2 Brodmann's cortical map, modified by Zilles and Amunts (2010).

Given the significance of minicolumns to cortical organization and circuitry, a fair amount of research has been devoted to investigating how minicolumnar organization is affected in different brain areas in ASD. Results from all of these studies show reduced minicolumn width throughout the brain in ASD subjects, which may result in reduced or impaired functional connectivity between cortical regions (Casanova et al. 2006). This difference is most notable in BA 9 (Casanova et al. 2006) and layers V/VI in BA 9, 10, and 11 (Casanova et al. 2010).

Dendritic morphology also appears to be altered in the PFC in both WS and ASD. Previous research demonstrates that branching of basal dendrites varies across functionally distinct cortical areas in TD subjects. In BA 10 dendrites are typically more branched with more dendritic spines compared to unimodal areas. On the other hand, dendrites in BA 11 are shorter, less branched, and less spinous than in BA 10 (Jacobs et al. 2001). In WS, neurons in BA 10 and BA 11 do not display longer and more branched dendrites than neurons in motor, sensory, and visual areas. Additionally, dendritic branching, dendritic length, and the number of dendritic spines differ little between BA 10 and BA 11 (Hrvoj-Mihic et al. 2017). In ASD dendritic spine density is increased in BA 9 (and temporal and parietal areas) compared to TD controls, and when analysis is restricted to apical dendrites alone, the most significant difference in spine density is in layer II in all three regions (Hutsler and Zhang 2010).

In addition to differences in neuronal organization and morphology, a few studies have investigated the role of glial cells in the PFC in WS and ASD. In WS there is increased glia density and an increased glia to neuron ratio in the ventromedial PFC, specifically in BA 25. This region is critically involved in social behaviors, inhibition, and

TABLE 15.1 Table Summarizing Microstructural Findings in WS

Age (range)	Number of subjects	Region(s) of interest	Findings (in WS compared to TD)	Methods	Reference
WS: 8 mo–53 yrs	WS: 4	BA 4, BA 17, BA 42, BA 6, BA 18, BA 22, BA 37, BA 39, BA 9	Increased cell size in BA 17 Decreased cell-packing density in BA 17	Nissl	Galaburda and Bellugi (2000)
WS: 44 yrs TD: 43 yrs	WS: 3 TD: 3	BA 17	Increased cell-packing density in left layer IVCβ Excess of small neurons in left layers IVA, IVCα, IVCβ, V, and VI	Nissl	Galaburda et al. (2002)
	WS: 2 TD: 4	BA 3, BA 4, BA 18	Larger total dendritic length in layer V/VI pyramidal neurons Higher number of dendritic spines in layer V/VI pyramidal neurons Increased numbers of dendritic segments and branching points in layer V/VI pyramidal neurons	Golgi	Chailangkarn et al. (2016)
WS: 18–45 yrs TD: 19–45 yrs	WS: 5 TD: 5	Striatum (dC, mC, aP, NA)	Increased glia density in dC Increased glia/neuron in dC and mC Increased oligodendrocyte density in mC	Nissl	Hanson et al. (2017)
WS: 31 and 47 yrs	WS: 2	BA 10, BA 11, (BA 4, BA 3, BA18)	Decreased branching in PFC basal dendrites	Golgi	Hrvoj-Mihic et al. (2017)
WS: 18–48 yrs TD: 1–951 yrs	WS: 6 TD: 6	BA 10, BA 11, (BA 4, BA 3, BA 18)	Decreased neuron density in PFC	Nissl	Lew et al. (2017)
WS: 26 days–48 yrs TD: 34 days–45 yrs	WS: 7 TD: 7	Amygdala (BLA and central nucleus)	Increased neuron number in lateral nucleus	Nissl	Lew et al. (2018)
WS: 114 days–43 yrs TD: 110 days–43 yrs	WS: 7 TD: 7	BA 25	Decreased neuron density Increased glia density Increased glia/neuron	Nissl	Wilder et al. (2018)
WS: 31, 43 yrs TD: 42 yrs	WS: 2 TD: 1	BA 10	Decreased density of SMI-32ir neurons	Immuno (SMI-32)	Hrvoj-Mihic and Semendeferi (2019)

(continued)

TABLE 15.1 Continued

Age (range)	Number of subjects	Region(s) of interest	Findings (in WS compared to TD)	Methods	Reference
WS: 18–69 yrs TD: 19–69 yrs	WS: 7 TD: 6	Striatum (dC, mC, aP, NA)	Reduced density of Ch+ interneurons in mC and NA	Immuno (PV+ and Ch+)	Hanson et al. (2020)
WS: 114 days–69 yrs TD: 107 days–69 yrs ASD: 20–64 yrs	WS: 7 TD: 7 ASD: 6	Amygdala (BLA and central nucleus)	Decreased density of SERT-ir axons in WS (compared to TD and ASD, only statistically significant compared to ASD)	Immuno (SERT)	Lew et al. (2020)

Galaburda and Bellugi (2000)—qualitative results. Hrvoj-Mihic and Semendeferi (2019)—pilot study.
aP = associative putamen; ASD = autism spectrum disorder; BA = Brodmann area; BLA = basolateral amygdala; Ch+ = cholinergic interneurons; dC = dorsal caudate; mC = medial caudate; NA = nucleus accumbens; PFC = prefrontal cortex; PV+ = parvalbumin-positive interneurons; SERT = serotonin transporter protein; SMI-32 = non-phosphorylated neurofilament protein; TD = typically developing; WS = Williams syndrome.

TABLE 15.2 Table Summarizing Microstructural Findings in ASD

Age (range)	Number of subjects	Region(s) of interest	Findings (in ASD compared to TD)	Methods	Reference
ASD: 5–28 yrs TD: NA; 15yrs	ASD: 9 TD: 4	BA 9, BA 22, BA 21	Reduced minicolumn width Increased number of minicolumns Less compact organization in minicolums Reduced neuropil space in periphery of minicolumns	Nissl	Casanova et al. (2002)
ASD: 3 yrs, 41 yrs TD: 2–44 yrs	ASD: 2 TD: 4	Dorsal, orbital, mesial frontal cortex, BA 17	Reduced minicolumn and neuropil spacing (dorsal, orbital, and mesial PFC in adults; dorsal and orbital PFC in child) Increased GLI (dorsal, orbital, and mesial PFC in adults; dorsal and orbital PFC in child)	Nissl	Buxhoeveden et al. (2006)
ASD: 4–24 yrs TD: 4–25 yrs	ASD: 6 TD: 6	BA 3b, BA 4, BA 17, BA 9	Reduced minicolumn width (greatest reduction in BA 9) Increased number of minicolumns Reduced neuron size within minicolumns	Nissl	Casanova et al. (2006)
ASD: 10–44 yrs TD: 11–44 yrs	ASD: 9 TD: 10	Amygdala (BLA, central, and other nuclei)	Reduced neuron number in total amygdala and lateral nucleus	Nissl	Schumann and Amaral (2006)

TABLE 15.2 Continued

Age (range)	Number of subjects	Region(s) of interest	Findings (in ASD compared to TD)	Methods	Reference
ASD: 3–41 yrs TD: 2–75 yrs	ASD: 4 TD: 5	FI	No reduction in number of spindle neurons Possible trend of increased number of spindle neurons in subset of ASD cases	Nissl	Kennedy et al. (2007)
ASD: 4–23 yrs TD: 4–65 yrs	ASD: 7 TD: 10	FG, BA 17, CGM	Reduced neuron density in layer III (FG) Reduced neuron numbers in layers III, V, and VI (FG) Reduced mean perikaryal volumes of neurons in layers V and VI (FG)	Nissl	van Kooten et al. (2008)
ASD: 15–54 yrs TD: 20–55 yrs	ASD: 9 TD: 4	ACC (BA 24)	Decrease in cell size in layers I–III and layers V–VI of area 24b Decrease in cell-packing density in layers V–VI of area 24c	Nissl	Simms et al. (2009)
ASD: 10–45 yrs TD: 11–51 yrs	ASD: 8 TD: 8	BA 21, BA 9, BA 7	Indistinct boundary between layer VI and subcortical WM	Nissl	Avino and Hutsler (2010)
ASD: 4–67 yrs TD: 4–65 yrs	ASD: 7 TD: 7	BA 9, BA 10, BA 11, BA 3b, BA 4, BA 17, BA 24, BA 43, BA 44	Reduced minicolumn width across all cortical areas (greatest in BA 44)	Nissl	Casanova et al. (2010)
ASD: 10–44 yrs TD: 11–46 yrs	ASD: 10 TD: 10	BA 7, BA 9, BA 21	Increased dendritic spine density (most significant in layer II in all three areas and layer V in BA 21)	Golgi	Hutsler and Zhang (2010)
ASD: 3–41 yrs TD: 1–44 yrs	ASD: 15 TD: 9	dlPFC	Increased WM microglial somal volume and GM microglial density Trend toward increased GM microglial somal volume Altered microglial morphology in some ASD cases	Immuno (iba-1)	Morgan et al. (2010)

(continued)

Age (range)	Number of subjects	Region(s) of interest	Findings (in ASD compared to TD)	Methods	Reference
ASD: 2–16 yrs TD: 2–16 yrs	ASD: 7 TD: 6	dlPFC mPFC	Increased neuron numbers in dlPFC and mPFC (greatest increase in dlPFC)	Nissl	Courchesne et al. (2011)
ASD: 4–11 yrs TD: 4–14 yrs	ASD: 4 TD: 3	Layer V of FI	Increased ratio of VENs to pyramidal neurons	Nissl	Santos et al. (2011)
ASD: 4–52 yrs TD: 4–48 yrs	ASD: 7 TD: 7	BA 44, BA 45	Reduced size of pyramidal neurons	Nissl	Jacot-Descombes et al. (2012)
ASD: 3–41 yrs TD: 1–44 yrs	ASD: 13 TD: 9	dlPFC (BA 9/46)	Increased microglia–neuron spatial clustering Increased neuron–neuron spatial clustering in older ASD subjects Normal microglia–microglia clustering	Immuno (iba-1)	Morgan et al. (2012)
ASD: 10–45 yrs TD: 11–59 yrs	ASD: 9 TD: 8	Layers II and III of BA 9, BA 21, BA 7	Increased cell density (greatest in layer II) Decreased cell size (layers II and III)	Nissl	Karst and Hutsler (2016)
ASD: 7–46 yrs TD: 7–44 yrs	ASD: 16 TD: 16	Amygdala (lateral nucleus)	Increased spine density in young ASD subjects Decreasing spine density with age in ASD subjects	Golgi	Weir et al. (2018)
ASD: 4–44 yrs TD: 2–48 yrs	ASD: 28 TD: 24	Amygdala (BLA and central nucleus)	Increased (excess) neuron number in childhood, followed by reduction in adulthood	Nissl Immuno (bcl-2)	Avino et al. (2019)
ASD: 20–64 yrs TD: 19–69 yrs WS: 17–69 yrs	ASD: 6 TD: 6 WS: 6	Amygdala (BLA and central nucleus)	Increased density of SERT-ir axons in ASD (compared to TD and WS, only statistically significant compared to WS)	Immuno (SERT)	Lew et al. (2020)

ACC = anterior cingulate cortex; ASD = autism spectrum disorder; BA = Brodmann area; BLA = basolateral amygdala; blc-2 = B cell lymphoma 2; CGM = cortical gray matter; dlPFC = dorsolateral prefrontal cortex; prefrontal cortex (PFC); FG = fusiform gyrus; FI = frontoinsular cortex; GLI = gray level index; GM = gray matter; iba-1 = ionized calcium binding adapter molecule 1; mPFC = mesial prefrontal cortex; NA = nucleus accumbens; PFC = prefrontal cortex; SERT = serotonin transporter; TD = typically developing; VENs = von Economo neurons; WM = white matter; WS = Williams syndrome.

decision-making and is heavily connected to several subcortical structures, most notably the amygdala, which is also significantly impacted in WS and ASD (Wilder et al. 2018). Non-neuronal cell types in prefrontal regions in ASD have been targeted with a focus on microglia (see Chapter 2). Increases have been reported in microglial somal volume in white matter and microglial density in gray matter in the dlPFC of autistic subjects compared to TD controls (Morgan et al. 2010). There is also a trend toward increased microglial somal volume in dorsolateral PFC gray matter in autistic subjects, although this did not reach statistical significance. Some ASD subjects also showed alterations in microglial morphology, including somal enlargement, process retraction and thickening, and extension of filopodia from processes. Microglia–neuron spatial clustering is also increased in older ASD subjects, suggesting a loss of typical neuronal organization with age. However, microglia–microglia spatial clustering was normal in ASD subjects, suggesting that the microglia–neuron associations are not the result of changes in microglial organization or distribution (Morgan et al. 2012).

Broca's Region, ACC, and Frontoinsular Cortex

While the primary focus of this section is on differences observed in the dlPFC over the course of human evolution and in WS and ASD, there are a few other frontal regions worth briefly discussing. These include Broca's area in the ventrolateral PFC, which is critically involved in language functioning; the ACC, which is involved in socioemotional processing and the management of social behaviors; and the frontoinsular cortex, which is also involved in socioemotional processing and body representation. The cytoarchitecture of all of these areas has been studied in humans and the great apes, as well as in TD and ASD subjects; but no reports exist yet on WS. BA 44 and BA 45 form Broca's region in humans and the homolog to Broca's region in nonhuman primates, although recent work suggests a more complex parcellation of this region in humans (reviewed in Amunts and Zilles 2012). Comparative analysis of BA 44 and 45 in humans and apes demonstrates that there are more and larger minicolumns in humans (reviewed in Semendeferi et al. 2010). In ASD minicolumn width appears to be decreased in cortical areas throughout the brain, and the greatest decrease is in BA 44 (Casanova et al. 2010). Additionally, stereological estimates of neuron size, neuron number, and regional volume of BA 44 and 45 in ASD reveal significantly smaller pyramidal neurons in autistic subjects compared to TD controls. However, there was no difference in the other parameters (Jacot-Descombes et al. 2012).

Von Economo neurons (VENs) are bipolar projection neurons with a restricted distribution. The ACC is one of two brain regions in which they can be found. While the exact function of these neurons is unknown, they are believed to be involved in socioemotional intelligence and are found in a number of social species. Among humans and apes the density of VENs increases with decreasing phylogenetic distance from humans, and humans have the greatest number and soma volume of VENs among the great apes (Allman et al. 2002). Analysis of VEN density in the ACC in ASD revealed no difference overall, but results could be divided into subsets, which may reflect the heterogeneity of the disorder. In three out of nine autistic subjects there was an increase in VEN density, while in the remaining six there was a decrease in VEN density. Neuron size and density are also altered in the subdomains of the ACC (areas 24a, 24b, and 24c) in ASD. In layers I–III and V/VI of area 24b there is a decrease in cell size, and in layers V/VI of area 24c there is a decrease in cell density (Simms et al. 2009). Additionally, and consistent with other brain regions in ASD, minicolumn width is decreased in the ACC (Casanova et al. 2010).

The frontoinsular cortex, a subdivision of the insula, is the second brain region where VENs can be found. As with the ACC, VEN density in the frontoinsular cortex has been investigated in a comparative evolutionary context, and humans have the highest density of VENs compared to other apes (Semendeferi 2018). A few studies of the frontoinsular cortex in ASD have investigated the number and density of VENs. Kennedy et al. (2007) undertook the first such investigation, hypothesizing that the number of VENs would be reduced in autistic subjects compared to TD controls. However, the results of their study showed no such reduction. They did, however, identify a possible trend toward an increase in VENs in a subset of ASD subjects. A further study focused specifically on autistic children reported an increased ratio of VENs to pyramidal neurons in ASD (Santos et al. 2011).

Cytoarchitectural findings in the frontal lobe and specifically in the PFC in WS and ASD reveal distinct and partly opposing phenotypes in each of these disorders. Neuron density is decreased throughout PFC areas in WS, while in ASD there is an increase in both number and density of neurons. Dendritic morphology is also altered in both disorders. In WS, there is a decrease in branching and spine density. In ASD, there is an increase in spine density, particularly in layer II. Moreover, many of these differences occur in areas which have undergone recent evolutionary change (namely, BA 10 and BA 11). Glial cells also are also impacted in these disorders. In WS there are increases in glia density and in the glia to neuron ratio in BA 25, while microglial morphology is altered in the dlPFC in ASD (Tables 15.1 and 15.2).

Temporal Lobe

Neuroimaging studies of the temporal lobe in humans, apes, and other primates have compared total temporal lobe and white matter volume. Unlike the frontal lobe, the human temporal lobe is larger than expected for an ape of human brain size. Total white matter volume is not larger than expected, but when subdivided into gyral and core white matter, the temporal lobe shows a similar pattern to the frontal lobe. That is to say, gyral white matter is also enlarged in the temporal lobe in humans, suggesting again an emphasis on short-range connections (Semendeferi et al. 2010).

Neuroimaging studies of WS and ASD reveal differences in volume and cortical surface area in temporal regions. Sampaio et al. (2008) report reduced absolute, although not relative, volume in the superior temporal gyrus (STG) in WS. They also found that WS subjects lack a typical left–right STG asymmetry. Reduced surface area has also been reported in the superior temporal sulcus in WS relative to TD controls (Ng et al. 2016). Several studies have also revealed a relative, but not absolute, increase in amygdala volume in WS (Martens et al. 2009; Capitão et al. 2011). Absolute amygdala volume is increased in children with ASD compared to TD controls, although this difference disappears by adolescence (Schumann et al. 2004). As previously stated, the early overgrowth typical of ASD also affects temporal regions. Moreover, regionally specific increases in gray matter volume have also been identified in ASD subjects in superior, middle, and inferior temporal gyri and the fusiform gyrus (reviewed in Rafiee et al. 2022; Figure 15.1).

Amygdala

The amygdala has previously been thought to be a relatively conserved structure in evolution, and several features do appear to be phylogenetically retained. However, there are also notable changes in size and cytoarchitecture of this region in humans compared to apes. In particular, four major amygdaloid nuclei (lateral, basal, accessory basal, and central nuclei) have been studied in an evolutionary context. The basolateral nuclei (lateral, basal, and accessory basal nuclei) have significant, bidirectional connections to the OFC and temporal association cortex and are considered part of the cognitive loop of the amygdala. The central nucleus receives significant input from the basolateral nuclei and serves as the primary output to regulatory regions (Lew et al. 2019). Among these four nuclei the lateral nucleus stands out in that it is largest in humans and larger than expected for an ape of human brain size. In other apes the basal nucleus is the largest of these nuclei. In humans the basal and central nuclei are smaller than expected. Volumetric differences in subdivisions of the amygdala are further reflected in neuron number in the major amygdaloid nuclei. In humans the greatest number of neurons is found in the lateral nucleus, while in apes the greatest number of neurons is found in the basal nucleus. Neuron numbers in the basal and central nuclei of humans are also lower than expected, although not significantly (Barger et al. 2012). Serotonergic innervation of the amygdala has also been considered in an evolutionary context. The serotonergic system plays a key role in both neurodevelopment and behavioral modulation, and the amygdala is heavily innervated by serotonergic projections. Serotonergic innervation of the amygdala varies between humans, chimpanzees, and bonobos. In humans, serotonin transporter immunoreactive (SERT-ir) axon density varies across the four major amygdaloid nuclei (central > accessory basal > basal > lateral), with the greatest difference found between the central and lateral nuclei. This pattern appears unique to humans. Additionally, SERT-ir axon density is greater in humans compared to chimpanzees in the basal, accessory basal, and central nuclei. Compared to bonobos, SERT-ir axon density in humans is greater in the accessory basal and central nuclei (Lew et al. 2019).

The amygdala is critically implicated in social behavior and cognition, and many of the same parameters discussed above have therefore been investigated in the amygdala in both WS and ASD given the atypical sociobehavioral phenotypes of these disorders. Total neuron numbers are increased in the lateral nucleus in WS (Lew et al. 2018). Additionally, there are several trends in the basolateral division of the amygdala that differentiate WS subjects and TD controls. WS subjects have a larger lateral nucleus with greater neuron density, a smaller basal nucleus with greater neuron number and density, and a smaller accessory basal nucleus with greater neuron density, though these trends did not reach statistical significance. Infants with WS followed the same general trend as adults in terms of neuron numbers and densities in the basolateral and central nuclei, and there was a statistically significant increase in neuron numbers in the lateral nucleus.

Interestingly, the first quantitative stereological study of the amygdala in ASD found no difference in volume of the total amygdala or any of its subdivisions in autistic subjects. In fact, this study revealed a reduced number of neurons in the total amygdala and in the lateral nucleus. It should, however, be noted that the subjects involved in this study were primarily adults (Schumann and Amaral 2006). Taken in conjunction with neuroimaging findings, the results of this study suggest an early overgrowth and ultimate reduction of amygdala neurons. Consistent with this suggestion, Avino et al. (2019) report an excess of neurons in the amygdala during childhood, which is followed by a reduction in adulthood. Additionally, dendritic spine density is increased in young ASD subjects and decreases with age (Weir et al. 2018).

Given the opposing sociobehavioral phenotypes of WS and ASD (hyper- and hyposociability, respectively) and the findings related to neuronal distribution of the amygdala in these disorders, Lew et al. (2020) hypothesized that WS subjects might exhibit a decrease in serotonergic innervation of the amygdala, while autistic subjects would demonstrate an increase. Analysis of serotonergic innervation of the amygdala in WS and ASD subjects, along with TD controls, did indeed reveal such a trend, with a statistically significant difference between the WS and ASD groups.

As was seen in the PFC, cytoarchitectural findings in the temporal lobe and specifically in the amygdala in WS and ASD reveal distinct phenotypes in each of these disorders. Moreover, some of these differences in the amygdala in WS and ASD appear to primarily affect the lateral nucleus, which seems to have been a target for recent evolutionary change. In WS there is an increase in neuron number and density in the lateral nucleus, while neuron number is reduced in the lateral nucleus in ASD. However, in WS the increase in number and density of neurons appears to present pre- or very early postnatally, whereas in ASD there is an early overabundance of neurons, followed by a period of reduction. Serotonergic innervation of the amygdala also varies in WS and ASD, with WS subjects demonstrating a decrease in SERT-ir density and autistic subjects demonstrating an increase (Tables 15.1 and 15.2).

Counterarguments and Evidence against Recent Evolutionary Influences

The findings discussed in the previous section support the idea of a possible link between recent evolutionary change and vulnerability to typical neurodevelopment. However, there are other brain regions which are significantly impacted by WS and ASD and for which no comparative studies are available to indicate whether or not they have been subject to recent evolutionary change. This leaves open the possibility that regions other than the ones discussed above are affected in the two disorders without having undergone recent changes during human evolution.

Neuroimaging studies in WS reveal a disproportionate decrease in parietal and occipital lobe volume in WS (Chiang et al. 2007) and an increased ratio of frontal to parietal and occipital tissue in WS subjects (Reiss et al. 2000). There are also reports of increased gyrification in the parietal and occipital lobes (Gaser et al. 2006) and reduced surface area in the inferior parietal lobule in WS subjects compared to TD controls (Ng et al. 2016). In ASD, regional increases in gray matter volume have been identified in the angular gyrus, precuneus, and inferior occipital gyrus. A regionally specific decrease in volume has also been identified in the postcentral gyrus in ASD (reviewed in Rafiee et al. 2022). Findings from neuroimaging studies in WS at the subcortical level reveal reduced volume in several basal ganglia nuclei, including the caudate and putamen, when compared to TD controls (Chiang et al. 2007), both of which have also been studied at the histological level (Figure 15.1).

Associative and limbic territories of the striatum (dorsal and medial caudate nucleus, associative putamen, and nucleus accumbens) are involved in learning, flexibility, and behavioral control. Hanson et al. (2017) hypothesized that the hypersociability characteristic of WS might be a result of deficiencies of inhibitory control of behavior, which might be reflected in these regions. Their results showed that average neuron densities are not affected by WS in associative and limbic territories of the striatum. However, glia densities are affected. They report an increase in glia density in the dorsal caudate, an increased glia to neuron ratio in the dorsal and medial caudate, and an increase in the density of oligodendrocytes in the medial caudate in WS. A further analysis reveals a reduction in the density of excitatory cholinergic interneurons in the medial caudate and nucleus accumbens in WS (Hanson et al. 2020). Excitatory cholinergic interneurons comprise less than 1% of the total neuronal population in the striatum. However, they do share extensive connections with cortical and subcortical projections and modulate inputs to inhibitory interneurons. Moreover, cholinergic interneurons may play a role in modulating social behaviors.

The findings from these neuroimaging and histological studies, including parietal and occipital regions and the basal ganglia, indicate the presence of differences in WS and ASD brain regions beyond areas that are known to have undergone recent evolutionary change, in particular parietal and occipital regions and the basal ganglia. These differences in WS and ASD combined with the lack, relative to other brain areas, of comparative data available might argue against recent evolutionary influences in the development of these two disorders and certainly highlight the significance of performing comparative primate studies.

It should further be noted here that a number of environmental influences have been identified as risk factors for ASD. These include prenatal viral infections, pre- and perinatal stress, zinc deficiency, and parental age. Maternal infections are thought to alter the immune system and immune status of the fetal brain, and this altered immune status has repeatedly been linked to ASD. Pre- and perinatal stress is also thought to impact immune function and increase risk of ASD. Additionally, there is a high incidence of zinc deficiency in ASD, and it has been suggested that maternal or early developmental zinc deficiency may provide a mechanism of gene/environment interactions. Finally, risk of development of ASD is associated with advancing age of either parent. In males increasing age leads to increased risk of genetic mutations, while in females increased age might contribute to the development of ASD as a result of increased pregnancy complications and maternal autoimmunity (Grabrucker 2013). The significant role played by these environmental factors might also argue against recent evolutionary influences in the development of ASD.

Clinical and Therapeutic Implications

Structural neuroimaging and postmortem evidence from WS and ASD subjects provide necessary information on the neuroanatomical phenotypes of these disorders in humans, a fundamental step that is needed in order to explore the underlying cellular and molecular mechanisms of WS and ASD. Use of animal and induced pluripotent stem cell (iPSC) models may, in turn, improve our clinical understanding of these disorders and aid in the development of novel therapeutics. The well-known and defined genetic etiology of WS makes this disorder particularly well suited to study in mouse and iPSC models. Moreover, evidence of WS-specific neuroanatomic features in infant subjects indicates a prenatal or early postnatal origin for the disorder, suggesting that genetic mechanisms play a significant role in the development of these features (Wilder et al. 2018). Mice with the complete human deletion, as well as partial deletions from *Gtf2i* to *Limk1* and *Limk1* to *Fkbp6*, replicate crucial aspects (increased sociability and acoustic startle response, cognitive defects, growth deficits, craniofacial abnormalities, impaired motor skills) of WS in humans (Li et al. 2009). Several genes in the WS critical region (*GTF2I*, *GTF2IRD1*, *LIMK1*, *CYLN2*, *FZD9*, *BAZ1B*, and *EIF4H*) are involved in brain development and are therefore of particular interest to this review.

> *GTF2I*. Mice with a deletion of *Gtf2i* have demonstrated anomalies in dendritic spines and synaptic plasticity in the hippocampus (Borralleras et al. 2015). This deletion has also been linked to white matter abnormalities. Deletion of *Gtf2i* results in reduced myelin-related gene transcripts, oligodendrocyte numbers, axon myelination, and neuronal function in mouse models (Barak et al. 2019).
> *LIMK1*. Deletions of *Limk1* appear to cause alterations in actin cytoskeleton in the hippocampus, which could impact spine morphogenesis, synaptic function, and synaptic plasticity (Hoogenraad et al. 2004).
> *CYLN2*. Deletion of *Cyln2* in mice also causes cytoskeletal alterations. In this case, proteins involved in regulating microtubule dynamics are affected (Hoogenraad

et al. 2002), which could have implications for axon guidance, along with synaptic morphology and plasticity (Hoogenraad et al. 2004).

FZD9. Mouse models with *Fzd9* deletions demonstrate an increase in apoptosis in the dentate gyrus beginning with the onset of development and continuing through the first postnatal week. This deletion is also associated with a potentially compensatory increase in dividing precursor cells (Zhao et al. 2005). The relationship of FZD9 to abnormal cell death has also been established using iPSCs. Fewer neural progenitor cells (NPC) were observed in a culture of WS iPSCs when compared to TD controls as a result of an increase in apoptosis. When FZD9 was restored in the WS population, comparable numbers of NPCs were observed in both WS and TD iPSCs, along with a reduction in apoptosis of NPCs in WS (Chailangkarn et al. 2016).

BAZ1B. An iPSC model has also been used to show that deletion of BAZ1B causes defects in cell differentiation (Lalli et al. 2016).

EIF4H. The deletion of *Eif4h* in mouse models results in reduced neuron numbers in the posterior cortex as well as alterations in morphology of these neurons (Capossela et al. 2012).

GTF2IRD1. Finally, there is evidence for alterations in serotonin metabolism in the amygdala and frontal and parietal cortices in mice with *Gtf2ird1* deletions (Capossela et al. 2012).

Taken together, findings from these studies indicate that these genes play an important role in neuronal morphology, synaptic function and plasticity, cell cycle regulation and offer insight into the mechanisms underlying some of the neuroanatomical features observed in WS.

Unlike WS, ASD does not have a well-defined genetic etiology. Rather, it seems likely that ASD is the result of a cascade of disrupted processes that results in the varying severities and heterogeneity characteristic of this disorder. That being said, the heritability of ASD does suggest a genetic component, and many genes have been identified as possible contributors to the development of ASD and investigated using animal and iPSC models (Courchesne et al. 2019). Moreover, mouse models of ASD have been shown to replicate characteristic features of ASD, including differences in social interactions and communication, restricted and repetitive behaviors, cognitive inflexibility, and anxiety (Pasciuto et al. 2015). While there are many genes which have been implicated in the development of ASD, a thorough review of them all is beyond the scope of this chapter. We will, however, review findings from animal and iPSC models for several of these ASD-risk genes below, many of which impact signaling pathways involved in different stages of brain development (Courchesne et al. 2019).

FMR1. Mice with mutations in *Fmr1* demonstrate alterations in synaptic plasticity (Gross et al. 2015a) and dendritic spine density (Gross et al. 2015b).

PTEN. Mutations in *Pten* enhanced proliferation of radial glia, cortical overgrowth (Chen et al. 2015), and hyperconnectivity of the PFC and amygdala (Huang et al. 2016).

SHANK3. *Shank3* mutations in mice result in deficits in synaptic function (Bozdagi et al. 2010).

MECP2. iPSC models of MECP2 mutations from Rett syndrome patients reveal excess proliferation of radial glia cells (Mellios et al. 2018) and abnormalities in soma size and spine density (Marchetto et al. 2010).

ARID1B. Mice with mutations in *Arid1b* demonstrate decreased dendritic arborization and alterations in dendritic spine morphology (Ka et al. 2016).

NF1. Mice with mutations in *Nf1* demonstrate altered spine morphology (Oliveira and Yasuda 2014).

16p11.2. Deletion of the *16p11.2* locus in mouse models results in abnormal proliferation and neurogenesis of neural progenitor cells (Pucilowska et al. 2015).

CHD8. Mice with mutations in *Chd8* demonstrate dysregulated proliferation and differentiation of neural progenitor cells (Cotney et al. 2015).

OXTR. OXTR is the oxytocin receptor gene and is of interest in studies of ASD given the significant role played by oxytocin in social cognition. *Oxrt* knockout mice lose the capacity to respond to social cues, and this capacity is fully restored by infusion of oxytocin to the amygdala (Ferguson et al. 2001). Interestingly, oxytocin levels are lower in autistic children, although this difference disappears in adulthood (John and Jaeggi 2021).

While the primary focus of this chapter is on the neuroanatomical profiles of WS and ASD in an evolutionary context and the findings discussed here have come primarily from studies involving structural neuroimaging techniques and analysis of postmortem tissue, findings from functional neuroimaging studies can also be very useful in a clinical and therapeutic setting. Results from functional imaging studies have shown a decreased interhemispheric functional connectivity in a number of regions, many of which are involved in social cognition, throughout the brain in ASD (Li et al. 2019). Additionally, evidence from functional neuroimaging studies indicates that the mirror neuron system is hyperactivated in ASD. The mirror neuron system plays an important role in the ability to imitate in humans, and imitation is thought to play a key role in social development (Chan and Han 2020). Combining lines of evidence from neuroanatomic studies, animal and stem cell models, and functional imaging studies may improve our understanding of structural and functional differences characteristic of WS and ASD, as well as the cellular and molecular mechanisms underlying these differences. This, in turn, may enhance our clinical understanding of these two disorders and aid in the development of novel therapeutics.

Unanswered Questions and Future Work

The human brain has undergone a significant amount of reorganization over the course of our evolution, and there are regions in the frontal and temporal lobes for which there is accumulating evidence of human-specific changes. The complexity of the human social environment is thought to be a driving force behind these changes, and many regions which have undergone a significant amount of change in human evolution are critically involved in social behaviors. A number of these regions also appear to be significantly impacted in certain neurodevelopmental disorders, such as WS and ASD, in which social behaviors are impacted. This overlap suggests a potential link between recent evolutionary change and susceptibility to atypical development.

In particular, the PFC and amygdala have been extensively studied in both an evolutionary context and the context of neurodevelopmental disorders. In the PFC there is evidence of human-specific microstructural changes in relative size of cortical areas, neuron density, minicolumnar organization, and dendritic morphology. Similarly, changes can be found in these parameters in WS and ASD. Neuron density is decreased throughout PFC areas in WS, while in ASD there is an increase in both number and density of neurons. Dendritic branching and spine density are decreased in WS. In ASD, there is an increase in spine density, with some layers more impacted than others. Glial cells are also impacted in these disorders. In WS there is an increase in glia density and the glia to neuron ratio in certain PFC areas, while microglial morphology is altered in ASD. In the amygdala, four major amygdaloid nuclei have been studied in an evolutionary context, and there is evidence of changes in volume, neuron number, and serotonergic innervation in some of these nuclei in humans. There are also changes in these parameters in these nuclei in WS and ASD. In WS there is an increase in neuron number and density in the lateral nucleus, while neuron number is reduced in the lateral nucleus in ASD. However, in WS the increase in number and density of neurons appears to present pre- or very early postnatally, whereas in ASD there is an early overabundance of neurons, followed by a period of reduction. Serotonergic innervation of the amygdala also varies in these disorders, with WS subjects demonstrating a decrease in SERT-ir density and ASD subjects demonstrating an increase. Given the opposing social behaviors characteristic of WS and ASD, it is interesting to note that these disorders also appear to be characterized by opposing neuroanatomical phenotypes.

Despite these interesting findings in the PFC and amygdala, there is more work to be done. As mentioned at the beginning of this chapter, ASD has been much more extensively studied than WS, and there are many brain regions which have been investigated in ASD but for which no comparable data are available in WS. Future work should seek to expand our knowledge of the neuroanatomy of WS outside of the regions discussed here. Moreover, of the studies discussed in this chapter, only one directly compared WS and ASD. More such direct comparisons would help elucidate the differences between these two disorders and might increase our understanding of both. It is also worth mentioning here that, while there is good evidence that some brain regions which have undergone recent evolutionary change are significantly impacted in WS and ASD, there are also brain regions impacted in these disorders for which little to no comparative primate data exist. Future work should also seek to expand our knowledge of human-specific brain changes in order to determine if other brain regions impacted in WS and ASD have also undergone recent evolutionary change.

Key Points to Remember

- The human brain has undergone significant reorganization over the course of evolution.
- The PFC and amygdala have undergone recent evolutionary changes and are also significantly impacted in WS and ASD.
- In WS and ASD, several brain regions are impacted for which no primate comparative data are available.
- Existing information suggests a possible link between recent evolutionary changes in parts of the brain and neurodevelopmental vulnerability in humans.
- iPSC and animal models can be useful in investigating cellular and molecular mechanisms underlying the neuroanatomical differences characteristic of ASD and WS.

References

Allman, J., Hakeem, A., & Watson, K. (2002). Two phylogenetic specializations in the human brain. *The Neuroscientist*, *8*, 335–346.

Amunts, K., & Zilles, K. (2012). Architecture and organizational principles of Broca's region. *Cell*, *16*, 4184–4226. https://doi.org/10.1016/j.tics.2012.06.005

Avino, T. A., Barger, N., Vargas, M. V., Carlson, E. L., Amaral, D. G., Bauman, M. D., & Schumann, C. M. (2019). Neuron numbers increase in the human amygdala from birth to adulthood, but not in autism. *Proceedings of the National Academy of Sciences of the United States of America*, *115*, 3710–3715. https://doi.org/10.1073/pnas.1801912115

Avino, T. A., & Hutsler, J. J. (2010). Abnormal cell patterning at the cortical gray-white matter boundary in autism spectrum disorders. *Brain Research*, *1360*, 138–146. http://doi.org/10.1016/j.brainres.2010.08.091

Badcock, C., & Crespi, B. (2006). Imbalanced genomic imprinting in brain development: An evolutionary basis for the aetiology of autism. *Evolutionary Biology*, *19*, 1007–1032. https://doi.org/10.1111/j.1420-9101.2006.01091.x

Barak, B., Zhang, Z., Liu, Y., Nir, A., Trangle, S. S., Ennis, M., Levandowski, K. M., Wang, D., Quast, K., Boulting, G. L., Li, Y., Bayarsaihan, D., He, Z., & Feng, G. (2019). Neuronal deletion of *Gtf2i*, associated with Williams syndrome, causes behavioral and myelin alterations rescuable by a remyelinating drug. *Nature Neuroscience*, *22*, 700–708. https://doi.org/10.1038/s41593-019-0380-9

Barger, N., Stefanacci, L., Schumann, C. M., Sherwood, C., Annese, J., Allman, J. M., Buckwalter, J. A., Hof, P. R., & Semendeferi, K. (2012). Neuronal populations in the basolateral nuclei of the amygdala are differentially increased in humans compared with apes: A stereological study. *Journal of Camparative Neurology*, *520*, 3035–3054. https://doi.org/10.1002/cne.23118

Bauman, M. L. (1991). Microscopic neuroanatomic abnormalities in autism. *Pediatrics*, *87*, 791–796.

Bellugi, U., Lichtenberger, L., Jones, W., Lai, Z., & St. George, M. (2000). I. The neurocognitive profile of Williams syndrome: A complex pattern of strengths and weaknesses. *Journal of Cognitive Neuroscience*, *12*, 7–29.

Bianchi, S., Stimpson, C. D., Bauernfeind, A. L., Schapiro, S. J., Baze, W. B., Mcarthur, M. J., Bronson, E., Hopkins, W. D., Semendeferi, K., Jacobs, B., Hof, P. R., & Sherwood, C. C. (2013). Dendritic morphology of pyramidal neurons in the chimpanzee neocortex: Regional specializations and comparison to humans. *Cerebral Cortex*, *23*, 2429–2436. https://doi.org/10.1093/cercor/bhs239

Borralleras, C., Sahun, I., Pérez-Jurado, L. A., & Campuzano, V. (2015). Intracisternal *Gtf2i* gene therapy ameliorates deficits in cognition and synaptic plasticity of a mouse model of Williams-Beuren syndrome. *Molecular Therapy*, *23*, 1691–1699. https://doi.org/10.1038/mt.2015.130

Bozdagi, O., Sakurai, T., Papapetrou, D., Wang, X., Dickstein, D. L., Takahashi, N., Kajiwara, Y., Yang, M., Katz, A. M., Scattoni, M. L., Harris, M. J., Saxena, R., Silverman, J. L., Crawley, J. N., Hof, P. R., & Buxbaum, J. D. (2010). Haploinsufficiency of the autism-associated *Shank3* gene leads to deficits in synaptic function, social interaction, and social communication. *Molecular Autism*, *1*, Article 15. https://doi.org/10.1186/2040-2392-1-15

Buxhoeveden, D. P., & Casanova, M. F. (2002). The minicolumn and evolution of the brain. *Brain, Behavior and Evolution*, *60*, 125–151. https://doi.org/10.1159/000065935

Buxhoeveden, D. P., Semendeferi, K., Buckwalter, J., Schenker, N., Switzer, R., & Courchesne, E. (2006). Reduced minicolumns in the frontal cortex of patients with autism. *Neuropathology and Applied Neurobiology*, *32*, 483–491. http://doi.org/10.1111/j.1365-2990.2006.00745.x

Campbell, L. E., Daly, E., Toal, F., Stevens, A., Azuma, R., Karmiloff-Smith, A., Murphy, D. G. M., & Murphy, K. C. (2009). Brain structural differences associated with behavioural phenotype in children with Williams syndrome. *Brain Research*, *1258*, 96–107. https://doi.org/10.1016/j.brainres.2008.11.101

Capitão, L., Sampaio, A., Sampaio, C., Vasconcelos, C., Férnandez, M., Garayzábal, E., Shenton, M. E., & Gonçalves, Ó. F. (2011). MRI amygdala volume in Williams syndrome. *Research in Developmental Disabilities*, *32*, 2767–2772. https://doi.org/10.1016/j.ridd.2011.05.033

Capossela, S., Muzio, L., Bertolo, A., Bianchi, V., Dati, G., Chaabane, L., Godi, C., Politi, L. S., Biffo, S., D'adamo, P., Mallamci, A., & Pannese, M. (2012). Growth defects and impaired cognitive-behavioral abilities in mice with knockout for *eif4h*, a gene located in the mouse homolog of the Williams-Beuren syndrome critical region. *American Journal of Pathology*, *180*, 1121–1135. https://doi.org/10.1016/j.ajpath.2011.12.008

Casanova, M. F., Buxhoeveden, D. P., Switala, A. E., & Roy, E. (2002). Minicolumnar pathology in autism. *Neurology*, *58*, 428–432.

Casanova, M. F., El-Baz, B., Vanbogaert, E., Narahari, P., & Switala, A. (2010). A topographic study of minicolumnar core width by lamina comparison between autistic subjects and controls: Possible minicolumnar disruption due to an anatomical element in-common to multiple laminae. *Brain Pathology*, *20*, 451–458.

Casanova, M. F., Van Kooten, I. a. J., Switala, A. E., Van Engeland, H., Heinsen, H., Steinbusch, H. W. M., Hof, P. R., Trippe, J., Stone, J., & Schmitz, C. (2006). Minicolumnar abnormalities in autism. *Acta Neuropathologica*, *112*, 287–303. https://doi.org/10.1007/s00401-006-0085-5

Cashin, A., & Barker, P. (2009). The triad of impairment in autism revisited. *Journal of Child and Adolescent Psychiatric Nursing*, *22*, 189–193. https://doi.org/10.1111/j.1744-6171.2009.00198.x

Chailangkarn, T., Trujillo, C. A., Freitas, B. C., Hrvoj-Mihic, B., Herai, R. H., Yu, D. X., Brown, T. T., Marchetto, M. C., Bardy, C., Mchenry, L., Stefanacci, L., Järvinen, A., Searcy, Y. M., Dewitt, M., Wong, W., Lai, P., Ard, M. C., Hanson, K. L., Romero, S., . . . Muotri, A. R. (2016). A human neurodevelopmental model for williams syndrome. *Nature*, *536*, 338–343. https://doi.org/10.1038/nature19067

Chan, M. M. Y., & Han, Y. M. Y. (2020). Differential mirror neuron system (MNS) activation during action observation with and without social-emotional components in autism: A meta-analysis of neuroimaging studies. *Molecular Autism*, *11*, Article 72. https://doi.org/10.1186/s13229-020-00374-x

Chen, Y., Huang, W., Séjourné, J., Clipperton, A. E., & Page, D. T. (2015). *Pten* mutations alter brain growth trajectory and allocation of cell types through elevated beta-catenin signaling. *Journal of Neuroscience*, *35*, 10252–10267. https://doi.org/10.1523/JNEUROSCI.5272-14.2015

Chiang, M., Reiss, A. L., Lee, A. D., Bellugi, U., Galaburda, A., Korenberg, J., Mills, D. L., Toga, A. W., & Thompson, P. M. (2007). 3D pattern of brain abnormalities in Williams syndrome visualized unsing tensor-based morphometry. *NeuroImage*, *36*, 1096–1109. https://doi.org/10.1016/j.neuroimage.2007.04.024

Cotney, J., Muhle, R. A., Sanders, S. J., Liu, L., Willsey, A. J., Niu, W., Liu, W., Klei, L., Lei, J., Yin, J., Reilly, S. K., Tebbenkamp, A. T., Bichsel, C., Pletikos, M., Sestan, N., Roeder, K., State, M. W., Devlin, B., & Noonan, J. P. (2015). The autism-associated chromatin modifier chd8 regulates other autism risk genes during human neurodevelopment. *Nature Communications*, *6*, Article 6404. https://doi.org/10.1038/ncomms7404

Courchesne, E., Mouton, P. R., Calhoun, M. E., Semendeferi, K., Ahrens-Barbeau, C., Hallet, M. J., Barnes, C. C., & Pierce, K. (2011). Neuron number and size in prefrontal cortex of children with autism. *Journal of the American Medical Association*, *306*, 201–210.

Courchesne, E., Pramparo, T., Gazestani, V. H., Lombardo, M. V., Pierce, K., & Lewis, N. E. (2019). The ASD living biology: From cell proliferation to clinical phenotype. *Molecular Psychiatry*, *24*, 88–107. https://doi.org/10.1038/s41380-018-0056-y

Crespi, B. J., & Procyshyn, T. L. (2017). Williams syndrome deletions and duplications: Genetic windows to understanding anxiety, sociality, autism, and schizophrenia. *Neuroscience & Biobehavioral Reviews*, *79*, 14–26. https://doi.org/10.1016/j.neubiorev.2017.05.004

Ferguson, J. N., Aldag, J. M., Insel, T. R., & Young, L. J. (2001). Oxytocin in the medial amygdala is essential for social recognition in the mouse. *Journal of Neuroscience*, *21*, 8278–8285. https://doi.org/10.1523/JNEUROSCI.21-20-08278.2001

Galaburda, A. M., & Bellugi, U. (2000). Multi-level analysis of cortical neuroanatomy in williams syndrome. *Journal of Cognitive Neuroscience*, *12*, 74–88. https://doi.org/10.1162/089892900561995

Galaburda, A. M., Holinger, D. P., Bellugi, U., & Sherman, G. F. (2002). Williams syndrome: Neuronal size and neuronal-packing density in primary visual cortex. *Archives of Neurology*, *59*, 1461–1467.

Galaburda, A. M., Wang, P. P., Bellugi, U., & Rossen, M. (1994). Cytoarchitectonic anomalies in a genetically based disorder—Williams syndrome. *Neuroreport*, *5*, 753–757. https://doi.org/10.1097/00001756-199403000-00004

Gaser, C., Luder, E., Thompson, P. M., Lee, A. D., Dutton, R. A., Geaga, J. A., Hayashi, K. M., Bellugi, U., Galaburda, A. M., Korenberg, J. R., Mills, D. L., Toga, A. W., & Reiss, A. L. (2006). Increased local gyrification mapped in Williams syndrome. *Neuroimage*, *33*, 46–54. https://doi.org/10.1016/j.neuroimage.2006.06.018

Gothelf, D., Searcy, Y. M., Reilly, J., Lai, P. T., Lanre-Amos, T., Mills, D., Korenberg, J. R., Galaburda, A., Bellugi, U., & Reiss, A. L. (2008). Association between cerebral shape and social use of language in Williams syndrome. *American Journal of Medical Genetics Part A*, *146A*, 2753–2761. https://doi.org/10.1002/ajmg.a.32507

Grabrucker, A. M. (2013). Environmental factors in autism. *Frontiers in Psychiatry*, *3*, Article 118. https://doi.org/10.3389/fpsyt.2012.00118

Gross, C., Chang, C., Kelly, S. M., Bhattacharya, A., Mcbride, S. M. J., Danielson, S. W., Jiang, M. Q., Chan, C. B., Ye, K., Gibson, J. R., Klann, E., Jongens, T. A., Moberg, K. H., Huber, K. M., & Bassell, G. J. (2015a). Increased expression of the PI3K enhancer PIKE mediates deficits in synaptic plasticity and behavior in fragile X syndrome. *Cell Reports*, *11*, 272–736. https://doi.org/10.1016/j.celrep.2015.03.060

Gross, C., Raj, N., Molinaro, G., Allen, A. G., Whyte, A. J., Gibson, J. R., Huber, K. M., Gourley, S. L., & Bassell, G. J. (2015b). Selective role of the catalytic PI3K subunit p110beta in impaired higher order cognition in fragile X syndrome. *Cell Reports*, *11*, 681–688. https://doi.org/10.1016/j.celrep.2015.03.065

Hanson, K. L., Lew, C. H., Hrvoj-Mihic, B., Cuevas, D., Greiner, D. M. Z., Groeniger, K. M., Edler, M. K., Halgren, E., Bellugi, U., Raghanti, M. A., & Semendeferi, K. (2020). Decreased density of cholinergic interneurons in striatal territories in Williams syndrome. *Brain Structure & Function*, *225*, 1019–1032. https://doi.org/10.1007/s00429-020-02055-0

Hanson, K. L., Lew, C. H., Hrvoj-Mihic, B., Groeniger, K. M., Halgren, E., Bellugi, U., & Semendeferi, K. (2017). Increased glia density in the caudate nucleus in Williams syndrome: Implications for frontostriatal dysfunction in autism. *Developmental Neurobiology*, *78*, 531–545. https://doi.org/10.1002/dneu.22554

Hoogenraad, C. C., Akhmanova, A., Galjart, N., & De Zeeuw, C. I. (2004). LIMK1 and CLIP-115: Linking cytoskeletal defects to Williams syndrome. *BioEssays, 26*, 141–150. https://doi.org/10.1002/bies.10402

Hoogenraad, C. C., Koekkoek, B., Akhmanova, A., Krugers, H., Dortland, B., Miedema, M., Van Aphen, A., Kistler, W. M., Jaegle, M., Koutsourakis, M., Van Camp, N., Verhoye, M., Van Der Linden, A., Kaverina, I., Grosveld, F., De Zeeuw, C. I., & Galjart, N. (2002). Targeted mutation of *cyln2* in the Williams syndrome critical region links CLIP-115 haploinsufficiency to neurodevelopmental abnormalities in mice. *Nature Genetics, 32*, 116–331. https://doi.org/10.1038/ng954

Hrvoj-Mihic, B., Hanson, K. L., Lew, C. H., Stefanacci, L., Jacobs, B., Bellugi, U., & Semendeferi, K. (2017). Basal dendritic morphology of cortical pyramidal neurons in Williams syndrome: Prefrontal cortex and beyond. *Frontiers in Neuroscience, 11*, Article 419. https://doi.org/10.3389/fnins.2017.00419

Hrvoj-Mihic, B., & Semendeferi, K. (2019). Neurodevelopmental disorders of the prefrontal cortex in an evolutionary context. *Progress in Brain Research, 250*, 109–127. https://doi.org/10.1016/bs.pbr.2019.05.003

Huang, W., Chen, Y., & Page, D. T. (2016). Hyperconnectivity of prefrontal cortex to amygdala projections in a mouse model of macrocephaly/autism syndrome. *Nature Communications, 7*, Article 13421. https://doi.org/10.1038/ncomms13421

Hutsler, J. J., & Zhang, H. (2010). Increased dendritic spine densities on cortical projection neurons in autism spectrum disorders. *Brain Research, 1309*, 83–94. https://doi.org/10.1016/j.brainres.2009.09.120

Jacobs, B., Schall, M., Prather, M., Kapler, E., Driscoll, L., Baca, S., Jacobs, J., Ford, K., Wainwright, M., & Treml, M. (2001). Regional dendritic and spine variation in human cerebral cortex: A quantitative Golgi study. *Cerebral Cortex, 11*, 558–571. https://doi.org/10.1093/cercor/11.6.558

Jacot-Descombes, S., Uppal, N., Wicinski, B., Santos, M., Schmeidler, J., Giannakopoulos, P., Heinsen, H., Schmitz, C., & Hof, P. R. (2012). Decreased pyramidal neuron size in brodmann areas 44 and 45 in patient with autism. *Acta Neuropathologica, 124*, 67–79. https://doi.org/10.1007/s00401-012-0976-6

John, S., & Jaeggi, A. V. (2021). Oxytocin levels tend to be lower in autistic children: A meta-analysis of 31 studies. *Autism, 25*, 2152–2161. https://doi.org/10.1177/13623613211034375

Ka, M., Chopra, D. A., Dravid, S. M., & Kim, W. (2016). Essential roles for ARID1B in dendritic arborization and spine morphology of developing pyramidal neurons. *Neurobiology of Disease, 36*, 2723–2742. https://doi.org/10.1523/JNEUROSCI.2321-15.2016

Karst, A. T., & Hutsler, J. J. (2016). Two-dimensional analysis of the supragranular layers in autism spectrum disorder. *Research in Autism Spectrum Disorders, 32*, 96–105. https://doi.org/10.1016/j.rasd.2016.09.004

Kennedy, D. N., Semendeferi, K., & Courchesne, E. (2007). No reduction of spindle neuron number in frontoinsular cortex in autism. *Brain and Cognition, 64*, 124–129. https://doi.org/10.1016/j.bandc.2007.01.007

Lalli, M. A., Jang, J., Park, J. C., Wang, Y., Guzman, E., Zhou, H., Audouard, M., Bridges, D., Tovar, K. R., Papuc, S. M., Tutulan-Cunita, A. C., Huang, Y., Budisteanu, M., Arghir, A., & Kosik, K. S. (2016). Haploinsufficiency of BAZ1B contributes to Williams syndrome through transcriptional dysregulation of neurodevelopmental pathways. *Human Molecular Genetics, 25*, 1294–1306. https://doi.org/10.1093/hmg/ddw010

Lew, C. H., Brown, C., Bellugi, U., & Semendeferi, K. (2017). Neuron density is decreased in the prefrontal cortex in Williams syndrome. *Autism Research, 10*, 99–112. https://doi.org/10.1002/aur.1677

Lew, C. H., Groeniger, K. M., Bellugi, U., Stefanacci, L., Schumann, C. M., & Semendeferi, K. (2018). A postmortem sterological study of the amygdala in Williams syndrome. *Brain Structure & Function, 223*, 1897–1907. https://doi.org/10.1007/s00429-017-1592-y

Lew, C. H., Groeniger, K. M., Hanson, K. L., Cuevas, D., Greiner, D. M. Z., Hrvoj-Mihic, B., Bellugi, U., Schumann, C. M., & Semendeferi, K. (2020). Serotonergic innervation of the amygdala is increased in autism spectrum disorder and decreased in Williams syndrome. *Molecular Autism, 11*. https://doi.org/10.1186/s13229-019-0302-4

Lew, C. H., Hanson, K. L., Groeniger, K. M., Greiner, D., Cuevas, D., Hrvoj-Mihic, B., Schumann, C. M., & Semendeferi, K. (2019). Serotonergic innervation of the human amygdala and evolutionary implications. *American Journal of Physical Anthropology, 170*, 351–360. https://doi.org/10.1002/ajpa.23896

Lew, C. H., & Semendeferi, K. (2017). Evolutionary specializations of the human limbic system. In J. Kaas (Ed.), *Evolution of nervous systems* (2nd ed., pp. 277–291). Elsevier.

Li, H. H., Roy, M., Kuscuoglu, U., Spencer, C. M., Halm, B., Harrison, K. C., Bayle, J. H., Splendore, A., Ding, F., Meltzer, L. A., Wright, E., Paylor, R., Deisseroth, K., & Francke, U. (2009). Induced chromosome deletions cause hypersociability and other features of Williams-Beuren syndrome in mice. *EMBO Molecular Medicine*, *1*, 50–65. https://doi.org/10.1002/emmm.200900003

Li, Q., Becker, B., Jiang, X., Zhao, Z., Zhang, Q., Yao, S., & Kendrick, K. M. (2019). Decreased interhemispheric functional connectivity rather than corpus callosum volume as a potential biomarker for autism spectrum disorder. *Cortex*, *119*, 258–266. https://doi.org/10.1016/j.cortex.2019.05.003

Marchetto, M. C., Carromeu, C., Acab, A., Yu, D., Yeo, G. W., Mu, Y., Chen, G., Gage, F. H., & Muotri, A. R. (2010). A model for neural development and treatment of Rett syndrome unsing human induced pluripotent stem cells. *Cell*, *143*, 527–539. https://doi.org/10.1016/j.cell.2010.10.016

Martens, M. A., Wilson, S. J., Dudgeon, P., & Reutens, D. C. (2009). Approachability and the amygdala: Insights from Williams syndroms. *Neuropsychologia*, *47*, 2446–2453. https://doi.org/10.1016/j.neuropsychologia.2009.04.017

Meda, S. A., Pryweller, J. R., & Thorton-Wells, T. A. (2012). Regional brain differences in cortical thickness, surface area, and subcortical volume in individuals with Williams syndrome. *PLOS ONE*, *7*, Article e31913. https://doi.org/10.1371/journal.pone.0031913

Mellios, N., Feldman, D. A., Sheridan, S. D., Ip, J. K. P., Kwok, S., Amoah, S. K., Rosen, B., Rodriguez, B. A., Crawford, B., Swaminathan, R., Chou, S., Li, Y., Ziats, M., Ernst, C., Jaenisch, R., Haggarty, S. J., & Sur, M. (2018). MeCP2-regulated mirnas control early human neurogenesis through differential effects on erk and akt signaling. *Molecular Psychiatry*, *23*, 1051–1065. https://doi.org/10.1038/mp.2017.86

Morgan, J. T., Chana, G., Abramson, I., Semendeferi, K., Courchesne, E., & Everall, I. P. (2012). Abnormal microglial–neuronal spatial organization in the dorsolateral prefrontal cortex in autism. *Brain Research*, *1456*, 72–81. https://doi.org/10.1016/j.brainres.2012.03.036

Morgan, J. T., Chana, G., Pardo, C. A., Achim, C., Semendeferi, K., Buckwalter, J., Courchesne, E., & Everall, I. P. (2010). Microglial activation and increased microglial density observed in the dorsolateral cortex in autism. *Biological Psychiatry*, *68*, 368–376. https://doi.org/10.1016/j.biopsych.2010.05.024

Ng, R., Brown, T. T., Erhart, M., Jarvinen, A. M., Korenberg, J. R., Bellugi, U., & Halgren, E. (2016). Morphological differences in the mirror neuron system in Williams syndrome. *Social Neuroscience*, *11*, 277–288. https://doi.org/10.1080/17470919.2015.1070746

Oliveira, A. F., & Yasuda, R. (2014). Neurofibromin is the major ras inactivator in dendritic spines. *Journal of Neuroscience*, *34*, 776–783. https://doi.org/10.1523/JNEUROSCI.3096-13.2014

Pasciuto, E., Borrie, S. C., Kanellopoulos, A. K., Santos, A. R., Cappuyns, E., D'Andrea, L., Pacini, L., & Bagni, C. (2015). Autism spectrum disorders: Translating human deficits into mouse behavior. *Neurobiology of Learning and Memory*, *124*, 71–87. https://doi.org/10.1016/j.nlm.2015.07.013

Pucilowska, J., Vithayathil, J., Tavares, E. J., Kelly, C., Colleen Karlo, J., & Landreth, G. E. (2015). The 16p11.2 deletion mouse model of autism exhibits altered cortical progenitor proliferation and brain cytoarchitecture linked to the ERK MAPK pathway. *Journal of Neuroscience*, *35*, 3190–3200. https://doi.org/10.1523/JNEUROSCI.4864-13.2015

Rafiee, F., Habibabadi, R. R., Motaghi, M., Yousem, D. M., & Yousem, I. J. (2022). Brain MRI in autism spectrum disorder: Narrative review and recent advances. *Journal of Magnetic Resonance Imaging*, *55*, 1613–1624. https://doi.org/10.1002/jmri.27949

Reiss, A. L., Eliez, S., Schmitt, E., Straus, E., Lai, Z., Jones, W., & Bellugi, U. (2000). Neuroanatomy of Williams syndrome: A high-resolution mri study. *Journal of Cognitive Neuroscience*, *12*, 65–73.

Reser, J. E. (2011). Conceptualizing the autism spectrum in terms of natural selection and behavioral ecology: The solitary forager hypothesis. *Evolutionary Psychology*, *9*, 207–238.

Sampaio, A., Sousa, N., Férnandez, M., Vasconcelos, C., Shenton, M. E., & Gonçalves, Ó. F. (2008). MRI assesment of superior temporal gyrus in Williams syndrome. *Cognitive and Behavioral Neurology*, *21*, 150–156. https://doi.org/10.1097/WNN.0b013e31817720e4

Santos, M., Uppal, N., Butti, C., Wicinski, B., Schmeidler, J., Giannakopoulos, P., Heinsen, H., Schmitz, C., & Hof, P. R. (2011). von Economo neurons in autism: A stereologic study of the frontoinsular cortex in children. *Brain Research*, *1380*, 206–217. https://doi.org/10.1016/j.brainres.2010.08.067

Schumann, C. M., & Amaral, D. G. (2006). Stereological analysis of amygdala neuron number in autism. *Journal of Neuroscience*, *26*, 7674–7679. https://doi.org/10.1523/JNEUROSCI.1285-06.2006

Schumann, C. M., Hamstra, J., Goodlin-Jones, B. L., Lotspeich, L. J., Kwon, H., Buonocore, M. H., Lammers, C. R., Reiss, A. L., & Amaral, D. G. (2004). The amygdala is enlarged in children but not adolescents with autism; the hippocampus is enlarged at all ages. *Neurobiology of Disease, 24*, 6392–6401. https://doi.org/10.1523/jneurosci.1297-04.2004

Semendeferi, K. (2018). Why do we want to talk? Evolution of neural substrates of emotion and social cognition, *Interaction Studies, 19*, 102–120. https://doi.org/10.1075/is.17046.sem

Semendeferi, K., Barger, N., & Schenker, N. (2010). Brain reorganization in humans and apes. In D. Broadfield, K. Yuan, K. Schick, & N. Toth (Eds.), *The human brain evolving: Paleoneurological studies in honor of Ralph L. Holloway* (pp. 119–149). Stone Age Institute Press.

Simms, M. L., Kemper, T. L., Timbie, C. M., Bauman, M. L., & Blatt, G. J. (2009). The anterior cingulate cortex in autism: Heterogeneity of qualitative and quantitative cytoarchitectonic features suggest possible subgroups. *Acta Neuropathologica, 118*, 673–684. https://doi.org/10.1007/s00401-009-0568-2

Tinbergen, E. A., & Tinbergen, N. (1972). *Early childhood autism: an ethological approach*. Parey.

Weir, R. K., Bauman, M. D., Jacobs, B., & Schumann, C. M. (2018). Protracted dendritic growth in the typically developing human amygdala and increased spine density in young ASD brains. *Journal of Camparative Neurology, 526*, 262–274. https://doi.org/10.1002/cne.24332

Van Kooten, I. a. J., Palmen, S. J. M. C., Von Cappeln, P., Steinbusch, H. W. M., Korr, H., Heinsen, H., Hof, P. R., Van Engeland, H., & Schmitz, C. (2008). Neurons in the fusiform gyrus are fewer and smaller in autism. *Brain, 131*, 987–999. https://doi.org/10.1093/brain/awn033

Wilder, L., Hanson, K. L., Lew, C. H., Bellugi, U., & Semendeferi, K. (2018). Decreased neuron density and increased glia density in the ventromedial prefrontal cortex (Brodmann area 25) in Williams syndrome. *Brain Sciences, 8*, Article 209. https://doi.org/10.3390/brainsci8120209

Zhao, C., Avil, C., Abel, R. A., Almli, C. R., Mcquillen, P., & Pleasure, S. J. (2005). Hippocampal and visuospatial learning defects in mice with a deletion of frizzled 9, a gene in the Williams syndrome deletion interval. *Development and Disease, 132*, 2917–2927. https://doi.org/10.1242/dev.01871

Zilles, K., & Amunts, K. (2010). Centenary of Brodmann's map—Conception and fate. *Nature Reviews Neuroscience, 11*, 139–145.

16

Attention-Deficit Hyperactivity Disorder

An Evolutionary View

Annie Swanepoel

Historical Background

Introduction

Attention-deficit hyperactivity disorder (ADHD) is a common neurodevelopmental disorder which is characterized by pervasive and persistent symptoms of hyperactivity/impulsivity and inattention that are extreme for age and impact on functioning (Dunn et al. 2019). The symptoms were first described more than 200 years ago (Faraone et al. 2021). The diagnostic criteria require that symptoms are present before age 12 and are present in more than one setting, usually at home and at school. The worldwide prevalence is about 5%, with boys being diagnosed at about 2–4 times the rate of girls (Dunn et al. 2019; Posner et al. 2020).

Despite all the evidence indicating that ADHD symptomatology occurs on a spectrum and is not a disorder that can be clearly delineated from those without ADHD, it continues to be seen as a categorical disorder even in the revised *Diagnostic and Statistical Manual of Mental Disorders*, Fifth Edition (DSM-5), and *International Classification of Diseases*, 11th Revision (ICD-11; Posner et al. 2020). We know that the cutoffs in terms of the number of symptoms are arbitrary and that the diagnosis depends on clinical experience of recognizing to what extent functioning is impaired in deciding whether intervention is warranted (Posner et al. 2020).

Prior to the DSM-5 and ICD-11, ADHD was considered to be a behavioral disorder and was not allowed to be diagnosed as a comorbid disorder with autism or learning disability (Swanepoel et al. 2017). This has now been rectified, recognizing the fact that ADHD is a neurodevelopmental disorder and that neurodevelopmental disorders commonly co-occur (Swanepoel et al. 2017). However, it remains the case that children with ADHD are more likely to develop oppositional defiant disorder and conduct disorder, particularly if the ADHD is not recognized and treated (Thapar and Cooper 2016).

Untreated ADHD has a range of unfavorable outcomes, including increased chances of school failure, serious accidental injury, illicit drug taking, problem gambling, involvement in the criminal justice system, and a decreased life expectancy (Bernardi et al. 2012; Thapar and Cooper 2016; Dunn et al. 2019). Although ADHD is usually thought of as a childhood disorder, half of children with ADHD continue to have impairing symptoms as adults, including poor work performance and unemployment (Dunn et al. 2019).

Etiology

ADHD can be caused by genetic as well as environmental factors and gene–environment interactions.

If we consider the genetic aspects first, it is well recognized that ADHD is highly heritable, with estimates of more than 70% (Dunn et al. 2019). Genome-wide association studies show that there is long-standing selection against ADHD-associated alleles since Palaeolithic times, which cannot be explained by African admixture or Neanderthal introgression (Esteller-Cucala et al. 2020). There are no genes of major effect, and the genetic liability is shared with several other common psychiatric disorders including anxiety and mood disorders (Dunn et al. 2019; Posner et al. 2020).

We know that many environmental factors also increase the chances of ADHD, particularly fetal exposure to increased maternal inflammation, for example, through maternal stress and maternal infection (Dunn et al. 2019). There is evidence for the hypothesis that prenatal exposure to inflammation leads to neuroinflammation in the fetus, which may predispose to ADHD (Dunn et al. 2019). Talge et al. (2007, p. 245) suggest that "extra vigilance or anxiety, readily distracted attention, or a hyper-responsive HPA axis may have been adaptive in a stressful environment during evolution, but exists today at the cost of vulnerability to neurodevelopmental disorders."

Interestingly, people with immune disorders such as eczema, asthma, rheumatoid arthritis, type 1 diabetes, and hypothyroidism are more likely to receive an ADHD diagnosis than those with normal immune systems (Dunn et al. 2019). In rare cases, ADHD can be caused by brain trauma, a single gene, or severe early deprivation (Posner et al. 2020). Environmental toxins, for example, lead, nitric oxide, and some medications in pregnancy, for example, valproate, lead to an increased risk of ADHD, as does prematurity and maternal obesity in pregnancy (Dunn et al. 2019). Indicators of deprivation have a cumulative effect. Once again, these effects are non-specific and are also related to other psychiatric disorders (Dunn et al. 2019).

It is not clear how much of the genetic effect is related to genotype-by-environment interactions (Dunn et al. 2019). What we do know is that genetically susceptible individuals are more likely to develop ADHD if they also experience particular environmental factors, although the precise mechanism of action is still unclear (Dunn et al. 2019) but may well involve epigenetic mechanisms (Hamza et al. 2019).

It is possible that children with ADHD, who are more difficult, have a greater ability to elicit maternal care and that this may improve survival in crises—however, in more settled environments maternal attention is more negative to children with ADHD than to children without ADHD, with resultant detrimental effects on cognitive and emotional development (Williams and Taylor 2006). It has been shown that parenting seems to be child-specific, in which children with ADHD elicit harsher parenting than their neurotypical siblings, adding an environmental risk to the genetic risk for ADHD (Swanepoel et al. 2017; Posner et al. 2020; Wren et al. 2021).

Neurotransmitters and Genetic Risk Factors

Children with ADHD have a slight reduction in overall gray matter and develop more slowly than typically developing children (Dunn et al. 2019). In line with the finding about slower maturation, several studies have found that children who are young in their school year are more likely to be diagnosed with ADHD than their older peers (Swanepoel et al. 2017). Whereas previously altered structural asymmetry was thought to be a potentially useful biomarker for ADHD, Postema et al. (2021) found that this is unlikely to be the case.

The main areas that are affected are important for the top-down regulation of attention, inhibition, emotion, and motivation, driven by the prefrontal cortex, orbitofrontal cortex, and anterior cingulate cortex (Dunn et al. 2019). People with ADHD particularly struggle with tasks that are long and repetitive, rather than interesting tasks with frequent changes (Posner et al. 2020).

In terms of neurotransmitters, a deficiency of dopamine is thought to play a role, mainly because medications that increase dopamine (e.g., methylphenidate) reduce ADHD-type symptoms (Dunn et al. 2019). Candidate gene studies identified polymorphisms in genes encoding for dopamine transporter (DAT1) and D4 and D5 dopamine receptors, as well as in the dopamine transporter gene (SLC6A3), in people with ADHD (Dunn et al. 2019). Furthermore, the FOXP2 gene, which regulates dopamine and neurodevelopment of frontal brain regions, was also found to be significantly associated with ADHD (Dunn et al. 2019).

However, we know that noradrenaline and serotonin also are important in ADHD. Serotonin is implicated in the hyperactive and impulsive symptoms of ADHD as it has an important role in regulating behavioral inhibition (Dunn et al. 2019). This fits with the observation that children with ADHD often have a comorbid anxiety disorder and are at twofold greater risk of future depression (Dunn et al.

2019). Both depression and anxiety are known to be associated with abnormalities in serotonin regulation. ADHD is therefore not purely a cognitive disorder and is associated with problems with emotional regulation (Dunn et al. 2019).

Once again, there is an interaction between genes and environment, and ADHD risk is increased by interaction between the type of polymorphism in the promoter region of the serotonin transporter encoding gene (5-HTTLPR) and environmental factors such as psychosocial stress (Dunn et al. 2019; Posner et al. 2020). Several other genes, including DRD4, have been implicated in meta-analyses but not yet proven in genome-wide association studies (Faraone et al. 2021).

Treatment

Medication (stimulant or non-stimulant) is the mainstay of treatment for children and adults with ADHD, with a significant positive impact on their ability to function at school and at work (Thapar and Cooper 2016). However, there are ethical concerns about whether to prescribe medication to children to make it easier for their parents and teachers (Swanepoel 2021).

In adults, Posner et al. (2020) summarized that a Swedish study found a significant reduction in criminality, a US-based insurance registry found reduced motor vehicle accidents, and a Taiwanese study found a lowering of depression risk during periods of stimulant use. There is a small risk of premature death in adults due to accidents and suicide, and reassuringly these risks have been shown to be reduced by medication (Faraone et al. 2021). Adults whose ADHD is treated also show improved occupational and social functioning (Sarkis 2014).

Evolutionary Evidence Specifically Related to Human Health and Disease

Each human alive today is the result of a continuous, unbroken line of ancestors stretching back 3.5 billion years. Genetic traits cannot survive across generations if the carriers of the genes do not reproduce. In evolutionary terms, it is better to survive and have the chance of reproducing, even if at considerable individual cost.

Traits that do not interfere with reproduction are not selected against—that is, any disease that originates in later life (i.e., after the reproductive age) is not eliminated through natural selection. The traits and behavioral strategies that have survived and that exist nowadays are likely to have done so for good reasons. This may include traits that have traditionally been seen as "pathological" and diagnosed as psychiatric disorders. We have previously suggested that an evolutionary understanding might be helpful in understanding the occurrence of psychopathology and might also help us to discriminate between adaptive responses and pathology in psychiatry more generally (Swanepoel et al. 2016).

In this chapter, we will attempt to answer the question of why ADHD exists and why it was not eliminated through natural selection.

An evolutionary perspective on ADHD can help us in the following three main ways:

1. Thinking about ADHD from an evolutionary perspective can help us see in what ways it can be adaptive for a child to develop ADHD if they experience maternal stress while in utero. This may result from intrauterine programming and may lead to a fast life history and potentially benefits for the group if not the individual (group selection).
2. Also, considering in which environments and situations ADHD traits offer a survival value and are adaptive can help us recognize also where this is not the case and where children with ADHD may be caught in an evolutionary mismatch between what they are adapted to cope with and the current environment.
3. Finally, people with ADHD may be better conceptualized as being at the extreme end of a spectrum, and taking their neurodiversity into account and making reasonable adjustments for them may benefit not just the people with ADHD but our whole society.

These aspects are discussed in more detail below.

Intrauterine Programming

We know that the developing fetus is very sensitive to maternal stress hormones. In our evolutionary history it is possible that some increase in the child's stress hormones in some individuals was adaptive in a stressful environment and that this type of intrauterine programming prepared the child for the environment in which they were going to find themselves (Glover 2011). If we consider the symptoms of what we now call ADHD, it is conceivable that those with the trait of "distractibility" may perceive a sudden danger promptly compared to those without the trait, who may be concentrating on something else and unaware of changes in the environment until it is too late. Also, impulsivity could lead to faster reactions and a willingness to take risks, which may have been adaptive in a dangerous environment (Talge et al. 2007).

Life History Theory

All organisms are faced with a range of decisions they need to make about the timing of reproduction, the number of offspring, and the amount invested in each. These decisions are usually unconscious and driven by the environment. Life history theory describes the different routes that individuals may take (Box 16.1).

In humans there is a spectrum too—those of us who live in safe and well-resourced circumstances can afford to wait before having children and then to invest a large amount into one or two children, therefore maximizing the chances of their future success. However, those of us who live in precarious, dangerous situations, where many children

> **BOX 16.1 Life History Theory and Relevance to ADHD**
>
> Individuals who follow a **slow life history** defer reproduction and tend to have fewer offspring, in whom they invest considerable resources. They can be described as following a biologically embodied unconscious strategy that prioritizes the quality of their offspring over their quantity. This is adaptive in benign environments in which offspring are expected to survive. For example, elephants, which live in a safe environment with few if any predators, have adapted to have few offspring, as each has a good chance to survive, and to invest a lot of time and energy into them.
>
> Individuals who follow a **fast life history**, begin reproducing at a young age and tend to have more offspring, each of whom gets little nurturance. They can be described as following a biologically embodied unconscious strategy that prioritizes quantity over quality. This is adaptive in harsh environments in which many offspring are expected to die young. Rabbits are an example of a species that follows a fast life history. Rabbits, which typically face a range of predators, leave more surviving offspring if they maximize the number of young ones they have, investing less time and energy in each.

die before reaching adulthood, will leave more surviving descendants if we have as many children as possible.

If a mother is stressed due to a harsh environment, her stress hormones will lead to an inflammatory state in her own body, possibly protecting her against infection but also causing neuroinflammation in the brain of her fetus if she is pregnant (Glover 2011). Maternal stress activates the mother's hypothalamopituitary axis (HPA) and increases cortisol in her blood circulation, which crosses the placenta and has a profound effect on the fetus's neurological development. Sapolsky (2017) suggests that maternal stress leads to neurodevelopmental changes in the baby, which may be evolutionarily adaptive as the baby is primed to survive in a dangerous environment. This means that the baby is adapted to cope with the environment the mother finds herself in—if it is stressful, both mother and newborn baby have activated stress systems. It is as if the fetus is prepared for the life it will likely be living. We know that a baby who is born to a stressed mother has a higher chance of ADHD and that this may well be adaptive in a harsh environment, where it may pay off to be hyperactive, impulsive, and easily distracted (Glover 2011; Swanepoel et al. 2017).

We know that people with ADHD are more inclined to follow a fast life history, with early pregnancy and many children with relatively little investment in each (Faraone and Larson 2019; Min et al. 2021). Studies from Scandinavia (Østergaard et al. 2017) and Taiwan (Hua et al. 2021) found teen pregnancies to be more common in those with ADHD and for the babies to also have an increased chance of having ADHD, leading to an intergenerational transmission of ADHD. Faraone and Larson (2019) collated evidence to show that ADHD is correlated with a younger parental age at birth, having more children, reduced years of schooling, reduced chances of college completion, and earlier death—all of which fits with a fast life history.

Interestingly teenage pregnancies reduce significantly if girls with ADHD are medicated, perhaps by reducing impulsivity and the risk of an unplanned pregnancy (Hua et al. 2021; Hinshaw et al. 2022). We also know that children and adults who are medicated for ADHD have a better chance of functioning well at school and work and that the chances of early death through accidents and suicide are reduced (Thapar and Cooper 2016). We can therefore hypothesize that effective treatment of ADHD helps people shift their life course from "fast" to "slow."

Group Selection

Apart from the potential survival value of the individual with ADHD in dangerous situations, there is some evidence to suggest that individuals with ADHD may have benefited the group they belonged to, although this is controversial. The evolutionary benefit to the group even if there is individual detriment is termed "group selection" (Rachlin 2019).

For example, in our hunter–gatherer past, if there were a drought and some individuals were more active and willing to take risks (the flip side of the coin of hyperactivity and impulsivity), they may explore further and potentially find areas with more resources that then benefit the rest of the group too. There is evidence that those of the Ariaal tribe with the DRD4 mutation, making ADHD more likely, were better nourished than those without—but only if they followed a nomadic lifestyle. This was not the case when they were farmers (Swanepoel et al. 2017). Ding et al. (2002) report that DRD4 receptor gene type 7 occurs in 0% of peoples who stay put, compared to 63% of nomads. It is conceivable that having the gene that predisposes toward novelty-seeking is adaptive in difficult circumstances (but not in easier times).

Williams and Taylor (2006) make the point that increased exploration of physical space, as happens with hyperactivity, may be of major benefit to the group not only through finding something useful but perhaps also through discovering a danger which others can then avoid. In that way ADHD genes may persist in the population even if they are of no use to the individual. Furthermore, impulsivity may test social limits, overcome superstitions, and perform dangerous experiments, which again may be of benefit for the group through gaining useful knowledge, even if only of what not to do (Williams and Taylor 2006).

However, Qi et al. (2021) found that an increase in novelty-seeking is not just a feature of ADHD but also a normal part of adolescent development and is furthermore associated with a variety of other psychiatric disorders, including bipolar disorder, major depressive disorder, and schizophrenia. Also, the role of DRD4 as a risk gene in ADHD is uncertain and will require further genome-wide association studies for confirmation (Faraone et al. 2021).

Evolutionary Mismatch

The environment that we live in now is very different from that which our ancestors lived in over tens of thousands of years. One example is that in the past humans had to suffer through food shortages and famine and that those who survived were skilled at eating lots of high-calorie food when they could find it and storing this as body fat for the leaner times. They survived, and we have inherited those genes. In the current environment where high-calorie food is cheap and ubiquitous, our innate tendency of wanting to stock up for the next famine leads to widespread obesity. The situation where the current environment differs in significant ways from the so-called environment of evolutionary adaptedness in the past is termed an "evolutionary mismatch."

We have previously argued that children with ADHD are also caught in an evolutionary mismatch and that this may partly account for the apparent increase in prevalence (Swanepoel et al. 2017). The current situation in which children need to sit still and concentrate for five hours a day, five days a week is clearly one which was not shared by most of our ancestors. Children used to be able to be a lot more active physically throughout the day, whether at play or at work. Children with ADHD struggle with the basic expectations of school, which are to sit still and pay attention.

Tegelbeckers et al. (2015) showed that children with ADHD engaged their frontal and temporal areas to a novel stimulus over and above what children without ADHD would do. They state that this indicates an inefficient use of neuronal resources in children with ADHD that could be closely linked to increased distractibility. This may well be the case in the safe, predictable environments many children experience in affluent countries. However, in a dangerous environment, it may have offered survival value if it meant recognizing camouflaged predators or poisonous snakes or insects.

Neurodiversity: Advantages of ADHD

In our hunter–gatherer past, the core symptoms of ADHD, namely hyperactivity/impulsivity and inattention, may have been evident as exploratory behavior, fast reactions, and fast attention to novel stimuli, which may all have been of benefit to prehistoric hunters.

Evolution does not just prepare us for the ideal. As there is a range of environments, the chances of success of a species are better if individuals are neurodiverse. Williams and Taylor (2006) propose that groups have a better chance of survival if a small percentage of the members are risk-taking. People with ADHD are typically more optimistic about taking risks (Shoham et al. 2021) and more creative (Girard-Joyal and Gauthier 2022). Many highly successful entrepreneurs credit their success to their ADHD and accompanying symptoms of having lots of energy and being willing to take risks (Sônego et al. 2021).

Nowadays, the neurodiversity movement challenges the concept of "disorder," and this fits with ADHD being seen as a "variation" instead (Sonuga-Barke and Thapar 2021).

Counterarguments and Evidence against Evolutionary Influences

The main argument against evolution would be if ADHD were a disease with no survival value. As discussed earlier in this chapter, we know that the outcomes for children and adults with ADHD are worse than for those without ADHD. However, evolution does not select for happiness or longevity—but only for survival and reproduction. People with ADHD tend to follow a fast life history, which is associated with earlier and more plentiful reproduction; and that may be why it still exists even in our modern society (Han and Chen 2020).

Furthermore, the fact that ADHD leads to worse outcomes can be explained as an evolutionary mismatch, in which the current environment is very different from that which evolution prepared us for. What is seen as pathological now may well have been adaptive in the past and may still be adaptive in particular environments where a willingness to take risks and high levels of energy are advantageous (Sedgwick et al. 2019).

Arildskov et al. (2022) have attempted to recreate a laboratory test for hunter–gatherer skills by comparing children with and without ADHD and seeing who was better able to find hidden coins. They rightly conclude that while they found no difference between the groups, this neither proves nor disproves that ADHD may have conferred an advantage in other ways. Thagaard and colleagues reviewed the evidence in 2016 and found that the natural selection–based accounts of ADHD had not been empirically tested.

Clinical and Therapeutic Implications

Children with ADHD are caught in an evolutionary mismatch in which their potential strengths of high activity levels and willingness to take risks are not valued in our modern schools (Swanepoel 2021). Schools currently expect that all children sit still and concentrate for long periods. This is not reasonable for those with ADHD, who learn much better if they are able to move and learn in a practical, hands-on way.

Ethical considerations also arise: is it acceptable to prescribe medication to change the biology of a child to fit with a system that is geared against them? The pressure to prescribe is usually from the parents or teachers, rather than the child. We know that medication for ADHD can help a child do better at school, and that is, of course, important. However, are we falling into the trap of prescribing medication instead of changing the way that schools are run? The Equality Act of 2010 in the United Kingdom mandates that "reasonable adjustments" are made to enable people with a disability to engage in school or work. This would include allowing movement breaks, making sure to get the child's attention before giving instructions, and seating the child close to the teacher to minimize other distractions.

Christiansen et al. (2019) summarizes the evidence and shows that 15 minutes of exercise at the beginning of the day can reduce ADHD-type behaviors and that this positive effect can be maintained by short bouts of moderate to vigorous activity every 90 minutes. Implementing a mandatory exercise regime would reduce the need for medication and would be beneficial also to those children who do not have ADHD.

One of the helpful insights that an evolutionary view provides is that the goodness of fit between an individual and the environment is important. The well-known maxim of "survival of the fittest" means that those individuals who have a good fit with their environment are those who thrive. By recognizing and valuing the unique talents that people with ADHD bring, we can help them find environments that play to their strengths. Schools may help these curious, active children by providing hands-on learning and self-paced computer assignments with lots of opportunities for physical exercise. Many children would benefit, not just those with ADHD.

Many adults with ADHD benefit from diagnosis and medication as that may enable them to function in a sedentary workplace that requires sustained concentration. However, other adults with ADHD prefer not to take medication as their hyperactivity may be useful in situations where a high level of physical activity is required. Also, as discussed previously, many entrepreneurs credit their ADHD for their willingness to take risks.

Unanswered Questions and Future Work

There are several new avenues that are being explored in terms of both prevention and treatment of ADHD:

- Addressing pre- and postnatal neuroinflammation (Dunn et al. 2019)
- Examining the gut–microbiota axis (Sukmajaya et al. 2021; Kalenik et al. 2021)
- Considering if emotional symptoms need to be incorporated into diagnostic criteria (Faraone and Larson 2019)
- Researching the benefits of exercise (Christiansen et al. 2019)

Key Points to Remember

- The evolutionary concepts of intrauterine programming, fast life history, and group selection may improve our understanding of ADHD over the life course.
- **ADHD favors novelty-seeking and risk-taking, which may have been adaptive in dangerous environments in our evolutionary past but are now deleterious in our modern school environments where children are expected to sit still and listen. This is called an evolutionary mismatch, in which evolution did not prepare us for the modern environment we find ourselves in.**

- Understanding ADHD in terms of an evolutionary mismatch raises significant issues regarding the management of childhood ADHD, including ethical ones. We should not see ADHD merely as a disorder that needs to be medicated—but rather as a variation which needs to be incorporated. Increasing opportunities for exercise and hands-on learning in schools would be of benefit.
- The neurodiversity movement correctly considers ADHD not as a disease but as a natural variation that can be of benefit in certain environments.

References

Arildskov, T. W., Virring, A., Thomsen, P. H., & Østergaard, S. D. (2022). Testing the evolutionary advantage theory of attention-deficit/hyperactivity disorder traits. *European Child and Adolescent Psychiatry*, *31*(2), 337–348. https://doi.org/10.1007/s00787-020-01692-4

Bernardi, S., Faraone, S. V., Cortese, S., Kerridge, B. T., Pallanti, S., Wang, S., & Blanco, C. (2012). The lifetime impact of attention deficit hyperactivity disorder: Results from the National Epidemiologic Survey on Alcohol and Related Conditions (NESARC). *Psychological Medicine*, *42*(4), 875–887. https://doi.org/10.1017/S003329171100153X

Christiansen, L., Beck, M. M., Bilenberg, N., Wienecke, J., Astrup, A., & Lundbye-Jensen, J. (2019). Effects of exercise on cognitive performance in children and adolescents with ADHD: Potential mechanisms and evidence-based recommendations. *Journal of Clinical Medicine*, *8*(6), Article 841. https://doi.org/10.3390/jcm8060841

Ding, Y. C., Chi, H. C., Grady, D. L., Morishima, A., Kidd, J. R., Kidd, K. K., Flodman, P., Spence, M. A., Schuck, S., Swanson, J. M., Zhang, Y. P., & Moyzis, R. K. (2002). Evidence of positive selection acting at the human dopamine receptor D4 gene locus. *Proceedings of the National Academy of Sciences of the United States of America*, *99*(1), 309–314. https://doi.org/10.1073/pnas.012464099

Dunn, G. A., Nigg, J. T., & Sullivan, E. L. (2019). Neuroinflammation as a risk factor for attention deficit hyperactivity disorder. *Pharmacology, Biochemistry, and Behavior*, *182*, 22–34. https://doi.org/10.1016/j.pbb.2019.05.005

Esteller-Cucala, P., Maceda, I., Børglum, A. D., Demontis, D., Faraone, S. V., Cormand, B., & Lao, O. (2020). Genomic analysis of the natural history of attention-deficit/hyperactivity disorder using Neanderthal and ancient *Homo sapiens* samples. *Scientific Reports*, *10*(1), Article 8622. https://doi.org/10.1038/s41598-020-65322-4

Faraone, S. V., Banaschewski, T., Coghill, D., Zheng, Y., Biederman, J., Bellgrove, M. A., Newcorn, J. H., Gignac, M., Al Saud, N. M., Manor, I., Rohde, L. A., Yang, L., Cortese, S., Almagor, D., Stein, M. A., Albatti, T. H., Aljoudi, H. F., Alqahtani, M. M. J., Asherson, P., . . . Wang, Y. (2021). The World Federation of ADHD international consensus statement: 208 Evidence-based conclusions about the disorder. *Neuroscience and Biobehavioral Reviews*, *128*, 789–818. https://doi.org/10.1016/j.neubiorev.2021.01.022

Faraone, S. V., & Larsson, H. (2019). Genetics of attention deficit hyperactivity disorder. *Molecular Psychiatry*, *24*(4), 562–575. https://doi.org/10.1038/s41380-018-0070-0

Girard-Joyal, O., & Gauthier, B. (2022). Creativity in the predominantly inattentive and combined presentations of ADHD in adults. *Journal of Attention Disorders*, *26*(9), 1187–1198. https://doi.org/10.1177/10870547211060547

Glover, V. (2011). Annual research review: Prenatal stress and the origins of psychopathology: An evolutionary perspective. *Journal of Child Psychology and Psychiatry*, *52*(4), 356–367. https://doi.org/10.1111/j.1469-7610.2011.02371.x

Hamza, M., Halayem, S., Bourgou, S., Daoud, M., Charfi, F., & Belhadj, A. (2019). Epigenetics and ADHD: Toward an integrative approach of the disorder pathogenesis. *Journal of Attention Disorders*, *23*(7), 655–664. https://doi.org/10.1177/1087054717696769

Han, W., & Chen, B. B. (2020). An evolutionary life history approach to understanding mental health. *General Psychiatry, 33*(6), Article e100113. https://doi.org/10.1136/gpsych-2019-100113

Hinshaw, S. P., Nguyen, P. T., O'Grady, S. M., & Rosenthal, E. A. (2022). Annual research review: Attention-deficit/hyperactivity disorder in girls and women: Underrepresentation, longitudinal processes, and key directions. *Journal of Child Psychology and Psychiatry, 63*(4), 484–496. https://doi.org/10.1111/jcpp.13480

Hua, M. H., Huang, K. L., Hsu, J. W., Bai, Y. M., Su, T. P., Tsai, S. J., Li, C. T., Lin, W. C., Chen, T. J., & Chen, M. H. (2021). Early pregnancy risk among adolescents with ADHD: A nationwide longitudinal study. *Journal of Attention Disorders, 25*(9), 1199–1206. https://doi.org/10.1177/1087054719900232

Kalenik, A., Kardaś, K., Rahnama, A., Sirojć, K., & Wolańczyk, T. (2021). Gut microbiota and probiotic therapy in ADHD: A review of current knowledge. *Progress in Neuro-Psychopharmacology and Biological Psychiatry, 110*, Article 110277. https://doi.org/10.1016/j.pnpbp.2021.110277

Min, X., Li, C., & Yan, Y. (2021). Parental age and the risk of ADHD in offspring: A systematic review and meta-analysis. *International Journal of Environmental Research and Public Health, 18*(9), Article 4939. https://doi.org/10.3390/ijerph18094939

Østergaard, S. D., Dalsgaard, S., Faraone, S. V., Munk-Olsen, T., & Laursen, T. M. (2017). Teenage parenthood and birth rates for individuals with and without attention-deficit/hyperactivity disorder: A nationwide cohort study. *Journal of the American Academy of Child and Adolescent Psychiatry, 56*(7), 578–584. e3. https://doi.org/10.1016/j.jaac.2017.05.003

Posner, J., Polanczyk, G. V., & Sonuga-Barke, E. (2020). Attention-deficit hyperactivity disorder. *Lancet, 395*(10222), 450–462. https://doi.org/10.1016/S0140-6736(19)33004-1

Postema, M. C., Hoogman, M., Ambrosino, S., Asherson, P., Banaschewski, T., Bandeira, C. E., Baranov, A., Bau, C., Baumeister, S., Baur-Streubel, R., Bellgrove, M. A., Biederman, J., Bralten, J., Brandeis, D., Brem, S., Buitelaar, J. K., Busatto, G. F., Castellanos, F. X., Cercignani, M., . . . Francks, C. (2021). Analysis of structural brain asymmetries in attention-deficit/hyperactivity disorder in 39 datasets. *Journal of Child Psychology and Psychiatry, 62*(10), 1202–1219. https://doi.org/10.1111/jcpp.13396

Qi, S., Schumann, G., Bustillo, J., Turner, J. A., Jiang, R., Zhi, D., Fu, Z., Mayer, A. R., Vergara, V. M., Silva, R. F., Iraji, A., Chen, J., Damaraju, E., Ma, X., Yang, X., Stevens, M., Mathalon, D. H., Ford, J. M., Voyvodic, J., . . . IMAGEN Consortium. (2021). Reward processing in novelty seekers: A transdiagnostic psychiatric imaging biomarker. *Biological Psychiatry, 90*(8), 529–539. https://doi.org/10.1016/j.biopsych.2021.01.011

Rachlin, H. (2019). Group selection in behavioral evolution. *Behavioural Processes, 161*, 65–72. https://doi.org/10.1016/j.beproc.2017.09.005

Sapolsky, R. M. (2017). *Behave: The biology of humans at our best and worst*. Penguin.

Sarkis, E. (2014). Addressing attention-deficit/hyperactivity disorder in the workplace. *Postgraduate Medicine, 126*(5), 25–30. https://doi.org/10.3810/pgm.2014.09.2797

Sedgwick, J. A., Merwood, A., & Asherson, P. (2019). The positive aspects of attention deficit hyperactivity disorder: A qualitative investigation of successful adults with ADHD. *Attention Deficit and Hyperactivity Disorders, 11*(3), 241–253. https://doi.org/10.1007/s12402-018-0277-6

Shoham, R., Sonuga-Barke, E., Yaniv, I., & Pollak, Y. (2021). ADHD is associated with a widespread pattern of risky behavior across activity domains. *Journal of Attention Disorders, 25*(7), 989–1000. doi:10.1177/1087054719875786

Sônego, M., Meller, M., Massuti, R., Campani, F., Amaro, J., Barbosa, C., & Rohde, L. A. (2021). Exploring the association between attention-deficit/hyperactivity disorder and entrepreneurship. *Revista Brasileira de Psiquiatria, 43*(2), 174–180. https://doi.org/10.1590/1516-4446-2020-0898

Sonuga-Barke, E., & Thapar, A. (2021). The neurodiversity concept: Is it helpful for clinicians and scientists? *Lancet Psychiatry, 8*(7), 559–561. https://doi.org/10.1016/S2215-0366(21)00167-X

Sukmajaya, A. C., Lusida, M. I., Soetjipto, & Setiawati, Y. (2021). Systematic review of gut microbiota and attention-deficit hyperactivity disorder (ADHD). *Annals of General Psychiatry, 20*(1), Article 12. https://doi.org/10.1186/s12991-021-00330-w

Swanepoel, A. (2021). Fifteen-minute consultation: To prescribe or not to prescribe in ADHD, that is the question. *Archives of Disease in Childhood: Education & Practice, 106*(6), 322–325. https://doi.org/10.1136/archdischild-2020-318866

Swanepoel, A., Music, G., Launer, J., & Reiss, M. J. (2017). How evolutionary thinking can help us understand ADHD. *BJPsych Advances, 23*(6), 410–418. https://doi.org/10.1192/apt.bp.116.016659

Swanepoel, A., Sieff, D. F., Music, G., Launer, J., Reiss, M. J., & Wren, B. (2016). How evolution can help us understand child development and behaviour. *BJPsych Advances*, *22*(1), 36–43. https://doi.org/10.1192/apt.bp.114.014043

Talge, N. M., Neal, C., Glover, V., & Early Stress, Translational Research and Prevention Science Network: Fetal and Neonatal Experience on Child and Adolescent Mental Health. (2007). Antenatal maternal stress and long-term effects on child neurodevelopment: how and why? *Journal of Child Psychology and Psychiatry*, *48*(3–4), 245–261. https://doi.org/10.1111/j.1469-7610.2006.01714.x

Tegelbeckers, J., Bunzeck, N., Duzel, E., Bonath, B., Flechtner, H. H., & Krauel, K. (2015). Altered salience processing in attention deficit hyperactivity disorder. *Human Brain Mapping*, *36*(6), 2049–2060. https://doi.org/10.1002/hbm.22755

Thagaard, M. S., Faraone, S. V., Sonuga-Barke, E. J., & Østergaard, S. D. (2016). Empirical tests of natural selection-based evolutionary accounts of ADHD: A systematic review. *Acta Neuropsychiatrica*, *28*(5), 249–256. https://doi.org/10.1017/neu.2016.14

Thapar, A., & Cooper, M. (2016). Attention deficit hyperactivity disorder. *Lancet*, *387*(10024), 1240–1250. https://doi.org/10.1016/S0140-6736(15)00238-X

Williams, J., & Taylor, E. (2006). The evolution of hyperactivity, impulsivity, and cognitive diversity. *Journal of the Royal Society Interface*, *3*(8), 399–413. https://doi.org/10.1098/rsif.2005.0102

Wren, B., Launer, J., Music, G., Reiss, M. J., & Swanepoel, A. (2021). Can an evolutionary perspective shed light on maternal abuse of children? *Clinical Child Psychology and Psychiatry*, *26*(1), 283–294. https://doi.org/10.1177/1359104520974418

17

Addiction
Diverted Reward and Motivation Principles

Roger Sullivan and Edward Hagen

Historical Background

One of the consequences of interdisciplinary academic boundaries is that most addiction scientists are unlikely to be aware of the importance of plant diets and plant-chemical defenses during the course of human evolution. Yet these constitute some of the key selection pressures affecting the evolution of the brain and nervous system in humans as well as in most other animals. This chapter seeks to highlight the implications of this insight for mainstream theories of addiction and to bring into question fundamental assumptions about the causes of drug abuse.

Domesticated plants transformed human lifeways a mere 10,000 years ago (on average). In comparison to wild plants, domesticated plants were artificially selected to reduce their toxic chemical defenses. This means that throughout the 6–7 million years of *hominin* (humans and extinct ancestor species) evolution, our ancestors obtained most of their calories from wild plants that were chemically toxic.

Plants are preyed upon by herbivores, which depend on the starches and other nutrients in plants for their survival and reproduction (in this treatment, the term "herbivore" includes any animal relatively dependent on plant foods). Plants do not "want" to be eaten and have evolved an array of adaptations to protect themselves from herbivores including both physical and chemical defenses (some plant secondary metabolites [PSMs], such as pigments and antioxidants, are not toxins; but we will use the term exclusively to refer to chemical defense). Plants redirect a substantial proportion of their energy and essential nutrients away from the production of primary metabolites required for their own growth and reproduction (like starches and proteins) and toward the synthesis of secondary metabolites for chemical defense. PSMs are mainly toxins that evolved to target the nervous systems and other organs of animal plant eaters, including humans and their ancestors. Fruiting plants that exploit animals for seed dispersal would seem to be an exception but actually maintain their chemical defenses except during the short period that their fruit is ripe (Nelson and Whitehead 2021). Plant chemical defenses were shaped by evolutionary forces based on how effectively they dissuaded herbivores. This evolutionary process favored the synthesis of toxins that could interfere with the functioning of the nervous systems of herbivores. The ideal form of a neurotoxic PSM was one that could bind at key receptor sites in the nervous systems of plant eaters, leading to the initially surprising phenomenon in which many of the main mammalian neurotransmitter systems are named after the plant chemicals that bind to them with uncanny accuracy (nicotinic, muscarinic, cannabinoid, opioid, etc.).

This evolutionary process, termed *evolutionary molecular modeling* (Wink 1998), indicates that humans and other animals have been subject to intense selection pressures from plant toxins targeting the brain and nervous system throughout the course of evolution. Note well that this dynamic does not occur in the case of ethanol and that this chapter is focused on drugs *excluding* alcoholic drinks. Ethanol is a yeast waste product, and its evolutionary dynamics are distinct from those of PSMs. See Dudley (2014) and Sullivan (2017) for a discussion of the evolutionary origins of alcohol consumption.

The Non-evolutionary History of Current Drug Reward Theory

With the exception of alcohol, almost all drugs of abuse are plant neurotoxins or their close chemical analogs. In our perspective, brain mechanisms mediating drug use and addiction cannot be understood without reference to evolutionary theory and the science of human evolutionary ecology. However, the fact that mammals, including human ancestors, evolved under constant selection from potent neurotoxins contradicts what we will call the "standard model" of addiction. In the standard model, all mammals are vulnerable to drugs of abuse because of a drug-sensitive neurobiological reward center that evolved to reinforce behaviors, such as obtaining food or sex, that increased fitness (Nesse and Berridge 1997; Wise 2000). Drugs of abuse "hijack" the natural functions of the reward center and produce pleasurable sensations independent of any natural cause or process. Mammals, including people, become addicted to drugs because of reinforcement of sensations of "liking" and "wanting" mediated by the reward center (Everitt and Robbins 2005; Nestler 2005). The contradiction is this—how could a reward center have evolved in mammals that is easily hijacked by plant neurotoxins (i.e., drugs of abuse) given millions of years of selection from plant neurotoxins targeting the nervous system? We have termed this the evolutionary "paradox of drug reward" (Sullivan et al., 2008).

The history of ideas underpinning the standard model is of particular interest here in the degree to which they have been insulated from evolutionary theory and the insights of evolutionary ecology. The two key "ideas" of the standard model are *behavioral* theory about reward and reinforcement and *neurobiological* theory about the putative reward center. Physiological evidence for a "reward center" in the brain was preceded by at least half a century by folk, educational, and psychological theories about learning. From the turn of the 20th century, the educational theorist E. L. Thorndike had already formalized the nascent ideas of a movement stressing reinforcement as the cause of behavioral learning with his *law of effect*: that learning is accelerated by positive rewards (Thorndike 1911). By the 1920s, Thorndike's work and other key influences such as Pavlov's classical conditioning experiments had matured into the behaviorism movement in US psychology. A key point is that behavioral psychology in this era emphasized the causal centrality of reward learning to the degree that biological factors, and therefore evolutionary causes, were excluded from the lexicon. Practitioners like Watson became anti-evolutionary, arguing that reward learning could explain all behavior and that innate predispositions or processes were irrelevant (as captured in his famous quotation "give me a child . . ."; Logue 1978). Although the influence of behaviorism was starting to wane by mid-century, one area of behaviorist research and theory that survived and thrived was reward theory applied to drug use by animals and people, which had been strongly boosted by 1950s research into the neurobiology of reward.

In a revolutionary 1954 research paper, Olds and Milner reported that rats will self-administer an electrical current to the septal region of the brain. They claimed that intracranial electrical stimulation was possibly the most potent reward stimulus used in animal experimentation to that date. The resemblance to the behavioral reward models explaining drug addiction was immediately evident, and the midbrain region became the candidate neurobiological "reward center" mediating the classical and operant conditioning observed by behaviorists, and the dysfunctional pleasurable reinforcement of drug-using behaviors (Olds and Milner 1954).

More than two decades of subsequent experimentation ensued to identify the neural mechanisms mediating the reinforcing effects of self-stimulation and drug reward. The neurons critical for both septal self-stimulation and the reinforcing properties of several classes of drugs turned out to be dopamine neurons in the midbrain (particularly the nucleus accumbens), commonly referred to as the mesolimbic dopamine system (MDS). With this discovery, the stimulus–response paradigm at the core of behaviorism was modified and merged with neurobiological insights into the dopamine theory of reinforcement learning and the dopamine theory of substance use and addiction (Hagen and Sullivan 2019; see Everitt and Robbins 2005 and Nestler 2005 for full reviews of the current model).

Despite the subsequent dominance of the MDS model, the old problem that emerged during the behaviorist era of natural factors affecting experiments of drug reinforcement in animals had never gone away. For example, Petrinovich and Bolles (1954) demonstrated that rats find it easier to learn relationships that are consistent with their natural ecology and will make such associations independently of experimentally induced motivational states such as hunger and thirst. Garcia and Ervin (1968) showed that rats will avoid novel foods paired with an aversive association after a single trial, but only if the aversive experience is nausea. Green et al. (2002) showed that rats in "enriched" environments (i.e., those that are relatively less artificial) are less inclined to lever-press for drug rewards. Ahmed has related that, when given a choice, rats prefer to self-administer saccharin (an analogue for natural sweetness) over cocaine, "suggesting that cocaine is less reinforcing than saccharin" (2012, p. 115). More generally, approximately 90% of rats "retain the ability to choose to abstain from cocaine for another non-drug pursuit when it is available" (Ahmed, 2012, p. 118), a pattern that is also observed in human drug users.

The common element in all of these studies is that experimental conditions are compromised by the intrusion of some element reflecting the natural world of the laboratory animal, whether it be an environmental defense mechanism like nausea, or a natural preference for sweet-tasting foods and fluids over neurotoxic plant defensive chemicals like cocaine. Studies that have been hugely influential on theories of drug reward, such as Olds and Milner's (1954) classic research on electrical brain stimulation in the rat, may have been successful only in so far as they could exclude natural confounding variables. Despite its dominance, therefore, the research outcomes that make the standard model credible are actually quite particular. The core experimental outcomes actually only "work" within quite narrow parameters that carefully avoid variables known to confound outcomes.

A key theme that we are developing here is that the a-evolutionary or even anti-evolutionary heritage of the science of reward is still with us to some degree. Although contemporary addiction science has come a long way since the behaviorist influences of the first half of the 20th century, it is still heavily dependent on notions of reward that emerged in the heyday of behaviorism, it still tends to minimize or ignore the problems of natural conditions, and it remains largely indifferent to evolutionary biological science.

Not all commentary has ignored evolution, however. As introduced previously, Nesse and Berridge (1997) focused on the hypothesis that both reward and aversion could be reinterpreted as evolved emotional faculties that encourage adaptive behaviors (we analyze the evolutionary significance of MDS-mediated aversion in a separate treatment; Hagen et al. 2009). In this perspective, drugs with novel purity and routes of administration "short-circuit" or "hijack" the evolved functions of the emotions and the brain's reward center. This then has become the default evolutionary interpretation of current reward theory—that drugs of abuse "hijack" natural brain functions and reward harmful repetitive drug use (see also Wise 2000). Our main disagreement with Nesse and Berridge's model (1997) is that it is more focused on making reward theory "work" in an evolutionary perspective than it is with approaching the problem using evolutionary ecological theory from *first principles*. We have called our own counter-perspective the "punishment" model and will develop it further here.

Evolutionary Evidence Specifically Related to Human Health and Disease

Punishment and Co-adaptation

Animals have been exposed to dietary toxins that affect the nervous system for hundreds of millions of years. Placental mammals have been exposed to dietary neurotoxins that cross the blood–brain barrier (BBB) and bind in the nervous system for ~60 million years. Humans and their hominin ancestors have been exposed to dietary neurotoxins throughout the 6–7-million-year course of their evolution.

Most PSMs may have evolved initially to protect plants from insect predators, but because of common neuronal signaling pathways in animals (e.g., the amine neurotransmitters dopamine, norepinephrine, and serotonin), PSMs also protect plants from vertebrate herbivores including mammals. We can get a picture of the scale of toxicity effects across different taxa by comparing the effects of common neurotoxins on well-studied insects and animals. For example, a 0.2 solution concentration of caffeine is lethal to 50% of exposed honeybees (LD50), whereas the LD50 of oral caffeine in mice is 127–137 mg/kg and 192 mg/kg in rats (a Starbucks "tall" coffee has about 235 mg of caffeine). Vincamine, a PSM in periwinkle garden plants, has an LD50 solution concentration of 0.4 in honeybees and an oral LD50 of 1 g/kg in mice (Wink et al. 1998). Nicotine has an LD50 solution concentration of 0.2 in honeybees (Wink et al. 1998), an oral LD50 in dogs of 9.2 mg/kg, 3.3 mg/kg in mice, 50 mg/kg in rats (National Institute for Occupational Safety and Health 1988), and an estimated 6.5–13 mg/kg in humans (Gosselin et al. 1984; Henstra et al. 2022). Consuming 1–5 g of PSMs in a single 100-g meal is an ecologically plausible range in a small herbivore, even though different metabolic parameters will affect actual toxicity in a living animal (Wink et al. 1998).

Every animal that extracts the energy it requires for survival and reproduction from plants (and/or fungi) is locked in a co-evolutionary relationship with the flora in its ecological niche. Another example animal, ideal because of its narrow diet breadth, is the koala bear. The koala has evolved to eat only the leaves of the eucalyptus tree. This tree produces phenolic compounds, terpenes, and other secondary metabolites to dissuade animal predators. The koala has been subject to the toxic effects of these PSMs during its evolution and has evolved its own counteradaptations to allow it to consume large quantities of eucalyptus leaves without being poisoned. A key adaptation in koalas is a unique evolution of the cytochrome P450 (CYP) pathway to allow the koala to detoxify a leaf diet that would be lethal to most other mammals. Under selection from toxins in the eucalyptus tree, the koala genome expanded to include no less than 31 CYP2C toxin-metabolizing genes; in comparison, humans have four CYP2C genes (Johnson et al. 2018). The koala and the eucalyptus plant, then, have co-evolved in an arms race, with each party subjecting the other to selection pressures leading to adaptation.

The herbivore strategy is often more complex than merely surviving the punishment from plant chemical defenses. In what is called a multitrophic interaction, many herbivores have evolved to exploit the chemical defenses of the plants that they eat and use them in some way to their own advantage. A good example is the tobacco hornworm (*Manduca sexta*), which has evolved to feed on the tobacco plant despite the toxic effects of the plant's secondary metabolite, nicotine. The tobacco hornworm has gone one step further than merely surviving the punishment from the plant's nicotine alkaloids; it sequesters the plant's nicotine in its body to dissuade parasitization from wasp larvae. Research has demonstrated that hornworms feeding on plants with nicotine are less parasitized than hornworms on a toxin-free diet (Bentz and Barbosa 1990). The strategy of consuming chemicals from the environment in order to make one's own body toxic to predators or parasites is known as *pharmacophagy*.

Large mammals also employ these strategies, but whereas insects often use pharmacophagy to avoid being eaten by their predators, larger animals are more likely to consume toxic plants to kill or control their parasites. In a study of sheep, Villalba and colleagues (2010) demonstrated that the animals are adept at avoiding dietary toxins but will selectively consume them when infected with parasites.

Phylogenetically closer to home, Krief et al. (2006) have described the chimpanzee use of toxic plants to "self-medicate" intestinal parasites and have assayed the anthelmintic chemical properties of the select plant species exploited. Similarly, Cousins and Huffman (2002) have proposed that gorillas strategically select plants for their anthelmintic and psychoactive properties. Animals also use plant toxins to protect against ectoparasites. Some mammal and bird species line nests with toxic leaves to suppress ectoparasites. Others wipe their fur with toxic leaves or chew leaves and then apply saliva to their fur as a strategy to discourage ectoparasites (Forbey et al. 2009).

The authors have promoted the hypothesis that humans may also engage in forms of pharmacophagy. They conceptualized a study of tobacco consumption and its effects on parasite load among Aka foragers in the Central African Republic. The study showed a relationship between tobacco consumption and reduced parasite load in the study population (Roulette et al. 2014).

How Are These Insights Relevant to Human Drug Use and Addiction?

Our hypothesis is that humans and other mammals have been subject to intense selection pressures from plant toxins targeting the brain throughout evolution. We have termed this the "punishment model" and contrasted it to the "hijack model" of the conventional perspective. What evidence is there to support the punishment model? One source is the suite of adaptations protecting mammals from plant chemical defenses that evolved over millions of years. These include the BBB, the chemical senses (taste and smell), the liver and specific chemical pathways for metabolizing environmental toxins (such as cytochrome P450), the vomiting reflex, sensory-specific satiety, neophobia, and other mechanisms of food selectivity and aversions.

Evolved Biological Defenses

A vertebrate's first line of defense against plant toxins is oral bitter taste receptors. Bitter taste, in contrast to the sweet, sour, salty, and umami receptors, must prevent the potential ingestion of tens of thousands of structurally diverse plant toxins. Not surprisingly, bitter taste is mediated by a large repertoire of 25 receptor genes, the TAS2Rs. Comparative genomics indicate that TAS2R genes evolved in vertebrates about 430 million years ago during the expansion of terrestrial vascular plants. The leading hypothesis explaining the evolution of this gene family is selection for an herbivore defense mechanism, although the TAS2R genes do have functions other than bitter taste detection (Wooding et al. 2021). Plant foods with PSMs taste bitter, allowing humans and other mammals to assess the degree of toxicity of plant foods before swallowing them or, in another sense, to assay the chemical environment using the palate.

Once a decision is made about the degree of toxicity of a plant food and it is swallowed, mammals and other vertebrates are protected from poisoning by several chemical pathways that metabolize PSMs. The most prominent of these is the CYP superfamily of genes coding for enzymes that catalyze the oxidation of a wide range of endogenous and exogenous chemicals in phase one metabolism. CYP enzymes are concentrated in the liver but are also found in the gut and other tissues. There are 57 CYP genes in humans. Whereas the majority of CYP genes metabolize endogenous chemicals, like steroids, the CYP 1, 2, and 3 genes evolved to specifically metabolize environmental chemicals (xenobiotics), although they also metabolize some endogenous substrates (Table 17.1; Nelson et al. 2013).

TABLE 17.1 Classes of Compounds Metabolized by Human Cytochrome P450 (CYP) and Major Gene/Isoenzymes Involved in Their Biotransformation: Xenobiotic-Metabolizing Gene/Isoenzymes in the CYP 1, 2, and 3 Families Are Highlighted

Classes of compounds	CYP enzymes [a]
Sterols	7A1, 7B1, 8B1, 11A1, 11B1, 11B2, 17A1, 19A1, 21A2, 27A1, 39A1, 46A1, 51A1
Xenobiotics	1A1, 1A2, 1B1, 2A6, 2A7, 2A13, 2B6, 2C8, 2C9, 2C18, 2C19, 2D6, 2E1, 2F1, 2J2, 2R1, 2S1, 2U1, 2W1, 3A4, 3A5, 3A7, 3A43
Fatty acids	4A11, 4B1, 4F11, 4F12, 4F22, 4V2, 4X1, 4Z1
Eicosanoids	4F2, 4F3, 4F8, 5A1, 8A1
Vitamins	24A1, 26A1, 26B1, 26C1, 27B1, 27C1
Other	4A22, 20

After Esteves et al. (2021) and Nelson et al. (2013).
[a] Note that CYP enzymes may metabolize substrates from more than one class.

Main Classes of Compounds Metabolized by Human CYPs

Our best examples of CYP substrates are from the pharmacology of medicines and recreational drugs, which are either PSMs or modified molecular products based on the chemical structure of PSMs. Substrates of the CYP 1 family include caffeine, theophylline, and theobromine (Figure 17.1; CYP1A2 gene); the CYP 2 family metabolizes >20% of commonly used pharmaceuticals and recreational drugs including nicotine (CYP 2A6 gene) and opiates (CYP 2D6 gene); and the CYP 3 family metabolizes around 50% of commonly used pharmaceuticals and recreational drugs including cocaine and quinine (CYP3A4 gene; Vuppalanchi, 2018). As is the case for the TAS2R bitter taste receptor genes, the xenobiotic metabolizing CYP genes evolved in vertebrates ~450 million years ago when the first land animals encountered terrestrial plants (Nelson et al. 2013).

A fundamental question is whether or not evolved defenses like TAS2R and CYP genes still "work" in domesticated *Homo sapiens* or whether they are an evolutionary legacy inherited from a common primate ancestor that is no longer required in the modern world. It became apparent early in the genomic era, however, that chemical perception and xenobiotic metabolism were genetic "hot spots" in the human genome.

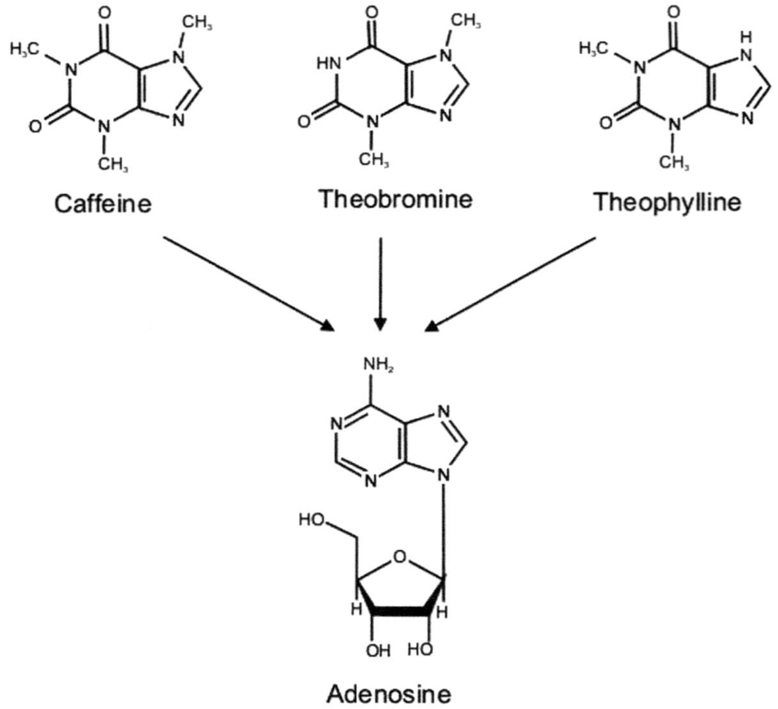

Figure 17.1 Evolutionary molecular modeling. Caffeine (coffee), theobromine (chocolate), and theophylline (tea and chocolate) mimic the structure and function of the adenosine neurotransmitter.

A key example is the Chimpanzee Sequencing and Analysis Consortium finding that the three most rapidly evolving (diverging) categories of genes in humans and chimpanzees were in "sensory perception of chemical stimulus," "perception of smell," and "xenobiotic metabolism" based on synonymous to nonsynonymous mutation rates in the respective genomes (Chimpanzee Sequencing and Analysis Consortium 2005). This is important for two reasons: it shows that environmental chemicals have been one of the primary selection pressures in human evolution and that selection from plant toxins has continued after the human/chimpanzee divergence into the evolutionary present.

In mature adults, one of the primary defense mechanisms from environmental toxins, along with CYP and other metabolic pathways, is the BBB. The BBB is a cluster of adaptations that restricts the passage of proteins and large molecules from the circulatory system into the brain and central nervous system (CNS). The main barrier is formed by tight-fitting endothelial cells in the microvessels surrounding the CNS. Not only is the BBB a physical barrier to toxic molecules, but it is also infused with xenobiotic metabolizing CYP enzymes (and other toxin binding and transport proteins) (Dauchy et al. 2008).

Evolved Behavioral Defenses

In addition to evolved physiological mechanisms, plant-eating animals have a range of evolved behavioral or biobehavioral adaptations to avoid dietary PSMs. Sensory-specific satiety (SSS) is a mammalian phenomenon whereby a food becomes less desirable the more it is eaten. Related to SSS, experiments with domesticated ruminants show that food selectivity is a key dietary strategy affected by food toxicity. Cows and sheep will graze on a range of plants but will consume them in an order reflecting their toxicity; they prefer to eat plants with the least toxins but will consume increasingly toxic plants as the quantity of available non-toxic plants diminishes (Provenza 1997). In a sense the animals appear to titrate their exposure to plant toxins as a function of nutritional need. Neophobia, or a reluctance to eat novel foods, is a further part of this dynamic of dietary selectivity. In a process that can also induce the production of toxin metabolizing CYP enzymes, mammals, ranging from cows and sheep to rodents, will carefully try new foods by tasting and consuming small quantities before committing to feeding extensively (Provenza 1997; Andrews and Horn 2006).

If an animal makes a mistake and commits to eating a food that is too toxic, the vomiting reflex has evolved to evacuate the stomach of dietary toxins before damage can occur. The vomiting reflex is present across multiple taxa including most mammalian orders and vertebrates including fish, amphibia, reptiles, and birds. Interestingly, mice, rats, and several other mammals do not vomit and appear to rely instead on acute selectivity and neophobia, as well as a large number of xenobiotic-metabolizing genes, to protect them from ingested toxins (Andrews and Horn 2006).

Are PSMs Still Relevant to People Today?

Humans are *not* cattle or rodents, but they do share a common mammalian ancestry with herbivorous mammals and persistent selection from plant toxins. Even though modern lifeways and domesticated foods mean that we are not as exposed to PSMs as our ancestors were, they are still ubiquitous in the environment; and we encounter them in foods, spices, recreational drugs, and commercial medicines. More than 50,000 different PSMs have been identified in foods, drugs, and medicines; and all the most commonly used recreational drugs, with the exception of alcohol, are PSMs (Table 17.2). Chemically, PSMs fall into a number of different categories, such as the nitrogen-containing alkaloids (the largest known category), amino acids, amines, peptides, lectins, cyanogenic glycosides, and compounds lacking nitrogen, such as monoterpenes, sesquiterpenes, steroids, tannins, and flavonoids (Wink 2006). Of most relevance here are the pharmacologically active PSMs in recreational and medical drugs. It is important to note that distinctions customarily drawn between PSMs as food toxins, recreational drugs, and pharmaceuticals are categorically and pharmacologically arbitrary and not very useful in assessing the ecological relationships between plants and humans. This arbitrariness is largely because many PSMs have evolved to interfere with basic cellular signaling systems in

TABLE 17.2 Relationships between Commonly Used Plant Drugs and Human (Mammalian) Nervous System Receptors

Plant drug	Plant toxin	Neurotransmitter/receptor
Tobacco, pituri	Nicotine [a]	Acetylcholine/nicotinic
Betel nut	Arecoline [a]	Acetylcholine/muscarinic
Coca	Cocaine [c]	Epinephrine, norepinephrine/adrenergic
Khat	Ephedrine, [c] Cathinone [a,c]	Epinephrine, norepinephrine/adrenergic
Cactus	Mescaline [a]	Serotonin
Coca	Cocaine [c]	Dopamine
Khat	Cathinone [a,c]	Dopamine
Kratom	Mitragynine [a]	Dopamine
Coffee, cola nut	Caffeine [b]	Adenosine
Tea	Caffeine,[b] theophylline, [b] theobromine [b]	Adenosine
Chocolate	Theobromine [b]	Adenosine
Opium	Codeine, [a] morphine [a]	Endorphins/MOP opioid
Salvia	Salvinorin A [a]	Dynorphins/KOP opioid
Cannabis	Δ9-THC [a]	Anandamide/cannabinoid

[a] Receptor agonist.
[b] Receptor antagonist.
[c] Reuptake inhibitor.
KOP = kappa opioid receptor; MOP = mu opioid receptor; THC = tetrahydrocannabinol.

animal herbivores, and the same chemical properties that render a PSM toxic also permit the manipulation of these very same signaling pathways for medical benefit or "recreation." It should not be too surprising, then, that numerous PSMs have been transformed into commercial pharmaceuticals. Sparteine, for example, is an alkaloid found in several legumes that has been synthesized in a range of pharmaceuticals for the treatment of cardiac arrythmias, and aconitine is another tuber alkaloid that is present in several pharmaceuticals used in the treatment of rheumatism and neuralgia (Schmeller and Wink 1998). Though it is beyond the scope of this chapter to review all the PSMs that have been exploited in commercial pharmaceuticals, one survey concluded that more than half of the new chemical compounds introduced as medicines between 1981 and 2002 were PSMs or their synthetic analogues (Newman et al. 2003).

Relationships between Commonly Used Plant Drugs

The differentiation of PSMs into non-psychoactive versus psychoactive, or even psychedelic, is equally arbitrary and is more accurately "dose-dependent." Nutmeg is a good example of the relationship between dosage and physiological/psychoactive effects; it is commonly used in low concentrations as a mundane food condiment but has also been used in high dosages by prison inmates and others as a hallucinogen (Rudgley 1998). Most people will think of nicotine as a mild stimulant that is not hallucinogenic, but Aboriginal Australians and Native American "tobacco shamans" used high dosages of nicotine from the indigenous plant *Duboisia hopwoodii* and various species of *Nicotiana*, respectively, to induce hallucinogenic states (Watson 1983); and like all PSMs, nicotine is fatal at a relative threshold of systemic concentration in animals. The advent of vaping products using vials of flavored nicotine solutions has caused the number of accidental nicotine poisonings in children to skyrocket by 1,500%, and suicide attempts using nicotine are also increasing (Henstra et al. 2022). Other examples of seemingly mundane plants that are psychoactive in a dose-dependent relationship are eucalyptus, lettuce seeds, rhododendron, rosemary, and saffron (Rudgley 1998).

To summarize this section on the evidence for punishment from PSMs, we have described some examples of mechanisms that have evolved in mammals and other animals to solve the problem of dietary plant poisons. These confirm that PSMs, including those that interfere with neural signaling, have been a strong selection pressure on animals for hundreds of millions of years. In our "punishment model" we are suggesting that the hominin ancestors of modern humans were subject to the same ecological conditions as any other animal dependent on a mainly plant diet. They were subject to selection from plant toxins and counterexploited those plant toxins to increase their fitness, perhaps by controlling their intestinal parasites. Our human ancestors then were chemical titrators—they were aware of the chemical environment and exploited that environment when possible.

Counterarguments and Evidence against Evolutionary Influences

The main objection to our punishment model of drug use in human evolution is the standard (reward) model itself. A century of psychological and neurobiological research has coalesced around a hypothesis that the mammalian brain has evolved with a reward center vulnerable to being "hijacked" by drugs of abuse. The rationale is that recreational drugs are evolutionarily novel in their concentrations and modes of delivery, and hence humans could not have evolved defenses against them (Wise 2000). Here, we discuss some of the problems of the "hijack hypothesis" when placed in an evolutionary ecological context.

Functions of the MDS

Despite more than a half-century of focused biological and behavioral research on the MDS and the nucleus accumbens (NAc), their functions remain elusive, and experimental outcomes have not always supported modeled predictions. For example, Salamone et al. (2005) relate that dopamine depletion to impair stimulant self-administration does not affect food-reinforced behaviors and that dosing with dopamine antagonists impairs learning but not appetite. If NAc dopamine mediates primary reinforcement for all natural stimuli, then modulation of learning and self-administration should also affect appetite and feeding behaviors. Although much research has supported the view that dopamine mediates reward prediction error, new results contradict that model and instead suggest that mesolimbic dopamine release mediates learning of causal associations (Jeong et al. 2022). How this alternative view of the MDS might explain drug use is not clear. As a recent review stated, "circuitry involving the NAc is far more heterogeneous and complicated than previously envisioned" (Klawonn and Malenka 2018, p. 125). In our view, no conclusions can yet be drawn regarding the evolved functions of the MDS and how these might explain drug use.

Sensitivity and Domains of Cognitive Functioning

Another potentially problematic aspect of the standard model is the generalized functioning of the putative MDS NAc pathway. In the arguments over whether the human brain is cognitively organized in a domain-general or domain-specific configuration, evolutionary theorists tend to favor domain-specific models because they more realistically reflect the likelihood of a brain evolving in response to a series of selection pressures over an extended evolutionary time frame (Barrett and Kurzban 2006). In contrast, the conventional MDS NAc model is an examplar generalized mechanism in that this one "center" mediates reinforcement of all of the natural rewards and their associated emotional reinforcers. One could well ask, though, under what conditions could such a generalized mechanism evolve, and what would be the selection pressures causing its evolution as a *single* device with so many functions?

What is arguably more likely from an evolutionary theoretical perspective is that each adaptation, or domain of adaptations, would evolve separately from different selection pressures in time and space. In this view, a suite of cognitive adaptations would evolve to mediate food-seeking and eating behaviors, distinct from a suite of adaptations that mediate sexual reproduction.

A remarkable aspect of the NAc is that it is activated by drugs with widely differing, or even opposing, mechanisms of action. Whereas stimulants such as amphetamines and cocaine directly increase dopaminergic transmission in the NAc, opiates inhibit gamma-aminobutyric acid signaling in the ventral tegmental area (VTA) by disinhibiting VTA dopamine neurons projecting to the NAc; and nicotine seems to activate VTA dopamine neurons both directly and indirectly via stimulation of glutamatergic terminals innervating dopamine cells (Nestler 2005; Lüscher and Ungless 2006). In other words, the increase or decrease of DA concentrations in the NAc is caused by most drugs of abuse but by distinct and often complex cascades of chemical pathways from other functional areas of the brain.

This model of cause and effect is questionable from an evolutionary perspective in that each of these chemical pathways represents a distinct co-evolved evolutionary dynamic. For example, the neurotoxin nicotine has evolved to bind with cholinergic receptors, not dopaminergic receptors. The same is true for most drugs of abuse: they have evolved from, or are similar to, naturally occurring plant toxins that substitute neurotransmitters and/or bind at mammalian nervous system receptors. As mentioned, this fact has led to the discovery and naming of several mammalian neurotransmitter systems. In all cases, the primary evolved dynamic is between the plant toxin and its target in the mammalian nervous system; the effects on NAc dopamine concentrations in most cases are many steps beyond this primary interaction, a fact raised by other critics of the dopamine model of addiction (Nutt et al. 2015).

From an evolutionary perspective, then, the standard "hijack" model is problematic on two levels. Firstly, the evolution of many distinct neurotoxins, separately in time and space, would function via distinct biochemical pathways, all leading back to the *single* mediating mechanism—the NAc. Secondly, the evolutionary paradox of drug reward is not that a mechanism merely vulnerable to plant toxins could evolve but that a mechanism so *exquisitely sensitive* to natural plant toxins could evolve. The standard model proposes that despite hundreds of millions of years of constant and strong selection from plant defensive toxins, mammals as a group evolved with an MDS that is easily subverted by a wide range of psychoactive toxins irrespective of their mode of action (i.e., stimulating, sedating, hallucinogenic, etc.). It is a given in evolutionary theory that strong selection leads to counteradaptation and the evolution of defense mechanisms, and yet the standard model describes a pan-mammalian mechanism that seems to have no evolved defenses against environmental toxins once they have entered the nervous system; moreover, this mechanism accidentally reinforces their consumption.

"Drugs" Are an Evolutionary Novelty

Some addiction scientists might question the relevance of an ancestral mammalian exposure to PSMs to addictions in modern people. Nesse and Berridge (1997) have asserted that the standard model still holds despite the evolution of PSMs because human exposure to drugs that are potent and widely available is an evolutionary *novelty*. This idea acknowledges that human ancestors were exposed to plant toxins to some degree but that exposure was qualitatively and quantitatively less than the exposure to potent drugs of abuse in modernity. Wise (2000), one of the developers of the dopamine theory of reward, makes a similar argument.

In terms of *availability*, ancestral humans were exposed to higher levels of "drugs" than modern people in the form of PSM toxins, including neurotoxins, consumed in most meals (Sullivan and Hagen 2002; Sullivan et al. 2008). To reiterate the point made at the start of this chapter, all plant foods prior to the agricultural revolution were wild varieties loaded with toxins to prevent herbivory. Each of these toxins evolved by evolutionary molecular modeling to disrupt critical functions in herbivores, including neural signaling; and this is true whether we are talking about the highly toxic solanine glycoalkaloids in wild potatoes or the nicotine alkaloid in tobacco (and other) plants. Even today, cassava, a staple food for hundreds of millions of people, contains dangerous levels of toxins (cyanide; Ferraro et al. 2016). Thus, our ancestors had to contend with the possibility of poisoning with every mouthful of food, and their total exposure to neurotoxic PSMs would have been far greater than for people in the present who eat detoxified domesticated plants and must seek out recreational neurotoxins in legal or illegal drugs.

The degree to which our ancestors consumed "drugs," that is, deliberate consumption of neurotoxins for purposes other than nutrition, is another topic we have addressed at length (Sullivan and Hagen 2002; Sullivan et al. 2008). Historical speculations about the antiquity of drug use by Indigenous peoples have always been vague and often assumed that the practices commenced shortly "before colonization." Our own research indicates that the prehistory of human drug use is greatly underestimated. For example, the most widely used (non-alcohol) drugs by people today are caffeine (coffee, tea), nicotine (tobacco products), and arecoline (betel nut). Two of these plant alkaloids are identifiable in the archaeological record before the emergence of agriculture in their respective geographies: betel nut circa 13,000 years ago in Southeast Asia and tobacco circa 8,000 years ago in the Americas (Sullivan and Hagen 2002). At the time of European contact, Australian Aborigines, hunter–gatherers who never developed agriculture, were apparently engaged in the long-distance trade of multiple tons of pituri, a native shrub (*Duboisia hopwoodii*) with higher nicotine content than tobacco (Silcock et al. 2012). The use of medicinal plants by Neandertals, including an analgesic, has been documented circa 48,000 years ago (Weyrich et al. 2017). These facts indicate that the actual time frame for the intentional use of drugs by our ancestors was, at minimum, much earlier than previously considered and, in reality, at an unknown point on an open-ended timescale (Sullivan and Hagen 2002).

The second argument supporting the notion that drugs are an evolutionary novelty is that substances available to recreational drug users today are far more *potent* and/or can be delivered directly to the bloodstream using novel technologies such as the hypodermic syringe (Wise 2000). If a drug is introduced directly into the bloodstream, it avoids the first-pass hepatic metabolism that occurs for any substance ingested through the gut. For medical treatment, this means a reduced dosage necessary to achieve a therapeutic effect, but for drug users it means a higher concentration of the drug reaching the brain and nervous system. There are several methods for avoiding first-pass metabolism employed by different indigenous groups. One is placing the plant drug next to the blood-rich buccal mucosa (of the cheek), or under the tongue, to facilitate the transport of alkaloids directly into the bloodstream (Sullivan and Hagen 2002). This method is used currently by millions of people on a daily basis throughout Asia who chew betel nut, by people who "dip" tobacco, and by indigenous Andeans who chew the coca leaf. In the case of betel and coca, lime, or a similar base, is added to change the alkaloid into its free base form in the buccal cavity. Many drugs are smoked today, as they were in antiquity. The obvious common examples are smoked tobacco and cannabis, but most recreational drugs are amenable to burning and smoking. Smoking is another simple method to introduce a drug into the body while avoiding first-pass metabolism. The main point we are trying to make here is that people in the past could use strategies to maximize the absorbable concentration of drugs in the same way that *most* people *still* do today.

What then of the potency of synthetic versions of natural opiates like fentanyl or oxycodone and the tragedy and high visibility of opioid overdoses? Our response to this sad phenomenon is to acknowledge that it is a real aspect of modern society and that it did not exist in the past but that *it occurs on too limited a scale* to be an explanatory model for human addiction or to contradict the evolutionary punishment model. If we look to the latest available federal data at the time of writing, we see that the largest proportion of recreational non-alcohol substance use in the United States is nicotine (tobacco products or vaping but excluding pipe tobacco; 26.9%) and cannabis (18.7%), followed by hallucinogens (2.7%) and cocaine (2.0%; population estimate ≥18 years in the last year). The storied targets of law enforcement, crack, methamphetamine, and heroin, were 1.0%, 0.3%, and 0.4%, respectively (Figure 17.2; US Department of Health and Human Services, Substance Abuse and Mental Health Services Administration, Center for Behavioral Health Statistics and Quality 2020).

Some of the federal data are in overlapping collective categories including "pain relievers" (3.5%) and "(all) opioids" (3.6% in the last year), which will reflect the prescription drug abuse evident in the current opioid crisis in the United States.

The point in presenting these data is not to minimize the scale of human tragedy represented in overdose deaths from fentanyl and oxycodone but to make clear that the majority of people who consume recreational drugs prefer natural PSMs in a form and manner that are similar to those available to our ancestors and that the hedonic model of drug reinforcement (emphasizing the use of potently pleasurable/reinforcing synthetic

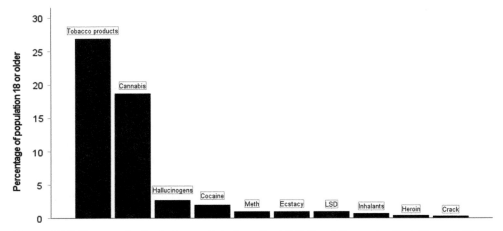

Figure 17.2 Drug use in the last year. Data from the US National Survey on Drug Use and Health 2020 (US Department of Health and Human Services, Substance Abuse and Mental Health Services Administration, Center for Behavioral Health Statistics and Quality 2020).

analogues of PSMs) can only be applied to a very small proportion of human populations, particularly given that the user data in Figure 17.2. will include individuals counted in multiple categories.

Clinical and Therapeutic Implications

Famously, a large proportion of US soldiers serving in Vietnam experimented with heroin use, and yet only a very small percentage went on to become habitual users (Robins et al. 1974). This anecdote is challenging for the standard reward model because of its assumption of vulnerability to addiction in the presence of drugs; when a large group of people are exposed to a potent drug, a relatively large proportion should become victims of it. In the ecological punishment model, however, the outcome of this natural experiment is not surprising because there is no assumption that people (or mammals in general) have evolved to be vulnerable to psychoactive drugs or that the drugs themselves are a novelty to the nervous system. The assumptions of the standard model are also noted by Ahmed (2012, p. 118), who points out the implications for drug policy and treatment:

> the prevailing default view in drug prevention and policy has always been to consider each one of us as a potential addict. Given sufficient drug exposure, each one of us could be turned into a cocaine addict. A biological resilience against cocaine addiction would not exist, or such resilience would exist only in rare individuals.

The ecological punishment model has implications for policy and therapy that are polar to the standard model. People are like all mammals and have been exposed to drugs in the form of PSMs throughout the course of evolution. Drugs are not a cultural novelty to people, and their evolved adaptations for dealing with PSMs (TCRs, CYP450, etc.) still work as designed to detect and identify "drugs" for what they are, toxic xenobiotics. From this perspective, there is no inherent vulnerability in the presence of drugs. If we now return to the people who are overdosing in the current opioid crisis, the issue is *not* that *all people* need to be isolated from fentanyl and oxycodone to protect them; it is why are *these particular individuals* overdosing? What is it about their individual biochemistry, genetics, or circumstances that makes *them*, not every human, vulnerable to opioid addiction? Recent research using genome-wide association analyses of large population cohorts reveals clues including heritable liabilities linking nicotine, alcohol, and cannabis dependence (Schoeler et al. 2023). Schoeler and colleagues (2023) identified liability genes in multiple neuronal signaling pathways and neurotransmitters including GABA, glutamate, and dopamine. Their analysis also identified overlap between heritable liability for dependence and complex traits including impulsivity and mood (Schoeler et al. 2023). This last finding overlaps with substantial research identifying links between addiction and heritable personality traits including impulsivity and novelty-seeking (Qi et al. 2021). It is beyond the scope of this chapter to fully review the literature of individual differences in drug use and addiction, but further research focused on individual biological differences, individual differences in heritable risk, and individual situational risk factors may yield new insights into the causes of drug use and addiction beyond those predicated on dopamine-mediated reward and reinforcement (Ahmed et al. 2020). The ecological punishment model also introduces an original approach for new research that follows from nature's example: that animals subject to plant toxins in their ecological niche will evolve counteradaptations. In many large species, that adaptation is pharmacophagy to control parasites, and we have pursued the question as to whether humans have also evolved to seek out PSMs for their therapeutic effects (Sullivan et al. 2008; Roulette et al. 2014).

There is also an implicit political dimension to the ecological punishment model. The "war on drugs" was predicated in part on the assumptions of the standard model: that the mammalian reward center is vulnerable to "hijack" in the presence of drugs; ergo a logical strategy is to isolate society from drugs of abuse. The "war on drugs" set out to do just that: to use the law and law enforcement to bring about a de facto prohibition of recreational psychoactive drugs to protect people from their own neurobiological vulnerabilities. With all animal life co-evolving with neurotoxic psychoactive plants, there can be no evolved vulnerability to psychoactive substances, and no *scientific* justification for prohibition (Figure 17.3).

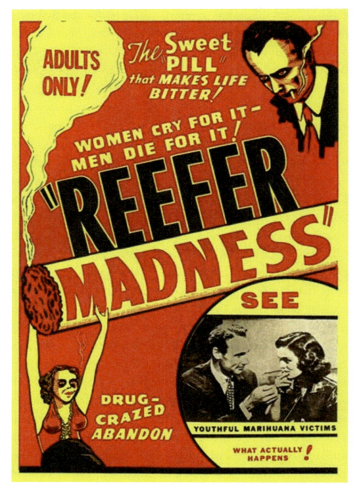

Figure 17.3 *Reefer Madness* movie poster 1936. The idea of a human weakness in the presence of mind-altering substances has legitimized several prohibition movements, including the "war on drugs."

Unanswered Questions and Future Work

If we can no longer model human (and animal) drug-seeking and abuse on the assumptions of the MDS drug reward pathway, then we need to allow more space for the considerable existing data explaining other dynamics affecting drug use behaviors. As stated above, these must emphasize dynamics affecting individuals rather than groups or species, such as individual differences in biology, heredity, and circumstances, and more investment in research on "self-medication."

Research on pharmacophagy and feeding strategies to avoid or exploit PSMs in animals has tended to be applied to and focused on possible effects on livestock and food products (Forbey et al. 2009). There is a small but growing interest in ecological factors affecting drug-seeking or PSM exploitation in animals and nonhuman primates in the wild, often found within the field of research called "zoopharmacognosy" (Krief et al. 2006; Neco et al. 2019). There are very few studies of human pharmacophagy, with our own central African project being one example (Roulette et al. 2014). More scientific research on ecological factors affecting human drug-seeking may be an important source of new data and insights into mechanisms of motivation for drug use and addiction.

A key problem in the history of animal research in addiction glossed above is the failure of experimental outcomes because of confounds from natural factors, for example, that rats in "enriched environments" are less motivated to lever-press (Green et al. 2002) or that rats will choose sweetened water over cocaine, when given the opportunity (Ahmed 2012). Future animal research on drug use and addiction that includes experiments that are *ecologically plausible* to the laboratory animal may avoid such confounds and may yield new insights (Sullivan et al. 2008).

It is our suggestion that the focus on the NAc and MDS drug–reward pathway has become a distraction in the neurobiology of drug-seeking and addiction. From an ecological perspective, the alkaloid caffeine evolved to mimic adenosine and bind at adenosine receptors in animal nervous systems (Figure 17.1). Nicotine evolved to mimic acetylcholine and to bind at cholinergic receptors. Tetrahydrocannabinol evolved to mimic the anandamide ligand and to bind at cannabinoid receptors. In none of these familiar cases has the molecule evolved to bind with dopaminergic neurons in the midbrain. The interaction between caffeine and adenosine or between nicotine and cholinergic binding sites is one that evolved for millions of years. It implies a specific ecological relationship between the environment and animal brains that may yield new insights. We suggest that research that focuses on the primary neurobiological target of the neurotoxin may generate new hypotheses about the unique biochemistry and use behaviors of *each* drug, independent of secondary or tertiary "downstream" interactions with dopamine in the NAc.

A prominent theme in this chapter has been that the notion of a neurobiological reward mechanism in the human/mammalian brain that is vulnerable to "hijacking" by diverse neurotoxic PSMs is at odds with the science of evolutionary ecology and that such a mechanism is unlikely to have evolved under constant selection pressures from plant chemical defenses. This is a bold claim that may be rejected out of hand by many addiction scientists. Yet consider for a moment that the science of plant/predator dynamics is relatively settled. There is argument about the details, but the hypothesis that plants have evolved chemical defenses targeting the nervous system of animals is a settled issue (Hartmann 2008). Similarly, the science of the ancestral diets of apes and hominins includes arguments about particulars of diet, but the fact that humans evolved from an *herbivorous* ape ancestor subsisting mainly on wild plants is not in question (Milton 1999). Finally, the forces of evolution themselves and how they work in nature is, obviously, settled scientific theory; and, as we have discussed in this chapter, the maturation of modern addiction theory has tended to steer away from any influence from evolutionary theory. How robust is the science of addiction in comparison? As we have related here, there are fundamental logical inconsistencies in the theory and practice of addiction science, and the science is arguably far less settled than the evolutionary ecological principles underpinning the "punishment model." If throwing out the mesolimbic dopamine reward model of addiction with the proverbial bathwater is too much for most readers, at least please consider that the science of evolutionary ecology has something substantial to offer addiction science.

Key Points to Remember

- The evolution of neurotoxic chemical defenses in plants to protect them from animal herbivores over hundreds of millions of years indicates that animals (including humans) are unlikely to be universally and inherently vulnerable to "drugs."
- If humans are like other animals, the seeking and use of "drugs" are explained by an evolutionary legacy of plant herbivory and adaptations to co-opt neurotoxic plant defenses.
- People who are at higher risk of abusing drugs have *individual* vulnerabilities, not a universal human/mammalian vulnerability to drugs. An ecological approach to treatment would focus on individual genetics, the effects of particular contexts, and evolved adaptations mediating drug-seeking.
- Animal experiments on drug-using behaviors need to be ecologically plausible. The science of human drug abuse and addiction is based in part on observations of animals. Changing the contexts of animal experimentation may change explanations of human drug-seeking behaviors.

References

Ahmed, S. H. (2012). The science of making drug-addicted animals. *Neuroscience, 211*, 107–125. https://doi.org/10.1016/j.neuroscience.2011.08.014

Ahmed, S. H., Badiani, A., Miczek, K. A., & Müller, C. P. (2020). Non-pharmacological factors that determine drug use and addiction. *Neuroscience and Biobehavioral Reviews, 110*, 3–27. https://doi.org/10.1016/j.neubiorev.2018.08.015

Andrews, P. L., & Horn, C. C. (2006). Signals for nausea and emesis: Implications for models of upper gastrointestinal diseases. *Autonomic Neuroscience, 125*(1–2), 100–115. https://doi.org/10.1016/j.autneu.2006.01.008

Barrett, H. C., & Kurzban, R. (2006). Modularity in cognition: Framing the debate. *Psychological Review, 113*(3), 628–647. https://doi.org/10.1037/0033-295X.113.3.628

Bentz, J., & Barbosa, P. (1990). Effects of dietary nicotine (0.1%) and parasitism by *Cotesia congregata* on the growth and food consumption and utilization of the tobacco hornworm, *Manduca sexta*. *Entomologia Experimentalis et Applicata, 57*, 1–8. https://doi.org/10.1111/j.1570-7458.1990.tb01409.x

Chimpanzee Sequencing and Analysis Consortium. Waterson Robert H. waterston@ gs. washington. edu Lander Eric S. lander@ broad. mit. edu Wilson Richard K. rwilson@ watson. wustl. edu. (2005). Initial sequence of the chimpanzee genome and comparison with the human genome. *Nature, 437*(7055), 69–87.

Cousins, D., & Huffman, M. A. (2002). Medicinal properties in the diet of gorillas—An ethnopharmacological evaluation. *African Study Monographs, 23*, 65–89. https://doi.org/10.14989/68214

Dauchy, S., Dutheil, F., Weaver, R. J., Chassoux, F., Daumas-Duport, C., Couraud, P. O., Scherrmann, J. M., De Waziers, I., & Declèves, X. (2008). ABC transporters, cytochromes P450 and their main transcription factors: Expression at the human blood–brain barrier. *Journal of Neurochemistry, 107*(6), 1518–1528. https://doi.org/10.1111/j.1471-4159.2008.05720.x

Dudley, R. (2014). *The drunken monkey: Why we drink and abuse alcohol*. University of California Press.

Esteves, F., Rueff, J., & Kranendonk, M. (2021). The central role of cytochrome P450 in xenobiotic metabolism—A brief review on a fascinating enzyme family. *Journal of Xenobiotics, 11*(3), 94–114.

Everitt, B. J., & Robbins, T. W. (2005). Neural systems of reinforcement for drug addiction: From actions to habits to compulsion. *Nature Neuroscience, 8*, 1481–1489. https://doi.org/10.1038/nn1579

Ferraro, V., Piccirillo, C., Tomlins, K., & Pintado, M. E. (2016). Cassava (*Manihot esculenta* Crantz) and yam (*Dioscorea* spp.) crops and their derived foodstuffs: Safety, security and nutritional value. *Critical Reviews in Food Science and Nutrition, 56*(16), 2714–2727. https://doi.org/10.1080/10408398.2014.922045

Forbey, J. S., Harvey, A. L., Huffman, M. A., Provenza, F. D., Sullivan, R., & Tasdemir, D. (2009). Exploitation of secondary metabolites by animals: A response to homeostatic challenges. *Integrative and Comparative Biology, 49*(3), 314–328. https://doi.org/10.1093/icb/icp046

Garcia, J., & Ervin, F. R. (1968). Gustatory–visceral and telereceptor–cutaneous conditioning: Adaptation in internal and external milieus. *Communications in Behavioral Biology, 1*, 389–415.

Gosselin, R. E., Smith, R. P., & Hodge, H. C. (1984). *Clinical toxicology of commercial products* (Vol. 2). Lippincott Williams and Wilkins.

Green, T., Gehrke, B., & Bardo, M. (2002). Environmental enrichment decreases intravenous amphetamine self-administration in rats: Dose–response functions for fixed- and progressive-ratio schedules. *Psychopharmacology, 162*, 373–378. https://doi.org/10.1007/s00213-002-1134-y

Hagen, E. H., & Sullivan, R. J. (2018). The evolutionary significance of drug toxicity over reward. In H. Pickard & S. Ahmed (Eds.), *The Routledge handbook of philosophy and science of addiction* (pp. 102–120). Routledge. https://doi.org/10.4324/9781315689197-10

Hagen, E. H., Sullivan, R. J., Schmidt, R., Morris, G., Kempter, R., & Hammerstein, P. (2009). Ecology and neurobiology of toxin avoidance and the paradox of drug reward. *Neuroscience, 160*(1), 69–84. https://doi.org/10.1016/j.neuroscience.2009.01.077

Hartmann, T. (2008). The lost origin of chemical ecology in the late 19th century. *Proceedings of the National Academy of Sciences of the United States of America, 105*(12), 4541–4546. https://doi.org/10.1073/pnas.0709231105

Henstra, C., Dekkers, B. G., Olgers, T. J., Ter Maaten, J. C., & Touw, D. J. (2022). Managing intoxications with nicotine-containing e-liquids. *Expert Opinion on Drug Metabolism & Toxicology, 18*(2), 115–121. https://doi.org/10.1080/17425255.2022.2058930

Jeong, H., Taylor, A., Floeder, J. R., Lohmann, M., Mihalas, S., Wu, B., Zhou, M., Burke, D. A., & Namboodiri, V. M. K. (2022). Mesolimbic dopamine release conveys causal associations. *Science, 378*(6626), Article eabq6740. https://doi.org/10.1126/science.abq6740

Johnson, R. N., O'Meally, D., Chen, Z., Etherington, G. J., Ho, S. Y., Nash, W. J., Grueber, C. E., Cheng, Y., Whittington, C. M., Dennison, S., & Peel, E. (2018). Adaptation and conservation insights from the koala genome. *Nature Genetics, 50*(8), 1102–1111. https://doi.org/10.1038/s41588-018-0153-5

Klawonn, A. M., & Malenka, R. C. (2018). Nucleus accumbens modulation in reward and aversion. *Cold Spring Harbor Symposia on Quantitative Biology, 83*, 119–129. https://doi.org/10.1101/sqb.2018.83.037457

Krief, S., Huffman, M. A., Sévenet, T., Hladik, C. M., Grellier, P., Loiseau, P. M., & Wrangham, R. W. (2006). Bioactive properties of plant species ingested by chimpanzees (*Pan troglodytes schweinfurthii*) in the Kibale National Park, Uganda. *American Journal of Primatology, 68*(1), 51–71. https://doi.org/10.1002/ajp.20206

Logue, A. W. (1978). Behaviorist John B. Watson and the continuity of the species. *Behaviorism, 6*(1), 71–79.

Lüscher, C., & Ungless, M. A. (2006). The mechanistic classification of addictive drugs. *PLOS Medicine, 3*(11), Article e437. https://doi.org/10.1371/journal.pmed.0030437

Milton, K. (1999). A hypothesis to explain the role of meat eating in human evolution. *Evolutionary Anthropology, 8*, 11–21. https://doi.org/10.1002/(SICI)1520-6505(1999)8:1%3C11::AID-EVAN6%3E3.0.CO;2-M

National Institute for Occupational Safety and Health. (1988). *RTECS: Registry of toxic effects of chemical substances, 1985–86*. US Department of Health and Human Services, Occupational Safety and Health.

Neco, L. C., Abelson E, S., Brown, A., Natterson-Horowitz, B., & Blumstein, D. T. (2019). The evolution of self-medication behaviour in mammals. *Biological Journal of the Linnean Society, 128*(2), 373–378. https://doi.org/10.1093/biolinnean/blz117

Nelson, A. S., & Whitehead, S. R. (2021). Fruit secondary metabolites shape seed dispersal effectiveness. *Trends in Ecology & Evolution, 36*(12), 1113–1123. https://doi.org/10.1016/j.tree.2021.08.005

Nelson, D. R., Goldstone, J. V., & Stegeman, J. J. (2013). The cytochrome P450 genesis locus: The origin and evolution of animal cytochrome P450s. *Philosophical Transactions of the Royal Society B: Biological Sciences, 368*(1612), Article 20120474. https://doi.org/10.1098/rstb.2012.0474

Nesse, R. M., & Berridge, K. C. (1997). Psychoactive drug use in evolutionary perspective. *Science, 278*, 63–66. https://doi.org/10.1126/science.278.5335.63

Nestler, E. J. (2005). Is there a common molecular pathway for addiction? *Nature Neuroscience, 8*(11), 1445–1449. https://doi.org/10.1038/nn1578

Newman, D. J., Cragg, G. M., & Snader, K. M. (2003). Natural products as sources of new drugs over the period 1981–2002. *Journal of Natural Products, 66*(7), 1022–1037. https://doi.org/10.1021/np030096l

Nutt, D. J., Lingford-Hughes, A., Erritzoe, D., & Stokes, P. R. (2015). The dopamine theory of addiction: 40 years of highs and lows. *Nature Reviews Neuroscience, 16*(5), 305–312. https://doi.org/10.1038/nrn3939

Olds, J., & Milner, P. (1954). Positive reinforcement produced by electrical stimulation of septal area and other regions of rat brain. *Journal of Comparative and Physiological Psychology, 47*(6), 419–427. https://doi.org/10.1037/h0058775

Petrinovich, L., & Bolles, R. (1954). Deprivation states and behavioral attributes. *Journal of Comparative and Physiological Psychology, 47*(6), 450–453. https://doi.org/10.1037/h0054731

Provenza, F. D. (1997). Feeding behavior of herbivores in response to plant toxicants. In J. P. Felix D'Mello (Ed.), *Handbook of plant and fungal toxicants* (pp. 231–242). CRC Press.

Qi, S., Schumann, G., Bustillo, J., Turner, J. A., Jiang, R., Zhi, D., Fu, Z., Mayer, A. R., Vergara, V. M., Silva, R. F., & Iraji, A. (2021). Reward processing in novelty seekers: A transdiagnostic psychiatric imaging biomarker. *Biological Psychiatry, 90*(8), 529–539. https://doi.org/10.1016/j.biopsych.2021.01.011

Robins, L. N. (1974). A follow-up study of Vietnam veterans' drug use. *Journal of Drug Issues, 4*(1), 61–63. https://doi.org/10.1177/002204267400400107

Roulette, C. J., Mann, H., Kemp, B. M., Remiker, M., Roulette, J. W., Hewlett, B. S., Kazanji, M., Breurec, S., Monchy, D., Sullivan, R. J., & Hagen, E. H. (2014). Tobacco use vs. helminths in Congo basin hunter-gatherers: Self-medication in humans? *Evolution and Human Behavior, 35*(5), 397–407. https://doi.org/10.1016/j.evolhumbehav.2014.05.005

Rudgley, R. (1998). *The encyclopedia of psychoactive substances*. Thomas Dunne Books.

Salamone, J. D., Correa, M., Mingote, S. M., & Weber, S. M. (2005). Beyond the reward hypothesis: Alternative functions of nucleus accumbens dopamine. *Current Opinion in Pharmacology, 5*, 34–41. https://doi.org/10.1016/j.coph.2004.09.004

Schmeller, T., & Wink, M. (1998). Utilization of alkaloids in modern medicine. In M. F. Roberts & M. Wink (Eds.), *Alkaloids: Biochemistry, ecology, and medicinal applications* (pp. 435–459). Plenum Press. https://doi.org/10.1007/978-1-4757-2905-4_18

Schoeler, T., Baldwin, J., Allegrini, A., Barkhuizen, W., McQuillin, A., Pirastu, N., Kutalik, Z., & Pingault, J. B. (2023). Novel biological insights into the common heritable liability to substance involvement: A multivariate genome-wide association study. *Biological Psychiatry, 93*(6), 524–535. https://doi.org/10.1016/j.biopsych.2022.07.027

Silcock, J. L, Tischler, M., & Smith, M. (2012). Quantifying the Mulligan River pituri trade of central Australia. *Ethnobotany Research and Applications, 10*, 37–44. https://ethnobotanyjournal.org/index.php/era/article/view/589

Sullivan, R. (2017). Toxin evolution for organismal defense: Is ethanol a special case? *American Journal of Physical Anthropology, 162*(S64), 374. https://doi.org/10.1002/ajpa.23210

Sullivan, R. J., & Hagen, E. H. (2002). Psychotropic substance-seeking: Evolutionary pathology or adaptation? *Addiction, 97*(4), 389–400. https://doi.org/10.1046/j.1360-0443.2002.00024.x

Sullivan, R. J., Hagen, E. H., & Hammerstein, P. (2008). Revealing the paradox of drug reward in human evolution. *Proceedings of the Royal Society B: Biological Sciences*, 275(1640), 1231–1241. https://doi.org/10.1098/rspb.2007.1673

Thorndike, E. L. (1911). *Animal intelligence: Experimental studies*. Macmillan. https://doi.org/10.5962/bhl.title.55072

US Department of Health and Human Services, Substance Abuse and Mental Health Services Administration, Center for Behavioral Health Statistics and Quality. (2020). https://www.datafiles.samhsa.gov/get-help/citations/why-and-how-should-i-cite-samhda-data

Villalba, J. J., Provenza, F. D., Hall, J. O., & Lisonbee, L. D. (2010). Selection of tannins by sheep in response to gastrointestinal nematode infection. *Journal of Animal Science*, 88(6), 2189–2198. https://doi.org/10.2527/jas.2009-2272

Vuppalanchi, R. (2018). Metabolism of drugs and xenobiotics. In R. Saxena (Ed.), *Practical hepatic pathology: A diagnostic approach* (2nd ed., pp. 319–326). Elsevier. https://doi.org/10.1016/B978-0-323-42873-6.00022-6

Watson, P. (1983). *This precious foliage: A study of the aboriginal psycho-active drug pituri*. University of Sydney.

Weyrich, L. S., Duchene, S., Soubrier, J., Arriola, L., Llamas, B., Breen, J., Morris, A. G., Alt, K. W., Caramelli, D., Dresely, V., & Farrell, M. (2017). Neanderthal behaviour, diet, and disease inferred from ancient DNA in dental calculus. *Nature*, 544(7650), 357–361. https://doi.org/10.1038/nature21674

Wink, M. (1998). Modes of action of alkaloids. In M. F. Roberts & M. Wink (Eds.), *Alkaloids: Biochemistry, ecology, and medicinal applications* (pp. 301–326). Plenum Press. https://doi.org/10.1007/978-1-4757-2905-4_12

Wink, M. (2006). Importance of plant secondary metabolites for protection against insects and microbial infections. In M. Rai & M. C. Carpinella (Eds.), *Naturally occurring bioactive compounds* (pp. 251–268). Elsevier. https://doi.org/10.1016/S1572-557X(06)03011-X

Wink, M., Schmeller, T., & Latz-Bruning, B. (1998). Modes of action of allelochemical alkaloids: Interactions with neuroreceptors, DNA, and other molecular targets. *Journal of Chemical Ecology*, 24, 1881–1937. https://doi.org/10.1023/A:1022315802264

Wise, R. A. (2000). Addiction becomes a brain disease. *Neuron*, 26, 27–33. https://doi.org/10.1016/s0896-6273(00)81134-4

Wooding, S. P., Ramirez, V. A., & Behrens, M. (2021). Bitter taste receptors: Genes, evolution and health. *Evolution, Medicine, and Public Health*, 9, 431–447. https://doi.org/10.1093/emph/eoab031

PART 3

Consequences and Perspectives on Research and Clinical Sciences

18

Conditions of Comparative Brain Connectomics

Kathleen S. Rockland, Daniel Zachlod, and Katrin Amunts

Introduction

The relevance of comparative brain studies has been recognized since the time of the early pioneer researchers: Cécile and Oskar Vogt and Brodmann in brain cartography and Cajal and many others at the more cellular level. Comparative studies have always been recognized as important for assessing conserved and divergent features, the potential and limits of different brain organizations, and, increasingly, brain models of disease and therapeutics. In the current era, a major emphasis of comparative studies has resulted in more detailed parcellations of the brain and better definition of cell types, especially with the relatively standardized approach of single-cell transcriptomics. Maps of gene expression profiles are becoming available at the systems level, combined with traditional neuroanatomic, neurochemical, and electrophysiological criteria. How cells and areas intercommunicate and what the mechanisms are to form networks are investigated in the living human brain by functional imaging or tractography. At the micro level, whole neuron detail (from soma through axon branches and synaptic terminations) is addressable by multiple light and electron microscopic techniques. Such diversity of information, which comes from research at different spatial scales and targets different levels of brain organization, is increasingly benefiting from integration of multimodal atlases that allow combined analysis. Together, these technical advances are inaugurating a new, high-resolution, high-throughput era of comparative connectomics. This can be expected to significantly improve our understanding of human-specific features, as well as inform our mode of interacting with other species, cultures, and the natural environment.

In this chapter, we selectively review several human-specific neuroanatomical features, how these differ from nonhuman brains, and how they are thought to be involved in vulnerabilities and disease. The first part of the chapter focuses primarily on macro-scale organization and the second on more micro-scale features. We discuss in more detail the evolution of connectivity for selected functional systems with high relevance for disease and the role of asymmetry of connectivity and provide an overview of web-based community resources of connectivity data.

Most of the comparisons are between human and nonhuman primates (NHPs), with extensive discussion of rodent brains but, in part for space constraints, less discussion of other species. We also include relatively little on development, although specific aspects of the proliferative cell cycle and metabolism of neural progenitor cells are widely recognized as having a fundamental role in shaping human cortical organization. Similarly, we opted for little comment on the important issue of environmental–genetic or –evolutionary interactions. Finally, it is acknowledged that the list of references is by far not complete; rather, it was intended to include some illustrative, recent examples.

The Macro Level

The large-scale network organization of the human brain (often referred to as the "connectome") has been extensively investigated by neuroimaging studies analyzing structural and functional connectivity. Tracer studies, common in animal models, play only a minor role in humans.

Tractography of fiber tracts in humans is mainly performed with diffusion tensor imaging (DTI) in the millimeter range in vivo and with a submillimeter resolution postmortem. With DTI it is possible to trace with high spatial resolution the large long-range fiber bundles connecting the different cortical lobes as well as single trajectories (e.g., the mammillothalamic tract of the Papez neuronal circuit). It is less successful in capturing the course of fibers in the cerebral cortex or subcortical nuclei. Tractography data exist not only for humans but for many different species (e.g., rabbit, prairie vole, marmoset, macaque). This makes DTI highly suitable for comparative studies, and reference data have been provided during the past years for several species (see below, "Access to Connectivity Data: Atlases and Databases").

The Default Mode Network

Coactivation studies are used to explore functional connectivity either during a specific task or in goal-free resting-state condition. Of special interest is the functionally interconnected default-mode network (DMN). The DMN seems to be evolutionarily conserved as it is found, with some modifications, from rodents (Whitesell et al. 2021) to NHPs (Garin et al. 2022). The DMN exhibits decreased neural activity during externally oriented tasks and increased activity during resting baseline and conditions with interoceptive-like features such as (in humans) autobiographical, episodic, and semantic memory; future thinking; or mind-wandering. In humans, this distributed network includes portions of the prefrontal cortex, inferior parietal lobule, and temporal cortex. Functional Magnetic Resonance Imaging (fMRI), tractography, and anatomical tracing studies in rodents provide evidence that these functionally related regions are densely interconnected but with major dissimilarities in the connectivity profiles across species. The connection of medial prefrontal cortex and posterior cingulate cortex seems to be a unique human feature (Garin et al. 2022), whereas in NHPs the medial prefrontal cortex is involved in a frontotemporal network. The alteration of the connectivity profile together with a higher network centrality (Ardesch et al. 2019) may boost the evolution of cognitive functions in humans.

This greater degree of connectional complexity is associated with cell class–specific connectivity, so far demonstrated in rodents. For example, in mice, supragranular neurons in identified DMN regions are reported to project mostly to other DMN domains, whereas layer V neurons project both within and without the DMN. Sublayers in layer V are populated with midline (in DMN targets) and visually projecting neurons (without DMN targets). Although the two neuronal types belong to the same transcriptomic cluster, they can be distinguished by marker genes. The neurons of layer V are considered as forming a distinct network connectivity and may be responsible for switching between the DMN and other brain networks. Cross-species comparisons of the DMN reveal a hierarchy-like organization from primary sensory to transmodal cortices, which is likely to be phylogenetically conserved from rodents to primates (Whitesell et al. 2021). It seems reasonable to suppose that the same laminar-differentiated general scaffolding will occur in humans.

Connectivity-Based Parcellations

fMRI and DTI are methods to investigate not only connectivity networks in different species but also the parcellation of the brain in rodents, monkeys, and humans. Connectivity-based parcellation (CBP) groups voxels with similar connectivity under the premise that function, connectivity, and cytoarchitectonic brain organization covary.

Functional neuroimaging can map coactivation patterns in task-based or task-free condition (resting state). A detailed map of the macaque resting-state connectome acquired with ultra-high-field MRI improved image quality and enabled identification of 30 resting-state networks also present in humans (Yacoub et al. 2020).

In humans, individual-specific mapping of resting-state fMRI data indicates a preferential association of the anterior hippocampal formation (HF) with the DMN, but the posterior HF was in contrast preferentially associated with the parietal memory network (Zheng et al. 2021). The differentiable neuroanatomical dichotomy is interpreted in the context of human memory and mental processes responsible for sense of self and raises addressable questions as to further substrates and hippocampal–cortical interactions.

To verify such parcellations, histology and/or physiology are mandatory. A recent study analyzed the relationship between connectivity- and cytoarchitecture-based parcellation as compared to classical tracing (Gao et al. 2018). While the patterns of parcellation appeared to be rather similar, differences between them were also observed that may be influenced partly by methodological biases (e.g., resulting from differences in spatial resolution) and/or true differences in parcellation. However, early studies demonstrated changes of electrophysiological properties precisely at the borders of two cytoarchitectonic areas (Luppino et al. 1991; i.e., showed that the two modalities covary). Along the same line of reasoning, fMRI showed area-specific activations during hand grasping and observation (Nelissen et al. 2005; i.e., a correlation of cytoarchitectonic organization with motor function).

Dynamic Causal Modeling

To find networks within fMRI data, dynamic causal modeling (DCM; Friston et al. 2003) estimates the coupling of brain regions and calculates the effective connectivity among them to build realistic models of interacting brain regions and how their activity changes in a task-specific way. DCM was applied in animal studies investigating induced epilepsy, like in a rat model or a zebrafish model. DCM was also applied to investigate drug-induced changes in excitatory and inhibitory synaptic transmission in the auditory region of rodents. The effective connectivity model of the auditory cortex appears to correspond to previous neurophysiological studies. However, pharmaceuticals can alter the cerebrovascular function and modulate communication within and between brain networks (reviewed for NHPs in Jovellar and Doudet 2019), which may affect direct cross-species comparisons.

Diffusion Tensor Imaging

DTI allows us to study large white matter fiber tracts and to elucidate the networks formed by these connections (Figure 18.1). DTI has the advantage of being applicable to the living human brain and providing homologous data in NHP brains. Analysis of fiber organization is broadly applicable across species; for example, in the HF, the segregation along the anterior–posterior long axis is conserved (homologous to ventral–dorsal in rodents).

Neuroimaging cross-species comparisons exist for humans and NHPs (macaque, chimpanzee). The evolutionary changes in the functional organization between macaques and humans are smaller for unimodal and sensory-motor areas but present even within the visual system and ventral visual pathway and larger for areas associated with attention and higher-order cognition. There is an evolutionary areal surface expansion in these areas, which seems to correspond to myelination mapping.

DTI-based tracing of white matter bundles has been carried out in chimpanzees (Mars et al. 2019). Interestingly, there are connectional features not present in macaques

CHAPTER 18 CONDITIONS OF COMPARATIVE BRAIN CONNECTOMICS | 403

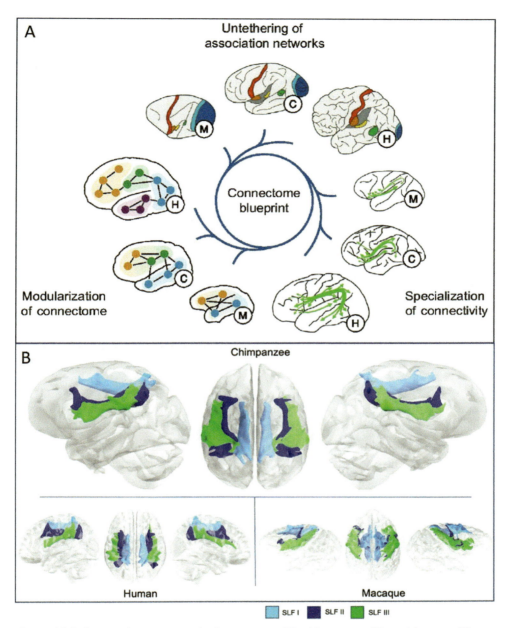

Figure 18.1 Comparative connectomics in macaques (M), chimpanzees (C), and humans (H) concerning network specialization (A) and long-range fiber tracts (B). Humans share evolutionarily conserved connectivity features ("connectivity blueprint") with adaptions in network specialization (i.e., untethering of association networks and an increase in modular organization). Tracing of the superior longitudinal fascicle (SLF), as an example of long-range fiber tracts, showed similar target areas in all three species for SLF I and III and a specialization of SLF II in hominoids, with subtle differentiation between humans and chimpanzees (in chimpanzees SLF II also reached the superior parietal lobe). (A) From Ardesch et al. (2019). (B) From Bryant et al. (2020).

and marmosets but resembling those in humans. Thus, chimpanzee connectivity data will be an important link in probing comparative brain evolution.

DTI-based tractography of human language regions and the putative homologues in chimpanzees revealed unique human features such as more connections of the posterior middle temporal cortex via the parietotemporal branch of the arcuate fascicle and fewer ventral connections. The anterior temporal lobe has unique expansions of the uncinate and middle and inferior longitudinal fascicles. A similar study in the auditory and visual cortices of monkeys and humanoids showed human-specific connections between the auditory core region and the anterior temporal lobe and major connectivity to the anterior superior temporal gyrus (Bryant et al. 2019). Thus, interspecies differences can be found in different functional systems.

A putative shift in the evolution of brain connectomics was revealed by a DTI-based study comparing the connectomes of chimpanzees and humans (Ardesch et al. 2019). In humans they found less interhemispheric connectivity and stronger connectivity between multimodal areas of the temporal, lateral parietal, and inferior frontal cortices, brain areas which are part of the language network. Furthermore, the human connectome showed a higher network centrality and an increased modular organization (i.e., area-specific specialization), prerequisites for the human abilities for language and cognition (Figure 18.1).

Network Theory

Several concepts have been formulated regarding cortical network organization. These all recognize the comparative aspect as an important source of insight, but the actual species coverage has remained rather limited. An ideal design, for example, might span rodents (both mouse and rat), at least macaques and marmoset monkeys (and maybe galago to represent nocturnal species), chimpanzees, and humans. At the macro scale, however, imaging studies tend to emphasize humans and only sometimes include (ex vivo) chimpanzees and macaques; and tracer studies, conversely, necessarily have emphasized mice, marmosets, and macaques.

In one of the older frameworks, cortical areas are ranked in an ordered hierarchy, relying on laminar connectivity patterns usually interpreted as progressing from primary sensory areas to higher-order, multimodal association areas (macaque: Felleman and Van Essen 1991). Especially over time, some categorization criteria have seemed less straightforward and open to multiple interpretations ("indeterminate": Hilgetag and Goulas 2020).

In the influential graph theoretical framework, neurons and connecting axons are conceived as "nodes and edges" (Bullmore and Sporns 2009). Key concepts include densely interconnected "hubs," reciprocity of connections, and economy of wiring or shortest pathway. This framework has been extensively applied to analyze imaging data sets of human brains, both event-driven and resting state (DMN). With continued investigations, however, further complexities are emerging. The DMN itself is recognized as divisible into multiple distinct networks. Distributed and parallel organizations in the mature brain have been attributed to fractionation and specialization of a simpler proto-organization by activity-driven processes during early development (Buckner and DiNicola 2019). Along different lines, arousal-related traveling waves are proposed as an important physiological phenomenon recapitulating the large-scale organization of the brain (Raut et al. 2021). These waves slowly propagate throughout the neocortex, striatum, and cerebellum in both humans and macaques in relation to autonomic infra-slow arousal fluctuations and may be a mechanism for spatiotemporal patterning of brain-wide excitability.

At a finer granularity, visualization of large numbers of single axons is providing novel results about quantitative and qualitative details of axon configuration, trajectory, and variability (frontal cortex in macaque; Xu et al. 2021; in mice: Gao et al. 2022). Results point to a previously unsuspected large number of long-range projecting axonal subtypes, within the classic cortico-cortical and cortico-subcortical networks (Gao et al. 2022). How this extremely heterogeneous and divergent topography maps onto previous proposals of brain organization remains to be established.

Asymmetries in Connectivity Patterns

A hallmark of human brain organization is anatomical asymmetry and functional laterality (see also Chapter 1). However, left–right differences of the brain have also been observed in a large number of mammalian species and in Amphibia, Reptilia, insects, fish, birds, and even invertebrates such as nematode (for a recent review on asymmetries, see Ocklenburg and Gunturkun 2024). Comparative studies on connectivity are focused on a few systems in relatively few species, in particular the motor system (e.g., in mammals the pyramidal tract, effective connectivity in humans; Pool et al. 2014), the language and vocalization system (the arcuate fascicle in particular), and the visual system (Graïc et al. 2022).

Asymmetries within the habenula nuclei of the epithalamus have been reported for all vertebrate species from fishes, amphibians, and reptiles to birds and mammals with respect to size, cytoarchitecture, neurochemistry, and connectivity.

An asymmetry in the functional connectivity of humans has been observed in the resting-state condition, which showed a left-lateralized language network and DMN, whereas visual, attentional (Liu et al. 2009), and face-processing networks (Badzakova-Trajkov et al. 2016) are right-lateralized. The left hemispheric bias of networks processing language and gesture prompted the hypothesis that the primate mirror neuron system became increasingly lateralized and then split up into subsystems (Häberling et al. 2016).

Cross-species comparisons have found a right hemispheric predominance for networks processing emotion in birds, dogs, primates, and humans. However, in humans, both hemispheres seem to contribute to emotion processing: either the emotions are positive or show approach motivational tendencies (left hemisphere) or the emotions are negative-valanced and show withdrawal tendencies (right hemisphere). This was formulated in the valence lateralization hypothesis. As both hemispheres seem to process emotions, the question arises if this is due to different network activations. In a dynamic model, proposed by Stanković (2021), the initial state of the emotional network is right-biased; but with increasing task demands, the network reconfigures with the inclusion of the left hemisphere.

There is ample evidence for neurochemical asymmetries. Neural signaling asymmetries in the dopaminergic nigrostriatal system are reported; they have direct effects on the motor system of mice for paw and rotation preference (Molochnikov and Cohen 2014) and seem to be crucial for motor preferences in humans. Changes in the dopamine neuroreceptor have also clinical relevance for Parkinson's disease (PD; see Chapter 11) and schizophrenia. PD patients show a reduced dopamine uptake in the posterior putamen and nigrostriatal neuronal loss contralateral to the symptom onset. Reduced hemispheric asymmetry has been identified as a risk factor in schizophrenia (see Chapter 14). Furthermore, neurodegenerative diseases show an asymmetric glucose metabolism (Alzheimer's disease [AD], multiple sclerosis [MS]), gray matter loss (AD, amyotrophic lateral sclerosis [ALS], MS), limb weakness (ALS, MS), depletion of neuronal connections (AD), and distribution patterns of proteins (frontotemporal dementia, primary progressive aphasia; reviewed in Lubben et al. 2021). As a particularly striking example, the brain seems to be organized through neurochemical asymmetries in the dopaminergic system for the left hemisphere (right manual preference and speech) and the noradrenergic arousal system in the right hemisphere, which maintains alertness, integrates bilateral information, and orients to new stimuli (Tucker and Williamson 1984).

The Micro Level

Bridging from Macro to Micro: Three-Dimensional Polarized Light Imaging

Postmortem three-Dimensional (3D) Polarized Light Imaging, based on the birefringence of myelin sheaths, is a technique that is capable of visualizing thin fibers and even axons at the whole-brain level in thin unstained brain sections (Figure 18.2; Axer et al. 2011). It can reliably visualize the fiber architecture of the cortex and fibers around and within the subcortical nuclei, thus bridging the scale between neuroimaging (macro level) and features of single-neuron morphology (micro level). In contrast to tracer injections, it is fully applicable in human and nonhuman brains, thus allowing us to perform interspecies comparison. Importantly, both short-range and long-range connections can be visualized, and 3D fiber distributions can be compared with classical myeloarchitecture as well as results obtained with DTI (Axer and Amunts 2022). A comparable approach is enabled by polarization-sensitive optical coherence tomography (Wang et al. 2018).

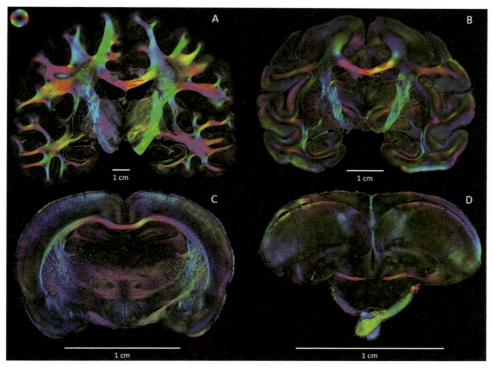

Figure 18.2 Whole-brain coronal sections showing fiber orientation maps based on three-dimensional polarized light imaging (Axer et al., 2011) of human (A), vervet monkey (B), rat (C), and pigeon (D). The sections show two organizational principles during evolution: upscaling in brain size (A) and surface expansion (gyrification, B]). This resulted in a higher demand for information transfer between brain regions (i.e., increased connectivity). Color wheel in upper left indicates fiber orientation. Scale bars, 1 cm as shown.

Cellular Circuits

Since the 1970s, tracer injections in experimental animals have mapped the general layout of brain network connectivity at the cellular level. Across species, this consists of ipsi- and contralateral cortical connections; multiple cortico-subcortical connections, often but not always reciprocal; neuromodulatory inputs; and, for the cerebral cortex, dense local intrinsic connections. Among the species-shared features (for mammals) is a recognizably conserved cortical laminar organization, where layer IV is a dominant input layer (but less so in limbic and motor areas), supragranular layers II and III are the dominant source of ipsi- and contralateral cortico-cortical output, and infragranular layers V and VI are the dominant source of cortico-subcortical connections. There are many area- and species-specific modifications; for example, inputs can target layers other than layer IV (layer I or VI or, in the case of "columnar" terminations, layers I–VI). In NHPs, a component of cortico-striatal output in some areas is known to originate from the upper layers, and feedback in the sensory cortical areas has a bistratified origin in the upper and lower layers, as do cortico-cortical projections among association areas (Rockland 2020).

How identified inputs are spatially distributed along identified dendrites is unknown for cortical neurons (for CA1, see Megias et al. 2001) but may influence patterns and degree of spine loss in the context of cellular vulnerabilities.

Given the degree of architectural commonality, despite variations, what can we say about "human-specific"? An obvious comment is that the human brain is large, with a large number of neurons, and, in contrast with rodents and other small animals, has an additional number of cortical association areas. The relative expansion of the supragranular layers in primates is thought to correlate with an enhanced prominence of cortico-cortical connections, originating from these layers. More particularly, one can assume there are conspicuously enhanced iterative cortico-cortical and cortico-subcortical loops. These are not hard-wired but dynamic and responsive to multiple intero- and exteroceptive stimuli. The complex dialogue within and between interoceptive and exteroceptive signals is generally discussed as basic to a sense of self and the adaptive relation of organisms to their environment (Panksepp and Northoff 2009). A robust hypothesis sees interoceptive processing as resulting from the interplay of two systems, that is, a basal subcortical system, likely to be widely present in all vertebrates, and a forebrain system including the medial and lateral cerebral hemispheres. The expansion of the forebrain system as well as its massive interconnections with the medial brainstem have been associated with human-specific attributes of self and world representation, in normal and pathological conditions (Craig 2002; Fabbro et al. 2015; Cheng et al. 2022). Also to be considered is the greater amount and variability of exteroceptive stimuli customary to humans.

This interactive "connectome" in itself constitutes a vulnerability, given the heightened opportunities for errors in synaptic convergence and orchestrated timing. Cell-specific loss or malfunction in any area would have a propagated effect throughout the network.

In the following, we briefly review the main cellular and subcellular connectional substrates.

Neurons

Basic neuron types are morphologically recognizable across species, for example, the excitatory pyramidal neurons in cortical areas and the cerebellar Purkinje cells. There are also, however, species-specific specializations even at the morphological level (Figure 18.3). Basal dendrites have more complex branching patterns and a greater number of spines in different areas and species, with the greatest number of both in the human association cortex (Luebke 2017). The protracted cycle of dendritic spinogenesis and pruning in higher order association areas of humans has been signaled out as a potential source of vulnerability. Spine loss is a prominent feature of pathological and psychiatric conditions (Figure 18.3).

Large brains have an increased cortical thickness and, in consequence, longer apical dendrites of pyramidal neurons, with more dendritic spines and greater segmental complexity (Schmidt et al. 2021; Figure 18.3). These features imply more inputs, more

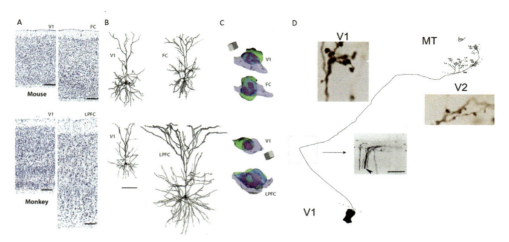

Figure 18.3 Morphological comparison of neurons in visual and prefrontal areas of mouse and monkey. (A) Comparative cytoarchitecture of V1 and frontal cortex in mouse (top) and monkey (bottom). (B) Dendritic arborization of reconstructed LIII pyramidal neurons in mouse (top) and monkey (bottom). (C) Three-dimensional reconstruction of axo-spinous synapses (spines in green, boutons in blue, perforated synapses in purple). Note the significantly larger perforated synapses between visual and prefrontal areas in monkeys (bottom) but not in mice (top). (D) Reconstruction of a single axon projecting from V1 (bottom left) to extrastriate mediotemporal cortex MT/V5 in macaque. Insets, clockwise from top: large V1 boutons in target area MT/V5, small V2 boutons in area MT/V5, and (from the boxed area) a bundle of mixed-diameter axons projecting from V1 to MT/V5. (C) From Luebke (2017). (D) From Rockland (1995), with permission.

integrative capacity, and greater computational power but are also associated with disease. The giant Betz cells in the human motor cortex undergo profound apical dendritic degeneration in cases of ALS (familial and sporadic) and cases with frontotemporal dementia (Genç et al. 2017).

Transcriptomic investigations are rapidly providing new evidence of a marked cell type diversity within populations previously considered as homogeneous. Transcriptomic signatures can also be expected to facilitate finer investigations of cell-specific vulnerabilities, serve as a gauge of experience-dependent gene expression, and provide clearer standards of species homologies and differences (see also Chapter 5). One recent study of the hippocampal–entorhinal system identified preferential expression of *METTL7B* (a gene which interacts with proteins associated with the endoplasmic reticulum, oxidative phosphorylation, and neurodegenerative diseases) in humans and macaques but not mice or pigs (Franjic et al. 2022). More broadly, from conserved or divergent molecular neuronal identities in a lizard and mouse brain, researchers have proposed the novel view that the mammalian brain is a mixture of ancient and novel neuron types in a distributed mosaic of new and ancestral traces (Hain et al. 2022).

Glia

Glia cells are phylogenetically ancient, but several uniquely hominid glia specializations have been described. For example, as compared with rodents, protoplasmic astrocytes are

2.6-fold larger in diameter, with 10-fold more GFAP+ primary processes; exhibit faster calcium waves in slice preparations; and comprise more subtypes, including interlaminar astrocytes in layer I with processes extending deeper toward layer IV. (See Chapter 2 for a comprehensive overview.)

Extracellular Space

Cortical cell-packing density varies in a species-specific progression, being less in humans than in monkeys, rats, and mice. Commensurately, there is a reverse species-specific "neuropil fraction" where a greater spatial volume in humans is apportioned to pre- and postsynaptic components. The cell-to-cell extracellular space (ECS) contains interstitial fluid, extracellular matrix, and secreted molecules. It is thought to comprise ~20% of brain space and is conserved across species. The ECS is now recognized as an intricate 3D structure with significant influence on multiple functional parameters, for example, interstitial viscosity, charge interactions, steric hindrance, and physical drag (Hrabetova et al. 2018).

Synapses

Glutamatergic synapses are a key component of cell–cell communication and are increasingly recognized as being highly diverse in molecular and morphological features. In electron microscopic tomography, surgical samples of synapses from the human temporal lobe were found to have larger active zones and a 2- to 3-fold larger total pool of synaptic vesicles as compared with rodents or macaque monkeys (Rollenhagen et al. 2020). Synaptic loss is a central feature of AD and is thought to correlate with molecular properties, still to be elucidated, of vulnerable and resilient synapses. Renormalization of synaptic weights during sleep is proposed to reset or weaken most synapses but afford protection to some, including those directly activated by learning (Cirelli and Tononi 2019). This occurs across species, but possible species- and area-specific variations remain to be determined.

Myelin

Myelination, important in timing and synchronization of brain function, is developmentally prolonged in humans, extending beyond late adolescence. This protracted developmental time course results in greater vulnerability to cultural and environmental effects. Waves of myelin gain and loss have been implicated in some types of learning and memory (de Faria et al. 2021). Transcriptomic profiles in humans and mice, an important model of human myelin disorders, have been reported to correlate well but with divergent expression of distinct genes.

General Comments on Architecture

The current view of cortical organization is to some extent "cortico-centric," tending to neglect the subcortical and neuromodulatory contributions. Further attention to whole-brain organization, however, can be expected from several new research directions. The elaboration of cell types, as described above, is being redefined by transcriptomic data, in combination with morphological, connectivity, and electrophysiological criteria. In accord, visualization of connections at the single-axon level reveals a previously unappreciated heterogeneity. Results in rodents demonstrate that neurons in one area can have "dedicated" projections to a single cortical target area or "broadcast" projections to a variable number of multiple areas (Han et al. 2018). On the basis of collateral patterns, 64 subtypes have been identified for four previously identified networks originating from the mouse frontal cortex (Gao et al. 2022). This architecture is not limited to cortical areas. Segregated subpopulations in the substantia nigra differentially branch to over 40 previously identified targets, with a 50-fold range of terminal density (in mouse: McElvain et al. 2021).

Origin and Impact of Collateralization

Relevant to whole-brain organization is the existence of extrinsic collateralization, especially from association areas (Rockland 2020). In addition, the intrinsic collaterals of pyramidal cells typically extend over a domain several millimeters in diameter. One idea has been that collateralization is a mechanism for corollary discharge (or efference copy). Interestingly, "reafferent sensing"—responses to the consequences of the animal's own actions as in the context of gravity-sensing, flow-sensing, or proprioception—has been proposed as an early phylogenetic precursor of corollary discharge (Jékely et al. 2021). A larger sample size is needed to determine how common this might be in large brains; and other techniques will be necessary to investigate long-distance extrinsic branching in humans, the extent to which branching is greater or less than in rodents, the structural details, and the functional processes.

The possible burden of long extended axons with multiple synaptic release sites is often discussed as a factor in circuit dysfunction. Of these, dopaminergic neurons are a prime example, with specific implications for PD; but, as just summarized, subpopulations of cortical neurons not uncommonly branch divergently to multiple, spatially distributed targets. In addition, since the intrinsic collaterals of pyramidal cells typically extend over several millimeters, there is abundant opportunity for intra- and interlaminar interactions.

A basic question about vulnerability and cell death is the relative influence of cell autonomous factors versus circuit "dysfunction." Cell autonomous factors include soma and dendritic morphology, connectional identity, and gene expression profiles, the latter including shifts over life span and species, as well as broader factors, such as metabolism. An exceptionally clear example of this is from a recent postmortem study of the substantia nigra in PD, which identified 10 spatially localized neuronal clusters, one of which, expressing the gene *AGTR1*, was highly susceptible to loss in PD (Kamath et al. 2022).

Evolution of Connectivity of Two Key Brain Regions Involved in Brain Diseases

Dorsal Thalamus—Gateway to the Cortex

The dorsal thalamus is a major input source to the cerebral cortex and, from the parafascicular and centromedian thalamic nuclei, to the basal ganglia. It is widely recognized across vertebrate species, with mammals having the greatest degree of cross-species homology in terms of nuclear subdivisions and their location, cell types, neurochemistry, and connectivity. The dorsal thalamus constitutes upward of 50 closely abutting distinguishable nuclei, broadly divisible into a small number of "first-order" sensory-recipient nuclei and a larger number of "higher-order" nuclei, synaptically more removed from the sensory periphery (retina, cochlea, and skin) and interconnected with cortical association areas. The higher-order nuclei are thought to have coevolved with cortical areas (Halley and Krubitzer 2019). Conspicuously conserved features include the cross-species persistence of the gamma-aminobutyric acid-ergic (GABAergic) thalamic reticular nucleus, with vital functions related to sleep and wakefulness, and, for mammals, the widely conserved bilaminar corticothalamic projections from layers V and VI (Rouiller and Welker 2000). Thalamic synapses express a particular isoform of vesicular glutamate transporter (VGLUT2), which facilitates their identification by immunocytochemistry and distinguishes them from cortical synapses (expressing VGLUT1).

Thalamic nuclei commonly project to multiple targets. In rodents (but unknown for humans), subpopulations of thalamic projection neurons can be differentiated by patterns of collateral branching. Reuniens and paraventricular midline nuclei, for example, have neuronal subpopulations that project to the frontal cortex only, the ventral hippocampus only, or both by branched collaterals (Viena et al. 2022). Similarly, from the centromedian nucleus, some neurons (approximately half) project densely to the striatum, some (approximately one-third) project diffusely to the cortex only, and others branch to both targets (Smith et al. 2009).

When comparing the dorsal thalamus in rodents and primates, it becomes evident that there are at least three notable functionally significant differences. First, in rodents, but not primates, the thalamus is largely free of inhibitory interneurons, except for the lateral geniculate nucleus. Second, in rodents, excitatory thalamocortical neurons do not express calcium binding proteins. Third, the extent and density of axons immunoreactive for the dopamine transporter are conspicuously greater in the primate dorsal thalamus (García-Cabezas et al. 2009).

Subdivisions of the pulvinar nucleus, in particular the multimodal medial subdivision (PM), are considered as recently evolved and expanded in humans and NHPs, in correlation with the evolutionary expansion of the parietal, temporal, and other association cortical areas with which these are reciprocally interconnected (Homman-Ludiye and Bourne 2019). In these areas and in the prefrontal cortex, multiple iterative cortico-pulvinar and pulvino-cortical loops partially overlap with inputs from the mediodorsal thalamus and basolateral amygdala. Pulvinar responses are reported to be sensitive to the temporal structure of visual stimuli and, in association with their cortical targets, to exhibit functional specializations related to attentional effects (dorsal PM) or visual recognition (ventral PM; Arcaro et al. 2018). Imaging studies in humans (reviewed in Homman-Ludiye and Bourne 2019) report volume changes in the PM in conditions of attention-deficit hyperactivity disorder (reduced volume), autism (increased volume, correlated with increased volume in the temporal and prefrontal cortex), and schizophrenia (reduced volume). (See Chapters 14–16.)

Thalamo-cortical terminations preferentially, but not exclusively, target cortical layer IV. Some geniculo-cortical axons to the primary visual cortex in macaques have collaterals in layer VI, and inputs from the visual pulvinar to the primary visual cortex target layer I, a zone of distal apical dendrites and inhibitory interneurons. Pulvino-cortical inputs, and those from the mediodorsal thalamus, typically have major input to layer IV but with additional input to layer I and other layers. In cetaceans and elephants, both of which species have pyramidalized cortices devoid of layer IV, thalamic input is likely to directly target pyramidal cell dendrites, without specialized interlaminar terminations through layer IV (Graïc et al. 2022).

Mapping thalamo-cortical interactions onto the architecture of human cognition and behavior is very much a work in progress. Recent views propose a dynamic perspective where internuclei thalamic coactivations are configured and reconfigured according to conditions of cognitive flexibility (Roy et al. 2022). Specific outcomes may result from thalamic subnetworks, as opposed to specific nuclei, interacting in different combinations and proportions.

The dorsal thalamus has an important place in human disease and therapeutic measures. The anterior thalamic (AT) nuclei are implicated in loss of episodic memory in early AD, possibly in consequence of an extended mnemonic system which critically involves the interdependent relationship of the AT with the retrosplenial cortex (Aggleton et al. 2016). Profound neurodegeneration in the centromedian/parafascicular complex occurs in PD and is thought to be associated with cognitive impairments in PD (Villalba et al. 2019). The midline thalamus is a target for deep brain stimulation for generalized epilepsy (centromedian nucleus; Agashe et al. 2022).

The Hippocampal Formation

The HF is recognizable, moderately accessible in experimental animals, and widely conserved across mammals in its several subdivisions (dentate gyrus, CA1–4, and subiculum), although with volumetric and cellular species-specific modifications. The human HF is reported as disproportionately large, with an especially large auto-associative CA3 among the primates (Schilder et al. 2020). There is evidence for a homologous nuclear structure in reptiles (by single-cell transcriptomics: Tosches et al. 2018) and birds (by receptor profiles: Herold et al. 2014). The three-layered hippocampus is part of the archicortex and has appeared earlier in phylogeny than the six-layered neocortex.

Early Golgi studies of the HF established the basic cellular architecture; and extensive tracer studies in NHPs and other species confirmed and extended these observations to include cortical, thalamic, and other connections. In rats, there are detailed quantitative data for a small number of CA3–CA3 and CA3–CA1 axons, showing distinct variability in this population (Ropireddy et al. 2011). This type of data, with 200–500-mm axon length distributions, will be hard to replicate in humans or NHPs but is becoming feasible by whole-cell visualization in mice. A combined light and electron microscopic study tabulated the number and distribution of excitatory and inhibitory synapses as mapped in relation to identified CA1 dendrites (Megias et al. 2001). Comparable data on synaptic density can be obtained from postmortem tissues of humans and other species and are important for further studies of synaptic dysfunction in AD and other pathologies and as constraints on modeling and simulation approaches.

As far as is known, the basic anatomical connections of the HF are widely conserved, for example, major cortical input from the entorhinal cortex, inputs from the medial septum and the supramammillary bodies to CA2, thalamic input from anterior and reuniens nuclei, reciprocal connections between the basolateral amygdala and anterior CA1 ("ventral" in rodents), and neuromodulatory inputs. A functional, genetic, and neuroanatomic dissociation has been consistently remarked between the anterior and posterior HF (or ventral and dorsal HF in rodents).

Reciprocal connections with the entorhinal cortex, a funnel-point for widespread cortical inputs, are considered a signature feature of HF; but in addition to entorhinal input, the HF has a small number of direct cortical connections from perirhinal areas, from parietal and temporal cortices (selectively to posterior HF in macaques), and, unidirectionally, to the frontal cortex (in NHPs and rodents). The thalamic connections with reuniens are complex and comprise a set of direct reciprocating connections (CA1-reuniens), direct reciprocating reuniens–frontal cortex connections, and a subpopulation of branched connections from reuniens to frontal cortex and CA1 (Hoover and Vertes 2012). The comparable "hippocampal connectome" is less well worked out in humans.

Species-specific modifications are prominent in the HF. For humans, these include greater thickness of the upper and lower layers of CA1 and likely further quantitative differences (for CA3 collaterals, etc.). The presence of adult neurogenesis in the human hippocampus has been a topic of intensive discussions. However, there is evidence that methodical constraints may limit the detection of the relevant markers and that adult neurogenesis plays a role in the pathomechanism of AD (Moreno-Jiménez et al. 2019). Differential vulnerabilities, perhaps associated with a greater functional burden, are reported for the human hippocampus and include neurodegenerative changes in entorhinal cortex and CA1 and volumetric changes in schizophrenia and depression.

Animal Models of Human Brain Diseases of Connectivity

It is widely acknowledged that animal models are important tools in research, but, at the same time, they are only partially recapitulating to human disease. Small animal models and transgenic mouse lines in particular, although effective in probing initial stages of proteinopathy and genetically based concepts of neurodegenerative disease, are more limited in the context of circuit dysfunction, vascular and immunological interactions, and cultural–genetic interactions, especially as proxies for these features in the human brain ("No animal model of AD, PD, FTD or ALS fully phenocopies human disease"; Dawson et al. 2018; p. 1370). Nevertheless, animal models provide an essential opportunity for invasive studies and experimental manipulation and are uniquely important for precise behavioral paradigms that might further elucidate the functional contributions of anatomical connections.

Macaque monkeys have provided an important model for PD, after chronic administration of 1-methyl-4-phenyl-1,2,3,6-tetrahydropyridine (MPTP; Masilamoni and Smith 2018). They also present as a promising model for AD, given the relatively large, gyrencephalic brain and the closer phylogenetic relation to humans, but so far, have not been extensively used. As a first shortcoming, the pathophysiological hallmarks of the disease in humans appear to be only incompletely recapitulated in monkeys. That is, amyloid plaques have been repeatedly reported in aged macaques, but the signature, naturally occurring tauopathies are more problematic (but see Beckman et al. 2021; Freire-Cobo et al. 2021). Importantly, the diagnostic cognitive decline of human AD has so far not been replicated in macaques; but among a variety of reasons for this is the small sample size, the difficulty of behavioral training of fragile aged macaques, and the lack of comparison points over a sufficiently long life span. The pathologic hallmarks of human AD (amyloid plaques, neurofibrillary tangles, and selective cell loss) have been reported in aged chimpanzees (Edler et al. 2017) but without documentation of cognitive decline. Thus, whether there is exceptional vulnerability of humans to AD, compared to NHPs, is still under discussion (Walker and Jucker 2017). Other gyrencephalic species develop plaque pathology (e.g., dogs; Mckean et al. 2021; aged pinnipeds: Takaichi et al. 2021), but correlation with cognitive decline is not established.

Small brains are eminently suited to powerful whole-brain analyses; and individual neurons are now routinely visualized through their complete, and usually extensive, axon arborization. More complete documentation, to include identified inputs, receptor profiles, and activity-related interactions, among other aspects, can be expected. Further, these and other detailed results from animal models provide a valuable source of comparison—validation or inspiration—with the rapidly advancing fields of artificial neuronal networks and machine learning.

A promising development is the renewed interest in broader comparative studies, including non-standard laboratory species such as pigs (having gyrencephalic brains and a longer gestational period) and an added variety of reptilian, amphibian, and avian brains. This is in part driven by several newer techniques. Postmortem single-cell transcriptomic investigations can address cell type– and species-specific comparisons and have been applied, for example, to investigations of neurogenic potential and cell diversity in the hippocampal–entorhinal system across the gyrencephalic brains of humans, macaques, and pigs (Franjic et al. 2022). Organoids offer another approach, especially suited to early developmental stages or as a source of humanoid implants in animal brains.

In another direction, comparative neuroimaging is well suited for repeated, longitudinal studies and, by facilitating larger sample sizes, is advantageous for achieving better standardization. Efficacious cross-species research has been outlined as a three-prong program: different tools within the same species for improved validation and interpretation, the same tools across multiple species, and different tools customized for direct comparison between human and animal research (Barron et al. 2021).

Access to Connectivity Data: Atlases and Databases

Connectivity data exist for different species, from nematodes to humans, either as atlases or single data sets (see Table 18.1). To provide some examples, atlases show anatomically delineated fiber bundles (e.g., zebrafish, rat, human) or provide connectivity data based on diffusion MRI (quail, macaque, chimpanzee, human), retrograde tracer studies (marmoset and macaque, among other species), electron microscopy (*Drosophila*), or by measurements of single-neuron activation (*Caenorhabditis elegans*). Data sets exist for the electrophysiology of single cells (Allen brain cell type database), DTI scans of different species (Digital Brain Zoo), or for the data sets of the multimodal atlases stored in the EBRAINS knowledge graph. These resources, all publicly available to the scientific community, enable cross-species comparisons and help to trace the evolution of the brain.

Future Directions

Each advance in neuroscience opens multiple new directions so that the number of "unanswered questions" is likely to keep increasing for the foreseeable future. Conspicuous among these, however, are several broad themes which apply to multiple levels of organization, from the subcellular to the supracellular. One is the issue of *variability*, including the molecular, neuronal, circuit, or network level. Insight as to factors and parameters in core variability has practical importance in the context of healthy lifestyle and precision medicine. This pertains to individual variability; variability of species, gender, and life span; and degree of variability across different brain regions. At a shorter timescale,

TABLE 18.1 Examples of Atlases and Databases Providing Connectivity Data in Different Species

Atlas	Description	Link or reference
Connectivity atlas of *Caenorhabditis elegans*	Functional connectivity atlas in the nematode *C. elegans*	Randi et al. (2022)
Connectome of *Drosophila*	Electron microscopy–based connectome of *Drosophila*	Hulse et al. (2021)
AZBA	Anatomy and fiber bundles in the zebrafish	http://azba.wayne.edu/; Kenney et al. (2021)
Atlas of the Japanese quail	DTI-based structural connectivity of the Japanese quail	Yebga Hot et al. (2022)
Waxholm Space Rat Brain Atlas	Anatomical delineations and fiber tracts in the adult rat brain	https://ebrains.eu/service/rat-brain-atlas
Marmoset brain connectivity atlas	3D connectivity atlas based on retrograde tracer injections	https://www.marmosetbrain.org/; Majka et al. (2020)
Macaque brain atlas	DTI-based atlas of 10 macaque brains in template space	www.brainmrimap.org; Feng et al. (2017)
SARM	Subcortical Atlas of the Rhesus Monkey	https://afni.nimh.nih.gov/pub/dist/doc/htmldoc/nonhuman/macaque_tempatl/; Hartig et al. (2021)
Chimpanzee atlas	DTI-based atlas of WM tracts in the chimpanzee	Bryant et al. (2020)
Multi-level human brain atlas of the HBP	Structural and functional connectivity data in different reference spaces	https://ebrains.eu/service/human-brain-atlas/
Allen cell-type database	Morphological, electrophysiological, and transcriptomic data of single cells	https://celltypes.brain-map.org/
Neurodata without Borders data sets	Data sharing platform for neurophysiological and anatomical data	https://www.nwb.org/
Digital Brain Bank	Digital Brain Zoo provides neuroimaging data of cetaceans, monkeys, prosimians, marupials, and apes	https://open.win.ox.ac.uk/DigitalBrainBank
Brain catalogue	Collaborative platform for comparative neuroanatomy	http://braincatalogue.pasteur.fr
EBRAINS knowledgegraph	Connectivity data sets of human, rat, mouse, macaque, and vervet monkey	https://search.kg.ebrains.eu/
Nonhuman Primate Neuroimaging and Neuroanatomy Project (NHP-NNP)	The NHP-NNP provides high-quality multimodal data on macaques, marmosets, and owl monkeys	Hayashi et al. (2021)

3D = three-dimensional; DTI = diffusion tensor imaging; WM = white matter; HBP = Human Brain Project.

there is state-dependent variability, as relating to internal (e.g., sleep) and external environments. Including parallel cross-species investigations, this is already a huge program entailing multiple and interdisciplinary collaborations.

A second broad theme is the need for more detailed and multidimensional investigation of neural substrates of *various behaviors*. For example,

"Detailed analysis of network level activity is needed to understand the dynamic interactions [. . .] that vary with time and task demand. The greatest insights will be obtained when complex circuit analysis and sophisticated behavioral paradigms are combined" (McNaughton and Vann 2022, p. 559). We can anticipate a shift to more naturalistic behaviors, for example, the emerging paradigms addressing two brains in action (two macaques: Lacal et al. 2022; two bats interacting in flight: Sarel et al. 2022). Dynamic and temporal dimensions are only beginning to be addressed ("little is known about how brain circuits rapidly switch between different natural behaviours"; Sarel et al. 2022, p. 119).

Third, we can expect continued and accelerating utilization of *multimodal atlases and 3D visualization*, for the human brain and, increasingly, across species. In vivo whole-brain visualization is increasingly feasible in rodents by wide-field imaging, in NHPs by fMRI or large-population neuropixel recordings, and to some extent in humans by selective in-depth electrodes in neurosurgical protocols. Standard non-invasive techniques such as high-density electroencephalographic (EEG) recordings will contribute toward a standardized atlas of effective connectivity.

Fourth, we can expect continued and accelerating progress in the elucidation of *neuronal cell types*, in the framework of transcriptomics and local and long-distance connectivity, as well as extended investigation of changes over the life span and across species. The neuron–glia–vascular "unit" has only recently been highlighted and will certainly recruit more attention, as pertaining to normal and pathological conditions.

Fifth, a new field of *nano-architectonics* can already be seen emerging from a growing database of ultrastructural results and will undoubtedly develop rapidly. This includes, among many other directions, synapse fine structure, heterogeneity, and specific pathologies in AD and other diseases (Griffiths and Grant 2022), as well as new findings on the intricate molecular organization of the axon cytoskeleton (Vassilopoulos et al. 2019). This level of data should be readily quantifiable and amenable to comparisons across species and conditions.

Finally, in any discussion of future directions, the interplay of *culture and genetics with neural systems* has to figure prominently: what prompts adaptations, and how may these penetrate the genetic level? These interactions have repeatedly been recognized, but a better understanding takes on renewed and even urgent relevance given the increasing number of stressors in the environment, with direct and indirect effects over individual life spans.

BOX 18.1 Three-Dimensional Polarized Light Imaging

Three-dimensional polarized light imaging (3D PLI) is a neuroimaging technique to explore structural connectivity in postmortem whole-brain sections. The visualization of fiber architecture is based on the birefringence of (myelinated and non-myelinated) nerve fibers. The interaction of polarized light with brain tissue leads to an alteration of the polarization state of light depending on the nerve fiber direction. This alteration is measured by a polarization microscope and can be used to calculate nerve fiber orientations in 3D space. The representation of these orientations is called a "fiber orientation map" (FOM), in which the orientation is encoded in color. The reconstruction of 3D PLI data of consecutive sections enables tracing the spatial orientation of nerve fibers and fiber tracts in the whole brain. As an example, the FOM of the human hippocampal region is shown in the figure. Nerve fiber orientation is indicated by the color sphere; that is, fiber inclinations are encoded by means of saturation value (i.e., the darker the pixel color, the more inclined the fibers). The figure shows the alveus (red/yellow) and bending nerve fibers entering the hippocampal formation in blue (see red box).

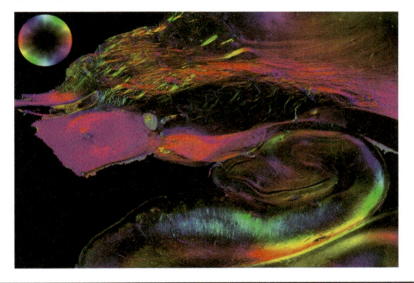

Key Points to Remember

- **Within the primate lineage and to some extent across mammals, the human brain shows evolutionarily conserved features with specialized adaptations.**
- **The DMN is the most salient conserved feature at the macro level; the cortical column organization is an illustrative example at the micro level.**

- Mapping network connectomics across species is challenging (e.g., defining homologies) but has the potential to provide new insights about coactivation patterns across distributed pathways, plasticity, and complex interactions of genetics and environment, among others.
- Although human-specific vulnerabilities to disease occur from the genetic to the cultural level, the interactive connectome in itself can be seen as a major vulnerability, carrying multiple opportunities for errors in synaptic convergence and finely orchestrated timing.
- The rapid tempo of results and the production of massive data sets are bringing about significant alterations in research style. In the future there will be greater emphasis on investigator cooperativity; on digital, brain modeling, and simulation; collaborative platforms; and reuse of existing data sets, based on FAIR (findable, accessible, interoperable, reusable) data.
- There is a greater perceived need for interdisciplinary approaches, extending to interactions of the central nervous system with metabolism, the immune system, and other major functional systems.

References

Agashe, S., Burkholder, D., Starnes, K., Van Gompel, J. J., Lundstrom, B. N., Worrell, G. A., & Gregg, N. M. (2022). Centromedian nucleus of the thalamus deep brain stimulation for genetic generalized epilepsy: A case report and review of literature. *Frontiers in Human Neuroscience, 16*, Article 858413. https://doi.org/10.3389/fnhum.2022.858413

Aggleton, J. P., Pralus, A., Nelson, A. J., & Hornberger, M. (2016). Thalamic pathology and memory loss in early Alzheimer's disease: Moving the focus from the medial temporal lobe to Papez circuit. *Brain, 139*, 1877–1890. https://doi.org/10.1093/brain/aww083

Arcaro, M. J., Pinsk, M. A., Chen, J., & Kastner, S. (2018). Organizing principles of pulvino-cortical functional coupling in humans. *Nature Communications, 9*(1), Article 5382. https://doi.org/10.1038/s41467-018-07725-6

Ardesch, D. J., Scholtens, L. H., & van den Heuvel, M. P. (2019). The human connectome from an evolutionary perspective. *Progress in Brain Research, 250*, 129–151. https://doi.org/10.1016/bs.pbr.2019.05.004

Axer, M., & Amunts, K. (2022). Scale matters: The nested human connectome. *Science, 378*, 500–504. https://doi.org/10.1126/science.abq2599

Axer, M., Amunts, K., Grässel, D., Palm, C., Dammers, J., Axer, H., Pietrzyk, U., & Zilles, K. (2011). A novel approach to the human connectome: Ultra-high resolution mapping of fiber tracts in the brain. *NeuroImage, 54*(2), 1091–1101. https://doi.org/10.1016/j.neuroimage.2010.08.075

Badzakova-Trajkov, G., Corballis, M. C., & Häberling, I. S. (2016). Complementarity or independence of hemispheric specializations? A brief review. *Neuropsychologia, 93*, 386–393. https://doi.org/10.1016/j.neuropsychologia.2015.12.018

Barron, H. C., Mars, R. B., Dupret, D., Lerch, J. P., & Sampaio-Baptista, C. (2021). Cross-species neuroscience: Closing the explanatory gap. *Philosophical Transactions of the Royal Society B: Biological Sciences, 376*(1815), Article 20190633. https://doi.org/10.1098/rstb.2019.0633

Beckman, D., Chakrabarty, P., Ott, S., Dao, A., Zhou, E., Janssen, W. G., Donis-Cox, K., Muller, S., Kordower, J. H., & Morrison, J. H. (2021). A novel tau-based rhesus monkey model of Alzheimer's pathogenesis. *Alzheimer's & Dementia, 17*(6), 933–945. https://doi.org/10.1002/alz.12318

Bryant, K. L., Glasser, M. F., Li, L., Jae-Cheol Bae, J., Jacquez, N. J., Alarcón, L., Fields, A., III, & Preuss, T. M. (2019). Organization of extrastriate and temporal cortex in chimpanzees compared to humans and macaques. *Cortex, 118*, 223–243. https://doi.org/10.1016/j.cortex.2019.02.010

Bryant, K. L., Li, L., Eichert, N., & Mars, R. B. (2020). A comprehensive atlas of white matter tracts in the chimpanzee. *PLOS Biology, 18*, Article e3000971. https://doi.org/10.1371/journal.pbio.3000971

Buckner, R. L., & Dinicola, L. M. (2019). The brain's default network: Updated anatomy, physiology and evolving insights. *Nature Reviews Neuroscience, 20*, 593–608. https://doi.org/10.1038/s41583-019-0212-7

Bullmore, E., & Sporns, O. (2009). Complex brain networks: Graph theoretical analysis of structural and functional systems. *Nature Reviews Neuroscience, 10*, 186–198. https://doi.org/10.1038/nrn2575

Cheng, S., Butrus, S., Tan, L., Xu, R., Sagireddy, S., Trachtenberg, J. T., Shekhar, K., & Zipursky, S. L. (2022). Vision-dependent specification of cell types and function in the developing cortex. *Cell, 185*, 311–327. https://doi.org/10.1016/j.cell.2021.12.022

Cirelli, C., & Tononi, G. (2019). Linking the need to sleep with synaptic function. *Science, 366*, 189–190. https://doi.org/10.1126/science.aay5304

Craig, A. D. (2002). How do you feel? Interoception: The sense of the physiological condition of the body. *Nature Reviews Neuroscience, 3*, 655–666. https://doi.org/10.1038/nrn894

Dawson, T. M., Golde, T. E., & Lagier-Tourenne, C. (2018). Animal models of neurodegenerative diseases. *Nature Neuroscience, 21*(10), 1370–1379. https://doi.org/10.1038/s41593-018-0236-8

de Faria, O., Jr., Pivonkova, H., Varga, B., Timmler, S., Evans, K. A., & Káradóttir, R. T. (2021). Periods of synchronized myelin changes shape brain function and plasticity. *Nature Neuroscience, 24*(11), 1508–1521. https://doi.org/10.1038/s41593-021-00917-2

Edler, M. K., Sherwood, C. C., Meindl, R. S., Hopkins, W. D., Ely, J. J., Erwin, J. M., Mufson, E. J., Hof, P. R., & Raghanti, M. A. (2017). Aged chimpanzees exhibit pathologic hallmarks of Alzheimer's disease. *Neurobiology of Aging, 59*, 107–120. https://doi.org/10.1016/j.neurobiolaging.2017.07.006

Fabbro, F., Aglioti, S. M., Bergamasco, M., Clarici, A., & Panksepp, J. (2015). Evolutionary aspects of self- and world consciousness in vertebrates. *Frontiers in Human Neuroscience, 9*, Article 157. https://doi.org/10.3389/fnhum.2015.00157

Felleman, D. J., & Van Essen, D. C. (1991). Distributed hierarchical processing in the primate cerebral cortex. *Cerebral Cortex, 1*(1), 1–47. https://doi.org/10.1093/cercor/1.1.1-a

Feng, L., Jeon, T., Yu, Q., Ouyang, M., Peng, Q., Mishra, V., Pletikos, M., Sestan, N., Miller, M. I., Mori, S., Hsiao, S., Liu, S., & Huang, H. (2017). Population-averaged macaque brain atlas with high-resolution ex vivo DTI integrated into in vivo space. *Brain Structure & Function, 222*(9), 4131–4147. https://doi.org/10.1007/s00429-017-1463-6

Franjic, D., Skarica, M., Ma, S., Arellano, J. I., Tebbenkamp, A. T. N., Choi, J., Xu, C., Li, Q., Morozov, Y. M., Andrijevic, D., Vrselja, Z., Spajic, A., Santpere, G., Li, M., Zhang, S., Liu, Y., Spurrier, J., Zhang, L., Gudelj, I., . . . Sestan, N. (2022). Transcriptomic taxonomy and neurogenic trajectories of adult human, macaque, and pig hippocampal and entorhinal cells. *Neuron, 110*(3), 452–469.e14. https://doi.org/10.1016/j.neuron.2021.10.036

Freire-Cobo, C., Edler, M. K., Varghese, M., Munger, E., Laffey, J., Raia, S., In, S. S., Wicinski, B., Medalla, M., Perez, S. E., Mufson, E. J., Erwin, J. M., Guevara, E. E., Sherwood, C. C., Luebke, J. I., Lacreuse, A., Raghanti, M. A., & Hof, P. R. (2021). Comparative neuropathology in aging primates: A perspective. *American Journal of Primatology, 83*(11), Article e23299. https://doi.org/10.1002/ajp.23299

Friston, K. J., Harrison, L., & Penny, W. (2003). Dynamic causal modelling. *NeuroImage, 19*, 1273–1302. https://doi.org/10.1016/s1053-8119(03)00202-7

Gao, L., Liu, S., Gou, L., Hu, Y., Liu, Y., Deng, L., Ma, D., Wang, H., Yang, Q., Chen, Z., Liu, D., Qiu, S., Wang, X., Wang, D., Wang, X., Ren, B., Liu, Q., Chen, T., Shi, X., . . . Yan, J. (2022). Single-neuron projectome of mouse prefrontal cortex. *Nature Neuroscience, 25*(4), 515–529. https://doi.org/10.1038/s41593-022-01041-5

Gao, Y., Schilling, K. G., Stepniewska, I., Plassard, A. J., Choe, A. S., Li, X., Landman, B. A., & Anderson, A. W. (2018). Tests of cortical parcellation based on white matter connectivity using diffusion tensor imaging. *NeuroImage, 170*, 321–331. https://doi.org/10.1016/j.neuroimage.2017.02.048

García-Cabezas, M. A., Martínez-Sánchez, P., Sánchez-González, M. A., Garzón, M., & Cavada, C. (2009). Dopamine innervation in the thalamus: Monkey versus rat. *Cerebral Cortex*, *19*(2), 424–434. https://doi.org/10.1093/cercor/bhn093

Garin, C. M., Hori, Y., Everling, S., Whitlow, C. T., Calabro, F. J., Luna, B., Froesel, M., Gacoin, M., Ben Hamed, S., Dhenain, M., & Constantinidis, C. (2022). An evolutionary gap in primate default mode network organization. *Cell Reports*, *39*(2), Article 110669. https://doi.org/10.1016/j.celrep.2022.110669

Genç, B., Jara, J. H., Lagrimas, A. K., Pytel, P., Roos, R. P., Mesulam, M. M., Geula, C., Bigio, E. H., & Özdinler, P. H. (2017). Apical dendrite degeneration, a novel cellular pathology for Betz cells in ALS. *Scientific Reports*, *7*, Article 41765. https://doi.org/10.1038/srep41765

Graïc, J. M., Peruffo, A., Corain, L., Finos, L., Grisan, E., & Cozzi, B. (2022). The primary visual cortex of cetartiodactyls: Organization, cytoarchitectonics and comparison with perissodactyls and primates. *Brain Structure & Function*, *227*(4), 1195–1225. https://doi.org/10.1007/s00429-021-02392-8

Griffiths, J., & Grant, S. G. N. (2022). Synapse pathology in Alzheimer's disease. *Seminars in Cell & Developmental Biology*, *139*, 13–23. https://doi.org/10.1016/j.semcdb.2022.05.028

Häberling, I. S., Corballis, P. M., & Corballis, M. C. (2016). Language, gesture, and handedness: Evidence for independent lateralized networks. *Cortex*, *82*, 72–85. https://doi.org/10.1016/j.cortex.2016.06.003

Hain, D., Gallego-Flores, T., Klinkmann, M., Macias, A., Ciirdaeva, E., Arends, A., Thum, C., Tushev, G., Kretschmer, F., Tosches, M. A., & Laurent, G. (2022). Molecular diversity and evolution of neuron types in the amniote brain. *Science*, *377*(6610), Article eabp8202. https://doi.org/10.1126/science.abp8202

Halley, A. C., & Krubitzer, L. (2019). Not all cortical expansions are the same: The coevolution of the neocortex and the dorsal thalamus in mammals. *Current Opinion in Neurobiology*, *56*, 78–86. https://doi.org/10.1016/j.conb.2018.12.003

Han, Y., Kebschull, J. M., Campbell, R. A. A., Cowan, D., Imhof, F., Zador, A. M., & Mrsic-Flogel, T. D. (2018). The logic of single-cell projections from visual cortex. *Nature*, *556*(7699), 51–56. https://doi.org/10.1038/nature26159

Hartig, R., Glen, D., Jung, B., Logothetis, N. K., Paxinos, G., Garza-Villarreal, E. A., Messinger, A., & Evrard, H. C. (2021). The Subcortical Atlas of the Rhesus Macaque (SARM) for neuroimaging. *NeuroImage*, *235*, Article 117996. https://doi.org/10.1016/j.neuroimage.2021.117996

Hayashi, T., Hou, Y., Glasser, M. F., Autio, J. A., Knoblauch, K., Inoue-Murayama, M., Coalson, T., Yacoub, E., Smith, S., Kennedy, H., & Van Essen, D. C. (2021). The nonhuman primate neuroimaging and neuroanatomy project. *NeuroImage*, *229*, Article 117726. https://doi.org/10.1016/j.neuroimage.2021.117726

Herold, C., Bingman, V. P., Ströckens, F., Letzner, S., Sauvage, M., Palomero-Gallagher, N., Zilles, K., & Güntürkün, O. (2014). Distribution of neurotransmitter receptors and zinc in the pigeon (*Columba livia*) hippocampal formation: A basis for further comparison with the mammalian hippocampus. *Journal of Comparative Neurology*, *522*(11), 2553–2575. https://doi.org/10.1002/cne.23549

Hilgetag, C. C., & Goulas, A. (2020). "Hierarchy" in the organization of brain networks. *Philosophical Transactions of the Royal Society B: Biological Sciences*, *375*, Article 20190319. https://doi.org/10.1098/rstb.2019.0319

Homman-Ludiye, J., & Bourne, J. A. (2019). The medial pulvinar: Function, origin and association with neurodevelopmental disorders. *Journal of Anatomy*, *235*(3), 507–520. https://doi.org/10.1111/joa.12932

Hoover, W. B., & Vertes, R. P. (2012). Collateral projections from nucleus reuniens of thalamus to hippocampus and medial prefrontal cortex in the rat: A single and double retrograde fluorescent labeling study. *Brain Structure & Function*, *217*(2), 191–209. https://doi.org/10.1007/s00429-011-0345-6

Hrabetova, S., Cognet, L., Rusakov, D. A., & Nägerl, U. V. (2018). Unveiling the extracellular space of the brain: From super-resolved microstructure to in vivo function. *Journal of Neuroscience*, *38*(44), 9355–9363. https://doi.org/10.1523/JNEUROSCI.1664-18.2018

Hulse, B. K., Haberkern, H., Franconville, R., Turner-Evans, D., Takemura, S. Y., Wolff, T., Noorman, M., Dreher, M., Dan, C., Parekh, R., Hermundstad, A. M., Rubin, G. M., & Jayaraman, V. (2021). A connectome of the *Drosophila* central complex reveals network motifs suitable for flexible navigation and context-dependent action selection. *eLife*, *10*, Article e66039. https://doi.org/10.7554/eLife.66039

Jékely, G., Godfrey-Smith, P., & Keijzer, F. (2021). Reafference and the origin of the self in early nervous system evolution. *Philosophical Transactions of the Royal Society B: Biological Sciences*, *376*(1821), Article 20190764. https://doi.org/10.1098/rstb.2019.0764

Jovellar, D. B., & Doudet, D. J. (2019). fMRI in non-human primate: A review on factors that can affect interpretation and dynamic causal modeling application. *Frontiers in Neuroscience, 13*, Article 973. https://doi.org/10.3389/fnins.2019.00973

Kamath, T., Abdulraouf, A., Burris, S. J., Langlieb, J., Gazestani, V., Nadaf, N. M., Balderrama, K., Vanderburg, C., & Macosko, E. Z. (2022). Single-cell genomic profiling of human dopamine neurons identifies a population that selectively degenerates in Parkinson's disease. *Nature Neuroscience, 25*(5), 588–595. https://doi.org/10.1038/s41593-022-01061-1

Kenney, J. W., Steadman, P. E., Young, O., Shi, M. T., Polanco, M., Dubaishi, S., Covert, K., Mueller, T., & Frankland, P. W. (2021). A 3D adult zebrafish brain atlas (AZBA) for the digital age. *eLife, 10*, Article e69988. https://doi.org/10.7554/eLife.69988

Lacal, I., Babicola, L., Caminiti, R., Ferrari-Toniolo, S., Schito, A., Nalbant, L. E., Gupta, R. K., & Battaglia-Mayer, A. (2022). Evidence for a we-representation in monkeys when acting together. *Cortex, 149*, 123–136. https://doi.org/10.1016/j.cortex.2021.12.012

Liu, H., Stufflebeam, S. M., Sepulcre, J., Hedden, T., & Buckner, R. L. (2009). Evidence from intrinsic activity that asymmetry of the human brain is controlled by multiple factors. *Proceedings of the National Academy of Sciences of the United States of America, 106*, 20499–20503. https://doi.org/10.1073/pnas.0908073106

Lubben, N., Ensink, E., Coetzee, G. A., & Labrie, V. (2021). The enigma and implications of brain hemispheric asymmetry in neurodegenerative diseases. *Brain Communications, 3*(3), Article fcab211. https://doi.org/10.1093/braincomms/fcab211

Luebke, J. I. (2017). Pyramidal neurons are not generalizable building blocks of cortical networks. *Frontiers in Neuroanatomy, 11*, Article 11. https://doi.org/10.3389/fnana.2017.00011

Luppino, G., Matelli, M., Camarda, R. M., Gallese, V., & Rizzolatti, G. (1991). Multiple representations of body movements in mesial area 6 and the adjacent cingulate cortex: An intracortical microstimulation study in the macaque monkey. *Journal of Comparative Neurology, 311*(4), 463–482. https://doi.org/10.1002/cne.903110403

Majka, P., Bai, S., Bakola, S., Bednarek, S., Chan, J. M., Jermakow, N., Passarelli, L., Reser, D. H., Theodoni, P., Worthy, K. H., Wang, X. J., Wójcik, D. K., Mitra, P. P., & Rosa, M. G. P. (2020). Open access resource for cellular-resolution analyses of corticocortical connectivity in the marmoset monkey. *Nature Communications, 11*(1), Article 1133. https://doi.org/10.1038/s41467-020-14858-0

Mars, R. B., O'Muircheartaigh, J., Folloni, D., Li, L., Glasser, M. F., Jbabdi, S., & Bryant, K. L. (2019). Concurrent analysis of white matter bundles and grey matter networks in the chimpanzee. *Brain Structure & Function, 224*(3), 1021–1033. https://doi.org/10.1007/s00429-018-1817-8

Masilamoni, G. J., & Smith, Y. (2018). Chronic MPTP administration regimen in monkeys: A model of dopaminergic and non-dopaminergic cell loss in Parkinson's disease. *Journal of Neural Transmission, 125*, 337–363. https://doi.org/10.1007/s00702-017-1774-z

McElvain, L. E., Chen, Y., Moore, J. D., Brigidi, G. S., Bloodgood, B. L., Lim, B. K., Costa, R. M., & Kleinfeld, D. (2021). Specific populations of basal ganglia output neurons target distinct brain stem areas while collateralizing throughout the diencephalon. *Neuron, 109*(10), 1721–1738.e4. https://doi.org/10.1016/j.neuron.2021.03.017

Mckean, N. E., Handley, R. R., & Snell, R. G. (2021). A review of the current mammalian models of Alzheimer's disease and challenges that need to be overcome. *International Journal of Molecular Sciences, 22*(23), Article 13168. https://doi.org/10.3390/ijms222313168

McNaughton, N., & Vann, S. D. (2022). Construction of complex memories via parallel distributed cortical–subcortical iterative integration. *Trends in Neurosciences, 45*(7), 550–562. https://doi.org/10.1016/j.tins.2022.04.006

Megías, M., Emri, Z., Freund, T. F., & Gulyás, A. I. (2001). Total number and distribution of inhibitory and excitatory synapses on hippocampal CA1 pyramidal cells. *Neuroscience, 102*(3), 527–540. https://doi.org/10.1016/s0306-4522(00)00496-6

Molochnikov, I., & Cohen, D. (2014). Hemispheric differences in the mesostriatal dopaminergic system. *Frontiers in Systems Neuroscience, 8*, Article 110. https://doi.org/10.3389/fnsys.2014.00110

Moreno-Jiménez, E. P., Flor-García, M., Terreros-Roncal, J., Rábano, A., Cafini, F., Pallas-Bazarra, N., Ávila, J., & Llorens-Martín, M. (2019). Adult hippocampal neurogenesis is abundant in neurologically healthy subjects and drops sharply in patients with Alzheimer's disease. *Nature Medicine, 25*(4), 554–560. https://doi.org/10.1038/s41591-019-0375-9

Nelissen, K., Luppino, G., Vanduffel, W., Rizzolatti, G., & Orban, G. A. (2005). Observing others: Multiple action representation in the frontal lobe. *Science, 310*(5746), 332–336. https://doi.org/10.1126/science.1115593

Ocklenburg, S., & Gunturkun, O. (2024). *The lateralized brain: The neuroscience and evolution of hemispheric asymmetries.* Elsevier.

Panksepp, J., & Northoff, G. (2009). The trans-species core SELF: The emergence of active cultural and neuro-ecological agents through self-related processing within subcortical–cortical midline networks. *Consciousness and Cognition, 18*, 193–215. https://doi.org/10.1016/j.concog.2008.03.002

Pool, E. M., Rehme, A. K., Fink, G. R., Eickhoff, S. B., & Grefkes, C. (2014). Handedness and effective connectivity of the motor system. *NeuroImage, 99*, 451–460. https://doi.org/10.1016/j.neuroimage.2014.05.048

Randi, F., Sharma, A. K., Dvali, S., & Leifer, A. M. (2022). A functional connectivity atlas of *C. elegans* measured by neural activation. arXiv, Article 2208.04790. https://doi.org/10.48550/arXiv.2208.04790

Raut, R. V., Snyder, A. Z., Mitra, A., Yellin, D., Fujii, N., Malach, R., & Raichle, M. E. (2021). Global waves synchronize the brain's functional systems with fluctuating arousal. *Science Advances, 7*(30), Article eabf2709. https://doi.org/10.1126/sciadv.abf2709

Rockland, K. S. (1995). Morphology of individual axons projecting from area V2 to MT in the macaque. *Journal of Comparative Neurology, 355*, 15–26. https://doi.org/10.1002/cne.903550105

Rockland, K. S. (2020). What we can learn from the complex architecture of single axons. *Brain Structure & Function, 225*, 1327–1347. https://doi.org/10.1007/s00429-019-02023-3

Rollenhagen, A., Walkenfort, B., Yakoubi, R., Klauke, S. A., Schmuhl-Giesen, S. F., Heinen-Weiler, J., Voortmann, S., Marshallsay, B., Palaz, T., Holz, U., Hasenberg, M., & Lübke, J. H. R. (2020). Synaptic organization of the human temporal lobe neocortex as revealed by high-resolution transmission, focused ion beam scanning, and electron microscopic tomography. *International Journal of Molecular Sciences, 21*(15), Article 5558. https://doi.org/10.3390/ijms21155558

Ropireddy, D., Scorcioni, R., Lasher, B., Buzsáki, G., & Ascoli, G. A. (2011). Axonal morphometry of hippocampal pyramidal neurons semi-automatically reconstructed after in vivo labeling in different CA3 locations. *Brain Structure & Function, 216*(1), 1–15. https://doi.org/10.1007/s00429-010-0291-8

Rouiller, E. M., & Welker, E. (2000). A comparative analysis of the morphology of corticothalamic projections in mammals. *Brain Research Bulletin, 53*, 727–741. https://doi.org/10.1016/s0361-9230(00)00364-6

Roy, D. S., Zhang, Y., Halassa, M. M., & Feng, G. (2022). Thalamic subnetworks as units of function. *Nature Neuroscience, 25*(2), 140–153. https://doi.org/10.1038/s41593-021-00996-1

Sarel, A., Palgi, S., Blum, D., Aljadeff, J., Las, L., & Ulanovsky, N. (2022). Natural switches in behaviour rapidly modulate hippocampal coding. *Nature, 609*(7925), 119–127. https://doi.org/10.1038/s41586-022-05112-2

Schilder, B. M., Petry, H. M., & Hof, P. R. (2020). Evolutionary shifts dramatically reorganized the human hippocampal complex. *Journal of Comparative Neurology, 528*(17), 3143–3170. https://doi.org/10.1002/cne.24822

Schmidt, E. R. E., Zhao, H. T., Park, J. M., Dipoppa, M., Monsalve-Mercado, M. M., Dahan, J. B., Rodgers, C. C., Lejeune, A., Hillman, E. M. C., Miller, K. D., Bruno, R. M., & Polleux, F. (2021). A human-specific modifier of cortical connectivity and circuit function. *Nature, 599*(7886), 640–644. https://doi.org/10.1038/s41586-021-04039-4

Smith, Y., Raju, D., Nanda, B., Pare, J. F., Galvan, A., & Wichmann, T. (2009). The thalamostriatal systems: Anatomical and functional organization in normal and parkinsonian states. *Brain Research Bulletin, 78*(2–3), 60–68. https://doi.org/10.1016/j.brainresbull.2008.08.015

Stanković, M. (2021). A conceptual critique of brain lateralization models in emotional face perception: Toward a hemispheric functional-equivalence (HFE) model. *International Journal of Psychophysiology, 160*, 57–70. https://doi.org/10.1016/j.ijpsycho.2020.11.001

Takaichi, Y., Chambers, J. K., Takahashi, K., Soeda, Y., Koike, R., Katsumata, E., Kita, C., Matsuda, F., Haritani, M., Takashima, A., Nakayama, H., & Uchida, K. (2021). Amyloid β and tau pathology in brains of aged pinniped species (sea lion, seal, and walrus). *Acta Neuropathologica Communications, 9*(1), 1–15. https://doi.org/10.1186/s40478-020-01104-3

Tosches, M. A., Yamawaki, T. M., Naumann, R. K., Jacobi, A. A., Tushev, G., & Laurent, G. (2018). Evolution of pallium, hippocampus, and cortical cell types revealed by single-cell transcriptomics in reptiles. *Science*, *360*(6391), 881–888. https://doi.org/10.1126/science.aar4237

Tucker, D. M., & Williamson, P. A. (1984). Asymmetric neural control systems in human self-regulation. *Psychological Review*, *91*(2), 185–215. https://doi.org/10.1037/0033-295X.91.2.185

Vassilopoulos, S., Gibaud, S., Jimenez, A., Caillol, G., & Leterrier, C. (2019). Ultrastructure of the axonal periodic scaffold reveals a braid-like organization of actin rings. *Nature Communications*, *10*(1), Article 5803. https://doi.org/10.1038/s41467-019-13835-6

Viena, T. D., Rasch, G. E., & Allen, T. A. (2022). Dual medial prefrontal cortex and hippocampus projecting neurons in the paraventricular nucleus of the thalamus. *Brain Structure & Function*, *227*, 1857–1869. https://doi.org/10.1007/s00429-022-02478-x

Villalba, R. M., Pare, J. F., Lee, S., Lee, S., & Smith, Y. (2019). Thalamic degeneration in MPTP-treated Parkinsonian monkeys: Impact upon glutamatergic innervation of striatal cholinergic interneurons. *Brain Structure & Function*, *224*(6), 3321–3338. https://doi.org/10.1007/s00429-019-01967-w

Walker, L. C., & Jucker, M. (2017). The exceptional vulnerability of humans to Alzheimer's disease. *Trends in Molecular Medicine*, *23*(6), 534–545. https://doi.org/10.1016/j.molmed.2017.04.001

Wang, H., Magnain, C., Wang, R., Dubb, J., Varjabedian, A., Tirrell, L. S., Stevens, A., Augustinack, J. C., Konukoglu, E., Aganj, I., Frosch, M. P., Schmahmann, J. D., Fischl, B., & Boas, D. A. (2018). as-PSOCT: Volumetric microscopic imaging of human brain architecture and connectivity. *NeuroImage*, *165*, 56–68. https://doi.org/10.1016/j.neuroimage.2017.10.012

Whitesell, J. D., Liska, A., Coletta, L., Hirokawa, K. E., Bohn, P., Williford, A., Groblewski, P. A., Graddis, N., Kuan, L., Knox, J. E., Ho, A., Wakeman, W., Nicovich, P. R., Nguyen, T. N., van Velthoven, C. T. J., Garren, E., Fong, O., Naeemi, M., Henry, A. M., . . . Harris, J. A. (2021). Regional, layer, and cell-type-specific connectivity of the mouse default mode network. *Neuron*, *109*, 545–559.e8. https://doi.org/10.1016/j.neuron.2020.11.011

Xu, F., Shen, Y., Ding, L., Yang, C. Y., Tan, H., Wang, H., Zhu, Q., Xu, R., Wu, F., Xiao, Y., Xu, C., Li, Q., Su, P., Zhang, L. I., Dong, H. W., Desimone, R., Xu, F., Hu, X., Lau, P. M., & Bi, G. Q. (2021). High-throughput mapping of a whole rhesus monkey brain at micrometer resolution. *Nature Biotechnology*, *39*(12), 1521–1528. https://doi.org/10.1038/s41587-021-00986-5

Yacoub, E., Grier, M. D., Auerbach, E. J., Lagore, R. L., Harel, N., Adriany, G., Zilverstand, A., Hayden, B. Y., Heilbronner, S. R., Uğurbil, K., & Zimmermann, J. (2020). Ultra-high field (10.5 T) resting state fMRI in the macaque. *NeuroImage*, *223*, Article 117349. https://doi.org/10.1016/j.neuroimage.2020.117349

Yebga Hot, R., Siwiaszczyk, M., Love, S. A., Andersson, F., Calandreau, L., Poupon, F., Beaujoin, J., Herlin, B., Boumezbeur, F., Mulot, B., Chaillou, E., Uszynski, I., & Poupon, C. (2022). A novel male Japanese quail structural connectivity atlas using ultra-high field diffusion MRI at 11.7 T. *Brain Structure and Function*, *227*(5), 1577–1597. https://doi.org/10.1007/s00429-022-02457-2

Zheng, A., Montez, D. F., Marek, S., Gilmore, A. W., Newbold, D. J., Laumann, T. O., Kay, B. P., Seider, N. A., Van, A. N., Hampton, J. M., Alexopoulos, D., Schlaggar, B. L., Sylvester, C. M., Greene, D. J., Shimony, J. S., Nelson, S. M., Wig, G. S., Gratton, C., McDermott, K. B., . . . Dosenbach, N. U. F. (2021). Parallel hippocampal–parietal circuits for self- and goal-oriented processing. *Proceedings of the National Academy of Sciences of the United States of America*, *118*(34), Article e2101743118. https://doi.org/10.1073/pnas.2101743118

19

Are Evolutionary Concepts Helpful in Designing Preventive Strategies for Brain Diseases?

Gilberto Levy and Bruce Levin

Introduction

Among the most prevalent diseases in advanced countries nowadays, brain diseases of complex etiology (i.e., caused by multiple genetic and environmental factors), in particular neurodegenerative diseases, have enjoyed relatively less progress in effective treatments or cures, compared with other chronic diseases including cardiovascular diseases and cancers (Cummings 2017; Sjögren et al. 2021). This is not surprising given that the predominant approach for achieving therapeutic progress in recent decades has relied on pathogenetic knowledge about specific diseases, but such brain diseases involve intricate and largely unknown pathogenetic mechanisms. While the further pursuit of therapeutic progress using this approach may eventually produce more effective treatments or cures for brain diseases, evolutionary concepts can contribute to preventive strategies that may reduce the incidence of these diseases (primary prevention) or slow their progression (secondary prevention). This chapter describes how two evolutionary concepts, as integrated into an overarching causation framework, may naturally lead to preventive strategies for brain diseases (Figure 19.1).

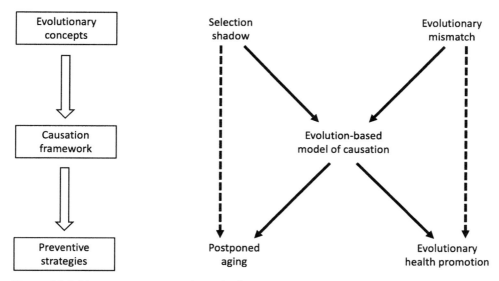

Figure 19.1 Diagram representing the paths from evolutionary concepts to preventive strategies, passing through a causation framework applicable to aging-related diseases. Both selection shadow and evolutionary mismatch inform the evolution-based model of causation, which in turn makes it possible to conceptualize how they are related to postponed aging and evolutionary health promotion. The dashed arrows from selection shadow and evolutionary mismatch to postponed aging and evolutionary health promotion, respectively, are intended to represent that they can be directly connected for those diseases that are not regarded as an aging-related disease (e.g., Huntington's disease and multiple sclerosis, respectively).

Evolutionary Concepts
Selection Shadow

Brain diseases including neurodegenerative diseases (e.g., Alzheimer's disease, Parkinson's disease, and amyotrophic lateral sclerosis) and stroke are aging-related diseases, characterized by complex etiology and increasing age-specific incidence rates with increasing age (Levy and Levin 2014). This group also includes ischemic heart disease and several types of cancers but excludes adult-onset brain diseases of Mendelian etiology (e.g., Huntington's disease) or without increasing incidence rates with age (e.g., multiple sclerosis). The relation of aging-related diseases with aging is not necessarily at the proximate or pathogenetic level. Rather, it is most satisfactorily conceptualized at the ultimate or evolutionary level through the evolutionary theory of aging. Some authors have favored the use of the term "senescence" for the deterioration accompanying aging, while the term "aging" would reflect the mere passage of time or "merely growing old" (Medawar 1952). In this chapter, we do not make such distinction and use "aging" as intrinsically associated with deterioration (Levy and Levin 2014).

The evolutionary theory of aging had its basic argument initially articulated by Sir Peter Medawar (1952). From an evolutionary perspective, aging seems paradoxical because natural selection acting on individuals supposedly causes the evolution of increased, not decreased, fitness. An earlier evolutionary theory of aging, proposed by the 19th-century German biologist August Weismann, was predicated on group selection. The theory considered that aging was somehow adaptive or beneficial in that the elderly would age and die, through a specific death mechanism designed by natural selection, in order to eliminate the old and leave room and resources for the young, who otherwise could not reproduce. Among other problems, Weismann's theory is not consistent with the fact that organisms in the wild are most often killed before aging becomes manifest, by extrinsic causes such as accident, predation, and infectious disease, so that there would be no opportunity for a mechanism to terminate life to evolve (Williams 1957; Rose 1991). As opposed to that, Medawar's (1952) reasoning started by considering a theoretical potentially immortal and ever-reproducing population. The older the members of this population are, the fewer there will be of them simply because they are exposed for a longer time to the hazard of death due to the extrinsic causes that prevailed throughout human evolutionary history. Thus, older individuals make progressively less contribution in terms of reproduction to the next generation, implying that "the force of natural selection" (a measure of the intensity of selection on genes according to their age of expression or "a measure of how effectively selection acts on survival rate or fecundity as a function of age"; Fabian and Flatt 2011, p. 2) weakens with increasing age. This thought experiment explains how, starting from an immortal and ever-reproducing population, evolution may lead to a population with increasing mortality rates with increasing age as well as reproductive senescence. In doing so, it reconciles individual selection with the fact that aging is non-adaptive.

If one now considers a population with a window of reproductive ages, a lethal mutation whose effect occurs before the earliest age of reproduction in the population is not passed to the next generation (meaning that the force of natural selection is maximum), while a lethal mutation whose effect occurs after the end of reproduction in the population would freely pass to the next generation (the force of natural selection is zero). In between, the force of natural selection declines with increasing age because of a decreasing contribution to reproductive output. By graphing the force of natural selection versus age, the age at which natural selection becomes all but ineffective to eliminate deleterious mutations marks the onset of a "selection shadow"; this would roughly correspond to the point at which the survival curve for a population starts to decline more precipitously (Figure 19.2). For deleterious mutations expressing themselves after the age given by this point, it is as if they are not "seen" by natural selection and are thus allowed to accumulate over generations with little or no check (Kirkwood and Austad 2000; Fabian and Flatt 2011). Selection shadow can thus be defined as a "blind spot" due to the ineffectiveness of natural selection to eliminate deleterious mutations expressing in the later period of life. In human evolutionary history, since reproductive output depends not only on fertility but also on survival, the effectiveness of natural selection was conditioned by lower life expectancies, such that diseases with onset in early or middle adult life may have become

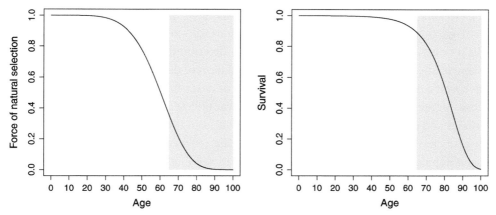

Figure 19.2 Hypothetical graphs of the force of natural selection versus age (left) and of the survival function for intrinsic mortality (right). The age at which natural selection becomes all but ineffective to eliminate deleterious mutations, arbitrarily chosen to be 65 years, marks the onset of the "selection shadow" (gray rectangles). This age corresponds to the point at which the survival curve starts to decline more precipitously. Selection shadow can be defined as a "blind spot" due to the ineffectiveness of natural selection to eliminate deleterious mutations expressing in the later period of life.

relatively frequent. Huntington's disease provides a prototypical example for that, as reflected in the following statement by Medawar (1952, p. 22): "It is only in the last century or so that selection has had a real chance to get a grip on it, for it is only within this period that the average expectation of life at birth has come to equal the average age of onset of the disease."

Population genetics has provided the evolutionary theory of aging with a mathematically explicit basis. Hamilton (1966) formally showed that the force of natural selection acting on a mutation that reduces survival decreases with age starting at the earliest age of reproduction in the population. In humans, there is an extreme dependence of offspring during infancy and early childhood (Williams 1957; Kirkwood and Austad 2000), such that the force of natural selection depends not only on reproductive output but also on transfers of food and care (e.g., parental care and help from others such as older siblings or grandparents; Lee 2003). A lethal mutation expressing itself just after, instead of before, the earliest age of reproduction in the population, by impeding the care and survival of the offspring, would similarly not pass to the next generation. This reasoning can be extended to the initial reproductive years, as stated by Williams (1957, p. 409): "In many primitive human societies the death of teen-age parents must have greatly reduced the survival prospects of any children they might have produced.... this factor should result in a very low rate of senescence during the first decade of man's reproductive life." Such "very low rate of senescence" would result from a very slowly declining force of natural selection during this period, which would also delay the onset of the selection shadow beyond what would be expected on the basis of reproduction alone and extend post-reproductive life (see Figure 19.2).

To sum up, the force of natural selection acting on genes whose effects occur at a given age (i.e., age-specific genetic effects) declines with increasing adult age, which entails a selection shadow. Then, the progressive deterioration of physiological function characteristic of aging results from the accumulation of deleterious mutations with late age-specific effects, which increases the hazard of death in the population as a function of age. Similarly, as initially suggested by Medawar (1952) with respect to cardiovascular diseases and cancers, aging-related diseases result from the accumulation of deleterious (disease-causing) mutations with late age-specific effects, which increases disease risk in the population as a function of age. This is to say that aging and aging-related diseases result from the same evolutionary process, but it does not necessarily follow that aging and aging-related diseases share proximate pathogenetic mechanisms. At the population genetics level, two mechanisms have been proposed to contribute to this evolutionary process: mutation accumulation (the passive accumulation of mutations with late-onset deleterious effects; Medawar 1952) and antagonistic pleiotropy (the active fixation of mutations with early beneficial effects and late deleterious effects; Williams 1957). Experimental and comparative biology studies have provided empirical evidence for the operation of both mechanisms in the evolution of aging (Rose 1991). Studies have also supported the operation of both population-genetics mechanisms in neurodegenerative diseases, including Huntington's disease and Alzheimer's disease (Albin 1993; Jasienska et al. 2015; Provenzano and Deleidi 2021).

Evolutionary Mismatch

Evolutionary mismatch results from the discordance between the human genetic makeup, which was selected in past environments, and the current environment of modern life, characterized by exposure to noxious substances and lifestyle conditions (e.g., diet and sedentarism) that represent a dramatic departure from those past environments (Eaton et al. 1988, 2002b). The relevance of evolutionary mismatch to disease causation is simply expressed as follows: our genome evolved to adapt to conditions that no longer exist, the lifestyle and environmental changes have occurred too rapidly for adequate genetic adaptation (i.e., cultural evolution has outpaced genetic evolution), and the resulting mismatch contributes to the causation of common chronic diseases (Konner and Eaton 2010). Nesse (2005, p. 67) put mismatch at the top of a list of "six evolutionary explanations for vulnerability to disease." So did Gluckman et al. (2011, p. 253), in an expanded list of eight "pathways that mediate the influence of evolutionary processes on disease vulnerability." While Nesse (2005, p. 67) emphasized the relative slowness of selection and "the time lag between environmental change and adaption by natural selection," Gluckman et al. (2011, p. 250) alluded to "the exceptional capacity of humans to alter their environment profoundly" and the challenge to adapt to an entirely novel environment. The concept of evolutionary mismatch is also central to the three key principles of evolutionary medicine proposed by these authors: "that selection acts on fitness, not health or longevity; that our evolutionary history does not cause disease, but rather impacts on our risk of disease in particular environments; and that we are now living in novel environments compared to those in which we evolved" (Gluckman et al. 2011, p. 249). As discussed below, evolutionary mismatch is relevant to brain diseases as diverse in their pathogenetic mechanisms as stroke, multiple sclerosis, and Alzheimer's disease.

The expression "environment of evolutionary adaptedness," developed within the field of evolutionary psychology in the late 1960s, has been used in connection with mismatch to denote that the relevant "past environments" (to which the modern human genome is adapted) are the hunter–gatherer environments of the Paleolithic (Williams and Nesse 1991; Eaton et al. 2002a; Rook 2009; Gluckman et al. 2011). Presumably undermining the application of the concept of mismatch, caveats have been raised that the Paleolithic environments were heterogeneous and that species can evolve quite rapidly, such that adaptations in the last 10,000 years since the end of the Paleolithic period are possible (Deaner and Winegard 2013). However, while the ancestral environments may have been heterogeneous, the way of living differed strikingly in its core essentials from the modern lifestyle. For example, as stated by Eaton et al. (2002a, p. 122), "Whether Stone Agers lived in the arctic or the tropics, vigorous physical exertion was essential; for foragers living 500,000 or 50,000 years ago food was derived from naturally occurring vegetation and wild game." Moreover, the application of mismatch to present human health relies mainly on the discordance that has occurred over the last 200 years since the Industrial Revolution, rather than the 10,000 years since the introduction of agriculture; and "No one proposes that genetic adaptations could have caught up with dietary and lifestyle changes over the past 2 centuries" (Konner and Eaton 2010, p. 595).

The diseases fostered by evolutionary mismatch are called "diseases of civilization" because they have emerged or their incidence has clearly increased in close temporal relation with the establishment of the modern Western lifestyle (Eaton et al. 1988). According to Williams and Nesse (1991, p. 14), diseases of civilization "are either caused by differences between our current environment and the environment we evolved to live in, or they are aspects of senescence that have been uncovered by preventing earlier causes of mortality." Chief among diseases of civilization are ischemic heart disease and diabetes type 2. Omran (1971, table 4) hinted at that in the early 1970s in his theory of epidemiologic transition, by noting that in "the age of degenerative and man-made diseases," there is "a tendency to overnutrition including consumption of rich and high-fat foods which may increase the risk of heart and metabolic diseases." Autoimmune diseases including multiple sclerosis are similarly regarded as diseases of civilization because of the well-established increase in their incidence in the 20th century (Bilbo et al. 2011; Murdaca et al. 2021). Alzheimer's disease may also be viewed in this way, even if its increasing prevalence in the 20th century may be primarily a function of the increased life expectancy and changing age distribution of the population (Fox 2018); another factor contributing to this observed increasing prevalence is increased ascertainment due to higher awareness about the disease.

The Hygiene or Old Friends Hypothesis

The hygiene hypothesis was initially developed with respect to allergies, another group of diseases presenting a major increase in incidence in the 20th century (Bloomfield et al. 2016). The origin of the hygiene hypothesis is attributed to a study by Strachan (1989) showing an inverse relation between the occurrence of hay fever in British children and the number of older siblings in the household. Strachan (1989, p. 1260) did not actually use the expression "hygiene hypothesis" but speculated that the observation could "be explained if allergic diseases were prevented by infection in early childhood, transmitted by unhygienic contact with older siblings." More than 10 years earlier, a study comparing atopic diseases in Caucasian and Indigenous people in Canada had already led to the suggestion that "atopic disease is the price paid by some members of the white community for their relative freedom from diseases due to viruses, bacteria and helminths" (Gerrard et al. 1976, p. 91). Since Strachan's (1989) study, numerous studies have shown an inverse relation between allergy prevalence and other "indicators of lower hygiene," such as growing up on a farm, use of early day care, and low socioeconomic status (Björkstén 2009). At the same time, the hygiene hypothesis was extended to autoimmune diseases as they shared with allergies an increasing incidence in the 20th century as well as analogous findings suggesting an association with lack of microbial exposures (Murdaca et al. 2021). With respect to brain diseases, the hygiene hypothesis is particularly relevant to multiple sclerosis, as will be discussed in the section "Evolutionary Health Promotion."

After its beginnings, some major developments have refined the hygiene hypothesis. Importantly, in contrast with the initial assumption that the relevant exposures were viruses causing childhood infections, the evidence now favors instead the "Old Friends," as proposed by Rook and Brunet (2005). From an evolutionary perspective, the likely candidates would be organisms that coevolved with humans in the Paleolithic period, in such a way that humans may have become adapted to their presence and can no longer perform well without them (evolved dependence). The circulation of many of the viruses causing childhood infections depends on crowding situations that did not exist during the Paleolithic, and they were acquired relatively recently from other animal species during the Neolithic. Furthermore, human populations were too small until recently to sustain these childhood infections as endemic, so it is theoretically improbable that exposures to the viruses causing them would become physiological necessities (Rook 2009). Indeed, evidence now indicates that the relevant microbial exposures are to those organisms most present during primate evolution and hunter–gatherer times, the so-called Old Friends including harmless commensal organisms in the gut microbiota, environmental saprophytic bacteria encountered in contact with mud and untreated water, and helminths that needed to be tolerated in the host organism (Rook 2009; Bloomfield et al. 2016). Diversity of exposure has also been shown to be crucial (von Hertzen et al. 2015). Although the most important time for exposure to Old Friends is apparently early development (so much so that cesarean section is linked to increased risk of allergies, and breast- versus bottle-feeding has a large influence on the gut microbiota), later contact with microbial diversity from the natural environment is also important. Factors that may impact the maintenance of a diverse gut microbiota include diet and antibiotic use (Bloomfield et al. 2016).

Regarding the immunological mechanism, it was initially thought that diminished microbial exposure failed to drive T helper 1 (Th1) cells (which promote cellular immune response), with a consequent overproduction of T helper 2 (Th2) cells (which promote humoral immune response), such that the mechanism underlying the hygiene hypothesis involved a Th1/Th2 balance (Rook and Brunet 2005; Guarner et al. 2006; Rook 2009). However, among three groups of chronic inflammatory diseases encompassed by the hygiene hypothesis, classified according to whether the immune system responds inappropriately to external stimulants (allergies), self-antigens (autoimmune diseases), or gut content (inflammatory bowel diseases), allergies are mediated by Th2 effectors, but autoimmune diseases are mediated by Th1 effectors and inflammatory bowel diseases also by Th1 effectors predominantly. It has now become clear that the critical balance is not Th1 versus Th2 but rather between regulatory and effector T cells. Diminished exposure to Old Friends fails to prime or "educate" regulatory T cells so that they are unable to stop inappropriate immunological responses (Rook and Brunet 2005; Bloomfield et al. 2016). Thus, in its current version, the hygiene hypothesis implies that dysregulation of the immune system has led to increased susceptibility to immunological and inflammatory diseases. In addition, the gut microbiota has come to play a central role in the hygiene hypothesis (the microbial flora of the skin and lung have received less attention), even if environmental organisms that do not necessarily colonize the host also play a role in driving immunoregulation (Rook 2009).

The Biome Depletion Paradigm

At least since the early 2000s, as a full-fledged "biome depletion paradigm" (Parker 2014) has taken shape, there has been a recognition that the "hygiene hypothesis" is a misnomer and should be replaced by a better alternative (Hadley 2004; Parker 2014; Bloomfield et al. 2016). The main reasons are as follows. First, while biome depletion in Western culture was indeed related to major advancements in sanitation (e.g., sewer systems and water treatment facilities), the use of the term "hygiene" today is much more associated with personal health habits such as hand washing (Parker 2014). Second, the use of "hygiene hypothesis" in connection with dreaded autoimmune diseases gives credence to the view that hygiene ought to be avoided, undermining public health efforts against the spread of infectious diseases (Hadley 2004; Parker 2014). Third, targeted hygiene, focused on the circumstances that matter most for infection transmission, is compatible with microbial exposures necessary for the establishment of the immune system in early life (Bloomfield et al. 2016; Rook and Bloomfield 2021); that is, "allergic diseases are not the price we have to pay for protection against infection" (Bloomfield et al. 2016, p. 218). Lastly, the term diverts attention from the apparent solution to the problem of biome depletion: biome restoration or reconstitution (Bilbo et al. 2011; Parker 2014). However, there has not been a consensus for a new term, as suggested by the large number of proposed alternatives, including among others the "microbial exposure hypothesis" (Hadley 2004), the "Old Friends mechanism" (Rook and Brunet 2005), the "microbial deprivation hypothesis" (Björkstén 2009), the "biome depletion theory" (Bilbo et al. 2011), and the "biodiversity hypothesis" (von Hertzen et al. 2015).

The term "Old Friends" has the virtue of alluding to the concept of evolutionary mismatch; it was chosen to emphasize humans' interaction extending back into the Paleolithic with the microorganisms potentially involved in the hygiene hypothesis, and Old Friends must satisfy the criterion that they are "virtually absent, and increasingly so over the last century, from the modern environment" (Rook 2009, p. 12). Indeed, biome depletion may be viewed as a special form of evolutionary mismatch. Maziak (2002, p. 416) drew a parallel between "the potential effect of lack of exposure to infectious agents on the function of the immune system" and "the effect of lack of physical activity on the function of the cardiovascular system" because, unlike in modern societies, microbial exposure and physical activity "were inherent features of life for the most part of human evolution." Yet physical activity also has an effect on chronic inflammatory diseases, and chronic inflammation plays a role in cardiovascular diseases (Parker 2016). Meanwhile, the relevance of the biome depletion paradigm has been extended beyond allergies, autoimmune diseases, and inflammatory bowel diseases to include diseases in which chronic inflammation plays a role such as stroke and neurodegenerative diseases (Rook 2009). This is to say that evolutionarily mismatched factors most likely interact with other mismatched factors to increase the risk of several chronic diseases in modern times, including brain diseases, which highlights the potential of the preventive strategy discussed below in the section "Evolutionary Health Promotion."

The Evolution-Based Model of Causation

The causation model presented here provides an overarching framework linking the evolutionary concepts discussed in the previous section and the preventive strategies to be discussed in the next section (Figure 19.1). The sufficient and component causes model of causation in epidemiology is particularly useful for complex diseases because it provides a convenient way of conceptualizing biological interactions (i.e., gene–gene, environment–environment, and gene–environment interactions). A sufficient cause constitutes a minimal set of conditions that produce disease. Each component cause (a genetic or environmental factor) is part of one or more sufficient causes, while each sufficient cause includes one or more component causes. The evolution-based model of causation (EBMC) takes the sufficient and component causes model as a starting point and is applicable to aging-related diseases and intrinsic mortality (i.e., excluding deaths due to extrinsic causes such as accidents and infections; Levy and Levin 2014, 2022).

In the EBMC, component causes are considered to occur as random events following time-to-event distributions. The time to event of the component cause is the age at which the component cause expresses its necessary causal role. Further, component causes are classified based on evolutionary reasoning. Genetic factors are classified in terms of the timing of their effects as "early-onset" or "late-onset." Early-onset genetic effects (EOGE) and late-onset genetic effects (LOGE) are defined according to whether the expression of their causal role occurs, respectively, before or after the earliest age of reproduction in the population (i.e., the age at which the force of natural selection starts to decline). In turn, environmental factors are classified as "evolutionarily conserved" or "evolutionarily recent." Evolutionarily conserved environmental factors (ECEF) are defined as having been present during enough time in evolutionary history for adaptation to their effect to take place. They are what is considered "part of nature," implying that members of the population are exposed to these environmental factors early in life, such that it can be reasonably assumed that they express themselves before the earliest age of reproduction in the population. Evolutionarily recent environmental factors (EREF) are recent enough on an evolutionary scale so that adaptation to their effects has not occurred. They have appeared mostly over the last 200 years since the Industrial Revolution. Importantly, environmental factors brought about by the modern lifestyle, such as changes in diet, sedentarism, and cigarette smoking, as well as exposures related to industrialization, such as pollution and toxic chemicals, fall under this category. These four categories of component causes can combine into sufficient causes involving one to four categories, as shown in Figure 19.3.

In relation to the declining force of natural selection with increasing age, the probability of fixation of a deleterious mutation, hence its ultimate frequency in the population under mutation-selection balance, depends on its age of expression. For example, a deleterious mutation that expresses its effect around age 40 years will be subjected to stronger negative selection and be present at lower frequency in the population than another that expresses its effect around age 60 years. Under the EBMC, it is appropriate to follow this reasoning at the level of the sufficient causes, rather than at the level of a single mutation or component cause, because the phenotypic expression necessary for the action of natural selection only occurs once a sufficient cause is complete. Sufficient causes containing LOGE (without EREF) components have been subjected to natural selection over evolutionary history as a consequence of LOGE being heritable genetic factors. Among these, sufficient causes with time-to-event distributions shifted closer to the earliest age at reproduction in the population would be present at low frequencies, while those with distributions shifted toward older ages of expression would be present at successively higher frequencies. This is not the case for sufficient causes containing EREF components, even if these sufficient causes may also contain heritable genetic factors, because the EREF components necessary for their phenotypic expression only came into play relatively recently on an evolutionary scale, such that natural selection hasn't acted on them long

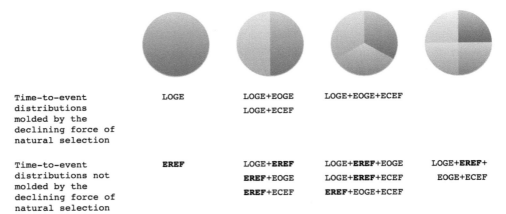

Figure 19.3 Schematic representation using "causal pies" of sufficient causes of aging-related diseases under the evolution-based model of causation (EBMC), grouped according to the number of categories of component causes. The EBMC is concerned with the evolutionarily defined aging phase of life, which starts at the earliest age of reproduction in the population. Thus, it is excluded from consideration the sufficient causes containing only early-onset genetic effects (EOGE) or evolutionarily conserved environmental factors (ECEF), or both EOGE and ECEF, because either by definition or by assumption they express themselves before the earliest age of reproduction in the population. If EOGE and/or ECEF participate in a sufficient cause also containing late-onset genetic effects (LOGE) and/or evolutionarily recent environmental factors (EREF), the expression of the sufficient cause occurs after the earliest age of reproduction because the sufficient cause only expresses itself once all its component causes do, and the age of expression of the sufficient cause is given by the maximum age of expression of its component causes. EREF component causes may express themselves before or after the earliest age of reproduction in the population, but the sufficient causes in which they participate are only taken into account in the EBMC when these sufficient causes express themselves after the earliest age of reproduction. The sufficient causes are also grouped according to whether their time-to-event distributions are molded by the declining force of natural selection. The distinguishing characteristic is the absence or presence of EREF, which is in bold typeface for that reason. The slices represent categories of component causes, and the presence of a category of component cause in a sufficient cause means that one or more genetic effects or environmental factors of that category are part of the sufficient cause. Each aging-related disease likely involves at least hundreds of sufficient causes; a monogenic form of an aging-related disease corresponds to a sufficient cause containing only one LOGE.

enough. Sufficient causes can thus be classified according to whether or not their time-to-event distributions are "molded" by the declining force of natural selection: sufficient causes containing LOGE (without EREF) and those containing EREF (with or without LOGE), respectively, have time-to-event distributions molded and not molded by the declining force of natural selection (Figure 19.3).

The EBMC takes account of both evolutionary concepts described in the previous section. The earliest age of reproduction in the population (the age at which the force of natural selection starts to decline), which provides the cutoff for early-onset versus late-onset in the classification of genetic component causes, has a reflection on the selection shadow because its onset is related or "calibrated" to the age at which the force of natural selection starts to decline, in addition to its rate of decline (e.g., a higher earliest age of reproduction in the population would postpone the onset of the selection shadow). The effect of the selection shadow on disease causation corresponds to the contribution of the group of sufficient causes containing LOGE (without EREF), whose time-to-event distributions are molded by the declining force of natural selection. On the other hand, evolutionary mismatch is reflected in the definition of EREF component causes, and the effect of mismatch on disease causation corresponds to the contribution of the group of sufficient causes containing EREF (with or without LOGE), whose time-to-event distributions are not molded by the declining force of natural selection. The EBMC can also accommodate the relevance of epigenetic mechanisms (e.g., DNA methylation and histone modifications) to disease causation, representing an alternative route for the action of environmental component causes through the modification of gene expression. Such epigenetic mediation of environmental effects can involve mitotic inheritance of epigenetic changes at any time during an individual's lifetime as well as transgenerational meiotic inheritance (for a more detailed discussion on this, see Levy and Levin 2014, 2022). As discussed below, through the EBMC it is then possible to conceptualize how selection shadow and evolutionary mismatch are related to the preventive strategies of postponed aging and evolutionary health promotion.

Preventive Strategies
Postponed Aging

As generally understood, postponed aging (also referred to as "delayed aging") involves a modification of the fundamental processes of aging or "manipulating 'aging' genes through techniques developed in molecular biology" (Olshansky et al. 1991, p. 204). The potential implications of this strategy for primary and secondary disease prevention are expressed in the "geroscience hypothesis," which "posits that manipulation of aging will delay (in parallel) the appearance or severity of many chronic diseases" (Sierra and Kohanski 2017, p. 1). Under the EBMC, the operation of sufficient causes containing LOGE (without EREF) underlies aging in its essential or evolutionarily derived form (i.e., as the ultimate result of selection shadow) and without the added contribution of environmental factors introduced by industrialization and the modern lifestyle. The manipulation of "aging genes" (e.g., through a hypothetical gene-editing intervention to prevent the expression of a LOGE component) would interfere with the contribution of these sufficient causes to intrinsic mortality, which can be regarded as a measure of aging at the population level (Levy and Levin 2014). The manipulation of aging genes might also interfere with the contribution of sufficient causes containing EREF to intrinsic mortality, as long as they also contain LOGE (i.e., LOGE component causes are also necessary for their expression). Thus, postponed aging might also indirectly prevent some of the effects of evolutionary mismatch. Despite this potential, there are three reasons to be cautious and not overly optimistic about the strategy of postponed aging.

First, it is unclear that postponed aging would reach the goal of reducing the burden of disease, as opposed to just increasing the length of life without decreasing the number of unhealthy years and disability at the end of life (i.e., without attaining a "compression of morbidity"; Fries 2016; Levy and Levin 2022). According to Brody (1985, p. 465), one of two promising vehicles for "reaching the potential of better health and quality of life in later years" is "major research efforts to understand and postpone the aging processes." At least since the late 1970s, the supposition that advanced countries might be approaching a biologic limit to life expectancy has motivated a call for research investment in postponed aging (Keyfitz 1978; Goldman et al. 2013). While proponents of postponed aging argue that, in addition to increasing life expectancy, it may decrease the duration of unhealthy years at the end of life, this is not a given from an evolutionary perspective because aging and aging-related diseases do not necessarily share proximate mechanisms, even though they result from the same evolutionary process, as noted previously. This is also to say that aging genes and genes involved in increased risk of aging-related diseases do not necessarily overlap, even though they result from the same population-genetic mechanisms of mutation accumulation and antagonistic pleiotropy.

Second, to the extent that genes with antagonistic-pleiotropic effects underlie aging, there is a risk that manipulating aging genes may have unintended adverse effects for early-life health. George Williams's seminal paper describing antagonistic pleiotropy contained nine "testable deductions from the theory," the last of which was that "Successful selection for increased longevity should result in decreased vigor in youth" (Williams 1957, p. 410). Referring to possible tests of this prediction, he argued that "if senescence results from genes that increase youthful vigor at the price of vigor later on, the loss of some of these genes through selection should result in decreased youthful vigor." While Williams (1957) conceded that not all genes involved in aging needed to have opposite effects on fitness, the prediction has generally been supported by laboratory studies (Gaillard and Lemaître 2017). Later on, Williams and Nesse (1991, p. 13) applied the same reasoning to specific diseases, in particular Alzheimer's disease: "If we naively assume that ridding the population of the genetic capability for Alzheimer's disease is an unconditionally desirable goal, we might incidentally eliminate unsuspected benefits." In support of this, studies have suggested that the *ApoE4* allele, a risk factor for Alzheimer's disease, confers a reproductive advantage in women (Jasienska et al. 2015). An antagonistic-pleiotropic effect for the *APOE* genotype has also been suggested in terms of a benefit early in life due to an immunological/inflammatory role leading to pathogen resistance (Provenzano and Deleidi 2021).

Third, evolutionary theory suggests that hundreds or thousands of genes and hundreds of biochemical pathways play a role in aging, which has been largely corroborated by experimental studies, thus indicating that postponed aging will remain a challenging and unfruitful strategy for many years to come (Rose 1991). Dating back to the beginnings of the evolutionary theory of aging, different authors have expressed a negative view of the prospects of postponed aging in more or less fatalistic terms. According to Williams (1957, p. 407), the evolutionary theory implication of a large number of physiological processes involved in aging "banishes the 'fountain of youth' to the limbo of scientific impossibilities where other human aspirations, like the perpetual motion machine and Laplace's 'superman', have already been placed by other theoretical considerations." More recently, Rose (2009, p. 448) considered that the evolutionary perspective that aging results from a failure of adaptation involving a large number of genetic variants and biochemical pathways may lead one to "regard the slowing of human aging as an essentially intractable problem."

Evolutionary Health Promotion

Evolutionary health promotion is a model for prevention research and potentially (if justified by experiments or epidemiological research) for health recommendations based on evolutionary principles (Eaton et al. 2002a, 2002b). Eaton et al. (1988, p. 739) initially considered the lifestyle factors relevant to evolutionary health promotion to include "nutrition, exercise, and exposure to harmful substances such as alcohol and tobacco." As to its potential impact, the authors noted that the "mismatch between our ancient, genetically controlled biology and certain important aspects of our daily lives . . . promotes chronic degenerative diseases that have their main clinical expression in the post-reproductive period, but that together account for nearly 75 percent of the deaths occurring in affluent Western nations." The goals of evolutionary health promotion have been described as follows: "(1) Better characterize differences between ancient and modern life patterns. (2) Identify which of these affect the development of disease. (3) Integrate epidemiological, mechanistic, and genetic data with evolutionary principles to create an overarching formulation upon which to base persuasive, consistent, and effective recommendations" (Eaton et al. 2002b, p. 109). This is to say that, even as it is neither possible nor desirable to revert back to the time of our ancestors, we can benefit from characterizing differences in lifestyle and exposures between past and modern times, identifying those factors that are associated with disease, and acting on them.

Evolutionary health promotion is fully consistent with Geoffrey Rose's epidemiological approach, within the "population strategy" of disease prevention, described as "the restoration of biological normality by the removal of an abnormal exposure (e.g., stopping smoking, controlling air pollution, moderating some of our recently acquired dietary deviations)" (Rose 1985, p. 432). On the evolutionary view, what makes modern life exposures "abnormal" is the lack of human adaptation to their effects or the mismatch between the human genetic makeup and modern life. Evolutionary health promotion also converges with the other of Brody's (1985, p. 465) promising vehicles for better health in later years: "health promotion through the improvement of personal health practices throughout life." There is evidence for a decrease in the duration of unhealthy years under health promotion (Fries 2016). Longitudinal studies have demonstrated the occurrence of fewer years of disability at the end of life for subjects with a lower number of lifestyle risk factors such as smoking, physical inactivity, and obesity (Vita et al. 1998; Hubert et al. 2002). In the EBMC, the implication of evolutionary health promotion that diseases can be prevented through changes in lifestyle and minimization of exposures corresponds to a reduction in the contribution of sufficient causes containing EREF to the incidence of aging-related diseases. The contribution of sufficient causes containing LOGE (without EREF) represents a sort of "evolutionary wall" beyond which health promotion cannot succeed in increasing health span.

The EBMC, as a causation framework, and evolutionary health promotion, as a preventive strategy (Figure 19.1), can accommodate the special form of evolutionary mismatch represented by biome depletion. Reduced exposure to Old Friends leading to biome depletion, with the ensuing immune dysregulation and/or chronic inflammation, can be represented as an EREF component cause in the EBMC; and different ways for achieving biome reconstitution can be tested and recommended under the prevention model of evolutionary health promotion. This is relevant to a wide range of diseases, either because they are primarily inflammatory or because chronic inflammation plays a role in their pathogenesis. Parker (2016, p. 195) noted that tallying "inflammation-related diseases," comprising not only chronic inflammatory diseases but also other common diseases not labeled as inflammatory but that are nevertheless associated with inflammation, "provides a list of the vast majority of disease seen in Western clinics and hospitals." Among brain diseases, given the role of inflammation in atherosclerosis (Ait-Oufella et al. 2009), this list must include stroke. In addition, it must include multiple sclerosis and neurodegenerative diseases, as further discussed below.

As the prototypical autoimmune disease affecting the brain, multiple sclerosis is the most studied brain disease in relation to the hygiene hypothesis (Wendel-Haga and Celius 2017). Even before the hygiene hypothesis was proposed, Leibowitz et al. (1966) showed an association between multiple sclerosis and higher levels of sanitation in Israel. Forty years later, compiling data from 35 countries around the world, Fleming and Cook (2006) showed a striking inverse correlation between the prevalence of multiple sclerosis and the prevalence of the helminth *Trichuris trichiura* (percentage of surveyed population infected). At about the same time, Correale and Farez (2007) presented the results of a prospective study in Argentina showing an association between parasite infection and more favorable clinical and neuroimaging course in patients with multiple sclerosis, as well as immune responses consistent with induction of regulatory T cells. However, phase 1 and 2 clinical trials involving oral inoculation of *Trichuris suis* ova have had conflicting results and do not support the use of helminth immunotherapy in multiple sclerosis (Wendel-Haga and Celius 2017; Charabati et al. 2020). Studies have also evidenced differences in the gut microbiota of patients with multiple sclerosis compared with healthy controls. This finding has supported the conduct of clinical trials involving either diets aimed at modifying the microbiota, use of probiotics, or fecal microbiota transplantation, with some promising preliminary results (Schepici et al. 2019; Cabré et al. 2022).

One may also reasonably ask, as did Griffin and Mrak (2009), if the hygiene hypothesis is relevant to Alzheimer's disease. It has long been recognized that inflammation plays a role in neurodegenerative diseases (Amor et al. 2014), and both the innate and adaptive immune systems have been implicated in Alzheimer's disease pathogenesis (Wu et al. 2021). A study showed an inverse correlation between measures of parasite burden and global prevalence of Alzheimer's disease (Fox et al. 2013), but this finding was not confirmed in a later analysis using an updated data set correcting for underdiagnosis and underreporting (Lathe and Lathe 2020). Although the evidence is still limited, studies of the gut microbiota in patients with Alzheimer's disease have shown different composition and lower microbial diversity compared with healthy controls (Cabré et al. 2022). Fox (2018) broadly discussed how evolutionary mismatch could be relevant to Alzheimer's disease based on five pathogenetic pathways or etiological factors that have been implicated in the disease: insulin resistance, estrogenic neuroprotection, inflammation, environmental toxins, and *APOE* genotype. The mismatch rationale for the *APOE* genotype relied on two lines of evidence (*APOE*'s effect on inflammation and the lack of association of the *ApoE4* allele with Alzheimer's disease in studies conducted in non-industrialized societies) and a hypothesized gene–environment interaction between *APOE* genotype and exposure to Old Friends, in order to suggest that the *ApoE4* allele may "not cause inflammation in non-industrialized environments in which individuals' immunoregulatory capacity is stronger" (Fox 2018, p. 145).

The importance of inflammation in aging-related diseases is highlighted by the term "inflammaging," proposed by Franceschi et al. (2000) to denote an increase in the pro-inflammatory status with advancing age. The authors alluded both to antagonistic pleiotropy and to the concept of selection shadow by considering inflammation in terms of beneficial effects in early life that become detrimental late in life, "in a period largely not foreseen by evolution" (Franceschi et al. 2000, p. 244). Among brain diseases, for example, inflammaging has been suggested to play a role in Parkinson's disease pathogenesis (Calabrese et al. 2018). In the context of evolutionary mismatch, the importance of inflammation is supported by a study that assessed the pattern of variability in C-reactive protein, a biomarker of chronic inflammation and cardiovascular disease risk, in a non-industrialized setting in Ecuador with relatively high burden of infectious disease. In contrast with environments with low burden of infectious disease, the results showed no evidence of chronic low-grade inflammation; in fact, the study failed to detect a single case of "high-risk" C-reactive protein among 52 adults according to established guidelines (McDade et al. 2012). Recently, other than the immune dysregulation/chronic inflammation mechanism, there has been some evidence for a more direct effect of the gut microbiota on brain diseases, through the so-called microbiota–gut–brain axis, a complex bidirectional pathway conveying immune, neural, endocrine, and metabolic signals (Cryan et al. 2020; Cabré et al. 2022). The preliminary evidence for a role of the microbiota–gut–brain axis in neurodegenerative diseases pertains mostly to Parkinson's disease but also to Alzheimer's disease and amyotrophic lateral sclerosis (Willyard 2021; Cabré et al. 2022).

The clinical trials aiming at biome reconstitution mentioned above were conducted with the purpose of impacting disease progression in patients with multiple sclerosis (i.e., secondary prevention). Similar trials have been conducted for

Alzheimer's disease and Parkinson's disease (Cabré et al. 2022). However, the gist of evolutionary health promotion is primary prevention. Current evidence is strong enough to encourage lifestyle choices that are already recommended because of other health benefits or are unlikely to have harmful effects and that may have the added value of promoting an adequate immunoregulation. These choices include natural childbirth, breastfeeding, and stimulating interaction among children (siblings and non-siblings). At the same time, it is important to avoid unnecessary antibiotic use and to eat a diet rich in fibers, which favor the maintenance of a diverse gut microbiota (Bloomfield et al. 2016). A case can also be made for lifelong contact with the natural environment (Rook 2013). In modern societies, widespread exposure to some Old Friends (e.g., helminths) was irreversibly lost, which reinforces the importance of contact with the remaining sources of biodiversity in the natural environment (animals, soil, air, plants) and in other humans. Toward that end, outdoor activities in green spaces, exposure to pets, social interaction, and practicing a team sport are likely to have beneficial effects (Rook 2013; Bloomfield et al. 2016). Under a "green space effect," there would be a win–win situation, in the sense of both an immunological benefit due to "the evolved need for the immune system to receive inputs provided by microbial biodiversity" and a psychological benefit due to "an evolved psychological reward from contemplating the ideal hunter–gatherer habitat" (Rook 2013, p. 18365). In this way, the enjoyable aspects of evolutionary health promotion might help make prevention less of "a hard-sell."

Conclusion

A decline in the influence of evolutionary thinking in medicine in the second half of the 20th century, other than being a reaction to eugenics, was linked to "the rise of a reductionist, molecular biological approach to disease" (Gluckman et al. 2011, p. 249). This approach has produced many successes for the management of prevalent diseases nowadays, including autoimmune and cardiovascular diseases; but progress has been limited for other diseases such as some types of cancers and neurodegenerative diseases. The approach afforded by evolutionary health promotion, by relying on mismatched lifestyle factors and environmental exposures, is distinct in that it involves prevention rather than treatment and potentially decreases the risk of several diseases associated with the modern Western lifestyle at the same time, rather than pursuing effective treatments or cures for one disease at a time. As put by Parker et al. (2012, p. 1199), "modern medical science is trying to stabilize an entire system that is inherently unstable (prone to pathologic over-reactivity) due to evolutionary mismatches," and it "is generally dependent on pharmaceuticals that are directed at a single target or at best a very limited number of targets." Evolutionary health promotion does not depend on detailed knowledge about (and ability to pharmacologically intervene in) the pathogenetic mechanisms of specific diseases.

Figure 19.4 Schematic drawing representing chronic diseases as "low-hanging fruits" or "higher-hanging fruits" in a tree, according to the complexity of their pathogenetic mechanisms. The predominant approach of biomedical research in recent decades, represented by the white-coated lab researcher on the ground, relies on detailed knowledge about (and ability to pharmacologically intervene in) the pathogenetic mechanisms of specific diseases, hence the limited progress in effective treatments or cures for higher-hanging fruits like some types of cancers and neurodegenerative diseases. The approach afforded by evolutionary health promotion, represented by a "flying Darwin," involves prevention rather than treatment and does not depend on the ability to pharmacologically intervene in pathogenetic mechanisms. Thus, it can reach with equal ease low-hanging and higher-hanging fruits. (Apple tree drawing downloaded from vecteezy.com under a Free License.)

Such dependence is one explanation for why therapeutic progress for chronic diseases in recent decades has mostly involved autoimmune and cardiovascular diseases (i.e., "low-hanging fruits" from a pathogenetic point of view, as compared with "higher-hanging fruits," such as cancers and neurodegenerative diseases; Figure 19.4). For cardiovascular diseases, therapeutic progress has been accompanied by successful preventive strategies, as shown by the decline in the incidence of ischemic heart disease starting in the 1960s when education efforts were directed at smoking and diet (Jones and Greene 2013). Similarly, there have been recent observations of declining incidence of dementia and Alzheimer's disease in a time frame inconsistent with broad genetic changes in the population (Wolters et al. 2020). If confirmed, this is likely due to lifestyle modifications, including changes in diet, physical exercise, and smoking, as well as control of environmental exposures such as air pollution (Livingston et al. 2020), all of which are consistent with evolutionary health promotion. As illustrated by the biome depletion paradigm, evolutionary health promotion can potentially contribute to a greater and broader impact on brain diseases. Indeed, the lifestyle choices discussed at the end of the previous section, by forestalling immune dysregulation and chronic inflammation, have the potential to impact the incidence of such brain diseases as stroke, multiple sclerosis, and neurodegenerative diseases including Alzheimer's disease and Parkinson's disease.

Key Points to Remember

- According to the evolutionary theory of aging, the force of natural selection acting on age-specific genetic effects declines with increasing adult age.
- The age at which natural selection becomes all but ineffective to eliminate deleterious mutations marks the onset of selection shadow.
- Evolutionary mismatch results from the discordance between the human genetic makeup, which was selected in past environments, and the environment of modern life.
- Postponed aging (or delayed aging) would involve a modification of the fundamental processes of aging through the manipulation of "aging genes."
- Evolutionary health promotion, a model for prevention research and potentially for health recommendations based on evolutionary principles, is a more promising preventive strategy for brain diseases in the short and mid-term.

References

Ait-Oufella, H., Tedgui, A., & Mallat, Z. (2009). Immune regulation in atherosclerosis and the hygiene hypothesis. In G. A. W. Rook (Ed.), *The hygiene hypothesis and Darwinian medicine* (pp. 221–238). Birkhäuser. https://doi.org/10.1007/978-3-7643-8903-1_12

Albin, R. L. (1993). Antagonistic pleiotropy, mutation accumulation, and human genetic disease. *Genetica*, *91*(1–3), 279–286. https://doi.org/10.1007/BF01436004

Amor, S., Peferoen, L. A., Vogel, D. Y., Breur, M., van der Valk, P., Baker, D., & van Noort, J. M. (2014). Inflammation in neurodegenerative diseases—An update. *Immunology*, *142*(2), 151–166. https://doi.org/10.1111/imm.12233

Bilbo, S. D., Wray, G. A., Perkins, S. E., & Parker, W. (2011). Reconstitution of the human biome as the most reasonable solution for epidemics of allergic and autoimmune diseases. *Medical Hypotheses*, *77*(4), 494–504. https://doi.org/10.1016/j.mehy.2011.06.019

Björkstén, B. (2009). The hygiene hypothesis: Do we still believe in it? In P. Brandtzaeg, E. Isolauri, & S. L. Prescott (Eds.), *Microbial–host interaction: Tolerance versus allergy* (pp. 11–18). Karger. https://doi.org/10.1159/000235780

Bloomfield, S. F., Rook, G. A., Scott, E. A., Shanahan, F., Stanwell-Smith, R., & Turner, P. (2016). Time to abandon the hygiene hypothesis: New perspectives on allergic disease, the human microbiome, infectious disease prevention and the role of targeted hygiene. *Perspectives in Public Health*, *136*(4), 213–224. https://doi.org/10.1177/1757913916650225

Brody, J. A. (1985). Prospects for an ageing population. *Nature*, *315*(6019), 463–466. https://doi.org/10.1038/315463a0

Cabré, S., O'Riordan, K. J., & Cryan, J. F. (2022). Neurodegenerative diseases and the gut microbiota. In G. A. W. Rook & C. A. Lowry (Eds.), *Evolution, biodiversity and a reassessment of the hygiene hypothesis* (pp. 339–392). Springer International Publishing. https://doi.org/10.1007/978-3-030-91051-8_11

Calabrese, V., Santoro, A., Monti, D., Crupi, R., Di Paola, R., Latteri, S., Cuzzocrea, S., Zappia, M., Giordano, J., Calabrese, E. J., & Franceschi, C. (2018). Aging and Parkinson's disease: Inflammaging, neuroinflammation and biological remodeling as key factors in pathogenesis. *Free Radical Biology & Medicine*, *115*, 80–91. https://doi.org/10.1016/j.freeradbiomed.2017.10.379

Charabati, M., Donkers, S. J., Kirkland, M. C., & Osborne, L. C. (2020). A critical analysis of helminth immunotherapy in multiple sclerosis. *Multiple Sclerosis*, *26*(12), 1448–1458. https://doi.org/10.1177/1352458519899040

Correale, J., & Farez, M. (2007). Association between parasite infection and immune responses in multiple sclerosis. *Annals of Neurology*, *61*(2), 97–108. https://doi.org/10.1002/ana.21067

Cryan, J. F., O'Riordan, K. J., Sandhu, K., Peterson, V., & Dinan, T. G. (2020). The gut microbiome in neurological disorders. *Lancet Neurology*, *19*(2), 179–194. https://doi.org/10.1016/S1474-4422(19)30356-4

Cummings, J. (2017). Disease modification and neuroprotection in neurodegenerative disorders. *Translational Neurodegeneration*, *6*, Article 25. https://doi.org/10.1186/s40035-017-0096-2

Deaner, R. O., & Winegard, B. M. (2013). Book review: Throwing out the mismatch baby with the paleobathwater. *Evolutionary Psychology*, *11*(1), 263–269. https://doi.org/10.1177/147470491301100123

Eaton, S. B., Cordain, L., & Lindeberg, S. (2002a). Evolutionary health promotion: A consideration of common counterarguments. *Preventive Medicine*, *34*(2), 119–123. https://doi.org/10.1006/pmed.2001.0966

Eaton, S. B., Konner, M., & Shostak, M. (1988). Stone agers in the fast lane: Chronic degenerative diseases in evolutionary perspective. *American Journal of Medicine*, *84*(4), 739–749. https://doi.org/10.1016/0002-9343(88)90113-1

Eaton, S. B., Strassman, B. I., Nesse, R. M., Neel, J. V., Ewald, P. W., Williams, G. C., Weder, A. B., Lindeberg, S., Konner, M. J., Mysterud, I., & Cordain, L. (2002b). Evolutionary health promotion. *Preventive Medicine*, *34*(2), 109–118. https://doi.org/10.1006/pmed.2001.0876

Fabian, D., & Flatt, T. (2011). The evolution of aging. *Nature Education Knowledge*, *3*(10), 1–10.

Fleming, J. O., & Cook, T. D. (2006). Multiple sclerosis and the hygiene hypothesis. *Neurology*, *67*(11), 2085–2086. https://doi.org/10.1212/01.wnl.0000247663.40297.2d

Fox, M. (2018). "Evolutionary medicine" perspectives on Alzheimer's disease: Review and new directions. *Ageing Research Reviews*, *47*, 140–148. https://doi.org/10.1016/j.arr.2018.07.008

Fox, M., Knapp, L. A., Andrews, P. W., & Fincher, C. L. (2013). Hygiene and the world distribution of Alzheimer's disease: Epidemiological evidence for a relationship between microbial environment and age-adjusted disease burden. *Evolution, Medicine, and Public Health*, *2013*(1), 173–186. https://doi.org/10.1093/emph/eot015

Franceschi, C., Bonafè, M., Valensin, S., Olivieri, F., De Luca, M., Ottaviani, E., & De Benedictis, G. (2000). Inflamm-aging: An evolutionary perspective on immunosenescence. *Annals of the New York Academy of Sciences*, *908*(1), 244–254. https://doi.org/10.1111/j.1749-6632.2000.tb06651.x

Fries, J. F. (2016). On the compression of morbidity: From 1980 to 2015 and beyond. In M. Kaeberlein & G. M. Martin (Eds.), *Handbook of the biology of aging* (8th ed., pp. 507–524). Elsevier.

Gaillard, J. M., & Lemaître, J. F. (2017). The Williams' legacy: A critical reappraisal of his nine predictions about the evolution of senescence. *Evolution*, *71*(12), 2768–2785. https://doi.org/10.1111/evo.13379

Gerrard, J. W., Geddes, C. A., Reggin, P. L., Gerrard, C. D., & Horne, S. (1976). Serum IgE levels in White and metis communities in Saskatchewan. *Annals of Allergy*, *37*(2), 91–100.

Gluckman, P. D., Low, F. M., Buklijas, T., Hanson, M. A., & Beedle, A. S. (2011). How evolutionary principles improve the understanding of human health and disease. *Evolutionary Applications*, *4*(2), 249–263. https://doi.org/10.1111/j.1752-4571.2010.00164.x

Goldman, D. P., Cutler, D., Rowe, J. W., Michaud, P. C., Sullivan, J., Peneva, D., & Olshansky, S. J. (2013). Substantial health and economic returns from delayed aging may warrant a new focus for medical research. *Health Affairs*, *32*(10), 1698–1705. https://doi.org/10.1377/hlthaff.2013.0052

Griffin, W. S. T., & Mrak, R. E. (2009). Is there room for Darwinian medicine and the hygiene hypothesis in Alzheimer pathogenesis? In G. A. W. Rook (Ed.), *The hygiene hypothesis and Darwinian medicine* (pp. 257–278). Birkhäuser. https://doi.org/10.1007/978-3-7643-8903-1_14

Guarner, F., Bourdet-Sicard, R., Brandtzaeg, P., Gill, H. S., McGuirk, P., van Eden, W., Versalovic, J., Weinstock, J. V., & Rook, G. A. (2006). Mechanisms of disease: The hygiene hypothesis revisited. *Nature Reviews: Gastroenterology & Hepatology*, *3*(5), 275–284. https://doi.org/10.1038/ncpgasthep0471

Hadley, C. (2004). Should auld acquaintance be forgot. *EMBO Reports*, *5*(12), 1122–1124. https://doi.org/10.1038/sj.embor.7400308

Hamilton, W. D. (1966). The moulding of senescence by natural selection. *Journal of Theoretical Biology*, *12*, 12–45. https://doi.org/10.1016/0022-5193(66)90184-6

Hubert, H. B., Bloch, D. A., Oehlert, J. W., & Fries, J. F. (2002). Lifestyle habits and compression of morbidity. *Journals of Gerontology: Series A*, *57*(6), M347–M351. https://doi.org/10.1093/gerona/57.6.m347

Jasienska, G., Ellison, P. T., Galbarczyk, A., Jasienski, M., Kalemba-Drozdz, M., Kapiszewska, M., Nenko, I., Thune, I., & Ziomkiewicz, A. (2015). Apolipoprotein E (ApoE) polymorphism is related to differences in potential fertility in women: A case of antagonistic pleiotropy?. *Proceedings of the Royal Society B: Biological Sciences*, *282*(1803), Article 20142395. https://doi.org/10.1098/rspb.2014.2395

Jones, D. S., & Greene, J. A. (2013). The decline and rise of coronary heart disease: Understanding public health catastrophism. *American Journal of Public Health*, *103*(7), 1207–1218. https://doi.org/10.2105/AJPH.2013.301226

Keyfitz, N. (1978). Improving life expectancy: An uphill road ahead. *American Journal of Public Health*, *68*(10), 954–956. https://doi.org/10.2105/ajph.68.10.654

Kirkwood, T. B., & Austad, S. N. (2000). Why do we age? *Nature*, *408*(6809), 233–238. https://doi.org/10.1038/35041682

Konner, M., & Eaton, S. B. (2010). Paleolithic nutrition: Twenty-five years later. *Nutrition in Clinical Practice*, *25*(6), 594–602. https://doi.org/10.1177/0884533610385702

Lathe, J. C., & Lathe, R. (2020). Evidence against a geographic gradient of Alzheimer's disease and the hygiene hypothesis. *Evolution, Medicine, and Public Health*, *2020*(1), 141–144. https://doi.org/10.1093/emph/eoaa023

Lee, R. D. (2003). Rethinking the evolutionary theory of aging: Transfers, not births, shape senescence in social species. *Proceedings of the National Academy of Sciences of the United States of America*, *100*(16), 9637–9642. https://doi.org/10.1073/pnas.1530303100

Leibowitz, U., Antonovsky, A., Medalie, J. M., Smith, H. A., Halpern, L., & Alter, M. (1966). Epidemiological study of multiple sclerosis in Israel. II. Multiple sclerosis and level of sanitation. *Journal of Neurology, Neurosurgery & Psychiatry*, *29*(1), 60–68. https://doi.org/10.1136/jnnp.29.1.60

Levy, G., & Levin, B. (2014). *The biostatistics of aging: From Gompertzian mortality to an index of aging-relatedness*. Wiley.

Levy, G., & Levin, B. (2022). An evolution-based model of causation for aging-related diseases and intrinsic mortality: Explanatory properties and implications for healthy aging. *Frontiers in Public Health*, *10*, Article 774668. https://doi.org/10.3389/fpubh.2022.774668

Livingston, G., Huntley, J., Sommerlad, A., Ames, D., Ballard, C., Banerjee, S., Brayne, C., Burns, A., Cohen-Mansfield, J., Cooper, C., Costafreda, S. G., Dias, A., Fox, N., Gitlin, L. N., Howard, R., Kales, H. C., Kivimäki, M., Larson, E. B., Ogunniyi, A., . . . Mukadam, N. (2020). Dementia prevention, intervention, and care: 2020 report of the *Lancet* Commission. *Lancet*, *396*(10248), 413–446. https://doi.org/10.1016/S0140-6736(20)30367-6

Maziak, W. (2002). The hygiene hypothesis and the evolutionary perspective of health. *Preventive Medicine*, *35*(4), 415–418. https://doi.org/10.1006/pmed.2002.1092

McDade, T. W., Tallman, P. S., Madimenos, F. C., Liebert, M. A., Cepon, T. J., Sugiyama, L. S., & Snodgrass, J. J. (2012). Analysis of variability of high sensitivity C-reactive protein in lowland Ecuador reveals no evidence of chronic low-grade inflammation. *American Journal of Human Biology*, *24*(5), 675–681. https://doi.org/10.1002/ajhb.22296

Medawar, P. B. (1952). *An unsolved problem of biology*. H. K. Lewis.

Murdaca, G., Greco, M., Borro, M., & Gangemi, S. (2021). Hygiene hypothesis and autoimmune diseases: A narrative review of clinical evidences and mechanisms. *Autoimmunity reviews*, *20*(7), Article 102845. https://doi.org/10.1016/j.autrev.2021.102845

Nesse, R. M. (2005). Maladaptation and natural selection. *Quarterly Review of Biology*, *80*(1), 62–70. https://doi.org/10.1086/431026

Olshansky, S. J., Rudberg, M. A., Carnes, B. A., Cassel, C. K., & Brody, J. A. (1991). Trading off longer life for worsening health: The expansion of morbidity hypothesis. *Journal of Aging and Health*, *3*, 194–216. https://doi.org/10.1177/089826439100300205

Omran, A. R. (1971). The epidemiologic transition. A theory of the epidemiology of population change. *Milbank Memorial Fund Quarterly*, *49*(4), 509–538. https://doi.org/10.1111/j.1468-0009.2005.00398.x

Parker, W. (2014). The "hygiene hypothesis" for allergic disease is a misnomer. *British Medical Journal*, *348*, Article g5267. https://doi.org/10.1136/bmj.g5267

Parker, W. (2016). Lessons learned from drug design and development. *Perspectives in Public Health*, *136*(4), 195–196. https://doi.org/10.1177/1757913916641589

Parker, W., Perkins, S. E., Harker, M., & Muehlenbein, M. P. (2012). A prescription for clinical immunology: The pills are available and ready for testing. A review. *Current Medical Research and Opinion*, *28*(7), 1193–1202. https://doi.org/10.1185/03007995.2012.695731

Provenzano, F., & Deleidi, M. (2021). Reassessing neurodegenerative disease: Immune protection pathways and antagonistic pleiotropy. *Trends in Neuroscience*, *44*(10), 771–780. https://doi.org/10.1016/j.tins.2021.06.006

Rook, G. A. W. (2009). Introduction: The changing microbial environment, Darwinian medicine and the hygiene hypothesis. In G. A. W. Rook (Ed.), *The hygiene hypothesis and Darwinian medicine* (pp. 1–27). Birkhäuser. https://doi.org/10.1007/978-3-7643-8903-1_1

Rook, G. A. W. (2013). Regulation of the immune system by biodiversity from the natural environment: an ecosystem service essential to health. *Proceedings of the National Academy of Sciences of the United States of America*, *110*(46), 18360–18367. https://doi.org/10.1073/pnas.1313731110

Rook, G. A. W., & Bloomfield, S. F. (2021). Microbial exposures that establish immunoregulation are compatible with targeted hygiene. *Journal of Allergy and Clinical Immunology*, *148*(1), 33–39. https://doi.org/10.1016/j.jaci.2021.05.008

Rook, G. A. W., & Brunet, L. R. (2005). Old friends for breakfast. *Clinical & Experimental Allergy*, *35*(7), 841–842. https://doi.org/10.1111/j.1365-2222.2005.02112.x

Rose, G. (1985). Sick individuals and sick populations. *International Journal of Epidemiology*, *14*(1), 32–38. https://doi.org/10.1093/ije/14.1.32

Rose, M. R. (1991). *Evolutionary biology of aging*. Oxford University Press.

Rose, M. R. (2009). Adaptation, aging, and genomic information. *Aging*, *1*(5), 444–450. https://doi.org/10.18632/aging.100053

Schepici, G., Silvestro, S., Bramanti, P., & Mazzon, E. (2019). The gut microbiota in multiple sclerosis: An overview of clinical trials. *Cell Transplant*, *28*(12), 1507–1527. https://doi.org/10.1177/0963689719873890

Sierra, F., & Kohanski, R. (2017). Geroscience and the trans-NIH Geroscience Interest Group, GSIG. *GeroScience*, *39*(1), 1–5. https://doi.org/10.1007/s11357-016-9954-6

Sjögren, M., Huttunen, H. J., Svenningsson, P., & Widner, H. (2021). Genetically targeted clinical trials in Parkinson's disease: Learning from the successes made in oncology. *Genes*, *12*(10), Article 1529. https://doi.org/10.3390/genes12101529

Strachan, D. P. (1989). Hay fever, hygiene, and household size. *British Medical Journal*, *299*(6710), 1259–1260. https://doi.org/10.1136/bmj.299.6710.1259

Vita, A. J., Terry, R. B., Hubert, H. B., & Fries, J. F. (1998). Aging, health risks, and cumulative disability. *New England Journal of Medicine*, *338*(15), 1035–1041. https://doi.org/10.1056/NEJM199804093381506

von Hertzen, L., Beutler, B., Bienenstock, J., Blaser, M., Cani, P. D., Eriksson, J., Färkkilä, M., Haahtela, T., Hanski, I., Jenmalm, M. C., Kere, J., Knip, M., Kontula, K., Koskenvuo, M., Ling, C., Mandrup-Poulsen, T., von Mutius, E., Mäkelä, M. J., Paunio, T., . . . de Vos, W. M. (2015). Helsinki alert of biodiversity and health. *Annals of Medicine*, *47*(3), 218–225. https://doi.org/10.3109/07853890.2015.1010226

Wendel-Haga, M., & Celius, E. G. (2017). Is the hygiene hypothesis relevant for the risk of multiple sclerosis? *Acta Neurologica Scandinavica*, *136*(201), 26–30. https://doi.org/10.1111/ane.12844

Williams, G. C. (1957). Pleiotropy, natural selection, and the evolution of senescence. *Evolution*, *11*, 398–411. https://doi.org/10.1126/sageke.2001.1.cp13

Williams, G. C., & Nesse, R. M. (1991). The dawn of Darwinian medicine. *Quarterly Review of Biology*, *66*(1), 1–22. https://doi.org/10.1086/417048

Willyard, C. (2021). How gut microbes could drive brain disorders. *Nature*, *590*(7844), 22–25. https://doi.org/10.1038/d41586-021-00260-3

Wolters, F. J., Chibnik, L. B., Waziry, R., Anderson, R., Berr, C., Beiser, A., Bis, J. C., Blacker, D., Bos, D., Brayne, C., Dartigues, J. F., Darweesh, S. K. L., Davis-Plourde, K. L., de Wolf, F., Debette, S., Dufouil, C., Fornage, M., Goudsmit, J., Grasset, L., . . . Hofman, A. (2020). Twenty-seven-year time trends in dementia incidence in Europe and the United States: The Alzheimer Cohorts Consortium. *Neurology*, *95*(5), e519–e531. https://doi.org/10.1212/WNL.0000000000010022

Wu, K. M., Zhang, Y. R., Huang, Y. Y., Dong, Q., Tan, L., & Yu, J. T. (2021). The role of the immune system in Alzheimer's disease. *Ageing Research Reviews*, *70*, Article 101409. https://doi.org/10.1016/j.arr.2021.101409

20

Evolutionary Aspects of Neuropsychopharmacology

Martin Brüne, Riadh Abed, and Paul St. John-Smith

Introduction

Neuropsychopharmacology is the field of medicine that deals with the neural mechanisms explaining how neurotropic molecules affect brain function. The mechanisms involved include neurological processes, such as the functioning of sensory and motor systems, as well as the functioning of the mind (i.e., the study of cognition, emotion, and behavior) in relation to the action of neurotransmitters, neuromodulators, and psychotropic drugs.[1] Accordingly, neuropsychopharmacology draws on concepts and knowledge from the fields of neuroanatomy, neuropathology, pharmacokinetics, and neuroscience. Neuropsychopharmacology is closely related to psychopharmacology but adds to the latter a more in-depth analysis of biochemical processes contributing to the modulation of receptor activity that are relevant for the understanding and treatment of psychiatric and neurological conditions.

In recent years, new task forces have been formed with the aim of replacing disorder- or disease-driven nomenclature (e.g., antidepressants, antipsychotics, antiepileptics, etc.) with a classification of drugs based on their pharmacological properties. This change helps shift focus from symptoms to the understanding of mechanisms and from disease (entity) to treatments. In addition, a new nomenclature may improve the integration of new findings from neuroscience research. More details on the goals of the task forces can be found on the homepages of the American College of Neuropsychopharmacology (ACNP), the European College of Neuropsychopharmacology, the International Union of Basic and Clinical Pharmacology, and others.

Critiques of these approaches include the claim that neuropsychopharmacology is overly reductionist. Accordingly, neuropsychopharmacology does reductionistically focus on the mechanistic understanding and therapeutic modification of neuronal activity. However, most clinicians recognize the complexity of the human brain and behavior, the countless number of possible interactions among different neurotransmitter and neuromodulator systems within the central nervous system (CNS), and interactions of the brain with other organs of the body, which make this only part of the endeavor that can never be totally explanatory in itself.

This critique notwithstanding, in the last century, huge progress has been made regarding the discovery of substances that help ameliorate psychological dysfunction, such as excessive states of anxiety, depression, psychosis, etc. This is also true concerning the treatment of neurological diseases, including epilepsy and movement disorders. However, many psychopharmacological discoveries were made by chance, not design. Prescription of medications for medical conditions serendipitously revealed neurotropic effects that improved psychiatric symptoms. Often, the underlying mechanisms involved were determined later on (a typical example is the discovery of antipsychotics). Contrastingly, in many medical and neurological diseases, the mechanisms of the identified dysfunction were known, at least to some extent, before the medication was developed. These differences in the drug discovery process occurred mainly because the precision of phenotypic description of "disease entities" with regard to psychiatric conditions (e.g., psychosis) is poor, whereas it is much more accurate in neurological diseases (e.g., diseases of the basal ganglia; see Chapter 9). Overall, the localization of central nervous dysfunction is much easier to characterize in neurological diseases, as compared to psychiatric disorders. This lack of consistent neuronal or pathophysiological correlates for any psychiatric condition is demonstrated by the common use of the term "disorder" over "disease" by psychiatric classifications. Furthermore, the distinction between normal function and disease/disorder is clearer in neurological conditions compared to psychiatric disorders. Nevertheless, medical models assume that mental states associated with low mood are disorders, whereas an evolutionary stance may also consider potential adaptive benefits of low mood (Nesse 2019).

Overall, neuropsychopharmacology has advanced over the years, not only regarding drug development and discovery but also concerning the understanding of the neuronal "machinery" (i.e., receptor activity and the modes of action of neuropsychopharmacological substances). The ACNP, for instance, provides many excellent publications on neurotransmitters and their roles in neuropsychiatric conditions (Davis et al. 2002). Surprisingly, however, there is very little, if any, reference to the evolution of those substances we call "neurotransmitters" or "biomodulators," their receptors, or underlying genetic mechanisms, let alone the selection processes involving their genes. Nor does this work refer to the evolutionary history of CNSs and how and why they came to utilize substances that existed even before the first synapses evolved to improve and accelerate the transmission of information (Emes and Grant 2012). We propose that this lack of understanding is a major shortcoming for the understanding of the evolved constraints, design flaws, and compromises involved in neurotransmission and, therefore, for the development of new neuropsychopharmacological agents.

The present chapter thus aims to address some of these knowledge gaps by highlighting the evolutionary history of neuropsychopharmacologically active substances and the evolution of CNSs. In addition, we seek to discuss potential implications of acknowledging the evolutionary processes involved in neuropsychopharmacology, including the development of new drugs. This approach is based on the work of Nicolaas Tinbergen (1963), who proposed that all biological phenomena must be studied on four different levels. He labeled these: "causation" (i.e., mechanism), "ontogeny" (individual development), "survival value" (biological fitness), and "phylogeny" (evolutionary history), whereby mechanisms and development are referred to as the "proximate" causes, whereas the other two levels are known as "ultimate" (or evolutionary) causes. Proximate causes are influenced by an organism's individual developmental life history and include effects of gene–environment interaction as well as epigenetic factors. In contrast, "ultimate" or evolutionary causes can only be studied on a vastly different timescale, namely changes occurring over evolutionary time. Accordingly, the study of the evolutionary causes entails similarities and dissimilarities between phylogenetically related species (comparative method) and an understanding of the adaptive value of a given trait. The biologist Ernst Mayr put the relevance of all four levels in the following words: "No biological problem is solved until both the proximate and the evolutionary causation has been elucidated. Furthermore, the study of evolutionary causes is as legitimate a part of biology as is the study of the usually physico-chemical proximate causes" (1982, p. 73).

An abridged overview of the evolutionary history of substances known as "neurotransmitters" or "biomediators" will be first presented. Aspects of the evolution and evolutionary constraints of cell communication, mainly concerning the communication among neurons and astroglia, will be outlined next. The evolutionary story of glutamate and its role in health and disease, with particular emphasis on neuropsychiatric diseases and disorders, will be presented in greater detail. Finally, the possible implications of an evolutionary perspective on neuropsychopharmacology for the understanding of the limitations of neuropsychotropic drug action and therefore for the development of new agents for the treatment of psychiatric and neurological conditions will be discussed.

Evolutionary History of Neurotransmitters and Biomediators

Substances referred to as "neurotransmitters" have been shown to be evolutionarily ancient molecules that likely have hundreds of millions of years of biological history. It appears that they were selected for and thus evolved for vastly different biological functions than the current use in neurotransmission (i.e., long before the first synaptic transmission emerged or CNSs were formed). Roshchina (2010) thus suggested that the term "biomodulators" would be more accurate for this group of substances. Using cross-species and even cross-kingdom comparisons, the utilization of catecholamines, dopamine, epinephrine (adrenaline), and norepinephrine (noradrenaline) appears to have evolved to promote bacterial growth. Other substances, including acetylcholine, inhibit chemotaxis in bacteria, while biogenic amines promote this process. In plants, such biomodulators foster growth by regulating membrane permeability and root shooting, whereas at higher concentrations, many biomodulators are involved in defense reactions against microbial colonization or feeding (Roshchina 2010). Plant toxicity to herbivores and parasites is widespread in nature, including the well-known examples of alkaloid- and serotonin-enriched plants. Paradoxically, while the ingestion of large quantities of these substances can be fatal to animals, including humans, some have therapeutic effects in low concentrations. This therapeutic or even protective consumption is known as "pharmacophagy" and occurs in many species (St. John-Smith et al. 2013).

Biomodulators, including the enzymes that are necessary to produce them, are evolutionarily highly conserved substances that over evolutionary time have been "co-opted" for many different functions. Interestingly, neurotransmission appears to occur quite late in evolutionary terms. Biomodulators are involved in the communication within and between different microorganisms, between microorganisms and plants and animals, between plants and animals, and between animals of different species, ranging from commensal or symbiotic purposes to deterrence (Roshchina 2010). In multicellular organisms, including higher vertebrates, many biogenic amines play a significant role in immunity and inflammatory responses (Franco et al. 2021). It is, however, important to consider the evolved adaptive properties of biomodulators with regard to CNS function, including the evolution of receptors that helped refine communication among neurons as well as the neuronal networks. It is also essential to recognize that many biomodulators utilized for neurotransmission are toxic in some situations, especially in high concentrations. Also, most biomediators are metabolically expensive in that they require a substantial amount of energy to be produced in CNSs; this is especially pertinent if blood–brain barriers preclude these substances from entering the CNS directly from the bloodstream. There is evidence to suggest that blood–brain barriers evolved multiple times and that their main function is to protect the neural tissue from being functionally disrupted or damaged by toxins. This protective function includes barriers to those substances that serve as biomodulators in CNSs or brains (Dunton et al. 2021). Both the toxic properties and the high energy expense characteristic of biomodulators have had an impact on the evolution of CNSs in terms of the evolution of reuptake mechanisms and the enzymatic degradation of these substances.

Acetylcholine

Acetylcholine has been found in bacteria, plants, and animals. It is synthesized from the nutrient choline and acetyl coenzyme A in the presence of choline acetyltransferase and degraded enzymatically by acetylcholinesterase. Acetylcholine binds to two types of receptors, namely nicotinic and muscarinic receptors. Both types of receptors appear to be present in unicellular organisms, and similar receptors also exist in plants. In higher vertebrates, nicotinic receptors are found in skeletal muscle cells and other tissue, whereas muscarinic receptors modify visceral parasympathetic activity (Roshchina 2010). In the mammalian brain, acetylcholine-producing nuclei are mainly found in the basal forebrain. Ontogenetically, cholinergic projections from the basal forebrain reach the cortex in the early prenatal period. Acetylcholine is synthesized at the synapse and stored in presynaptic vesicles. After release by a depolarization signal, acetylcholine is rapidly degraded by acetylcholinesterase.

Clinically, acetylcholine is involved in several neuropsychiatric diseases and disorders (e.g., myasthenia gravis and Alzheimer's disease) where a cholinergic deficit arises from a progressive degeneration of the nucleus basalis of Meynert. Muscarinic receptor alterations have also been found in schizophrenia (Dean 2009).

Catecholamines

Catecholamines (dopamine, norepinephrine, epinephrine) have been identified in unicellular organisms, plants, and multicellular animals. Interestingly, the concentration of catecholamines is higher in bacteria than in other organisms. Catecholamines are synthesized from the amino acid phenylalanine by hydroxylation, through which tyrosine is produced. Tyrosine is then dehydroxylated to dehydroxyphenylalanine (DOPA; note that animals synthesize only levo-DOPA [L-DOPA] but not its counterpart dextro-DOPA). DOPA is transformed into dopamine by decarboxylation (bacteria use tyrosine decarboxylase for this metabolic step), and another hydroxylation produces norepinephrine (noradrenaline). Epinephrine (adrenaline) results from methylation of norepinephrine. Tyrosine hydroxylase is the limiting step in the production of dopamine. Interestingly, gnathostomes (jawed vertebrates) have two types of tyrosine hydroxylase, one of which was lost in the evolution of placental mammals (platypuses and opossums possess two types of tyrosine hydroxylase). However, the functional consequences of this gene loss remain unclear (Yamamoto and Vernier 2011).

Importantly, in contrast to dopamine, L-DOPA can cross the blood–brain barrier, such that centrally active dopamine has to be synthesized in the CNS. It has been speculated that the reduction of tyrosine to dopamine may cause oxidative stress, which can be attenuated by reuptake mechanisms. Reuptake of transmitters can thus be seen as an evolved mechanism to save energy for the manufacturing of transmitters and reduce "side effects" of the required metabolic steps (Franco et al. 2021). The evolutionary history of reuptake mechanisms is highly interesting. Reuptake is not specific to dopamine as in vertebrates there are other transporter proteins that regulate the reuptake of noradrenaline and serotonin. The transporter proteins appear to be phylogenetically related, and therefore not specific for one neurotransmitter or biomodulator. Hence, the reuptake of dopamine and noradrenaline can be executed by the serotonin transporter, such that reuptake of a serotonergic drug like sertraline is possible via the dopamine or noradrenaline transporters (Yamamoto and Vernier 2011).

The activity of catecholamines is terminated intracellularly by catechol-*O*-methyltransferase (COMT) and by the monoamine-oxidase enzymes (MAO-A and MAO-B). The evolutionary history of COMT is not entirely clear. There seems to be just one isoform in jawed vertebrates, which derived from one of numerous methyltransferases in living organisms. Similarly, the origin of the two MAO paralogues found in vertebrates is uncertain. Teleost fish (the largest group of bony fish species), for instance, possess only one MAO. The two MAO versions of vertebrates probably evolved by gene duplication. MAO-A and MAO-B have different functions in the brain. MAO-A is mainly found in catecholamine-containing cells, whereas MAO-B is more abundant in serotonergic cells (Yamamoto and Vernier 2011).

In the brain, dopamine synthesis takes place in the pars compacta of the substantia nigra. In the vertebrate brain, dopaminergic neurons are abundant in the ventral tegmental area connecting to the nucleus accumbens in the ventral striatum and the cortex via mesolimbic and mesocortical pathways. Norepinephrine-synthesizing cells are abundant in the locus coeruleus, from which pathways connect with the neocortex, hypothalamus, cerebellum, brainstem, and spinal cord.

Clinically, the catecholamines are involved in many neurological and psychiatric conditions. Dopamine deficiency is the emblematic biochemical signature or fingerprint of Parkinson's disease. This biomodulator also plays a role in other basal ganglia diseases, including Huntington's disease and Tourette syndrome. Moreover, dopamine overactivity in the mesolimbic system is part of the pathogenesis of schizophrenia. A prefrontal dopaminergic deficit is believed to play a role in negative symptoms associated with psychosis. Norepinephrine deficiency is involved in attention-deficit hyperactivity disorder.

Serotonin

Serotonin utilization by organisms probably occurred at least 500 million years ago and is another example of a biogenic amine that is highly conserved over evolutionary time. Serotonin exists in bacteria, plants, and multicellular animals. It is synthesized in a two-step process from the amino acid tryptophan. Importantly, animals have lost the ability to produce tryptophan, in contrast to plants. In bacteria serotonin has growth-stimulating properties but may also be involved as a defense agent in higher concentrations. In animals possessing a CNS, 5-hydroxytryptophan can cross the blood–brain barrier, where it is transformed into serotonin by decarboxylation. A derivative of serotonin is melatonin, which in plants acts as an antioxidant and in animals is involved in sleep regulation (Azmitia 2010). In the brain, serotonin is produced in the raphe nuclei of the brainstem. Serotonergic pathways project to the lower brainstem and spinal cord and to the striatum, hypothalamus, amygdala, hippocampus, and cortex.

In mammals, more than 90% of the serotonin is produced and stored in the guts. In the brain, at least 16 types of serotonin receptors exist that probably evolved by gene duplication and subsequent selection for differential properties of the phenotypes produced. The different serotonin receptors are distributed across different brain regions in different densities. The reuptake of serotonin from the synaptic cleft into the terminal axon is under control of a serotonin transporter gene.

Serotonin deficiency is commonly thought to be involved in the pathogenesis of mood disorders, pathological anxiety, obsessive-compulsive disorder, and eating disorders, even though its role in depression has recently been disputed (Moncrieff et al. 2023).

Glutamate and GABA

The evolutionary history of the amino acid glutamate is particularly interesting for several reasons. First, there is evidence to suggest that the existence of glutamate is older than life on earth itself, and its non-biological synthesis has been recreated by several abiotic experiments. Both L-glutamate and D-glutamate were produced abiotically perhaps as early as 4 billion years ago by photochemical reaction, ionizing radiation, volcanic activity, and even meteoritic impact, rendering glutamate as perhaps the most important "proto-molecule" (overview in Moroz et al. 2021). Second, its abundance and biochemical properties made glutamate a perfect candidate to become one of the first energy sources utilized in cell metabolism. Glutamate is a product of the oxidative Krebs cycle (tricarboxylic acid cycle), whereby the intermediate substance alpha-ketoglutarate is reduced with ammonium or glutamine acting as nitrogen sources (Moroz et al. 2021). Glutamate is rich in energy and can serve as a donor of carbon and nitrogen; it is thus involved in over 200 metabolic pathways. Biochemically, L-glutamate seems to have advantages over D-glutamate because the L-enantiomer is structurally more flexible than the D-enantiomer, and thus more "reactive."

Glutamate is the most abundant metabolite in bacteria and in mammals. Glutamate also plays an important role in plant metabolism, including the regulation of osmolarity, growth, photosynthesis, and reproduction. Remarkably, when leaves are damaged by herbivores, glutamate is released at the injured site; and within minutes, undamaged leaves activate defense responses to deter herbivores from feeding (Moroz et al. 2021).

In bilaterians (animals that are symmetrical along a vertical axis of the body), glutamate is present in high intracellular concentrations, even though its role differs vastly between invertebrates and vertebrates. For example, glutamate can have inhibitory functions in the CNSs of invertebrates but acts exclusively as an excitatory transmitter in vertebrate CNSs, suggesting that its inhibitory function was lost over evolutionary time. Moreover, glutamate serves as neuromuscular transmitter in arthropods, while in vertebrates this function is carried out by acetylcholine.

In the vertebrate brain, glutamate binds to four major receptors: *N*-methyl-D-aspartic acid (NMDA), kainate, alpha-amino-3-hydroxy-5-methylisoxazole-4-proprionic acid (AMPA), and delta. Due to its excitatory properties and high concentrations in neuronal tissue, extracellular glutamate, when released by pathological processes including neurodegeneration, stress, or ischemia, is toxic to the CNS or brain itself and is involved in the pathogenesis of several neurodegenerative diseases, stroke, as well as affective disorders (Haroon et al. 2017; Fu et al. 2018).

Gamma-amino butyric acid (GABA) is the most abundant inhibitory neurotransmitter in the brain. It is synthesized from glutamate in a single step through decarboxylation. GABA binds to $GABA_{A-C}$ receptors which are mainly located on GABAergic interneurons. GABA is therefore part of the evolutionarily conserved solution to control the excitatory effects of glutamate in highly efficient ways.

Oxytocin and Vasopressin

Oxytocin and vasopressin are nonapeptides, in contrast to the single amino acids glutamate, glutamine, and glycine. Oxytocin and vasopressin differ in only two amino acids. Evolutionarily, they evolved by gene duplication from isotocin, which has been found in bony fish species. Similar peptides exist in invertebrates, including arthropods (where the molecule is called "vasotocin"), nematodes ("nemotocin"), and other phyla (overview in Carter 2014).

Oxytocin and its evolutionary precursors have been involved in associative learning, sensory processing, and various roles in reproduction, including egg laying, courting, and, in mammals, parturition and lactation. Vasopressin, in contrast, is more involved in water conservation and regulation of blood pressure (in vertebrates) and may have facilitated the transition from water-bound to terrestrial life. Together, given their role in CNSs and brains, oxytocin and vasopressin are rightly called "neuropeptides." Oxytocin and vasopressin are synthesized in large cells (magnocellular cells) in the paraventricular nucleus and the supraoptic nuclei of the hypothalamus. Oxytocinergic pathways reach the amygdala and the pituitary gland.

In recent decades, the role of oxytocin in mammalian, including human, social behavior has been unveiled in countless studies. For example, oxytocin promotes parental bonding and partner bonding in nonhuman animals and in humans (Donaldson and Young 2008). In humans, oxytocin can improve social cognitive functions such as empathy and trust, although such prosocial effects very likely depend on the quality of early attachment experiences and experience of childhood adversity (Domes et al. 2007; Brüne 2016). Oxytocin is more abundant in female brains and is released upon infant suckling, a process that strengthens the mother–child dyad (Feldman 2012). Oxytocin also interacts with dopamine, serotonin, and the opioidergic system in manifold ways (Walker and McGlone 2013).

The synthesis of vasopressin, in contrast, seems to be in part under the control of testosterone. Vasopressin promotes male sexuality and aggression in humans and nonhuman mammals.

Our understanding of the role of these and other neuropeptides in psychopathology has gradually improved. For example, it has been suggested that disruption of early infant attachment can have profound long-term consequences on the social affiliative role of oxytocin and vasopressin (Carter 2014).

Due to constraints of space, the preceding list of biomodulators is necessarily incomplete. In addition to the substances mentioned, there are many more bioactive molecules, including neuropeptides, endocannabinoids, nitric oxide, etc., that have a long evolutionary history and have been "co-opted" for the purpose of neuronal cell communication. In the next section, we will explore some aspects explaining how and why these substances came to be utilized for neurotransmission. This is mainly a story of the evolution of synaptic transmission and the evolution of receptors.

Evolution of Cell Communication, Synaptic Transmission, and Receptors

Cell Communication

Communication between cells appears to have been selected for very early in the evolution of life. Accordingly, relatively simple forms of cell communication exist in unicellular organisms including the extant prokaryotes, bacteria, and archaea. During the evolution of prokaryotes, transmembrane ion exchange became regulated by molecules that bind to the ion channel and modulate its permeability (Le Duc and Schöneberg 2019).

In contrast to prokaryotes (lacking a nucleus), eukaryote cells possess internal organization. Eukaryotic cells likely evolved as a symbiotic cooperation between archaeal species and alpha-protobacteria, which were the precursors of intracellularly located mitochondria (Sagan 1967). Mitochondria still possess their own RNA-based genome. It is not entirely clear when this "fusion" happened, but it could have been related to the increase of oxygen in the earth's atmosphere by the activity of cyanobacteria about 2 billion years ago (Moroz et al. 2021). In other words, the symbiotic fusion of cells could have served, in the first place, as an adaptation to protect the cells against the toxic effects of oxygen, whereby later on the oxidative Krebs cycle was formed, becoming an important method of energy production (Moroz et al. 2021).

In multicellular organisms, communication among cells utilizes multiple routes of information exchange. Endocrine and paracrine signals come from the cell's environment, whereas autocrine signals come from within the cell. Cell communication is energetically expensive, so there must have been a selective advantage for the communicative pathways to evolve. One such advantage is certainly the maintenance of homoeostasis (Le Duc and Schöneberg 2019).

Aside from glutamate, nucleotides and their phosphates (e.g., adenosine triphosphate [ATP] and guanosine triphosphate [GTP]) are also rich in energy and among the oldest biomolecules. They form the structural basis of RNA and DNA and serve as metabotropic molecules in signaling pathways. Nucleotide triphosphate cyclases are membrane-anchored, and they are soluble catalytic enzymes producing cyclic adenosine monophosphate (cAMP) and cyclic guanosine monophosphate (cGMP), which serve as intracellular signaling molecules, or second messengers. Cyclases are regulated by a variety of molecules, including calmodulin, guanine nucleotide-binding proteins (G proteins), nitric oxide, calcium, and bicarbonate. Adenyl cyclase, for example, has been found in prokaryotic and eukaryotic cells. cAMP and cGMP can also be targeted by drugs and toxins, including methylxanthines (e.g., caffeine) that increase their availability by inhibiting the phosphodiesterase. In addition, catecholamines and their agonists and blockers can modulate cAMP or cGMP availability (overview in Le Duc and Schöneberg 2019).

Over evolutionary time, the signaling cascades became increasingly complex and specific, and signal specificity was mainly achieved at the receptor level. The increasing complexity of cell communication seems to be associated with genome size in both prokaryotes and eukaryotes.

Synaptic Transmission

Intercellular junctions between neurons are traditionally called "synapses." Comparative analyses of the proteomic composition of synapses suggest that proto-synapses evolved earlier than the first primitive nervous systems. Some protein families found in synapses, such as calcium transporters and protein kinases, also exist in unicellular organisms. Synaptic protein families are present in unicellular eukaryotes such as yeast and amoebae, and many proteins found in synapses probably served intracellular communicative purposes in the first place (Ryan and Grant 2009; Emes and Grant 2012). After multicellularity emerged, genome duplication led to an increase in the number of synaptic gene families, including a variety of genes coding for glutamate and GABA receptors, and postsynaptic membrane proteins (Emes and Grant 2012). Cell communication in primitive neuronal systems might not have been synaptic but paracrine or hormonal instead (Moroz et al. 2021), supporting the idea that neurons and synapses evolved independently (Moroz and Kohn 2016). The first synaptic transmission, as found, for example, in the nerve net (not a real CNS) of primitive comb jellyfish (Ctenophora), was probably glutamatergic, signaling stress or injury (Moroz 2009). Ctenophores, interestingly, lack the genes coding for the enzymes that are required to synthesize catecholamines and other biomodulators. Together, evidence suggests that the *ur*-synapse evolved at some point between 750 and 1,000 million years ago (Emes and Grant 2012).

In any event, the evolution of cell–cell communication via synapses was not just adaptive (and thus advantageous); synaptic communication also co-opted intracellular communicative pathways to exchange information within the first neuronal networks. Indeed, the "synapse first" hypothesis suggests that the evolution of the synapse was a major driving force in the evolution of CNSs and, eventually, brains (Emes and Grant 2012).

Brains possess cell types that became specialized in the transmission of information over long distances. Axonal impulses, for example, can be transmitted at a speed of 100 meters per second, whereby the evolution of myelin sheaths greatly accelerated the axonal transport (Hofman 2001). In general, axosomatic synapses are largely inhibitory, while axodendritic synapses are mainly excitatory, and axo-axonal synapses are double inhibitory, thus disinhibitory (Zilles 2005).

Over evolutionary time, the number of synapses per neuron has substantially increased. Nonhuman primates have an average number of synapses per neuron which increased from 2,000 to 5,600, whereas the figure for neurons of the human brain is 6,800–10,000 synapses per neuron (Changeux 2005; see Chapter 22 in this volume).

Information transfer in brains, including the human brain, does not solely depend on neuronal activity. Research has revealed the influences of neuroglia on brain function (i.e., oligodendrocytes and astrocytes). For example, each oligodendrocyte produces the myelin sheaths for up to 60 axons. This cell type is essential in the repair of damaged myelin sheaths after stroke (Edgar and Sibille 2012). In addition, oligodendrocytes play a role in neuronal plasticity but are also vulnerable to oxidative stress and to glutamatergic toxicity (Haroon et al. 2017).

Astrocytes are involved in maintaining local ion concentrations and pH homoeostasis. A single astrocyte cell can interact with up to 2 million synapses (Fields et al. 2014). Astrocytes also deliver metabolites such as glucose, cholesterol, glutamate, steroids, lipids, neuropeptides, and growth factors. Astrocytes produce the extracellular matrix, mainly proteoglycans (Maeda 2015), and remove toxic metabolic products, including excess glutamate, from the synaptic cleft (Nedergaard et al. 2003; Faissner et al. 2010). Furthermore, astrocytes are important in maintaining the blood–brain barrier. The role of astroglia as part of the "tripartite synapse" (Faissner et al. 2010) is also reflected in the increase in number of this cell type relative to neurons. In nematodes, for example, the ratio of the number of neurons to astroglia is about 6:1. This ratio has constantly decreased, reaching about 3:1 in small rodents, and has been inversed in humans down to about 1:1.4 (i.e., the human brain contains more astroglia than neurons). This evolutionary change cannot be explained by the metabolic demands of neurons alone (which are relatively similar among higher vertebrates); its adaptive advantage probably resides in the regulatory function of astroglia concerning synaptic transmission, synaptogenesis, neurogenesis, and detoxification (Nedergaard et al. 2003).

Evolution of Neurotransmitter Receptors

Synapses consist of a presynaptic neuron and a postsynaptic neuron, which are separated by a synaptic cleft. In the presynaptic neuron, neurotransmitters (biomodulators) are stored in vesicles. Upon an electrical signal, the transmitters are released into the synaptic cleft and diffuse across the synaptic cleft to bind to specific receptors on the surface of the postsynaptic neuron, which include ionotropic receptors (also termed "ligand-gated ion channels"), and G-protein coupled receptors (GPCRs; also termed "metabotropic receptors"). Generally speaking, postsynaptic receptor proteins can excite, inhibit, or otherwise modify the excitability of the postsynaptic neuron (Guyton and Hall 2006; Raven et al. 2017), whereby excitation occurs directly via cation channels, whereas inhibition occurs by opening of anion channels. Some neurotransmitters produce rather short, sharply peaking excitatory action potentials at ionotropic receptors, for example, nicotinic acetylcholine receptors, whereas other compounds produce much slower alterations at metabotropic receptors, for example, metabotropic glutamate receptors.

Many neurotransmitters can reach high concentrations in the synaptic cleft, while their affinity to corresponding postsynaptic receptors is relatively low. Accordingly, the neuronal signal is often terminated by the dissolution of the neurotransmitter from the receptor. Alternative ways of terminating the action of neurotransmitters include their enzymatic degradation or reuptake in the presynaptic vesicles. These pathways have been fine-tuned over hundreds of millions of years of evolution (Ryan and Grant 2009).

GPCR Evolution

The evolution of GPCRs has led to one of the most successful and versatile signaling systems. GPCRs are present in all organs, and humans possess about 800 GPCR-coding genes. Mutations in GPCRs are involved in a plethora of human diseases, including retinitis pigmentosa, diabetes insipidus, and cancers. Toxins and drugs such as amphetamine, cannabinoids, psilocybin, opiates, and LSD utilize GPCR pathways (Le Duc and Schöneberg 2019).

The dopamine receptors also belong to the GPCR superfamily. However, the evolution of these receptors is quite complex. The D1 and D2 receptors have evolved independently from one another and are thus only distantly related (Opazo et al. 2018). This has functional consequences as the group of D1 receptors activate the production of cAMP, while D2-type receptors reduce cAMP. Moreover, D1 receptors are present only on the postsynaptic membrane, whereas D2 receptors can also be found presynaptically (Opazo et al. 2018). In the last common ancestor of vertebrates, only D1-type and D2-type receptors were present. Subsequent gene duplications led to the evolution of four different receptor types within the D1 family and the D2 family, respectively. Mammals have retained only five types of dopamine receptors (D1–D5), whereas other vertebrate clades have many other dopamine receptors that originated from either D1- or D2-type receptors but were lost during mammalian evolution (Opazo et al. 2018). Phylogenetic analysis suggests that D1 and D5 originated from the D1-type receptor, whereas D2, D3, and D4 derived from the ancestral D2-type receptor. In summary, while the dopaminergic pathways seem to be highly conserved over evolutionary time, there is also evidence for a refinement of ligand specificity and high genetic variability among GPCRs, which may have important ramifications for future drug development (Opazo et al. 2018).

The evolutionary history of metabotropic glutamate receptors is also highly complex and characterized by multiple gene duplication events that eventually led to a greater diversity of receptors in vertebrates, compared to invertebrates. One classic function of a metabotropic glutamate receptor is the sensation of the "umami" taste (Moroz et al. 2021).

Evolution of Ligand-Gated Ion Channels

The exchange of ions with the interior cell milieu was probably one of the first metabolic challenges for early life forms. The exchange was made possible by the evolution of transmembrane-spanning proteins, to which certain molecules bound in order to open or close the barrier between the cell interior and the environment. The molecules that bind to the transmembranous proteins are called "ligands" (Le Duc and Schöneberg 2019). The nicotinic acetylcholine receptor is probably the most widely studied ionotropic receptor. Ionotropic acetylcholine receptors and serotonin receptors are excitatory because they open cation channels to depolarize the postsynaptic neuron. In contrast, ionotropic GABA and glycine receptors are inhibitory because they are selective for anions and thus hyperpolarize (inhibit) the postsynaptic neuron (Ortells and Lunt 1995; Guyton and Hall 2006). Both the excitatory and the inhibitory pathways have a common evolutionary origin that may date back up to 2.5 billion years (Ortells and Lunt 1995). Within the excitatory (cationic) pathways it seems that the acetylcholinergic receptor split early from the serotonergic pathway. The anionic receptor molecules diverged somewhat later. Interestingly, it seems that the glycine receptor derived from the GABA receptor, which is counterintuitive as GABA is a more complex molecule compared to the simple amino acid glycine (Ortells and Lunt 1995).

Ionotropic transmission is also known for glutamatergic synapses. Ionotropic glutamatergic transmission is evolutionarily younger than metabotropic glutamate receptors. The first ionotropic glutamate receptors were discovered in cnidarians (an evolutionarily ancient aquatic animal phylum comprising sea anemones and different species of jellyfish that appeared about 1 billion years ago), which include postsynaptic NMDA and AMPA receptors (Ryan and Grant 2009). With the evolution of vertebrates, NMDA receptor complexity increased significantly. Ryan and Grant (2009) point out that the protein sequences of the NMDA NR2 subunit are several times larger than those of its homolog in invertebrates, yielding many more protein interaction and phosphorylation sites. In vertebrates, gene duplications led to further diversification of the NR2 subunit (NR2A to NR2D). Interestingly, the NR2A subunit seems to have undergone positive selection in primates (Ryan and Grant 2009). On the other hand, the impression of increasing receptor complexity emerging over evolutionary time is misleading. In fact, as vertebrates possess only four glutamate receptor types (i.e., AMPA, NMDA, kainate, and delta, the distribution of which in the human brain can be made visible by receptor autoradiography; Zilles and Amunts 2010; Amunts and Zilles 2015; see also Figure 20.1), the number of receptor types can be considerably larger in invertebrates. For example, in *Aplysia* (sea slugs), there are 14 different glutamatergic receptor types,

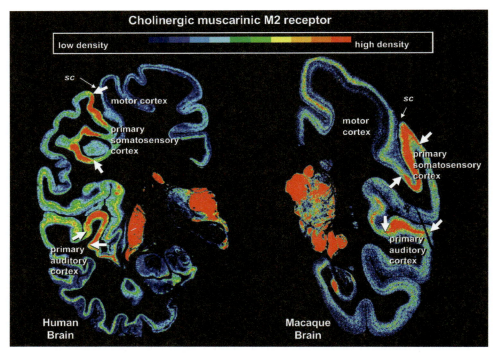

Figure 20.1 Receptor distributions of the muscarinic M2 receptors (labeled with [^{3}H]oxotremorine-M) for acetylcholine in coronal sections of a (left) human and a (right) macaque brain hemisphere. The densities of the receptors are color-coded from blue (low) to red (high). The receptor densities differ between brain areas, with high values in the primary somatosensory and auditory cortices and low values in the motor cortex in both species. Also note differences in the densities between cortical layers. cs = central sulcus. Image adapted from Zilles and Amunts (2010). Reprinted with permission. Image courtesy of Dr. Katrin Amunts.

which interact with neuropeptide co-transmitters in highly complex and insufficiently understood ways (Moroz et al. 2021). This clearly suggests that some receptor types were not preserved in the evolution of vertebrates. Glutamate is now regarded as the major excitatory neurotransmitter in vertebrates, and there is evidence to suggest that the bioenergetic properties of glutamate explain why this molecule prevailed in the competition with other biomodulators in becoming so central to neurotransmission in large and energetically costly brains (Moroz et al. 2021). Glutamate is also intricately linked to brain disease and deserves special attention in a chapter about evolutionary insights in neuropsychopharmacology.

Glutamate in Brain Health and Disease

A Very Short Story of Human Brain Evolution

The human brain has considerably increased in size, compared to the brain sizes of our closest extant relatives, the great apes. With its 1.300 or so ccm, it is about three times larger than expected for a primate of human body weight. The neocortex is also about three times more convoluted, even though the absolute number of neurons is not larger than in other large-brained animals such as whales or elephants (Herculano-Houzel et al. 2014). The most likely explanation for why such a large organ could evolve resides in the complexity of *Homo sapiens*' social environments (Dunbar and Shultz 2007).

A large human brain grows in a large braincase. According to a widespread hypothesis, this fact, together with bipedalism, led to the evolutionary necessity to antedate parturition. Indeed, compared to other great apes, human birth seems to be considerably anticipated. Put differently, a human baby as mature as a chimpanzee newborn would leave the womb after 22 months of gestation, way too long to allow for natural birth-giving. Aside from the "obstetrical dilemma" hypothesis (recently updated in Häusler et al. 2021), however, there may be energetical factors contributing to antedating human birth; and these factors may, in part, relate to the brain's hunger for glutamate (Campbell 2010).

Energy-Relevant Facts about the Human Brain

The adult human brain is energetically costly, consuming about 20% of the total energy ingestion, but is only 2% of the human body weight. Increases in brain size over evolutionary time have, in part, been constrained by energy supply (Herculano-Houzel 2011). A number of adaptations have evolved in humans that maintained the provisioning of energy. First is the digestion of food that is particularly rich in calories and protein, while, at the same time, the size of another energy-expensive organ decreased, namely the intestines (Aiello and Wheeler 1995). The preprocessing and cooking of food (Furness and Bravo (2015) created the term "cucinivore" for this evolutionary and cultural development) was crucial to human brain evolution. The consumption of a raw diet such as fruit or leaves, typical for great apes, could never energetically maintain a human-size brain. Second, the basal metabolic rate of humans is much higher than that of any other hominoid species. This is unusual because high basal metabolic rates are mainly found in small animals that reproduce early with large litter sizes, at the expense of longevity. In contrast, life expectancy is higher in humans compared to other apes, while our reproduction rate is also higher than in other apes. Interestingly, humans have more body fat compared to other great ape species, which may serve as an energy storage for the brain (Pontzer et al. 2016). Finally, the width of the carotid canal has also increased over evolutionary time in relation to brain size (Braga and Hublin 1998). Accordingly, the perfusion rate of the brain via the carotid arteries has disproportionately increased. This observation sheds additional light on the brain's energy consumption because the lumen radius of the internal carotid artery has been shown to be a reliable estimate of cerebral blood flow and metabolic rate (Seymour et al. 2016).

Developmentally, a large proportion of energy for the developing human fetus comes from glutamine and glutamate, which is produced in the fetal liver and partially recycled in the placenta. Fetal growth restriction is often due to placental dysfunction associated with reduced activity of the glutamine and glutamate transporter (McIntyre et al. 2020). During normal pregnancy, maternal metabolic rate doubles at 6 months of gestation and does not increase further, even in populations where mothers do physical work during pregnancy (Dunsworth et al. 2012). In other words, aside from anatomical constraints (i.e., an evolved trade-off between maternal pelvis size and fetal head circumference), it could well be that the fetal energy demand requires the antedating of human birth as it seems the maternal body cannot provide sufficient energy beyond 40 weeks of gestation (Campbell 2010; Dunsworth et al. 2012). This interpretation is indirectly further supported by the fact that postnatally human milk contains much higher levels of glutamate than the milk of other primates, suggesting that the human baby continues to be hungry for glutamate (Campbell 2010).

Why is the selection of glutamate apparently so ideal for the brain? Notably the brain as an organ requires disproportionally large quantities of energy, even "at rest." That is, cognitive task performance adds very little to the basal energy consumption of the brain. About three-quarters to 80% is used for the production of action potentials and postsynaptic activity, whereas the synthesis of glutamine and the uptake of glutamate consume only 5% (reviewed in Moroz et al. 2021). This is surprisingly "thrifty" for the production of a neurotransmitter. In addition, as Moroz et al. (2021) point out, one ATP molecule is sufficient for the uptake of one glutamate molecule. Conversely, the oxidation of one glutamate molecule can produce 24 to 27 ATP molecules. Moroz et al. (2021) further calculate that one cortex neuron, with its up to 17,500 synaptic contacts can produce, in theory, up to 20 billion ATP molecules, even if only 10% of the neuron's synapses released glutamate. In addition, the presence of glutamate may conserve energy by reducing glucose oxidation by up to 75%. This makes sense from a bioenergetic point of view; therefore, glutamate outcompeted other neurotransmitters to become the main excitatory neurotransmitter in the vertebrate brain (Moroz et al. 2021). This evolutionary success story comes at a cost, however; and this is dealt with in the next section.

The Role of Glutamate in Brain Disease

Extracellular glutamate is highly toxic. Given its abundance in the brain, there are a number of metabolic strategies which have evolved to buffer the brain against the potential excitotoxicity of glutamate. Some astrocytes synthesize lactate, which can ameliorate the excitotoxic effects of glutamate. Protoplasmic astrocytes protect the extrasynaptic space from a "spillover" of excess glutamate by converting it into glutamine. The brain also possesses a specialized lymphatic system (often referred to as "glymphatic" because astrocytes play a central role), through which toxic substances are eliminated from the brain. The glymphatic system is particularly active during sleep, while it is down-regulated during wakefulness. However, the clearance of toxic metabolic products may be compromised by several pathological conditions, including inflammation, degeneration, stroke, and traumatic brain injury. In addition, aging reduces the production of cerebrospinal fluid and is often associated with poor quality of sleep (thus impairing the functioning of the glymphatic system), which negatively influences the brain's capacity to eliminate toxic metabolic products (reviewed in Brüne 2019).

In health, excess glutamate is adequately removed from the extracellular space by the action of excitatory amino acid transporters. In young infants, glutamate transport into astrocytes is underdeveloped, which may account for the higher risks of epileptic seizures early in life, particularly when associated with fever (Campbell 2010). Generally, oxidative stress and inflammatory processes can exceed the capacity of excitatory acid transporters. Oligodendrocytes and neurons are particularly vulnerable to the toxic effects of glutamate. Oligodendrocyte damage may also result from an overstimulation of AMPA and kainate receptors, which is believed to be part of the pathophysiology of demyelinating diseases, whereas toxic effects from NMDA receptors on oligodendrocytes are possibly involved in the pathogenesis of mood disorders (Haroon et al. 2017).

Both aberrant metabotropic and ionotropic glutamate receptor functioning may be involved in a variety of degenerative brain disorders. Ionotropic glutamate receptor dysfunction has been described in Alzheimer's disease, whereas differential expression of metabotropic glutamate receptors on vulnerable motor neurons may play a role in amyotrophic lateral sclerosis. NMDA and AMPA overactivation seems to selectively affect vulnerable neurons in the striatum in Huntington's disease (Fu et al. 2018).

Finally, the glutamatergic system appears crucially involved in diseases such as neurodegeneration and stroke as well as in some psychiatric disorders. It may be difficult to disentangle its primary or secondary involvement (Fu et al. 2018). Aging is a risk factor for many of these illnesses, possibly because the metabolic pathways contributing to the clearance of excess glutamate become more evident only with increasing age. Such design flaws may escape purifying selection[2] because after peak reproductive age selection processes become less powerful.

Future Directions

This chapter has explored the long evolutionary history of biomolecules now known as "neurotransmitters" (see Box 20.1 for key evolutionary messages). These substances were utilized and/or synthesized by simple life forms for chemotaxis and primitive ways of cell communication. In multicellular organisms, many of these molecules were co-opted to signal injury or were toxic to other organisms.

When the first neuronal networks evolved, neurotransmitters were co-opted for communication between neurons. The term "co-optation" is somewhat unfortunate because it suggests that something in nature was predesigned for later purposes, which has a tinge of goal-directedness, which never occurs in evolution by natural selection The substances that were selected to be neurotransmitters or biomodulators in synaptic communication may have adopted this function because they also served as donors of carbon, nitrogen, and other reactive ions providing energy to the cell. Glutamate, which

> **BOX 20.1 Evolution, Neurotransmitters, Phylogenetics, Function, and Phenotypes**
>
> - The molecular basis of neurotransmitters has been highly conserved over evolutionary history.
> - A relatively small number of molecules have been recycled by evolution and have been found across kingdoms, phyla, etc.
> - While the molecules have been highly conserved, their phenotypic effects have, by and large, not been, the exception being effects at the cellular level.
> - It may be that natural selection conserved certain molecules because of their efficacy as messengers or chemical signalers through the evolution of receptors. Whether some molecules lend themselves more readily to engage with receptors compared to others or the evolution of receptors is the limiting process remains an open question. It appears that once a receptor evolves it is conserved by selection at all costs. This may be the most difficult and expensive step for selection to achieve.
> - While the chemical messengers (neurotransmitters/biomodulators) are highly conserved, the content of the messages/signals is not. These can alter radically from one species to the next.
> - Hence, few generalizations can be made regarding the function or phenotypic effects of a given neurotransmitter molecule across species.
> - Therefore, the evolutionary study of neurotransmitters should start with phenotypic effect(s) or function in a given species and proceed downward to the molecular level, not the other way around. As it is the phenotypic effect/end product that is the target of selection, importantly, this end product/function can be multiply realized at the molecular level (i.e., the same effect at the level of the phenotype) through diverse neurotransmitter routes and configurations across species and even within the same species.

existed before the emergence of life on earth, is the index example in this respect (Moroz et al. 2021).

As CNSs and brains became more complex and compartmentalized, neurotransmitter receptor systems diversified, often by gene duplication. However, the evolution of neurotransmitter receptors was not a linear process from "primitive" to "complex." Receptors arose, duplicated, mutated, and disappeared in different evolutionary selection processes in different phyla, through independent events. Consequently, some systems may be actually even more diverse or complex in arthropods (insects) than in vertebrate species. In human brain evolution, evidence suggests that positive selection was involved in the evolution of acetylcholine, dopamine, GABA, oxytocin, and glutamate systems, especially the receptors and their genes, which resulted in improved behavioral flexibility, though possibly at the cost of increased risk for neuropsychiatric disease (Ryan and Grant 2009).

Furthermore in order to reduce the potentially toxic effects of neurotransmitters, evolution has resulted in the selection of protective systems for organisms that posess brains, giving rise to a range of metabolic and detoxifaction mechanisms. One such system is the blood–brain barrier, which probably evolved multiple times in different species (Dunton et al. 2021). Another mechanism concerns the rapid reuptake of neurotransmitters in presynaptic vesicles. A third is enzymatic degradation. There is evidence for positive selection processes being involved in the evolution of both genes coding for reuptake transporters and genes encoding catabolizing enzymes such as MAO-A in humans, which may be associated with heightened risk for neuropsychiatric disease in combination with adverse life events (Ryan and Grant 2009).

Under normative physiological conditions, the detoxification of neurotransmitters (biomodulators) works sufficiently well. However, there are circumstances that highlight the deficiencies in the systems and can cause failure of the detoxification mechanisms. This includes problems arising from infectious disease and inflammation, very young age (immaturity of mechanisms) and old age (senescence of mechanisms), neurodegeneration (the predisposing genes of which often escape natural selection) and stroke, and exotoxins or arising from externally synthesized chemicals (e.g., the inadvertent or intentional consumption of toxins). Alkaloids such as caffeine, nicotine, and arecoline, for example, act on the acetylcholinergic neurotransmission and may improve cognitive functioning at low dose, with detrimental, even lethal outcomes when consumed in higher dosages. Other substances such as alcohol, opiates, or cocaine "usurp" the brain's reward system and can cause untold damage in the biological as well as psychological and social domains.

Interestingly, "pharmacophagy" of some toxins in limited amounts for various reasons is widespread in nature. As regards our own species, paleoanthropological studies have demonstrated both pharmacophagy and geophagy in our ancestors and in our cousins, the Neanderthals (Huffman 1997; St. John-Smith et al. 2013).

An important message conveyed by evolutionary perspectives on neuropsychopharmacology includes the observation that biomodulators are not specific to CNSs or brains. They are abundant in peripheral organs, which makes it extremely difficult to target brain circuits selectively for the treatment of neuropsychiatric disease, without undesirable effects in the "periphery." This is amply demonstrated by observations that most psychoactive drugs have cardiac, respiratory, or intestinal side effects. Similarly, dopamine receptors are highly expressed in immunocompetent cells and are involved in the expression of antigens, T-cell activation, and inflammatory processes. Dopamine seems to play a role in the gut–brain axis, yet the role of the intestines (and microbiota) in the pathogenesis of Parkinson's disease has only just begun to be uncovered (Franco et al. 2021).

Ultimately, we have an opportunity to creatively utilize the insights from an evolutionary perspective on neuropsychopharmacology for the discovery of new drugs. Many psychoactive substances have plant origins. Carefully exploring the range of natural bioactive compounds from plant and animal sources and even other kingdoms has great promise regarding the development of medical products. Even microorganisms provide a rich source of bioactive molecules. Castaneda et al. (2021) have suggested that Cyanobacteria are excellent candidates for the discovery of new drugs for the treatment of neurodegenerative diseases such as Alzheimer's disease. Cyanobacteria is an evolutionarily old phylum of photosynthesizing bacteria that probably contributed to the oxygenation of the earth's atmosphere. Cyanobacteria evolved in extreme adverse conditions; they can tolerate very high temperatures and high salinity. This group of bacteria is of great interest to biotechnology because Cyanobacteria synthesize exopolysaccharides against desiccation, trehalose against freezing, and substances that act as sunscreens or anti-inflammatory and antioxidant agents. Among other substances, Cyanobacteria produce tasiamide B, which has the potential to inhibit beta-secretase, an enzyme that is considered central in the pathogenesis of Alzheimer's disease (Castaneda et al. 2021). This brief example may illustrate the enormous potential of natural compounds for the treatment of medical conditions. It will be catastrophic if, among other reasons, humankind would deprive itself of these opportunities by ongoing habitat destruction and impoverishment of biodiversity. After all, *Homo sapiens* needs nature, but nature does not need us.

Key Points to Remember

- **Neurotransmitters are evolutionarily ancient molecules that originally served a range of biological functions, including chemotaxis, defense against pathogens, and hormonal activity.**
- **Neurotransmitters, better termed "biomodulators," are older than the first nervous systems and were metabolically utilized by single-cell organisms long before the first synapse evolved.**
- **The main excitatory neurotransmitter in the brain is glutamate. The evolutionary history of this molecule is particularly interesting from a neuropsychiatric point of view because glutamate is not only indispensable for normal brain development but paradoxically also potentially highly toxic.**
- **Mammalian, including human, brains possess several evolved strategies that buffer against the toxic effects of glutamate. If these mechanisms are overstretched, the resulting excess glutamate plays a significant role in the pathophysiology of brain diseases, including neurodegeneration, stroke, and affective disorders.**
- **An evolutionary framework may inform research into the search for new drugs.**

Notes

1. Note that the term "drug" has a double meaning in medicine and embraces therapeutic substances as well as molecules used for recreational purposes and abuse. In this chapter, we use the term "drug" or "drugs" only in its therapeutic meaning.
2. Purifying selection removes deleterious variants, whereas positive selection fixes beneficial variants in the population and promotes the emergence of new phenotypes.

References

Aiello, L. C., & Wheeler, P. (1995). The expensive-tissue hypothesis—The brain and the digestive-system in human and primate evolution. *Current Anthropology, 36*(2), 199–221.

Amunts, K., & Zilles, K. (2015). Architectonic mapping of the human brain beyond Brodmann. *Neuron, 88*, 1086–1107. https://doi.org/10.1016/j.neuron.2015.12.001

Azmitia, E. C. (2010). Evolution of serotonin: Sunlight to suicide. In C. P. Müller & B. L. Jacobs (Eds.), *Handbook of the behavioral neurobiology of serotonin* (pp. 3–22). Academic Press.

Braga, J., & Hublin, J. J. (1998). What do carotid canals tell us about human brain evolution? *American Journal of Biological Anthropology, 105*(S26), 112.

Brüne, M. (2016). On the role of oxytocin in borderline personality disorder. *British Journal of Clinical Psychology, 55*(3), 287–304. https://doi.org/10.1111/bjc.12100

Brüne, M. (2019). Brain, spinal cord, and sensory systems. In M. Brüne & W. Schiefenhövel (Eds.), *The Oxford handbook of evolutionary medicine* (pp. 739–814). Oxford University Press.

Campbell, B. C. (2010). Human biology, energetics, and the human brain. In M. P. Muehlenbein (Ed.), *Human evolutionary biology* (pp. 425–438). Cambridge University Press.

Carter, C. S. (2014). Oxytocin pathways and the evolution of human behavior. *Annual Review of Psychology, 65*, 10.1–10.23. https://doi.org/10.1146/annurev-psych-010213-115110

Castaneda, A., Ferraz, R., Vieira, M., Cardoso, I., Vasconcelos, V., & Martins, R. (2021). Bridging cyanobacteria to neurodegenerative diseases: A new potential source of bioactive compounds against Alzheimer's disease. *Marine Drugs, 19*(6), Article 343. https://doi.org/10.3390/md19060343

Changeux, J. P. (2005). Genes, brains, and culture: From monkey to human. In S. Dehaene, J. R. Duhamel, M. Hauser, & G. Rizzolatti (Eds.), *From monkey brain to human brain* (pp. 73–94). MIT Press.

Davis, K. L, Charney, D., Coyle, J. T., & Nemeroff, C. (Eds.). (2002). *Neuropsychopharmacology—The fifth generation of progress*. Lippincott, Williams and Wilkins.

Dean, B. (2009). Evolution of the human CNS cholinergic system: Has this resulted in the emergence of psychiatric disease? *Australian and New Zealand Journal of Psychiatry, 43*(11), 1016–1028. https://doi.org/10.3109/00048670903270431

Domes, G., Heinrichs, M., Michel, A., Berger, C., & Herpertz, S. C. (2007). Oxytocin improves "mind-reading" in humans. *Biological Psychiatry, 61*(6), 731–733. https://doi.org/10.1016/j.biopsych.2006.07.015

Donaldson, Z. R., & Young, L. J. (2008). Oxytocin, vasopressin, and the neurogenetics of sociality. *Science, 322*(5903), 900–904. https://doi.org/10.1126/science.1158668

Dunbar, R. I., & Shultz, S. (2007). Evolution in the social brain. *Science, 317*(5843), 1344–1347. https://doi.org/10.1126/science.1145463

Dunsworth, H. M., Warrener, A. G., Deacon, T., Ellison, P. T., & Pontzer, H. (2012). Metabolic hypothesis for human altriciality. *Proceedings of the National Academy of Sciences of the United States of America, 109*(38), 15212–15216. https://doi.org/10.1073/pnas.1205282109

Dunton, A. D., Göpel, T., Ho, D. H., & Burggren, W. (2021). Form and function of the vertebrate and invertebrate blood–brain barriers. *International Journal of Molecular Sciences, 22*(22), Article 12111. https://doi.org/10.3390/ijms222212111

Edgar, N., & Sibille, E. (2012). A putative functional role for oligodendrocytes in mood regulation. *Translational Psychiatry, 2*(5), Article e109. https://doi.org/10.1038/tp.2012.34

Emes, R. D., & Grant, S. G. N. (2012). Evolution of synapse complexity and diversity. *Annual Review of Neuroscience, 35*, 111–131. https://doi.org/10.1146/annurev-neuro-062111-150433

Faissner, A., Pyka, M., Geissler, M., Sobik, T., Frischknecht, R., Gundelfinger, E. D., & Seidenbecher, C. (2010). Contributions of astrocytes to synapse formation and maturation—Potential functions of the perisynaptic extracellular matrix. *Brain Research Reviews, 63*(1–2), 26–38. https://doi.org/10.1016/j.brainresrev.2010.01.001

Feldman, R. (2012). Oxytocin and social affiliation in humans. *Hormones and Behavior, 61*(3), 380–391. https://doi.org/10.1016/j.yhbeh.2012.01.008

Fields, R. D., Araque, A., Johansen-Berg, H., Lim, S. S., Lynch, G., Nave, K. A., Nedergaard, M., Perez, R., Sejnowski, T., & Wake, H. (2014). Glial biology in learning and cognition. *The Neuroscientist, 20*(5), 426–431. https://doi.org/10.1177/1073858413504465

Franco, R., Reyes-Resina, I., & Navarro, G. (2021). Dopamine in health and disease: Much more than a neurotransmitter. *Biomedicines, 9*(2), Article 109. https://doi.org/10.3390/biomedicines9020109

Fu, H., Hardy, J., & Duff, K. E. (2018). Selective vulnerability in neurodegenerative diseases. *Nature Neuroscience, 21*, 1350–1358. https://doi.org/10.1038/s41593-018-0221-2

Furness, J. B., & Bravo, D. M. (2015). Humans as cucinivores: Comparisons with other species. *Journal of Comparative Physiology, 185*(8), 825–834. https://doi.org/10.1007/s00360-015-0919-3

Guyton, A. C., & Hall, J. E. (2006). *Textbook of medical physiology.* Elsevier.

Haroon, E., Miller, A. H., & Sanacora, G. (2017). Inflammation, glutamate, and glia: A trio of trouble in mood disorders. *Neuropsychopharmacology, 42*(1), 193–215. https://doi.org/10.1038/npp.2016.199

Häusler, M., Grunstra, N. D. S., Martin, R. D., Krenn, V. A., Fornai, C., & Webb, N. M. (2021). The obstetrical dilemma hypothesis: There's life in the old dog yet. *Biological Reviews of the Cambridge Philosophical Society, 96*(5), 2031–2057. https://doi.org/10.1111/brv.12744

Herculano-Houzel, S. (2011). Scaling of brain metabolism with a fixed energy budget per neuron: Implications for neuronal activity, plasticity, and evolution. *PLOS ONE, 6*(3), Article e17514. https://doi.org/10.1371/journal.pone.0017514

Herculano-Houzel, S., Avelino-de-Souza, K., Neves, K., Porfírio, J., Messeder, D., Mattos Feijó, L., Maldonado, J., & Manger, P. R. (2014). The elephant brain in numbers. *Frontiers in Neuroanatomy, 8*, Article 46. https://doi.org/10.3389/fnana.2014.00046

Hofman, M. A. (2001). Evolution and complexity of the human brain: Some organizing principles. In G. Roth & M. F. Wullimann (Eds.), *Brain evolution and cognition.* Wiley.

Huffman, M. A. (1997). Current evidence for self-medication in primates: A multidisciplinary perspective. *American Journal of Physical Anthropology, 104*(S25), 171–200. https://doi.org/10.1002/(SICI)1096-8644(1997)25+<171::AID-AJPA7>3.0.CO;2-7

Le Duc, D., & Schöneberg, T. (2019). Cellular signaling systems. In M. Brüne & W. Schiefenhövel (Eds.), *The Oxford handbook of evolutionary medicine* (pp. 45–75). Oxford University Press.

Maeda, N. (2015). Proteoglycans and neuronal migration in the cerebral cortex during development and disease. *Frontiers in Neuroscience, 9*, Article 98. https://doi.org/10.3389/fnins.2015.00098

Mayr, E. (1982). *The growth of biological thought: Diversity, evolution, and inheritance.* Belknap Press of Harvard University Press.

McIntyre, K. R., Vincent, K. M. M., Hayward, C. E., Li, X., Sibley, C. P., Desforges, M., Greenwood, S. L., & Dilworth, M. R. (2020). Human placental uptake of glutamine and glutamate is reduced in fetal growth restriction. *Scientific Reports, 10*(1), Article 16197. https://doi.org/10.1038/s41598-020-72930-7

Moncrieff, J., Cooper, R. E., Stockmann, T., Amendola, S., Hengartner, M. P., & Horowitz, M. A. (2023). The serotonin theory of depression: A systematic umbrella review of the evidence. *Molecular Psychiatry, 28*, 3243–3256. https://doi.org/10.1038/s41380-022-01661-0

Moroz, L. L. (2009). On the independent origins of complex brains and neurons. *Brain, Behavior and Evolution, 74*(3), 177–190. https://doi.org/10.1159/000258665

Moroz, L. L., & Kohn, A. B. (2016). Independent origins of neurons and synapses: Insights from ctenophores. *Philosophical Transactions of the Royal Society B: Biological Sciences, 371*(1685), Article 20150041. https://doi.org/10.1098/rstb.2015.0041

Moroz, L. L., Nikitin, M. A., Poličar, P. G., Kohn, A. B., & Romanova, D. Y. (2021). Evolution of glutamatergic signaling and synapses. *Neuropharmacology, 199*, Article 108740. https://doi.org/10.1016/j.neuropharm.2021.108740

Nedergaard, M., Ransom, B., & Goldman, S. A. (2003). New roles for astrocytes: Redefining the functional architecture of the brain. *Trends in Neurosciences, 26*(10), 523–530. https://doi.org/10.1016/j.tins.2003.08.008

Nesse, R. M. (2019). *Good reasons for bad feelings: Insights from the frontiers of evolutionary psychiatry*. Penguin Books.

Opazo, J. C., Zavala, K., Miranda-Rottmann, S., & Araya, R. (2018). Evolution of dopamine receptors: Phylogenetic evidence suggests a later origin of the DRD2l and DRD4rs dopamine receptor gene lineages. *PeerJ, 6*, Article e4593. https://doi.org/10.7717/peerj.4593

Ortells, M. O., & Lunt, G. G. (1995). Evolutionary history of the ligand-gated ion-channel superfamily of receptors. *Trends in Neurosciences, 18*(3), 121–127. https://doi.org/10.1016/0166-2236(95)93887-4

Pontzer, H., Brown, M. H., Raichlen, D. A., Dunsworth, H., Hare, B., Walker, K., Luke, A., Dugas, L. R., Durazo-Arvizu, R., Schoeller, D., Plange-Rhule, J., Bovet, P., Forrester, T. E., Lambert, E. V., Thompson, M. E., Shumaker, R. W., & Ross, S. R. (2016). Metabolic acceleration and the evolution of human brain size and life history. *Nature, 533*(7603), 390–392. https://doi.org/10.1038/nature17654

Raven, P., Johnson, G., Mason, K., Losos, J., & Duncan, T. (2017). The nervous system. In *Biology* (Chapter 42). McGraw-Hill Education.

Roshchina, V. V. (2010). Evolutionary considerations of neurotransmitters in microbial, plant, and animal cells. In M. Lyte & P. Freestone (Eds.), *Microbial endocrinology* (pp. 17–52). Springer. https://doi.org/10.1007/978-1-4419-5576-0_2

Ryan, T. J., & Grant, S. G. (2009). The origin and evolution of synapses. *Nature Reviews Neuroscience, 10*(10), 701–712. https://doi.org/10.1038/nrn2717

Sagan, L. (1967). On the origin of mitosing cells. *Journal of Theoretical Biology, 14*(3), 225–274. https://doi.org/10.1016/0022-5193(67)90079-3

Seymour, R. S., Bosiocic, V., & Snelling, E. P. (2016). Fossil skulls reveal that blood flow rate to the brain increased faster than brain volume during human evolution. *Royal Society Open Science, 3*(8), Article 160305. https://doi.org/10.1098/rsos.160305

St. John-Smith, P., McQueen, D., Edwards, L., & Schifano, F. (2013). Classical and novel psychoactive substances: Rethinking drug misuse from an evolutionary psychiatric perspective. *Human Psychopharmacology, 28*(4), 394–401. https://doi.org/10.1002/hup.2303

Tinbergen, N. (1963). On aims and methods of ethology. *Zeitschrift für Tierpsychologie, 20*(4), 410–433. https://doi.org/10.1111/j.1439-0310.1963.tb01161.x

Walker, S. C., & McGlone, F. P. (2013). The social brain: Neurobiological basis of affiliative behaviours and psychological well-being. *Neuropeptides, 47*(6), 379–393. https://doi.org/10.1016/j.npep.2013.10.008

Yamamoto, K., & Vernier, P. (2011). The evolution of dopamine systems in chordates. *Frontiers in Neuroanatomy, 5*, Article 21. https://doi.org/10.3389/fnana.2011.00021

Zilles, K. (2005). Evolution of the human brain and comparative cyto- and receptor architecture. In S. Dehaene, J. R. Duhamel, M. Hauser, & G. Rizzolatti (Eds.), *From monkey brain to human brain* (pp. 41–56). MIT Press.

Zilles, K., & Amunts, K. (2010). Centenary of Brodmann's map—Conception and fate. *Nature Reviews Neuroscience, 11*(2), 139–145.

21

Ongoing Human Evolution?

Nicole Bender, Frank Rühli, and Maciej Henneberg

Historical/Evolutionary Background

Our evolutionary lineage separated from that of great apes some 7–6 million years ago (Henneberg and Saniotis 2016).[1] In our earliest ancestors, an erect posture of the trunk and bipedal gait emerged (Galik et al. 2004), and relationships between the sexes changed from male competition for access to females to transitional monogamy and collaborative parenting (Lovejoy 2009). This latter change, possibly together with the change in diet and balancing of the head on the erect neck, resulted in the loss of large canines used in male-to-male competition, and the shortening of the jaws. The new facial morphology became conducive to production of vowels in the upper respiratory tract, leading to the beginning of speech and singing (Clark and Henneberg 2017). Erect bipedalism freed the upper limbs from locomotor activities, thus allowing their use for manipulation and transport of objects.

Cooperative social relations, biparental care for children, object manipulation, and ability to communicate with a large variety of sounds initiated the development of a series of feedbacks between human biological characteristics, behaviors including technology and social organization, and the environment. The nature of these feedbacks is positive, in that they are self-amplifying (Henneberg and Eckhardt 2022). Operation of these feedbacks increased efficiency of the central nervous system. A simple measure of brain size—cranial capacity—shows a double exponential change from the earliest human ancestors some 5 million years ago to the end of the Pleistocene, the Ice Age (De Miguel and Henneberg 2001). This increase in brain size turned out to be parallel to an increase in hominin body size (Henneberg 1998). Given the decrease in human brain size over the past 10,000 years by about 10% (i.e., one standard deviation; Henneberg 1988) and the lack of any substantial correlation between brain size and mental aptitude of modern people (Henneberg et al. 1985), it is obvious that modifications of mental abilities in the human lineage must have been facilitated not so much by anatomical changes but by alterations in neurohormonal regulation and levels and distribution of neurotransmitters (Previc 2009; Saniotis et al. 2021), rather than by increased brain size. These changes—as is always the case with the results of natural selection—were balanced against physiological requirements of the entire body (evolutionary trade-offs). As long as they performed their task of increasing reproductive fitness of individuals, they were promoted. However, they had side effects. Neurotransmitter changes, notably increased levels of pleasure-causing dopamine (Previc 2009), resulted in humans developing natural cravings for various substances that produced pleasant effects in their brains (Saniotis and Henneberg 2012) and facilitated the occurrence of schizophrenia (Brisch et al. 2014). Relatively fast evolution of brain physiology in the series of feedback loops driving human evolution is likely to have produced a number of side effects resulting in mental instability and psychiatric pathologies.

Most studies of human evolution focus on changes in the hominin lineage over the last 6 million years that led to the appearance of "our species"—*Homo sapiens*. It is tacitly assumed that, once individuals belonging to this species had emerged in the evolutionary process, their biological traits subsequently remained unchanged. This is, of course, a macroevolutionary interpretation. However, the process of macroevolution takes place through the same mechanisms that still occur from generation to generation: mutations, natural selection, assortative mating, genetic drift, and gene flow. It is thus inevitable that since the "emergence" of *H. sapiens* its various populations have undergone some changes labeled "microevolutionary," although they are actually nothing other than normal evolutionary processes that have not so far resulted in recognizable speciation. Those changes did, however, produce morphological—and probably physiological and genetic—shifts that distinguish present-day humans from the earliest populations of *H. sapiens*. Recognition and classification of species are based on expert opinion involving arbitrary boundaries, with some 23 different definitions of species in current use (Henneberg 1997). As a result, it is possible that we are now members of a species different from what paleoanthropologists described as the first representatives of *H. sapiens* living over 200,000 years ago (Vidal et al. 2022). Conversely, our species may have existed continuously from over 1 million years ago (Clark and Henneberg 2021). A suggestion now entertained in paleontology is that the epoch that started after the last glaciation—the Holocene—has ended and that we are now living in a new, different epoch—the Anthropocene (Tong et al. 2022).

Theoretical discussions of epochs and species being what they may be, during the past few centuries in which modern science emerged, changes in human biological characteristics were clearly observed. Reflecting methods and opportunities for observation, most of them are morphological changes, although some physiological and genetic changes have also been documented. All these changes altered relationships between humans and their environments, including pathogens and causative agents of non-infectious diseases. These altered relationships yielded new adaptive challenges to humans, influencing ongoing evolution.

Evolutionary Evidence Specifically Related to Human Health and Disease

Since the Late Pleistocene, about 30,000 years ago, human body size and shape have changed (Figure 21.1). Skeletal evidence indicates that body height became reduced by about 10 cm prior to 1900 (Mathers and Henneberg 1995). By the early 20th century, bodily robusticity, including muscle mass, was reduced and brain size shrank by about 100–150 ml, representing a decrease of ~10% (Henneberg 1988). Subsequently, in developed countries, there has been an increase in both body height (Henneberg 2001) and brain size (Henneberg et al. 1985), along with a striking and much-discussed increase in adiposity (Budnik and Henneberg 2017). These changes reflected altered energy balance at the transition from gathering/hunting economies to agricultural food production. This transition involved reduction of animal meat in the diet and a shift toward reliance on carbohydrates, often without adequate variety of other plant foods, resulting in mineral and vitamin deficiencies that have accelerated over recent decades. The transition to reliance on agriculture resulted in profound changes in human social organization because structured food production, unlike foraging on natural resources, required division of

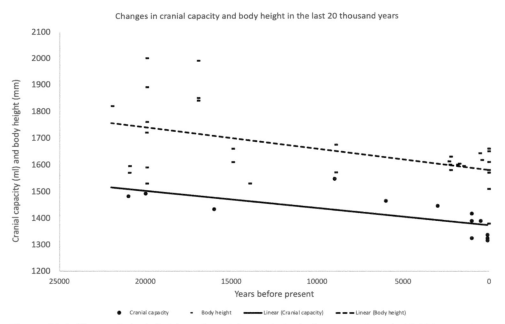

Figure 21.1 Changes in body height and cranial capacity in the last approximately 20,000 years. Data sources as provided by Saniotis et al. (2022).

labor, storage of products, their subsequent distribution, and formation of permanent settlements with large population sizes. There are also suggestions of other reasons for the decrease of brain size in the Holocene (DeSilva et al. 2021) that relate this change to alterations in the use of human brains in social systems of distributed cognition that produced the externalization of knowledge and advantaged group-level decision-making over individual decision-making. Such an explanation, based on a limited sample of relevant braincase sizes, still argues for the doubtful link between brain size and human cognition and does not relate it to ecological conditions.

Conditions of daily existence and of relationships among larger numbers of individuals, beyond a family or a local group, changed relations between pathogens and human organisms. They introduced psychological stressors and altered patterns of gene exchange since numbers of adults in local populations (i.e., effective population sizes) were increased, reducing the need to search for mates in other populations that was previously required to avoid the effects of inbreeding. Altered pressures of natural selection affected the mutation/selection balance. Concerning host–pathogen interactions, local pathogen pressure led to high frequencies of heterogenetic constellations of disease genes that, among other things, offered a partial resistance against the infection in question. Examples are high frequencies of sickle-cell gene variants in regions with high levels of malaria or high frequencies of Tay-Sachs gene variants in populations with a strong environmental pressure imposed by tuberculosis (Withrock et al. 2015). At the physiological level, it led, for example, to an increase in the number of copies of amylase genes (Rossi et al. 2021) and to an increase in the frequency of genes determining lactose tolerance among adults in populations relying on the consumption of unfermented milk (Holden and Mace 2009). Spread of pandemics in the last millennium in densely populated parts of Eurasia resulted in selection effects on immune system genes (Klunk et al. 2022).

The case of lactase persistence is, however, a good example of how complex adaptive mechanisms can be. Some years ago, it had already been observed that the overall picture of convergent evolution of high lactase persistence in areas of dairy consumption was not always consistent. It was noted, for instance, that lactase persistence frequencies were low in central Asian herders, despite high dairy consumption, while at the same time frequencies were high in certain African hunter–gatherers who did not consume milk. The relevance of the microbiome for milk digestion has been discussed in this context. Furthermore, it was observed that lactase persistence was also selected against in areas where dairy products were traditionally fermented, such that the lactose content was strongly reduced. It was therefore concluded that there might be alternative selective forces at work, other than the advantages of the energy content of the dairy products themselves (Segurel and Bon 2017).

Possible explanations were recently proposed following a large-scale archaeological investigation in which pottery fat residues from more than 550 European archaeological sites spanning a time frame of 9,000 years were analyzed. The resulting data were linked to UK Biobank data on the association between lactase persistence, milk consumption, and health indicators of 500,000 people. The results indicated that milk was already being consumed before the onset of large-scale agriculture but was probably not of major disadvantage for healthy individuals. In the case of famine or heavy pathogen load, however, lactase persistence might have been advantageous. As such unfavorable environmental conditions arose with the spread of farming, lactase persistence could have been selected for (Evershed et al. 2022).

The general phenomena producing the complex of changes described are known collectively as "cultural evolution" and "human auto-domestication" (Saniotis et al. 2022). This latter term should be used with caution (Brüne 2007). Like the domestication of animals, auto-domestication reduces individuals' ability to cope with environmental challenges, introduces metabolic deficiencies, leads to dental deficiencies, and results in an array of mental abnormalities. These changes are the consequence of domesticators' care that prevents detrimental results that could have occurred in natural environments. For instance, an individual in a state of depression would not be able to react efficiently to a threat to life or to self-access food but, given care by others, may survive with sufficient nutrition. Auto-domestication and its consequences were also discussed in connection with ancient selection for less social aggression, especially among males (Hare 2017), or in connection with the evolution of a sophisticated language communication and language-based conspiracy (Wrangham 2019). In these evolutionary frameworks, self-domestication is seen as one of the possible mechanisms leading to more peacefulness in the human species as compared to other hominid species. However, it was also hypothesized that self-domestication was involved in the development of targeted conspiratorial killing of antisocial individuals, a behavior unknown in other hominid species (Wrangham 2021).

Food production and increasing size of populations over the last few thousand years stimulated various technological advances. These were required for more comprehensive control of the environment and developments in social structures needed to manage labor forces, organization of housing, food distribution, defense against natural forces, and defense against antagonistic human groups. These developments resulted in social stratification. The complex of technological and social adaptations also included health care. Before the advent of scientific medicine related to industrialization, the effects of health care, apart from treatment of traumatic injuries and general hospitalization of some sick individuals, were limited by the lack of knowledge of bacteriology and constrained by the lack of anesthetics, surgical techniques, and availability of effective medicines. The situation changed with the introduction of antiseptics, anesthetics and chemical pharmacotherapies in the 19th century, followed by widespread effective birth control. Since that time, the operation of natural selection through premature mortality and differential fertility has become significantly reduced (Figure 21.2), altering the mutation/selection balance—a phenomenon known as "relaxed natural selection." Increased numbers of deleterious mutations, especially mildly deleterious kinds, were propagated. Paradoxically, with the decreasing mortality due to efficient medical interventions and

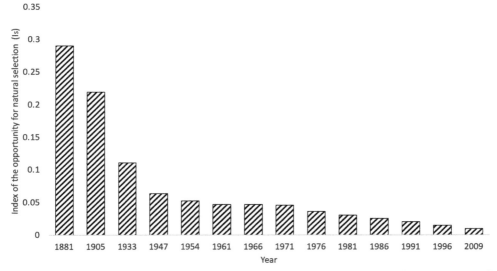

Figure 21.2 Changes in opportunities for natural selection in Australia. Details of the calculation of the index I$_s$ are provided by Stephan and Henneberg (2001) and You and Henneberg (2018).

improvements in sanitation and public health activities, incidence of various non-communicable diseases, such as diabetes and cancers, increased and required in turn more medical interventions (Rühli and Henneberg 2016; You and Henneberg 2018). Those ailments include dementia (You et al. 2022), which suggests that an altered mutation/selection balance may be affecting other functions of brains of modern humans (Keller and Miller 2006) and other aging phenomena escaping the purview of natural selection (see Chapter 8).

Counterarguments and Evidence against Evolutionary Influences

In basic scientific discourse, leaving aside any theological considerations (Chaberek and Carleial 2022), the origin of human brain diseases may be seen as not so much a result of evolutionary processes as of present-day living conditions and disease definitions that inform diagnostic criteria (Ehrenreich 2017). In this view, the brain formed by evolutionary processes is undergoing pathological changes due to the influence of previously unencountered stresses and challenges, including greater aging due to expansion of life expectancy by medical interventions, a phenomenon known as "evolutionary mismatch." Among these challenges may be stresses caused by somatic diseases, such as coronary heart disease (De Hert et al. 2018). What remains to be explained in such cases is whether the heart disease itself is a result of evolution. A widespread use of illegally trafficked psychotropic substances ("drugs") may be seen as yet another immediate, rather than evolutionary, cause of mental disorders (Rehm and Shield 2019). However, considering an argument that a human propensity for mind-altering substances may be a result of the rapid physiological evolution of the brain (Saniotis and Henneberg 2012), the link between "drug" use and mental disorders can be seen as yet another result of evolution. In a general sense, the fact that modern life has significantly altered daily relationships among individuals and between them and the environment, relative to the pre-industrial era, has presented new challenges for the workings of human brains. It responds to these challenges in ways sometimes exceeding ranges of what is considered "normal behavior" and altering baseline levels of physiological responses of the central nervous system that are an outcome of prior evolution.

Another group of biological mechanisms linking rapid responses to the environment to human biology, which are not directly evolutionary mechanisms, is summarized as "biological plasticity." This term encompasses mechanisms such as ontogenetic adaptability during growth and development (e.g., to food availability or parasite load), pre- and postpartum epigenetic mechanisms, maternal effects (e.g., hormonal changes) during pregnancy and lactation, and resulting secular trends.

These are the trends in phenotypic characters that are observed over periods of, approximately, a century (i.e., *seculum* in Latin). Most of them, especially those concerning quick increases in body size, have been seen as adaptive responses to improved living conditions (Henneberg 1999, 2001). They are regarded as important mosaic stones in the picture of ongoing evolution as plastic changes can subsequently be evolutionarily fixed by genetic changes. As with the evolutionary mismatch effects described above, plastic responses emerge out of ecologically adaptive contexts but may attain pathologic conditions in fast-changing or unstable environments. For instance, while the secular trends in human body height described above are mostly explicable through plasticity effects during growth; other phenomena—such as the ongoing and increasing obesity pandemic—are partially attributable to an epigenetically driven mismatch state (Gluckman et al. 2011).

Biological explanations of epigenetic mismatch conditions are evolutionarily based. According to these explanations, during pregnancy the fetus is epigenetically prepared for expected environmental conditions, in order to be ontogenetically optimally suited for the environment after birth. The biochemical mechanisms of this process encompass histone modifications, DNA methylations, and others. If the environment changes rapidly, faulty epigenetic preparation might occur, resulting in a mismatch to the environment. For instance, if the fetus is prepared for an environment of food scarcity through epigenetically modified physiological conditions that facilitate the accumulation of fat deposits, while the environment after birth is calorie-rich, a mismatch arises as the risk of obesity (Gluckman et al. 2011).

When traits change rapidly it is hence important to discern the mechanisms by which the modifications are implemented as changes due to plasticity do not normally involve genetic changes and are therefore more easily influenceable by therapeutic or preventive measures. Many complex traits might incorporate both genetic and plastic components in their responses to the environment. If we learn more about the underlying mechanisms involved, we can use this knowledge for improved treatments or prevention of diseases.

One of the first questions emerging in the debate about the underlying principles of human biology is often "nature or nurture?" From what has been discussed so far, we can conclude that this question is misleading as human biology is influenced both by evolutionary biology (nature) and by the environment (nurture). However, one aspect that should be stressed in more detail before moving on to the health implications of this discussion is the role of cultural evolution. Cultural evolution brings additional factors into the picture that are especially relevant to the human being because we rely most heavily on cultural capital such as language, cumulatively learned skills and knowledge, sophisticated tool use, and highly complex communication and social systems (Stanford 2020). While none of these features is per se a unique feature in the animal kingdom, their degree of complexity is particularly high in humans and allowed our species not only to populate large parts of the planet but also to dominate in many ecosystems. This development was accompanied by stunning cultural and technological advances but also recent downsides such as climate change, loss of biodiversity, and pollution of water, soil, and air.

As cultural evolution is also observed to a certain degree in other animals and especially in apes (Castro and Toro 2004), the interplay between natural selection and cultural advancement probably started very early in human evolution. Witnesses are the earliest known stone tools or signs of early fire use that predate the emergence of modern *H. sapiens* by millions of years (de la Torre 2019; see Chapter 22). This means that, especially for brain and higher cognitive evolution, it is difficult to disentangle effects of natural selection and culturally acquired skills. Furthermore, cultural practices in recent years influenced the environment at greater speeds than those occurring in changes of the genetic endowment.

Cultural evolution proceeds much more quickly than genetic evolution, because individuals can change behaviours within one lifetime, and there is no need to wait for death or reproduction for the distribution of cultural traits to change. (Stanford 2020)

Clinical and Therapeutic Implications

The evolutionary and plastic effects of the human–environmental interaction, and in particular the mismatch with the modern environment of the Anthropocene described in this chapter, are not only important aspects of ongoing human evolution but also the fundaments of many human pathologies, including both physical and psychiatric diseases. In 1994, a new field of research was established under the name "evolutionary medicine" (Williams and Nesse 1994), leading to a dedicated journal (*Evolution, Medicine & Public*

Health: https://academic.oup.com/emph) and an international society (International Society for Evolution, Medicine & Public Health: https://isemph.org/) that stages annual conferences. A body of literature exists within this field exploring the evolutionary background and interaction with human diseases and the possible ways in which therapies and prevention can be improved using this knowledge (Gluckman et al. 2009; Stearns and Medzhitov 2015; Brüne and Schiefenhövel 2019). Here, we review some of the most commonly cited examples of applications of evolutionary thinking in medicine and public health, specifically in brain diseases and mental health.

Sex Differences in Psychiatric Diseases

An interesting aspect of human biology in relation to health is sex. In recent years it has been observed that many disease symptoms differ between men and women. However, many medical professionals are not yet aware of such differences, and medical training is commonly focused largely or exclusively on male symptoms. As a result, women are underdiagnosed for many diseases, such as cardiovascular diseases and myocardial insults, which have a higher specific mortality rate in females (Mauvais-Jarvis et al. 2020). Biological sex differences apply not only to physical diseases but also to mental health. Likewise, prevalence and symptoms of many brain diseases and psychiatric disorders differ between men and women (Mauvais-Jarvis et al. 2020). For instance, autism spectrum disorders and schizophrenia are more common in men, while depression and anxiety disorder are more common in women (Brown et al. 2022). Another very common disease with a clearly higher prevalence in men is Parkinson's (Russillo et al. 2022).

Explanations for sex differences in mental health are diverse and range from sex differences in brain anatomy (DeCasien et al. 2022) to theoretical explanations relying on sexual selection, sexual hormones, and sex differences in social stress (Hagen and Rosenstrom 2016; Brown et al. 2022) to the interplay between sex and gender as modifiers of health (Mauvais-Jarvis et al. 2020). Brown et al. (2022), for instance, suggest the sexual selection–sex hormone–psychosocial stress theory to explain why women show a higher prevalence of stress-related mental health disorders, such as post-traumatic stress disorder, depression, and anxiety. In this theoretical construct, genetic, hormonal, and environmental influences are combined to explain the worldwide differences in prevalence of mental health disorders between the sexes. Specifically, sexual antagonistic behavior shaped by sexual selection modifies the propensity for mental disorders in different ways in men than in women. Similarly, sexual hormones influence health differently between the sexes, leading to more inflammation and stress propensity in women (Brown et al. 2022).

Another approach was taken by Hagen and Rosenström (2016), who correlated physical upper body strength with propensity to depression or anger in men and in women. They found that the stronger the grip strength, the smaller the odds for depression in their sample, and the difference in upper body strength was the most important explanatory factor for the sex difference in depression prevalence. The only other relevant factor modifying this relation was age (Hagen and Rosenstrom 2016).

As discussed, many sex differences in mental health prevalence can be explained by biological factors, such as genetics or hormonal influences, and are therefore evolved traits. One example is provided by differences in immunity modulated by sex hormones, leading to a generally stronger immune system in women than in men. The consequences are a higher prevalence of many autoimmune diseases in women, while men show a higher prevalence of several non-sex-specific cancers (Mauvais-Jarvis et al. 2020). One common autoimmune disease in the Northern Hemisphere is multiple sclerosis (MS). From the observation of a higher prevalence of MS in women with the peak onset at young adulthood and the observation that other hominids (apes) do not seem to spontaneously develop MS, Bove (2018) proposed a life history -based hypothesis that MS emerged in humans because of a disruption of myelin production and maintenance mechanisms during our uniquely long brain development in childhood and due to the modern westernized lifestyle.

Aging and Pleiotropy

One of the most relevant topics of ongoing human evolution in relation to health is aging and the pleiotropic effects of certain disease-risk genes. A basic principle of evolutionary medicine is that natural selection favors the dissemination of genes through reproductive success and not primarily health and longevity. This principle can explain the existence of genetic variants that offer an advantage early in life but a disadvantage at a later age, a phenomenon called "antagonistic pleiotropy." Many such genetic variants have been found in animals, where experiments and genetic manipulations are possible. In humans the evidence is predominantly correlational, and therefore less conclusive; but several diseases show a clear pattern of a possible antagonistic pleiotropy, such as a disease onset later in life, a correlation with reproductive performance, and a genetic basis with a signature of positive selection. Among these diseases are several cancers, cardiovascular disease, cystic fibrosis, Huntington's disease, and Alzheimer's disease. In the example of Alzheimer's disease, the risk gene variant is *APOE ε4*, and the potential early life benefits are increased fecundity and fertility (Byars and Voskarides 2020). Interestingly, the fitness advantage of this gene variant was found in pre-industrial and non-Western societies in environments with a high infectious disease load. It therefore seems that this variant not only is pleiotropic but also represents a mismatch in modern environments.

In this chapter it was already mentioned that in populations with a strong environmental pressure imposed by tuberculosis a high frequency of Tay-Sachs gene variants evolved (Withrock et al. 2015), suggesting another example of evolutionary pleiotropy. Interestingly, the high frequency of two recent Tay-Sachs gene variants among French Canadians in eastern Quebec was explained by convergent evolution due to natural selection for heterogeneous advantage. Tay-Sachs disease leads to an excessive storage of sphingolipids, a lipid needed for the development of neurons in the brain. While the homozygous state leads to a debilitating disease and early death, the heterozygous state allows a normal life and is even linked to an increased learning capacity. Similarly to the sickle-cell disease in malaria regions, the heterozygous state of Tay-Sachs disease can represent a selective advantage in certain environments and can therefore be selected for. In the case of the French Canadians of eastern Quebec, it was assumed that the enhanced intellectual capabilities offered a selective advantage during the time of immigration (Frost 2012). It remains an open question if the resistance against tuberculosis or the enhanced cognitive capabilities, or a combination of both, represented the primary selective advantage for Tay-Sachs genetic variants in different regions and populations.

Regional Differences in Gene Frequencies

Due to genetic drift, the founder effect, migration, or local adaptation, genetic variants can differ in frequency between populations of different regions of the world. While for many gene variants these different frequencies do not have a major impact on health, some variants can show associations with health outcomes under specific circumstances. Famous examples are the already mentioned adaptations to milk consumption and the associated selection of lactase gene variants or the copy number variants of the amylase gene associated with starch consumption. Other well-known examples are local adaptations to high altitude or regional adaptations in skin pigmentation. These examples demonstrate past evolutionary adaptations to local environmental conditions that only show health implications under specific circumstances, such as increased skin cancer risk in people of European descent living at latitudes close to the equator (Fan et al. 2016).

There are also more subtle examples that are of medical, neurological, and psychiatric relevance and that become of increasing importance with increasing worldwide migration movements.

One disease affecting the brain and showing a clear geographic pattern is MS. The Global Burden of Disease Collaboration for MS confirmed previous research that there was a strong gradient for the prevalence of MS in latitude, with a prevalence increase of 1.03 times per degree of latitude. This gradient was found in the Northern as well as in the Southern Hemisphere and seems to correlate with the socioeconomic status of the countries. Relevant for the acquisition of MS seems to be the geographical location before the onset of the disease (GBD 2016 Multiple Sclerosis Collaborators 2019).

These patterns clearly point to an environmental influence on the development of MS which seems to play a role apart from genetic predisposition. In fact, MS was discussed in the framework of the hygiene hypothesis, which states that a disruption of the microbial environment and/or the microbiome can lead to a mistuning of the immune system early in life, increasing the risk for allergies or autoimmune diseases. Such microbial disruptions are more common in Westernized countries with lower parasite loads, higher prevalence of cesarean sections, and lower rates of breastfeeding (Alexandre-Silva et al. 2018).

Unanswered Questions and Future Work

The examples discussed in the preceding sections illustrate the potential for an evolutionary application to medical and public health problems. Many more such examples could be listed, and a substantial body of literature has in fact been published on the subject (Gluckman et al. 2009; Stearns and Medzhitov 2015; Brüne and Schiefenhövel 2019). Thus far, however, the impact on medical or public health practice seems to be limited, possibly due to the theoretical nature of the field and to the fact that evolutionary biology is not taught in most medical schools. As human evolution is not only ongoing at present, but will also continue into the future, evolutionary medicine will surely remain a relevant and topical field in research, practice, and teaching also in the future. The evolutionary approach will be relevant in future, for example, to disentangle the complex interplay between the many influences on human health, such as genetics, epigenetics, early childhood environment, nutrition, physical activity, microbiota, pathogens, environmental toxins and pollutants, social interactions, etc. The evolutionary approach offers the necessary framework to analyze, compare, and interpret data and observations in a comprehensive way.

One main message is that the prevailing one-size-fits-all preventive approach adopted for most health issues has, as of yet, not generated many positive results. Evolutionary biology teaches us that individual, intergroup, and intergenerational variations are the norm rather than the exception. Accordingly, a much more differentiated approach is needed that respects in each case underlying specific causes of disease as well as environmental risk factors. Furthermore, preventive interventions should follow a life-course approach and adapt to different needs of different people at different stages in their lives. We will discuss several examples of possible future research fields in these directions applicable specifically to brain disorders.

Mental Health

Especially complex diseases and mental health are easier to understand if the evolutionary perspective is considered. For instance, variations in human personality that relate to mental health disorders, such as attention-deficit hyperactivity disorder, borderline personality, or broad autism phenotypes, could represent traits that were once selected under very specific conditions (social niches), resulting in "specialized minds" with specific advantages in their primary contexts. Once genetic predispositions of such traits reached a minimal threshold, they were able to spread through the population. As traits normally show a Gaussian distribution, this would lead to extreme cases at the edges of the distribution, with a low degree of adaptability (autism, schizophrenia, etc.). Today, differentiated personalities and specializations of behavioral traits might explain the existing variation in modern life paths (Hunt and Jaeggi 2022).

Since the early beginnings of evolutionary medicine, psychiatric diseases were described as cases of evolutionary mismatches in modern environments but with a certain potential adaptive value in the past. For instance, mild depression could have been adaptive to save energy in certain ancient circumstances, and anxiety could be a part of an ancient psychological defense mechanism (Nesse 1999). Similarly, borderline personality disorder is dysfunctional in the modern, safe environment but might have been functional in ancient, more dangerous, and unpredictable environments. In fact, borderline personality disorder shows certain similarities to a fast life history style, such as increased risk-taking behavior and impulsivity, and might therefore have been adaptive in such a context (Brüne 2016).

With such novel views on and explanations for psychiatric diseases it is hoped, on the one hand, to remove the stigma from psychiatric patients and, on the other, to help them to deal more effectively with their disease or personality disorder.

Brain Developmental Dilemma

One of the most frequently asked questions concerns the future development of the human brain and mind. Can we further increase our brain size and our intellectual capacities? In this respect humans have long been trapped in an evolutionary trade-off since the emergence of erect bipedalism and the accompanying expansion of brain size. On the one hand, the female pelvic canal must be wide enough to permit the passage of a large-brained infant at birth and yet narrow enough to enable economical bipedal locomotion. As a result, for millions of years the female pelvis has been under opposing selective pressures, resulting in a trade-off. This compromise has been given the name "obstetric dilemma" in the literature (Haeusler et al. 2021). A second player in this dilemma is the human neonate, who is born in a more dependent state compared to the other apes, with part of early brain development and expansion postponed until the postnatal stage. At the same time the human neonate exhibits large deposits of subcutaneous fat, probably to buffer nutrient and energy needs for the large and fast-developing brain. Any further evolutionary increase in human neonate size is therefore constrained by the dimensions of the adult female pelvis (Cunnane and Crawford 2003). Today, the incidence of obstructed labor is 3%–6% worldwide, resulting in an increasing number of cesarean sections. Indeed, in certain regions of the world, cesarean sections have reached 20%–50% of all births. It was calculated that if the development of medical intervention were to continue at this pace over many generations, it could disrupt the subtle balance in the obstetric dilemma trade-off, leading to an additional increase in neonatal body size or a narrowing of the female pelvis. This effect would in turn also increase the incidence of obstructed labor in the future (Mitteroecker et al. 2016).

As a solution to this dilemma, an evolutionary approach was proposed. It was assumed that midwifery is an antique invention and socially assisted birth is an antique human feature as this cultural development can be observed all over the planet. If people consequently adopt a supporting and caring environment for the delivering mothers, many of the unnecessary cesarean sections might be avoided (Rosenberg and Trevathan 2018).

Ethical Considerations

With increasing medical–technological advancement in the future, many ethical issues concerning health will be accentuated, such as questions about survival at the beginning and end of life, quality of life, optimization of health benefits versus costs, and global dissemination of resources, including in medical care. Ethical issues in medicine can be addressed only in international ethical consortia including a large number of stakeholders. There is, however, a danger inherent in applying ethical standards to human health: historically, ethical standards were heavily based on religious beliefs, and they continue to be influenced by religions, political ideologies, and cultural practices. Such foundations, though deserving their place in human culture, are not derived from objective observations of facts. They may therefore lead to detrimental health effects and, on an individual basis, deny principal human rights. One example that we discussed was the case of sex differences in brain anatomy that should avoid any sexistic notion (DeCasien et al. 2022). Another example was the term "auto-domestication" or "self-domestication," which was abused by social Darwinists in the last century (Brüne 2007). What is therefore needed is caution with the use of critical terms, application of the best available objective knowledge, and a cautious approach to medical decision-making, both for individual patients and for public health policies. As in the evolutionary process itself, involving a constantly changing web of complex feedbacks in nature, medicine requires an approach drawing upon the entire nexus of constantly changing knowledge—moderated by a profound concern for human well-being and unimpeded by rigid ideologies.

Concluding Remarks

In this chapter we have demonstrated that human evolution is not a phenomenon from the remote past with little impact on today's human health. The evolution of humans is an ongoing process, modified in its mechanisms by cultural, technological, and rapid environmental changes. In this regard it is important to understand that today human evolution follows somewhat different rules than in the past, such as the increased relevance of relaxed natural selection in modern environments, the impact of cultural evolution, the role of large migration movements, medical advances, and the like. It can be expected that human evolution will continue in the future as well and that its pace and rules will also continue to change over time.

Since human activity changes the environment, the relationship between humans and the environment becomes bidirectional. We are active ecological niche constructors and influence our own evolution. In this chapter we discussed how this feedback influence accelerated over time and led to a mismatch between the evolved human genome and the modern physical and social environment. We discussed several examples of this evolutionary mismatch and how it can increase the risk for several disease conditions, with a focus on brain and mental diseases. We therefore argued for the study of evolutionary processes in human health and disease, as well as of non-evolutionary mechanisms, such as epigenetics, secular trends, ontogenetic plasticity, the microbiome, etc., to better understand these interactions.

With several examples we demonstrated the value of the evolutionary approach to human health and disease. Because the human brain is among the most specialized organs of humans, human behaviors, psychiatric diseases, and mental health are of particular interest for the evolutionary medicine approach. In the future, evolutionary medicine might contribute further to the emergence of novel insights into therapeutic or preventive strategies for human diseases.

Key Points to Remember

- Human evolution is an ongoing process, modified in its mechanisms by cultural and rapid environmental changes. This process therefore follows slightly different rules than evolution of other animals and is expected to do so increasingly in the future.
- Humans actively change their environments and ecological niches and interfere with evolutionary processes more than other animals do.
- Cultural evolution and niche construction led to relaxed natural selection in several traits.
- These processes contribute to a mismatch between the human genome and the modern physical and social environment, increasing the risk for several disease conditions.
- The study not only of evolutionary processes but also of non-evolutionary mechanisms, such as epigenetics, secular trends, and ontogenetic plasticity, might promote the emergence of novel insights into therapeutic or preventive strategies for human diseases.
- In the future, global health challenges such as climate change, pandemics, or new zoonoses may increase. Such threats can best be faced with evolutionarily informed strategies, such as One Health, Global Health, and Planetary Health.

Acknowledgments

We thank Robert D. Martin for valuable comments on the original manuscript, which improved its final version.

Note

1. For the sake of brevity, the following description of the evolution of main human characteristics may create an impression of teleological changes toward the modern human state; however, it is made with the full understanding of the purely natural forces shaping the origins of humans.

References

Alexandre-Silva, G. M., Brito-Souza, P. A., Oliveira, A. C. S, Cerni, F. A., Zottich, U., & Pucca, M. B. (2018). The hygiene hypothesis at a glance: Early exposures, immune mechanism and novel therapies. *Acta Tropica, 188*, 16–26. https://doi.org/10.1016/j.actatropica.2018.08.032

Bove, R. M. (2018). Why monkeys do not get multiple sclerosis (spontaneously): An evolutionary approach. *Evolution, Medicine, and Public Health, 2018*(1), 43–59. https://doi.org/10.1093/emph/eoy002

Brisch, R., Saniotis, A., Wolf, R., Bielau, H., Bernstein, H. G., Steiner, J., Bogerts, B., Braun, K., Jankowski, Z., Kumaratilake, J., Henneberg, M., & Gos, T. (2014). The role of dopamine in schizophrenia from a neurobiological and evolutionary perspective: Old fashioned, but still in vogue. *Frontiers in Psychiatry, 5*, Article 47. https://doi.org/10.3389/fpsyt.2014.00047

Brown, C. M., Wong, Q., Thakur, A., Singh, K., & Singh, R. S. (2022). Origin of sex-biased mental disorders: Do males and females experience different selective regimes? *Journal of Molecular Evolution, 90*(6), 401–417. https://doi.org/10.1007/s00239-022-10072-2

Brüne, M. (2007). On human self-domestication, psychiatry, and eugenics. *Philosophy, Ethics, and Humanities in Medicine, 2*, Article 21. https://doi.org/10.1186/1747-5341-2-21

Brüne, M. (2016). Borderline personality disorder: Why "fast and furious"? *Evolution, Medicine, and Public Health, 2016*(1), 52–66. https://doi.org/10.1093/emph/eow002

Brüne, M., & Schiefenhövel, W. (2019). *Oxford handbook of evolutionary medicine*. Oxford University Press.

Budnik, A., & Henneberg, M. (2017). Worldwide increase of obesity is related to the reduced opportunity for natural selection. *PLOS ONE, 12*(1), Article e0170098. https://doi.org/10.1371/journal.pone.0170098

Byars, S. G., & Voskarides, K. (2020). Antagonistic pleiotropy in human disease. *Journal of Molecular Evolution, 88*(1), 12–25. https://doi.org/10.1007/s00239-019-09923-2

Castro, L., & Toro, M. A. (2004). The evolution of culture: From primate social learning to human culture. *Proceedings of the National Academy of Sciences of the United States of America, 101*(27), 10235–10240. https://doi.org/10.1073/pnas.0400156101

Chaberek, M., & Carleial, R. (2022). Human origins revisited: On the recognition of rationality and the antiquity of the human race. *Studia Gilsoniana, 2*, 249–287. https://doi.org/10.26385/SG.110210

Clark, G., & Henneberg, M. (2017). *Ardipithecus ramidus* and the evolution of language and singing: An early origin for hominin vocal capability. *Homo, 68*(2), 101–121. https://doi.org/10.1016/j.jchb.2017.03.001

Clark, G., & Henneberg, M. (2021). Cognitive and behavioral modernity in *Homo erectus*: Skull globularity and hominin brain evolution. *Anthropological Review, 84*(4), 485–503. https://doi.org/10.2478/anre-2021-0030

Cunnane, S. C., & Crawford, M. A. (2003). Survival of the fattest: Fat babies were the key to evolution of the large human brain. *Comparative Biochemistry and Physiology, 136*(1), 17–26. https://doi.org/10.1016/s1095-6433(03)00048-5

DeCasien, A. R., Guma, E., Liu, S., & Raznahan, A. (2022). Sex differences in the human brain: A roadmap for more careful analysis and interpretation of a biological reality. *Biology of Sex Differences, 13*(1), Article 43. https://doi.org/10.1186/s13293-022-00448-w

De Hert, M., Detraux, J., & Vancampfort, D. (2018). The intriguing relationship between coronary heart disease and mental disorders. *Dialogues in Clinical Neuroscience, 20*(1), 31–40. https://doi.org/10.31887/DCNS.2018.20.1/mdehert

de la Torre, I. (2019). Searching for the emergence of stone tool making in eastern Africa. *Proceedings of the National Academy of Sciences of the United States of America, 116*(24), 11567–11569. https://doi.org/10.1073/pnas.1906926116

De Miguel, C., & Henneberg, M. (2001). Variation in hominid brain size: How much is due to method? *Homo, 52*(1), 3–58. https://doi.org/10.1078/0018-442x-00019

DeSilva, J. M., Traniello, J. F. A., Claxton, A. G., & Fannin, L. D. (2021). When and why did human brains decrease in size? A new change-point analysis and insights from brain evolution in ants. *Frontiers in Ecology and Evolution, 9*, Article 742639. https://doi.org/10.3389/fevo.2021.742639

Ehrenreich, H. (2017). The impact of environment on abnormal behavior and mental disease: To alleviate the prevalence of mental disorders, we need to phenotype the environment for risk factors. *EMBO Reports*, *18*(5), 661–665. https://doi.org/10.15252/embr.201744197

Evershed, R. P., Davey Smith, G., Roffet-Salque, M., Timpson, A., Diekmann, Y., Lyon, M. S., Cramp, L. J. E., Casanova, E., Smyth, J., Whelton, H. L., Dunne, J., Brychova, V., Šoberl, L., Gerbault, P., Gillis, R. E., Heyd, V., Johnson, E., Kendall, I., Manning, K., . . . Thomas, M. G. (2022). Dairying, diseases and the evolution of lactase persistence in Europe. *Nature*, *608*(7922), 336–345. https://doi.org/10.1038/s41586-022-05010-7

Fan, S., Hansen, M. E., Lo, Y., & Tishkoff, S. A. (2016). Going global by adapting local: A review of recent human adaptation. *Science*, *354*(6308), 54–59. https://doi.org/10.1126/science.aaf5098

Frost, P. (2012). Tay-Sachs and French Canadians: A case of gene–culture co-evolution? *Advances in Anthropology*, *2*(3), 132–138. https://doi.org/10.4236/aa.2012.23016

Galik, K., Senut, B., Pickford, M., Gommery, D., Treil, J., Kuperavage, A. J., & Eckhardt, R. B. (2004). External and internal morphology of the BAR 1002'00 *Orrorin tugenensis* femur. *Science*, *305*(5689), 1450–1453. https://doi.org/10.1126/science.1098807

GBD 2016 Multiple Sclerosis Collaborators. (2019). Global, regional, and national burden of multiple sclerosis 1990–2016: A systematic analysis for the Global Burden of Disease Study 2016. *Lancet Neurology*, *18*(3), 269–285. https://doi.org/10.1016/S1474-4422(18)30443-5

Gluckman, P. D., Beedle, A., & Hanson, M. A. (2009). *Principles of evolutionary medicine*. Oxford University Press.

Gluckman, P. D., Hanson, M. A., & Low, F. M. (2011). The role of developmental plasticity and epigenetics in human health. *Birth Defects Research*, *93*(1), 12–18. https://doi.org/10.1002/bdrc.20198

Haeusler, M., Grunstra, N. D. S., Martin, R. D., Krenn, V. A., Fornai, C., & Webb, N. M. (2021). The obstetrical dilemma hypothesis: There's life in the old dog yet. *Biological Reviews of the Cambridge Philosophical Society*, *96*(5), 2031–2057. https://doi.org/10.1111/brv.12744

Hagen, E. H., & Rosenström, T. (2016). Explaining the sex difference in depression with a unified bargaining model of anger and depression. *Evolution, Medicine, and Public Health*, *2016*(1), 117–132. https://doi.org/10.1093/emph/eow006

Hare, B. (2017). Survival of the friendliest: *Homo sapiens* evolved via selection for prosociality. *Annual Review of Psychology*, *68*, 155–186. https://doi.org/10.1146/annurev-psych-010416-044201

Henneberg, M. (1988). Decrease of human skull size in the Holocene. *Human Biology*, *60*(3), 395–405.

Henneberg, M. (1997). The problem of species in hominid evolution. *Perspectives in Human Biology*, *3*, 21–31. https://doi.org/10.1142/9789812816603_0003

Henneberg, M. (1998). Evolution of the human brain: Is bigger better? *Clinical and Experimental Pharmacology and Physiology*, *25*(9), 745–749. https://doi.org/10.1111/j.1440-1681.1998.tb02289.x

Henneberg, M. (Ed.) (1999). Child growth, secular trends and continuing human evolution. *Perspectives in Human Biology*, *4*(2), Centre for Human Biology, University of Western Australia, Perth.

Henneberg, M. (2001). Secular trends in body height—Indicator of general improvement in living conditions or of a change in specific factors? In P. Dasgupta & R. Hauspie (Eds.), *Perspectives in human growth, development and maturation* (pp. 159–168). Kluwer Academic Publishers.

Henneberg, M., Budnik, A., Pezacka, M., & Puch, A. E. (1985). Head size, body size, and intelligence—Intraspecific correlations in *Homo sapiens sapiens*. *Homo*, *36*(4), 207–218.

Henneberg, M., & Eckhardt, R. B. (2022). Evolution of modern humans is a result of self-amplifying feedbacks beginning in the Miocene and continuing without interruption until now. *Anthropological Review*, *85*(1), 77–83. https://doi.org/10.18778/1898-6773.85.1.05

Henneberg, M., & Saniotis, A. (2016). *The dynamic human*. Bentham Science.

Holden, C., & Mace, R. (2009). Phylogenetic analysis of the evolution of lactose digestion in adults. *Human Biology*, *81*(5–6), 597–619. https://doi.org/10.3378/027.081.0609

Hunt, A. D., & Jaeggi, A. V. (2022). Specialised minds: Extending adaptive explanations of personality to the evolution of psychopathology. *Evolutionary Human Sciences*, *4*, Article e26. https://doi.org/10.1017/ehs.2022.23

Keller, M. C., & Miller, G. (2006). Resolving the paradox of common, harmful, heritable mental disorders: Which evolutionary genetic models work best? *Behavioral and Brain Sciences*, *29*(4), 385–452. https://doi.org/10.1017/S0140525X06009095

Klunk, J., Vilgalys, T. P., Demeure, C. E., Cheng, X., Shiratori, M., Madej, J., Beau, R., Elli, D., Patino, M. I., Redfern, R., DeWitte, S. N., Gamble, J. A., Boldsen, J. L., Carmichael, A., Varlik, N., Eaton, K., Grenier, J. C., Golding, G. B., Devault, A., . . . Barreiro, L. B. (2022). Evolution of immune genes is associated with the Black Death. *Nature*, *611*(7935), 312–319. https://doi.org/10.1038/s41586-022-05349-x

Lovejoy, C. O. (2009). Reexamining human origins in light of *Ardipithecus ramidus*. *Science*, *326*(5949), 74–74e8. https://doi.org/10.1126/science.1175834

Mathers, K., & Henneberg, M. (1995). Were we ever that big? Gradual increase in hominid body size over time. *Homo*, *46*, 141–173.

Mauvais-Jarvis, F., Bairey Merz, N., Barnes, P. J., Brinton, R. D., Carrero, J. J., DeMeo, D. L., De Vries, G. J., Epperson, C. N., Govindan, R., Klein, S. L., Lonardo, A., Maki, P. M., McCullough, L. D., Regitz-Zagrosek, V., Regensteiner, J. G., Rubin, J. B., Sandberg, K., & Suzuki, A. (2020). Sex and gender: Modifiers of health, disease, and medicine. *Lancet*, *396*(10250), 565–582. https://doi.org/10.1016/S0140-6736(20)31561-0

Mitteroecker, P., Huttegger, S. M., Fischer, B., & Pavlicev, M. (2016). Cliff-edge model of obstetric selection in humans. *Proceedings of the National Academy of Sciences of the United States of America*, *113*(51), 14680–14685. https://doi.org/10.1073/pnas.1612410113

Nesse, R. M. (1999). What Darwinian medicine offers psychiatry. In W. R. Trevathan, E. O. Smith, & J. J. McKenna (Eds.), Evolutionary medicine (pp. 351–375). Oxford University Press.

Previc, F. (2009). *The dopaminergic mind in human evolution and history*. Cambridge University Press.

Rehm, J., & Shield, K. D. (2019). Global burden of disease and the impact of mental and addictive disorders. *Current Psychiatry Reports*, *21*(2), Article 10. https://doi.org/10.1007/s11920-019-0997-0

Rosenberg, K. R., & Trevathan, W. R. (2018). Evolutionary perspectives on cesarean section. *Evolution, Medicine, and Public Health*, *2018*(1), 67–81. https://doi.org/10.1093/emph/eoy006

Rossi, N., Aliyev, E., Visconti, A., Akil, A. S. A., Syed, N., Aamer, W., Padmajeya, S. S., Falchi, M., & Fakhro, K. A. (2021). Ethnic-specific association of amylase gene copy number with adiposity traits in a large Middle Eastern biobank. *NPJ Genomic Medicine*, *6*(1), Article 8. https://doi.org/10.1038/s41525-021-00170-3

Rühli, F., & Henneberg, M. (2016). Biological future of humankind: Ongoing evolution and the impact of recognition of human biological variation. In M. Tibayrenc & F. J. Ayala (Eds.), *On human nature* (pp. 263–275). Academic Press.

Russillo, M. C., Andreozzi, V., Erro, R., Picillo, M., Amboni, M., Cuoco, S., Barone, P., & Pellecchia, M. T. (2022). Sex differences in Parkinson's disease: From bench to bedside. *Brain Sciences*, *12*(7), Article 917. https://doi.org/10.3390/brainsci12070917

Saniotis, A., Bednarik, R. G., & Henneberg, M. (2022). Auto-domestication hypothesis and the rise in mental disorders in modern humans. *Medical Hypotheses*, *164*, Article 110874. https://doi.org/10.1016/j.mehy.2022.110874

Saniotis, A., Grantham, J. P., Kumaratilake, J., Henneberg, M., & Mohammadi, K. (2021). Going beyond brain size: An evolutionary overview of serotonergic regulation in human higher cortical functions. *Anthropologie*, *59*(1), 101–105. https://doi.org/10.26720/anthro.20.08.10.1

Saniotis, A., & Henneberg, M. (2012). Craving for drugs is a consequence of evolution. *Anthropos*, *107*(2), 571–578. https://doi.org/10.5771/0257-9774-2012-2-571

Segurel, L., & Bon, C. (2017). On the evolution of lactase persistence in humans. *Annual Review of Genomics and Human Genetics*, *18*, 297–319. https://doi.org/10.1146/annurev-genom-091416-035340

Stanford, M. (2020). The cultural evolution of human nature. *Acta Biotheoretica*, *68*(2), 275–285. https://doi.org/10.1007/s10441-019-09367-7

Stearns, S. C., & Medzhitov, R. (2015). *Evolutionary medicine*. Oxford University Press.

Stephan, C. N., & Henneberg, M. (2001). Medicine may be reducing the human capacity to survive. *Medical Hypotheses*, *57*(5), 633–637. https://doi.org/10.1054/mehy.2001.1431

Tong, S. L., Bambrick, H., Beggs, P. J., Chen, L. M., Hu, Y. B., Ma, W. J., Steffen, W., Tan, J. G. (2022). Current and future threats to human health in the Anthropocene. *Environment International*, *158*, Article 106892. https://doi.org/10.1016/j.envint.2021.106892

Vidal, C. M., Lane, C. S., Asrat, A., Barfod, D. N., Mark, D. F., Tomlinson, E. L., Tadesse, A. Z., Yirgu, G., Deino, A., Hutchison, W., Mounier, A., & Oppenheimer, C. (2022). Age of the oldest known *Homo sapiens* from eastern Africa. *Nature*, *601*(7894), 579–583. https://doi.org/10.1038/s41586-021-04275-8

Williams, G. C., & Nesse, R. M. (1994). *Why we get sick*. Times Books.
Withrock, I. C., Anderson, S. J., Jefferson, M. A., McCormack, G. R., Mlynarczyk, G. S. A., Nakama, A., Lange, J. K., Berg, C. A., Acharya, S., Stock, M. L., Lind, M. S., Luna, K. C., Kondru, N. C., Manne, S., Patel, B. B., de la Rosa, B. M., Huang, K. P., Sharma, S., Hu, H. Z., . . . Carlson, S. A. (2015). Genetic diseases conferring resistance to infectious diseases. *Genes & Diseases*, *2*(3), 247–254. https://doi.org/10.1016/j.gendis.2015.02.008
Wrangham, R. W. (2019). Hypotheses for the evolution of reduced reactive aggression in the context of human self-domestication. *Frontiers in Psychology*, *10*, Article 1914. https://doi.org/10.3389/fpsyg.2019.01914
Wrangham, R. W. (2021). Targeted conspiratorial killing, human self-domestication and the evolution of groupishness. *Evolutionary Human Sciences*, *3*, Article e26. https://doi.org/10.1017/ehs.2021.20
You, W., & Henneberg, M. (2018). Cancer incidence increasing globally: The role of relaxed natural selection. *Evolutionary Applications*, *11*(2), 140–152. https://doi.org/10.1111/eva.12523
You, W., Henneberg, R., & Henneberg, M. (2022). Healthcare services relaxing natural selection may contribute to increase of dementia incidence. *Science Reports*, *12*(1), Article 8873. https://doi.org/10.1038/s41598-022-12678-4

22

Human Cultural Evolution Outpaces Biological Evolution
A Brain Connectomic Approach

Jean-Pierre Changeux

Introduction

Although rudimentary culture has been reported in nonhuman species (chimpanzees cracking nuts, songbirds developing local dialects, humpback whales perfecting hunting techniques), *Homo sapiens'* "culture" has developed to such an extent that it may be labeled a species-specific attribute. Right away it must be underscored—in the context of the Human Brain Project—that culture proceeds not from a single individual but from a community of individual human brains that epigenetically collect, maintain, transmit, and accumulate information in its network (Changeux et al. 2021). As defined, culture is grounded in the fact that humans form social groups. Their brains must possess connectomic features that make them social, able to establish common and reciprocal relationships. In other words, there exists for neurobiologists a "social brain" which, progressively over the past millennia, became a "cultural brain" carrying connectomic signatures of the extraordinary expansion and diversification of culture. All intervening developments are epigenetically stored in the brain (Changeux 2021) but remain strictly constrained by the human genome envelope. Adult cognition and brain architecture result from the interplay between phylogeny and genetics, on the one hand, and ontogeny and individuals' interactions with physical and social environments, on the other. Timescales of cultural evolution range from psychological time—the genesis of mental objects and their communication—to millennia—conservation of extracerebral artifacts, tools, sculptures, architectures, etc. The *Homo sapiens* genome itself is the outcome of a long Darwinian evolution—here referred to as "hominization"—which one may arbitrarily view as starting with *Australopithecus* about 3 million years ago and reaching *Homo sapiens* about 300,000 years ago (Changeux et al. 2021; Hublin and Changeux 2022). This is relatively fast in terms of paleontological time but slow compared to the recent expansion and diversification of culture as already 35,000 years ago art approached its contemporary quality with, for instance, the Chauvet cave paintings. Certainly, cultural evolution outpaces genetic evolution. In this review, the contribution of genetic versus epigenetic mechanisms that account for this timing difference shall be evaluated, in terms of genome structure and brain connectomics.

Hominization of the Brain

Brain hominization—with the genetic and connectomic aspects involved—is a fascinating topic in contemporary neuroscience (Changeux et al. 2021). One of its most evident manifestations is the dramatic increase in brain connectivity and higher functions that occurred at a paleontological pace. The simplest marker of this evolution is the increase in brain size (Hublin and Changeux 2022), starting about 4 million years ago with *Australopithecus* (400–550 ml, 30% larger than chimpanzees) and then "early *Homo*" including *H. habilis* (510–850 ml) about 2 million years ago, *H. erectus* (546–1250 ml) from 1.9 million to 125,000 years ago, and *H. neanderthalensis* (1300–1700 ml) from 650,000 to 40,000 years ago. Then, amazingly, about 300,000 years ago brain size did not fluctuate much, with *H. sapiens* staying around 1350 ml (on average). This evolution in size reflects a parallel increase in neuron number as seen in the primate lineage: 3.3 billion in the squirrel monkey brain and 6.4 billion in the macaque compared to at least 87 billion in the human brain (Herculano-Houzel, 2009). Yet the human brain is not simply a homothetically enlarged squirrel monkey brain. Considerable changes have taken place in brain organization, macroscopic as well as microscopic. These aspects of hominization have recently been reviewed, in close collaboration with Claus Hilgetag and Alexandros Goulas (Changeux et al. 2021). Following are the most critical elements, or "fundamentals," of the proposed "connectomic hypothesis."

Expansion of the Neocortex

A major feature of brain hominization is the differential expansion of the neocortex (see also Chapter 1). For instance, the surface area of a cortical hemisphere increased from ~105 cm^2 in the macaque (Van Essen et al. 2012) to 973 ± 88 cm^2 in humans. This expansion is due to enhanced proliferation of neuron precursors during the prolonged prenatal development, which takes 35 weeks in chimpanzees, 37 weeks in gorilla and orangutan, and 38 weeks in humans (Finlay and Darlington 1995). The expansion is accompanied by an increase in the number of cortical columns (Rakic 2009) and an abundant gyrification of the cerebral cortex (Rash et al. 2019). All primates exhibit a striking enlargement of the neocortex relative to total brain volume. Yet the increased surface area is not uniformly distributed throughout the cortical sheet. The "association centers"—prefrontal, temporal, and parietal cortices—undergo preferential enlargement. In contrast to unimodal sensory areas, the association cortex is marked by a complex non-canonical circuit organization (Changeux et al. 2021) that supports the parallel and re-entrant processing necessary for the development of cognitive functions. Preferential expansion of the neocortex, especially prefrontal cortices, is undoubtedly associated with "higher-order" executive, social, and cultural capabilities.

Relevant to our discussion of biology and culture, we note that the expansion of the neocortex does not simply parallel the evolution of total brain weight. All primates exhibit preferential neocortex enlargement relative to total brain volume. This "unbalanced" evolution becomes important with the increased cerebral cortical surface area in humans (Finlay and Darlington 1995). Allometric scaling profiles for each distinct brain structure reveal that the human neocortex undergoes the steepest expansion (Finlay and Darlington 1995). During hominization, expansion of the "rational" cortex and associated executive functions acquired a selective advantage over subcortical territories such as the "emotional and social" limbic system.

In the course of brain development from newborn to adult a similar evolution takes place. While the cerebral isocortical surface area expands dramatically from 7.0 to 20.9 cm^2 (the greatest change occurring in the frontal lobes), here again subcortical structures change relatively little (Sowell et al. 2002). Between ages 8 and 30, the allocortical hippocampus, thalamus, and amygdala change the least, as do basal ganglia and nucleus accumbens structures.

In brief, while the main signature of brain hominization is a striking expansion of the neocortex, an "imbalance"—sometimes qualified as a "mismatch" or "disharmony"—between cortical areas and subcortical nuclei and diverse brain territories develops during biological evolution as well as in the course of development. I shall return to this issue in the second part of this review.

Increased Number of Cortical Areas

With the expansion of the cerebral cortex, the number of specialized cortical areas augments (e.g., from 129 areas in macaque parceling [Van Essen et al. 2012] to over 180 areas per hemisphere in humans [Amunts et al. 2020]). These areas are cytoarchitectonically diverse, with individual connectional "fingerprints" of their specialized functions. This architectonic diversity is manifest in both the biochemistry and the distribution of neurotransmitter receptors (Zilles et al. 2002; Goulas et al. 2021). Receptor densities in human cerebral cortex have been found to form a natural axis with progressively increasing ratios of excitatory to inhibitory and ionotropic to metabotropic receptor densities (Goulas et al. 2021). Such diversity is of particular interest in the case of areas involved in language processing, considering the fact that multiple (over 10) brain areas may contribute to language generation and production in adults, including the left frontal lobe, left temporal/

parietal lobes, right temporal lobe, cerebellum, hippocampus, basal ganglia (Fedorenko et al. 2011), and especially the prefrontal cortex (Amunts and Zilles 2012).

Differences in microarchitecture have been identified between Broca's region in humans and the homologous areas 44 and 45 in ape and macaque brains (Zilles et al. 2002) and hypothesized—together with other anatomical features—to account for the unique human ability to produce language, an essential component of cultural evolution. Specifically, in all layers of areas 44 and 45, neuropil volume relative to cell body increases from macaque to great apes to *H. sapiens* (Zilles et al. 2002) and is also found to vary in human subjects with different verbal IQ scores (Heyer et al. 2022). Brains of subjects with higher general and verbal IQ scores show a thicker left temporal cortex (Brodmann area 21 [BA21]), due to a selective thickening of layers 2 and 3 accompanied by lower pyramidal neuron density but larger cell body size and dendrite proliferation (Heyer et al. 2022). Enhanced dendritic arborization associated with neuropil enlargement might allow for increased integration in both local and long-range cortical circuits (Changeux et al. 2021).

Another consequence of cortical expansion, noted by Changeux et al. (2021), is that the human brain network is *sparser* and favors local connections over global ones. Such increased sparsity has functional benefits, serving to separate and stabilize local pattern representations and helping create functional specialization. These features may contribute to the diversification of brain representations, of primary importance for cultural evolution.

Network sparsity at the global level is counteracted locally by connections organized into *modules*, that is, communities of nodes that have more connections within their home community than with nodes elsewhere. These modules with sparse intermodular connections may have important implications for cultural evolution. They allow locally sustained activity while simultaneously preventing global network overexcitation. Networks able to self-sustain activation patterns are a necessary precondition to the ability to maintain the dynamic representations underlying short-term or working memory such as may play a critical role in social interaction (Changeux et al. 2021).

In sum, throughout the hominization lineage over a few million years, the increased number of cortical areas, together with the appearance of sparsity and modularity, directly enhanced the representational capacity of the human brain, its diversification and epigenetic storage, and its acquisition of a wide range of patterns (Kaiser et al. 2007). These features have contributed to the emergence of human-specific cognitive and social functions such as language. The biological evolution of the human brain thus predisposes it to the genesis and evolution of culture, which itself, as we shall see, will then progress at a faster pace.

Multilevel Processing and Global Neuronal Workspace

Differentiation of multiple cortical areas together with increased sparsity and modularity resulted in a more complex connectome which (at variance with artificial intelligence developments) required some spatial reorganization. An architecture with *multiple nested levels of organization* developed, As seen in the schemes of Van Essen et al. (2012; Markov et al. 2013), which illustrate the superimposition—and nesting—of levels of organization from the primary area (representation of simple lines) to the parietotemporal cortex (representation of hands, face recognition, etc.). Thus, a serial, convergent signal processing allows the development of increasingly sophisticated representations. The larger number of hierarchical levels of organization may enable the human brain to form and process more refined, diverse, and abstract representations (Changeux et al. 2021; Volzhenin et al. 2022; Axer and Amunts 2022).

Also, at each level of organization, the neuronal composition and microscopic circuit organization underlie the capacity for particularly elaborate cognitive tasks. For example, a computational analysis reveals that the ability to successfully perform "trace conditioning" requires cooperation of inhibitory neurons, dopamine neurons, and spontaneous activity among other components (Volzhenin et al. 2022). Indeed, a fivefold hierarchical increase in the proportion of inhibitory GABA (gamma-aminobutyric acid; calretinin) neurons—in higher-order associative areas (such as the prefrontal cortex)—is found between non-human primates and humans (Dzaja et al. 2014).

In the context of cultural evolution, one may speculate that mental representations mobilizing "vertically" several superimposed levels of organization give access to new forms of semantic innovation, including, for instance, the social environment (like theory of mind) and highly structured "blending" of items, which ultimately lead to the open-ended recursive and hierarchical organization of language (Changeux et al. 2021).

At the highest level of organization, conscious processing arises. Various theories have been proposed regarding the neuronal bases of the content of conscious access (Tononi et al. 2016; Mashour et al. 2020) that differ from the mechanisms known to govern the states of consciousness (wakefulness, sleep, anesthesia; Llinas and Paré 1991). Among these theories, the "global neuronal workspace" (GNW) hypothesis (Dehaene et al. 1998; Mashour et al. 2020) offers a simple connectomic scheme based upon the contribution of neurons with long-range axons, which would form a global workspace (Baars 1988), broadcasting signals from the sensory periphery to the entire brain, thereby yielding conscious experience. The GNW hypothesis relies on the presence of a "horizontal" and reciprocally connected set of brain areas (Goldman-Rakic 1995) in monkeys, sometimes referred to as "core–periphery" or "rich club." This tightly interconnected set of areas in primates (marmoset and macaque monkeys, humans) encompasses areas of the prefrontal, temporal, and parietal (i.e., association) cortices. The core–periphery network architecture of the primate brain develops in humans to support the core long-range connectivity of the GNW (Dehaene et al.1998; Mashour et al. 2020). It was initially proposed, in agreement with von Economo and Koskinas (1925), that the pyramidal cells with long-range horizontal axons linking multiple cortical areas originate from cortical layers II–III. This view is supported by the observation of Goulas et al. (2021) that, in the course of biological evolution, a shift of cortical layer reafference from lower (V–VI) to upper (II–III) layers selectively takes place in humans (Changeux et al. 2021). Since pyramidal cells with long-range axons mostly originate from layers II–III, such a cortical layer shift further contributes to populating the GNW network connectivity (Markov et al. 2013).

The contribution of long-range connections to the brain's higher functions has been debated (Rosen and Halgren 2022). The former compared human postmortem corpus callosum long-range axon density to that of the short-range connections between neighboring areas. The number of long-axon connections is surprisingly small compared to the short-range ones, on average only about 6,000 long axons connect any two areas *in the same cortical hemisphere*. The average number of long-axon connections between areas in *different hemispheres* is even smaller, fewer than 1,500. Yet, potential mechanisms by which long-range projection signals could be amplified have been proposed, including formation of multiple/highly effective synapses and subsequent amplification of incoming signals by local cortical circuitry (Changeux et al. 2021).

In any case, this developmental evolution of the neural architectures that give access and frame conscious processing offers extraordinary opportunities for generating new forms of abstract representations. They offer the capacity to build up the open-ended recursive and hierarchical possibilities of language efficiently, thus contributing to cultural evolution on psychological timescales.

Extended Postnatal Development of the Human Brain

In addition to the prolongation, already mentioned, of prenatal development from apes to humans, a unique feature of the human brain is the extension of its postnatal development for up to 15 years (about half the life span of early *H. sapiens*) and perhaps longer, into the third decade (Petanjek et al. 2011). For rhesus monkeys, the brain at birth is 48%–68% of the adult size; for chimpanzees, 36%–46%; and for humans, only 25%–29%. Humans experience much more postnatal brain growth than chimpanzees or rhesus monkeys. The postnatal ratio of brain growth to body growth increases between chimpanzees and humans and in the latter continues postnatally at a rate similar to the prenatal one until the adult brain size is reached. Humans reach their adult brain size more slowly than other primates, thus allowing for a much longer period of "experience-expectant information storage" (Greenough et al. 2002) during which synaptic proliferation, competition, and selective stabilization can take place. During this period about half of all adult synaptic connections are formed at a very fast pace (averaging half a million per second). This long period of extreme plasticity in human children directly contributes to the formation and shaping of synaptic architecture and the development of "higher-order" cognitive functions in close interaction with the physical, social, and cultural environment. As we shall see, the germination and growth of culture are major outcomes of this extension of postnatal development.

Cultural Evolution Takes Over
Synaptic Epigenesis and the Origin of Culture

As mentioned in the introduction, culture may be viewed, in its present form, as a species-specific attribute of modern *H. sapiens*. A "cultural brain" develops, which stores the connectomic signatures of the extraordinary expansion and diversification of culture. Its origin and evolution are predicated on the long period of extreme neural plasticity of developing human children, which directly contributes to the genesis of higher brain functions. The fundamental mechanism of plasticity involved in the assembly of billions of synapses in the adult human brain relies upon the variability of developing interneuronal connections and the progressive consolidation of robust circuits by trial and error (Changeux Courrège Danchin 1973 (CCD model); Rakic 1976; Shatz and Stryker 1978; Purves and Lichtman 1980; Changeux 2021). This process formally resembles an evolutionary one by variation selection yet without major changes in genome sequence. It was therefore referred to as an *epigenesis by selective stabilization of synapses* (Changeux et al. 1973) (Figure 22.1). The CCD mechanism is radically distinct from the concept of DNA epigenetics subsequently hijacked by molecular biologists to refer to unrelated mechanisms of DNA covalent modifications such as methylation and/or chromatin remodeling that affect gene regulation without altering the DNA sequence. By formal analogy, the original concept has been extended to include "neural Darwinism" (Edelman and Mountcastle 1978), the group-selection theory of higher brain functions (Edelman 1987), and more recently the theory of symmetry breaking in space–time hierarchies, among others.

The CCD model is based on the fact that most connections in the brain do not form via precise one-to-one cell–cell recognition. At critical periods of development, for a given circuit, axons grow toward their targets in a rather random manner, their axon terminals branching exuberantly at first. This transient diversity is then reduced by selective stabilization of some contacts and elimination (or retraction and/or pruning) of others, depending on the activity state of the target neuron. Both excitatory and inhibitory synapses exist in at least three possible connective states: labile (L), stable (S), and degenerate (D). Only states L and S transmit impulses, and state transitions are limited to L→S, L→D, and S→L. A critical element of the model is that evolution of the connective state of a given synaptic contact is governed by *global activity* (i.e., by the total afferent input onto the postsynaptic soma) within the time window of the synapse's firing (evolutive power of the soma). It includes the standard Hebbian time–coincidence relationship and the popular phrase "cells that fire together wire together." Activity of the postsynaptic cell regulates, in retrograde fashion, the stabilization and/or elimination of afferent nerve endings. The associative property that results from this learning process is structurally laid down as a particular pattern of synaptic connections. In the course of development an immense cascade of such nested elementary epigenetic steps takes place until the adult state is reached. This is, of course, of fundamental importance for cultural evolution since

Growth

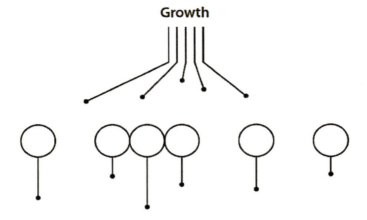

Maximum variability

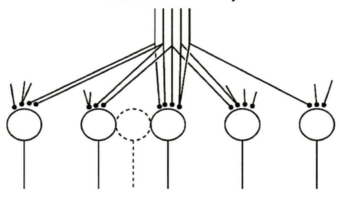

Selective stabilization

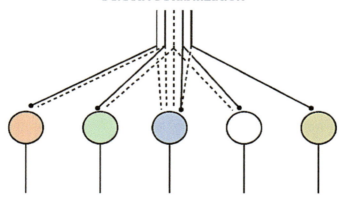

Figure 22.1 Model of the epigenesis of neuronal networks by selective stabilization of synapses as proposed by Changeux et al. (1973). Adapted with permission from Changeux (1983).

the intervening state of activity of the developing circuits is directly controlled by various social and cultural interactions occurring during postnatal development. The timescales for this "printing" process range from psychological time (0.1 sec) to seconds and days, in any case a much faster pace than biological (genetic) evolution, which takes paleontological times.

Epigenetic Variability of Neural Circuits

An unexpected but critical feature of the theory was originally stated as the *variability theorem* (Changeux et al. 1973) that "different learning inputs may produce different connective organizations and neuronal functioning abilities, but the same behavioral abilities." Thus, the neuronal connectivity code exhibits "degeneracy" (Edelman and Mountcastle 1978); that is, different code words (connection patterns) may carry the same meaning (function). One prediction of this theorem is that the synaptic connectivity of monozygotic twins may differ. This was initially demonstrated in the small invertebrate *Daphnia magna* by electron microscopy of genetically identical individuals. It was found that, while the number of cells and main categories of contacts are preserved, the exact number of synapses and precise form of axonal branches vary between pairs of identical twins (Macagno et al. 1973). Similar findings have been reported by the same team for the Müller cells of a parthenogenetic fish (*Poecilia formosa*) and confirmed in the nervous system of isogenic *Caenorhabditis*, a species, until recently, viewed as wired reproducibly, down to the individual synapse (Oren-Suissa et al. 2016). Yet, individual *Daphnia*, worms, or fish all move/swim similarly. Lastly, magnetic resonance imaging of human monozygotic twins reveals images of the planum temporale that differ between left- and right-handed twins (Steinmetz et al. 1995). Wide phenotypic variance of brain connectivity is quite general (Lu et al. 2009). It was—until recently—underestimated in the many attempts worldwide to establish high-resolution brain maps. Such attempts should now include the cultural background and actual social environment of the subject—in other words, their "habitus" (Bourdieu 2018).

Early Evidence for Synapse Selection

The abundant experimental evidence supporting the CCD model has recently been reviewed (Changeux 2021); a few examples shall suffice. The classical Hubel and Wiesel experiments on how monocular deprivation affects binocular vision in cats are consistent with the selective stabilization hypothesis (Hubel and Wiesel 1965; Rakic 1976; Shatz and Stryker 1978; LeVay et al. 1980). Occlusion of one eye causes the relevant cortical cells to lose their ability to respond, and in one week a dramatic retraction of geniculocortical arbor branches takes place (Sretavan et al. 1988). An early demonstration of active synapse stabilization accompanied by elimination of supernumerary nerve synapses was obtained on skeletal muscle motor neurons (Benoit and Changeux 1975, 1978). In adults each muscle fiber has a single motor nerve ending, while at birth several active motor nerve endings converge on a common endplate (Redfern 1970). All except one are then eliminated, the state of activity of the neuromuscular junction playing a decisive role (Benoit and Changeux 1975, 1978). Similarly, in the adult cerebellar cortex, there is a single climbing fiber synapse on each Purkinje cell. In rats this mature single innervation is preceded by a transient stage of synapse exuberance, with up to five climbing fibers per Purkinje cell (Mariani and Changeux 1980). Genetic mutations have also been shown to alter synaptic epigenesis in the mouse. Last, the innervation of sympathetic submandibular ganglion cells undergoes profound reorganization during postnatal development, leaving adult individual ganglion cells innervated by many fewer axons than at birth (Lichtman 1977). This widespread occurrence of synapse selection during development results in a significant change in the total number of cortical synapses that become very important in humans. Whereas in rats maximal synaptic number is reached a few weeks after birth, in humans it takes over three years. Moreover, rats show little loss of synapses after maximal density is reached, whereas in humans there is a steady decline until the total number stabilizes around puberty (Huttenlocher and Dabholkar 1997; Bourgeois 1997; Petanjek et al. 2011). The major decline in synapse number in humans plausibly results from a nested cascade of multiple epigenetic steps that takes place until the adult state is reached. In contemporary humans, synaptic refinement continues far beyond puberty: epigenetic learning is lifelong (Petanjek et al. 2011).

There are abundant examples supporting synapse selection in the nervous system, from the simplest to the most elaborate cognitive functions. In all cases, the network's state of activity regulates synaptic elimination. This validates the quote "to learn is to eliminate" (Changeux 1985).

A Tentative Model for the Origin and Evolution of Oral Language

The synapse selection model has been extended to the acquisition of higher brain functions, in particular the development of newborns and their interaction with the social environment, thereby contributing to the genesis and internalization of culture, as well as its transgenerational transmission (Vygotsky 1926/1992; Changeux et al. 2021). Among the many manifestations of cultural evolution, *oral language* stands out.

The origin of language is a hotly contested topic, with some languages tentatively traced back to ancestors of H. sapiens (Hublin and Changeux 2022). Early signs of the relevant cortical organization appear between 1.7 and 1.5 million years ago with "early Homo" species (in particular H. habilis and early H. erectus), after the first dispersals of Homo from Africa. They include the differential enlargement and reorganization of the frontal lobe, especially the inferior prefrontal cortex and Brodmann areas (BA) 44/45/47 (which make up Broca's area), in tandem with the posterior parietal cortex, the temporal lobe, and to some extent the occipital lobe. Such frontal lobe innovations might have constituted the foundation of the "language-ready" brain of later Homo species. They might also represent some still unknown form of communication intermediate between chimp grunts, calls, or signs and human language proper.

Along these lines, Fitch (2017), on the basis of empirical linguistic data, has delineated some elementary "derived components of language" that are unique to humans. First is phonology, the ability to acquire—epigenetically—a basic lexicon, including symbols, that maps signals in particular sounds to concepts and dramatically develops during human evolution. Early Homo species (Hublin and Changeux 2022) exhibit expansion of the cerebral cortex, enlargement and reorganization of the frontal lobe, together with the already mentioned increased sparsity and modularity, which directly enhance the epigenetic representational capacity of their brain.

Subsequent changes in the endocranial organization through paleontological time are best documented in Neanderthals and in H. sapiens (Hublin and Changeux 2022). Both lineages, though more marked in the latter, display a broadening of the frontal lobes. Moreover, H. sapiens underwent a more modest increase in brain size than Neanderthals but a significant reorganization into a more globular shape, together with a striking development of the cerebellum. A plausible connectomic backbone, which would make the brain "more globular" or compact—"linking together" these widely distributed territories of the brain—has been tentatively interpreted (Hublin and Changeux 2022) as an enhanced development of the long-range—horizontal—connectivity of the brain (Changeux et al. 2021) postulated to contribute to the GNW (Dehaene et al. 1998; Mashour et al. 2020). One may further speculate that theory of mind and high-order language processing depend on the epigenetic fine-tuning of these long-range connections and of their hierarchical organization, and thus possibly contribute to the observed difference from the Neanderthal brain. This might relate to Fitch's (2017) second important component of human language: the unique ability to produce an unlimited variety of linear signal strings that communicate complex semantic messages in a recursive and hierarchical manner, named "dendrophilia" (Fitch 2017) or "merge" (Chomsky 1957). The neuronal origin of dendrophilia is still largely controversial. Yet, one may note that between ages 2 and 4, children automatically develop syntactic rules *without explicit instruction* (Dehaene-Lambertz and Spelke 2015). But interestingly, they concomitantly develop syntax, the capacity of recursive thought, and another distinct cognitive ability referred to as "theory of mind" (Fitch 2017). Theory of mind is defined in the field as a cognitive science that investigates how we ascribe mental states to other persons and how we explain and predict the actions of those other persons using these states (Frith & Frith 2003; Miller 2006; Völlm 2006; Decety and Lamm 2007). It is often also called "mindreading" (Premack and Premack 2003; Petanjek et al. 2019) and may be viewed as a basic component of social interaction, with the actual evaluation of the relationships between individuals in the group. One may speculate that, as a cognitive operation, its "difficulty" or "complexity" is akin to that of understanding (or producing) a sentence (i.e., to assign a global meaning to an ordered string of words). It thus legitimizes the further development of the hypothesis that dendrophilia and theory of mind are both produced by the emergence of a common higher-new-level of the GNW organization in postnatal development. Indeed, around 2 years of age, characteristic changes in the postnatal maturation of pyramidal projection neurons from the prefrontal cortex have been shown to take place (Petanjek et al. 2019; Changeux et al. 2021). It happens at the upper layers—specifically layer IIIc—in the human prefrontal cortex where pyramidal neurons exhibit, between 16 months and 2.5 years, a unique differential increase in the number of segments and length of their basal dendrites (contrasting with deep layer V projection neurons). Furthermore, differential epigenetic elimination (pruning) of supernumerary dendritic spines is most pronounced and protracted in layer IIIc neurons (Petanjek et al. 2011), especially in the prefrontal cortex (Petanjek et al. 2019). Also, the local axonal collaterals of layer IIIc are in control of the prefrontal cortico-cortical output, while their long projections modulating interareal processing. They are the major integrative element of cortical processing and regulate global cortical—GNW—functional organization. This new level of conscious processing, as mentioned in the preceding sections, would then become a major integrative element of cortical processing fine-tuned via epigenetic processing. It would in particular confer mental-tree reading ability, which would mobilize in a concomitant top-down and bottom-up manner the multiple levels of GNW organization from the prefrontal cortex. The outcome of the hypothesis would be that thanks to such vertical and horizontal interconnectivity and their relationship, a recursive- or center-embedding, global cortical synthesis takes place and is at the origin of language dendrophilia. The hypothesis, of course, remains to be demonstrated.

Lastly, Fitch (2017) has called "glossogeny" the human-specific proclivity to communicate socially, which is of special importance for the epigenetic diversification and perpetuation of sociocultural aspects of language. It is associated with the ability to share culturally acquired knowledge with close kin during the prolonged postnatal synaptic epigenesis through teaching or "pedagogy" (Premack and Premack 2003). The epigenetic evolution of culture needs relearning with each new generation (without requiring any change in the genome). This transmission is not always faithful, and alterations may contribute to the diversification of cultures, especially languages.

During the past roughly 2 million years, most of the basic neurobiological steps—genetic and epigenetic—underpinning human language acquisition and formation have progressively occurred, dramatically enhancing cultural evolution (Fitch 2017; Changeux et al. 2021). As a result, oral languages have been generated. Research on their identity and distribution has revealed substantial changes over time. According to Greenberg (1963), highly diverse Nilo-Saharan languages originated 200,000–20,000 years ago in the Sahara during the "wet period" of subpluvial Neolithic times. Population genetics further suggests that the earliest precursors of Dravidian languages have been spoken in southwest Iran between 15,000 and 10,000 years ago before much later spreading to India. The Eastern Sudanic group of Nilo-Saharan languages may have merged around

7,000 years ago. Since then, a broad diversity of languages has exploded throughout the world (Cavalli-Sforza 1997) at a pace much faster than genetic evolution.

Epigenetic Signature of Writing and Reading

Writing and reading can be traced back to abstract cave drawings, dated around 30,000 BCE. Clay counting tokens are known from Mesopotamia (9,000 BCE), and the first pictograms from Ur are from around 4,000 BCE. Writing and reading are recent cultural inventions that evolved—at a cultural-historical pace—into distinct subsystems and may be viewed as a typical signature of epigenetically laid down "cultural circuits." Historically, the first evidence for acquired brain circuits underpinning writing and reading was the early discovery, in 1914, by French neurologist Joseph Jules Dejerine of pure alexia, also known as alexia without agraphia. Individuals with pure alexia have severe reading problems, while other language-related skills are typically intact. Alexia results from lesions in the supramarginal and angular gyri. An acquired set of new connections has thus been selectively stabilized during written language learning. Recent brain imaging studies on illiterate versus literate subjects by Castro-Caldas et al. (1998) confirm Dejerine's pioneering work. Comparison of positron-emission tomography (PET) scans from illiterate and literate groups shows considerably more activation, for the literate group, in the right frontal opercular–anterior insular region, the left anterior cingulate, the left lentiform nucleus, and the anterior thalamus/hypothalamus (Castro-Caldas et al. 1998). These conclusions have been strengthened and further expanded by functional magnetic resonance imaging (fMRI) scans from illiterate and literate groups (Dehaene et al. 2010). Acquisition of reading and writing takes place at a time of still very rapid synaptogenesis and persists into adulthood (Dehaene et al. 2010). It may be viewed as an example of epigenetically laid-down "cultural circuits." The neural mechanism involved in laying down the epigenetic neural signature of reading and writing is still debated. The evolution of reading and face recognition during the first three years of reading acquisition has been followed by fMRI in three groups of French children: 6-year-olds, pre- and beginning readers, and 9-year-old advanced readers (Feng et al. 2022). The results showed that specific responses to written words are absent prior to reading but appear in beginning readers irrespective of age. Crucially, the sectors of ventral visual cortex that become specialized for words and faces harbored their own functional connectivity prior to actual reading acquisition, which would argue (according to Feng et al. 2022) in favor of a hypothetical "recycling" of already differentiated territories. Yet these authors overlook the fact that new sets of connections become established during this period and might then be selectively stabilized without any need to recycle existing circuits. The selective stabilization hypothesis is not excluded by these low-resolution data, but its demonstration requires new quantitative investigations at the synaptic level.

The brain interprets Western alphabetic writing using circuits that differ in part from those used for Chinese ideography. Japanese writing uses two kinds of signs, one phonetic and combinatory (Kana) and the other ideographic (Kanji), borrowed from Chinese writing. In Kanji, each individual written sign has a meaning. Following Dejerine's tradition, Japanese neurologist T. Imura, in 1943 (Yamadori 2019), described an aphasic syndrome called Gogi's aphasia. The main characteristic is a selective alteration of Kanji processing. Lesions of Broca's or Wernicke's area in the left hemisphere profoundly affect the use of Kana, less so of Kanji. Brief presentation of Kana characters and Kanji signs to the two eyes independently suggests left hemisphere superiority for Kana and right hemisphere superiority for Kanji, especially when it comes to nouns. The results suggest that Kana and Kanji are processed differently by the right and left hemispheres.

A similar type of synaptic epigenesis mechanism may possibly account for abundant acquired sociocultural elements still not examined in a "neuro-cultural" context, ranging from language diversity, symbolic and social representations, and institutions.

The connectivity of the adult human brain might then be viewed as a complex tangle of cognitive, social, and cultural circuitry, epigenetically stabilized during development, nevertheless limited within the species-specific envelope of the human genome.

Conclusions

The question of the non-linear evolution of the complexity of the human brain connectome versus the apparent "simplicity" of the genome has been recently reviewed and is still challenging (Changeux et al. 2021). Many of the genes involved in the brain's hominization are part of its genetic envelope already functional in the primate lineage. Few, if any, are unique to *H. sapiens*; and there appears to be no single "intelligence," "math," or "art" gene! The most likely innovation underlying hominization is differences in gene regulation. Indeed, gene duplications (Samonte and Eichler 2002) have been shown to occur in the human lineage, some exclusively in humans. Also, non-transcribed regulatory sequences may control entire sections of the connectome in ways proper to humans. The connectomic hypothesis (Changeux et al. 2021) seeks to define a minimal set of principles that accounts for the uniquely human brain architecture. Its goal is to characterize larger connectomic "fundamentals" that may serve as intermediate between the genome and higher cognitive functions. As mentioned, the "single structural genes" current approach may not suffice to account for brain hominization (Changeux et al. 2021). The hypothesis is that these connectomic fundamental are under the regulation of wider patterns of genes—or a gene community—themselves determined by a small number of genetic regulatory events. Accordingly, "gene networks would encode neuronal networks," (Changeux et al. 2021) which ultimately determine the fundamental connectomic organization of the human brain (Changeux et al. 2021; Hublin and Changeux 2022).

The pace of genetic changes involved in the brain's hominization is that of paleontological time, while cultural evolution occurs several orders of magnitude faster. Examples of the latter are the invention of writing thousands of years ago, closer to the use of printing, and especially the new informatic means due to the advent of computers. The pace of standard cultural evolution is limited by the direct contribution of the human brain and the so-called psychological (fractional second) time. Recall that our brain operates at the speed of sound, while computers process information at the speed of light, a million times faster. This creates an additional boost in the pace of cultural evolution. Most importantly, many "cultural inventions" took place in the course of human cultural evolution in addition to those mentioned, including ethical regulation and artistic activity (which have been discussed elsewhere [Changeux and L'Yvonnet 2023]). It has been argued (see above Expansion of the neocortex) that, during hominization, expansion of the "rational" cortex and associated executive functions acquired a selective advantage over subcortical territories such as the "emotional" limbic system. As a consequence, highly cognitive processes such as science and technologies outpaced cultural evolution to the detriment of more "emotional" disciplines such as art and ethics. An imbalance—sometimes qualified as "disharmony" or "mismatch"—between cortical and non-cortical territories—develops during biological evolution as well as in the course of development. Cultural evolution would have then introduced "epigenetic rules" (Evers and Changeux 2016)—ethical and artistic—aiming, from the start, to overcome this disharmony and, for instance, prevent the self-destruction of the species and the ruin of its environment. Unambiguously, after two world wars, Hiroshima, Afghanistan, Ukraine—the list is long—we have to conclude that "peace is not hardwired in Homo sapiens brain" but has to be epigenetically reinvented constantly at the world scale. The recurrent confrontation with the diversity of cultures (and the relevant epigenetic traces, languages, territories, beliefs, etc.) challenges it over and over. Institutions have to be invented at the world scale to urgently enforce the universal ethical rules needed (Changeux and L'Yvonnet 2023).

Key Points to Remember

- The non-linear evolution of genomic complexity versus brain connectivity is hypothesized to originate from a unique relationship between genetic events and characteristic large-scale network organization of the brain referred to as "connectomic fundamentals," thus the statement "gene networks encode neuronal networks."
- Brain evolution is mostly under genetic control, while cultural evolution largely relies upon non-genetic selection of developing synapses during postnatal development.
- The pace of the genetic changes involved in the hominization of the brain is much slower (millions to tens of thousands of years) than that of cultural evolution (hundreds of milliseconds to years), which, as a consequence, outpaces biological evolution.
- The pace of cultural diversification and evolution is limited by the direct contribution of the human brain and the so-called psychological times of hundreds of milliseconds, by its intergenerational transmission, and by the storage of extracerebral memories.
- An important interindividual variability of brain connectomics results from the essential contribution of synaptic epigenesis in the course of postnatal development.

Acknowledgments

The author gratefully acknowledges Claus Hilgetag for discussions and Paul De Weere for the careful editing of the manuscript and the European Union's Horizon 2020 Framework Program for Research and Innovation under Specific Grant Agreement No. 945539 (Human Brain Project SGA3).

References

Amunts, K., Mohlberg, H., Bludau, S., & Zilles, K. (2020). Julich-Brain: A 3D probabilistic atlas of the human brain's cytoarchitecture. *Science, 369*(6506), 988–992. https://doi.org/10.1126/science.abb4588

Amunts, K., & Zilles, K. (2012). Architecture and organizational principles of Broca's region. *Trends in Cognitive Sciences, 16*(8), 418–426. https://doi.org/10.1016/j.tics.2012.06.005

Axer, M., & Amunts, K. (2022). Scale matter: The nested human connectome. *Science, 378*, 500–504. https://doi.org/10.1126/science.abq2599

Baars, B. J. (1988). *A cognitive theory of consciousness*. Cambridge University Press.

Benoit, P., & Changeux, J. P. (1975). Consequences of tenotomy on the evolution of multiinnervation in developing rat soleus muscle. *Brain Research, 99*, 354–358. https://doi.org/10.1016/0006-8993(75)90036-0

Benoit, P., & Changeux, J. P. (1978). Consequences of blocking the nerve with a local anesthetic on the evolution of multiinnervation at the regenerating neuromuscular junction of the rat. *Brain Research, 149*(1), 89–96. https://doi.org/10.1016/0006-8993(78)90589-9

Bourdieu, P. (2018). Structures, habitus, practices. In J. Faubion (Ed.), *Rethinking the subject* (pp. 31–45). Routledge. https://doi.org/10.4324/9780429497643

Bourgeois, J. P. (1997). Synaptogenesis, heterochrony, and epigenesis in the mammalian neocortex. *Acta Paediatrica, 422*, 27–33. https://doi.org/10.1111/j.1651-2227.1997.tb18340.x

Castro-Caldas, A., Petersson, K. M., Reis, A., Stone-Elander, S., & Ingvar, M. (1998). The illiterate brain. Learning to read and write during childhood influences the functional organization of the adult brain. *Brain, 121*(Pt. 6), 1053–1063. https://doi.org/10.1093/brain/121.6.1053

Cavalli-Sforza, L. L. (1997). Genes, peoples, and languages. *Proceedings of the National Academy of Sciences of the United States of America, 94*(15), 7719–7724. https://doi.org/10.1073/pnas.94.15.7719

Changeux, J. P. (1983). *L'homme neuronal*. Fayard.

Changeux, J. P. (1985). *Neuronal man: The biology of mind*. Pantheon Books.

Changeux, J. P. (2021). Epigenesis, synapse selection, cultural imprints, and brain development: From molecules to cognition. In Houdé & G. Borst (Eds.), *The Cambridge handbook of cognitive development* (pp. 27–49). Cambridge University Press.

Changeux, J. P., Courrege, P., & Danchin, A. (1973). A theory of the epigenesis of neuronal networks by selective stabilization of synapses. *Proceedings of the National Academy of Sciences of the United States of America, 70*, 2974–2978. https://doi.org/10.1073/pnas.70.10.2974

Changeux, J. P., Goulas, A., & Hilgetag, C. C. (2021). A connectomic hypothesis for the hominization of the brain. *Cerebral Cortex, 31*(5), 2425–2449. https://doi.org/10.1093/cercor/bhaa365

Changeux, J. P., & L'Yvonnet, F. (2023). *Le beau et la splendeur du vrai*. Albin Michel.

Chomsky, N. (1957). *Syntactic structures*. Gruyter Mouton.

Decety, J., & Lamm, C. (2007). The role of the right temporoparietal junction in social interaction: How low-level computational processes contribute to meta-cognition. *The Neuroscientist, 13*(6), 580–593.

Dehaene, S., Kersberg, M., & Changeux, J. P. (1998). A neuronal model of a global workspace in effortful cognitive tasks. *Proceedings of the National Academy of Sciences of the United States of America, 95*, 14529–14534. https://doi.org/10.1073/pnas.95.24.14529

Dehaene, S., Pegado, F., Braga, L. W., Ventura, P., Filho, G. N., Jobert, A., Dehaene-Lambertz, G., Kolinsky, R., Morais, J., & Cohen, L. (2010). How learning to read changes the cortical networks for vision and language. *Science, 330*, 1359–1364. https://doi.org/10.1126/science.1194140

Dehaene-Lambertz, G., & Spelke, E. S. (2015). The infancy of the human brain. *Neuron, 88*, 93–109. https://doi.org/10.1016/j.neuron.2015.09.026

Dejerine, J. (1914). *Sémiologie des affections du système nerveux*. Masson.

Dzaja, D., Hladnik, A., Bičanić, I., Baković, M., & Petanjek, Z. (2014). Neo-cortical calretinin neurons in primates: Increase in proportion and microcircuitry structure. *Frontiers in Neuroanatomy, 8*, Article 103. https://doi.org/10.3389/fnana.2014.00103

Edelman, G. M. (1987). *Neural Darwinism: The theory of neuronal group selection*. Basic Books.

Edelman, G. M., & Mountcastle, V. B. (1978). Group selection and phasic reentrant signaling: A theory of higher brain function. In *The mindful brain: Cortical organization and the group-selective theory of higher brain function* (pp. 51–98). MIT Press.

Evers, K., & Changeux, J. P. (2016). Proactive epigenesis and ethical innovation: A neuronal hypothesis for the genesis of ethical rules. *EMBO Reports, 17*(10), 1361–1364. https://doi.org/10.15252/embr.201642783

Fedorenko, E., Behr, M. K., & Kanwisher, N. (2011). Functional specificity for high-level linguistic processing in the human brain. *Proceedings of the National Academy of Sciences of the United States of America, 108*, 16428–16433. https://doi.org/10.1073/pnas.1112937108

Feng, X., Monzalvo, K., Dehaene, S., & Dehaene-Lambertz, G. (2022). Evolution of reading and face circuits during the first three years of reading acquisition. *NeuroImage, 259*, Article 119394. https://doi.org/10.1016/j.neuroimage.2022.119394

Finlay, B., & Darlington, R. (1995). Linked regularities in the development and evolution of mammalian brains. *Science, 268*, 1578–1584. https://doi.org/10.1126/science.7777856

Fitch, W. T. (2017). Empirical approaches to the study of language evolution. *Psychonomic Bulletin & Review, 24*, 3–33. https://doi.org/10.3758/s13423-017-1236-5

Frith, U., & Frith, C. D. (2003). Development and neurophysiology of mentalizing. *Philosophical Transactions of the Royal Society B: Biological Sciences, 358*(1431), 459–473.

Goldman-Rakic, P. S. (1995). Cellular basis of working memory. *Neuron, 14*(3), 477–485. https://doi.org/10.1016/0896-6273(95)90304-6

Goulas, A., Changeux, J. P., Wagstyl, K., Amunts, K., Palomero-Gallagher, N., & Hilgetag, C. C. (2021). The natural axis of transmitter receptor distribution in the human cerebral cortex. *Proceedings of the National Academy of Sciences of the United States of America, 118*(3), Article e2020574118. https://doi.org/10.1073/pnas.2020574118

Greenberg, J. H. (1963). The languages of Africa. *International Journal of American Linguistics, 29*(1), 1–171.

Greenough, W. T., Black, J. E., & Wallace, C. S. (2002). Experience and brain development. In M. H. Johnson, Y. Munakata, & R. O. Gilmore (Eds.), *Brain development and cognition* (pp. 539–559). Blackwell Publishing.

Herculano-Houzel, S. (2009). The human brain in numbers: A linearly scaled-up primate brain. *Frontiers in Human Neuroscience, 3*(31), 1–11. https://doi.org/10.3389/neuro.09.031.2009

Heyer, D. B., Wilbers, R., Galakhova, A. A., Hartsema, E., Braak, S., Hunt, S., Verhoog, M. B., Muijtjens, M. L., Mertens, E. J., Idema, S., & Baayen, J. C. (2022). Verbal and general IQ associate with supragranular layer thickness and cell properties of the left temporal cortex. *Cerebral Cortex, 32*(11), 2343–2357. https://doi.org/10.1093/cercor/bhab330

Hubel, D. H., & Wiesel, T. N. (1965). Receptive fields and functional architecture in two nonstriate visual areas (18 and 19) of the cat. *Journal of Neurophysiology, 28*(2), 229–289. https://doi.org/10.1152/jn.1965.28.2.229

Hublin, J. J., & Changeux, J. P. (2022). Paleoanthropology of cognition: An overview on hominins brain evolution. *Comptes Rendus: Biologies, 345*(2), 57–75. https://doi.org/10.5802/crbiol.92

Huttenlocher, P., & Dabholkar, A. (1997). Regional differences in synaptogenesis in human cerebral cortex. *Journal of Comparative Neurology, 387*, 167–178. https://doi.org/10.1002/(sici)1096-9861(19971020)387:2<167::aid-cne1>3.0.co;2-z

Kaiser, M., Görner, M., & Hilgetag, C. (2007). Criticality of spreading dynamics in hierarchical cluster networks without inhibition. *New Journal of Physics, 9*, Article 110. https://doi.org/10.1088/1367-2630/9/5/110

LeVay, S., Wiesel, T., & Hubel, D. (1980). The development of ocular dominance columns in normal and visually deprived monkeys. *Journal of Comparative Neurology, 191*, 1–51. https://doi.org/10.1002/cne.901910102

Lichtman, J. (1977). The reorganization of synaptic connexions in the rat submandibular ganglion during post-natal development. *Journal of Physiology, 273*(1), 155–177. https://doi.org/10.1113/jphysiol.1977.sp012087

Llinás, R. R., & Paré, D. (1991). Of dreaming and wakefulness. *Neuroscience, 44*(3), 521–535. https://doi.org/10.1016/0306-4522(91)90075-y

Lu, J., Tapia, J. C., White, O. L., & Lichtman, J. W. (2009). The interscutularis muscle connectome. *PLOS Biology, 7*(2), Article e1000108. https://doi.org/10.1371/journal.pbio.1000032

Macagno, E. R., Lopresti, V., & Levinthal, C. (1973). Structure and development of neuronal connections in isogenic organisms: Variations and similarities in the optic system of *Daphnia magna*. *Proceedings of the National Academy of Sciences of the United States of America, 70*(1), 57–61. https://doi.org/10.1073/pnas.70.1.57

Mariani, J., & Changeux, J. P. (1980). Multiple innervations of Purkinje cells by climbing fibers in the cerebellum of the adult staggerer mutant mouse. *Journal of Neurobiology, 11*(1), 41–50. https://doi.org/10.1002/neu.480110106

Markov, N. T., Ercsey-Ravasz, M., Van Essen, D. C., Knoblauch, K., Toroczkai, Z., & Kennedy, H. (2013). Cortical high-density counter stream architectures. *Science, 342*(6158), Article 1238406. https://doi.org/10.1126/science.1238406

Mashour, G. A., Roelfsema, P., Changeux, J. P., & Dehaene, S. (2020). Conscious processing and the global neuronal workspace hypothesis. *Neuron, 105*, 776–798. https://doi.org/10.1016/j.neuron.2020.01.026

Miller, C. (2006). Developmental relationships between language and theory of mind. *American Journal of Speech-Language Pathology, 15*(2), 142–154.

Oren-Suissa, M., Bayer, E., & Hobert, O. (2016). Sex-specific pruning of neuronal synapses in *Caenorhabditis elegans*. *Nature, 533*, 206–211. https://doi.org/10.1038/nature17977

Petanjek, Z., Judaš, M., Šimic, G., Rašin, M. R., Uylings, H. B. M., Rakic, P., & Kostovic, I. (2011). Extraordinary neoteny of synaptic spines in the human prefrontal cortex. *Proceedings of the National Academy of Sciences of the United States of America, 108*(32), 13281–13286. https://doi.org/10.1073/pnas.1105108108

Petanjek, Z., Sedmak, D., Džaja, D., Hladnik, A., Rašin, M. R., & Jovanov-Milosevic, N. (2019). The protracted maturation of associative layer IIIC pyramidal neurons in the human prefrontal cortex during childhood: A major role in cognitive development and selective alteration in autism. *Frontiers in Psychiatry, 10*, Article 122. https://doi.org/10.3389/fpsyt.2019.00122

Premack, D., & Premack, A. (2003). *Original intelligence: Unlocking the mystery of who we are*. McGraw-Hill.

Purves, D., & Lichtman, J. W. (1980). Elimination of synapses in the developing nervous system. *Science, 210*(4466), 153–157. https://doi.org/10.1126/science.7414326

Rakic, P. (1976). Prenatal genesis of connections subserving ocular dominance in the rhesus monkey. *Nature, 261*(5560), 467–471. https://doi.org/10.1038/261467a0

Rakic, P. (2009). Evolution of the neocortex: A perspective from developmental biology. *Nature Reviews Neuroscience, 10*(10), 724–735.

Rash, B. G., Duque, A., Morozov, Y. M., Arellano, J. I., Micali, N., & Rakic, P. (2019). Gliogenesis in the outer subventricular zone promotes enlargement and gyrification of the primate cerebrum. *Proceedings of the National Academy of Sciences of the United States of America, 116*(14), 7089–7094. https://doi.org/10.1073/pnas.1822169116

Redfern, P. (1970). Neuromuscular transmission in new-born rats. *Journal of Physiology, 209*(3), 701–709. https://doi.org/10.1113/jphysiol.1970.sp009187

Rosen, B. Q., & Halgren, E. (2022). An estimation of the absolute number of axons indicates that human cortical areas are sparsely connected. *PLOS Biology, 20*(3), Article e3001575. https://doi.org/10.1371/journal.pbio.3001575

Samonte, R. V., & Eichler, E. E. (2002). Segmental duplications and the evolution of the primate genome. *Nature Reviews Genetics, 3*(1), 65–72. https://doi.org/10.1038/nrg705

Shatz, C. J., & Stryker, M. P. (1978). Ocular dominance in layer IV of the cat's visual cortex and the effects of monocular deprivation. *Journal of Physiology, 281*, 267–283. https://doi.org/10.1113/jphysiol.1978.sp012421

Sowell, E. R., Trauner, D. A., Gamst, A., & Jernigan, T. L. (2002). Development of cortical and subcortical brain structures in childhood and adolescence: A structural MRI study. *Developmental Medicine and Child Neurology, 44*(1), 4–16. https://doi.org/10.1017/s0012162201001591

Sretavan, W., Shatz, C. J., & Stryker, M. P. (1988). Modification of retinal ganglion cell morphology by prenatal infusion of tetrodotoxin. *Nature, 336*(6198), 468–471. https://doi.org/10.1038/336468a0

Steinmetz, H., Herzog, A., Schlaug, G., Huang, Y., & Jäncke, L. (1995). Brain (a)symmetry in monozygotic twins. *Cerebral Cortex, 5*(4), 296–300. https://doi.org/10.1093/cercor/5.4.296

Tononi, G., Boly, M., Massimini, M., & Koch, C. (2016). Integrated information theory: From consciousness to its physical substrate. *Nature Reviews Neuroscience, 17*(7), 450–461. https://doi.org/10.1038/nrn.2016.44

Van Essen, D. C., Glasser, M. F., Dierker, D. L., & Harwell, J. (2012). Cortical parcellations of the macaque monkey analyzed on surface-based atlases. *Cerebral Cortex, 22*, 2227–2240. https://doi.org/10.1093/cercor/bhr290

Völlm, B. A. (2006). Neuronal correlates of theory of mind and empathy: A functional magnetic resonance imaging study in a nonverbal task. *NeuroImage, 29*(1), 90–98.

Volzhenin, K., Changeux, J. P., & Dumas, G. (2022). Multilevel development of cognitive abilities in an artificial neural network. *Proceedings of the National Academy of Sciences of the United States of America, 119*(39), Article e2201304119. https://doi.org/10.1073/pnas.2201304119

von Economo, C., & Koskinas, G. N. (1925). *Die zytoarchitektonik der hirnrinde des erwachsenen menschen*. Springer-Verlag.

Vygotsky, L. (1992). *Educational psychology*. Saint Lucie Press. (Original work published 1926)

Yamadori A. (2019). Gogi (word meaning) aphasia and its relation with semantic dementia. *Frontiers of Neurology and Neuroscience, 44*, 30–38.

Zilles, K., Palomero-Gallagher, N., Grefkes, C., Scheperjans, F., Boy, C., Amunts, K., & Schleicher, A. (2002). Architectonics of the human cerebral cortex and transmitter receptor fingerprints: Reconciling functional neuroanatomy and neurochemistry. *European Neuropsychopharmacology, 12*(6), 587–599. https://doi.org/10.1016/s0924-977x(02)00108-6

23

Concluding Remarks and Future Directions

Martin Brüne, Nico J. Diederich,
Christopher G. Goetz, and Katrin Amunts

The human brain is an extraordinary organ. It is considered to be one of the most complex biological structures that evolved by natural and sexual selection. Given that the first primitive nervous systems emerged to maintain homoeostasis of the organism, the contemporary human brain's evolution has involved an increasingly large, convoluted, and nutritionally "expensive" design that not only keeps threats to metabolic equilibrium at bay but permits active exploration and manipulation of the environment (Allman 1999). In fact, this evolutionary journey has lasted billions of years, with no end in sight.

With respect to the human brain, it is still not entirely clear which driving forces have been most relevant in giving the brain its particular shape and functional capacities. The most plausible scenarios include selection pressures from the social environment (i.e., "social brain hypothesis"; Brothers, 1990) but also tool use and foresight (i.e., "mental time travel"; Suddendorf and Corballis 1997; "predictive brain"; Brown and Brüne 2012), all of which were important for survival and reproductive fitness. While these factors in combination sound like a straightforward evolutionary success story, we should not overlook the fact that, anatomically, modern humans have been more than once on the brink of extinction (Shultz & Maslin 2013), something that is admittedly difficult to imagine, in light of the current 8 billion people on the planet.

The human brain not only yields the cognitive and behavioral capacities to manipulate the environment of its carrier but has also evolved mechanisms to manipulate other organs of the body to its own advantage through neuronal and humoral (endocrinological) pathways. For example, in times of food shortage, the brain manages to extract the resources that it requires to maintain its function, at the expense of tissue loss of other organs. Only in very rare circumstances does the brain digest its own tissue, as is the case, for instance, in severe anorexia nervosa, but only after all other organs have been exploited (Hitze et al. 2010).

A brain of the human size is difficult to maintain nutritionally because it largely lacks the capacity to store energy. The human brain runs on 20 to 30 W, which seems not much but is in line with estimates of approximately 20% of the body's energy production of around 100 W at rest. In relation to body weight, however, the brain's energy consumption is fairly high, as the adult human brain makes up only 2% of the total body weight (Magistretti and Allaman 2015). Notably, at birth the human brain is barely myelinated and continues to grow postnatally at the same pace as before birth for another 13 months (Jones et al. 1992). Moreover, myelination of the human brain is accomplished as late as early adulthood (e.g., Yakovlev and Lecours 1967), which together renders the human brain vulnerable to insults during a greatly expanded growth period and functional maturation.

The present multi-authored volume covers a lot of ground in regard of healthy brain function and mechanisms of breakdown from an evolutionary point of view. We recognize that we have focused on a select group of prototypic illnesses. For example, our book lacks chapters on mechanical trauma (concussion), epilepsy, acute and chronic inflammatory diseases such as meningitis and encephalitis, as well as multiple sclerosis. Stroke and other diseases associated with a dysfunction of the brain's vasculature have also not been addressed. However, these diseases are seen in other mammals as well or can be experimentally modeled. This is not the case for most of the diseases discussed in this volume, their being psychiatric or neurological. Similarly, in the domain of psychiatry the coverage of topics is incomplete. We view this volume as a compendium of relevant and archetypal examples of the role of evolutionary science in a modern discussion of health and brain disease, but we admittedly had to limit ourselves. These neglected issues may be addressed in future editions. Our major premise stressed throughout the book, even with our restricted examples, is that the brain, even though meticulously adapted to the environmental conditions of our evolutionary past, is vulnerable to dysfunction and disease. This vulnerability is, in part, related to "modern" lifestyles, which include (in industrialized countries) an oversupply of energy, lack of physical activity, and heightened exposure to stress, foremost social stress. This particular form of stress is fostered by the sheer number of social encounters with unfamiliar people, something that was absent in our evolutionary past. Another factor may reside in the exceptional human longevity, at least in industrialized countries. As the volume editors, we feel the chapters have admirably highlighted these tenets. Before closing and in the contemporary context of concerns regarding the health impacts of heat, waste control, infections, and inflammation, we add a few final reflections of our own.

Example 1: Heat Regulation

The human brain is extremely heat-sensitive. An increase in body temperature by 4°C may lead to states of confusion, convulsions, or epileptic seizures, whereby young children are particularly vulnerable to fever convulsions (febrile seizures) due to their immature thermoregulation. Accordingly, protection of the brain from overheating is critical and was a significant driving force for a couple of adaptations to emerge during human evolution (Wang et al. 2016). These adaptive processes were spurred by changes in nutrition as early humans adopted hunting skills requiring long-distance running in arid environments (Hublin 2005). Even in moderate climatic conditions, the body temperature during a marathon can rise to 40°C, yet the brain is protected by a cooling system that operates similarly to a radiator (Falk 1990).

Compared to nonhuman apes, the human venous sinuses in the dura mater are largely drained via the internal jugular vein and by the vertebral venous plexus (Kunz and Iliadis 2007). Blood temperature in the internal jugular vein is higher than that in the carotid artery, indicating that the removal of excess heat from the brain is an important task of the venous draining system.

Interestingly, bipedalism has apparently shifted blood drainage to the vertebral venous plexus (Falk 1990). In addition, an upright body position itself reduces the exposure of body surface to the heat of the day, and thus prevents hyperthermia (Wheeler 1988).

The evolution of this peculiar cooling system of the brain probably preceded the enlargement of the brain. In fact, the number of emissary veins connecting intracranial veins and sinuses with extracranial blood vessels increased already in early hominids. Moreover, in anatomically modern humans, small cortical capillaries cross the subarachnoid space and allow thermal exchange with the cerebrospinal fluid (CSF), which adds to the venous system's cooling properties (Bruner et al. 2011; Wang et al. 2016). Other anatomical adaptations involved in thermoregulation of the brain include the nasal cavity, and forehead sweating is particularly effective at cooling the cerebrum.

Clinically, it is conceivable that the human brain's radiator system is more vulnerable to thrombosis of the large sinuses, with critical life events such as pregnancy posing specific risks for this potentially life-threatening disease (Silvis et al. 2017). In addition, Rzechorzek et al. (2022) recently demonstrated that the brain's nocturnal cooling mechanisms functionally decline with age. Nevertheless, there are also profound regional differences in temperature within the brain, with temperature in medial brain regions such as the thalamus and hypothalamus rising to over 40°C in clinically healthy individuals. Clinically, it has been discussed that a disruption of the circadian variation in brain temperature may be linked to sundowning and disturbed sleep patterns in degenerative brain diseases (Rzechorzek et al. 2022).

Example 2: Waste Clearance and Protection against Oxidative Stress: Glymphatic Function

Aside from thermoregulation, a brain consuming large amounts of energetically highly potential substances needs to effectively eliminate metabolic waste products and, ideally, utilize agents with detoxification properties. One effective way of clearing the brain of metabolic products takes the route of perivascular tunnels. The brain glymphatic system (so called due to the role of glia cells, which is where the "g" comes from) is composed of a vast network of perivascular tunnels formed by astrocytic endfeet which surround the entire vasculature. The perivascular space (the macroscopically visible part of it called the "Virchow-Robin space") between the wall of the vasculature and the astrocytic endfeet is filled with CSF. Its flow into the perivascular space is supported by arterial pulsatility and respiration. Metabolic and toxic substances are enriched in the interstitial fluid and cleared by convective movement of CSF into the brain parenchyma. As the total volume of the CSF of about 150 ml is renewed every 6 hours, the clearance of metabolic products is highly effective (Jessen et al. 2015).

The elimination of metabolic waste products is intimately linked to the circadian rhythm of sleep and wakefulness. For example, the brain's metabolic rate decreases only slightly during sleep, yet the convective flux of interstitial and cerebrovascular fluid is greatly enhanced (Herculano-Houzel 2013). During wakefulness, the action of the brain's glymphatic system is rather suppressed (Jessen et al. 2015).

The duration of sleep (i.e., how much sleep an organism can afford to eliminate toxic substances) is traded off against the time spent foraging, at least in mammals. Larger animals require relatively more feeding time to supply their bodies (and brains) with energy, which thus needs to be balanced against the duration of sleep (Herculano-Houzel 2015). However, large animals often have large brains, which produce more metabolic waste than smaller brains. This, in turn, could be associated with selection pressures toward more time for sleep. Sleep requirements seem to be related to neural density, that is, lower neuronal density in cortical areas is associated with shorter sleep duration, possibly due to less metabolic waste production (Herculano-Houzel 2015). Sleep is also involved in inflammatory activity of interleukins and other immunologically active substances (Irwin and Opp 2017), whereby sleep deprivation is associated with an increase of tumor necrosis factor in the blood and may cause dysfunction of the brain's glymphatic system. However, we are far from understanding the optimal ratio between sleep and wakefulness. For example, quality of sleep is influenced by many factors, including lifestyle and traumatic experiences (Reddy and van der Werf 2020). Moreover, both shorter and longer sleep duration are risk factors for dementia and are also observed in depressive states.

The production of CSF declines with age, as does the quality and duration of sleep. Together, this may influence the brain's capacity to eliminate toxic metabolic substances such as ß-amyloid, tau protein, and glutamate (Jessen et al. 2015; Irwin and Opp 2017).

In addition, degenerative brain diseases, stroke, intracerebral hemorrhage, and traumatic brain injury can disrupt the convective flow into the interstitial compartment and the CSF, thus potentiating the detrimental effects of metabolic waste clearance in the brain.

Aside from the elimination of potentially toxic substances such as misfolded ß-amyloid protein, oxidative stress is also a threat to normal brain function. Indeed, the metabolic problem of how to deal with free radicals has confronted eukaryotes since the emergence of aerobic metabolism utilizing oxygen to produce energy-rich molecules. For example, research has shown that, when the ancestors of cyanobacteria evolved to produce molecular oxygen on a large scale that eventually filled the atmosphere with a large amount of free oxygen, their descendants soon adapted to this situation. In fact, the first superoxide dismutase (SOD) enzymes probably evolved some 3.3 to 3.6 billion years ago. The first SOD used copper and zinc as cofactors, whereas later on, iron, nickel, and manganese served as cofactors to reduce free oxygen radicals. These radicals are produced during photosynthesis but also during respiratory activity and extracellular processes. Free radicals have beneficial effects, to some degree, by helping the organism acquire iron. However, when present in excess, they impair protein synthesis and may damage DNA (Boden et al. 2021). SODs are potential metalloenzymes that catalyze the superoxide anion free radical to hydrogen peroxide, which is further enzymatically degraded to water. SOD can act as an antioxidant, when its activity is orchestrated with other enzymes, but also as a pro-oxidant, leading to an accumulation of hydrogen peroxide (Rosa et al. 2021).

SOD supplementation could be beneficial in diseases associated with high oxidative stress, including cardiovascular disease, diabetes, and hypercholesterolemia. Oral SOD supplementation in the form of plant extracts is, however, fraught with low bioavailability, and parenteral application of SODs can be associated with allergic reactions (Rosa et al. 2021). A few randomized placebo-controlled studies suggest that the administration of SOD may be helpful in neurodegenerative diseases such as Alzheimer's disease (AD), Parkinson's disease (PD), and amyotrophic lateral sclerosis (ALS), as well as stroke. SOD supplementation may also reduce the subjective experience of distress, which could be interesting for a range of psychiatric conditions, including schizophrenia, autism, depression, and anxiety disorders (reviewed in Rosa et al. 2021). It would be premature, however, to recommend SOD supplementation as a general treatment strategy for these conditions.

Example 3: Infectious Diseases—Fast and Slow

The human brain is highly susceptible to infection with neurotropic agents such as viruses, bacteria, and other single-cell organisms. While the blood–brain barrier (BBB) partly protects the brain from infection, this barrier is far from being impermeable. For example, the BBB is under the continuous surveillance of immune-competent cells such as macrophages and leucocytes. Paradoxically, however, these cells may—under specific circumstances—serve as Trojan horses for pathogenic agents (Salinas et al. 2010). In addition, viruses and bacterial toxins may utilize retrograde axonal transport to enter the central nervous system. Other agents enter the brain via neuromuscular junctions, where they bind to receptors on the terminal synapse (Salinas et al. 2010). However, there is also evidence for pathogen–host coevolution, and some retroviruses are known to be permanently (dormant) in the host genome. It is estimated that about 8% of the human genome originated from retroviruses, some of which are non-defective or may be even beneficial to the host (Pellett et al. 2014). Acute infection with SARS-CoV-2, causing COVID-19, is associated with neurological and/or psychiatric symptoms in about 35% of the affected individuals. The infection of central nervous structures triggers reactive astrogliosis, an evolutionarily conserved defense mechanism associated with pro-inflammatory reactions (Tavčar et al. 2021). However, long-term effects of infections with RNA viruses are not yet fully understood. For example, there is mounting evidence to suggest that SARS-CoV-2 infection in aging populations may foster the development of neurodegenerative diseases, even though the mechanisms are not entirely clear (Strong 2023).

Aside from acute infection, "slow virus" diseases caused by prions affect many mammalian species, with misfolding of the prion and protein aggregation as key characteristics. AD, PD, ALS, and other neurodegenerative diseases are conceptualized as proteinopathies with potentially similar characteristics to prionopathies including the propagation mechanism. For classic prionopathies it has been proposed that the still highly "prevalent human prion protein 129M/V mutation could be a living fossil from a Palaeolithic panzootic superprion pandemic" (Nyström and Hammarström 2014). This provocative hypothesis remains unproven however, and there is also no comparable concept explaining the evolutionary emergence and persistence of proteins acting like prions in neurodegenerative diseases. It is established that formation of amyloid fibrils is a phylogenetically ancient process that can be observed in some of the "oldest inhabitants of the earth, bacteria and fungi," showing here protective and detoxification properties (de Eguileor et al. 2022). Also, human amyloid β evidences characteristics of mammalian prions including the property to form functionally variant strains. While the conceptual frameworks of protein aggregation—being pathogenic, protective, or just epiphenomenal—are still heavily debated (Espay et al. 2019), they seem to be remarkably species-specific. For instance, in various animals mutant α-synuclein (such as the A53T mutation) does not develop the same level of cytotoxicity as in humans. It has been shown that this missing pathogenicity could be due to reduced fibrillation propensity. There is also weaker cellular membrane disruption due to soluble α-synuclein oligomers (Sahin et al. 2018). These findings suggest that if downstream effects are missing, the misfolding process per se is not a sufficient condition. In humans, the propagation speed of neurodegenerative diseases is highly variable from one subject to the next, and it remains poorly understood. Beside numerous lifestyle factors and old-age comorbidities, the rate of progression may depend on subtle differences of the structural conformers of the misfolded proteins with variable potency to speed up the pathological process. These could also be subliminal local variations of the intrinsic template, as proposed here. As with classical prionopathies, it remains unknown why and how the degenerative process starts in a specific brain region, if it may start almost synchronously in different regions, how it is transferred from one cell to the next, or why brain tissue affected by proteinopathies does not have the cross-subject contagious properties of brain tissue affected by classical prionopathies. In the case of AD, as a prime example of a progressive neurodegenerative disease with a focal pathological onset and spread within the cortex, it remains terra incognita how the β-sheet misfolding is linked ultimately to a tauopathy (Kabir and Safir 2014; Rasmussen et al. 2017). In PD and other diseases with highly branched axons, it is unknown how the process is linked to mitochondrial deficiencies in distant axonal parts. These are key issues that impact disease progression as well as therapeutic strategies to interrupt this process, and evolutionary studies may be important to unveiling those relationships.

Added Challenges and Future Directions

We have covered both psychiatric and neurological disorders, but one fundamental difference between them is, with few exceptions, their average onset of signs and symptoms. By and large, psychiatric disorders usually manifest clinically in adolescence or early adulthood, whereas neurological diseases usually become clinically apparent much later in life, often after the peak of reproductive activity (or fecundity). Nevertheless, we do not know if in neurological diseases there is a long silent period possibly extending over decades, as discussed for AD or in Huntington's disease. For example, in carriers of the Huntington gene, neurodevelopmental abnormalities have already been observed in embryonic states of development (Barnat et al. 2020). From a genetic point of view, pleiotropic antagonistic effects underlying neurological diseases have to be considered as well. Despite these restrictions, at the front line, the differences in clinical manifestation of psychiatric and neurological diseases mean that the asymptomatic period of life is, on average, much longer in neurological diseases than in psychiatric disorders, suggesting that the options for neurological preventive or compensatory measures and prodromal recognition are much better than in psychiatry. This difference could pertain to the accumulation of toxic substances that may promote the risk for neurodegenerative diseases but also the control of "lifestyle" factors contributing to the risk of disease, including diet and physical activity. These factors seem to be much less relevant for psychiatric conditions that strike during the peak of reproductive activity. Accordingly, selection pressure must be higher in psychiatric conditions compared to neurological diseases because the latter may escape selection, especially when associated with aging or, paradoxically enough, even associated with some early advantage. This difference also suggests that the concept of "evolutionary mismatch," whereby evolved functions adaptive to ancestral conditions are no longer adapted to modern environments, is far more relevant for the understanding, recognition, and prevention of psychiatric, as opposed to neurological, disease. A better understanding of evolutionary genetics and deep phenotyping of both archetypal disease manifestations as well as variants and *formes frustes* may permit evolutionary perspectives to be directly integrated into diagnosing and treating individual patients. In this regard, a better understanding of the adaptability/flexibility of epigenetic mechanisms, sometimes with transgenerational impact as well as the forward and backward interplay/complementarity of inheritance, environment, and mechanisms may help us to project and model how human brains may adapt—or fail to adapt—to precipitating changes in external conditions dictated by climate changes, mass migrations, emergence of new anthropo-zoonoses, etc. Whatever route this knowledge base may take, an essential conclusion from the contributions to this volume is that, in spite of differences in onset and manifestation, the medical fields of neurology and psychiatry are much more intimately linked to one another than is currently taught at medical schools. It is the hope of the editors that the present book can help build bridges between the two disciplines in the interest of medical science.

References

Allman, J. M. (1999). *Evolving brains*. W. H. Freeman and Co.

Barnat, M., Capizzi, M., Aparicio, E., Boluda, S., Wennagel, D., Kacher, R., Kassem, R., Lenoir, S., Agasse, F., Braz, B. Y., & Liu, J. P. (2020). Huntington's disease alters human neurodevelopment. *Science*, *369*(6505), 787–793. https://doi.org/10.1126/science.aax3338

Boden, J. S., Konhauser, K. O., Robbins, L. J., & Sánchez-Baracaldo, P. (2021). Timing the evolution of antioxidant enzymes in Cyanobacteria. *Nature Communications*, *12*(1), Article 4742. https://doi.org/10.1038/s41467-021-24396-y

Brothers, L. (1990). The social brain: A project for integrating primate behavior and neurophysiology in a new domain. *Concepts in Neuroscience*, *1*, 27–51. https://doi.org/10.7551/mitpress/3077.003.0029

Brown, E. C., & Brüne, M. (2012). Evolution of social predictive brains? *Frontiers in Psychology*, *3*, Article 414. https://doi.org/10.3389/fpsyg.2012.00414

Bruner, E., Mantini, S., Musso, F., De La Cuétara, J. M., Ripani, M., & Sherkat, S. (2011). The evolution of the meningeal vascular system in the human genus: From brain shape to thermoregulation. *American Journal of Human Biology*, *23*, 35–43. https://doi.org/10.1002/ajhb.21123

de Eguileor, M., Grimaldi, A., Pulze, L., Acquati, F., Morsiani, C., & Capri, M. (2022). Amyloid *fil rouge* from invertebrate up to human ageing: A focus on Alzheimer disease. *Mechanisms of Ageing and Development*, *206*, Article 111705. https://doi.org/10.1016/j.mad.2022.111705

Espay, A. J., Vizcarra, J. A., Marsili, L., Lang, A. E., Simon, D. K., Merola, A., Josephs, K. A., Fasano, A., Morgante, F., Savica, R., & Greenamyre, J. T. (2019). Revisiting protein aggregation as pathogenic in sporadic Parkinson and Alzheimer diseases. *Neurology*, *92*(7), 329–337. https://doi.org/10.1212/WNL.0000000000006926

Falk, D. (1990). Brain evolution in *Homo*: The "radiator" theory. *Behavioral and Brain Science*, *13*(2), 333–381.

Herculano-Houzel, S. (2013). Sleep it out. *Science*, *342*(6156), 316–317. https://doi.org/10.1126/science.1245798

Herculano-Houzel, S. (2015). Decreasing sleep requirement with increasing numbers of neurons as a driver for bigger brains and bodies in mammalian evolution. *Proceedings of the Royal Society B: Biological Sciences, 282*(1816), Article 20151853. https://doi.org/10.1098/rspb.2015.1853

Hitze, B., Hubold, C., van Dyken, R., Schlichting, K., Lehnert, H., Entringer, S., & Peters, A. (2010). How the selfish brain organizes its supply and demand. *Frontiers in Neuroenergetics, 2*, Article 7. https://doi.org/10.3389/fnene.2010.00007

Hublin, J. J. (2005). Evolution of the human brain and comparative paleoanthropology. In S. Dehaene, J. R. Duhamel, M. D. Hauser, & G. Rizzolatti (Eds.), *From monkey brain to human brain* (pp. 57–72). MIT Press.

Irwin, M. R., & Opp, M. R. (2017). Sleep health: Reciprocal regulation of sleep and innate immunity. *Neuropsychopharmacology, 42*(1), 129–155. https://doi.org/10.1038/npp.2016.148

Jessen, N. A., Munk, A. S., Lundgaard, I., & Nedergaard, M. (2015). The glymphatic system: A beginner's guide. *Neurochemical Research, 40*(12), 2583–2599. https://doi.org/10.1007/s11064-015-1581-6

Jones, S., Martin, R. D., & Pilbeam, D. R. (Eds.). (1992). *The Cambridge encyclopedia of human evolution.* Cambridge University Press.

Kabir, M. E., & Safar, J. G. (2014). Implications of prion adaptation and evolution paradigm for human neurodegenerative diseases. *Prion, 8*(1), 111–116. https://doi.org/10.4161/pri.27661

Kunz, A. R., & Iliadis, C. (2007). Hominid evolution of the arteriovenous system through the cranial base and its relevance for craniosynostosis. *Child's Nervous System, 23*(12), 1367–1377. https://doi.org/10.1007/s00381-007-0468-5

Magistretti, P. J., & Allaman, I. (2015). A cellular perspective on brain energy metabolism and functional imaging. *Neuron, 86*(4), 883–901. https://doi.org/10.1016/j.neuron.2015.03.035

Nyström, S., & Hammarström, P. (2014). Is the prevalent human prion protein 129M/V mutation a living fossil from a Paleolithic panzootic superprion pandemic? *Prion, 8*(1), 2–10. https://doi.org/10.4161/pri.27601

Pellett, P. E., Mitra, S., & Holland, T. C. (2014). Basics of virology. In A. C. Tselis & J. Booss (Eds.), *Handbook of clinical neurology* (pp. 45–66). Elsevier Science.

Rasmussen, J., Jucker, M., & Walker, L. C. (2017). Aβ seeds and prions: How close the fit? *Prion, 11*(4), 215–225. https://doi.org/10.1080/19336896.2017.1334029

Reddy, O. C., & van der Werf, Y. D. (2020). The sleeping brain: Harnessing the power of the glymphatic system through lifestyle choices. *Brain Sciences, 10*(11), Article 868. https://doi.org/10.3390/brainsci10110868

Rosa, A. C., Corsi, D., Cavi, N., Bruni, N., & Dosio, F. (2021). Superoxide dismutase administration: A review of proposed human uses. *Molecules, 26*(7), Article 1844. https://doi.org/10.3390/molecules26071844

Rzechorzek, N. M., Thrippleton, M. J., Chappell, F. M., Mair, G., Ercole, A., Cabeleira, M., CENTER-TBI High Resolution ICU (HR ICU) Sub-Study Participants and Investigators, Rhodes, J., Marshall, I., & O'Neill, J. S. (2022). A daily temperature rhythm in the human brain predicts survival after brain injury. *Brain, 145*(6), 2031–2048. https://doi.org/10.1093/brain/awab466

Sahin, C., Kjær, L., Christensen, M. S., Pedersen, J. N., Christiansen, G., Pérez, A. M. W., Møller, I. A., Enghild, J. J., Pedersen, J. S., Larsen, K., & Otzen, D. E. (2018). α-Synucleins from animal species show low fibrillation propensities and weak oligomer membrane disruption. *Biochemistry, 57*(34), 5145–5158. https://doi.org/10.1021/acs.biochem.8b00627

Salinas, S., Schiavo, G., & Kremer, E. J. (2010). A hitchhiker's guide to the nervous system: The complex journey of viruses and toxins. *Nature Reviews Microbiology, 8*(9), 645–655. https://doi.org/10.1038/nrmicro2395

Shultz, S., & Maslin, M. (2013). Early human speciation, brain expansion and dispersal influenced by African climate pulses. *PLOS ONE, 8*(10), Article e76750. https://doi.org/10.1371/journal.pone.0076750

Silvis, S. M., de Sousa, D. A., Ferro, J. M., & Coutinho, J. M. (2017). Cerebral venous thrombosis. *Nature Reviews Neurology, 13*(9), 555–565. https://doi.org/10.1038/nrneurol.2017.104

Strong, M. J. (2023). SARS-CoV-2, aging, and post-COVID-19 neurodegeneration. *Journal of Neurochemistry, 165*(2), 115–130. https://doi.org/10.1111/jnc.15736

Suddendorf, T., & Corballis, M. C. (1997). Mental time travel and the evolution of the human mind. *Genetic, Social, and General Psychology Monographs, 123*(2), 133–167.

Tavčar, P., Potokar, M., Kolenc, M., Korva, M., Avšič-Županc, T., Zorec, R., & Jorgačevski, J. (2021). Neurotropic viruses, astrocytes, and COVID-19. *Frontiers in Cellular Neuroscience, 15*, Article 662578. https://doi.org/10.3389/fncel.2021.662578

Wang, H., Kim, M., Normoyle, K. P., & Llano, D. (2016). Thermal regulation of the brain—An anatomical and physiological review for clinical neuroscientists. *Frontiers in Neuroscience, 9*, Article 528. https://doi.org/10.3389/fnins.2015.00528

Wheeler, P. (1988). Stand tall and stay cool. *New Scientist, 118*(1612), 62–65.

Yakovlev, P. I., & Lecours, A. R. (1967). The myelogenetic cycles of regional maturation of the brain. In A. Minkowski (Ed.), Regional development of the brain in early life (pp. 3–70). Blackwell.

Glossary

Active sleep Part of the sleep cycle of newborns, the other part being quiet sleep. During active sleep the newborn may move or cry. Considered to be the predecessor of REM sleep seen later in life.

Adaptive introgression Advantageous archaic genetic variants adapted to local environments as the result of interbreeding between anatomically modern humans (see definition) and archaic human species in Eurasia.

Allostatic load Accumulation of dysfunctional physiological pathways secondary to repetitive exposure to stressful life events and exceeding the body's capacity to maintain equilibrium or allostasis.

Anatomically modern humans Humans that are anatomically identical with the phenotypes seen in modern humans (*Homo sapiens*); to be distinguished from extinct and anatomically distinguishable archaic human species such as *Homo erectus*.

Anhedonia Loss of ability to feel pleasure. Characteristic sign of depressive disorders. Can hamper execution of routine daily activities, induce social withdrawal, etc.

Antagonistic pleiotropy Concept explaining how an ancestral feature or mechanism (old tool) is reused by evolution for newer functions, mostly by exchange of the afferent input and the efferent output.

Apomorphy Novel trait, characteristic of a newly evolved group or species, not present in the species ancestry.

ARTAG Aging-related tau astrogliopathy is a phenomenon associated with a morphological range of tau accumulation found in astrocytes, which is not necessarily associated with cognitive decline typical of Alzheimer's disease.

Astrogliosis Spectrum of changes in astrocyte morphology and function due to multiple types of insults, including inflammation, mechanical or metabolic stress.

AUG Most common start codon encoding for methionine in eukaryotes.

Auto-domestication Self-imposed domestication reducing individuals' ability to cope with environmental challenges, possibly introducing metabolic deficiencies and even mental abnormalities. However, the concept has been criticized as serving also as an explanation for alleged genetic degeneration of pollutions.

Autophagy Natural, biological process involving the selective degradation of excessive or dysfunctional cellular components by lysosomes.

Axodendritic synapse Synapse that a neuron makes onto the dendrite of another neuron (see also **Axosomatic synapse**).

Axosomatic synapse Axon synapse directly connected to the soma (cell body) of another cell (see also **Axodendritic synapse**).

Bauplan "Building plan" or morphoplan common to the cerebral organization in all vertebrates. Identification of homologous brain regions is needed to detect the common architecture.

Belt and parabelt organization of the auditory system Cerebral areas surrounding the core region of the auditory system. The belt immediately surrounds the core, whereas the parabelt is adjacent to the lateral side of the belt. These areas are tonotopically and hierarchically organized, with areas on higher hierarchy levels processing more complex stimulus properties.

Beta-amyloid Extraneuronal protein formed from the breakdown of the larger amyloid precursor protein. Pathological misfolding and aggregation of this protein are pathological hallmarks of Alzheimer's dementia.

Betweenness centrality Concept of graph theory defining the amount of influence a node has over the flow of information in a graph. It identifies nodes serving as a bridge from one part of a graph to another. It is calculated based on the shortest path between all pairs of nodes in a graph.

Betz cells Very large or giant pyramidal cells in layer V of the primary motor cortex.

Biomodulator Evolutionarily highly conserved substances that over evolutionary time have been "co-opted" for different functions, most recently as neurotransmitters.

Bipedalism Modus of locomotion using exclusively two (lower) limbs. Characteristic feature of hominins, in contrast to other members of the larger family of hominids.

Blindsight Ability to adequately respond to visual stimuli that are not consciously perceived (e.g., due to lesions of the primary visual cortex). However, the fast and evolutionarily old pathways of blindsight running through the superior colliculi also operate in parallel in visually unimpaired subjects.

Blood oxygen level–dependent imaging (BOLD) Technique used in functional MRI studies taking into account local fluctuations of oxygenated versus deoxygenated hemoglobin as the latter one is a paramagnetic signal, thus changing the MRI signal. Considered to be an indirect measure of neuronal activity. Mostly used with the abbreviation "BOLD."

Braak's prion hypothesis Hypothesis proposing a progressive neuron-to-neuron, prion-like propagation of aggregated misfolded intracellular proteins, thus acting as seeds in the next neurons. First proposed as an ascending degeneration pathway of misfolded α-synuclein disease, later also applied to Alzheimer's dementia and other neurodegenerative diseases.

Bradykinesia Slowness of movements.

Brain globularity Round shape of the brain of anatomically modern humans partially due to large and bulging parietal lobes. This brain globularity contrasts with the elongated brain shape of the extinct human species of Neanderthals.

CCD model Theory elaborated by Changeux, Corrège, and Danchin proposing that, based on high variability of developing interneuronal connections, there is progressive consolidation of robust circuits by trial and error; it implies selective stabilization of synapses.

Central pattern generators Phylogenetically preserved cellular networks in invertebrates and vertebrates located in the medulla and the brainstem allowing automatic, although complex, behaviors (deambulation, mastication, feeding).

CG methylation DNA methylation on CpG sites.

CH methylation DNA methylation on CH sites. H indicates A, C, or T nucleotides. This methylation type is relatively abundant in the brain.

Chemokine receptor (CCR) Receptors on cells interacting with chemokines, a family of small signaling proteins involved, for instance, in immune responses, morphogenesis, and pathogenesis of cancer.

Circadian clock Internally driven rhythm governed by a circadian biochemical clock or oscillator located in the suprachiasmatic nucleus and intertwined with numerous metabolic processes. While genetically set at roughly 24 hours, it also depends on environmental cues (see also **Internal clock**).

***Cis* and *trans* effects** Gene expression differences can be driven either by the sequences on the same DNA molecule as the target gene and acting on the gene in *cis* (i.e., *cis* effect) or by the variations in diffusible molecules that are coded elsewhere in the genome and act on the target gene in *trans* (i.e., *trans* effect).

Clade Group of organisms that includes all species having the same common ancestor.

Cognitive reappraisal Cognitive behavioral method to tackle unwanted repetitive thoughts with the aim of transforming their emotional impact and thus controlling them.

Connectivity-based parcellation Method of parcellation of brain structures based on similar connectivity as shown by functional magnetic resonance imaging or diffusion tensor imaging.

Constitutionality One of three principles coined by Crow in order to describe the pathogenesis of schizophrenia. It means in this context that schizophrenia is a neurodevelopmental condition that does not depend on an exogenous or environmental factor to trigger its expression (see also **Continuity** and **Universality**).

Contextual cues Environmental stimuli repeatedly associated with a planned action and thus able to elicit bottom-up automatic stimulus processing associated with a pre-defined response. Can also aid in memory retrieval, when there is regular association between these cues and a certain memory.

Continuity One of three principles coined by Crow to describe the pathogenesis of schizophrenia. It means here that schizophrenia is not a disease entity distinct from other conditions but that states described as schizophrenic merge imperceptibly into affective states and non-psychotic conditions (see also **Constitutionality** and **Universality**).

Continuity hypothesis Dream hypothesis proposing that dreams are embodied simulations that dramatize our concerns and conceptions.

Continuous variation Term describing that different types of variations are distributed on a continuum; used here in the context of schizophrenia meaning pathogenetic incorporation of multiple genetic systems, neurocognitive processes, and psychopathologies.

Copy-number variation Fragments of the genome that vary in the number of times they are repeated in different people.

Corollary discharge Neuronal signal encoding a blueprint of a planned motor command to a non-motor structure of the brain, able to compare the intended movement with the sensory feedback. For instance, based on this information, the world does not appear to move when the eyes move.

Cortical parcellation Division of the human cortex into meaningful anatomical units allowing comparability between different individuals. Based on cortical landmarks and atlases, this parcellation is achieved by highly automatized processing of neuroimaging.

Corticobasal degeneration Neurodegenerative disease caused by accumulation of aggregated, misfolded tau proteins and causing impairment of movements, speech, and memory (see also **Progressive supranuclear palsy**).

Corticospinal tract Major neuronal pathway connecting the cortex to the spinal cord and enabling voluntary movements of the extremities.

CpG sites Indicates a cytosine nucleotide followed by a guanine nucleotide on a DNA sequence. CpG sites are the prime target for DNA methylation.

Cucinovore Species eating essentially cooked or in another way processed food.

Cyanobacteria Single-celled organisms living in water; some species produce toxins affecting the hepatic or renal system, others the central nervous system.

Cybernetic Research field dealing with circular causal processes such as feedbacks or self-regulating loops.

Cynomolgus monkey Synonym for the long-tailed or crab-eating macaque.

Cytoarchitectonics Method for studying the structure of cortical brain regions.

Cytoarchitecture Arrangement, density, and distribution of cells in the brain.

Cytopathology Pathological study and diagnosis of diseases at the cellular level.

Damage-associated molecular pattern (DAMPs) Molecules released from senescent or dying cells able to trigger an immune response. DAMPs are a warning sign for the organism to inform it of infection or damage at the cellular level. DAMPs include parts of the cell like organelles, including mitochondria, and other proteins; and they are released into the extracellular matrix. In particular, mitochondria hold the potential to be large players in inflammaging caused by this type of molecule, due to their endosymbiont origin.

Default mode network (DMN) Network of functionally coactivated brain areas with decreased activity during externally oriented tasks and increased activity during goal-free resting state conditions. The DMN seems to be evolutionary conserved but with major dissimilarities across species.

Degree centrality values In graph theory, the degree centrality is a simple basic numeric centrality measure. For instance, when a node (or brain region) has 10 connections or edges, it has a degree centrality value of 10.

Dendrophilia Human ability to produce an unlimited variety of linear signal strings that communicate complex semantic messages in a recursive and hierarchical manner.

Diffusion tensor imaging (DTI) Neuroimaging technique that enables the study of large white matter tracts in vivo, measuring the diffusion of water molecules along axons. It is applicable across species to trace the evolutionary changes in their functional organization.

Diseases of civilization Chronic diseases in highly industrialized countries and in relation to the lifestyle in these countries, including cardiovascular diseases, obesity, diabetes type 2, asthma, and osteoporosis. Hypothetically characterized by systemic low-grade inflammation and quasi-absent in hunter–gatherer societies. Similar expression: "diseases of modernity."

Diseases of modernity See **Diseases of civilization.**

Dream–action isomorphism Dream action corresponds to externally observable sleep behavior.

Dysbiosis State of imbalanced microbiota characterized by decrease in microbial diversity and increase in proinflammatory species.

Ectoderm The outermost of the three germ layers, giving rise, among other organs, to the central and peripheral nervous systems.

Emergent phenotype Phenotype not easily explicable by underlying genetic or environmental causes but rather due to complex interactions of numerous factors, thus inducing the emergence of this phenotype. Used, for instance, when discussing the etiologies of autistic or schizophrenic syndromes.

Endocranial cast Mold of the endocranial cavity.

Endogenous template Term used to describe the selective disease-specific vulnerability for various protein deposits and propagation routes seen in neurodegenerative diseases.

Endosymbiont theory Theory proposing that mitochondria—and chloroplasts—are descendants of formally free-living bacteria. According to the theory, roughly 2 billion years ago, a larger cell engulfed the bacteria without digesting it, and the bacteria did not kill the host cell. This conferred advantages to both parties; the host cell gained a new source of energy and provided an extra layer of protection for the bacteria. Over billions of years, this symbiotic relationship led to the integration of the bacterial ancestor into the host cell, including the transfer of its genes.

Environment of evolutionary adaptedness Ancestral environment to which a species has primarily been adapted by selection pressures. Has shaped certain traits persisting in a new environment, for instance, in human food preferences including meat, social interactions, and responses to stress.

Ependyma Thin neuroepithelial lining within the ventricles of the brain and the central canal of the spinal cord. Cells are glial (express GFAP), and produce the cerebrospinal fluid in the ventricular system.

Evolutionary mismatch Fact that a species or genotype formerly well adapted to the environment has become maladapted to the present environment, possibly due to a change of this environment (migration, climate change, evolution of other species, etc.).

Evolutionary molecular modeling Evolutionary process that describes the selection pressures on animals including humans from, for example, plant toxins targeting the central nervous system.

Evolution-based model of causation Based on the sufficient and component causes model, model using statistical and evolutionary theories to better explain variable penetrance of genetic variants, age as a risk factor, emergence of diseases associated with the Western lifestyle, etc.

Exaptation Concept explaining that an evolutionarily ancestral mechanism can be reused for newer functions, mostly by exchanging the input and output targets.

Fight or flight Automated, immediate, and probably evolutionarily old reaction to a frightening event.

Fitness (evolutionary sense) Concept explaining how well a species is able to reproduce and survive in a specific environment. According to Darwinism, due to natural and sexual selection. In a wider sense, developmental plasticity, synaptic plasticity, and life history are contributive as well.

Fluorodeoxyglucose Also known as ^{18}F or FDG, a radiopharmaceutical compound homologous of glucose, used in combination with PET as a radiotracer to detect metabolic activity. It is used to study malignant lesions, including cancer, and in research to study the uptake and consumption of glucose.

FOXP2 gene Coding for the protein forkhead box P2, a transcription factor; variants in its coding region linked with language disorders.

Fragile X syndrome Genetic disorder caused by a mutation in the fragile X messenger ribonucleoprotein 1 (FMR1) gene in terms of excessive repeats of the triplet CGG. As males only have one copy of this gene, the clinical manifestations of developmental delays and learning disabilities are more pronounced in males than females.

Free-radical aging theory Theory hypothesizing that aging can be linked to the accumulation of oxidative cellular damage by aerobic respiration. Mitochondria are specialized organelles that create ATP through aerobic respiration. A normal byproduct of that process includes reactive oxygen species (ROS)—highly reactive and unstable chemicals formed from diatomic oxygen such as peroxide and superoxide. ROS can cause cellular damage and induce mutations and deletions of mitochondrial DNA.

Friedreich ataxia Autosomal recessive disease characterized by multiple coordination and balance symptoms. Mostly due to an excessive number of the trinucleotide repeat expansion GAA, causing reduced expression of the gene and therefore reduced production of the protein frataxin.

Gene regulatory element Non-coding genomic elements regulating gene expression; most important types are promotors, enhancers, and silencers.

Genetic drift Variation in the relative frequency of different genotypes due to chance disappearance of particular genes rather than evolutionary selection processes.

Geophagy Practice of eating earthy substances, can supplement a mineral-deficient diet.

Geroscience hypothesis Hypothesis claiming that the course of aging can be retarded and simultaneously onset of diseases related to old age can be delayed, based on the fact that deterioration mechanisms common to these processes are known.

Ghost population Hypothetical archaic human population based on statistical analyses on genetic material but without formal proof by bones or ancient DNA.

Ghost tangles Extracellular remnants of formerly intracellular tau seen in brain sections after neuronal death as sparsely spread dots composed of non-fibrillary hyperphosphorylated misfolded tau aggregates.

Gliopathy Pathological condition specifically affecting glial cells.

Glossogeny Human-specific proclivity to communicate socially.

Glycolysis Metabolic pathway converting glucose into pyruvate, occurring in the cytosol and resulting in quickly producing a small amount of ATP by reducing NADH, and generating lactate as one of the byproducts of the reaction.

Glymphatic system Vast network of perivascular tunnels formed by astrocytic endfeet which surround the entire vasculature and essential for waste clearance.

Goal-directed system Actions are selected to obtain a specific outcome or goal. This system contrasts with the habitual system (see **Habitual system**).

G protein Membrane-bound protein that binds GTP when activated by a membrane receptor.

G protein–coupled receptor Group of membrane receptors detecting molecules outside the cell and activating internal signal transduction pathways by coupling with G proteins. In this way they stimulate or inhibit intracellular biochemical responses (see also **Metabotropic receptor**).

Graded action potential Local changes in the membrane potential, mostly occurring in postsynaptic dendrites and dependent on the size of the stimulus.

Group selection Mechanism of evolution with natural selection impacting fitness (survival and reproduction) of groups but not individuals. For instance, altruistic behavior rather than competitive behavior can enhance fitness of the group.

Gyrification Formation of folds or gyri on the surface of the brain.

Habit Individual task component that is automatically selected.

Habitual system System allowing execution of routine tasks under automatic habitual control. These are mostly motor responses to stimuli performed due to repeated reinforcement. This system contrasts with the goal-directed system (see **Goal-directed system**).

Heterochrony Alterations in the developmental timing or rate of one species compared to another species.

Heteroplasmy Two or even more variants of mitochondrial DNA coexist in the same cell.

Hominin Group consisting of all human species (extinct human species, such as Neanderthals and Denisovans), unknown human ancestors, and modern humans. Bipedalism is a common denominator.

Hub Concept explaining how well a species is able to reproduce and survive in a specific environment. According to Darwinism, due to natural and sexual selection. In a wider sense developmental plasticity, synaptic plasticity, and life history are contributive as well. A cerebral region with numerous afferent or efferent connections with other cerebral regions. Also called "a node with a large degree."

Human accelerated regions (HARs) DNA sequences that are conserved throughout mammalian evolution but have accumulated high numbers of substitutions in the human lineage.

Hybridization (genetic) Process of interbreeding subjects from genetically different populations, here the interbreeding of anatomically modern humans with humans from various early human species.

Hygiene hypothesis Hypothesis establishing an inverse relation between allergy prevalence and indicators of lower hygiene, such as growing up on a farm, lower socioeconomic status, etc.

Hyperfrontality Excessively frontal activation during a task (see also contrasting **Hypofrontality**).

Hypersociability Behavior characterized by inappropriate and undistinguished behavior when approaching familiar or unfamiliar persons, leading to over-friendliness.

Hypofrontality Reduced or dysfunctional activation of the frontal lobes inducing failure or dysregulation of emotion, attention, and working memory. Part of the clinical syndrome in schizophrenia, depression, and attention-deficit disorder.

Hypomimia Reduction of facial expression and common symptom in Parkinson's disease.

Hyposmia Reduced faculty of smelling.

Hyposociability Abnormal social behavior characterized by inappropriate dismissive behavior.

Index of dexterity Quantification of the ability to effectively use the hands with precision.

Induced pluripotent stem cells (iPSCs) Skin or blood cells which have been reprogrammed back into an embryonic-like pluripotent state that can then be directed to become any type of cell. iPSC-derived human neural progenitor cells and neurons can be used to study cellular and molecular mechanisms underlying neurodevelopmental disorders in vitro.

Inflammaging Importance of chronic low-grade inflammation for the aging process.

Inflammasome Sensors and receptors of the innate immune system that trigger inflammation in response to an infection by secretion of inflammatory cytokines, etc.

Internal clock Internally driven rhythm governed by a circadian biochemical clock or oscillator located in the suprachiasmatic nucleus and intertwined with numerous metabolic processes. While genetically set at roughly 24 hours, it also depends on environmental cues as well as internal inputs (see also **Circadian clock**).

Intrauterine programming Genetic programming occurring during embryonic and fetal phases. Stimuli or insults at critical periods can have definite and lifelong consequences on structure and functions of organs.

Introgression desert Genetic sequences devoid of introgressed variation (see also **Adaptive introgression**).

Ionotropic receptors Ion channels opening by the binding of neurotransmitters and allowing ions to diffuse through the membrane following their electrochemical gradient. Synonym: "ligand-gated ion channels."

Lamelliform Processes showing the morphology of a thin plate, with one size (thickness) significantly smaller than the other two.

Law of effect Term used in education theory and later in behavioral psychology to describe the fact that learning can be accelerated by positive rewards or, in terms of behavioral psychology, that there is reinforcement of this behavior.

Lewy body Intracellular inclusion of protein aggregates, among them especially aggregated α-synuclein (see definition) and a pathological hallmark of so-called Lewy body diseases, among them Parkinson's disease and Lewy body dementia.

Life history Expression used in psychology and psychiatry to globally apprehend behavioral adaptations to social and ecological conditions as encountered during (early) lifetime with impact on survival, reproduction, and growth.

Ligand-gated ion channels See **Ionotropic receptors**.

Maladaptation Phenotypic trait that has become more harmful than helpful. Probably senseless co-evolution of another—fully adaptive—trait.

Massively parallel reporter assays (MPRAs) A reporter assay that is used to simultaneously measure the effect of thousands of sequences for regulatory activity. In contrast to traditional reporter assays that measure the protein level of the reporter gene, MPRAs are sequencing-based and measure the effect at the level of barcoded transcripts, allowing them to be multiplexed.

Mesolimbic dopamine system Dopamine networks in the midbrain, particularly the ventral tegmental area, considered to be part of the reward system of the brain, involved in motivational and affective functions, and innervating, for instance, the nucleus accumbens, amygdala, prefrontal cortex, etc.

Metabolome Total quantitative collection of metabolites present in a cell.

Metabotropic receptors See **G protein–coupled receptor**.

Metazoans Multicellular eukaryotic organisms deriving from an embryo with two or three tissue layers, developing tissues and organs, consuming organic material, reproducing sexually, and requiring oxygen. Humans belong to metazoans.

Microbiome Collection of all bacteria, viruses, and fungi living in direct contact with the human body (inside or outside).

Minicolumn Radially oriented network of neurons and smallest cerebral module for information processing. Constituted of around 100 neurons, all highly interconnected and with identical inputs and outputs.

Mitochondrial haplotype Population sharing similar sequences of mitochondrial DNA, thus unraveling the phylogenetic origins of maternal lineages.

Module Connectomic term describing communities of nodes that have more connections within their home community than with nodes elsewhere, thus only sparse intermodular connections.

mtDNA heteroplasmy Within any given mitochondria, there can be up to 10 copies of mitochondrial DNA (mtDNA). Within any given cell, there can be several hundred to 1,000 mitochondria. Thus, thousands of copies of mtDNA can exist within one cell. With only one DNA polymerase (the DNA "proofreader") and malfunctions in nuclear–mitochondrial intergenomic signaling, the error rate of mtDNA replication is far higher than that of nuclear DNA, leading to mtDNA depletion and the accumulation of mutations and deletions. Errors do not happen consistently, thereby causing differences in mtDNA copies between cells and within mitochondria. mtDNA heteroplasmy describes the state in which different variants of mtDNA copies coexist within a single cell or individual mitochondrion.

Myeloarchitecture Feature of myelination across the cerebral cortex, used to understand and visualize the cortical organization.

Natural selection Reproduction and survival of a species as dictated by random mutational gene changes with resulting impact on protein production and expression. Some of these genes are better adapted to a certain environment, and their carriers have therefore higher chances of survival and reproduction.

Negative selection Purifying natural selection hindering the spread of alleles detrimental in a certain environment.

Neontology Research domain dealing with living species. It contrasts with paleontology (see **Paleontology**).

Neophobia In general, fear of anything new or unfamiliar. Food neophobia means reluctance to eat novel food.

Neoteny A special type of heterochrony that is used to describe deceleration of developmental rate in one species compared to another.

Network theory Application of graph theory to the human brain when focusing on neuronal connectivity—distant or close—and by using concepts of graph theory, such as edges, nodes, etc.

Neural Darwinism Neurobiological theory applying the Darwinian concept of natural selection to higher brain functions, thus proposing selectionist pruning or strengthening of synapses and neurons, etc.

Neuroanatomical profile Suite of macro- and microstructural brain traits characteristic of a particular disorder, like Williams syndrome and autism spectrum disorders.

Neurodiversity movement Movement coined by the disability rights activist Judy Singer to sustain people who by the way their brain is organized, and functions are different from the majority of the population. They are neurodivergent. For instance, people with Tourette syndrome, autism, or attention-deficit disorder could be ranged here.

Neuroinflammation In response to an inflammatory challenge with impact on the brain, there is activation of the local, mostly innate immune system (microglia), which triggers a cascade of cellular and molecular reactions.

Neuron doctrine Concept considering the neurons as the fundamental functional units of the nervous systems as being polarized structures highly interconnected through synapses.

Neuronal workspace hypothesis Hypothesis proposing links between widely distributed territories of the human brain and thus enhanced development of the long-range—horizontal—cerebral connectivity.

Novelty seeking Personality trait expressing an increased tendency to actively search for new, thrilling, although potentially risky experiences, not considering the risk of deception or unforeseen dangerous situations.

Nychthemeral rhythm Natural biological oscillation that repeats approximately every 24 hours (see also **Circadian rhythm**). There are, for instance, nychthemeral rhythms in core temperature, heart rate, and endocrine secretions.

Old friends Organisms that coevolved with humans in the Paleolithic period, in such a way that humans became adapted to their presence and even can no longer perform well without them. The early development is the most important time for exposure to "old friends."

One Health Concept proposing a unified approach to the health of humans, animals, plants, and ecosystems, as interconnected at numerous levels.

Glossary | 529

Ontogeny Development of an organism from egg fertilization or a single cell to an adult.

Onufrowicz nucleus Also called Onuf's nucleus. Motor nucleus composed of motor neurons involved in the maintenance of urethral and rectal continence and thus controlling, for instance, the external anal sphincter muscle tone; located in the anterior horn of the sacral region.

Orthologous Two different genes present in different species, deriving from a common ancestor, exerting the same function.

Out of Africa model Model proposing that anatomically modern humans first appeared 100,000–200,000 years ago in East Africa and spread throughout the world from there around 50,000 years ago. Also called "African replacement" model.

Paleoneurology Research domain investigating brain evolution in extinct species.

Paleontology Research domain dealing with extinct species (fossils). It contrasts with neontology (see **Neontology**).

Perivascular space Anatomical space surrounding cerebral blood vessels. If macroscopically visible, called also "Virchow-Robin space."

PGO wave Phasic bursts of activation from the pons (P) to the lateral geniculate nucleus (G) and ending in the occipital cortex (O), supposed to trigger eye movements and imaging during REM sleep, as seen (mostly) before REM sleep.

Pharmacophagy Strategy of consuming chemicals from the environment in order to make one's own body toxic to predators or parasites.

Phonology Human ability to acquire—epigenetically—a basic lexicon, including symbols that map signals in particular sounds to concepts.

Phylogenesis Biological sequence of events by which a taxon appears during evolutionary development.

Phylogeny History of descent of a group of species from their common ancestor through evolution; can be applied also to the genealogy of genes (see also **Ontogeny**).

Plant secondary metabolites Toxins but also antioxidants derived from plant metabolism and targeting the nervous system of animal plant eaters.

Plexin protein Family of receptor proteins for signaling proteins of the semaphorin class. Involved, for instance, in axonal growth and guidance or signal transduction.

Polyalanine diseases (polyA) Diseases due to expansions of the polyalanine repeats. While occurring often in transcription factors, polyalanine tract expansions have been seen in congenital malformation syndromes (see also **Polyglutamine diseases**).

Polyglutamine diseases (polyQ) Diseases due to expansions of the trinucleotide polyglutamine repeats. Relatively common neurological diseases such as Huntington's disease and spinocerebellar ataxias are due to expansions of these repeats. Most often causing detrimental nuclear aggregates beyond a certain threshold (see also **Polyalanine diseases**).

Positive selection Also called "Darwinian selection," as favoring the spread of beneficial alleles for survival and reproduction.

Post-traumatic stress disorder (PTSD) Mental health disorder due to the exposure to or remembrance by specific cues of a threatening life event. Beyond physiological stress reactions, can trigger anxiety attacks, flashbacks, and involuntary memory intrusions and induce abnormal avoidance strategies.

Postural instability Falling tendency when standing or arising from a chair or bed (especially backward: retropulsion or sideward: lateropulsion). Testing of postural instability is part of the standard neurological examination of a patient at risk for Parkinson's disease. If present, indicates a higher risk for sudden falls.

Prionopathy Brain disease caused by accumulation of abnormally aggregated prion proteins. Prion proteins are proteins able to trigger misfolding of normally healthy proteins.

Progressive supranuclear palsy (PSP) Neurodegenerative disease caused by accumulation of aggregated, misfolded tau proteins and causing impairment of eye movements, postural stability, gait, speech, and cognition (see also **Corticobasal degeneration**).

Protein aggregation Pathological process seen in neurodegenerative diseases in which erroneously assembled or misfolded proteins assemble and form insoluble aggregates.

Proximate causes Direct physiological, biochemical, or genetic explanations of a behavior ("how"). Remains muted on the phylogenetic background (see **Ultimate causes**).

Pruning Epigenetic elimination of supernumerary dendritic spines.

Random gene flow Gene flow where all possible genotypes are equally likely to disperse in the same manner.

Reafference Fact that an organism is able to predict the sensory feedback of its actions or an effect of an organism's sensory mechanisms that is due to the animal's own actions.

Recent evolutionary change Changes in the human brain after the split from the last common ancestor with the chimpanzee.

Relaxed natural selection Reduced impact of natural selection through premature mortality and differential fertility, thus changing the mutation/selection balance.

REM atonia Characteristic loss of muscle tone during REM sleep, essentially due to inhibition of motor neurons in the spinal cord.

Repeat-associated non-AUG translation (RAN) Aberrant translation mediated by expanded repeats; enables the ribosome machinery to utilize a near-AUG codon for initiation.

Reporter activity Expression of a reporter gene (usually a fluorescent protein) placed downstream of a non-coding DNA sequence with unknown function. The activity of a reporter gene reflects the level and pattern of the non-coding DNA sequence activity.

Resting-state functional magnetic resonance imaging Neuroimaging method measuring spontaneous low-frequency fluctuations in the BOLD signal (see **BOLD**), thus providing a window into the functional architecture of the brain. Different resting-state networks have been defined, among them the default mode network (see **Default mode network**).

Reticulation Applied to human evolution, this term means that evolution is tree-like, not linear.

Reward theory Psychological theory describing the fact that incentives and reinforcement ("reward") can motivate people to perform a certain action.

Rigidity Increased muscle tone, characterized by an increased resistance to passive movements.

Secular trends Trends in phenotypic characters that are observed over periods of approximately a century, especially those concerning quick increases in body size; seen as adaptive responses to improved living conditions.

Selection shadow "Blind spot" due to ineffectiveness to eliminate deleterious mutations expressed in later life periods. In other words, selection pressures become ineffective, once sexual maturity has been passed, as they do not consider deleterious mutations expressing their effect only with old age.

Sensory-specific satiety Phenomenon seen, for instance, in mammalians whereby a food becomes less desirable the more it is eaten (decline in pleasantness), in comparison to a food that has not yet been eaten. However, not restricted to mammalians and to gustatory perceptions.

Serotonin transporter immunoreactivity Histological technique to establish the density of serotonin transporters in the brain. Widely used in animal modeling for evaluation of the effect on various drugs inhibiting the transport of serotonin.

Sickness behavior Summarizing term for numerous behavioral changes during an infectious or autoimmune syndrome, all expressing the host's adaptation to this syndrome by reorganization of the priorities. Immobilization and sleepiness are main features. Depression may share similar immunoinflammatory pathways.

Slow virus disease Viral disease with long latency period and, mostly, slowly progressing clinical phase. Misleading term, as also in use for prionopathies (see **Prionopathy**).

Social brain Concept describing an assembly of brain regions governing social interactions. Classically the amygdala, the temporal cortex, and the orbital frontal cortex are considered to be part of the social brain.

Southern Dispersal theory Theory proposing a coastal dispersal of modern humans from the Horn of Africa crossing the Bab-el-Mandeb to Yemen at a lower sea level. These early modern humans may have populated Southeast Asia and Oceania. This theory could explain the discovery of early human sites in these areas much earlier than those in the Levant.

Split-hand syndrome Description of the fact that in amyotrophic lateral sclerosis there is preferential muscle wasting of the thenar, while the hypothenar muscles are relatively spared. Thought to be caused by particular vulnerability of the cortico-motoneuronal projections to the thenar motoneurons over and above a more general lower motoneuron disorder.

Synapse first hypothesis Hypothesis suggesting that the evolution of the synapse was a major driving force in the evolution of the central nervous system.

α-Synuclein Neuronal protein regulating synaptic vesicle trafficking. It is encoded by the gene SNCA. Misfolded and aggregated α-synuclein is a major component of Lewy bodies (see **Lewy body**), found in patients suffering from Parkinson's disease or from Lewy body dementia.

Tandem repeats "Tandem" means located directly one after another. Up to six repeats one after another are called "short tandem repeats."

Tau Abbreviation for tubulin-associated unit; group of six soluble intracellular protein isomers maintaining the stability of the axonal microtubules; produced by alternative splicing of the gene MATP.

TDP43 TAR DNA–binding protein of 43 kDa.

Tectum opticum Phylogenetic correspondent of superior colliculus.

Telencephalization Progressive migration of regulatory cerebral functions from lower brain regions to the cortex inducing growth and higher complexity of the cortex.

Theory of mind Cognitive ability to infer mental states of other persons and, in this way, to explain and predict the actions of these other persons.

Therian Subclass of mammals (Theria) including marsupials and placental mammals.

Threat simulation theory Theory claiming that dream contents repetitively simulate threatening events and that this biological function has been selected by evolution as a defense mechanism. According to its proponents, threat simulation during sleep probes the mechanisms necessary for successful perception and avoidance of threats, and thus indirectly increases the likelihood for reproduction and survival during human evolution.

Three-dimensional polarized light imaging (3D-PLI) Neuroimaging technique showing fiber architecture at micrometer resolution in postmortem brains. It uses the birefringence of the myelin sheath to visualize fiber bundles as well as single axons.

Trade-off Observation that an evolutionarily beneficial change in one trait is associated with a detrimental change in another trait. When considering human brain evolution and the high energy expenditure of neurons, the "expensive tissue" hypothesis has proposed a trade-off between the size of the human brain (expansion) and the size of the human digestive tract (retraction). However, the trade-off concepts are often oversimplifying. For instance, retraction of the human gastric tube was also a consequence of the switch from starch-rich vegetarian to calory-rich cooked protein nutrition.

Tripartite synapse Functional unit composed of neuronal presynapse and postsynapse and astrocytic synapse, allowing bidirectional communication between neurons and astrocytes.

Ultimate (or evolutionary) causes Evolutionary explanations for phenotypic traits or behaviors as well as genotypic characteristics such as allele frequencies. Ultimate causes complete comprehension of a specific feature incompletely apprehended by "proximate" causes only.

Universality One of three principles coined by Crow to describe the pathogenesis of schizophrenia. It means here that schizophrenia has been identified in biologically and culturally diverse populations and societies across the globe (see also **Constitutionality** and **Continuity**).

Valence lateralization hypothesis Describes the concept of valence-specific organization of emotion processing in both hemispheres; that is, positive emotions are processed in the left hemisphere and negative emotions in the right one; in contrast to the right hemispheric hypothesis, which postulates a general right hemispheric dominance for all emotions.

Variable number tandem repeat Location in DNA where a short nucleotide sequence is seen as a tandem repeat. However, there may be length variations between different subjects. Usually, the number of repeats varies between 10 and 100.

Vomeronasal organ An accessory olfactory organ situated on the anteroinferior third of the nasal septum, playing only an accessory role or no role at all in adult human olfaction, as rudimentary development in humans. It is involved with pheromone detection in several tetrapode species.

von Economo neuron Projection neuron located in anterior parts of the insula and cingular cortex. It has a characteristic bipolar spindle configuration and allows rapid communication in large brains (also seen in elephants and great apes). Abnormal development linked to psychotic disorders.

What stream Also called the "ventral stream." It leads to the temporal lobe, where there is identification and recognition of objects. Is part of the two-streams hypothesis, proposing segregation of visual and auditory inputs (see also **Where stream**).

Where stream Also called "dorsal stream." It leads to the parietal lobe, where there is processing of the spatial localization of an object (in relation to the viewer). It is part of the two-streams hypothesis (see also **What stream**).

Zeitgeber External visual cues occurring at regular intervals and helping the organism to maintain each day a rhythm of 24 hours. These can be biological cues, such as sunlight, or social cues, such as regularity of working or eating hours.

Index

For the benefit of digital users, indexed terms that span two pages (e.g., 52–53) may, on occasion, appear on only one of those pages.

Tables and figures are indicated by an italic *t* and, *f* following the page/paragraph number.

acetylcholine, evolutionary history of, 452–53
"acting out of dreams," 270–71
   See also rapid eye movement (REM) sleep behavior disorder (RBD)
active sleep, 271, 285
   definition, 521
   early REM sleep and, 273–74
AD. *See* Alzheimer's disease (AD)
adaptation, concept of, 194–96
adaptive archaic introgression
   functional consequences in human cranium and brain-related traits, 143–47
   functional consequences of adaptive, from Denisovan, 142–43
   functional consequences of adaptive, from Neanderthal, 140–42
   functional value of introgressed regions, 136–47
   future work and unanswered questions, 147–48
   genomic signatures of adaptive, 138–47
   key points to remember, 149
   phenotypic impact of introgressed variants from ancient genomes on modern human systems, 137*f*
   selection against introgressed regions at genome and loci levels, 137–38
adaptive introgression
   definition, 521
   genomic signatures of, 138–47
   process, 138–39

   See also adaptive archaic introgression
addiction
   classes of compounds metabolised by cytochrome P450 (CYP) enzymes, 379*t*
   clinical and therapeutic implications, 388
   counterarguments and evidence against evolutionary influences, 384–88
   "drugs" as evolutionary novelty, 386–88
   evolutionary evidence relating to human health and disease, 377–83
   evolutionary molecular modeling, 380*f*
   evolved behavioral defenses, 381
   evolved biological defenses, 379
   functions of mesolimbic dopamine system (MDS), 384
   future work and unanswered questions, 390
   goal-directed and habitual control, 168*f*
   goal-directed behavior, 172
   habitual behaviors of drug, 170–71
   historical background, 373–76
   human drug use and, 378
   impact of abused drugs, 170
   key points to remember, 392
   main classes compounds metabolized by human CYPs, 380–81
   non-evolutionary history of current drug reward theory, 374–76
   plant secondary metabolites (PSM) and, 373–74, 382–83

534 | INDEX

addiction (*cont.*)
    punishment and co-adaptation, 377–78
    Reefer Madness movie poster (1936), 390*f*
    relationships between commonly used plant drugs and human nervous system receptors, 382*t*, 383
    sensitivity and domains of cognitive functioning, 384–85
    standard model of, 374–75
adenosine triphosphate (ATP), 58
    oxidative phosphorylation for, 60–61
ADHD. *See* attention-deficit hyperactivity disorder (ADHD)
admixture, 133
adult humans
    rapid eye movement (REM) sleep and dreaming, 274–75
    regulation of emotions, 274–75
    rehearsing instinctual behaviors, 275
    simulating threats, 275
    thermoregulation, 274
African populations, genetic variation, 148
aging
    epigenetic modulation of mtDNA, 71–72
    evolutionary theory of, 427–30, 438–39
    evolution-based model of causation (EBMC), 434–37, 436*f*
    free-radical aging theory, 61
    inflammation and, 441
    manipulating, 437
    mitochondria-derived damage-associated molecular patterns (DAMPs), 62
    mitochondrial contribution to, 60–62
    natural selection and, 435–37
    pleiotropy and, 481–82
    postponed, 437–39
aging-related tau astrogliopathy (ARTAG), 48
    definition, 521
alcoholism, goal-directed behavior, 172
Allen brain cell type database, 416, 417
allocortex, cerebral cortex, 15–16
allostatic load, definition, 521
ALS. *See* amyotrophic lateral sclerosis (ALS)
Alzheimer's dementia, 3, 4
Alzheimer's disease (AD), 12–15
    aggregation of ß-amyloid plaques, 190–91, 193
    amyloid deposition, 48
    antagonistic pleiotropy in late-onset, 184–86
    ApoE4 allele as risk factor, 438, 481
    apolipoprotein ε (APOE-ε) gene, 184–86
    clinical and therapeutic implications, 195
    counterarguments and evidence against evolutionary influences, 194–95
    depression, 304
    disease-specific neuronal groups, 81–82
    DNA methylation, 122
    downside of complex parietal cortex, 192–94
    future work and unanswered questions, 195–96
    glial cells, 49–50
    glial tauopathy, 47–48
    human and nonhuman primates, 105
    human-specific DNA evolution in, 111
    hygiene hypothesis, 440–41
    induced pluripotent stem cells (iPSCs), 48–50
    insulin resistance, 67
    key points to remember, 198
    limbic neurons as internal templates, 91*t*
    long silent period of, 516–17
    memory and its disruption by tau deposition, 93–96
    mitochondrial dysfunction, 62, 73–100
    neurite architecture, 82
    parietal cortex and, 190–91
    past, present, and future of, 97–99
    reactive oxygen species (ROS) in, 44–45
    shift from four-repeat (4R) to three-repeat (3R) tau, 93–96, 94*f*
    social brain and dementia, 197
    structural and functional brain changes in parietal lobe in, 191–92
    superoxide dismutase (SOD) supplementation, 514
    tau accumulation, 99
    testing the hypothesis, 195–96
    von Economo neurons (VENs), 22
American College of Neuropsychopharmacology (ACNP), 449, 450–51
American Psychiatric Association, 304
amygdala
    multimodal dysfunction in Parkinson's disease, 211*f*
    Parkinson's disease symptoms, 211–12
    William's syndrome and autism, 347–48
amyloid deposition, 48
amyotrophic lateral sclerosis (ALS)
    animal model for ALS, 260–63, 261*b*
    Betz cells, 259
    clinical and therapeutic implications, 263
    CM connections and evolution of dexterity, 256–57
    cortico-mononeuronal (CM) cells, 254*f*
    counterarguments and evidence against evolution, 263
    disease-specific neuronal groups, 81–82
    evolutionary evidence related to, 255–60
    evolution of primate motor cortex, 257–58
    fast-conducting CM system, 259
    future work and unanswered questions, 263–64
    glial cells, 49–50

historical background of corticospinal tract
(CST) in, 252–55
key points to remember, 264
mitochondrial dysfunction, 62
motor neurons as internal template for, 91t
neurite architecture, 82
past, present, and future of, 97–99
pattern of weakness in upper and lower limb
muscles, 255f
primate features of CST, 257–58
range of axon sizes within human CST, 258
repeat expansion, 235
selecting animal models to study, 260–63
skilled hand movements and, 260, 262–63
species differences in corticospinal
connectome, 259–60
straight axons for, 87–89, 88f
straight myelinated axons of, 88f, 88–89
superoxide dismutase (SOD)
supplementation, 514
anal sphincter function, motor neurons, 88–89
anatomically modern humans (AMH)
definition, 521
future work and unanswered questions, 147–48
interbreeding between Denisovans and, 135–36
interbreeding between Neanderthals
and, 133–35
interbreeding of modern humans and, 133–36
interbreeding with "ghost" populations, 136
out of Africa (OOA) model, 132–33
See also archaic introgression
anatomical tracing, 400–1
anhedonia
definition, 521
major depression disorder (MDD)
symptom, 307
animal models
brain research, 10b
human brain diseases of connectivity, 415–16
animals and REM sleep behavior disorder (RBD)
comparing movement phenomenology in
different species, 277t
Jouvet's pioneer studies in cats, 276
RBD-like behaviors in dogs, 276
REM sleep in different, 272f
rodent models of RBD, 277–78
See also rapid eye movement (REM) sleep
behavior disorder (RBD)
anorexia nervosa, 510–11
schizophrenia and, 323
antagonistic pleiotropy, 184–86, 195–96, 481
definition, 521
anterior cingulate cortex (ACC), 335–36
William's syndrome and autism, 345–46
Anthropocene, 479–80

apathy, Parkinson's disease, 218
aphasic syndrome, 503–4
apomorphy, 40–42
definition, 521
appetite changes, major depression disorder
(MDD) symptom, 306–7
archaic hominins, 131, 132f, 147–48
Aristotle, 154
ARTAG (aging-related tau astrogliopathy), 48
definition, 521
association centers, 492–93
astrocyte(s), 38–40
aging-related tau astrogliopathy, 48
comparison between murine and
human, 41f
coupling of neurons and, 45
emergence of, 40–42
lactate cycle, 45–47
mammalian species, 43
phylogenesis of, 47
three-dimensional models of, 46f
vertebrates, 42
astrocyte-neuron-lactate shuttle (ANLS)
model, 45
astrogliosis, 47–48
definition, 521
astron, 39–40
attention-deficit hyperactivity disorder
(ADHD), 483
clinical and therapeutic implications, 368–69
counterarguments and evidence against
evolutionary influences, 368
etiology, 361–62
evolutionary evidence related to human health
and disease, 363–67
evolutionary mismatch, 367
future work and unanswered questions, 369
glial cells, 49–50
group selection, 366
historical background, 360–63
intrauterine programming, 364
key points to remember, 369–70
life history theory, 364–66, 365b
neurodevelopmental disorder, 360
neurodiversity and advantages of, 367
neurotransmitters and genetic risk
factors, 362–63
treatment, 363
auditory areas, mammalian and primate
brains, 24–25
AUG, 231, 232f, 238
definition, 521
Australopithecus, 491–92
autism, 4, 483
autism phenotypes, 483

536 | INDEX

autism spectrum disorder (ASD), 334
   amygdala, 347–48
   Broca's region, anterior cingulate cortex (ACC), and frontoinsular cortex, 345–46
   clinical and therapeutic implications, 350–52
   counterarguments and evidence against evolutionary influences, 348–50
   evolutionary evidence related to neurodevelopmental disorders, 336–48
   frontal lobe, 337–46
   future work and unanswered questions, 353–54
   glial cells, 49–50
   heterogenous group of disorders, 335
   historical background of, 335–36
   human-specific DNA evolution in, 110f, 111
   human-specific gene regulatory changes, 119–20
   human-specific regulatory changes, 121f
   induced pluripotent stem cells (iPSCs), 48–50
   key points to remember, 354
   microstructural findings in, 342t
   neuroimaging findings in, 339f
   prefrontal cortex (PFC), 338–45
   schizophrenia and, 323–24
   social skills and, 105
   temporal lobe, 346–48
   von Economo neurons (VENs), 22
autobiographical frame, precuneus as "eye of the self," 197
auto-domestication
   definition, 521
   humans, 476
   term, 485
autophagy, 65, 73, 83–84, 233–34
   definition, 521
autosomal dominant diseases, 229
availability, drugs, 386
avian species, brain layouts in, 19f–20
axodendritic synapse, 93–94
   definition, 521
axosomatic synapse, definition, 522
axons. *See* highly branched axons

Bain, Alexander, 160–61
Baratela-Scott syndrome (BSS), 229
basal ganglia (BG), 205–6
   goal-directed to habitual behavior, 162–63
   micro- and macro-anatomy of, 208
   Parkinson's disease, 210
bat, brain layout, 19f–20
bauplan
   definition, 522
   human brain, 12–18
behavioral theory, standard model, 374–75
behaviors. *See* goal-directed and habitual behaviors

belt and parabelt organization of the auditory system, 24–25
   definition, 522
Bernard, Claude, 98–99, 214
beta-amyloid, 190–91, 193
   definition, 522
betweenness centrality, 188f
   definition, 522
betz cells, 259
   definition, 522
Bichat, Xavier, 38–39
biocultural research, schizophrenia, 328
bioenergetics, mitochondrial, 66–70
   dysfunctional mitochondrial metabolism, 67–69
   hypoperfusion (reduced blood flow) of brain, 67
   insulin resistance, 67
   mitochondrial calcium dyshomeostasis, 69–70
biological normalcy, schizophrenia, 326
biological plasticity, 478
biomediators, 451
biomodulators
   adaptive properties of, 452
   definition, 522
   term, 451–52
bipedal gait, Parkinson's disease, 216–17
bipedalism, 98
   definition, 522
bipolar disorder, DNA methylation, 122
blindsight, 219
   definition, 522
   Parkinson's disease, 218–20
blood-brain barrier (BBB), 43–44
   formation, 22
   infection protection, 514–15
blood oxygen level-dependent imaging (BOLD), 191–92
   definition, 522
Boas, Franz, 317–18
body movements, rapid eye movement (REM) sleep, 273–74
body temperature, brain and heat regulation, 512
borderline personality disorder, 483–84
Braak-prion hypothesis, 94–95
   definition, 522
bradykinesia, 205–7
   definition, 522
   Parkinson's disease, 215
brain
   asymmetries, 27–29
   cartography, 399
   hypoperfusion (reduced blood flow) of, 67
   size and number of neurons, 20–21
brain connectivity, behavioral control options, 165–66
brain disease(s)

glial cells in, 47–48
role of glutamate in, 464–65
See also designing strategies for brain diseases
brain diseases with unstable repeats
   clinical and therapeutic implications, 240–41
   computational algorithms for evaluating repeat expansions, 243*t*
   counterarguments against evolutionary influences, 239–40
   evolutionary evidence related to, 236–39
   future work and unanswered questions, 241–44
   historical background, 228–36
   identifying pathological repeat expansions, 241
   key points to remember, 244–45
   modifiers of expanded alleles, 240
   need for temporal and spatial control, 240–41
   PCR-free library preparation and instrumentation, 242
   pipelines for informatic analyses, 242–43
   range of repeat expansion disorders, 229, 235–36
   therapeutic interventions, 240–41
   understanding neurological function and evolution, 244
   3' untranslated regions (UTRs), 229, 235–36
brain evolution
   glial metabolism supporting, 43–47
   human natural history, 181–82
brain globularity, parietal development and, 187–90
   definition, 522
brain hominization, 492–96, 504
   expansion of neocortex, 492–93
   extended postnatal development of human, 496
   global neuronal workspace, 494–96
   increased number of cortical areas, 493–94
   multilevel processing, 494–96
   pace of genetic changes, 504–5
brain layouts, mammalian, avian, and reptilian species, 19*f*–20
brain physiology, 472–73
brain research, mammalian model species in neuroscience, 10*b*
brainstem nuclei, Parkinson's disease, 213
brain studies. *See* comparative brain connectomics
Broca's region, 17, 25, 26–28
   William's syndrome and autism, 345–46
Brodmann's map, 16, 340*f*
Buffon, 271–72
burnout, depression induced by long-term stress, 300–1
Burr, Harold Saxton, 39–40

Caenorhabditis, 500
Caenorhabditis elegans, 11*b*, 416
   connectivity data, 417*t*

evolutionary evidence of, 40–42
schizophrenia-related genes, 325
caffeine, 380
   evolutionary molecular modeling, 380*f*
Cajal, Santiago Ramón y, 22
calcium dyshomeostasis, mitochondrial, 69–70
cancer, depression, 304
cannabis, 387
   drug use in US, 388*f*
cascading loop, dopamine-related, 169
cat(s)
   brain layout, 19*f*–20
   studies of rapid eye movement (REM) sleep behavior disorder in, 276, 277*t*
catatonia, schizophrenia subtype, 314–15
catecholamines, evolutionary history of, 453–54
CCD model, 497–500, 498*f*
   definition, 522
cell communication, evolution of, 457–58
cellular circuits
   brain network connectivity, 407–10
   extracellular space, 410
   glia, 409–10
   myelin, 410
   neurons, 408–9, 409*f*
   synapses, 410
central nervous system (CNS), 228, 450
   diseases targeting, 255–56
   evolution of, 451, 452
   glial cells in, 39–40
central pattern generators, 208, 214, 216, 222*t*, 283, 285–86
   definition, 522
cerebral cortex
   cytoarchitecture, 15–16
   frontal lobe, 17
   insular lobe, 18
   limbic lobe, 18
   lobes of, 16–18
   occipital lobe, 17
   parietal lobe, 17
   specialized cortical areas, 493–94
   subdivision, 16
   temporal lobe, 18
cerebrospinal fluid (CSF)
   production with age, 513–14
   thermal exchange with, 512
   waste clearance, 513
CG methylation, 119
   definition, 522
CH methylation, 119
   definition, 522
Changeux Courrège Danchin 1973 (CCD model), 497–500, 498*f*
   definition, 522

Chaplin, Charlie, 284
chemically induced depression, 304
chemokine receptor (CCR), 142
   definition, 522
chicken, brain layout, 19f–20
chimpanzee(s)
   brain layout, 19f–20
   selection of neuroanatomical measures for, 14t
   white matter fiber tracts and network connectivity, 403f
Chimpanzee Sequencing and Analysis Consortium, 380–81
Chinese ideography, 503–4
Chinese writing, 503–4
chromatin immunoprecipitation followed by sequencing (ChIP-seq), 106–7
circadian clock, 221
   definition, 522
cis and *trans* effects, 108–9, 113
   definition, 523
clade, 132
   definition, 523
classical neuropathology, terms, 209
clinical psychologists, 5–6
clinical syndromes, 209
   Parkinson's disease, 215–21
clustered regularly interspersed short palindromic repeats (CRISPR)-CRISPR-associate 9 (Cas9), 240–41
cocaine, drug use in US, 388f
cognitive reappraisal, 170–71
   definition, 523
collateralization, whole-grain organization, 411
comparative brain connectomics, 399–400
   access to connectivity data, 416, 417t
   animal models of human brain diseases of connectivity, 415–16
   association networks for macaques, chimpanzees and humans, 403f
   asymmetries in connectivity patterns, 405–6
   brain connectivity evolution and brain disease, 412–14
   bridging from macro to micro, 406
   cellular circuits, 407–10
   connectivity-based parcellations, 401–2
   default mode network (DMN), 400–1
   diffusion tensor imaging (DTI), 400, 402–4
   dorsal thalamus as gateway to cortex, 412–13
   dynamic causal modeling, 402
   extracellular space, 410
   future directions, 416–18
   general comments on architecture, 410–11
   glia, 409–10
   hippocampal formation, 413–14
   key points to remember, 419–20
   macro level, 400–5
   micro level, 406–10
   myelin, 410
   network theory, 404–5
   neurons, 408–9, 409f
   origin and impact of collateralization, 411
   synapses, 410
   three-dimensional polarized light imaging (3D PLI), 406, 407f, 419b
   white matter fiber tracts for macaques, chimpanzees and humans, 403f
compulsivity, goal-directed behavior, 172, 173
computational algorithms, evaluating repeat expansions, 243t
connectivity-based parcellations (CBP)
   brain in rodents, monkeys, and humans, 401–2
   definition, 523
connectivity data, atlases and databases, 416, 417t
connectivity patterns, asymmetries in, 405–6
connectomes
   term, 400
   See also comparative brain connectomics
connectomic hypothesis, 492
constitutionality
   Crow's principle of, 321
   definition, 523
contextual cues, 159, 165–66
   definition, 523
continuity
   Crow's principle of, 321
   definition, 523
continuity hypothesis, 287
   definition, 523
continuous variation, 323–24
   definition, 523
controversies, 5
co-optation, term, 465–66
copy-number variation, 111, 140
   definition, 523
corollary discharge, 273–74, 411
   definition, 523
cortical areas
   auditory/vocalization areas, 24–25
   brain hominization, 493–94
   differentiation of multiple, 494–95
   olfactory areas, 25–26
   parcellation, 23–27
   prefrontal areas, 26–27
   somatosensory areas, 26
   visual areas, 23–24
cortical mapping
   goal-directed and habitual behavior, 163–65
   infralimbic cortex (ILC), 165
   orbitofrontal cortex (OFC), 164–65
cortical parcellation, 23–27
   definition, 523
corticobasal degeneration, 95–96, 281t

definition, 523
cortico-cortical connections, human cerebral connectome, 317f
cortico-motoneuronal (CM) connections, 252
　Betz cells and fast-conducting CM system, 259
　brain's motor network for complex movements, 256
　evolution of dexterity, 256–57
　importance of system, 252–54
　species differences in, 259–60
　synapse of CM cells, 254f
corticospinal tracts (CSTs)
　definition, 523
　involvement in amyotrophic lateral sclerosis/motor neuron disease (ALS/MND), 252
　primate features of, 257–58
　range of axon sizes within human, 258
corticostriatal relationship, during habits, 166
COVID-19, 514–15
CpG sites, 119
　definition, 523
cristae, term, 59–60
crocodile, brain layout, 19f–20
cross-cultural research, schizophrenia, 328
Crow, Timothy, 320
cucinivore, 463
　definition, 523
cultural brain, 491–92
cultural circuits, 503
cultural evolution, 476, 479
　early evidence for synapse selection, 499–500
　epigenesis by selective stabilization of synapses, 497, 498f
　epigenetic rules, 504–5
　epigenetic signature of writing and reading, 503–4
　epigenetic variability of neural circuits, 499
　key points to remember, 505
　synaptic epigenesis and origin of culture, 497–99
　tentative model for origin and evolution of oral language, 500–3
cyanobacteria, 468
　definition, 523
cybernetic, 91t
　definition, 523
cynomolgus monkey, 95–96
　definition, 523
cytoarchitectonic maps, Julich-Brain, 16
cytoarchitectonics, 16, 25
　definition, 523
cytoarchitecture, 15–16
　definition, 523
cytochrome P450 (CYP) pathways
　classes of compounds metabolized by, 379t, 380–81

evolution of, 377
cytopathology, 93–94
　definition, 523

damage-associated molecular patterns (DAMPs), 62, 73–74
　definition, 524
Daphnia magna, 500
Darwin, Charles, 2
Darwinian evolution, 491–92
Darwinism, 9, 497
deer, brain layout, 19f–20
default mode network (DMN)
　brain connectomics, 400–1
　definition, 524
deficient blindsight, 219–20, 222t
　affected areas and nuclei in, 220f
　Parkinson's disease, 218–20
deficient gait automatisms, Parkinson's disease, 216–17, 222t
deficient nictemeral regulation, Parkinson's disease, 221, 222t
deficient olfactory function, Parkinson's disease, 210–11, 222t
degree centrality values, 188f
　definition, 524
Dejerine, Joseph Jules, 503
delayed aging, 437
del Río-Hortega, Pío, 39–40
dementia
　social brain and, 197
　See also Alzheimer's disease
dementia with Lewy bodies (DLB), cognitive function for, 90
dendritic morphology, pyramidal neurons, 21f, 21–22
dendritic tau deposits, Alzheimer's disease, 93–96
dendrophilia, 501–2
　definition, 524
Denisova Cave, fossil discovery of, 135
Denisova human, 11–12
Denisovans, 2–3, 11–12
　adaptive introgression from, 140
　archaic introgression, 134f
　functional consequences of adaptive introgression from, 142–43
　genomic studies of introgression in early Eurasians, 147
　interbreeding between anatomically modern humans (AMH) and, 135–36
　interbreeding of Eurasians with, 142
　introgressed variation, 137–38
deoxyribonucleic acid (DNA)
　human-specific DNA sequences, 106–7
　mitochondrial DNA (mtDNA), 59–60, 61

deoxyribonucleic acid (DNA) sequences
  comparing human and nonhuman primates (NHP), 122–23
  evolution and function of human-specific, 108–9
depression, 4, 12–15
  attention deficit hyperactivity disorder (ADHD), 362–63
  chemically induced, 304
  Parkinsons' disease and, 15
  somatic diseases inducing, 304
  spring, 304
  starvation-induced, 304–5
  winter, 303–4
  See also major depressive disorder (MDD)
designing strategies for brain diseases
  animal models of, 415–16
  biome depletion paradigm, 433–34
  chronic diseases as "low-hanging fruits" or "higher-hanging fruits," 443*f*, 443–44
  evolutionary concepts, 427–34
  evolutionary health promotion, 439–42
  evolutionary mismatch, 430–34
  evolutionary thinking in medicine, 442–44
  evolution-based model of causation (EBMC), 434–37, 436*f*
  human-specific gene regulatory changes, 119–20
  hygiene hypothesis, 432–33
  key points to remember, 444
  natural selection versus age, 429*f*
  Old Friends hypothesis, 432–33
  postponed aging, 437–39
  preventive strategies, 426, 427*f*, 437–42
  selection shadow, 427–30, 429*f*
  survival versus age, 429*f*
dexterity, cortico-motoneuronal connections and evolution of, 256–57
diabetes, insulin resistance, 67
Diagnostic and Statistical Manual of Mental Disorders Fifth Edition (DSM-5)
  attention deficit hyperactivity disorder (ADHD), 360–61
  major depressive disorder (MDD), 294
diffusion tensor imaging (DTI)
  definition, 524
  schizophrenia brain imaging, 316
  tractography of fiber tracts in humans, 400
  white matter fiber tracts, 402–4, 403*f*
Digital Brain Bank, 416, 417
diseases of civilization
  definition, 524
  evolutionary mismatch, 431
distractibility, trait of, 364
DNA epigenetics, 497
DNA methylation profile, brain disorders, 122

dogs, rapid eye movement (REM) sleep behavior disorder (RBD) in, 276, 277*t*
dopamine
  addiction, 377, 389
  attention deficit hyperactivity disorder (ADHD), 362
  cascading loop, 169
  evolutionary history of catecholamines, 453–54
  goal-directed and habitual control, 167–68
  schizophrenia, 169
dopaminergic (DA) neurons
  death, 62, 63
  depletion of, for motor symptoms of PD, 90
  firing of, 83
  levodopa-responsive parkinsonism, 83–84
  Lewy body formation, 82
dopaminergic disease. *See* Parkinson's disease
dopamine theory, reinforcement learning, 375
dorsal thalamus, brain disease and, 412–13
dorsolateral striatum (DLS), habitual response with goal-oriented response, 163
dorsomedial striatum (DMS), goal-directed behavioral control, 163
dream action isomorphism, 281–82
  definition, 524
dream content, rapid eye movement (REM) sleep behavior disorder (RBD) and, 281–83
dreamlike behaviors, 276
Drosophila, 416
  connectivity data, 417*t*
  connectome project, 11*b*
  evolutionary evidence of, 40–42
  nervous system, 45
  schizophrenia-related genes, 325
drug reward theory, non-evolutionary history of, 374–76
drugs/drug use
  abusers and Parkinson-like symptoms, 62–63
  discovery of new, 468
  human, and addiction, 378
  US National Survey on Drug Use and Health (2020), 388*f*
  See also addiction
Duboisia hopwoodii (native shrub), 383, 386
Dunbar, Robin, 6
Durant, W., 154
dynamic causal modeling (DCM), connectivity, 402
dysbiosis, 306
  definition, 524
dysfunctional gait, Parkinson's disease, 215

EBMC. *See* evolution-based model of causation (EBMC)
EBRAINS, 416, 417

ecological punishment model
  addiction, 389
  "war on drugs" and, 389
ectoderm, 42
  definition, 524
Edinburgh Postnatal Depression Scale, 296
electroencephalographic (EEG) rhythm, rapid eye movement (REM) sleep, 272–73
electron microscopy, high-resolution three-dimensional, 46f
electron transport chain (ETC), 59
  cytochrome C of, 73
  efficiency of components, 59–60
emergent phenotype, 316
  definition, 524
emotional dysregulation, Parkinson's disease, 217–18
emotional pain, major depression disorder (MDD) symptom, 305
emotions, emergence of self, 82–83
encephalization, 29
endocranial cast, 183–84, 184f
  definition, 524
endocranial cavity, paleoneurology and casts of, 183–84, 184f
endogenous template, 379
  definition, 524
endosymbiont theory, 58
  definition, 524
endosymbiosis, 58–60
energy utilization, brain, 44–45
enteric nervous system (ENS)
  glial cells, 50
  spinal cord of Parkinson's disease, 214
environmental toxins, axon terminals as entry portal, 83–84
environment of evolutionary adaptedness, 209, 367
  definition, 524
ependyma, 42
  definition, 524
epidemic, major depression disorder, 295
epigenetic mismatch, biological explanations, 478
epigenetic modulation, mitochondrial DNA (mtDNA), 71–72
epilepsy, depression, 304
epinephrine, evolutionary history of catecholamines, 453–54
Equality Act of 2010 (UK), attention deficit hyperactivity disorder (ADHD), 368
Erlenmeyer-Kimling, L., 318–19
ethical issues, medicine, 485
European College of Neuropsychopharmacology, 449
evolution
  comparing human and chimpanzee proteins, 105–6
  counterevidence of novelty in human brain disorders, 120–22
  course of, 9
  human brain, 60
  See also ongoing human evolution
evolutionary concept, selection shadow, 427–28, 429f
evolutionary evidence
  Alzheimer's disease (AD), 111
  autism, 111
  counterarguments, 194–95
  human health and disease, 184–94
  human-specific DNA in human brain disorders, 110–11
  human-specific DNA sequences, 108–9
  overview of human-specific genomic changes, 110f
  schizophrenia, 110–11
evolutionary health promotion, prevention research, 439–42
evolutionary medicine, 479–80
  natural selection, 481
evolutionary mismatch, 477–78
  attention deficit hyperactivity disorder (ADHD), 367, 368
  biome depletion paradigm, 433–34
  brain diseases, 430–34
  concept of, 516–17
  definition, 524
  hygiene hypothesis, 432–33
  Old Friends hypothesis, 432–33
evolutionary molecular modeling
  definition, 525
  term, 374
evolutionary point of view, goal-directed and habitual behaviors, 157–58
evolutionary science, 1–2
evolutionary theory, aging, 427–30
evolutionary tree
  archaic and modern hominins, 132f
  See also human evolutionary history
evolution-based model of causation (EBMC), 434–37
  definition, 525
  early-onset genetic effects (EOGE), 435, 436f
  evolutionary health promotion, 439–42
  evolutionarily conserved environmental factors (ECEF), 435, 436f
  evolutionarily recent environmental factors (EREF), 435, 436f
  late-onset genetic effects (LOGE), 435–37, 436f
exaptation, 210
  definition, 525

Expression of the Emotions in Man and Animals, The (Darwin and Prodger), 2
extracellular space, cellular circuits, 410

facial expression, Parkinson's disease, 217–18
fatigue, major depression disorder (MDD) symptom, 307
feelings, emergence of self, 82–83
fentanyl, 387–88
    addiction, 389
"fight or flight"
    definition, 525
    during REM sleep, 212–13
    program, 3, 208–9
fight or flight reactions, 82–83
    dream content of RBD behaviors, 283
    RBD and threat simulation theory (TST), 288
    rehearsal of, for survival, 286
first brain, term, 214
fitness (evolutionary sense), 182, 193–94
    definition, 525
fluorodeoxyglucose
    definition, 525
    positron emission tomography (^{18}F-FDG PET), 47
focal Lewy body disease, 86–87
folate-sensitive fragile site 12 (FRA12A), 229
folate-sensitive fragile site on chromosome 2 (FRA2A), 229
folate-sensitive fragile site XE (FRAXE), 229
Folley, Bradley, 322–23
FOXP2 gene, 112–13, 124
    definition, 525
fragile X-associated disorders, prototypic, 229–31
fragile X-associated primary ovarian insufficiency (FXPOI), 231, 237
fragile X-associated tremor/ataxia syndrome (FXTAS), 231, 237
fragile X- mental retardation protein (FMRP), 237
    repeat-associated non-AUG (RAN) translation, 238–39
fragile X-related disorders
    fragile X messenger ribonucleoprotein (*FMR 1*) genes, 228, 244
    understanding neurological function and evolution, 244
    See also brain diseases with unstable repeats
fragile X syndrome (FXS), 229
    definition, 525
    induced pluripotent stem cells (iPSCs), 48–50
free-radical aging theory, 61
    definition, 525
Friedrich's ataxia
    definition, 525
    dynamic repeats in introns, 234–35

frontal lobe, 17
    neuroimaging studies of William's syndrome and ASD, 337–46
    prefrontal cortex (PFC), 338–45
frontoinsular cortex, William's syndrome and autism, 345–46
frontotemporal dementia
    behavior of, 170
    repeats in, 237
frontotemporal dementia (FTD), repeat expansion, 235
Fuchs' endothelial corneal dystrophy type 3 (FECD3), 235
functional diversity, glial cells, 49
functional magnetic resonance imaging (fMRI), 47, 207, 209
    connectivity profiles, 400–1
    illiterate and literate groups, 503
    parietal lobe in AD, 191–92
    Parkinson's disease, 207, 224
functional neuroimaging, 401

gamma-amino butyric acid (GABA)
    addiction, 389
    evolutionary history of, 456
GDS. *See* goal-directed and habitual behaviors
gene frequencies, regional differences in, 482
gene-profiling studies, 44–45
gene regulatory changes. *See* human-specific gene regulatory changes
gene regulatory elements (GREs)
    comparisons in adult brains, 118–19
    definition, 525
genetic drift, 138, 473, 482
    definition, 525
genetic expression, 2–3
genetic manipulation, glial cells, 51
genetic morphism, schizophrenia as, 318–19
genetic risk factors, attention deficit hyperactivity disorder (ADHD), 362–63
genome-wide association studies (GWAS), 107
    attention deficit hyperactivity disorder (ADHD), 361
    mental health phenotypes, 145
    schizophrenia, 322–24, 328, 329
genomic features, identifying human-specific brain disease, 107
genomic sequences, comparing human and nonhuman primates, 106–7
geophagy, 467
    definition, 525
geroscience hypothesis, 437
    definition, 525
"ghost" populations
    archaic introgression, 134*f*

definition, 525
  interbreeding between anatomically modern humans (AMH) and, 136
ghost tangles, 94f, 94–95
  definition, 525
glia, 38
  cellular circuits, 409–10
glial biomarkers, 50
glial cells
  disease, 49–50
  enteric nervous system, 50
  evolutionary evidence, 40–43
  functional studies, 51
  functions of, 40
  future of biology of, 50–51
  genetic manipulation, 51
  in health and disease, 47–50
  heterogeneity, 49
  historical background, 38–40
  history of research, 40
  key points, 51
  metabolism supporting brain evolution, 43–47
  morphology of, 39–40
  neuron-glia interactions, 49
  therapeutic interventions, 51
glial cells missing (gcm), 40–42
gliopathy, 47–48
  definition, 525
global activity, synaptic epigenesis, 497–99
Global Burden of Disease Collaboration, multiple sclerosis, 482
global developmental delay, progressive ataxia and elevated glutamine (GDPAG), 229
Global Health, 486
global neuronal workspace (GNW)
  brain hominization, 494–96
  connectivity of brain and language, 501–2
  functional organization, 501–2
  hypothesis, 495–96
  organization in postnatal development, 501–2
glossogeny, 502
  definition, 525
glutamate
  addiction, 389
  evolutionary history of, 455–56
  excitatory neurotransmitter, 468
  role in brain disease, 464–65
glutamate-glutamine cycle, 45–47
glutamate in brain health and disease
  energy-relevant facts about human brain, 463–64
  short story on human brain evolution, 462–63
glycogen distribution, analysis of, 46f
glycolysis, 60–61, 66–69
  definition, 525

glymphatic function, human brain, 513–14
glymphatic system, 199
  definition, 525
GNW. *See* global neuronal workspace (GNW)
goal-directed and habitual behaviors
  basal ganglia, 162–63
  behavioral stages across lifetime and cell loss in Parkinson's disease, 158f
  brain connectivity in control options, 165–66
  characteristics of, 156t
  clinical and therapeutic implications, 169–70
  cortical mapping of, 163–65
  corticostriatal relationship during habits, 166
  depictions of, 155f
  evolutionary point of view, 157–58
  fundamental behavioral selection mechanisms, 155–57
  future work and unanswered questions, 172
  habitual overload hypothesis, 157f
  historical background of, 154–55
  importance of training, 160–61
  infralimbic cortex (ILC), 165
  key points to remember, 172–73
  neural networks from evolutionary perspective, 162–65
  neurotransmitters involved in, 167–68
  orbitofrontal cortex (OFC), 164–65
  practice makes perfect, 156
  stimulus-response (S-R) associations, 155f
  summary of, 171–72
  theories of, 159
  therapeutic lessons and opportunities, 170–71
  transmission between, 168f
goal-directed system (GDS), 155
  characteristics of, 156t
  definition, 526
  See also goal-directed and habitual behaviors
Gogi's aphasia, 503–4
Golgi, Camillo, 39f, 39
Goulas, Alexandros, 492
G protein, 457–58
  definition, 526
G protein-coupled receptors (GPCRs), 459–60
  definition, 526
  evolution of, 460–61
graded action potential, 21–22
  definition, 526
grief, depression induced by, 302
Grillner, Sten, 208
group selection, 366
  definition, 526
guilt, major depression disorder (MDD) symptom, 308
gut microbiota, mood disorders and, 308–9

gyrification, 22–23
  definition, 526

habit, 167
  definition, 526
habitual behaviors. *See* goal-directed and habitual behaviors
habitual overload hypothesis, 157*f*
habitual system (HS), 155–56
  characteristics of, 156*t*
  definition, 526
  See also goal-directed and habitual behaviors
hallucinogens drug use in US, 388*f*
haplogroup, mitochondrial, 71
haplotypes, mitochondrial, 71
Harmon, Denham, 61
HARs. *See* human accelerated regions (HARs)
heat regulation, human brain, 512
hedonic model, drug reinforcement, 387–88
heterochrony, 113, 114
  definition, 526
heterogeneity, glial cells, 49
heteroplasmy, 61
  definition, 526
hierarchy conflict, depression induced by, 301–2
highly branched axons
  axon-first pathogenesis by, in Lewy body disease, 83–84
  closed-loop feedback, 89*f*, 90
  distribution of ALS-affected neurons and Lewy-prone neurons, 88*f*, 88–89
  fluctuation of symptoms of, without real-time feedback, 90–92
  hyperbranching axons in Lewy pathology, 83–84, 85*f*
  Lewy-prone neurons, 97
  motor, nigrostriatal, and limbic neurons as internal templates, 91*t*
  multifocal Lewy body disease with capricious spread, 84–87
  open-look feedback, 89*f*, 90
  output through hyperbranching axon without feedback, 89*f*, 90
  parallel involvement of Lewy-prone systems, 83–84, 85*f*
  straight axons for ALS versus, for Lewy pathology, 87–89
  See also intrinsic templates
high-throughput transcriptomic profiling assays, 106–7
hijack hypothesis, 384
hijack model
  addiction, 378
  evolutionary perspective, 385
Hilgetag, Claus, 492

hippocampal connectome, 414
hippocampal formation, brain disease and, 413–14
hippocampus, 15–16
Hoffer, Abram, 318–19
Holocene, 474–75
hominin, 131, 132*f*, 147–48
  definition, 526
hominization, 491–92
  brain, 492–96
Homo erectus, 492
Homo habilis, 492
Homo neanderthalensis, 11–12, 492
  cortical elements of parietal lobe, 184*f*
Homo sapiens, 2–3
  cortical elements of parietal lobe, 184*f*
  culture, 491–92
  dementia and social brain, 197
  emergence of, 131, 473
  emergence of modern, 479
  origin of language in early, 500–2
  parietal lobes in, 187–89
  prevalence of Alzheimer's disease in, 192
  selection of neuroanatomical measures for, 14*t*
Homo species
  evolution, 11–12
  origin of language in early, 500–2
hub, 18, 116–17, 162
  definition, 526
human(s)
  brain asymmetries, 27–29
  brain layout, 19*f*–20
  brain size, 44
  changes in body size and shape, 474*f*, 474–75
  comparing astrocytes of rodent and, 41*f*
  comparing movement phenomenology of animals and, 277*t*
  corticospinal tracts (CSTs), 258
  dendritic spines of neurons, 105
  dispersal of, 132–33, 133*f*
  environment and, 485–86
  evolutionary constraints, 283–84
  milk consumption and digestion, 475–76
  origin of, 131, 132*f*
  selection of neuroanatomical measures for, 14*t*
  white matter fiber tracts and network connectivity, 403*f*
  See also adult humans
human accelerated regions (HARs)
  autism and, 111
  definition, 526
  evolution, 105–6
  human-specific DNA sequences, 108–9
  schizophrenia and, 110–11, 324
human auto-domestication, 476
human biology, nature or nurture, 479

human brain(s), 9
  bauplan of, 12–18
  connectome of, 400, 504
  cortical parcellation, 23–27
  digital reconstruction, 26–27, 28f
  disease-specific proteins, 98
  energy-relevant facts about, 463–64
  energy storage and consumption, 511
  evolution, 2, 60
  extended postnatal development, 496
  function and mechanisms of, 511
  future development of, 484
  gyrification, 22–23
  heat regulation, 512
  infectious diseases, 514–16
  Lewy pathology formation in, 96–97
  pallium, 15
  protection against oxidative stress, 513–14
  pyramidal neurons, 21f, 21–22
  size and number of neurons, 20–21
  story of evolution, 462–63
  telencephalon of, 15
  waste clearance, 513–14
human brain disorders, counterevidence of evolutionary novelty in, 120–22
Human Brain Project, 491–92
human cerebral connectome, cortico-cortical connections of, 317f
human evolution
  anatomically modern humans (AMH), 132
  identification of genomic sequences, 106–7
  key points to remember, 486
  summary of, 11–12
  See also ongoing human evolution
human evolutionary history
  evolutionary tree showing phylogenetic relationships, 132f
  human dispersal, 132–33, 133f
  human origin, 131, 132f
  inferred pulses of archaic introgression between modern and archaic humans, 134f
  interbreeding between archaic and modern humans, 133–36
  interbreeding between Neanderthals and modern humans, 133–35
  interbreeding with archaic "ghost" populations, 136
  interbreeding with Denisovans, 135–36
human genomes, schizophrenia and modern, 324–25
human health, ethical issues, 485
human health and disease
  Alzheimer's disease (AD) and parietal cortex, 190–91
  antagonistic pleiotropy in late-onset Alzheimer's disease, 184–86
  biomechanical effects of tentorium cerebelli, 187, 188f
  brain globularity and parietal development, 187–90
  downside of complex parietal cortex, 192–94
  evolutionary evidence, 184–94
  parietal lobes and functional morphology, 186f, 186–87
  structural and functional brain changes in parietal lobe in AD, 191–92
human language, origin and evolution of, 500–3
human model species, pallial layout and architecture, 18–29
human natural history, brain evolution and, 181–82
human-specific gene regulatory changes
  epigenomic comparisons in adult brain, 118–19, 121f
  future work and unanswered questions, 122–24
  implications in brain diseases, 119–20
  key points to remember, 124
  overview of, 121f
  single-gene comparisons, 112–13
  transcriptomic changes, 121f
  transcriptomic comparisons in adult brain, 115–18
  transcriptomic comparisons in developing brain, 113–15
Huntington's disease, 3
  Huntington gene (HTT), 232–33
  long silent period of, 516–17
  mitochondrial dysfunction, 62
  monkey strains with nutant genes, 10b
  mutant Huntington gene (mHTT), 232–33
  neural atrophy and, 169
  repeats in coding regions, 231–34
  role of biomodulator, 454
  understanding neurological function and evolution, 244
  See also brain diseases with unstable repeats
Huxley, Julian, 318–19
hybridization (genetic), 133
  definition, 526
hybrid-origin theory, 133
Hydra, evolutionary evidence of, 40–42
hygiene, term, 433–34
hygiene hypothesis, 432–34
  Alzheimer's disease, 440–41
  definition, 526
  multiple sclerosis, 440
hyperbranching axons, parallel involvement of Lewy-prone systems, 83–84, 85f
hyperfrontality, 526

hypersociability, 334, 349
    definition, 526
hypervigilance cycle, schizophrenia, 326
hypofrontality, 526
hypomimia
    definition, 526
    Parkinson's disease, 215
hypoperfusion, reduced blood flow of brain, 67
hyposmia, 205–6, 219, 222t
    definition, 526
    dysfunction in Parkinson's disease, 211f
    olfactory loss, 210
    Parkinson's disease, 218–20
hyposociability, 334, 348
    definition, 526

Ice Age, 472–73
ideography, Chinese, 503–4
imaging techniques, advanced, 50
Imura, T., 503–4
index of dexterity, 256–57
    definition, 526
induced pluripotent stem cells (iPSCs), 48–50
    definition, 526
infection-induced depression, sickness behavior, 299–300
infectious diseases, fast and slow, 514–16
inflammaging
    condition of, 62
    definition, 526
inflammasome, 72–73
    definition, 527
inflammation
    aging-related diseases, 441
    major depressive disorder (MDD) and, 297–98
    mitochondria and, 72–73
inflammatory diseases, 511
infralimbic cortex (ILC), mapping goal-directed and habitual behavior, 165
inner carotid artery (ICA), 29
instinctual behaviors, rapid eye movement (REM) sleep and rehearsing, 275
Institute of Neurology, 208–9
insular lobe, 18
insulin resistance, mitochondrial bioenergetics, 67
intellectual disability, schizophrenia and, 323
interbreeding
    archaic and modern humans, 133–36
    archaic "ghost" populations, 136
    Denisovans and modern humans, 135–36
    hybridization, 133
    Neanderthals and modern humans, 133–35
internal clock, 221, 307
    definition, 527

International Classification of Diseases 11th Revision (ICD-11), attention deficit hyperactivity disorder (ADHD), 360–61
International Union of Basic and Clinical Pharmacology, 449
intralaminar astrocyte (ILA), 43, 51
intrauterine programming
    attention deficit hyperactivity disorder (ADHD), 364
    definition, 527
intrinsic apoptotic pathway, Parkinson's disease (PD)-associated proteins in, 73
intrinsic template(s)
    architecture of, 82
    axon-first pathogenesis by highly branched axons, 83–84
    goal-directed motor control system vs. goal-searching reward estimation system, 92–93
    highly branched axons, 83–92
    key points to remember, 99
    Lewy pathology only in human brain, 96–97
    "machineries" of brain, 81–82
    memory and its disruption by tau deposition, 93–96, 94f
    motor, nitrostriatal and limbic neurons as, 91t
    multifocal Lewy body disease, 84–87
    past, present, and future of neurodegenerative disorders, 97–99
    straight axons for ALS vs. highly branched axons for Lewy pathology, 87–89, 88f
    See also highly branched axons
introgression desert, 137–38
    definition, 527
invertebrates, glia-to-neuron ratio in phylogeny, 41f
ionotropic receptors, 459–60, 461
    definition, 527
ionotropic transmission, 461–62
isocortex, cerebral cortex, 15–16

Jackson, John Hughlings, 2, 255–56
James, William, 154, 160–61
Japanese writing, 503–4
Jouvet, Michel, 276
Julich-Brain cytoarchitectonic maps, 16

Kana characters, 503–4
Krebs cycle, 59

lactate dehydrogenase (LDH), comparing isoform levels, 45
lamelliform, 43–44
    definition, 527
Langston, J. William, 62–63
language

Dravidian, 502–3
epigenetic signature of writing and reading, 503–4
glossogeny, 502
neurobiological steps, 502–3
tentative model for origin and evolution of oral, 500–3
law of effect
definition, 527
drug reward theory, 374–75
lemur, brain layout, 19f–20
Lewy bodies (LB), 205–6, 209
definition, 527
Lewy body disease
alpha-synuclein in branched non-myelinated axons of, 88f
axonal degeneration of, 99
axon-first pathogenesis by highly branched axons in, 83–84
clinical manifestations of, 97
focal, 86–87
multifocal, with capricious spread, 84–87
REM sleep behavior disorder (RBD), 270–71
Lewy pathology
formation in human brain, 96–97
highly branched axons for, 98
hyperbranching axons as structural template, 83–84, 85f
Lewy-prone systems
highly branched axons for, 87–89
hyperbranching axons of, 83–84, 85f
life history
definition, 527
ligand-gated ion channels
definition, 527
evolution of, 461–62
ligands, 461
limbic lobe, 18
limbic neurons, Alzheimer's disease, 91t
limbic system(s), 18
locus coeruleus (LC), Parkinson's disease, 213
loneliness, depression induced by, 301
long-read sequencing, 242–43
long-term stress, depression induced by, 300–1
lori, brain layout, 19f–20
Louisiana State University, 9

Macaca mulatta, selection of neuroanatomical measures for, 14t
macaque, white matter fiber tracts and network connectivity, 403f
McGrath, John, 320
Magnetic Resonance Imaging (MRI), 28f
multimodal, 16
parietal lobes of humans and primates, 186f

major depressive disorder (MDD)
adaptive function of symptoms, 305–9
diagnosis of, 294
as disease of modern lifestyles, 295–96
evolution-based treatment for, 309
inflammation and, 297–98
key points to remember, 310
leading source of disability worldwide, 294
modern lifestyle and risk of, 296–97
pathways to MDD and adaptive coping, 299f
schizophrenia and, 323
subtypes, 299–305
major depressive disorder (MDD) subtypes, 299–305
chemically induced depression, 304
depression induced by grief, 302
depression induced by hierarchy conflict, 301–2
depression induced by loneliness, 301
depression induced by long-term stress (burnout), 300–1
depression induced by romantic rejection, 302
depression induced by somatic diseases, 304
depression induced by traumatic events, 301
infection-induced (sickness behavior), 299–300
postpartum depression, 303
season-related depression, 303–4
spring depression, 304
starvation-induced depression, 304–5
winter depression, 303–4
major depressive disorder (MDD) symptoms, 305–8
anhedonia, 307
appetite changes, 306–7
emotional pain, 305
fatigue, 307
gut microbiota and mood disorders, 308–9
loss of self-confidence, 308
pessimism, 308
psychomotor agitation or retardation, 307–8
rumination, 306
self-accusations and guilt, 308
sleep problems, 307
suicide proneness, 305–6
maladaptation, 156–57
definition, 527
mammalian brain(s)
bauplan of, 29, 30
brain size and number of neurons, 20–21
role of gyrification, 22–23
mammalian class, simplified phylogenetic trees for, 13f
mammalian evolution, summary of, 11–12
mammalian model species
neuroscience, 10b
selection of neuroanatomical measures for, 14t

mammalian species
  astrocytes of, 43
  brain layouts in, 19f–20
  pyramidal neurons, 21f, 21–22
mammals
  glia-to-neuron ratio in phylogeny, 41f
  pallium, 15
Manduca sexta (tobacco hornworm), 377–78
Marsden, David, 208–9
massively parallel reporter assays (MPRAs), 108–9
  definition, 527
Mayr, Ernst, 2, 318–19, 451
MDD. *See* major depressive disorder (MDD)
Mead, Margaret, 317–18
Medawar, Sir Peter, 427–28
medication, attention deficit hyperactivity disorder (ADHD), 363
medulla oblongata, 12–15
Megginson, Leon C., 9
mental health
  evolutionary perspective, 483–84
  sex differences in, 480–81
mental time travel, 510
mesencephalon, 12–15
mesolimbic dopamine system (MDS)
  definition, 527
  dominance of model, 375–76
  drug reward, 375, 391–92
  functions of, 384
  sensitivity and domains of cognitive functioning, 384–85
metabolic waste, elimination of, 513–14
metabolome, 44–45, 74
  definition, 527
metabotropic receptors, 459–60
  definition, 527
metazoans, 40–42
  definition, 527
methamphetamine, 387
  drug use in US, 388f
1-methyl-4-phenyl-1,2,3,6-tetrahydropyridine (MPTP), toxicity of, 62–63, 83–84
microbiome, 309, 475, 482, 485–86
  definition, 527
microglia, 39–40
migraine, depression, 304
migration paths, humans, 132–33, 133f
mindreading, 501–2
minicolumn, 338, 340, 342t, 345–46, 353
  definition, 527
mitochondria
  central factor in pathogenesis of PD, 62–66
  components of, 59–60
  contribution to aging, 60–62
  description, 58–60
  DNA, 59
  dysfunction, 73–74
  epigenetic modulation of genome, 71–72
  genome alterations, 70–72
  haplotypes and neurodegeneration, 71
  inflammation and, 72–73
  key points to remember, 74–75
  mtDNA epigenetic modulation, 71–72
  organelles and interconnected networks, 60
  PD-associated proteins in resolution of apoptosis, 73
  powerhouses of the cell, 59
mitochondria-associated membranes (MAMs), 69
mitochondrial, endosymbiotic theory, 58
mitochondrial apoptotic pathway, Parkinson's disease (PD)-associated proteins in, 73
mitochondrial dysfunction
  alpha-synucleic aggregation and, 64–65
  metabolism, 67–69
mitochondrial haplotype, 71
  definition, 527
mitochondrial metabolism, dysfunctional, 67–69
mitochondrial quality control, regulation of, 65–66
mitochondrial signaling, impairments in Parkinson's disease, 63f
mitophagy, process of, 65
modern lifestyles
  depression as disease, 295–96
  vulnerability of brain to disease, 511
module, 116–17, 187, 188f, 210
  definition, 527
monkey, brain layout, 19f–20
Montaigne, 271–72
mood disorders, gut microbiota and, 308–9
Moreau, Clara, 328
mosaicism, 2–3
motor control system
  goal-directed, 92–93
  outcome of, 93
motor neuron disease (MND), 252
motor neurons, internal template for ALS, 91t
mouse
  brain layout, 19f–20
  comparing astrocytes of humans and, 41f
  pyramidal neurons, 21f
  selection of neuroanatomical measures for, 14t
mtDNA heteroplasmy, 59–60, 61, 71–72
  definition, 528
multifocal Lewy body disease, spread of, 84–87
multiple nested levels or organization, architecture, 494–95
multiple sclerosis, 511
  environmental influence on development of, 482
  glial cells, 49–50
  Global Burden of Disease Collaboration, 482

hygiene hypothesis, 440
nervous system deregulation, 47–48
sex differences, 481
multiple system atrophy, REM sleep behavior disorder (RBD), 270–71
muscle atonia, loss of, in REM sleep, 278–80
Mus musculus, selection of neuroanatomical measures for, 14*t*
myelencephalon, 12–15
myelin, cellular circuits, 410
myeloarchitecture, 16, 406
definition, 528

Nakamura, Karen, 326–27
narcolepsy, movement phenomenology in humans, 277*t*
National Institutes of Health, 62–63
NatoSatellite, 243, 243*t*
natural selection
aging and, 435–37
changes in opportunities in Australia, 477*f*
Darwinian interpretation of, 92
definition, 528
neurotransmitters and, 472–73
relaxed, 476–77
sexual and, 510
Neanderthal genomes, schizophrenia and, 324–25, 328
Neanderthals, 2–3, 11–12
adaptive introgression from, 140
admixture, 146–47
alleles, 144–45
archaic introgression, 134*f*
comparing Europeans with Altai Neanderthal genome, 143–44
discovery of remains, 133
DNA contribution, 144–45
endocranial organization, 501–2
functional consequences of adaptive introgression from, 140–42
genomes, 134–35
genome-wide measure of burden of introgression, 146
genomic studies of introgression in early Eurasians, 147
interbreeding between anatomically modern humans (AMH) and, 133–35
interbreeding of Eurasians with, 142
introgressed variation, 137–38
parietal development, 187–90
negative selection, 146–47
definition, 528
neocortex, 15–16, 30
expansion in humans, 44–45
expansion of, in brain hominization, 492–93

neomammalian complex, 207–8
neontological study, species diversity, 182–83
neontology, 182–83, 192
definition, 528
neophobia, 378, 381
definition, 528
neoteny, 108–9, 123
definition, 528
network theory
cortical organization, 404–5
definition, 528
neural circuits, epigenetic variability of, 499
neural Darwinism, 497
definition, 528
neuroanatomical profile, 337, 352
definition, 528
neurobiological reward mechanism, human/mammalian brain, 445
neurobiological theory
reward center, 375
standard model, 374–75
Neurodata without Borders data sets, 417
neurodegeneration, mitochondrial haplotypes and, 71
neurodegenerative conditions, insulin resistance, 67
neurodegenerative diseases
asymmetric glucose metabolism, 406
misfolding of prion and protein aggregation, 515–16
selection shadow, 427–30
social impact of, 197
superoxide dismutase (SOD) enzymes, 514
See also designing strategies for brain diseases
neurodegenerative disorders, vulnerability of neuronal groups, 81–82
neurodiversity, attention deficit hyperactivity disorder (ADHD), 367
neurodiversity movement, 367, 370
definition, 528
neuroglia
glial cells, 38
phylogenetic advance of, 41*f*
neuroimaging, amygdala, 212
neuroimaging studies, goal-directed and habitual control, 169
neuroinflammation, 297–98
definition, 528
neurologists, 5–6
neuronal intranuclear inclusion disease (NIID), 229
neuronal networks, model of epigenesis of, 497–99, 498*f*
neuronal workspace hypothesis, 494–96
definition, 528

neuron doctrine
  definition, 528
  Ramon y Cajal, 87–88
neuron-glia interactions, 49
neurons
  brain size and, 44
  cellular circuits of brain, 408–9
  coupling astrocytes and, 45
  morphological comparison of, in mouse and monkey, 409f
  pyramidal, 21–22
  von Economo (VENs), 22
neuropsychiatric diseases, role of gyrification, 22–23
neuropsychologists, 5–6
neuropsychology, brain and behavior, 197
neuropsychopharmacology, 449–51
  evolutionary history of neurotransmitters and biomediators, 451–57
  evolutionary perspectives on, 467–68
  evolution of cell communication, 457–58
  evolution of ligand-gated ion channels, 461–62
  evolution of neurotransmitter receptors, 459–60
  evolution of synaptic transmission, 458–59
  as field of medicine, 449
  future directions of, 465–68
  glutamate in brain health and disease, 462–65
  GPCR (G-protein coupled receptors) evolution, 460–61
  key points to remember, 468
  receptor distribution of muscarinic M2 receptors for acetylcholine, 462f
  See also glutamate in brain health and disease
neuroscience
  contemporary, 492
  mammalian model species in, 10b
neurotransmitters, 451
  acetylcholine, 452–53
  addiction, 377, 389
  attention deficit hyperactivity disorder (ADHD), 362–63
  catecholamines, 453–54
  detoxification of, 467
  evolutionary history of, 465
  evolutionary history of biomediators and, 451–57
  evolution of, 466b
  evolution of receptors, 459–60
  glutamate and GABA, 455–56
  goal-directed and habitual behaviors, 167–68
  key points to remember, 468
  oxytocin and vasopressin, 456–57
  serotonin, 454–55
nexopathies, disease-specific neurite lesions, 82
Nicotiana, 383

nicotinamide adenine dinucleotide (NADH), 59
nicotine, 377–78, 387
nigrostriatal neurons, Parkinson's disease, 91t
N-methyl-D-aspartate (NMDA) receptors, 45–47
nonhuman model species, pallial layout and architecture, 18–29
nonhuman primates (NHPs)
  comparing genomic sequences of humans and, 106–7
  comparing protein-coding genes of humans and, 105–6
  dendritic spines of neurons, 105
  future work on, 122–24
nonmammalian species, lessons from, 11b
norepinephrine
  addiction, 377
  evolutionary history of catecholamines, 453–54
novelty-seeking, 3
  definition, 528
nychthemeral rhythm, 528

obsessive-compulsive disorder (OCD)
  goal-directed behavior, 172, 173
  schizophrenia and, 323
occipital lobe, 17
oculopharyndistal myopathy type 1 (OPDM1), 229
oculopharyngeal muscular dystrophy (OPMD), 233
Old Friends
  definition, 528
  term, 434
Old Friends hypothesis, 432–33
olfactory areas, mammalian and primate brains, 25–26
olfactory pathways, Parkinson's disease, 210–11
One Health, 486
  definition, 528
oneiric behaviors, 276
ongoing human evolution, 472–74, 485–86
  aging and pleiotropy, 481–82
  body height changes and cranial capacity in last 20,000 years, 474f
  brain developmental dilemma, 484
  changes in opportunities for natural selection in Australia, 477f
  clinical and therapeutic implications, 479–82
  counterarguments and evidence against evolutionary influences, 477–79
  ethical considerations, 485
  evolutionary evidence related to human health and disease, 474–77
  future work and unanswered questions, 483–85
  historical/evolutionary background, 472–74
  key points to remember, 486
  mental health, 483–84

regional differences in gene frequencies, 482
sex differences in psychiatric diseases, 480–81
ontogeny, 2
   definition, 529
Onufrowicz nucleus
   definition, 529
   motor neurons, 88–89
opioid(s), 387
   addiction, 389
   goal-directed and habitual control, 167–68
oral language, tentative model for origin and evolution of, 500–3
orbitofrontal cortex (OFC), mapping goal-directed and habitual behavior, 163–65
orthologous, 42
   definition, 529
Osmond, Humphry, 318–19
otherness, schizophrenia, 314
out of Africa (OOA) model
   anatomically modern humans (AMH), 132–33
   definition, 529
Oxford Nanopore, minion sequencing, 242
oxidative phosphorylation, dysfunctional mitochondrial, 67–68
oxidative stress, human brain and protection against, 513–14
oxycodone, 387–88
   addiction, 389
oxytocin, evolutionary history of, 456–57

Pacific Biosciences (PacBio), 242
   single-molecule real-time (SMRT) sequencing, 242
pain relievers, 387
paleomammalian complex, 207–8
paleoneurology
   brain evolution in extinct species, 183–84
   definition, 529
   endocranial casts and, 183–84, 184f
paleontology, 182–83, 473
   definition, 529
pallium, 15
pangolin, brain layout, 19f–20
Pan troglodytes
   cortical elements of parietal lobe, 184f
   selection of neuroanatomical measures for, 14t
paraquat, environmental toxin, 83–84
parasomnia
   REM sleep behavior disorder, 270–72, 281t
   youth adults with NREM, 286
   See also rapid eye movement (REM) sleep behavior disorder (RBD)
Parent, André, 208
parietal cortex, Alzheimer's disease and, 190–91
parietal lobe(s), 17
   development of, 187–90
   functional morphology of, 186f, 186–87
Parkinson, James, 271–72
Parkinsonian syndromes, 206, 271–72, 283
Parkinsonism, 206, 208–9
Parkinson's disease (PD), 3, 4
   amygdala, 211f, 211–12
   apathy, 218
   basal ganglia (BG), 205–6, 210
   behavioral stages across timeline and cell loss in, 158f
   bipedal gait, 216–17, 222t
   brainstem nuclei, 213
   clinical and therapeutic implications, 223–24
   commonalities between clinical syndromes in, 221, 222t
   commonalities between nuclei and pathways in, 214–15
   counterarguments and related evidence against evolutionary influences, 222–23
   deficient behaviors and affected areas, 222t
   deficient blindsight, 219–20, 220f, 222t
   deficient colon motility, 222t
   deficient nictemeral rhythm or missing light entrainment, 221, 222t
   degenerative cell loss, 169
   depression, 15, 304
   disease-specific neuronal groups, 81–82
   dopamine neuroreceptor changes, 406
   dopaminergic disease, 205–6
   dopaminergic neurons in substantia nigra, 172
   dysfunctional mitochondria, 67–69
   emotional dysregulation, 217–18
   enteric nervous system (ENS), 207, 222t
   evolutionary evidence related to, 209–15
   evolutionary view of clinical syndromes of, 215–21
   future research proposals, 224
   historical background of, 207–9
   human-specific gene regulatory changes, 119–20
   hyposmia and deficient blindsight examples, 218–20
   induced pluripotent stem cells (iPSCs), 48–50
   inflammaging in, 441
   insulin resistance, 67
   key points to remember, 224
   locus coeruleus (LC), 213
   mesencephalic and thalamic nuclei, 212–13
   mesencephalon for, 12–15
   mitochondria and inflammation, 72–73
   mitochondria as central factor in pathogenesis of, 62–66
   mitochondrial bioenergetics, 66–70
   mitochondrial dysfunction, 62, 73–100

552 | INDEX

Parkinson's disease (PD) (cont.)
    mitochondrial dysfunction and alpha-synucleic aggregation, 64–65
    mitochondrial genome alterations in, 70–72
    mitochondrial signaling impairments in, 63f
    monkey strains with nutant genes, 10b
    movement phenomenology in humans, 277t
    neurite architecture, 82
    nigrostriatal neurons as internal templates, 91t
    olfactory and visual pathways, 210–11
    past, present, and future of, 97–99
    PD-associated proteins in resolution of apoptosis, 73
    pedunculopontine nucleus (PPN), 213
    phenotype diversities of, 206–7
    from poor perception to blunted expression, 217–18
    prevalence of, 205–6
    reactive oxygen species (ROS) in, 44–45
    regulation of mitochondrial quality control, 65–66
    REM sleep behavior disorder (RBD), 270–71
    search for evolutionary grounding of human, 206
    sensory deficits, 218–20
    from spinal cord to ENS, 214
    superior colliculi and the pulvinar, 212–13
    superoxide dismutase (SOD) supplementation, 514
    von Economo neurons (VENs), 22
passerines, songs of, 11b
pathology
    axon-first pathogenesis in Lewy body disease, 83–84
    multifocal Lewy body disease, 84–87
pathophysiology, 4–5
    Parkinson's disease, 206, 223–24
    rapid eye movement (REM) sleep behavior disorder (RBD) in humans, 278–83
PD. See Parkinson's disease (PD)
Pearlson, Godfrey, 322–23
pedunculopontine nucleus (PPN), Parkinson's disease, 213
perivascular space, 513
    definition, 529
pessimism, major depression disorder (MDD) symptom, 308
PGO wave, 273–74
    definition, 529
pharmacophagy, 377–78, 451–52, 467
    definition, 529
    plant secondary metabolites (PSMs), 390–91
phenotypes, 466b
phonology, 501
    definition, 529

phylogenesis, 47
    definition, 529
phylogenetics, 466b
phylogenetic trees, mammalian class and primate order, 13f
phylogeny, 2
    definition, 529
Planetary Health, 486
plant secondary metabolites (PSM), 373–74
    definition, 529
    protecting plants from insect predators, 377
    relationships with human nervous system receptors, 382t, 383
    relevance of, 382–83
    See also addiction
Pleistocene, 472–73, 474–75
plexin protein, 257
    definition, 529
Pliny the Elder, 271–72
Poecilia formosa (parthenogenetic fish), 500
polyalanine diseases (polyA), 231–32, 233, 237
    definition, 529
polyglutamine diseases (polyQ), 231–34, 235–36, 237
    definition, 529
population strategy, disease prevention, 439–40
positive selection, 105–6, 110–11, 112–13, 135, 139, 140–41, 142, 324–25
    aging, 481
    definition, 529
    neurotransmitters, 466–67
positron-emission tomography (PET) scans, illiterate and literate groups, 503
postmortem transcriptomic comparisons, mouse model of Alzheimer's disease, 120–22
postpartum depression, 303
post-traumatic stress disorder (PTSD), 270–71
    definition, 529
    depression induced by, 301
postural instability, 205–7
    definition, 529
precuneus, as "eye of the self," 197
predictive brain, 510
prefrontal areas, mammalian and primate brains, 26–27
prefrontal cortex (PFC), goal-directed and habitual behavior, 162
preventative strategies, brain diseases, 426, 427f
prevention research, evolutionary health promotion, 439–42
preventive strategies
    brain diseases, 437–42
    evolutionary health promotion, 439–42
    postponed aging, 437–39
"price to pay," term, 3

primate motor cortex, evolution of, 257–58
primate order, simplified phylogenetic trees for, 13f
primates
　brain size and number of neurons, 20–21
　corticospinal tracts (CSTs), 257–58
　evolution of motor cortex, 257–58
　research in, 10b
primate species, 30
prion misfolding, neurodegenerative
　　diseases, 515–16
prionopathy, 515–16
　definition, 529
progressive myoclonic epilepsy of Unverricht-
　　Lundborg type 1A (EPM1), 229
progressive supranuclear palsy (PSP), 95–96
　definition, 530
protein aggregation
　definition, 530
　neurodegenerative diseases, 515–16
protein-coding genes, comparing humans and
　　nonhuman primates (NHPs), 105–6
protoplasmic astrocytes, 39
proximate
　concept of, 2
　definition, 530
　influences on disease, 4–5
pruning, 337–38, 408, 497–99, 501–2
　definition, 530
psychiatric diseases, sex differences in, 480–81
psychiatrists, 5–6
psychomotor agitation or retardation, major
　　depression disorder (MDD) symptom, 307–8
punishment model
　addiction, 378
　evolutionary, 387
　plant drugs, 383
pyramidal neurons, 30
　mammalian, 21f, 21–22

rabbit, brain layout, 19f–20
Ramon y Cajal, Santiago, 39f, 39
random gene flow, 134f, 473
　definition, 530
rapid eye movement (REM) sleep
　behavior disorder, 205–6, 215
　definition of, 272–73
　fight or flight during, 212–13
　functions of dreaming and, in adult
　　life, 274–75
　normal functional pathways of, 279f
　ontogenetic "maturation" of, 273–74
　regulation of emotions, 274–75
　rehearsing instinctual behaviors, 275
　simulating threats, 275
　thermoregulation, 274

rapid eye movement (REM) sleep behavior
　　disorder (RBD)
　appearance of complex movements and
　　behaviors, 280–81
　characteristics in different
　　species, 272f
　clinical and therapeutic implications, 287–88
　considering evolutionary basis of, 271
　counterarguments and evidence against
　　evolutionary influences, 286–87
　description of, 270–71
　dream content corresponding to RBD
　　behaviors, 281–83
　evolutionary constraints in humans, 283–84
　evolutionary evidence relating to human
　　disease, 276–86
　evolutionary evidence relating to human
　　health, 272–75
　future work and unanswered questions, 288
　historical background of, 271–72
　human RBD may be evolutionarily
　　impregnated, 285–86
　key points to remember, 288–89
　as "late" phylogenetic acquisition, 272–73
　loss of muscle atonia in REM sleep, 278–
　　80, 279f
　most frequent causes of, in humans, 281t
　normal REM sleep vs., 279f
　observations on RBD-like syndromes in
　　animals, 276–78
　ontogenetic "maturation" of, 273–74
　pathophysiology in humans, 278–83
　proposed functional pathways in human
　　RBD, 279f
　rehearsals of "fight or flight" survival
　　behavior, 286
　specificities of human RBD
　　phenomenology, 284–85
　See also animals and REM sleep behavior
　　disorder (RBD)
rat
　brain layout, 19f–20
　rapid eye movement (REM) sleep behavior
　　disorder (RBD), 277–78, 277t
　selection of neuroanatomical measures for, 14t
Rattus norvegicus, selection of neuroanatomical
　　measures for, 14t
reactive oxygen species (ROS), 44–45, 60–61
reading, epigenetic signature of, 503–4
reafference, 495–96
　definition, 530
recent evolutionary change, 110, 334, 337, 346,
　　348–49, 353–54
　definition, 530
regional differences, gene frequencies, 482

regulation of emotions, rapid eye movement (REM) sleep and, 274–75
relaxed natural selection, 476–77, 485, 486
   definition, 530
REM atonia, 278–80, 279f
   definition, 530
repeat-associated non-AUG (RAN) translation, 231, 232f
   definition, 530
repeat expansion disorders
   range of, 229, 230f
   untranslated regions (UTRs), 229
repeat expansions
   computational algorithms evaluating, 243t
   range of disorders, 229
reporter activity, 108–9
   definition, 530
reptilian complex (R-complex), 207–8
reptilian species, brain layouts in, 19f–20
resting-state functional magnetic resonance imaging, 191–92, 401
   definition, 530
reticulation, 187–89
   definition, 530
retino-colliculo-thalamo-amygdala (RCTA) pathway, 212–13
Rett syndrome, induced pluripotent stem cells (iPSCs), 48–50
reward estimation, autonomic independence of, 97
reward estimation system, goal-searching, 92–93
reward theory, 374–76
   definition, 530
rhesus macaque/monkey, selection of neuroanatomical measures for, 14t
rigidity, 205–7, 284
   definition, 530
rodents, rapid eye movement (REM) sleep behavior disorder (RBD), 277–78, 277t
romantic rejection, depression induced by, 302
Rose, Geoffrey, 439–40
rotenone, environmental toxin, 83–84
rumination, major depression disorder (MDD) symptom, 306

Sahelanthropus tchadensis, 11–12
SARS-CoV-2 infection, 514–15
schizophrenia, 4, 12–15
   catatonia, 314–15
   clinical and therapeutic implications, 325–27
   cortico-cortical connections of human cerebral connectome, 317f
   counterarguments and evidence against evolutionary influences, 322–25
   diagnosis of, 314–16
   DNA methylation, 122
   evolutionary evidence related to human health and disease, 319–22
   future work and unanswered questions, 327–28
   genome-wide association studies (GWAS), 322–24
   glial cells, 49–50
   goal-directed behavior, 172
   historical background, 317–19
   human condition of, 314–16
   human-specific DNA evolution in, 110f, 110–11
   human-specific gene regulatory changes, 119–20
   human-specific regulatory changes, 121f
   key points to remember, 329
   monkey strains with nutant genes, 10b
   motivational deficits, 169
   neurotransmitters, 472–73
   social skills and, 105
schizophrenia-shaman connection, 318
Schizophrenia Working Group of the Psychiatric Genomics Consortium, 322–23
seasonal affective disorder (SAD), 303
   spring, 304
   winter, 303–4
season-related depression, 303–4
secular trends, 478, 485–86
   definition, 530
selection shadow
   as "blind spot," 428–29
   definition, 530
   evolutionary concept, 427–28
selective stabilization, epigenesis of neuronal networks by, 497–99, 498f
"self," emergence of, 82–83
self-accusations, major depression disorder (MDD) symptom, 308
self-confidence loss, major depression disorder (MDD) symptom, 308
self-domestication, term, 485
self-medication, addiction, 378
sensations de passage, 219–20
sensory deficits, Parkinson's disease, 218–20
sensory-specific satiety (SSS), 381
   behavioral defenses, 381
   definition, 530
serotonin
   addiction, 377
   attention deficit hyperactivity disorder (ADHD), 362–63
   evolutionary history of, 454–55
   goal-directed and habitual control, 167–68
serotonin transporter immunoreactivity, 347
   definition, 530
sex differences, psychiatric diseases, 480–81
shaman hypothesis, 318
short tandem repeats (STRs), 228, 236–37

polymorphic discovery, 244
sickness behavior
  definition, 530
  infection-induced depression, 299–300
simulating threats, rapid eye movement (REM) sleep, 275
single-cell transcriptomics, glial cells, 50
single-cell transcriptome profiling, 106–7
single nucleotide polymorphisms (SNPs), haplotypes, 71
skilled hand movements, amyotrophic lateral sclerosis (ALS), 260, 262–63
sleep duration, human brain and, 513
sleep problems, major depression disorder (MDD) symptom, 307
sleep-wake dysregulation, Parkinson's disease, 221
slow virus diseases
  definition, 531
  infection, 515–16
social brain, 491–92
  definition, 531
  dementia and, 197
social brain hypothesis, 510
social Darwinists, 485
social zeitgebers, 221
somatic diseases, depression induced by, 304
somatosensory areas, mammalian and primate brains, 26
songbirds, 11b
Southern Dispersal theory, 133
  definition, 531
species diversity, evolutionary anthropology and, 182–83
spinocerebellar ataxia type 12 (SCA12), 229
split ankle, 254–55
"split elbow," 254–55
split food, 254–55
split-hand syndrome, 253–54
  definition, 531
spring depression, 304
stages of life, behavior in Parkinson's disease, 158f
starvation-induced depression, 304–5
statistical norms, schizophrenia, 326
stellate cells, 38–39
stigma, reducing cost of, for schizophrenia, 327
stimulus-response control, behavior and brain regions, 171
stimulus-response structure, goal-directed and habitual control, 168
striatum, 15
stroke, depression, 304
subpallium, 15
substantia nigra (SN), 205–6
suicide proneness, major depression disorder (MDD) symptom, 305–6

superior colliculi (SC), coordinating alarm system, 212–13
superoxide dismutase (SOD) enzymes, neurodegenerative diseases, 514
synapse first hypothesis, 531
synapses
  cellular circuits, 410
  intercellular junctions between neurons, 458
synapse selection, early evidence for, 499–500
synaptic epigenesis
  mouse, 499–500
  selective stabilization, 497–99, 498f
synaptic transmission, evolution of, 458–59
synthetic heroin, Parkinson-like symptoms, 62–63
α synuclein (αSyn), 205–6
  definition, 531
  mitochondrial dysfunction and, 64–65

tandem repeats (TRs), 228
  definition, 531
  See also brain diseases with unstable repeats
tau
  definition, 531
  memory disruption by deposition of, 93–96
Tay-Sachs disease, 481–82
TDP43, 240, 260, 262, 263
  definition, 531
tectum opticum, 212–13, 222t, 283–84
  definition, 531
telencephalization, 9, 29
  definition, 531
  selective vulnerability enhanced by, 97
  size and complexity of cerebral cortex, 96–97
telencephalon (endbrain), 9, 15
temporal lobe, 18
  William's syndrome and autism, 346–48
Ten Country Study, World Health Organization, 320
tentorium cerebelli, developmental effect of, 187, 188f, 189–90
theobromine (chocolate), 380
  evolutionary molecular modeling, 380f
theophylline (tea and chocolate), 380
  evolutionary molecular modeling, 380f
theory of mind, 105, 283–84, 495, 501–2
  definition, 531
therapeutic interventions, glial cells, 51
therian, 43
  definition, 531
thermoregulation, rapid eye movement (REM) sleep and, 274
Thorndike, E. L., 374–75
threat simulation theory (TST)
  definition, 531
  REM sleep behavior disorder (RBD), 271, 275

three-dimensional polarized light imaging (3D PLI)
  bridging from macro to micro, 406, 419b
  definition, 531
tobacco, drug use in US, 388f
tobacco hornworm (*Manduca sexta*), 377–78
tobacco shamans, 383
Tourette syndrome, role of biomodulator, 454
tractography
  comparative brain studies, 399
  connectivity profiles, 400–1
tractography data, fiber tracts in humans, 400
trade-off, 2–3, 110–11, 143, 182, 472–73
  definition, 531
training, transitioning from goal-directed to habitual behavior, 160–61
transcription factor FOXP2, single-gene comparisons, 112–13, 124
transcriptomes
  comparisons in adult human and NHP brains, 115–18
  comparisons in developing human and NHP brains, 113–15
transposase-accessible chromatin using sequencing (ATAC-seq), 106–7
traumatic brain injury (TBI), depression, 304
traumatic events, depression induced by, 301
Trichuris suis, 440
Trichuris trichiura, 440
tripartite synapse, 43–44
  definition, 531
triune brain, 2
Triune Brain in Evolution, The (MacLean), 207–8

UK Biobank data, 475–76
ultimate
  concept of, 2
  definition of cause, 531
  roots of disease, 4–5
Unified Parkinson's Disease Rating Scale, 282
universality
  Crow's principle of, 320
  definition, 532
unstable repeats. *See* brain diseases with unstable repeats

valence lateralization hypothesis, 405–6
  definition, 532
variability theory, neural circuits, 499
variable number tandem repeats (VNTRs), 228, 236
  definition, 532
vasopressin, evolutionary history of, 456–57
vertebrates, astrocytes of, 42
Virchow, Rudolf, 38–39, 39f

visual areas, mammalian and primate brains, 23–24
visual imaging, parietal cortex, 193–94
visual pathways, Parkinson's disease, 210–11
vocalization areas, mammalian and primate brains, 24–25
vomeronasal organ, 25–26, 218–19
  definition, 532
von Economo, Constantin, 22
von Economo neurons (VENs), 22
  definition, 532
  William's syndrome and autism, 345–46
von Lenhossek, Michel, 39–40

Wallace, Alfred Russel, 2
waste clearance, human brain, 513–14
watershed border zones, 189–90
Waxholm Space Rat Brain Atlas, 417
Weismann, August, 427–28
Wernicke's region, 18, 25
Western lifestyles, depression, 295–96, 301–2, 307–8, 309, 310
what stream (ventral stream), 24
  definition, 532
where stream (dorsal visual stream), 24
  definition, 532
Williams, George, 438
William's syndrome (WS), 334
  amygdala, 347–48
  Broca's region, anterior cingulate cortex (ACC), and frontoinsular cortex, 345–46
  clinical and therapeutic implications, 350–52
  counterarguments and evidence against evolutionary influences, 348–50
  evolutionary evidence related to neurodevelopmental disorders, 336–48
  frontal lobe, 337–46
  future work and unanswered questions, 353–54
  historical background of, 335–36
  key points to remember, 354
  microstructural findings in, 341t
  neuroimaging findings in, 339f
  prefrontal cortex (PFC), 338–45
  rare neurodevelopmental disorder, 334–35
  temporal lobe, 346–48
Winston, Ellen, 317–18
winter depression, 303–4
World Health Organization (WHO), 295, 320
Writing, epigenetic signature of, 503–4
WS. *See* William's syndrome (WS)

zebrafish, schizophrenia-related genes, 325
zeitgebers, 221
  definition, 532
zoopharmacognosy, 390–91